# Industrial Electricity

## Eighth Edition

Michael E. Brumbach

DELMAR
CENGAGE Learning™

Australia • Brazil • Japan • Korea • Mexico • Singapore • Spain • United Kingdom • United States

**Industrial Electricity, 8th Edition**
**Michael Brumbach**

Vice President, Career and Professional
    Editorial: Dave Garza

Director of Learning Solutions: Sandy Clark

Acquisitions Editor: Stacy Masucci

Managing Editor: Larry Main

Senior Product Manager: John Fisher

Senior Editorial Assistant: Dawn Daugherty

Vice President, Career and Professional
    Marketing: Jennifer Baker

Marketing Director: Deborah Yarnell

Marketing Manager: Jimmy Stephens

Associate Marketing Manager: Scott A. Chrysler

Production Director: Wendy Troeger

Production Manager: Mark Bernard

Senior Art Director: David Arsenault

Technology Project Manager:
    Christopher Catalina

Production Technology Analyst: Thomas Stover

© 2011, 2005 Delmar, Cengage Learning

For product information and technology assistance, contact us at
**Professional Group Cengage Learning Customer & Sales Support, 1-800-354-9706**

For permission to use material from this text or product,
submit all requests online at **cengage.com/permissions**
Further permissions questions can be e-mailed to
**permissionrequest@cengage.com**

Library of Congress Control Number: 2009938506

ISBN-13: 978-1-4354-8374-3

ISBN-10: 1-4354-8374-X

**Delmar**
5 Maxwell Drive
Clifton Park, NY 12065-2919
USA

Cengage Learning is a leading provider of customized learning solutions with office locations around the globe, including Singapore, the United Kingdom, Australia, Mexico, Brazil, and Japan. Locate your local office at:
**international.cengage.com/region**

Cengage Learning products are represented in Canada by Nelson Education, Ltd.

For your lifelong learning solutions, visit **delmar.cengage.com**

Visit our corporate website at **cengage.com.**

**Notice to the Reader**
Publisher does not warrant or guarantee any of the products described herein or perform any independent analysis in connection with any of the product information contained herein. Publisher does not assume, and expressly disclaims, any obligation to obtain and include information other than that provided to it by the manufacturer. The reader is expressly warned to consider and adopt all safety precautions that might be indicated by the activities described herein and to avoid all potential hazards. By following the instructions contained herein, the reader willingly assumes all risks in connection with such instructions. The publisher makes no representations or warranties of any kind, including but not limited to, the warranties of fitness for particular purpose or merchantability, nor are any such representations implied with respect to the material set forth herein, and the publisher takes no responsibility with respect to such material. The publisher shall not be liable for any special, consequential, or exemplary damages resulting, in whole or part, from the readers' use of, or reliance upon, this material.

Printed in the United States
1 2 3 4 5 12 11 10 09

*"Never stop a pulling horse."*

CHARLES D. PEEK

*A friend and mentor, Charlie oozes wisdom and common sense. I am proud to have worked for him and know him for over twenty years. Oh yea, he was SAC's best navigator, too!*

# CONTENTS

The continued growth and technological advances of the electrical and electronics industry result in many challenges for those entering or working in the field of industrial electrical maintenance. Not only must you be knowledgeable in these fields, you must stay abreast of changes and advancements.

A thorough knowledge of electrical theory is not enough to be successful as an electrical technician. You must also learn to apply the knowledge that you have gained. Since *Industrial Electricity* was first published, it has proven to be a valuable tool for beginners and experienced technicians alike. This text covers the theory of electricity and various applications. In addition, this text goes beyond the basic theories by introducing installation, maintenance, and troubleshooting applications as well.

The eighth edition of *Industrial Electricity* represents a refinement to the seventh edition. The text remains well illustrated. Terms are defined when introduced and a glossary is included as well. Each chapter begins with objectives and ends with review questions. To aid the instructor, an *Instructor's Guide*, detailing answers to the review questions, is available to course instructors.

Michael E. Brumbach has taught industrial electricity/electronics courses at the technical college level continuously since 1986. He has taught both credit and noncredit courses and has served as the Industrial Maintenance Department manager since 1994. He is responsible for programs in industrial electricity/electronics, industrial mechanics, and welding. Mr. Brumbach holds an associate degree in electronics technology and has eleven years' experience in the electrical/electronics industry. Mr. Brumbach is the author of *Electronic Variable Speed Drives* and coauthor of *Industrial Maintenance*.

## NEW FOR THE EIGHTH EDITION

New coverage is included on programmable logic controllers (PLCs) and motors. Chapter 10, Wiring Methods, is the most revised and there are extensive changes to Chapters 16 through 19. Any reference to the *National Electric Code* has been updated to the 2008 NEC. There are sixteen new photographs and twenty-two new drawings designed to better convey important concepts and information. In addition, this text now has an updated Instructor Guide, chapter presentations in PowerPoint, and a computerized test bank as explained directly below.

## SUPPLEMENT TO THIS BOOK

Also available to the instructor is the *Instructor Resource to Accompany Industrial Electricity*, eighth edition. Thoroughly updated to reflect changes to the seventh edition book, the *Instructor's Resource* contains:

- Instructor Guide with answers to the questions in the book
- Chapter Presentations in PowerPoint
- Test bank

(Order #: 1435483731)

Visit us now at our newly designed Web site *www.delmarelectric.cengage.com* featuring other titles available from Delmar Cengage Learning, industry links, career profiles, job opportunities, and more!

## Michael E. Brumbach

Mike is the Industrial Maintenance Department manager as well as an instructor in the Industrial Electricity/Electronics program at York Technical College in Rock Hill, S.C. He has been employed by York Technical College since 1986, having spent the previous eleven years working in industry. Mike possesses an AS degree and is a member of the National Fire Protection Association (NFPA) and the American Society for Nondestructive Testing (ASNT).

You may e-mail Mike at mike@mikebrumbach.com.

# A C K N O W L E D G M E N T S

I would like to thank my wife Kathy, my mother Doris, my daughter Jennifer, and her husband Jason. Working on a project such as this is so much easier when you have the support and love of a great family.

I would also like to thank my friends and colleagues who have been so supportive, helpful, and encouraging during this time. In addition, I wish to thank my students who inspire me and remind me why I teach.

A very big thank you to Mr. Steve Crowley, V.P. Sales and Marketing and Northeast Region manager for Rome Cable Corporation, and to Mr. Jim Tehan, senior marketing manager for EGS Electrical Group, for granting permission to use their company's information in this text. An equally big thank you to those at the National Fire Protection Association (NFPA) for their permission to reprint the *National Electrical Code (NEC)* tables, which appear in the appendixes.

I wish to thank Mark Huth and Dawn Daugherty, both of whom retired recently from Delmar Learning. My thanks are also due to the rest of the team at Delmar Learning. Every one of you is tops in my book.

Finally, I wish to thank my reviewers:

**Recayi Pecen**
University of Northern Iowa
Cedar Falls, Iowa

**James Blackett**
Thomas Nelson Community College
Chesapeake, Virginia

**James Medlin**
Richmond Community College
Hamlet, North Carolina

**Javier Lara**
Flint Hills Technical College
Emporia, Kansas

# Language of Electricity

## OBJECTIVES

After studying this chapter, the student will be able to:

■ Explain the purpose of electrical symbols and drawings.

■ List three main categories of electrical symbols and drawings.

■ Identify and draw from memory some common electrical symbols.

■ Use electrical symbols to construct electrical drawings.

■ Use scientific and engineering notation.

One of the greatest accomplishments of humanity is the development of systems of symbols for communication. The modern English alphabet is a set of 26 symbols, called letters, which convey information when combined in the proper order. Our number system is a set of 10 symbols, which can be used individually or grouped to represent small quantities or large quantities. Symbols used in arithmetic indicate the operation to be performed. For example, the symbol + indicates addition and the symbol × indicates multiplication. Another set of symbols known as shorthand enables stenographers to write words at a rapid pace.

# ELECTRICAL SYMBOLS

There are many different types of symbols used in electrical drawings. These symbols are used to convey meaning, while conserving drawing space. It is important for you to become familiar with these symbols so that you can read and interpret electrical drawings.

Table 1–1A shows some of the typical symbols used to represent power sources and grounds. Table 1–1B shows the symbols used to represent overcurrent protective devices such as fuses and circuit breakers. Notice that there are two symbols

shown for a fuse. This is one of the challenges that you will face in the electrical industry. There is often more than one recognized symbol for a device. You must become familiar with all forms of symbols in use at your facility. This is especially important when dealing with equipment or machinery manufactured in other countries. The symbols used will generally be different from the symbols used within the United States.

Various types of switches are shown in Tables 1–1C, 1–1D, and 1–1E. Table 1–1C shows various types of switch arrangements. Table 1–1D shows manually operated switches, while Table 1–1E shows switches that are operated automatically. Please note that these are not all of the possible switch arrangements in existence today. Often, the symbols are combined so that a more complex switch can be represented. For example, a four-pole double-throw normally open switch symbol would resemble two two-pole double-throw normally open switch symbols combined.

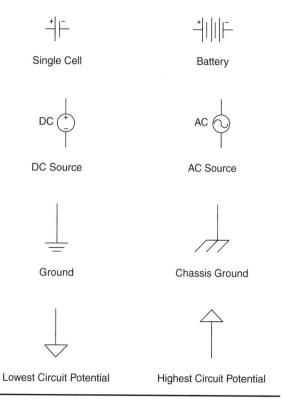

**TABLE 1–1A** Power sources and ground symbols.

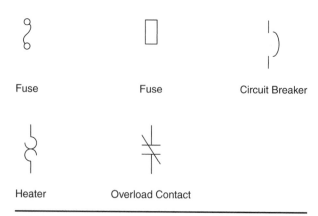

**TABLE 1–1B** Overcurrent protective device symbols.

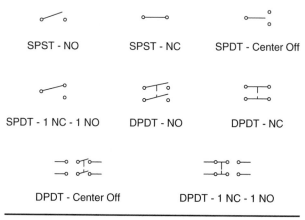

**TABLE 1–1C** Symbols of various types of switch arrangements.

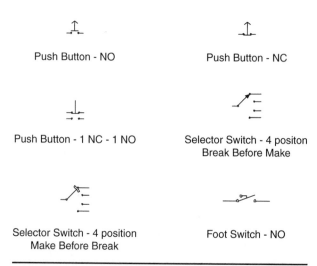

**TABLE 1–1D** Manually operated switch symbols.

Table 1–1F shows the symbols used to represent various types of contacts, relays, motor starters, motors, and generators. Notice that the symbol used to represent motors and generators is the same, the main difference being the qualifier, that is, the flat, horizontal line used to denote DC, or the

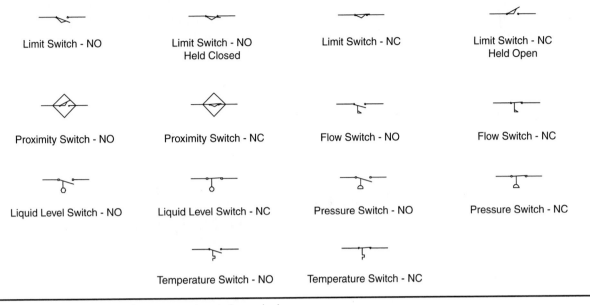

Limit Switch - NO    Limit Switch - NO Held Closed    Limit Switch - NC    Limit Switch - NC Held Open

Proximity Switch - NO    Proximity Switch - NC    Flow Switch - NO    Flow Switch - NC

Liquid Level Switch - NO    Liquid Level Switch - NC    Pressure Switch - NO    Pressure Switch - NC

Temperature Switch - NO    Temperature Switch - NC

**TABLE 1–1E**  Automatically operated switch symbols.

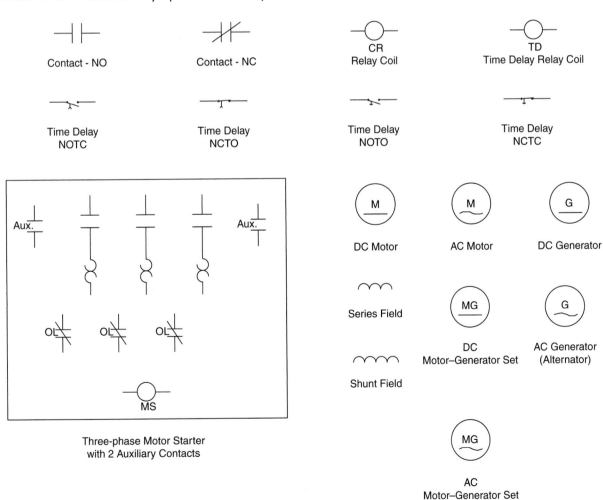

Contact - NO    Contact - NC    CR Relay Coil    TD Time Delay Relay Coil

Time Delay NOTC    Time Delay NCTO    Time Delay NOTO    Time Delay NCTC

Three-phase Motor Starter with 2 Auxiliary Contacts

DC Motor    AC Motor    DC Generator

Series Field    DC Motor–Generator Set    AC Generator (Alternator)

Shunt Field

AC Motor–Generator Set

**TABLE 1–1F**  Contact, relay, motor starter, motor, and generator symbols.

sine wave used to denote AC. The letter M is used to represent a motor, a G is used to represent a generator, and the letters MG are used to represent a motor–generator set.

Various components are shown in Tables 1–1G, 1–1H, 1–1I, 1–1J, and 1–1K. Resistors and capacitors are shown in Table 1–1G. Notice the alternate symbols for some of these devices. Inductors and transformers are shown in Table 1–1H. Again, notice the alternate symbols. Tables 1–1I and 1–1J show the symbols used to represent various solid-state devices. Finally, integrated circuits and digital logic gates are shown in Table 1–1K.

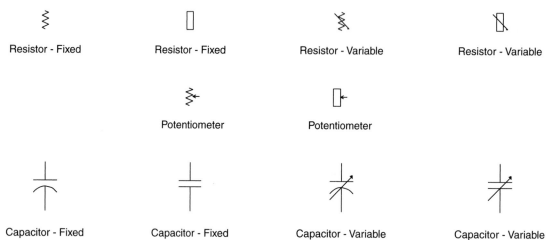

TABLE 1–1G   Resistor and capacitor symbols.

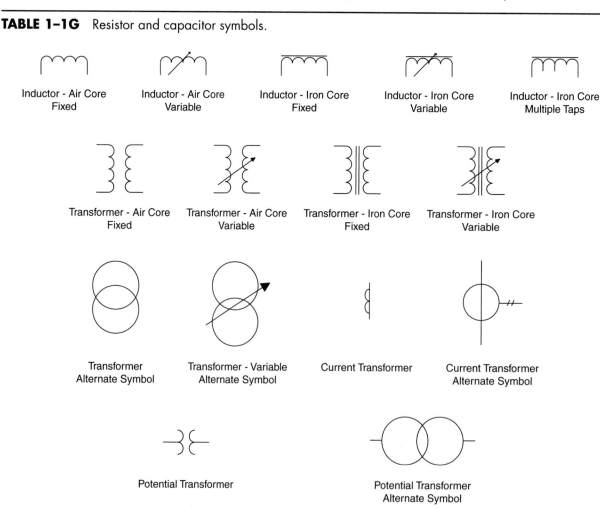

TABLE 1–1H   Inductor and transformer symbols.

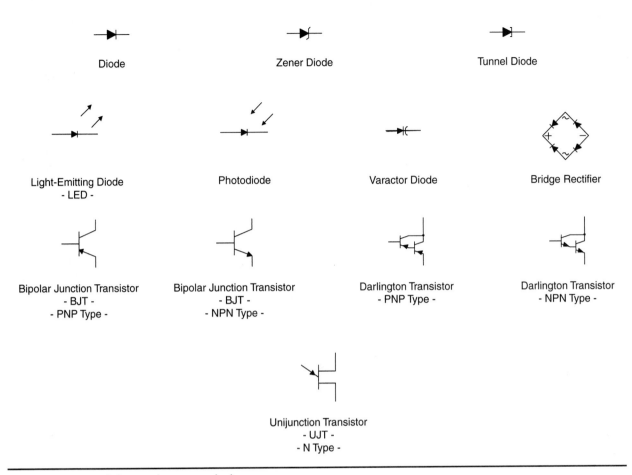

**TABLE 1-1I** Solid-state device symbols.

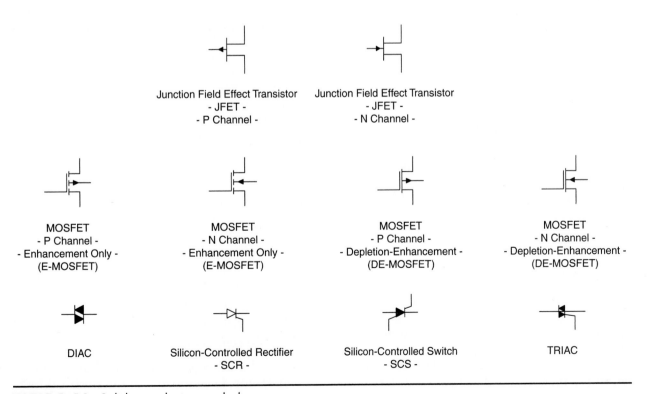

**TABLE 1-1J** Solid-state device symbols.

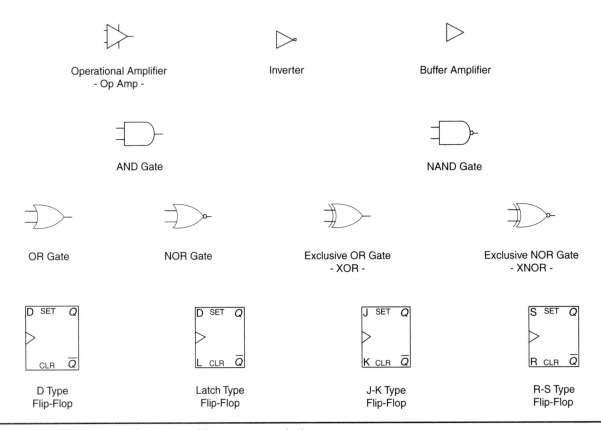

**TABLE 1–1K**  Integrated circuit and logic gate symbols.

Remember, these figures show the essential symbols used in the electrical field. There are many more symbols in use. You will need to become familiar with the symbols used on the machinery and equipment within your facility. Should you have any questions about a particular symbol, ask your facility's supervisor or the equipment manufacturer's representative.

## Architectural Symbols

Architectural symbols, shown in Tables 1–2A and 1–2B, are used in conjunction with architectural drawings (for buildings and structures). These symbols indicate the type of device or equipment and its location as well as the number of electrical conductors necessary for proper operation.

## Pictorial Symbols

Pictorial symbols designate the component, its physical location with respect to other components, and the electrical connections. These symbols are frequently used in panel diagrams (which illustrate the location and connections of electrical and mechanical components within a panel). Pictorial diagrams frequently incorporate schematic symbols within the diagram. Some common pictorial symbols are listed in Table 1–3.

## ELECTRICAL DRAWINGS

Electrical symbols, by themselves, convey little information. Appropriate symbols are combined to form a drawing. An electrical drawing can provide the following information:

- Circuit operation
- Component location
- Electrical connections
- Component function or purpose
- Manufacturer's information
- Wire gauge
- Wire length
- Component specifications
- Circuit specifications
- Motor specifications
- Power specifications

Electrical drawings are typically drawn with all components de-energized. In the electrical field, there are four basic drawings used to convey

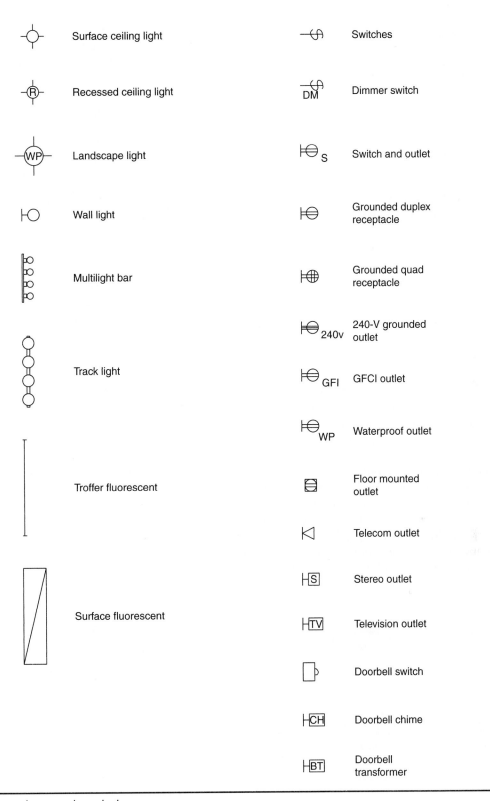

| | | | |
|---|---|---|---|
| | Surface ceiling light | | Switches |
| | Recessed ceiling light | | Dimmer switch |
| | Landscape light | | Switch and outlet |
| | Wall light | | Grounded duplex receptacle |
| | Multilight bar | | Grounded quad receptacle |
| | Track light | | 240-V grounded outlet |
| | | | GFCI outlet |
| | | | Waterproof outlet |
| | Troffer fluorescent | | Floor mounted outlet |
| | | | Telecom outlet |
| | | | Stereo outlet |
| | Surface fluorescent | | Television outlet |
| | | | Doorbell switch |
| | | | Doorbell chime |
| | | | Doorbell transformer |

**TABLE 1–2A** Architectural symbols.

information. They are the single-line drawing, the pictorial diagram, the schematic diagram, and the ladder diagram. We will look at each of these drawings in more detail.

## The Single-Line Drawing

The first drawing that we will study is the **single-line drawing,** as seen in Figure 1–1A. This type of drawing is generally used to convey an overview

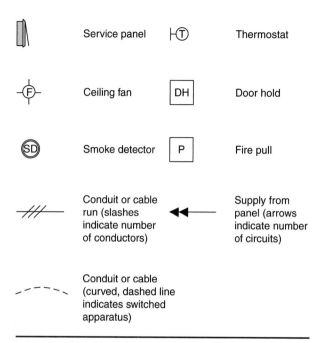

| | | | |
|---|---|---|---|
| Service panel | | Thermostat | |
| Ceiling fan | | DH Door hold | |
| Smoke detector | | P Fire pull | |
| Conduit or cable run (slashes indicate number of conductors) | | Supply from panel (arrows indicate number of circuits) | |
| Conduit or cable (curved, dashed line indicates switched apparatus) | | | |

**TABLE 1–2B** Architectural symbols.

of information but not a lot of detail. A single-line drawing does not show the actual electrical connections or the actual physical location of the devices. Single-line drawings show that some type of connection exists between components.

Probably the most common use for single-line drawings is to show the power distribution within a facility. This is helpful in the event that a particular section of the plant must be isolated for maintenance or repair. Using a single-line drawing, the technician can determine where to electrically isolate one particular section for maintenance without interrupting other portions of the facility.

The single-line drawing in Figure 1–1A shows the power distribution grid of a manufacturing facility. Notice that all the wires are shown as a single line, hence the name single-line drawing. Notice also that there is not a lot of detail shown on this drawing. The components have been arranged so that the highest voltage rating is located at either the top or the left side of the drawing. Lower voltage ratings are then located below or to the right. This means that the highest voltage rated devices will be at the top or left, and the lowest voltage rated devices will be at the bottom or right.

The simple nature of a single-line drawing allows you to easily trace the flow of power through the drawing. However, you can also use a single-line drawing in reverse. Imagine that you wish to determine the source of power for a specific component. Begin at that component and trace upwards (or to the left) until you arrive at the source of power. This helps you identify the power source and any disconnects that can be used to isolate the component from the remainder of the circuit.

## The Pictorial Diagram

The **pictorial diagram** is also called a wiring diagram. A pictorial diagram shows the relative physical location of the components, wires, and termination points. A pictorial diagram will not tell you how the circuit operates.

Pictorial diagrams are valuable because they help you locate the specific component or termination point that you may be looking for. Figures 1–1B and 1–1C show an example of a pictorial diagram. Figure 1–1B shows the front view of the cover of a

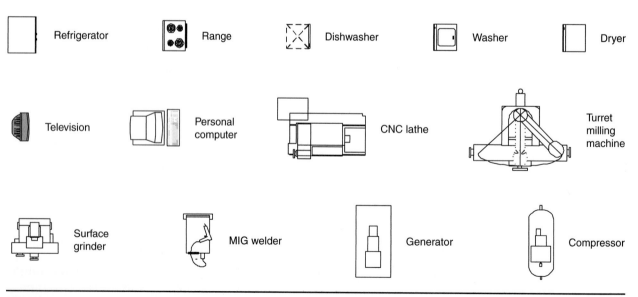

**TABLE 1–3** Pictorial symbols.

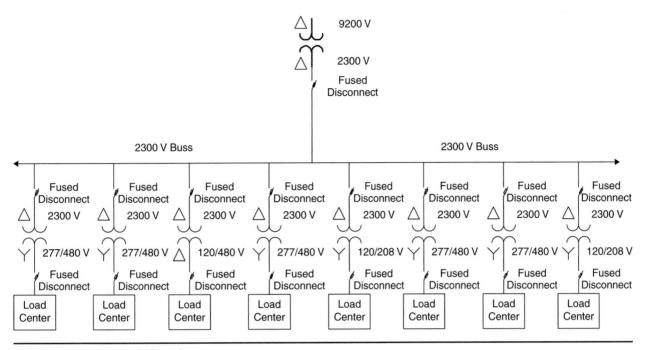

**FIGURE 1-1A** Single-line drawing.

control box. Figure 1–1C shows the same control box with the cover opened. You can now see the backside of the control box cover as well as the inside of the control box.

The circuit shown in Figures 1–1B and 1–1C is a forward–reverse control with push-button and electrical interlocks. There are also indicator lights to tell you that power is applied to the circuit, the motor is running forward, or the motor is running in reverse. You might have a difficult, if not impossible, time trying to determine the function of this circuit or the components by looking at the pictorial diagram. However, if you wanted to know the physical location of control relay CR1 within the panel, for example, the pictorial diagram would help you locate it.

## The Schematic Diagram

**Schematic diagrams** show the actual electrical connections, but not the actual physical location of components. Schematics are useful in determining how a circuit is designed to operate. Components on a schematic are generally arranged in their approximate location, relative to each other. However, the arrangement on a schematic may be very different from reality.

Figure 1–1D shows a schematic diagram of the forward–reverse control circuit seen in Figures 1–1B and 1–1C. Notice that the power source is located at the upper left side of the drawing.

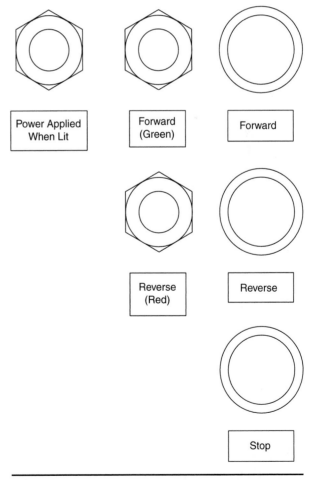

**FIGURE 1-1B** Pictorial diagram—control box—front view (cover closed).

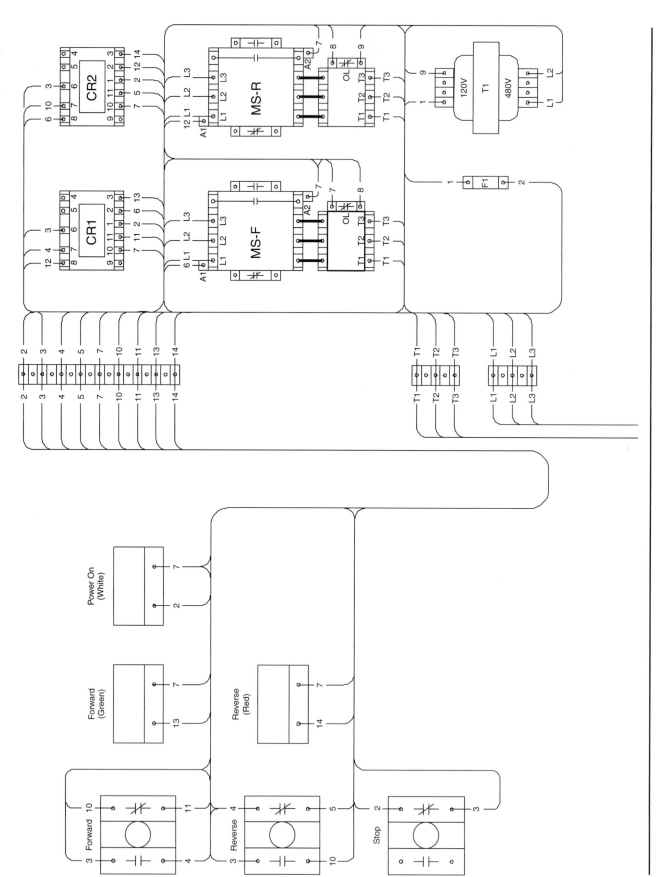

**FIGURE 1-1C** Pictorial diagram—control box—inside view (cover open).

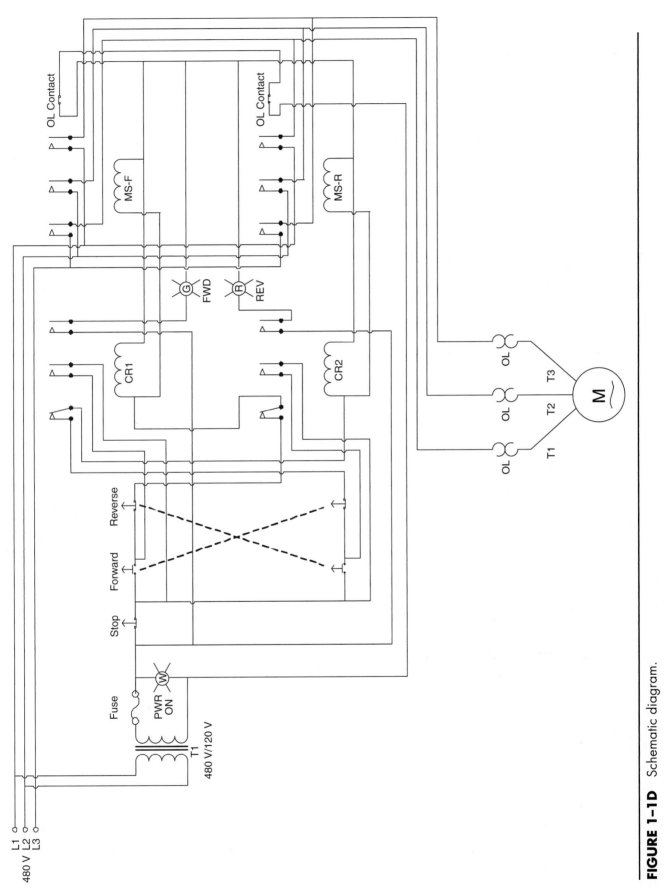

**FIGURE 1-1D**  Schematic diagram.

(Typically, the power source is located at the top or left side.) Notice also, that the motor (load) is located at the bottom of the drawing. (Typically, the load is located at the bottom or right side.) Schematics help you understand the action of the circuit. By reading a schematic, you can gain an understanding of the function of the various components within the circuit. This is useful when troubleshooting, modifying, or performing maintenance on the equipment.

## Ladder Diagram

The **ladder diagram** is a variation of the schematic diagram. Ladder diagrams show the actual electrical connections, but not the actual physical location of components. Ladder diagrams are useful in determining how a circuit is designed to operate in a more logical fashion than the traditional schematic diagram. Compare Figure 1–1D with Figure 1–1E. Which figure looks simpler and easier to follow?

Refer to Figure 1–1E. Two power rails are located on either side of the diagram. The control logic is then placed on rungs between the two power rails. Notice how the drawing resembles a ladder, hence the name, ladder diagram. Numbers along the left power rail identify rung numbers. Numbers along the right power rail, next to coils, identify the rung number of the contacts that are controlled by that particular coil. (A line under the number denotes a normally closed contact. The absence of the line indicates a normally open contact.)

Components on a ladder diagram are generally arranged in a logical order. Typically, the first component to operate (manually or automatically) is located at the top of the diagram. This component may then control other components on the same rung or on other rungs below. Sometimes, components control other components located on rungs above. This may seem confusing, but, with practice, the logic is apparent and ladder diagrams can be quite easy to use.

## USING THE DRAWINGS/ DIAGRAMS

Now imagine you are a maintenance technician. You must troubleshoot a defective machine on the plant floor. You know where the machine is located, but you must determine how to remove the power from the machine so that you can troubleshoot it. You need to understand how a portion of the machine operates. You must identify a possibly defective component and then locate the component. How do you do this? You could proceed as follows:

1. Use a single-line drawing to determine the source of power to the machine.
   A. Use the single-line drawing to identify the disconnect that can be used to remove power from the machine.
2. Use a schematic or ladder diagram to understand the control logic for the machine.
   A. Use the schematic or ladder diagram to identify the voltages and connections to check.
3. Use a pictorial diagram to locate the component and wiring.
   A. Use the pictorial diagram to help you determine where to place your meter leads for voltage/current measurements.
   B. Use the pictorial diagram to help you determine connection points that should be checked.

As you can see, you will need several types of drawings to help you to understand the details of operating, maintaining, and troubleshooting machinery and equipment at your facility.

## Architectural Drawings

Architectural drawings are used in conjunction with the plans of buildings and structures. One type of architectural drawing is the floor plan. The floor plan is a sort of bird's eye view of the building or a section of the building. It notes the location of structural elements such as walls, doors, windows, rooms, and stairs. The electrical floor plan also indicates the types and locations of the electrical equipment, as shown in Figure 1–2.

## SCIENTIFIC AND ENGINEERING NOTATION

Scientific and engineering notation uses the powers of 10 to simplify working with large and small numbers. Engineering notation uses increments of 1000 as well as word abbreviations to make large and small numbers even less confusing.

### Scientific Notation

When dealing with measurements and quantities in the field of electricity, it is not uncommon to encounter values that are quite large as well as quite

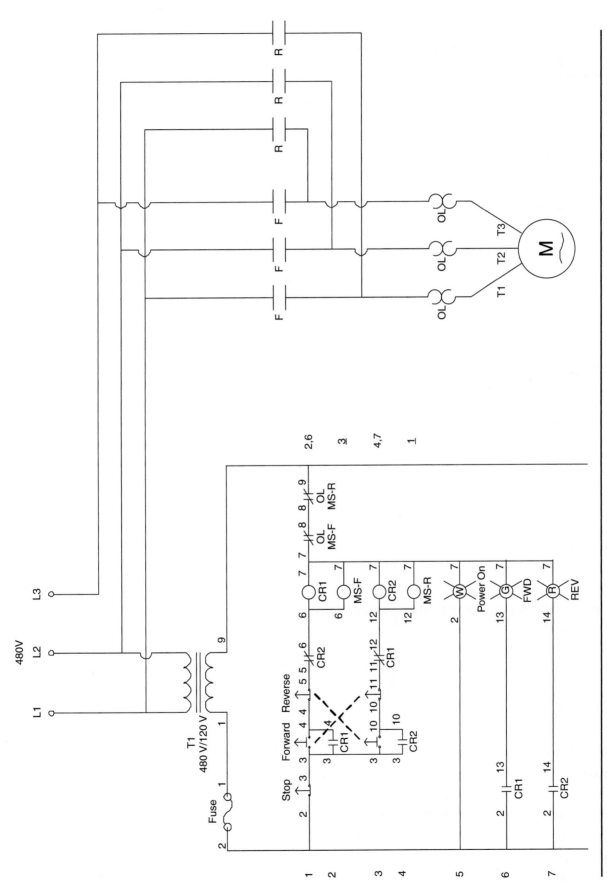

**FIGURE 1–1E** Ladder diagram.

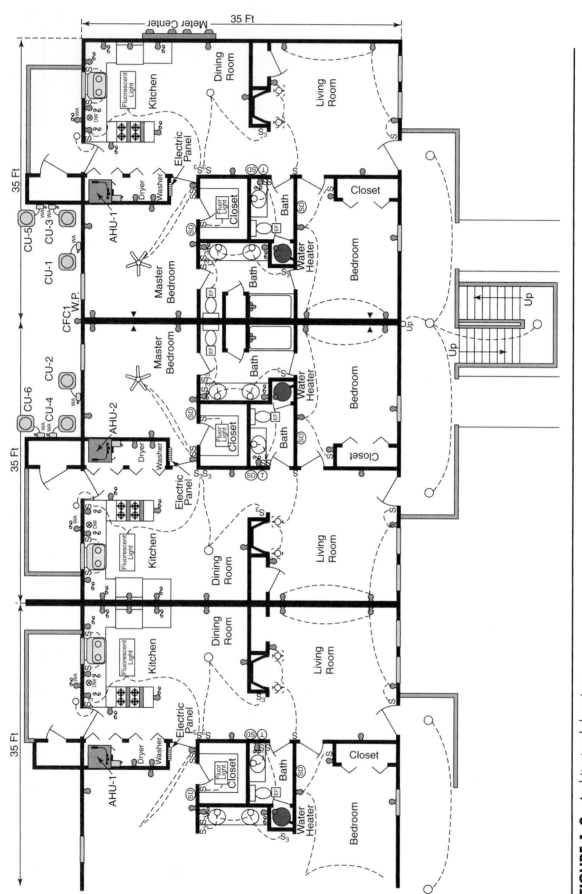

**FIGURE 1-2** Architectural drawing.

small. An example would be a current of 10,000 amperes or a voltage of 0.000006 volts. It can be seen that these numbers can be somewhat difficult to handle accurately. How many zeros appear to the right of the decimal point in the voltage example? It might be necessary to double-check to be certain that there are actually five.

Fortunately, there is an easier and more accurate method of writing and even speaking very large and very small numbers like these. The method is known as **scientific notation**.

Scientific notation is a type of shorthand that is used to make large and small numbers more easily manageable. Powers of 10 are used to abbreviate the number. For example, 10,000 amperes can be rewritten as $1.0 \times 10^4$ amperes. Since 10 raised to the fourth power is 10,000, $1.0 \times 10^4$ is $1 \times 10,000$ or 10,000.

Here is another example. The number 3,404,655 can be rewritten using scientific notation. The number becomes $3.404655 \times 10^6$. Ten raised to the sixth power is 1,000,000; therefore, $3.404655 \times 1,000,000$ is 3,404,655.

The same process is used for very small numbers. The only difference is that a negative (−) sign is placed in front of the power of 10 exponent to indicate that the zeros are to the right of the decimal point. The voltage example used earlier will illustrate this. The voltage was 0.000006 volts. Rewritten in scientific notation, the new number becomes $6.0 \times 10^{-6}$. Ten raised to the −6 power is 0.000001; therefore, $6.0 \times 0.000001$ becomes 0.000006.

Here is another example: 0.0234. Using scientific notation, 0.0234 becomes $2.34 \times 10^{-2}$. Ten raised to the −2 power is 0.01; therefore, $2.34 \times 0.01$ becomes 0.0234. Note that this example could also be written as $23.4 \times 10^{-3}$ or $234 \times 10^{-4}$. These are all correct; however, it is preferred to rewrite a number in scientific notation with one whole number to the left of the decimal point. Therefore, $2.34 \times 10^{-2}$ is the preferred method.

## Engineering Notation

A variation on scientific notation is **engineering notation**. Engineering notation works in a similar manner to scientific notation except that engineering notation moves in steps of one thousand, not ten.

The first example of 10,000 amperes will illustrate this. Using engineering notation, 10,000 amperes can be rewritten as $10.0 \times 10^3$. Ten raised to the third power is 1000; therefore, $10.0 \times 1000$ becomes 10,000. However, this is taken one step further in engineering notation. Words are substituted for the power of 10. For instance, $10^3$ becomes

kilo, which means "times 1000." So, 10,000 amperes can be rewritten as 10 kiloamperes. This is the same as saying $10 \times 10^3$ amperes. The number can be shortened even further by abbreviating the word *kilo* to a k and writing it as 10 k amperes. Since this is new, it might seem confusing, but actually this method can be more accurate, since it is not necessary to get the correct number of zeros in the number.

Remember the number 3,404,655? If it is rewritten in engineering notation, 3,404,655 becomes $3.404,655 \times 10^6$. This is the same as the scientific notation example. The reason is that engineering notation moves in steps of thousands. From one thousand, the next step is one thousand thousand or one million. This means that the power of 10 increases by threes, from $10^3$ to $10^6$ to $10^9$ and so on.

But remember that in engineering notation, words are used to replace the powers of 10; therefore, $10^3$ was called kilo or k and $10^6$ is called *mega* or M. "Mega" means "times 1,000,000." Now $3.404,655 \times 10^6$ can be rewritten as 3.404,655 megavolts or 3.404,655 M volts.

Figure 1–3 shows the standard units of engineering notation as well as the associated scientific notation values. Refer to Figure 1–3 in examining the next example.

In an earlier example, the voltage was 0.000006 volts. How would this value be expressed in engineering notation? 0.000006 becomes $6.0 \times 10^{-6}$ volts. This is the same as the scientific notation example. The reason is that engineering notation moves in steps of thousandths. From one one-thousandth, the next step is one thousand one-thousandths or one millionth. This means that the power of 10 increases by threes, from $10^{-3}$ to $10^{-6}$ to $10^{-9}$ and so on. Remember that in engineering notation, words are used to replace the powers of 10; therefore, $10^{-3}$ is called *milli* or m. "Milli" means "one thousandth." Consequently, $10^{-6}$ is called *micro* or $\mu$ (this is a Greek letter called mu). "Micro" means "one millionth." Now $6.0 \times 10^{-6}$ can be rewritten as 6.0 microvolts or 6.0 $\mu$ volts.

Here is one final example: 0.0234. Using engineering notation, 0.0234 becomes $23.4 \times 10^{-3}$ or 23.4 milli (23.4 m). Ten raised to the −3 power is 0.001. Therefore, $23.4 \times 0.001$ becomes 0.0234.

Notice that unlike the example in scientific notation, this is the only way to write 0.0234 in engineering notation. This number could be written as $23,400 \times 10^{-6}$, or 23,400 $\mu$, but notice how large the number to the left of the decimal point becomes. Likewise, 0.0234 could be written as $0.0000234 \times 10^3$, but look at all the zeros! 23.4 m is the easiest number to use and understand.

| Scientific Notation | Engineering Notation | Symbol | Multiply By |
|---|---|---|---|
| $1 \times 10^{12}$ | Tera | T | 1,000,000,000,000 |
| $1 \times 10^{9}$ | Giga | G | 1,000,000,000 |
| $1 \times 10^{6}$ | Mega | M | 1,000,000 |
| $1 \times 10^{3}$ | Kilo | K | 1000 |
| $1 \times 10^{0}$ | Base Unit | – | 1 |
| $1 \times 10^{-3}$ | milli | m | 0.001 |
| $1 \times 10^{-6}$ | micro | μ | 0.000,001 |
| $1 \times 10^{-9}$ | nano | n | 0.000,000,001 |
| $1 \times 10^{-12}$ | pico | p | 0.000,000,000,001 |

**FIGURE 1–3** Standard units of scientific and engineering notation.

# REVIEW QUESTIONS

*Multiple Choice*

1. The many different types of electrical symbols used in electrical drawings are:
   a. highly detailed and informative.
   b. vendor/manufacturer specific.
   c. used to convey meaning while conserving space.
   d. miniaturized, exact look-a-likes of the actual component.

2. How can you tell if an electrical symbol represents a motor or a generator?
   a. The symbols are different.
   b. The symbols are identical, but they have a qualifier that indicates the difference.
   c. The symbols are identical and there is no way to tell the difference.
   d. The symbols are drawn in different colors.

3. Single-line drawings show:
   a. an overview of information but not a lot of detail.
   b. very detailed information.
   c. the actual electrical connections for all devices.
   d. the physical location of all of the devices.

4. Architectural electrical symbols are used in conjunction with:
   a. architectural drawings.
   b. schematic drawings.
   c. pictorial drawings.
   d. ladder diagrams.

5. Pictorial electrical symbols are used in conjunction with:
   a. architectural drawings.
   b. panel diagrams.
   c. ladder diagrams.
   d. schematic drawings.

*Give Complete Answers*

1. List three types of diagrams.

2. A ladder diagram is a variation of what type of diagram/drawing?

3. What information is conveyed in schematic diagrams?

4. Describe a ladder diagram.

5. How is a ladder diagram useful?

6. When would you use a single-line drawing?

7. Describe an electrical floor plan.

8. What is a pictorial diagram?

9. What information is and is not found on a pictorial diagram?

10. Where is the power source located on a typical schematic diagram? Where is the load typically located?

11. Draw the architectural electrical symbols for the following:
    a. Incandescent ceiling outlet
    b. Surface-type fluorescent ceiling fixture
    c. Waterproof wall fixture
    d. Grounded duplex receptacle outlet
    e. 240-V grounded outlet

12. Draw the schematic symbols for the following:
    a. DC motor
    b. AC generator
    c. Push-button—NO
    d. Limit switch—NC
    e. Circuit breaker

13. Draw the pictorial symbols for the following:
    a. Motor
    b. Dishwasher
    c. Television
    d. MIG welder

14. Name these architectural electrical symbols.

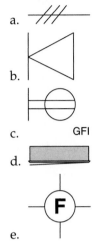

    a.

    b.

    c.          GFI

    d.

    e.

15. This is a pictorial diagram of a single-pole toggle switch controlling an incandescent lamp. Draw a schematic diagram of the same circuit.

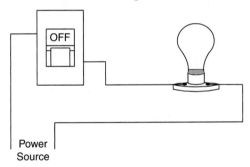

16. Express the following numbers in scientific notation format:
    a. 5280
    b. 12,000
    c. 1,600,000
    d. 0.012
    e. 0.000,324
    f. 0.000,000,007

17. Express the following as whole numbers:
    a. $9.75 \times 10^3$
    b. $1.6 \times 10^1$
    c. $2.637 \times 10^5$
    d. $3.54 \times 10^{-2}$
    e. $8.67 \times 10^{-5}$
    f. $0.043 \times 10^{-9}$

18. Express the following numbers in engineering notation format:
    a. 14,200
    b. 7600
    c. 1,700,000
    d. 0.002
    e. 0.000,100
    f. 0.000,000,078

19. Express the following as whole numbers:
    a. 3.3 kilo
    b. 470 kilo
    c. 6.8 mega
    d. 37 milli
    e. 454 micro
    f. 2 nano

# Electrical Fundamentals

## OBJECTIVES

After studying this chapter, the student will be able to:

■ Describe the structure of matter.

■ Define static electricity and explain its effect.

■ Define the unit of electricity and describe current flow.

■ List and define the three electrical quantities that are present in all energized electrical circuits.

■ Describe six methods used to produce electricity.

■ State and apply Ohm's law.

# STRUCTURE OF MATTER

It is necessary to study the structure of matter in order to develop a thorough understanding of electricity. **Matter** is anything that occupies space and has mass. Some examples of matter are wood, air, metal, and water. The smallest particle of matter that retains the same chemical properties is the **molecule**. A molecule can be divided into smaller parts called **atoms**. Dividing a molecule into atoms creates a chemical change. For example, a molecule of water undergoes a chemical change to become two parts of hydrogen (two hydrogen atoms) and one part oxygen (one oxygen atom).

Currently we know of more than 100 kinds of atoms, or elements, as seen in Table 2–1. *Elements* are frequently referred to as the "building blocks" of nature. Singly or in combination, they are the materials that constitute all matter. Each element is composed solely of one type of atom. Some examples of elements are iron, copper, aluminum, lead, carbon, and hydrogen.

The structure of an atom can be compared to the sun's planetary system. The nucleus, like the sun, is at the center, with tiny particles called **electrons** orbiting around it (Figure 2–1). Electrons have a negative electrical charge. The nucleus of the atom consists of **protons**, which have a positive electrical charge, and **neutrons**, which have no electrical charge.

The hydrogen atom (Figure 2–2) is the simplest of all atoms. The hydrogen atom contains 1 proton in the nucleus, with 1 electron revolving around the nucleus. An atom of aluminum has 13 protons and 14 neutrons in the nucleus, with 13 electrons orbiting around the nucleus (Figure 2–3).

Atoms in their natural state contain an equal number of electrons and protons. In both the aluminum atom and the copper atom (Figure 2–4), as in all atoms, the electrons orbit in arranged rings around the nucleus. These rings are called *shells*. The maximum number of electrons in any shell is determined by the "number" of the shell. The first shell is the one nearest to the nucleus, and the numbers of the shells increase consecutively as their distance from the nucleus increases. The maximum number of electrons

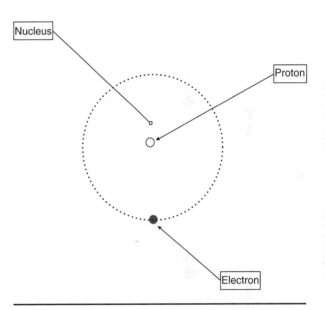

**FIGURE 2–2**  Atom of hydrogen.

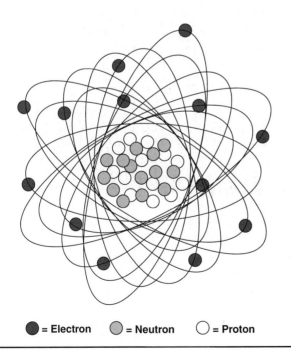

● = Electron  ● = Neutron  ○ = Proton

**FIGURE 2–1**  Atom containing protons and neutrons in the nucleus, with electrons orbiting around the nucleus.

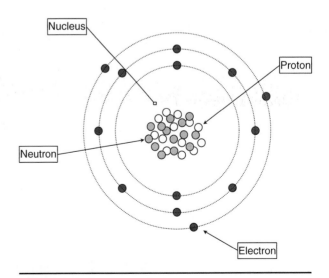

**FIGURE 2–3**  Atom of aluminum.

| 1 | 2 | 3 | 4 | 5 | 6 | 7 | 8 | 9 | 10 | 11 | 12 | 13 | 14 | 15 | 16 | 17 | 18 |
|---|---|---|---|---|---|---|---|---|---|---|---|---|---|---|---|---|---|
| hydrogen 1 **H** | | | | | | | | | | | | | | | | | helium 2 **He** |
| lithium 3 **Li** | beryllium 4 **Be** | | | | | | | | | | | boron 5 **B** | carbon 6 **C** | nitrogen 7 **N** | oxygen 8 **O** | fluorine 9 **F** | neon 10 **Ne** |
| sodium 11 **Na** | magnesium 12 **Mg** | | | | | | | | | | | aluminum 13 **Al** | silicon 14 **Si** | phosphorus 15 **P** | sulfur 16 **S** | chlorine 17 **Cl** | argon 18 **Ar** |
| potassium 19 **K** | calcium 20 **Ca** | scandium 21 **Sc** | titanium 22 **Ti** | vanadium 23 **V** | chromium 24 **Cr** | manganese 25 **Mn** | iron 26 **Fe** | cobalt 27 **Co** | nickel 28 **Ni** | copper 29 **Cu** | zinc 30 **Zn** | gallium 31 **Ga** | germanium 32 **Ge** | arsenic 33 **As** | selenium 34 **Se** | bromine 35 **Br** | krypton 36 **Kr** |
| rubidium 37 **Rb** | strontium 38 **Sr** | yttrium 39 **Y** | zirconium 40 **Zr** | niobium 41 **Nb** | molybdenum 42 **Mo** | technetium 43 **Tc** | ruthenium 44 **Ru** | rhodium 45 **Rh** | palladium 46 **Pd** | silver 47 **Ag** | cadmium 48 **Cd** | indium 49 **In** | tin 50 **Sn** | antimony 51 **Sb** | tellurium 52 **Te** | iodine 53 **I** | xenon 54 **Xe** |
| caesium 55 **Cs** | barium 56 **Ba** | lanthanium 57 **La\*** | hafnium 72 **Hf** | tantalum 73 **Ta** | tungsten 74 **W** | rhenium 75 **Re** | osmium 76 **Os** | iridium 77 **Ir** | platinum 78 **Pt** | gold 79 **Au** | mercury 80 **Hg** | thallium 81 **Tl** | lead 82 **Pb** | bismuth 83 **Bi** | polonium 84 **Po** | astatine 85 **At** | radon 86 **Rn** |
| francium 87 **Fr** | radium 88 **Ra** | actinium 89 **Ac#** | rutherfordium 104 **Rf** | dubnium 105 **Db** | seaborgium 106 **Sg** | bohrium 107 **Bh** | hassium 108 **Hs** | meitnerium 109 **Mt** | darmstadtium 110 **Ds** | 111 | 112 | | 114 | | 116 | | 118 |

**TABLE 2–1**  Periodic table of the elements.

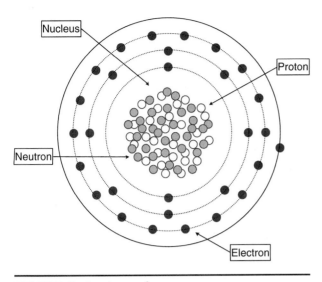

**FIGURE 2–4** Atom of copper.

contained in any shell of any atom is as follows: the first shell, 2 electrons; the second shell, 8; the third shell, 18; the fourth shell, 32; the fifth and sixth shells, 18 each; and the seventh shell, 2 electrons.

Atoms that have from 1 to 4 electrons in their last, or outer, shell are generally good conductors of electricity. For example, aluminum and copper are good conductors. (The aluminum atom has 3 electrons in its last shell; the copper atom has 1 electron in its last shell.) Atoms with 5, 6, or 7 electrons in their outer shell are classified as nonmetals and are poor conductors. Atoms with 8 electrons in their outer shell are insulators.

## STATIC ELECTRICITY

*Static electricity* is an electric charge accumulated on an object. One method of building up an electrical charge is by friction. Rubbing a hard rubber rod with fur causes both the rubber and the fur to become electrified. When a glass rod is rubbed with silk, both the silk and the glass become electrified. Further experiments demonstrate that when two charged rubber rods are brought near each other, a repelling force exists. Two charged glass rods also repel each other. However, when a charged glass rod is brought near a charged rubber rod, the rods attract each other. This attraction and repulsion indicate that there must be a difference between the charge on the rubber and the charge on the glass.

An object in its natural state contains an equal number of electrons and protons. Therefore, it is *uncharged*, or *neutral*. In order to be in a state of *charge*, the object must contain more electrons than protons or more protons then electrons. Removing

electrons from an object causes it to take on a positive charge. Adding electrons to an object causes it to become negatively charged. Note that electrons are the particles that are added to or removed from the object. The protons are held firm in the nucleus, but the electrons, being in the outer portion of the atom, are more easily removed from one object and deposited on another.

When a rubber rod is rubbed with fur, electrons are transferred from the fur to the rod. This transfer causes the rod to become negatively charged and the fur to become positively charged. Rubbing glass with silk removes electrons from the glass and deposits them on the silk. Two charged glass rods repel each other because they contain like charges. If a charged rubber rod is placed near a charged glass rod, the rods are attracted to each other because they contain opposite charges. Objects charged electrically by friction retain their charge for an indefinite period of time. This storage of an electrical charge is called "electricity at rest," or static electricity.

Two rules for electric charges are:

1. Like charges repel each other, and opposite charges attract each other.
2. The strength of the attraction or repulsion is directly proportional to the strength of the electric charge and inversely proportional to the square of the distance between the charged objects.

Two unlike charged objects are attracted to each other because of the nature of electrical balance. If the two objects touch, the electrons in the negatively charged object will move into the positively charged object until there is a balance between them. This is nature's way of restoring objects to their natural state. If the difference in potential (the surplus of electrons as opposed to the deficit of electrons) is great enough, it is not necessary for the objects to touch. They need only come close to each other and the electrons will move through the air to the object containing the positive charge. This discharge through the air is called *static discharge*. Lightning is a common form of static discharge.

## ELECTRIC CURRENT

An **electric current** is electricity in motion. It is the movement of electrons from a negatively charged object to a positively charged object. The direction of current flow is the same as the direction of electron flow.

In nonmetallic materials such as rubber and glass, the electrons are held firmly to their parent atom, and it is difficult to cause them to move

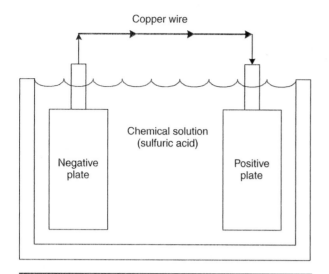

Copper wire

Chemical solution
(sulfuric acid)

Negative
plate

Positive
plate

**FIGURE 2–5** Battery with terminals connected by a length of copper wire. The arrows indicate the direction of electron flow.

through the material. In metals, however, some electrons are free to flow. These are called *free electrons*. Materials that contain free electrons are called **conductors**. Materials that do not contain free electrons are called **insulators**. Rubber and glass are good insulators; copper and aluminum are good conductors. If a copper wire is connected between two unlike charged bodies, electrons will flow freely through the wire, attempting to restore balance.

A **battery** is an apparatus that uses chemicals to cause an electron unbalance between its terminals. If a copper wire is connected to the terminals of a battery, the free electrons are attracted toward the positive terminal and repelled from the negative terminal (Figure 2–5). The difference in potential between the battery terminals causes this electron flow. As electrons arrive at the positive side of the battery, the chemical action forces them through the battery to the other terminal, thus maintaining an electrical unbalance. This movement of electrons continues as long as the battery chemicals are active. The flow of electrons through a circuit constitutes an electric current. It can be said, then, that current flows from the negative terminal of the battery, through the circuit, and back to the positive terminal.

## CURRENT MEASUREMENT

It is necessary to measure the rate of flow of electrons in order to determine circuit characteristics. The current (flow of electrons) through a 100-watt (W) lamp is approximately $6 \times 10^{18}$ electrons per

second. Because working with such large numbers is cumbersome, a unit much larger than the electron has been established. This unit is the **coulomb** (C). One coulomb is equal to $628 \times 10^{16}$ electrons. A motor that permits 5 coulombs per second to flow through its windings is rated at 5 amperes. The term **ampere** (A) means coulombs per second (C/s). A 10-ampere heater permits 10 coulombs per second to flow through it. The relationship between the coulomb and the ampere can be expressed mathematically as follows:

$$Q = It \qquad \text{(Eq. 2.1)}$$

where $Q$ = quantity of coulombs (C)
$I$ = current, in amperes (A)
$t$ = time, in second(s)

### Example 1

A battery forces a current of 10 A through a circuit for 1 hour (h). (a) How many coulombs will flow through the circuit? (b) How many electrons per second will flow through the circuit?

a. $Q = It$
   $Q = 10\,A \times 3600\,s$
   $Q = 36,000\,C$

b. $10 \times 628 \times 10^{16} = 628 \times 10^{17}$ electrons/s

To measure current flow through a circuit, a *current meter* must be inserted into the circuit. It must be connected so that all the current that flows in the circuit will flow through the meter. A current meter is called an **ammeter** because it measures the rate of current, or coulombs per second, flow. Figure 2–6 shows

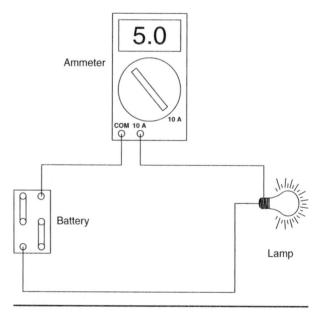

5.0

Ammeter

COM  10 A

10 A

Battery

Lamp

**FIGURE 2–6** Ammeter connected to measure the current flowing through the filament of a lamp.

an ammeter connected to measure the current flowing through the lamp. Because the ammeter must be inserted into the circuit, it must have a very low resistance so it will not hinder the flow of current and thereby alter the characteristics of the circuit.

# VOLTAGE

The flow of electrons through a circuit can be compared to the flow of water through a pipe. A good comparison is a closed-loop system similar to that used on a forced hot water heating system. As shown in Figure 2–7A, the pump blades cause the water to enter the pump at point D at a low pressure and leave at point A at a high pressure. The section of pipe between A and B causes very little friction; therefore, the pressure at B is only slightly less than at A. Section BC causes considerable friction, so the pressure at point C is much less than at point B. The section of pipe from C to D causes the same amount of friction as from A to B; therefore, the pressure drop is the same as from A to B.

Pressure is measured in *pounds per square inch* (psi). If the drop in pressure from A to B is 5 pounds per square inch (0.35 kilogram per square centimeter), from B to C is 100 pounds per square inch (7.0 kilograms per square centimeter), and from C to D is 5 pounds per square inch (0.35 kilogram per square centimeter), then the total drop is 5 + 100 + 5 = 110 pounds per square inch (7.7 kilograms

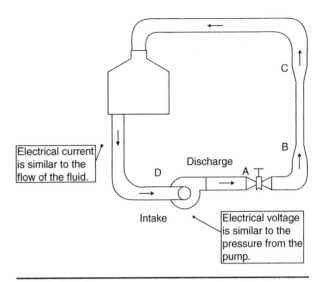

FIGURE 2–7A   Electrical current and voltage is analogous to a closed-loop hydraulic system with current equivalent to flow and voltage equivalent to pressure.

per square centimeter). This means that the pump must develop a pressure of 110 pounds per square inch (7.7 kilograms per square centimeter) between the intake at D and the discharge at A.

A similar condition exists in an electrical circuit, but the pressure drop is expressed in units called *volts* (V). Volt is used in an electric circuit in the same manner as the pounds per square inch unit is used in a hydraulic circuit. Figure 2–7B illustrates

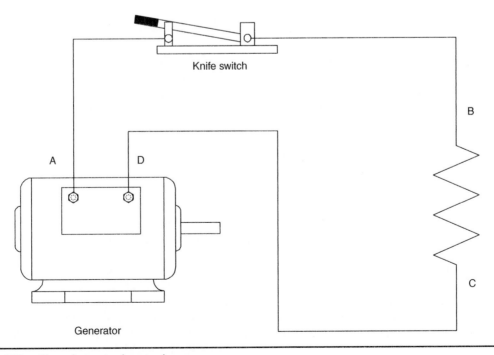

FIGURE 2–7B   Closed-circuit electrical system.

an electric circuit comparable to the hydraulic circuit in Figure 2–7A. The section of the wire from A to B has low resistance (electrical friction) and the pressure drop is low. Section BC has high resistance, causing a large drop in pressure. Section CD has the same resistance as AB and, therefore, has the same pressure drop. If the drop in pressure from A to B is 5 volts, B to C is 100 volts, and C to D is 5 volts, the total drop is 5 + 100 + 5 = 110 volts. This means that the generator, an electric pump, must develop a pressure of 110 volts between its terminals.

In an electric circuit, the term *electromotive force* (emf) is used to indicate the voltage generated by the source, such as a battery or generator. The term *voltage drop* is used to denote the drop in pressure in a particular part of a circuit. In circuits like that in Figure 2–7B, the sum of the voltage drops is equal to the emf.

## VOLTAGE SOURCES

The methods used to produce voltage include the following:

- Mechanical magnetic (induction)
- Chemical
- Thermoelectric
- Photovoltaic
- Piezoelectric
- Friction

Generators are used to produce a voltage by induction. There are two types of generators: the direct current (DC) generator and the alternating current (AC) generator. The *direct current generator* produces a voltage that acts in one direction only. In other words, it produces an emf that forces the current to flow in one direction only. The *alternating current generator* (sometimes called an **alternator**) produces a voltage that alternates back and forth. That is, it produces a voltage first in one direction and then in the opposite direction. The current flowing as a result of this alternating emf flows first in one direction and then in the opposite direction. This type of current is called **alternating current** (AC).

Generators produce a voltage by rotating coils of wire through a **magnetic field**, an invisible force produced by a magnet. As the coils move through the magnetic field, a voltage develops. It is not necessary to move the coils if the magnetic field is moved across the coils. Either method produces a voltage.

Various chemicals mixed in proper proportions produce a voltage in batteries. The chemical action removes electrons from one terminal and forces them to the other terminal, causing a difference in potential between the terminals.

A thermocouple is a device made by welding together two dissimilar metals at one end. The welded junction is heated to produce a voltage across the cold ends. This phenomenon is called the *thermoelectric effect*.

The *photovoltaic effect* is the production of a voltage in a material, such as copper oxide, in contact with copper. Light striking the copper-oxide surface causes electrons to flow from the copper to the copper oxide, thus developing a potential difference between the two metals. The amount of emf produced is proportional to the intensity of the light.

A voltage can be produced by applying pressure to certain crystals, such as Rochelle salts. When pressure is applied, a small emf is produced. The amount of voltage varies with the pressure. This phenomenon is known as the *piezoelectric effect*.

Electrostatic generators are used to develop a voltage by friction. The *electrostatic generator* consists of a latex belt that revolves over a metal pulley. The friction of the belt moving over the pulley builds up an electrostatic charge. A hollow metal ball surrounding part of the belt attracts the electrons. An emf of many thousands of volts can be produced with this machine.

## VOLTAGE MEASUREMENTS

The amount of water flowing in a pipe depends upon the difference in pressure between the two ends of the pipe. Figure 2–8A shows one method used to measure this difference. A pressure gauge is tapped (connected) into each end of the pipe at points X and Y. The difference between the two readings gives the amount of pressure necessary to force the water through the pipe.

The amount of current flowing through a conductor depends upon the difference in pressure between the two ends of the conductor. This difference in pressure is called the **potential difference** or **voltage**. To measure the potential difference between the two ends X and Y (Figure 2–8B), the voltmeter leads must be tapped to the ends of the wire. The voltmeter is an instrument that measures the difference in electrical pressure between two points. It will, therefore, indicate the voltage across the wire XY.

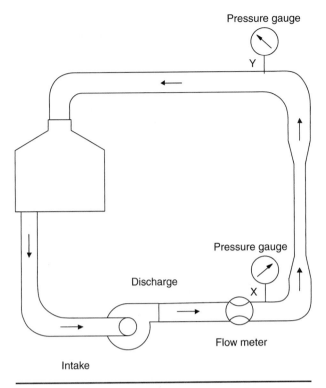

**FIGURE 2–8A** Pressure gauges tapped into a pipe indicate water pressure at points X and Y.

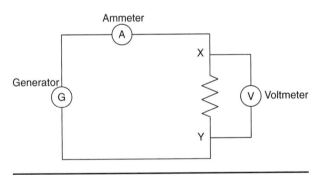

**FIGURE 2–8B** A voltmeter connected across points X and Y indicates the voltage between the two points.

# RESISTANCE

Some materials have many free electrons. These materials require only a comparatively low voltage to force the electrons to flow from atom to atom, which establishes a high current flow. These materials are good conductors. Other materials have only a few free electrons, thus permitting only a low current flow when the same voltage is applied. These materials are poor conductors.

All materials have some opposition to current flow. This opposition is caused by the type of material and by the friction produced when the electrons move through the material. Opposition to the flow of electrons is called **resistance**. The unit of measurement for resistance is the **ohm** ($\Omega$). The instrument for measuring resistance is called the **ohmmeter** (Figure 2–9).

> ⚠ **CAUTION** Never connect an ohmmeter to a circuit until the circuit has been disconnected from the power source. The ohmmeter has its own source of power, and connecting it across another power source can cause serious damage.

# OHM'S LAW

The amount of current flowing through a circuit is directly proportional to the **electromotive force** (voltage) and inversely proportional to the resistance of the circuit. In other words, if the voltage applied to a circuit is increased and the resistance remains the same, the current will increase in proportion to the increase in voltage. If the voltage remains the same but the resistance is increased, the current will decrease in proportion to the increase

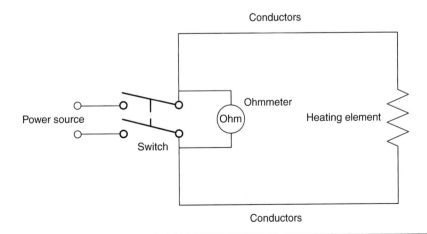

**FIGURE 2–9** Ohmmeter connected to measure circuit resistance.

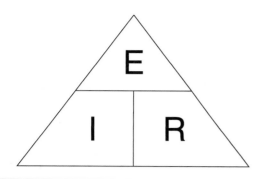

**FIGURE 2–10**   The Ohm's law triangle.

in resistance. This relationship between current, voltage, and resistance is called **Ohm's law**. The mathematical formula for Ohm's law is

$$I = \frac{E}{R} \left( \text{or } E = IR, \text{ or } R = \frac{E}{I} \right)$$   (Eq. 2.2)

where $I$ = current, in amperes (A)
$E$ = electromotive force, in volts (V)
$R$ = resistance, in ohms ($\Omega$)

If you have difficulty manipulating these three Ohm's law formulas, a simple device can help. This device is called the *Ohm's law triangle*. Refer to Figure 2–10. Notice the triangle and how it is divided. Notice also that the letter $E$ is located in the top compartment, the letter $I$ is located in the lower left compartment, and the letter $R$ is located in the lower right compartment. To use the Ohm's law triangle, you simply cover up the compartment of the variable that you are trying to find. The remaining compartments show you the formula that you should use. For example, if you needed to find the current, you would cover the $I$. This leaves the $E$ and the $R$ uncovered. However, notice the arrangement of the $E$ and the $R$. They appear as the formula $\frac{E}{R}$. Therefore, the triangle shows you the formula for finding current, $I = \frac{E}{R}$.

**Example 2**

How much current flows through a 30-$\Omega$ heating element when it is connected to a 120-V source?

$$I = \frac{E}{R}$$
$$I = \frac{120}{30}$$
$$I = 4A$$

**Example 3**

What is the resistance of an electric iron when it is connected to a 120-V source, and a current of 3 A flows?

$$R = \frac{E}{I}$$
$$R = \frac{120}{3}$$
$$R = 40\Omega$$

**Example 4**

What is the emf necessary to force 6 A through a circuit having a resistance of 20 $\Omega$?

$$E = IR$$
$$E = 6 \times 20$$
$$E = 120 \text{ V}$$

Ohm's law may be applied to an entire circuit or any part of the circuit. This may be stated as follows:

■ The total current flowing in a circuit is equal to the total voltage applied to the circuit divided by the total resistance of the circuit.

■ The current flowing in any part of a circuit is equal to the voltage across that part of the circuit divided by the resistance of that part of the circuit.

## REVIEW QUESTIONS

*Multiple Choice*

1. The nucleus of an atom is surrounded by
   a. neutrons.
   b. electrons.
   c. protons.
   d. positrons.

2. A positively charged body has a
   a. deficiency of electrons.
   b. surplus of electrons.
   c. surplus of protons.
   d. deficiency of protons.

3. Electrical pressure is measured with
   a. a voltmeter.
   b. an ammeter.
   c. an ohmmeter.
   d. a wattmeter.

4. Electric current is the movement of
   a. positrons.
   b. protons.
   c. neutrons.
   d. electrons.

5. Good conductors contain many free
   a. protons.
   b. neutrons.
   c. electrons.
   d. positrons.

6. Without changing it chemically, the smallest particle of matter is
   a. an element.
   b. a molecule.
   c. a proton.
   d. an atom.

7. Static electricity is generally produced by
   a. friction.
   b. induction.
   c. heat.
   d. light.

8. One coulomb is equal to
   a. 628 electrons.
   b. 628 × 108 electrons.
   c. 628 × 1016 electrons.
   d. 628 × 1018 electrons.

9. Voltage is the
   a. movement of electrons through a circuit.
   b. opposition to current flow.
   c. electrical pressure that causes current flow.
   d. electrical resistance.

10. Resistance is
    a. the movement of protons through a circuit.
    b. the opposition to current flow.
    c. an immovable force.
    d. the movement of electrons through a circuit.

*Give Complete Answers*

1. What elements are present in a molecule of water?

2. Describe the structure of an atom.

3. What is static electricity?

4. Write two rules related to static electricity.

5. What is electric current?

6. Define resistance and identify the cause of resistance in electric circuits.

7. Describe six methods used to produce electricity.

8. What instrument is used to measure electrical current?

9. How is a voltmeter connected in a circuit? (Use a diagram if necessary.)

10. Write Ohm's law.

*Solve each problem, showing the method used to arrive at the solution.*

1. It takes 360,000 C to charge a certain storage battery. If the charging rate is 15 A, how long will it take to fully charge the battery?

2. A voltmeter, which has a resistance of 10,000 $\Omega$, indicates 120 V. How much current is flowing through it?

3. A current of 2.5 A is flowing through the heating element of a soldering iron. If the iron is connected to a 120-V line, what is the resistance of the element?

4. What is the voltage drop across a resistor if its resistance is 100 $\Omega$ and 0.2 A flows through it?

5. How much current flows through a 144-$\Omega$ lamp connected to a 120-V line?

6. A circuit has a resistance of 480 ohms and is connected across 120 volts. How much current flows through the circuit?

7. If the circuit in Problem 6 is connected across 240 volts, how much current will flow?

8. If the resistance of the circuit in Problem 6 is reduced to 240 ohms, how much current will flow?

9. If the circuit in Problem 6 requires 1 ampere, what voltage must be applied to the circuit?

10. What is the resistance of a circuit if an ammeter in the line indicates 4 amperes and the circuit is connected across 480 volts?

# Electrical Power and Energy

## OBJECTIVES

After studying this chapter, the student will be able to:

- Define the terms relating to electrical power and energy.
- Calculate the power necessary to perform various electrical jobs.
- Calculate the amount of electrical energy required to perform various electrical jobs.
- Determine the horsepower necessary to drive various machines.
- Calculate the efficiency of electrical equipment.
- Calculate the cost of operating electrical equipment.

# WORK

**Work** may be defined as the overcoming of resistance through a distance. The unit of measurement of work is the *foot-pound* (kilogram-meter), abbreviated ft lb, which is the work performed when a force of 1 pound (0.4536 kilogram) acts through a distance of 1 foot (0.3048 meter). The unit of measurement of work in the metric system is the *joule*. One joule (J) is equal to 0.7376 foot-pound (0.102 kilogram-meter).

# POWER

Machines were invented to assist in doing work. The amount of work a machine can do in a specific time determines its power. **Power** is a measurement of the rate of doing work. The amount of work can be calculated by the formula Work = Force × Distance (W = FD), where the force is measured in pounds (or kilograms) and the distance is measured in feet (or meters).

To determine the power a machine must deliver, it is necessary to know the speed at which the work must be accomplished. The formula for power is Power = Work/Time (P = W/T), where work is measured in foot-pounds and time is measured in minutes (min).

## Horsepower

**Horsepower** (hp) is the common unit of measurement of mechanical power in the English system. In early times, water was pumped from mines by horses pulling and turning a wheel on the end of a shaft to drive a pump. Later, horses were replaced by the steam engine. Steam engines are rated as having the power of a certain number of horses (horsepower). It was determined that an average horse could do work at the rate of 33,000 foot-pounds (44,740 joules) per minute. This means that the average horse can move 1000 pounds (453.6 kilograms) through 33 feet (10.058 meters) in 1 minute. The mathematical formula for horsepower is

$$hp = \frac{ft\ 1b/min}{33,000} \qquad \text{(Eq. 3.1)}$$

### Example 1
A machine can place 50,000 lb (22,680 kg) of scrap metal onto a truck 10 ft (3.048 m) high in 5 min. What horsepower is the machine capable of delivering?

$$(50,000 \times 10) \div 5 = 100,000\ ft\ 1b/min$$

$$hp = \frac{ft\ 1b/min}{33,000}$$

$$hp = \frac{100,000}{33,000}$$

$$hp = 3$$

## Electric Power

Electric power is measured in **watts** (W) or kilowatts (kW). One thousand watts equal one kilowatt. It is relatively simple to convert from mechanical power to electrical power: 1 horsepower is equal to 746 watts. (One watt per second equals one joule.)

One watt of power is developed when one volt forces one ampere through a circuit. This can be expressed mathematically by the formula

$$P = IE \left( I = \frac{P}{E} \text{ or } E = \frac{P}{I} \right) \qquad \text{(Eq. 3.2)}$$

where  $P$ = power, in watts (W)
$I$ = current, in amperes (A)
$E$ = electrical pressure, in volts (V)

As was the situation with Ohm's law, if you have difficulty manipulating these three power law formulas, a simple device can help. This device is called the *power law triangle* and is illustrated in Figure 3–1. Notice the triangle and how it is divided with the letter $P$ located in the top compartment, the letter $I$ in the lower left compartment, and the letter $E$ in the lower right compartment. To use the power law triangle, you simply cover up the compartment of the variable that you are trying to find. The remaining compartments show you the formula that you should use. For example, if you needed to find the current, you would cover the $I$. This leaves the $E$ and the $P$ uncovered. Notice the arrangement of the $E$ and the $P$: they appear as the formula $\frac{P}{E}$. Therefore, the triangle shows you the formula for finding current, $I = \frac{P}{E}$.

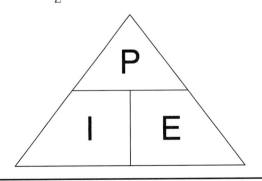

**FIGURE 3–1**  Power law triangle.

Suppose you need to find the voltage when the current and the power are known. Simply cover the $E$ (since you are trying to find the voltage), and you have the formula $\frac{P}{I}$. Therefore, the triangle shows you the formula for finding voltage, $E = \frac{P}{I}$. Finally, if you need to determine the power when you know the voltage and current, cover the $P$. The remaining uncovered variables give you the formula $I \times E$; therefore, the triangle shows you the formula for finding power, $P = I \times E$.

You now know six formulas for determining an unknown value when two values are known. However, suppose you had a circuit in which you knew the voltage and the power but needed to find the resistance. You do not have a formula that allows you to find $R$ when you know $E$ and $P$. Likewise, suppose you wanted to find the circuit current but only knew the circuit power and resistance. Again, you do not have a formula for finding $I$ when you know $P$ and $R$. Well, guess what? There are six more formulas that will enable you to solve for any unknown when two items are known. Let us look at the formulas first, and then we will see how they were developed.

$$P = \frac{E^2}{R} \qquad \text{(Eq. 3.3)}$$

$$P = I^2 \times R \qquad \text{(Eq. 3.4)}$$

$$R = \frac{E^2}{P} \qquad \text{(Eq. 3.5)}$$

$$R = \frac{P}{I^2} \qquad \text{(Eq. 3.6)}$$

$$R = \sqrt{P \times R} \qquad \text{(Eq. 3.7)}$$

$$I = \sqrt{\frac{P}{R}} \qquad \text{(Eq. 3.8)}$$

Let us begin by looking at equation 3.3, $P = \frac{E^2}{R}$. This formula is a combination of two formulas: (A) $P = I \times E$ and (B) $I = \frac{E}{R}$. Here is how these two formulas were combined. Begin with formula A:

$$P = I \times E$$

From formula B, we know that:

$$I = \frac{E}{R}$$

When you substitute formula B for the variable $I$ in formula A, the formula becomes:

$$P = \frac{E \times E}{R}$$

Combining like terms gives:

$$P = \frac{E^2}{R}$$

This results in equation 3.3.

Equation 3.4 is also a combination of two formulas. It is a combination of (A) $P = I \times E$ and (B) $E = I \times R$. Here is how these two formulas were combined. Begin with formula A:

$$P = I \times E$$

From formula B, we know that:

$$E = I \times R$$

When you substitute formula B for the variable E in formula A, the formula becomes:

$$P = I \times I \times R$$

Combining like terms gives:

$$P = I^2 \times R$$

This results in equation 3.4.

Equation 3.5, $R = \frac{E^2}{P}$, is found by rearranging equation 3.3:

$$P = \frac{E^2}{R}$$

Multiply both sides by $R$:

$$R \times P = \frac{E^2}{R} \times R$$

Cancel like terms:

$$R \times P = E^2$$

Then, divide both sides of the equation by $P$:

$$\frac{R \times P}{P} = \frac{E^2}{P}$$

Cancel like terms:

$$R = \frac{E^2}{P}$$

This is equation 3.5.

Equation 3.6 is derived from equation 3.4:

$$P = I^2 \times R$$

Divide both sides of the equation by $I^2$:

$$\frac{P}{I^2} = \frac{I^2 \times R}{I^2}$$

Cancel the like terms:

$$\frac{P}{I^2} = R \text{ or } R = \frac{P}{I^2}$$

This results in equation 3.6.

Equation 3.7, $E = \sqrt{P \times R}$, is also found by rearranging equation 3.3:

$$P = \frac{E^2}{R}$$

Multiply both sides by $R$:

$$R \times P = \frac{E^2}{R} \times R$$

Cancel the like terms:

$$R \times P = E^2$$

Take the square root of both sides of the equation:

$$\sqrt{R \times P} = \sqrt{E^2}$$

Simplify both sides of the equation:

$$\sqrt{R \times P} = E \text{ or } E = \sqrt{R \times P} \text{ or } E = \sqrt{P \times R}$$

This is equation 3.7.

Finally, we look to equation 3.8, which is also derived from equation 3.4:

$$P = I^2 \times R$$

Divide both sides of the equation by $R$:

$$\frac{P}{R} = \frac{I^2 \times R}{R}$$

Cancel the like terms:

$$\frac{P}{R} = I^2$$

Take the square root of both sides of the equation:

$$\sqrt{\frac{P}{R}} = \sqrt{I^2}$$

Simplify both sides of the equation:

$$\sqrt{\frac{P}{R}} = I \text{ or } I = \sqrt{\frac{P}{R}}$$

This results in equation 3.8.

### Example 2

How much electrical power is required to operate a water heater that is rated at 240 V, if its resistance (R) is 24 Ω?

### Solution

$$P = \frac{E^2}{R}$$
$$P = \frac{57,600}{24}$$
$$P = 2400 \text{ W, or } 2.4 \text{ kW}$$

### Example 3

How much power is dissipated by an electric oven if 30 A flows through a resistance of 10 Ω?

### Solution

$$P = I^2 \times R$$
$$P = 900 \times 10$$
$$P = 9000 \text{ W or } 9 \text{ kW}$$

By now, you are probably wondering how you will ever remember all of this. There is a device called a *PIRE wheel* that may help you remember all twelve formulas. The word PIRE is from the letters used to represent (P)ower, (I)ntensity, (R)esistance, and (E)lectromotive force. Figure 3–2 shows a PIRE wheel with all twelve formulas. Notice also that the formulas are grouped by the variable you are trying to find. The wheel is divided into four quadrants, and each quadrant contains the three formulas that can be used to find the same unknown variable. For instance, in the upper left quadrant are the three

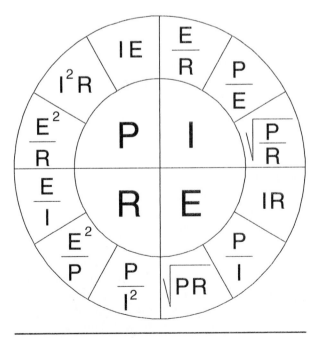

**FIGURE 3–2** PIRE wheel.

formulas that can be used to find power. You simply choose the formula that contains the known variables available to you.

# ENERGY

**Energy** is the ability to do work. Energy cannot be destroyed or consumed. However, it can be converted from one form to another. Some forms of energy are heat, light, mechanical, electrical, and chemical energy.

Many methods for converting energy from one form to another have been developed through scientific research. Energy conversion can be used to do work. For example, the electric motor converts electrical energy into mechanical energy. In the process, the motor shaft rotates to drive a machine.

The term *energy loss*, which is frequently used in the study of electricity, can be misleading. As just stated, energy cannot be destroyed or consumed. Energy loss means that during the process of energy conversion, some energy is converted into a form *other* than that which is desired. This so-called loss takes place in the electric motor. During the process of converting electrical energy into mechanical energy, some of the electrical energy is converted into heat energy. Because the heat energy is not desired and does not serve a useful purpose, it is considered a loss.

## Mechanical Energy

*Mechanical energy* is energy that causes motion. It is measured in the same units as work (foot-pounds or joules).

## Heat Energy

*Heat energy* is energy that causes substances to rise in temperature. It may be expressed in either the British unit or the metric unit. In the British system, heat is measured in **British thermal units** (Btu). One Btu is the amount of heat required to raise the temperature of 1 pound of water 1 degree Fahrenheit (F). In the metric system, heat energy is measured in units called *calories* (cal). One calorie is the amount of heat required to raise the temperature of 1 gram (g) of water 1 degree Celsius (C).

## Electrical Energy

*Electrical energy* is the product of power and time. Therefore, its unit of measurement is the *watt-hour* (Wh), or the *kilowatt-hour* (kW h). A 100-watt incandescent lamp will utilize 1 kilowatt-hour of energy if it is operated for 10 hours (100 watts × 10 hours = 1000 watt-hours = 1 kilowatt-hour).

Electrical consumers are billed by the utility company according to the amount of energy utilized. In other words, they are billed for the amount of power expended in a specific period of time. Billing is usually based on a sliding scale according to the number of kilowatt-hours indicated for the billing period.

## Example 4

The customer, an industrial plant, expends 5000 kW·h of electrical energy each working day. The plant is in operation 5 days per week. If the utility company charges $0.03 per kW·h for the first 50,000 kW·h and $0.015 per kW·h for all energy above 50,000 kW·h, how much does the customer pay for a 10-week period?

5000 kW·h/day × 5 days/week × 10 weeks = 250,000 kW·h

50,000 kW·h × $0.03/kW·h = $1500.00

250,000 kW·h − 50,000 kW·h = 200,000 kW·h

200,000 kW·h × $0.015/kW·h = $3000.00

$1500.00 + $3000.00 = $4500.00 paid for the 10-week period

# EFFICIENCY

**Efficiency** is the ratio of the useful power output to the total power input. Efficiency is generally stated in percent (%). No machine is 100 percent efficient. Not all the energy delivered to a machine serves the purpose for which it was intended. A motor is designed to produce motion (mechanical energy); however, some of the electrical energy received by the motor produces heat. The energy that is converted into heat is considered a loss. In order to operate machines at minimum cost, it is necessary to keep all losses to a minimum. A machine with very low energy loss is considered to be very efficient. The nearer the efficiency is to 100 percent, the more efficient the machine is. Efficiency can be calculated by using one of the following formulas:

$$\% \text{ Eff} = \frac{\text{Useful energy output}}{\text{Total energy input}} \times 100 \quad \text{(Eq. 3.9)}$$

or

$$\% \text{ Eff} = \frac{\text{Power output}}{\text{Power input}} \times 100 \qquad \text{(Eq. 3.10)}$$

# MECHANICAL TRANSMISSION OF POWER

Machines may be divided into two classes, the *driving machine* and the *driven machine*. The driving machine delivers the power to the machine that is being driven. Some types of driving machines are gasoline engines, steam turbines, and electric motors. Some examples of driven machines are presses, lathes, elevators, pumps, and saws.

# DRIVES

The usual types of connections between driving machines and driven machines are belts on pulleys, chains on sprockets, gear assemblies, and direct drives. It is very important that the proper drive be used to meet the needs of the job. Consideration must be given to such factors as speed requirements, direction of rotation, starting torque (twisting effort when the machine is starting), full-load torque, and starting current.

## Speed Requirements

If the speeds of both machines are the same, it may be possible to use a direct drive. There are two methods used to obtain direct drive. The machine and the motor may be mounted on the same shaft. This method is frequently used with motor–generator sets. A second method is to use a shaft coupling, a device that fastens two shafts together. This causes the entire assembly to operate as one machine.

If the speed of the driven machine differs from the speed of the motor, pulleys or gears are usually used. When pulleys are used, the speed of the machine is determined by the sizes of the pulleys. The speed relationship of the two machines is inversely proportional to the diameter relationship of their pulleys. The formula for calculating pulley sizes is

$$\frac{N_1}{N_2} = \frac{D_2}{D_1} \qquad \text{(Eq. 3.11)}$$

where $N_1$ = speed of the motor, in revolutions per minute (r/min)
$N_2$ = speed of the driven machine, in revolutions per minute (r/min)

$D_1$ = diameter of the motor pulley, in inches (in.) or centimeters (cm)
$D_2$ = diameter of the driven machine pulley, in inches (in.) or centimeters (cm)

## Example 5

The nameplate of a motor indicates that it rotates at a speed of 1700 r/min. The diameter of the pulley on the motor shaft is 6 in. (15.24 cm). If the machine must be driven at a speed of 850 r/min, what size pulley must be used on the machine?

$$\frac{N_1}{N_2} = \frac{D_2}{D_1}$$
$$\frac{1700}{850} = \frac{D_2}{6}$$
$$85 \, D_2 = 1020$$
$$D_2 = 12 \text{ in.}$$

## Example 6

A motor operates at a speed of 3250 r/min. The machine it is driving requires a speed of 650 r/min. If the machine pulley is 20 in. (50.8 cm), what size pulley must be installed on the motor?

$$\frac{N_1}{N_2} = \frac{D_2}{D_1}$$
$$\frac{3250}{650} = \frac{20}{D_1}$$
$$5 \, D_1 = 20$$
$$D_1 = 4 \text{ in.}$$

## Example 7

A motor operates at a speed of 500 r/min. The pulley attached to its shaft has a diameter of 20 in. (50.8 cm). The machine that the motor is driving has a pulley that is 8 in. (20.32 cm) in diameter. At what speed will the machine operate?

$$\frac{N_1}{N_2} = \frac{D_2}{D_1}$$
$$\frac{500}{N_2} = \frac{8}{20}$$
$$8 N_2 = 10,000$$
$$N_2 = 1250 \text{ r/min}$$

A gear drive is generally used where slippage is a factor. Although the gear drive is much more expensive than belts and pulleys, it is a positive drive and will not slip.

When using gears, the speed ratio is determined by the number of teeth in each gear. If both gears have the same number of teeth, the machine and the motor will operate at the same speed. The speed relationship is inversely proportional to the relationship of the number of teeth. With only two gears, the maximum speed ratio is generally 10 to 1. More than one set of gears is usually required to obtain greater speed ratios.

The formula for calculating speed and/or the number of teeth is

$$\frac{N_1}{N_2} = \frac{T_2}{T_1} \qquad \text{(Eq. 3.12)}$$

where $N_1$ = speed of the driving gear, in revolutions per minute (r/min)
$N_2$ = speed of the driven gear, in revolutions per minute (r/min)
$T_1$ = number of teeth in the driving gear
$T_2$ = number of teeth in the driven gear

### Example 8

The driving gear on a motor has 60 teeth and the driven gear has 90 teeth. If the driven gear revolves at 100 r/min, what is the speed of the driving gear?

$$\frac{N_1}{N_2} = \frac{T_2}{T_1}$$
$$\frac{N_1}{100} = \frac{90}{60}$$
$$6N_1 = 900$$
$$N_1 = 150 \text{ r/min}$$

### Example 9

The driving gear on a motor revolves at a speed of 1150 r/min and has 25 teeth. The speed of the machine must be 115 r/min. How many teeth must there be on the driven gear?

$$\frac{N_1}{N_2} = \frac{T_2}{T_1}$$
$$\frac{1150}{115} = \frac{T_2}{25}$$
$$T_2 = 250 \text{ teeth}$$

### Example 10

Drive gear A on a motor has 20 teeth and revolves at a speed of 2250 r/min. Gear A drives a second gear, gear B, which has 30 teeth and meshes with gear C, which has 40 teeth. What is the speed of gear C?

| Solution | Proof |
|----------|-------|
| $\dfrac{N_A}{N_C} = \dfrac{T_C}{T_A}$ | $\dfrac{N_A}{N_B} = \dfrac{T_B}{T_A}$ |

$$\frac{2250}{N_C} = \frac{40}{20}$$
$$N_C = 1125 \text{ r/min}$$

$$\frac{2250}{N_B} = \frac{30}{20}$$
$$3N_B = 4500$$
$$N_B = 1500 \text{ r/min}$$
$$\frac{N_B}{N_C} = \frac{T_C}{T_B}$$
$$\frac{1500}{N_C} = \frac{40}{30}$$
$$4N_C = 4500$$
$$N_C = 1125 \text{ r/min}$$

## Direction of Rotation

The directions of rotation of the motor and the machine must be considered when selecting the type of drive. When a direct drive, belt driven (Figure 3–3) or chain driven, is used, the machine rotor (revolving part) will rotate in the same direction as the motor rotor. Reversing the direction of rotation of the driven pulley necessitates reversing the direction of rotation of the motor rotor or crossing the belt (Figure 3–4). Crossing the belt is not generally recommended because of the extra wear on the belt.

When a two-gear drive is used, the driven gear will rotate in the opposite direction from the driving gear (Figure 3–5). Three gears must be used in order to have the driven gear rotate in the same direction as the driving gear (Figure 3–6).

A motor can deliver only the amount of horsepower indicated on the nameplate. No arrangement of gears and/or pulleys will change the horsepower delivered to the machine.

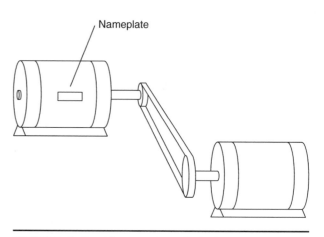

**FIGURE 3–3**  Motor using a belt drive.

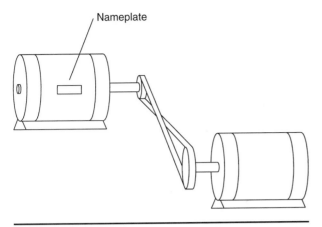

**FIGURE 3–4** Crossed belt drive for reversing the direction of the driven machine rotor.

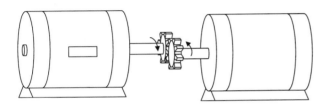

**FIGURE 3–5** Two-gear drive. The arrows indicate the direction of gear rotation.

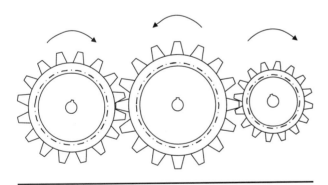

**FIGURE 3–6** Three-gear drive. The arrows indicate the direction of gear rotation.

## Torque

It can be observed that horsepower is a combination of **torque** (a rotating force) and speed. At a specific horsepower, the torque is inversely proportional to the speed. Any change in speed will cause a change in torque. An increase in speed will result in a decrease in torque, and vice versa. One must ensure adequate torque to drive the machine

at the required speed. The value of torque can be calculated for various speeds by the formula

$$hp = \frac{TN}{5252} \qquad \text{(Eq. 3.13)}$$

where hp = horsepower
  $T$ = torque, in pound-feet (lb ft)
  $N$ = speed, in revolutions per
    minute (r/min)

### Example 11

How much torque is required to drive a machine rated at 10 hp? The speed of the machine is 500 r/min.

$$T = \frac{5252\,hp}{N}$$
$$T = \frac{5252 \times 10}{500}$$
$$T = 105.04 \text{ lb ft}$$

## Starting Torque

The **starting torque** of a motor is the amount of torque developed at the instant the motor is energized. This value may be more or less than the torque developed when the motor is running at full load. The value depends upon the design of the motor.

Some machines, such as electric vehicles, elevators, and presses, require high starting torque. Other types of equipment may be damaged by sudden application of a large amount of torque. In the selection of motors, consideration must be given to the type of load and the maximum torque required to produce motion.

## Starting Current

**Starting current** is the value of current required by the motor from the instant it is energized until it reaches its rated speed. The largest amount of current flows at the instant the motor is energized. A motor that produces a high starting torque requires a high starting current.

The value of current available is limited by the size of the circuit conductors and the overcurrent devices protecting the circuit. With some motors, the starting current may be great enough to cause the overcurrent devices to open. To eliminate this problem, special motor-starting equipment is connected to the circuit. This equipment limits the maximum current to a specific value during the starting cycle. Many sophisticated devices have been designed for this purpose. A very simple device to use, however, is the rheostat. A rheostat connected in series with the motor will both limit and control the starting current.

## Other Factors in Selecting Drive

Other factors that must be considered in the selection of the type of drive include the following:

- Safety requirements
- Space requirements
- Size and type of shafts
- Horizontal or vertical shafts
- Distance from the motor to the machine

# SIZING MOTORS

The size of a motor for driving a machine depends upon the speed of the machine, the torque necessary to drive the machine at the rated speed, and the efficiency of the machine. For the most efficient operating conditions, motors should be sized as accurately as possible. Oversizing results in wasted power and high operating costs. Undersized motors cause excessive heating, insulation breakdown, and frequent operation of the overcurrent protection devices. The power required to drive a machine depends upon the rate that the machine does work and the efficiency of the machine. The power may be calculated by using variations of the following formula:

$$hp = \frac{Wh}{33,000 \times Eff} \qquad \text{(Eq. 3.14)}$$

where hp = mechanical power, in horsepower
$W$ = weight lifted, in pounds (lb)
$h$ = height lifted, in feet per minute (ft/min)
Eff = efficiency, in percent (%)

## Example 12

What size motor is required to drive a hoist if it must lift 1000 lb (453.6 kg) to a height of 30 ft (9.144 m) in 1 minute? The hoist is 90% efficient.

$$hp = \frac{Wh}{33,000 \times Eff}$$
$$hp = \frac{1000 \times 30}{33,000 \times 0.9}$$
$$hp = \frac{30,000}{29,700}$$
$$hp = 1$$

Manufacturers' literature and machine reference books provide information to calculate the horsepower of the driving motor for other types of driven machines, such as pumps, compressors, fans, and presses. Basically, the equations used are adaptations of equation 3.14.

# REVIEW QUESTIONS

*Multiple Choice*

1. Energy is the
   a. overcoming of resistance through a specific distance.
   b. rate of doing work.
   c. ability to do work.
   d. power.

2. The unit of measurement of work is the
   a. foot-pound.
   b. watt.
   c. kilowatt-hour.
   d. horsepower.

3. A common unit of measurement of mechanical power is the
   a. watt.
   b. horsepower.
   c. kilowatt-hour.
   d. volt.

4. One kilowatt is equal to
   a. 0.001 watt.
   b. 1000 watts.
   c. 100 watts.
   d. 1,000,000 watts.

5. One horsepower is equal to
   a. 746 watts.
   b. 986 watts.
   c. 1000 watts.
   d. 0.5 watt.

6. An incandescent lamp converts electrical energy into
   a. light energy.
   b. heat energy.
   c. both a and b.
   d. chemical energy.

7. Energy loss means that
   a. energy has been consumed.
   b. energy has been destroyed.
   c. one form of energy has been converted into an unwanted form of energy.
   d. energy has dissipated.

8. The electric motor converts
   a. light energy into mechanical energy.
   b. electrical energy into mechanical and heat energy.

c. mechanical energy into electrical and heat energy.

d. chemical energy into mechanical energy.

9. Horsepower is a measurement of
   a. mechanical power.
   b. electrical power.
   c. useful power.
   d. atomic power.

10. The percent efficiency of a motor is:
    a. the ratio of the total energy input to the total energy output.
    b. the ratio of the total power input to the total power output.
    c. the ratio of the power output to the power input.
    d. always 100% or more.

*Give Complete Answers*

1. Define *power*.

2. What is meant by the term *watt-hour*?

3. Write three formulas, in equation form, for calculating electrical power.

4. What is the unit of measurement of electrical energy?

5. Explain how to produce 1 watt of electrical power.

6. Define *work*.

7. Write the formula, in equation form, for calculating work.

8. Name the most common unit of measurement of mechanical power.

9. How many watts are there in 1 kilowatt?

10. List five forms of energy.

11. Name the two classes of machines.

12. List four types of machine drives.

13. Identify two methods used to obtain direct drive.

14. What method of drive is used if the speed of the driven machine differs from the speed of the motor?

15. List two factors that determine the speed of a machine that is belt driven.

16. If the direction of rotation of the machine rotor is opposite to the direction of the rotation of the motor rotor, what type of drive is preferred?

17. Name two types of driving machines.

18. Name three types of driven machines.

19. Write the formula (in equation form) for horsepower when the torque and speed are known.

20. List seven factors that must be considered when selecting the type of drive to be used on a machine.

*Solve each problem, showing the method used to arrive at the solution.*

1. A person raised 450 lb of shingles 100 ft in 20 min. Calculate the average horsepower expended.

2. A current of 20 A flows through a heating element when a pressure of 110 V is applied. What power does it dissipate?

3. A current of 50 A flows through a circuit having a resistance of 10 Ω. What power is dissipated by the circuit?

4. A bank of heating lamps requires 230 V for full heat. If the lamps' (hot) resistance is 5 Ω, what power do they require?

5. What horsepower is developed by the heating element in Problem 2?

6. How much electrical energy is utilized in 1 week by the circuit in Problem 3 if it is in operation for 8 hours a day, 6 days per week?

7. How much will it cost to operate the lamps in Problem 4 if they are to be used for 10 hours? The cost of energy is $0.05 per kW h.

8. A machine motor requires 20 A at 120 V. If the motor's output is 2300 W, what is the efficiency of the motor?

9. Motors are usually rated in terms of their horsepower output. In order to calculate the efficiency, the power (output and input) must be in the same unit of measurement. An electric motor on a forklift is rated at 10 hp. The input current to the motor is 70 A at 120 V. What is the efficiency of the motor?

10. A compressor motor delivers 20 hp to the pump. If the motor is 75% efficient, what is the kW input to the motor?

11. The driving gear on a motor has 30 teeth. If it revolves at a speed of 1150 r/min, how many teeth must the driven gear have if it is to revolve at a speed of 575 r/min?

12. A machine must be driven by a gear that rotates at a speed of 875 r/min and has 100 teeth. If the rotor of the driving motor rotates at a speed of 1750 r/min, how many teeth must the driving gear contain?

13. The revolving component of a machine rotates at a speed of 100 r/min and is driven by a gear containing 100 teeth. If the driving gear on the motor has 25 teeth, what is the speed of the motor?

14. A motor shaft contains a gear that has 10 teeth. The shaft rotates at a speed of 2350 r/min. If the machine that is being driven by the motor has a gear containing 200 teeth, what is the speed of the machine?

15. A machine rotor is belt driven and must revolve at a speed of 345 r/min. If the driving motor shaft rotates at a speed of 1725 r/min, determine the size of pulleys for both the motor and the machine.

16. The nameplate of a motor indicates that it rotates at a speed of 3550 r/min. The diameter of the pulley on the motor shaft is 4 in. (10.16 cm). If the machine must be driven at a speed of 710 r/min, what size pulley must be used on the machine?

17. A motor operates at a speed of 2950 r/min. The machine it is driving requires a speed of 295 r/min. If the machine has a pulley 30 in. (76.2 cm) in diameter, what size pulley must be installed on the motor?

18. A motor operates at a speed of 250 r/min. The pulley attached to its shaft has a diameter of 30 in. (76.2 cm). The machine the motor is driving has a pulley that is 4 in. (10.16 cm) in diameter. What is the speed of the machine?

19. How much torque is required to drive a machine rated at 50 hp? The machine must operate at a speed of 250 r/min.

20. What size motor is required to operate an elevator if its maximum capacity is 3000 lb (1360.8 kg)? The elevator travels at a speed of 10 ft (3.048 m) per minute. The motor is 80% efficient.

# Test Equipment

## OBJECTIVES

After studying this chapter, the student will be able to:

- Correctly set up and use a digital multimeter for the measurement of voltage, current, and resistance.
- Correctly set up and use a voltage tester.
- Correctly set up and use a clamp-on ammeter.
- Correctly set up and use a megohmmeter for the measurement of high resistances.
- Correctly set up and use an oscilloscope for the measurement of voltage, time, and frequency.

In this chapter, we will learn about the five most common pieces of test equipment used in industry today: the digital multimeter, the voltage tester, the clamp-on ammeter, the megohmmeter, and the oscilloscope. It is very important that you gain an understanding of these pieces of test equipment. You must know how to use them properly and safely, and you must understand their limitations. These pieces of test equipment will provide you with valuable information that will aid in maintaining and troubleshooting various pieces of electrical equipment.

# THE DIGITAL MULTIMETER

The most common multimeter in use today is the *digital multimeter* (Figure 4–1). The digital multimeter is so named because the measured value is displayed in a digital readout. This is an improvement over the older analog style of display in which you had to match a moving indicator (vane or needle) to a fixed scale. These analog indicators were prone to inaccuracies and difficulties in obtaining correct measurements. The digital display has made using a multimeter much easier.

The digital multimeter, or DMM, may be bench or panel mounted, or more commonly in maintenance, handheld. The operation of these DMMs is essentially the same. The major differences are in their functions and accuracies. Most bench or panel-type DMMs have more ranges, or more specific ranges, with higher accuracies. However, this is not a problem for the maintenance technician, who needs a portable, handheld unit. The handheld DMMs have the function and accuracy necessary to get the job done.

## Premeasurement Inspection

Prior to using your multimeter, you should visually inspect the meter and test leads thoroughly. Take your time and fully examine both the red and black test leads from the probe end to the plug end. You are looking for cracks or breaks in the probe, insulation, or plug. If you find any damage to either test lead, *replace the set of leads!* Test leads are very inexpensive and it is far safer to replace a damaged test lead set than to try to repair it with electrical tape. Remember, the voltage or current that you are measuring must be sensed by your meter by way of the test leads. The insulation quality of these leads is all that stands between your hands and the voltage or current that you are measuring. If your leads are damaged, replace them with a new set. The life you save may be your own!

Also make a thorough inspection of the multimeter itself. If you find any cracks or missing pieces from the case, have the meter repaired. *Do not use it!* The case is your protection from the voltages and currents that you will be measuring. Cracked or missing pieces of the case may provide an opportunity for you to come into contact with dangerous levels of voltage or current.

If the meter and test leads check out visually OK, you should perform an operational test. When preparing to measure resistance, you should *short* your leads together to verify that your meter is responding properly when set to measure resistance, as seen in Figure 4–2. When preparing to measure voltage or current, you should verify the operation of your meter on a known, good source of voltage or current. You should test the meter on the function and range that you expect to use. Many a technician has been injured or killed as a result of trusting a faulty meter.

Imagine that you are preparing to work on a circuit. You need to know if the circuit is energized or not, so you connect your meter into the circuit to measure voltage, and you read 0 V. You assume that the circuit is de-energized and begin to work. Upon coming into contact with the circuit, you receive an electrical shock! How can this happen if the meter indicates 0 V? Your meter is defective and cannot measure voltage.

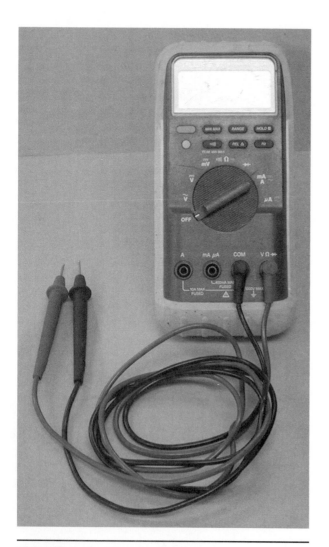

**FIGURE 4–1** Digital multimeter.

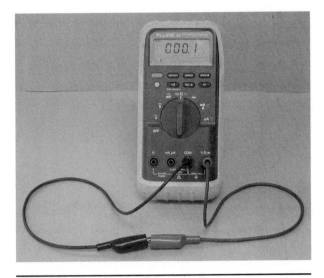

**FIGURE 4–2** DMM set to measure ohms with leads shorted.

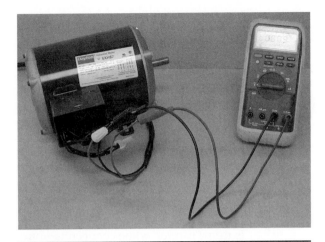

**FIGURE 4–3** DMM measuring voltage applied to a motor.

Prior to measuring the circuit voltage, you should have tested your meter on a known, good source of voltage. You should have set your meter to the voltage function and range that you expected to use, and then measured the voltage of a known energized circuit. If the meter indicated the proper type and amount of voltage, the meter would be safe to use on the circuit that you are troubleshooting. If not, the meter is in need of repair and should not be used until fixed.

## Premeasurement Setup

Before you even think about connecting your meter into a circuit to be tested, you must set up your meter for the measurement that you expect to make. For example, if you are going to measure the voltage of a circuit, you must connect the test leads to the appropriate jacks. You must then set the function and/or range switch to measure the correct type of voltage and the proper range for the expected amount. Only after you double-check your setup are you ready to connect your meter to the circuit to be tested.

We will now look at the correct procedures to follow for setting up and using your meter for measuring voltage, current, and resistance. Please keep in mind that there are many different manufacturers and models of digital multimeters. The information presented here will be generic in nature. You should familiarize yourself with the operating manual for the particular meter that you will be using. Should you have any questions or need additional help, check with your supervisor or the manufacturer of the meter.

## Measuring Voltage

Refer to Figure 4–3. The procedure for measuring voltage is as follows:

1. Perform a visual inspection of your meter and test leads.
2. Determine the amount of voltage that you expect to measure and verify that the voltage rating of your meter and test leads will not be exceeded.
3. Set up your meter to measure voltage as follows:
   A. Insert the black test lead into the COM test lead jack.
      1. Make certain the test lead is fully and securely inserted into the COM jack.
   B. Insert the red test lead into the Volts or V test lead jack.
      1. Make certain the test lead is fully and securely inserted into the Volts or V jack.
   C. Determine the type of voltage (AC or DC) and set the voltage selection switch (if available) to the proper voltage type.
4. Determine the approximate amount of voltage to be measured and set the range switch to the proper voltage range. Again, verify that the voltage rating of your meter and test leads will not be exceeded.
   A. If you are unsure of the amount of voltage that you will be measuring, set the range switch to the highest voltage range. If you have an autoranging meter, you will not need to set the voltage range. The meter will determine the correct range automatically!

5. Double-check your meter setup.

6. Connect the red and black test probes to a working circuit with a similar known amount of voltage present.

   A. When measuring voltage, your test leads will be connected across, or in parallel with, the portion of the circuit that you are testing.

      1. The safest method for making the meter connections is to de-energize the circuit to be tested, connect your meter into the circuit, and then re-energize the circuit. Unfortunately, in reality, this is impractical; therefore, exercise extreme caution when connecting your meter to an energized circuit. Be sure to wear appropriate personal protective equipment, as your facility requires.

7. Read the indicated voltage from the digital display of your meter.

   A. If the indicated voltage is correct for the amount of voltage that should be present in the circuit, your meter is functioning normally and you may disconnect your test probes from the test circuit and proceed to Step 8.

   B. If the indicated voltage is not correct for the amount of voltage that should be present in the circuit, your meter is faulty and should not be used until repaired. Disconnect your test probes from the test circuit.

8. Connect the red and black test probes to the circuit that you are troubleshooting.

   A. Remember that when measuring voltage, your test leads will be connected across, or in parallel with, the portion of the circuit that you are testing.

9. Read the indicated voltage from the digital display of your meter.

10. Carefully disconnect your meter from the circuit that you are testing.

11. Reconnect your meter to the working circuit with the similar known amount of voltage present that you previously used. This will verify that your meter is still functioning properly.

## Measuring Current

Refer to Figure 4–4. The procedure for measuring current is as follows:

1. Perform a visual inspection of your meter and test leads.

2. Determine the amount of current you expect to measure and verify that the current rating of your meter and test leads will not be exceeded.

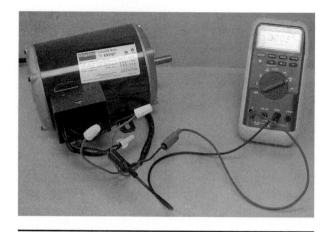

**FIGURE 4–4**  DMM measuring the current draw of a motor.

3. Set up your meter to measure current as follows:

   A. Insert the black test lead into the COM test lead jack.

      1. Make certain the test lead is fully and securely inserted into the COM jack.

   B. Insert the red test lead into the mA or A test lead jack depending on the amount of current that you expect to measure. If you are uncertain as to the amount of current that you will be measuring, insert the red test lead into the A jack. You can move it to the mA jack later if you discover that the current is less than 1 ampere.

      1. Make certain the test lead is fully and securely inserted into the mA or A jack.

   C. Determine the type of current (AC or DC) and set the current selection switch (if available) to the proper current type.

   D. Determine the approximate amount of current to be measured and set the range switch to the proper current range. If you are unsure of the amount of current that you will be measuring, set the range switch to the highest current range. If you have an autoranging meter, you will not need to set the current range. The meter will determine the correct range automatically.

4. Double-check your meter setup.

5. Connect the red and black test probes to a working circuit with a known amount of current present.

   A. When measuring current, your test leads will be connected in series with the portion of the circuit that you are testing.

      1. This means that you will need to remove power from the circuit.

2. Break or open the circuit at the point at which you wish to measure the current.

3. Insert your meter in series with the break so that one test lead is connected to one side of the break, and the other test lead is connected to the remaining side of the break. Now reapply power. Connecting your meter in this fashion causes the circuit current to flow from the circuit, into your meter, through your meter, and then return to the circuit. (This can be very difficult to do in an operating facility; therefore, the measurement of current in this manner is not widely performed. A special type of current meter, called a clamp-on ammeter, will be discussed later. We will continue to show you the correct procedure to follow so that you will be aware should the opportunity present itself.) Be sure to wear appropriate personal protective equipment, as your facility requires.

6. Read the indicated current from the digital display of your meter.

A. If the indicated current is correct for the amount of current that should be present in the circuit, your meter is functioning normally and you may proceed with your test. Remove power from the circuit, disconnect your test probes from the test circuit, and proceed to Step 7. (Don't forget to reconnect the circuit at the break and reapply power.)

B. If the indicated current is not correct for the amount of current that should be present in the circuit, your meter is faulty and should not be used until repaired. Remove power from the circuit and disconnect your test probes from the test circuit. (Don't forget to reconnect the circuit at the break and reapply power.)

7. Connect the red and black test probes to the circuit that you are troubleshooting.

A. Remember that when measuring current, you will need to remove power from the circuit. Break or open the circuit at the point at which you wish to measure the current. Insert your meter in series with the break so that one test lead is connected to one side of the break and the other test lead is connected to the remaining side of the break. Reapply power.

8. Read the indicated current from the digital display of your meter.

9. Remove power from the circuit.

10. Carefully disconnect your meter from the circuit that you are testing. Remember

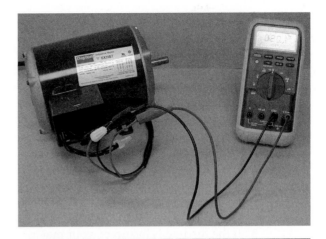

**FIGURE 4-5** DMM measuring the winding resistance of a motor.

to reconnect the circuit at the break and reapply power.

11. Reconnect your meter to the working circuit with the similar known amount of current present that you previously used. This will verify that your meter is still functioning properly.

## Measuring Resistance

Refer to Figure 4–5. The procedure for measuring resistance is as follows:

1. Perform a visual inspection of your meter and test leads.

2. Set up your meter to measure resistance as follows:

A. Insert the black test lead into the COM test lead jack.

　1. Make certain the test lead is fully and securely inserted into the COM jack.

B. Insert the red test lead into the Ohms or $\Omega$ test lead jack.

　1. Make certain the test lead is fully and securely inserted into the Ohms or $\Omega$ jack.

C. *Verify that all voltage has been removed from the circuit or component that you will be testing.*

D. Determine the approximate amount of resistance to be measured and set the range switch to the proper resistance range. If you are unsure of the amount of resistance that you will be measuring, *set the range switch to the highest resistance range.* If you have an autoranging meter, you will not need to set the resistance range. The meter will determine the correct range automatically.

E. Touch or short the red and black probe tips together. A near-zero reading on the display verifies that your meter is properly set up to measure resistance, your test leads are making good connection, and your meter is in working order. In addition, this verifies that the internal battery of your meter is capable of supplying enough energy to perform the resistance measurement.

1. You should see a reading of almost 0 Ω of resistance on your meter's display.

2. If the reading is correct, you may proceed to Step 3.

3. If the reading is incorrect, your meter is faulty and should not be used until repaired. This may be as simple as replacing the internal battery.

3. Double-check your meter setup, *and double-check that all power has been removed from the circuit or component that you are testing.*

4. Connect the red and black test probes to the circuit or component that you are testing.

A. When measuring resistance, your test leads will be connected across, or in parallel with, the component or portion of the circuit that you are testing.

5. Read the indicated resistance from the digital display of your meter.

6. Disconnect your meter from the circuit or component that you are testing.

## THE VOLTAGE TESTER

The voltage tester is a simple device used for making quick, ballpark voltage measurements. You may have heard a voltage tester referred to as a Wiggy. (Wiggy is a registered trademark of Schneider Electric.) Voltage testers typically measure AC and DC voltages. Most voltage testers will indicate AC voltages (50 Hz or 60 Hz) of 115/120 VAC, 220/240 VAC, 440/480 VAC, and 550/600 VAC. DC voltages of 115/120 VDC, 230/240 VDC, and 600 VDC can typically be measured as well.

Some voltage testers simply indicate the voltage by illuminating an LED (light emitting diode) located next to the voltage marking. Other voltage testers use small neon bulbs to indicate the measured voltage. Figure 4–6 shows a picture of a voltage tester. Notice the illuminated indicator next to the 120 V marking. Other voltage testers are designed to vibrate when voltage is sensed. The intensity of the vibration is in relation to the amount of voltage being measured. This means that the voltage

**FIGURE 4–6** Voltage tester measuring 120 V.

tester vibrates more violently when a high voltage is measured as opposed to when a low voltage is measured.

In addition to using a voltage tester to check for the presence of voltage, a voltage tester can also be used to identify the grounded conductor, check for blown fuses, and distinguish between AC and DC voltages.

To measure the voltage between two conductors, simply connect the voltage tester between the two conductors, as shown in Figure 4–7. If a neon-bulb-type voltage tester is used on an AC circuit, both plates or sides of the neon bulb will glow. Should DC voltage be tested with a neon-bulb-type voltage tester, only the negative plate or side of the neon bulb will glow.

To identify the grounded conductor of a circuit or system, connect the voltage tester between one conductor and a good ground, as shown in Figure 4–8. If the voltage tester indicates a voltage, the conductor is not grounded. Repeat this procedure with each conductor in the circuit or system. A measurement of 0 V from any conductor to ground is an indication that conductor is grounded.

**FIGURE 4–7** Voltage tester measuring the voltage across two conductors.

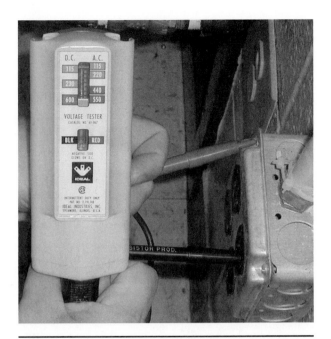

**FIGURE 4–8** Voltage tester used to test for a grounded conductor.

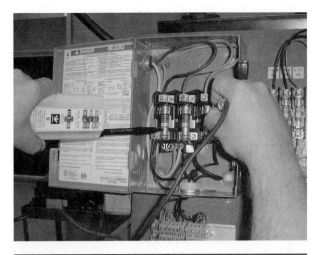

**FIGURE 4–9A** Voltage tester used to test fuses. The left-hand fuse is good.

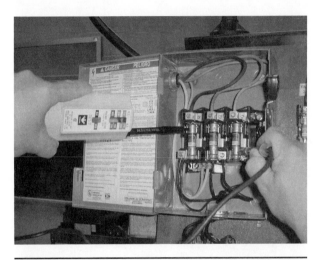

**FIGURE 4–9B** Voltage tester used to test fuses. The right-hand fuse is open.

To check for blown fuses, connect the voltage tester as shown in Figure 4–9A. In Figure 4–9A, the panel is fed from a 220 V source, and the voltage tester indicates "220 V." This means that the left-hand fuse, is good. Now, refer to Figure 4–9B. The voltage-tester connections allow for the testing of the right-hand fuse. The voltage tester indicates "0 V." This means that the right-hand fuse is blown and needs to be replaced. When performing this test on a three-phase circuit, lower-than-normal line voltage or no voltage indicates a blown fuse.

## THE CLAMP-ON AMMETER

When measuring current with a multimeter, you must remove power, break the circuit, insert your meter into the break, and reapply power. This is almost always impractical to do in an operating manufacturing facility. Power cannot easily be removed without disrupting other processes. The *clamp-on ammeter* allows the measurement of current without the need to de-energize the circuit.

Figure 4–10 shows a clamp-on ammeter. Notice the lack of test leads and the clamp at the top of the meter. The clamp is opened by depressing a button located on the side of the meter housing. Opening the clamp allows the clamp to be placed around a conductor, as seen in Figure 4–11. In order for the clamp-on ammeter to function properly, the clamp must be fully closed around the conductor. As current flows through the conductor, a magnetic field is produced. The clamp-on ammeter senses the strength of the magnetic field and translates that strength into a corresponding amount of current. The amount of current is then displayed by either an analog meter or a digital readout.

While clamp-on ammeters are typically used to measure significant amounts of current, they can be used to measure smaller amounts of current as well. There is, however, a technique that is used to aid in the measurement of smaller currents. Figure 4–12 shows the technique used to measure small values of current with a clamp-on ammeter.

Notice that the conductor has been wound around one of the jaws of the clamp. In fact, the

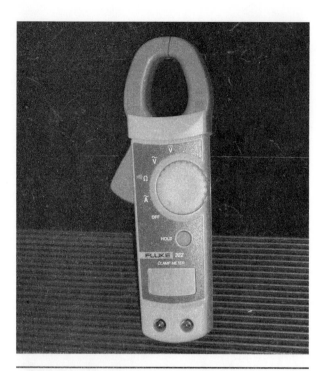

**FIGURE 4–10**   The clamp-on ammeter.

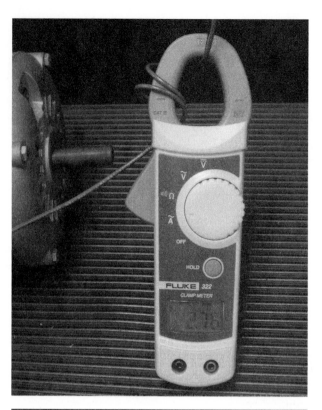

**FIGURE 4–12**   Winding the conductor around one jaw of a clamp-on ammeter in order to measure small amounts of current.

will now display a current reading. However, the amount of current displayed is artificially three times higher than the actual value as a result of winding the conductor around the jaw. Now, simply read the indicated amount of current, divide by 3, and you have a fairly accurate indication of the amount of current flow in the conductor. You will need to experiment to determine how many times to wind the conductor around the jaw to obtain a measurable amount of current. Just remember to divide the indicated current by the number of times the conductor is wound around the jaw.

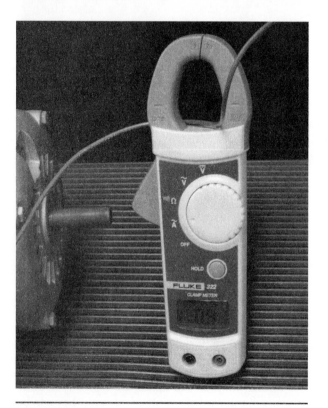

**FIGURE 4–11**   Using the clamp-on ammeter to measure the current flowing through a conductor.

conductor has been wound around the jaw a total of three times. This has the effect of tripling the strength of the magnetic field created by the current flowing through the conductor. The ammeter

# THE MEGOHMMETER

Figure 4–13 shows a **megohmmeter.** The megohmmeter, more commonly referred to as a Megger, allows the measurement of very high resistance values. (Megger is a registered trademark of AVO International Limited.) This test device is useful in determining the quality of insulation of wires, cables, transformers, motors, and generators. There are many different types, makes, and models of megohmmeters on the market today. Some models

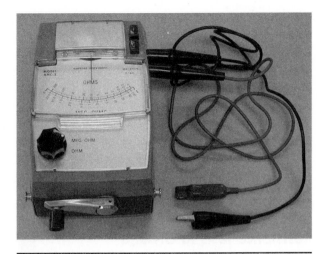

**FIGURE 4–13** A crank-type megohmmeter.

simply measure high resistance values, while others incorporate additional features that allow them to measure continuity and voltage as well. Some models use a hand crank to operate an internal DC generator. This DC generator is what produces the high voltages needed to test high resistance values. Other megohmmeters use internal electronic circuitry to develop the high test voltages and, therefore, do not have a hand crank. We will focus our attention on the measurement of high resistance values using a hand-crank model.

Next, we look at the correct procedures to follow for setting up and using your megohmmeter for measuring high resistance values. The information presented here will be generic in nature. You should familiarize yourself with the operating manual for the particular megohmmeter that you will be using. Should you have any questions or need additional help, check with your supervisor or the manufacturer of the megohmmeter that you are using.

In order to measure high values of resistance, a megohmmeter must deliver high test voltages to the circuit that is being tested. It is not uncommon for a megohmmeter to operate with 500 V or even 1000 V at its test terminals. Therefore, prior to using the megohmmeter, we will discuss some safety precautions that *must* be observed.

### Safety Precautions

1. Do not use the megohmmeter unless you have been trained in its proper use.

2. Wear your facility's approved and appropriate safety equipment when performing tests with a megohmmeter.

3. You must de-energize the circuit that you will be testing. Be absolutely certain that power

is removed and the circuit is isolated before performing any measurements with the megohmmeter.

4. Perform a visual inspection of the megohmmeter. Prior to using your megohmmeter, you should visually inspect the meter and test leads thoroughly. Take your time and fully examine the red, black, and guard or ground test leads from the clip end to the plug end. You are looking for cracks or breaks in the clip, insulation, or plug. If you find any damage to either test lead, *replace the set of leads!* Test leads are very inexpensive, and it is far safer to replace a damaged test lead set than to try to repair it with electrical tape. Remember, the megohmmeter operates with very high test voltages. If your leads are damaged, *replace them* with a new set. The life you save may be your own. Also give a thorough inspection to the megohmmeter itself. If you find any cracks or missing pieces from the case, have the meter repaired. *Do not use it!* Cracked or missing pieces of the case may provide an opportunity for you to come into contact with dangerous levels of voltage. The case is your protection from the voltages and currents that you will be measuring.

5. *While the test is in process, you must not touch or come into contact with the megohmmeter connections to the circuit that you are testing. High voltages will be present!*

6. *Discharge the circuit that has been tested before disconnecting the megohmmeter leads.* This not only applies to circuits containing capacitors, but also to circuits that become capacitive due to long lengths of cable. Most megohmmeters contain an internal discharge circuit to perform a circuit discharge automatically. However, be certain that your megohmmeter is equipped with this circuit and that it is operating properly. An automatic discharge circuit should not be taken for granted or viewed as a substitute for safe working practices.

7. *Never have someone hold the test leads while operating the megohmmeter! This is an irresponsible and very dangerous prank that can cause injury or death!*

## Preparing the Megohmmeter for Resistance Tests

1. Place the megohmmeter on a flat, level, solid surface.

2. Prior to connecting the test leads, set the test voltage selector switch to the highest test

voltage position (500 V or 1000 V depending on your unit).

3. Hold down the *test* button while turning the generator hand crank.
   A. The crank should be turned at a constant rate according to the manufacturer's specifications.
   B. The meter movement indicator should remain at ∞ (infinity).
   C. This test verifies that your megohmmeter does not have any internal leakage.

4. Stop turning the generator hand crank.

5. Release the test button.
   A. When the test button is released, any voltage stored by the circuit will be indicated on the voltage scale of the meter movement. Wait a few seconds for the voltage to completely discharge to 0 V before proceeding further.

6. Insert the red test lead into the + terminal and the black test lead into the − terminal. Verify that the clip ends of the test leads are not in contact with anything.

7. Hold down the test button while turning the generator hand crank.
   A. The crank should be turned at a constant rate according to the manufacturer's specifications.
   B. The meter movement indicator should remain at ∞ (infinity).
   C. This verifies the quality of the insulation of the test leads.

8. Stop turning the generator hand crank.

9. Release the test button.
   A. When the test button is released, any voltage stored by the circuit will be indicated on the voltage scale of the meter movement. Wait a few seconds for the voltage to completely discharge to 0 V before handling the test leads.

10. Connect the clip ends of the test leads together.

11. Hold down the test button while turning the generator hand crank.
    A. The crank should be turned at a constant rate according to the manufacturer's specifications.
    B. The meter movement indicator should now read approximately 0 Ω.
    C. A high or infinite reading means that one or both test leads are open and will need replacement. This may also indicate a problem with the megohmmeter. If the same

reading is obtained after repeating the test with replacement leads, the megohmmeter is faulty and must be repaired.

12. Stop turning the generator hand crank.

13. Release the test button.
    A. When the test button is released, any voltage stored by the circuit will be indicated on the voltage scale of the meter movement. Wait a few seconds for the voltage to completely discharge to 0 V before handling or disconnecting the test leads.

## Resistance Tests to Ground

The connections for using a megohmmeter to perform resistance tests to ground are shown in Figure 4–14. The procedures to perform these tests are as follows:

1. Place the test voltage selector switch to the required test voltage position.

2. Connect the red (+) test lead to a good ground or the metal frame of the equipment under test.

3. Connect the black (−) test lead to the portion of the circuit that is being tested.

4. Hold down the test button while turning the generator hand crank.
   A. The crank should be turned at a constant rate according to the manufacturer's specifications.
   B. The meter movement indicator will indicate the amount of insulation resistance on the MΩ scale.

5. Stop turning the generator hand crank.

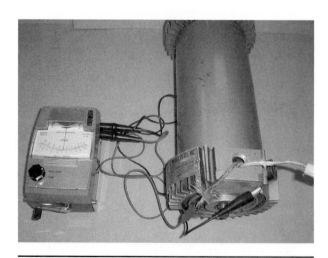

**FIGURE 4–14** Using a megohmmeter to check for a grounded motor winding.

6. Release the test button.

   A. When the test button is released, any voltage stored by the circuit will be indicated on the voltage scale of the meter movement. Wait a few seconds for the voltage to completely discharge to 0 V before handling or disconnecting the test leads.

## Insulation Test Between Two Insulated Wires

Refer to Figure 4–15. Procedures for the insulation test are as follows:

1. Place the test voltage selector switch to the required test voltage position.

2. Connect the red (+) test lead to the conductor of one of the wires.

3. Connect the black (−) test lead to the conductor of the remaining wire.

4. Hold down the test button while turning the generator hand crank.

   A. The crank should be turned at a constant rate according to the manufacturer's specifications.

   B. The meter movement indicator will indicate the amount of insulation resistance on the MΩ scale.

5. Stop turning the generator hand crank.

6. Release the test button.

   A. When the test button is released, any voltage stored by the circuit will be indicated on the voltage scale of the meter movement. Wait a few seconds for the voltage to completely discharge to 0 V before handling or disconnecting the test leads.

7. If a reading of ∞ (infinity) is obtained, the opposite ends (not the ends connected to the megohmmeter) of the conductors being tested should be connected together.

8. Hold down the test button while turning the generator hand crank.

   A. The crank should be turned at a constant rate according to the manufacturer's specifications.

9. The meter movement should indicate approximately 0 Ω.

   A. This verifies the measurement of the insulation as infinite and that the test leads are not disconnected or broken.

10. Stop turning the generator hand crank.

11. Release the test button.

   A. When the test button is released, any voltage stored by the circuit will be indicated on the voltage scale of the meter movement. Wait a few seconds for the voltage to completely discharge to 0 V before handling or disconnecting the test leads.

## THE OSCILLOSCOPE

As industrial controls and devices become more "intelligent" and advanced, the technician will need more advanced test equipment as well. One piece of test equipment that is becoming more and more common in the industrial environment is the **oscilloscope** (or o'scope or scope). The oscilloscope, as seen in Figure 4–16, is being used more frequently in the maintenance field. The oscilloscope not only measures voltage but also allows the user

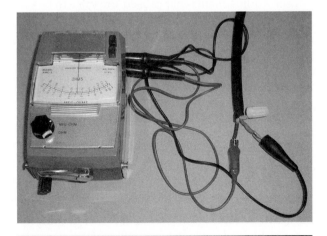

**FIGURE 4–15** Using a megohmmeter to check the quality of insulation between two conductors.

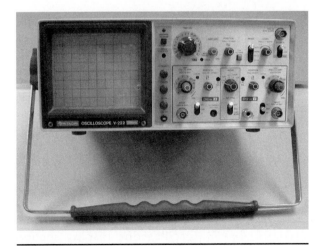

**FIGURE 4–16** An oscilloscope.

to *see* a representation of the voltage. This is quite helpful when diagnosing problems caused by voltage spikes, dirty power, or simply as a way to find failed components more rapidly.

At first glance, an oscilloscope may seem rather intimidating. There are many knobs and adjustments on the front of the scope. Rather than a digital readout or moving pointer, an oscilloscope displays an image of the voltage that you are measuring. This is what makes the scope so attractive for troubleshooting. Now the voltage can not only be measured but *seen* as well. So that you will have a basic understanding of the setup, operation, and interpretation of an oscilloscope, we will explore the function of the different controls found on a typical scope. Once a scope is properly set up, we will learn how to make and interpret the scope display. Please be aware that there are many different manufacturers of oscilloscopes. Each manufacturer may produce several different models. It is, therefore, impossible to present all of the possible features and functions in this text. The most common controls and their function will be presented here. You will need to familiarize yourself with your facility's oscilloscope.

## The Display Section

Refer to Figure 4–17. The section of an oscilloscope on which measurements are made is called the *cathode ray tube* or CRT. The CRT is part of the display section. There are several controls that allow you to adjust the display or **trace** on the CRT. These controls are called *intensity*, *trace rotation*, *beam find*, and *focus*.

■ Intensity—This allows you to adjust the brightness or intensity of the trace in various lighting conditions. For example, if you were to use

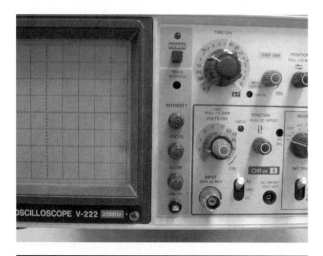

**FIGURE 4–17** Oscilloscope display section controls.

the scope in a brightly lit area, you may need to increase the intensity of the trace so that the ambient light does not wash out the display. On the other hand, if the scope is used in a dimly lit area, the intensity of the trace would be set to a lower level so that the trace does not *bloom* or *blossom*. Blooming or blossoming causes inaccuracies in your measurements. You should always adjust the Intensity control for the *minimum* brightness of the trace for a comfortable measurement. It is common practice to leave a scope turned on when not making measurements. This allows the scope to be ready for use at a moment's notice. However, leaving a trace displayed on the CRT for extended periods of time will damage the CRT permanently and lead to a very costly repair. It is good practice to turn the intensity of the trace down to a point where the trace is no longer visible. You can still leave the scope on, but turn the intensity down. Then, when you need to take a measurement, simply turn up the intensity to the required level. This avoids damage to the CRT.

■ Trace rotation—The CRT display is divided into grids by vertical and horizontal lines. It is important, for accurate measurements, that the trace be perfectly horizontal. That is, the trace must run parallel to the horizontal grid lines. Moving a scope from place to place will cause the trace to *tilt*. The trace is no longer parallel to a horizontal grid line but is now at an angle. This is a result of the variations in the earth's magnetic field as the scope is repositioned. To compensate for these variations, simply take a small screwdriver and adjust the Trace Rotation until the display trace is aligned horizontally with the horizontal grid lines.

■ Beam find—Sometimes you turn on your scope and wait for the display to appear, only to be disappointed. This may be a result of a previous user's adjustments that caused the trace to be off-screen. To quickly locate the trace, press the Beam Find button. You will then see an intensified spot on the CRT display. The spot will appear in one of the four quadrants on the CRT. This helps you adjust the Position controls to quickly bring the trace back on-screen. Do not depress the Beam Find button for extended periods of time. Due to the intensity of the spot, permanent damage to the CRT may result.

■ Focus—The Focus control is used to adjust the sharpness or crispness of the trace on the display. You will get a more accurate measurement using a finely focused trace than with a fuzzy trace.

The final element of the display section is the CRT itself. As already mentioned, the CRT is divided by vertical and horizontal lines that form a grid pattern on the face of the CRT. This grid is called a *graticule*. The horizontal lines will be used to measure the amount of voltage present. The vertical lines will be used to measure the time. Oscilloscopes measure voltage with respect to time. Notice that our CRT has 8 horizontal lines and 10 vertical lines (see Figure 4–16). These form the *major* divisions of the display. (Most oscilloscopes use an 8 × 10 grid.) Notice also that the center horizontal line and the center vertical line have smaller graduations. These are called *minor divisions* or *subdivisions*. There are four minor divisions between two major divisions. We will use these major and minor divisions to measure voltage and time.

## The Vertical Section

Refer to Figure 4–18. One of the reasons that the oscilloscope seems so intimidating is due to the number of knobs or controls that are found on the front of the scope. The vertical section is partly to blame for this. Most oscilloscopes are *two-channel* or *dual-trace* scopes. This means that the scope has the ability to display two different waveforms at the same time. In order to have control over each of these displays, you must have two separate sets of controls.

When you look at the vertical section of the scope, you will see that some of the controls have been duplicated. This accounts for many of the knobs or controls that make the scope intimidating. But remember, they are duplicate controls. If you understand how one set of controls works, you will understand the operation of the other set.

The vertical section consists of the VERTICAL POSITION, VERTICAL OPERATING MODE, VARIABLE VOLTS/DIV., INPUT SENSITIVITY,

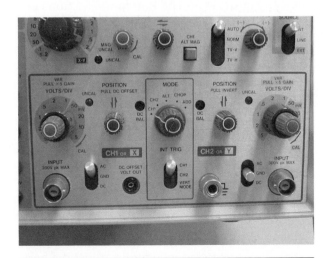

**FIGURE 4–18**  Oscilloscope vertical section controls.

and INPUT COUPLING controls. In addition, the INPUT JACKS for the *test probes* are a part of this section as well. Let's look at each of these controls in more detail:

- VERTICAL POSITION—This control allows you to move the trace displayed on the CRT up or down. In other words, if the displayed waveform is near the bottom of the CRT, you can use the VERTICAL POSITION control to reposition the trace to the center or top of the CRT. Do not be afraid to do this. You will make measurements more easily and accurately if you reposition the trace for the optimum viewing position.

- VERTICAL OPERATING MODE—This is actually several controls. These are CH 1–BOTH–CH 2, CH 2 INVERT, and ADD–ALT–CHOP.

  - CH 1–BOTH–CH 2—This control allows you to determine which trace is displayed on the CRT. You may wish to view only the trace from channel 1. If so, place this switch in the CH 1 position. Likewise, should you wish to view only the trace for channel 2, you would place the switch in the CH 2 position. However, if you wish to view both channels simultaneously, place the switch in the BOTH position. Now you will be able to view channel 1 and channel 2 on the CRT at the same time.

  - CH 2 INVERT—When this control is depressed, the signal displayed on channel 2 will be inverted. Using this switch, in conjunction with the ADD position of the ADD–ALT–CHOP switch, allows you to make *differential measurements*. For example, we will assume that the input signal on channel 1 is 20 V p-p, and the input signal on channel 2 is 7 V p-p. We wish to display the difference between these two input signals. Place the ADD–ALT–CHOP switch in the ADD position and depress the CH 2 INVERT switch. The signal for channel 2 is now inverted, and when added to the signal from channel 1, will yield the difference of the two input signals. Unless you are making differential measurements, always verify that the CH 2 INVERT switch is in the normal position.

  - ADD–ALT–CHOP—As we have seen previously, the ADD position allows us to algebraically add the signals from channel 1 and channel 2 together. This will give us the sum of the two input signals or the difference between the two input signals depending upon the setting of the CH 2 INVERT switch. The ALT or *alternate* position allows the scope to show the channel 1 signal, then the channel 2 signal, then the channel 1 signal, and so forth. When the scope is set to display higher frequencies, the ALT position will allow you to see both input signals at the same time. The scope will alternate between channel 1 and channel 2 at such a high speed, your eyes will not see the switching and the display will appear to show both input signals simultaneously. When the scope is set to display lower frequencies, you must use the CHOP position.

- VARIABLE VOLTS/DIV—A dual-trace scope will have two of these controls: one for channel 1 and one for channel 2. This is a variable control that allows us to independently adjust the vertical height of the displayed signal for channel 1 or channel 2. It is helpful in making quick comparisons of signals. For example, imagine that you only wanted to see if a signal was amplified, but you really did not care by how much it was amplified. You would measure the input signal to the circuit. While displaying the input signal, you adjust the VARIABLE VOLTS/DIV control until you fit the displayed waveform within the horizontal major division graticule lines. Now, without adjusting the setting, move your probe to the output of the circuit. If the displayed waveform extends beyond the graticule lines that you used for the input, the signal has been amplified. If the displayed waveform does not fill in the area between the graticule lines that you used for the input, the signal has not been amplified and is, in fact, smaller than the input signal. Notice that this control has a curved arrow and the word "CAL" on it. On this particular scope, you turn this control fully clockwise. There is actually a detent in the switch that provides a click when the control is turned fully clockwise. The VARIABLE VOLTS/DIV control must be in the detent or CAL position in order to make use of the values on the VOLTS/DIV switch behind.

- INPUT SENSITIVITY—This control is actually labeled CH 1 VOLTS/DIV or CH 2 VOLTS/DIV. Think of this as a range switch. You will adjust this control until the displayed waveform is the largest it can be, vertically, and still fit within the graticule of the CRT. This will provide the highest accuracy for your measurements. Notice there are two windows on this control. The 1X window is located at the ten o'clock position, and the 10X window is located at the two o'clock position. These windows work in conjunction with the type of oscilloscope probes that you use. If you use a 1X scope probe, you will make your measurements using the 1X window. If you use a

10X scope probe, you will make your measurements using the 10X window. The numbers on this control represent the *volts per division* for the graticule. For example, suppose you are using a 1X probe and displaying a waveform that is three major divisions high. If the VOLTS/DIV switch is set to 0.2 under the 1X window, the waveform measures 0.6 V p-p. (Three divisions times 0.2 volts per division equals 0.6 V p-p.)

- INPUT COUPLING—There are two of these switches, one for each channel. Each switch has three positions: AC–GND–DC. The DC position represents *direct coupled*. This allows the scope to display all components of the input signal. The AC position inserts a capacitor into the input circuit. This blocks any DC component from the signal that you are measuring. The scope will display only the AC component of the signal. The GND position disconnects the input signal from the scope and connects the scope input to the chassis ground of the scope. This prevents unwanted signals from being displayed while the trace is adjusted for a reference on the CRT. For example, by placing the input coupling switch in the GND position, the trace can be adjusted vertically (using the VERTICAL POSITION control) until it exactly lines up with the horizontal grid line that is second from the bottom of the graticule. Now, the input coupling switch is placed in the DC position. When a measurement is made, the trace moves vertically upward to the fourth horizontal graticule line from the bottom. We now know that a DC component is present because the trace is at some positive voltage above ground.

## The Horizontal Section

Refer to Figure 4–19. The horizontal section consists of the HORIZONTAL POSITION, VARIABLE SEC/DIV., HORIZONTAL MAGNIFICATION, and SWEEP SPEED controls. Let's look at each of these controls in more detail.

- HORIZONTAL POSITION—This control allows you to move the trace displayed on the CRT left or right. In other words, if the displayed waveform is near the left side of the CRT, you can use the HORIZONTAL POSITION control to reposition the trace to the center or right side of the CRT. Do not be afraid to do this. You will make measurements more easily and more accurately if you reposition the trace for the optimum viewing position.

- VARIABLE SEC/DIV—This is a variable control that allows us to independently adjust the horizontal width of the displayed signal for channel 1

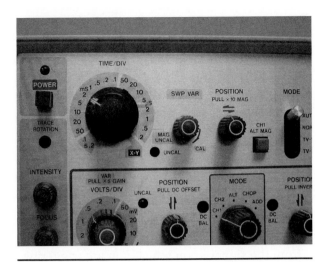

**FIGURE 4–19** Oscilloscope horizontal section controls.

and channel 2. This is helpful for making quick comparisons of signals. For example, imagine that you only wanted to compare the period of one waveform to another. You would display one input signal. While displaying the input signal, you adjust the VARIABLE SEC/DIV control until you fit one complete cycle of the displayed waveform within the vertical major division graticule lines. Now, without adjusting the setting, move your probe to the signal with which you wish to compare. If the one complete cycle of the displayed waveform extends beyond the graticule lines that you used for the input, the signal has a longer period. If the displayed waveform does not fill the area between the graticule lines that you used for the input, the signal has a shorter period. Notice that this control has a curved arrow and the word "CAL" on it. On this particular scope, you turn this control fully clockwise. There is actually a detent in the switch that provides a click when the control is turned fully clockwise. The VARIABLE SEC/DIV control must be in the detent or CAL position in order to make use of the values on the SEC/DIV switch behind.

- HORIZONTAL MAGNIFICATION—By pulling the VARIABLE SEC/DIV knob gently, you would be able to magnify the displayed waveform by a factor of 10. This allows you to examine a waveform more closely. The HORIZONTAL MAGNIFICATION control must be depressed in order to make use of the SEC/DIV switch behind.

- SWEEP SPEED—Think of this as a range switch. You will adjust this control until the displayed waveform is the largest it can be, horizontally, and still fit within the graticule of the CRT. This will provide the highest accuracy for your

measurements. Notice the settings are divided into S (seconds), ms (milliseconds), and μs (microseconds). After adjusting this control, you count the number of major and minor divisions, horizontally, on the graticule for one complete cycle of the measured signal. For example, imagine a signal that is displayed on the CRT. From the beginning to the end of one complete cycle, you count five major divisions. You look at the SEC/DIV control and find it set on 5 ms. This means that there are 5 ms per division. Therefore, the time of one cycle is 25 ms (5 major divisions times 5 ms per division equals 25 ms).

## The Trigger System

Refer to Figure 4–20. The *trigger system* is the portion of the scope that controls *when* the waveform is displayed on the CRT. If the trigger system controls are not adjusted properly, the waveform may drift slowly across the screen or may be a jumble of images. In any event, the display will be impossible to read. Proper adjustment of the trigger system controls is critical in displaying a usable image on the CRT.

Typical controls found in the trigger system are: VARIABLE TRIGGER HOLDOFF, TRIGGER OPERATING MODES, TRIGGER SLOPE, TRIGGER LEVEL CONTROL, and TRIGGER COUPLING. Let's look at these adjustments and controls in more detail:

- VARIABLE TRIGGER HOLDOFF—This control adjusts the delay time before the waveform is drawn on the CRT. For example, imagine you are trying to display a waveform that is rather complex. The scope will have difficulty displaying

**FIGURE 4–20** Oscilloscope trigger system controls.

a stable waveform because there are many points on the complex wave that would cause the scope to begin drawing the pattern. By adjusting the VARIABLE TRIGGER HOLDOFF, you cause the scope to wait for a particular portion of the waveform to be present before the waveform is drawn. This provides a stable display. Typically, this control is set to the NORM position, but don't be afraid to adjust this control if the display is not stable.

- TRIGGER OPERATING MODES—This is actually several controls. These are NORMAL, P-P AUTO, and TELEVISION modes and TRIGGER SOURCE switches.

- NORMAL—This control allows the scope to be triggered when a waveform is detected. When a signal is not present, the scope is not triggered, and the CRT is blank.

- P-P AUTO—This control allows the scope to be triggered when a waveform is detected. When a signal is not present, a trigger signal is internally sent to the scope causing a bright baseline to appear on the CRT. The scope is typically set for P-P AUTO mode.

- TELEVISION—This control allows the scope to trigger on TV fields or lines. This function is not applicable in the industrial maintenance field.

- TRIGGER SOURCE SWITCHES—This is comprised of two switches, SOURCE and INT, that work in conjunction with one another.

  - SOURCE—The SOURCE switch allows the user to choose the trigger source. INT, or internal, means that the scope will trigger on the signal applied to either channel 1 or channel 2. LINE means that the scope will use the frequency of the AC power to the scope for the trigger signal. EXT, or external, means that the scope will trigger on a separate trigger signal that is applied to the EXT INPUT jack. Usually, the SOURCE switch is set to the INT position.

  - INT—In the CH 1 position, the scope will use the signal from CH 1 for triggering. In the CH 2 position, the scope will use the signal from CH 2 as the trigger signal. Setting this switch to the VERT MODE position tells the scope to use whichever signal is present. This switch is used when the SOURCE switch is set to the INT position. Typically, the INT switch is set to CH 1; however, many individuals find it more convenient to set the INT switch to VERT MODE.

- TRIGGER SLOPE—This controls whether the scope triggers on the rising or falling edge of a

signal. Usually, the TRIGGER SLOPE switch is set for a rising slope.

- TRIGGER LEVEL CONTROL—This controls *where* the trigger point occurs on the signal. For most measurements, you will set this control to its mid position. However, do not be afraid to adjust this control if your waveform is unstable.

- TRIGGER COUPLING—This switch is used when the SOURCE switch is set to the EXT position. Trigger coupling functions similarly to the coupling switches for the vertical section. This switch has three positions, AC − DC − DC ÷ 10. The AC position inserts a capacitor into the external trigger input circuit. This blocks any DC component from the signal upon which you are triggering. The scope will trigger on only the AC component of the signal. The DC position represents "direct coupled." This allows the scope to trigger on all components of the input signal. The DC ÷ 10 position also uses direct coupling, but attenuates the signal by a factor of 10. This is useful if the external trigger signal is too large.

## OSCILLOSCOPE PROBES

Refer to Figure 4–21. In order for you to make any measurements with an oscilloscope, you must connect the circuit to be tested to the channel 1 or channel 2 input of your scope. While it is possible to use an ordinary set of wires to do the job, you will get better, more accurate measurements by using oscilloscope probes.

There are three basic types of oscilloscope probes. The first type is called a *direct measurement* or 1X probe. This probe supplies an input signal

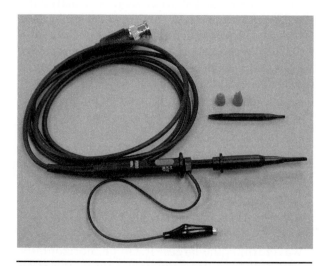

**FIGURE 4–21** A switchable oscilloscope probe.

that is of the same amplitude as the signal being measured. The second type of probe is called a 10X probe. This probe provides an input signal that is attenuated by a factor of 10. In addition, a 10X probe (due to its design) introduces less loading effect to the circuit being measured. For this reason, most users of oscilloscopes use a 10X probe at all times. The third type of oscilloscope probe combines the first two types. This probe has a small switch that allows the user to switch between the 1X setting and the 10X setting. This is the most convenient type of probe to use. When using this type of probe, always double-check the position of the switch so that your measurements are not off by a factor of 10.

The 10X probes must be matched to the oscilloscope input being used. This is called *compensating the probe*. To compensate the probe, connect the probe tip to the *probe adjust* test point on the oscilloscope. The probe adjust test point provides a test signal for compensating the probe. Adjust the scope for a stable display. You should see a square wave. Some probes are adjusted by rotating a collar near the probe tip. Other probes are compensated by adjusting a small screw with a nonmetallic screwdriver. The adjustment may be located in the probe or at the connector end. While adjusting the probe, look at the square wave on the CRT. Adjust the probe so that the tops and bottoms of the square wave are flat and the corners are sharp. Once your probe is compensated, you do not have to readjust it unless you move the probe to a different input or use it on a different scope. The probe is always compensated to the input that you are using.

The oscilloscope probes also contain a ground lead or clip. When making measurements, the probe tip is attached to the point of the circuit where the measurement is taken. The ground lead is attached to the reference point for the circuit. Be very careful when using the ground lead. You should be aware that on many dual channel scopes, the ground lead of channel 1 is internally connected to the ground lead of channel 2. This can cause problems. For example, suppose you wanted to compare the signal found on the primary of a transformer with the signal found on the secondary of the same transformer. You might be tempted to connect the channel 1 probe and ground lead across the primary and the channel 2 probe and ground lead across the secondary. *Don't do it!* You would create a short circuit from the primary to the secondary through the ground leads. Remember that the ground leads are internally connected. You should use an oscilloscope that has *isolated inputs*. This means that the ground for channel 1 is electrically isolated from the ground for channel 2.

Often, there is another precaution that you must observe when using ground leads. The ground leads may also be connected to the scope ground (the grounding conductor of the three-wire AC plug). This means that when you make a measurement, it is possible that you will connect the reference point in the circuit to AC ground through the ground lead and ground prong of the scope AC plug. To prevent this from occurring, some individuals may attach a three-prong to two-prong adapter to the AC plug of the scope. This is not the safest procedure to follow because it will defeat the equipment ground for the scope and may create a shock hazard. The ideal fix is to use a scope with isolated inputs.

# MAKING MEASUREMENTS

## Initializing the Scope

Before you begin to make your measurement, you should initialize the scope and check your setup. To initialize the scope, do the following:

1. Display System Controls
   A. Set the INTENSITY control to the mid-range position.
   B. Turn the FOCUS control to the mid-range position.
2. Vertical System Controls
   A. Turn the channel 1 POSITION control to the mid-range position.
   B. Turn the channel 2 POSITION control to the mid-range position.
   C. Place the VERTICAL MODE switch in the CH 1 position (if you are only using CH 1), the BOTH position (if you will be using CH 1 and CH 2 simultaneously), or CH 2 position (if you are only using CH 2).
   D. Turn both CH 1 and CH 2 VOLTS/DIV switches to their highest setting.
   E. Verify that the CH 1 and CH 2 VARIABLE VOLTS/DIV switch is in the CAL position or detent.
   F. Place the CH 1 and CH 2 INPUT COUPLING switches to DC (if you wish to view the DC and AC components of the signal), AC (if you only wish to view the AC component of the signal), or GND (if you wish to establish a reference).
3. Horizontal System Controls
   A. Turn the horizontal POSITION control to the mid-range position.

B. Set the SEC/DIV switch to the 0.5 ms position.

C. Verify that the VARIABLE SEC/DIV switch is in the CAL position or detent.

D. Verify that HORIZONTAL MAGNIFICATION is not selected by pushing in on the VARIABLE SEC/DIV switch.

4. Trigger System Controls

A. Set the VAR HOLDOFF control to the NORM position.

B. Set the TRIGGER OPERATING MODE switches as follows:

1. P-P AUTO on.
2. SOURCE to INT.
3. INT to VERT MODE.

C. Set the TRIGGER SLOPE switch to positive (if you wish to trigger on the leading edge of the signal) or negative (if you wish to trigger on the falling edge of the signal).

D. Set the TRIGGER LEVEL control to the midrange position.

Your scope is now initialized. Plug your scope into a properly grounded outlet and power up the scope. Connect an oscilloscope probe to the CH 1 input jack (or CH 2 if you are using channel 2 for your measurements, or connect a probe to both CH 1 and CH 2 if you will be measuring two different input signals).

5. Compensate the probe(s).

6. Verify that the ground of the circuit you will be testing is at the same potential as the oscilloscope ground.

A. Touch the probe tip to the circuit ground. If no difference of potential is detected, you may connect the oscilloscope ground lead to the circuit ground.

## Making Voltage Measurements

Refer to Figure 4–22. Voltage is measured vertically on the oscilloscope graticule. This means that when a waveform is displayed on the CRT, you will measure the height or amplitude of the waveform to determine the amount of voltage present. Following are the steps you should take to make a voltage measurement:

1. Connect the ground lead of the oscilloscope probe to the circuit ground.

2. Connect the probe tip of the oscilloscope probe to the point of the circuit where the voltage is to be measured.

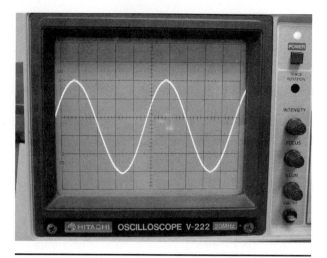

**FIGURE 4–22**   A 2.6 V peak-to-peak sine wave.

3. Adjust the SEC/DIV switch until a minimum of one complete cycle of the waveform is displayed horizontally across the CRT.

4. Adjust the TRIGGER LEVEL control as necessary to obtain a stable display.

5. Adjust the VERTICAL POSITION control until the waveform is centered on the CRT graticule.

6. Adjust the VOLTS/DIV switch until the displayed waveform fills the CRT from top to bottom. If the waveform extends beyond the top or bottom of the CRT, you will need to use the next larger VOLTS/DIV setting. (If you cannot adjust the waveform to fit vertically within the CRT, you will need to use a 10X probe instead of a 1X probe.)

7. Use the VERTICAL POSITION control to reposition the displayed waveform so that the bottom peak of the waveform is just touching the bottommost horizontal line (major division) on the CRT graticule.

8. Use the HORIZONTAL POSITION control to reposition the displayed waveform so that the top peak of the waveform is centered over the center vertical graticule line (the one with the minor divisions).

9. Count the major and minor divisions from the bottom of the negative peak to the top of the positive peak.

10. Multiply this number by the VOLTS/DIV setting. (Be sure to use the correct value of VOLTS/DIV for the type of probe that you are using.)

11. The product is the amount of peak-to-peak voltage for the displayed waveform.

Figure 4–22 shows an oscilloscope with the display properly adjusted for voltage measurement. If we count the number of major and minor divisions from the bottom of the negative peak to the top of the positive peak, we will have 5.2 divisions. Now, we multiply the number of divisions (5.2) by the VOLTS/DIV setting (0.5 V) and the product is 2.6 volts peak-to-peak.

## Making Time Measurements

Refer to Figure 4–23. Time is measured horizontally on the oscilloscope graticule. This means that when a waveform is displayed on the CRT, you will measure the width of one complete cycle of the waveform to determine the time of one complete cycle. Following are the steps you should take to make time measurements:

1. Connect the ground lead of the oscilloscope probe to the circuit ground.
2. Connect the probe tip of the oscilloscope probe to the point of the circuit where the time is to be measured.
3. Adjust the SEC/DIV switch until a minimum of one complete cycle of the waveform is displayed horizontally across the CRT.
4. Adjust the TRIGGER LEVEL control as necessary to obtain a stable display.
5. Adjust the VERTICAL POSITION control until the waveform is centered on the CRT graticule.
6. Adjust the VOLTS/DIV switch until the displayed waveform fills the CRT from top to bottom. If the waveform extends beyond the top or bottom of the CRT, you will need to use

the next larger VOLTS/DIV setting. (If you cannot adjust the waveform to fit vertically within the CRT, you will need to use a 10X probe instead of a 1X probe.)

7. Use the HORIZONTAL POSITION control to reposition the displayed waveform so that an easily identified key portion of the waveform is just touching the left-most vertical line (major division) of the CRT graticule.
8. Use the VERTICAL POSITION control to reposition the displayed waveform so that the key portion of the waveform is located on the center horizontal graticule line (the one with the minor divisions).
9. Count the major and minor divisions from the key portion of the left side of the waveform to the same key portion at the right side of the waveform (where the cycle repeats).
10. Multiply this number by the SEC/DIV setting.
11. The product is the amount of time (in sec, ms, or µs) of one complete cycle of the displayed waveform.

Figure 4–23 shows an oscilloscope with the display properly adjusted for time measurement. If we count the number of major and minor divisions from the key portion of the left side of the waveform to the same key portion of the right side of the waveform (where the cycle repeats), we will have 5.1 divisions. Now we multiply the number of divisions (5.1) by the SEC/DIV setting (2 ms) and the product is 10.2 ms.

## Frequency Measurement

Once the time it takes to complete one cycle is known, it is possible to compute the frequency of the waveform. We use a simple formula to determine frequency when time is known. The formula is:

$$F = \frac{1}{T}$$

where  $F$ = the frequency in Hz
$T$ = the time for one complete cycle, in seconds

As an example, we will use the values from the previous time measurement exercise. In this example:

$$T = 10.2 \text{ ms}$$
$$F = \frac{1}{T} = \frac{1}{10.2 \text{ ms}} = 98.04 \text{ Hz}$$

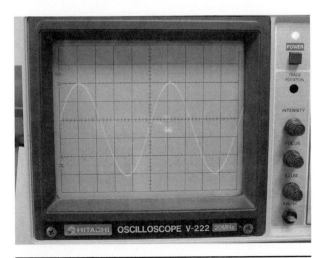

**FIGURE 4–23** A 10.2 ms sine wave thereby having a frequency of 98.04 Hz.

# REVIEW QUESTIONS

*Multiple Choice*

1. A voltage tester is
   a. good for ballpark measurements.
   b. very accurate.
   c. used to measure resistance.
   d. capable of indicating the exact amount of voltage measured.

2. Voltmeters are always connected
   a. in parallel with the component.
   b. in series with the component.
   c. next to the component.
   d. across the line.

3. Ammeters are always connected
   a. in series with the component.
   b. in parallel with the component.
   c. to the positive terminal.
   d. across the line.

4. Megohmmeters are used to measure
   a. high values of current.
   b. high values of voltage.
   c. high values of resistance.
   d. high values of frequency.

5. Oscilloscopes are useful because they allow you to
   a. measure and see the current.
   b. measure and see the resistance.
   c. measure and see the wattage.
   d. measure and see the voltage.

*Give Complete Answers*

1. List the steps that you would follow in preparing to make a voltage measurement with a digital multimeter.

2. How is a digital multimeter connected in a circuit when measuring current?

3. What precaution must you observe when measuring the resistance of a circuit with a digital multimeter?

4. List the steps you would follow to check for a blown fuse in a disconnect when using a voltage tester. Tell how you would know if the fuse was good or blown.

5. Name three uses for a voltage tester.

6. Describe the technique you would use that would allow you to measure small amounts of current with a clamp-on ammeter.

7. When would you use a megohmmeter?

8. The vertical sensitivity adjustment of an oscilloscope is set to 5 V/div, and the horizontal sensitivity adjustment is set to 2 ms/div. A displayed waveform is 6.2 divisions from peak to peak. What is the peak-to-peak voltage of the displayed waveform?

9. One cycle of the waveform in Problem 8 spans 4.7 divisions. What is the frequency of this waveform?

# Basic Resistive Electrical Circuits

## OBJECTIVES

After studying this chapter, the student will be able to:

■ Identify a series resistive circuit.

■ Identify a parallel resistive circuit.

■ Identify a combination resistive circuit.

■ Perform all necessary calculations to analyze a resistive electrical circuit.

Electrical circuits are the building blocks that comprise all electrical devices, no matter how simple or complex. A technician must gain a solid understanding of the basics if he or she ever hopes to understand more complicated devices. Regardless of the complexity, all electrical circuits consist of series, parallel, or combination circuits. This chapter will introduce you to these fundamental circuits upon which more complex circuits are built.

## SERIES CIRCUITS

Before we can begin our study of series, parallel, and combination circuits, we must first understand what comprises a *circuit*. A circuit must contain a minimum of three different elements. These are a *power source*, a *complete path* for current flow, and a *load*. Some examples of a power source would be a battery, a power supply, and a generator. To provide a path for the current flow, conductors are used. Conductors are generally in the form of wires, which connect the power source to the load. Conductors may also be the copper traces, or tracks, on a printed circuit board. A load is any device that draws current from the power source. A load could be a resistor, a light bulb, a motor, or other such devices.

In addition to these three items, some circuits also contain a form of *control*. A control device is used to switch the current flow on or off. Some examples of control devices would be a switch, a thermostat, and a relay. Figure 5–1 shows a complete circuit. Notice the power source (the battery), the conductors (wires), the load (light bulb), and the control (switch).

The first type of circuit that we will study is the **series circuit**. A series circuit is shown in Figure 5–2. Notice that this circuit is simple in appearance. The

circuit consists of a power source (indicated by the AC sine wave), conductors, loads (the three light bulbs), and a control device (the switch). When you look at this circuit, imagine yourself as the current flowing from the power source. As you leave the power source, you flow through the conductor to the switch. In Figure 5–2, the switch is open. This means that the current flow cannot continue to the loads. There is not a complete path because of the open switch. Therefore, the light bulbs will not light.

Now look at Figure 5–3. The switch has now been closed. The current can now flow from the power source, through the conductor to the switch, through the now-closed switch, through the conductor to light bulb 1, through light bulb 1, through the conductor to light bulb 2, through light bulb 2, through the conductor to light bulb 3, through light bulb 3, and through the conductor, back to the power source. Since there is a complete path for current flow, the light bulbs will light. Notice that the current could also flow in the opposite direction. That is, from the power source, through the conductor to light bulb 3, through light bulb 3, through the conductor to light bulb 2, through light bulb 2, through the conductor to light bulb 1, through light bulb 1, through the conductor to the switch, through the closed switch, and through the conductor back to

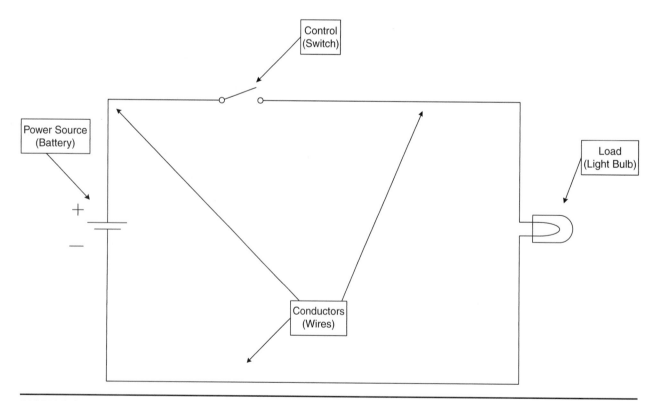

**FIGURE 5–1**  A complete electrical circuit containing a power source, complete path for current flow, a load, and a control.

the power source. Regardless of the direction of the current flow, there is no other path that the current could travel. This is an important characteristic of a series circuit. *A series circuit will have only one path for current flow.*

Now imagine that the filament of light bulb 2 has burned open, as seen in Figure 5–4. It is obvious that light bulb 2 will not light, but what about

light bulbs 1 and 3? Again, imagine yourself as the current flowing from the power source. As you leave the power source, you flow through the conductor to the closed switch. The current will flow through the closed switch through the conductor to light bulb 1, through light bulb 1, and through the conductor to light bulb 2. Since the filament of light bulb 2 has burned open, the current flow stops at

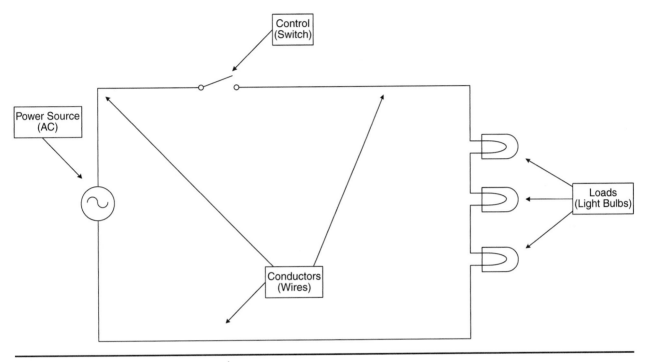

**FIGURE 5–2**   A series circuit—switch open.

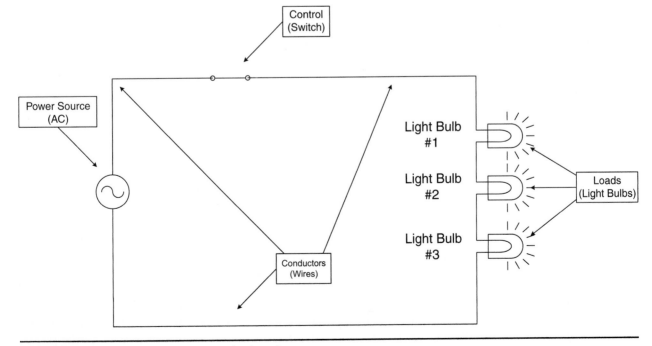

**FIGURE 5–3**   A series circuit—switch closed.

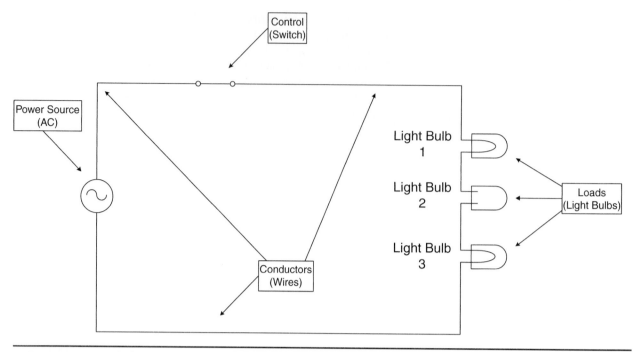

**FIGURE 5–4**   Light bulb 2 filament open.

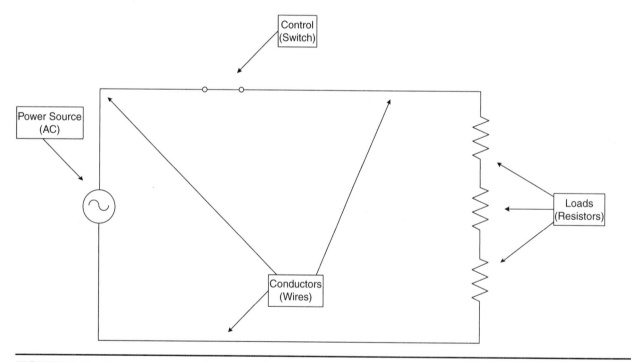

**FIGURE 5–5**   A series circuit with resistive loads.

this point. Since there is not a complete path for current flow, *none of the light bulbs will light*. In a circuit with light bulbs connected in series, if one light bulb burns out, all of the light bulbs will fail to light.

Figure 5–5 shows a circuit in which the light bulbs have been replaced with three resistors, $R_1$, $R_2$, and $R_3$. Is this a series circuit? Follow the path

for current flow. Is there only one path? Current will flow from the power source, through the conductor, through the closed switch, through the conductor to $R_1$, through $R_1$, through the conductor to $R_2$, through $R_2$, through the conductor to $R_3$, through $R_3$, and through the conductor back to the power source. Is this the only path that current can follow?

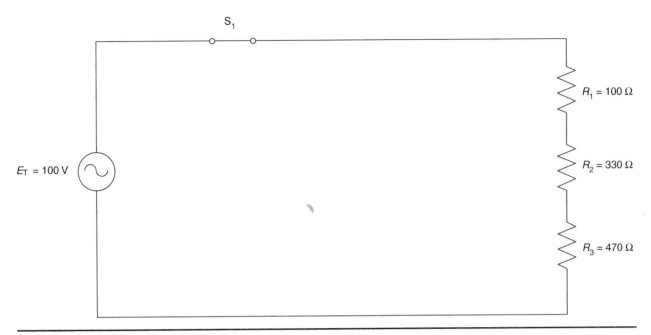

**FIGURE 5–6** A series circuit with values added.

Yes. The current could flow in the opposite direction, but the current must still follow the same path. There is only one current path in this circuit. Therefore, this is also a series circuit. In fact, resistors $R_1$, $R_2$, and $R_3$ are said to be connected in series.

Once we have identified the circuit as being a series circuit, we can determine other useful information about the circuit. Figure 5–6 is the same circuit as seen in Figure 5–5, except we have added values for the amount of voltage present as well as the values of $R_1$, $R_2$, and $R_3$. Since we know the amount of voltage present, can we determine the amount of current flowing in the circuit? In Chapter 2, we studied Ohm's law. We learned that by using Ohm's law, we could determine the amount of current if we know the amount of voltage and resistance. Recall the formula:

$$I = \frac{E}{R}$$

where       $I$ = current in amperes
              $E$ = voltage in volts
              $R$ = resistance in ohms

If we were to try to determine the amount of current flowing in this circuit, we would have some difficulty. What do we use for the value of $R$? Do we use the value of $R_1$, the value of $R_2$, or do we use the value of $R_3$? Actually, we will use the combined value of $R_1 + R_2 + R_3$. If we are to find the total circuit current, we must use the total circuit resistance as well as the total circuit voltage. This is another important characteristic about a series circuit. *The*

*total resistance of a series circuit is equal to the sum of the individual resistances.* Therefore, the formula for finding the total resistance of a series circuit is:

$$R_T = R_1 + R_2 + R_3 + \dots \qquad \text{(Eq. 5.1)}$$

What is the total resistance of the circuit shown in Figure 5–6?

What do we know?

$$R_1 = 100 \ \Omega$$
$$R_2 = 330 \ \Omega$$
$$R_3 = 470 \ \Omega$$

What do we not know?

$$R_T = ?$$

What formula could we use?

$$R_T = R_1 + R_2 + R_3 + \dots$$

Substitute the known values and solve for $R_T$:

$$R_T = R_1 + R_2 + R_3 + \dots$$
$$R_T = 100 \ \Omega + 330 \ \Omega + 470 \ \Omega$$
$$R_T = 900 \ \Omega$$

Therefore, the total resistance of the circuit shown in Figure 5–6 is 900 Ω. Now that we know the total resistance of the circuit, and we know the total applied voltage to the circuit, we can determine the total circuit current.

What do we know?

$$E_T = 100 \ V$$
$$R_T = 900 \ \Omega$$

What do we not know?

$$I_T = ?$$

What formula could we use?

$$I_T = \frac{E_T}{R_T}$$

Substitute the known values and solve for $I_T$:

$$I_T = \frac{E_T}{R_T}$$
$$I_T = \frac{100 \text{ V}}{900 \text{ }\Omega}$$
$$I_T = 111.11 \text{ mA}$$

Therefore, the total current flowing in the circuit in Figure 5–6 is 111.11 mA.

If the total current flowing in the circuit of Figure 5–6 is 111.11 mA, how much current is flowing through $R_1$? The answer is 111.11 mA. How much current is flowing through $R_2$? Again, the answer is 111.11 mA. The current flowing through $R_3$ will also be 111.11 mA. Since there is only one path for current flow in a series circuit, the current must be the same throughout a series circuit. This means that no matter where you measure the current in a series circuit, the amount of current will always be equal to the total circuit current. This is another important characteristic of a series circuit. *In a series circuit, the current is the same value throughout the entire circuit. The value of the current flowing through each individual component will be equal to the total circuit current.* This can be expressed in the formula:

$$I_T = I_1 = I_2 = I_3 = \dots \quad \text{(Eq. 5.2)}$$

Now we will turn our attention to the voltage. If we were to place a digital voltmeter across resistor $R_1$, how much voltage would we measure? How much voltage would we measure across resistor $R_2$? What about resistor $R_3$? Before we answer these questions, let us analyze this. There is 100 V applied to the circuit shown in Figure 5–6. We have determined that there is 111.11 mA of current flowing in this circuit. The power source is supplying 100 V of pressure to push 111.11 mA of current through $R_1$, $R_2$, and $R_3$. What happens to the pressure when the current reaches $R_1$? There will be a pressure drop. How much will the pressure drop? We can determine the amount of pressure, or voltage, drop by using an Ohm's law formula. Recall the formula for finding voltage when current and resistance are known:

$$E = I \times R$$

Since we are interested in knowing the voltage drop across resistor $R_1$, we will use the current flowing through $R_1$ and the resistance value of $R_1$.

What do we know?

$$I_1 = 111.11 \text{ mA}$$
$$R_1 = 100 \text{ }\Omega$$

What do we not know?

$$E_1 = ?$$

What formula could we use?

$$E_1 = I_1 \times R_1$$

Substitute the known values and solve for $E_1$:

$$E_1 = I_1 \times R_1$$
$$E_1 = 111.11 \text{ mA} \times 100 \text{ }\Omega$$
$$E_1 = 11.11 \text{ V}$$

Therefore, the voltage dropped across $R_1$ is 11.11 V. How much voltage is dropped across resistor $R_2$? We will use the same value for current, but we must use the resistance value for $R_2$.

What do we know?

$$I_2 = 111.11 \text{ mA}$$
$$R_2 = 330 \text{ }\Omega$$

What do we not know?

$$E_2 = ?$$

What formula could we use?

$$E_2 = I_2 \times R_2$$

Substitute the known values and solve for $E_2$:

$$E_2 = I_2 \times R_2$$
$$E_2 = 111.11 \text{ mA} \times 330 \text{ }\Omega$$
$$E_2 = 36.67 \text{ V}$$

Therefore, the voltage dropped across $R_2$ is 36.67 V. How much voltage is dropped across resistor $R_3$? We will use the same value for current, but we must use the resistance value for $R_3$.

What do we know?

$$I_3 = 111.11 \text{ mA}$$
$$R_3 = 470 \text{ }\Omega$$

What do we not know?

$$E_3 = ?$$

What formula could we use?

$$E_3 = I_3 \times R_3$$

Substitute the known values and solve for $E_3$:

$$E_3 = I_3 \times R_3$$
$$E_3 = 111.11 \text{ mA} \times 470 \text{ }\Omega$$
$$E_3 = 52.22 \text{ V}$$

Therefore, the voltage dropped across $R_3$ is 52.22 V. Now notice something interesting. If we take the voltage drop across $R_1$, add it to the voltage drop across $R_2$, and add the voltage dropped across $R_3$, we will find the total applied voltage.

$$E_1 = 11.11 \text{ V}$$
$$E_2 = 36.67 \text{ V}$$
$$E_3 = 52.22 \text{ V}$$
$$E_T = E_1 + E_2 + E_3$$
$$E_T = 11.11 \text{ V} + 36.67 \text{ V} + 52.22 \text{ V}$$
$$E_T = 100 \text{ V}$$

This is another important characteristic of a series circuit: *The sum of all individual voltage drops will equal the total applied voltage.* You should notice something else about the voltage drops in this circuit. Compare the voltage drop across resistor $R_1$ with the voltage drop across resistor $R_2$ and the voltage drop across resistor $R_3$. Which voltage drop is larger? The 52.22 V voltage drop across $R_3$ is larger. Which resistor is larger? $R_3$ at 470 Ω is larger than $R_1$ at 100 Ω or $R_2$ at 330 Ω. *The largest resistor will have the largest voltage drop.*

There is one other parameter of a circuit with which we must be concerned, that is, the amount of power consumed by the circuit and the individual components of the circuit. Circuit power is measured in *watts* (W). With the individual components, watts are a measure of the amount of power (in the form of heat) that the component must dissipate. Let us see how this works by again referring to Figure 5–6. We now know the total circuit voltage, the total circuit current, and the total circuit resistance. We do not know the total circuit power. We also know the voltage drops across $R_1$, $R_2$, and $R_3$, as well as the current flowing through each resistor, and the resistive value of each resistor. We do not know how much power each resistor must dissipate. We will begin by finding the total power, $P_T$. Recall from Chapter 3 the three formulas for finding power:

$$P = I \times E$$
$$P = \frac{E^2}{R}$$
$$P = I^2 \times R$$

We can use any of these three formulas to determine the total circuit power. We know $I_T$ and $E_T$; therefore, we could use the first formula. We know $E_T$ and $R_T$; therefore, we could use the second formula. Finally, we know $I_T$ and $R_T$; therefore, we could use the third formula. For this example, we will use the second formula. You may wish to try to solve the example using either or both of the remaining formulas. Your results should be the same. Now let us find $P_T$.

What do we know?

$$E_T = 100 \text{ V}$$
$$R_T = 900 \text{ Ω}$$

What do we not know?

$$P_T = ?$$

What formula could we use?

$$P_T = \frac{E_T{}^2}{R_T}$$

Substitute the known values and solve for $P_T$:

$$P_T = \frac{E^2}{R}$$
$$P_T = \frac{100\,^2\text{V}}{900\,\text{Ω}}$$
$$P_T = 11.11 \text{ W}$$

Therefore, this circuit will consume 11.11 W of power. Now, let us turn our attention to the individual resistors, $R_1$, $R_2$, and $R_3$. How much power must each resistor dissipate? We will begin with $R_1$. What do we know?

$$E_1 = 11.11 \text{ V}$$
$$I_1 = 111.11 \text{ mA}$$
$$R_1 = 100 \text{ Ω}$$

What do we not know?

$$P_1 = ?$$

What formula could we use? (We could use any one of the following.)

$$P_1 = I_1 \times E_1$$
$$P_1 = \frac{E_1{}^2}{R_1}$$
$$P_1 = I_1{}^2 \times R_1$$

Substitute the known values and solve for $P_1$. (We will use the first equation.)

$$P_1 = I_1 \times E_1$$
$$P_1 = 111.11 \text{ mA} \times 11.11 \text{ V}$$
$$P_1 = 1.23 \text{ W}$$

Therefore, resistor $R_1$ must dissipate 1.23 W of heat. In the actual circuit, the minimum wattage rating for resistor $R_1$ should be 2.46 W. A good rule of thumb is that the resistor wattage rating should be twice the actual wattage that must be handled. Now, let us determine how much power resistor $R_2$ must dissipate.

What do we know?

$$E_2 = 36.67 \text{ V}$$
$$I_2 = 111.11 \text{ mA}$$
$$R_2 = 330 \text{ Ω}$$

What do we not know?

$$P_2 = ?$$

What formula could we use? (We could use any one of the following.)

$$P_2 = I_2 \times E_2$$
$$P_2 = \frac{E_2^2}{R_2}$$
$$P_2 = I_2^2 \times R_2$$

Substitute the known values and solve for $P_2$. (We will use the third equation.)

$$P_2 = I_2^2 \times R_2$$
$$P_2 = 111.11^2 \text{ mA} \times 330 \text{ }\Omega$$
$$P_2 = 4.07 \text{ W}$$

Therefore, resistor $R_2$ must dissipate 4.07 W of heat. In the actual circuit, resistor $R_2$ should be rated at a minimum of 8.14 W. Now, let us determine how much power resistor $R_3$ must dissipate.

What do we know?

$$E_3 = 52.22 \text{ V}$$
$$I_3 = 111.11 \text{ mA}$$
$$R_3 = 470 \text{ }\Omega$$

What do we not know?

$$P_3 = ?$$

What formula could we use? (We could use any one of the following.)

$$P_3 = I_3 \times E_3$$
$$P_3 = \frac{E_3^2}{R_3}$$
$$P_3 = I_3^2 \times R_3$$

Substitute the known values and solve for $P_3$. (We will use the second equation.)

$$P_3 = \frac{E_3^2}{R_3}$$
$$P_3 = \frac{52.22^2 \text{V}}{470 \text{ }\Omega}$$
$$P_3 = 5.80 \text{ W}$$

Therefore, resistor $R_3$ must dissipate 5.80 W of heat. In the actual circuit, resistor $R_3$ should be rated at a minimum of 11.60 W.

It is interesting that if we take the actual wattage dissipated by $R_1$, add it to the actual wattage dissipated by $R_2$, and add the actual wattage dissipated by $R_3$, we will find the total wattage consumed by the circuit.

$$P_1 = 1.23 \text{ W}$$
$$P_2 = 4.07 \text{ W}$$

$$P_3 = 5.80 \text{ W}$$
$$P_T = P_1 + P_2 + P_3$$
$$P_T = 1.23 \text{ W} + 4.07 \text{ W} + 5.80 \text{ W}$$
$$P_T = 11.10 \text{ W}$$

(Do not be concerned with the 0.01 W difference between the two $P_T$ answers. This is a result of the rounding of numbers to two decimal places during calculations.) This is another important characteristic of a series circuit: *The sum of all individual wattages will equal the total circuit wattage.* You should notice something else about the wattage in this circuit. Compare the power dissipated by the three resistors, $R_1$, $R_2$, and $R_3$. Which wattage is larger? The 5.80 W of $R_3$ is larger. Which resistor is larger? $R_3$ at 470 $\Omega$ is larger than $R_1$ at 100 $\Omega$ or $R_2$ at 330 $\Omega$. *The largest resistor must dissipate the most wattage.*

## PARALLEL CIRCUITS

The second type of circuit we will study is the **parallel circuit** (see Figure 5–7). The circuit consists of a power source (indicated by the AC sine wave), conductors, three loads (the light bulbs), and a control device (the switch). When you look at this circuit, imagine yourself as the current flowing from the power source. As you leave the power source, you flow through the conductor, to the switch. In Figure 5–7, the switch is open. This means that the current flow cannot continue to the load. There is not a complete path because of the open switch; therefore, the light bulbs will not light.

Now look at Figure 5–8. The switch has been closed. The current can now flow from the power source, through the conductor to the switch, through the now-closed switch, and through the conductor to junction A. At junction A the current can take one of two paths. The current can flow through light bulb 1 or continue toward light bulb 2 and light bulb 3. Actually, the current will do both. The current will *divide*, with some of the current flowing through light bulb 1, and some of the current flowing toward light bulbs 2 and 3. If you have difficulty understanding this, think of connecting two garden hoses to a single faucet with a "Y" connector, as seen in Figure 5–9. Some of the water will flow through one hose, while some of the water will flow through the second hose. The same principle applies to our circuit. Returning to Figure 5–8, let us leave the current flowing through light bulb 1 for a moment and turn our attention to the current flowing toward light bulb 2 and light bulb 3. At junction B, the current will again divide. Some of the current will flow through light bulb 2, while some of the current will flow through light

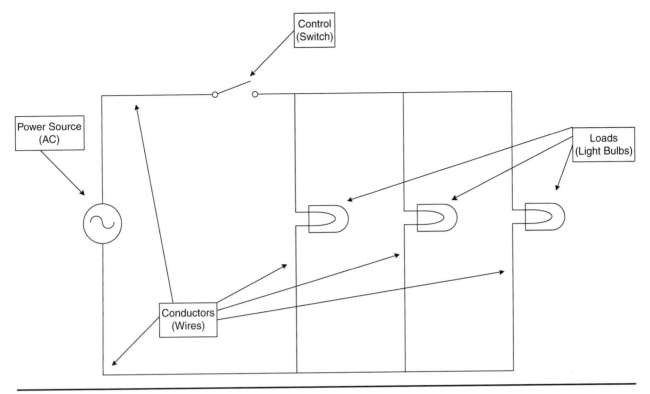

**FIGURE 5–7** A parallel circuit—switch open.

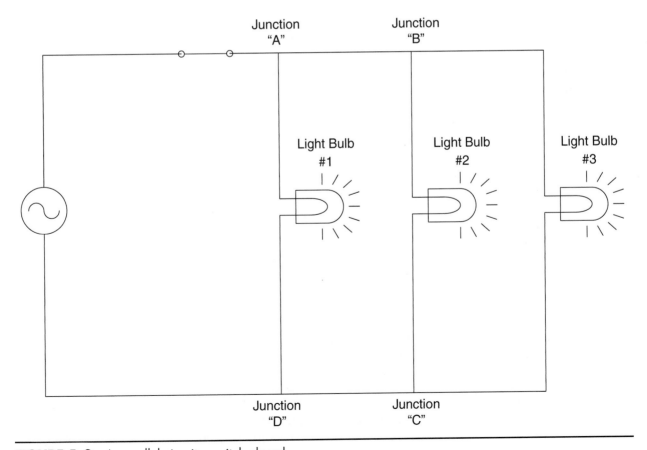

**FIGURE 5–8** A parallel circuit—switch closed.

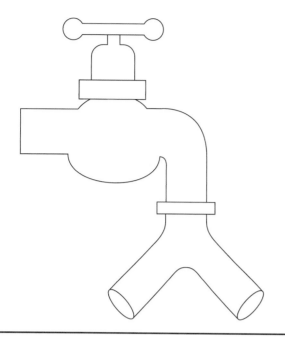

**FIGURE 5–9** Current flow in a parallel circuit is analogous to water flow through a Y-pipe.

bulb 3. The current flowing through the light bulbs must return to the power source. Therefore, after the current flows through light bulb 3, it will combine at junction C with the current flowing through light bulb 2. The combined currents (from light bulbs 2 and 3) will combine at junction D with the current flowing through light bulb 1. The total recombined current will then flow through the conductor back to the power source. Since there is a complete path for current flow, the light bulbs will light. Notice that the current could also flow in the opposite direction, that is, from the power source, through the conductor to junction D where the current will divide. Some of the current will flow through light bulb 1 and some of the current will flow toward light bulb 2 and light bulb 3. At junction C the current will divide. Some of the current will flow through light bulb 2, and some of the current will flow through light bulb 3. The current from light bulb 3 will combine with the current from light bulb 2 at junction B. This current will then combine at junction "A" with the current from light bulb 1. The total recombined current then flows through the conductor to the closed switch, through the switch, and back to the power source. Notice, in this circuit, there is more than one path for current flow. Current can flow from the source through light bulb 1, or current can flow from the source through light bulb 2, or current can flow from the source through light bulb 3. In fact, should the filament of light bulb 2 burn open, light bulb 1 and light bulb 3 would still light. Likewise, if light bulb 1 burned out, light bulb 2 and light bulb 3 would still light. Finally,

if light bulb 3 burned out, light bulb 1 and light bulb 2 would still light. This means that there are three paths for current flow in this circuit.

Now look at Figure 5–10. This circuit contains five light bulbs. Notice that each light bulb will work independently of the other. This circuit also has more than one path for current flow. In fact, there are five current paths in this circuit; therefore, this circuit is also a parallel circuit. This is an important characteristic of a parallel circuit. *A parallel circuit will have two or more paths for current flow.*

Figure 5–11 shows a circuit in which the light bulbs have been replaced with three resistors, $R_1$, $R_2$, and $R_3$. Is this a parallel circuit? Follow the path for current flow. Is there only one path or are there two or more paths? Current will flow from the power source, through the conductor, and through the closed switch to junction A. At junction A the current will divide. Some of the current will flow through $R_1$, and some of the current will flow toward $R_2$ and $R_3$. At junction B, the current will divide again. Some of the current will flow through $R_2$, and some of the current will flow through $R_3$. The currents from $R_2$ and $R_3$ will recombine at junction C. Next, the current will recombine with the current for resistor $R_1$ at junction A. This current will then flow through the conductor back to the power source. How many paths exist for current flow? There are three paths. Therefore, this is a parallel circuit. In fact, resistors $R_1$, $R_2$, and $R_3$ are said to be connected in parallel.

Once we have identified the circuit as being a parallel circuit, we can determine other useful information about the circuit. Figure 5–12 is the same circuit as seen in Figure 5–11, except we have now added values for the amount of voltage present as well as the values of $R_1$, $R_2$, and $R_3$. We will use the same values used in our previous example of a series circuit so that you can see the effects of connecting components in series or parallel. Since we know the amount of voltage present, can we determine the amount of current flowing in the circuit?

If we were to try to determine the amount of current flowing in this circuit, we would have some difficulty. What do we use for the value of $R$? Do we use the value of $R_1$, or do we use the value of $R_2$ or $R_3$? Actually, the total circuit resistance will be *less* than any individual resistor. How can this be? Think about our earlier example of the garden hose. Suppose you could connect three, four, or more hoses to that single faucet. Would you agree that you have provided more paths for the water to flow? If there are more paths for the water to flow, then the overall opposition to the flow of the water must be less. The same applies to our circuit. If more paths are provided for current flow, then the overall opposition

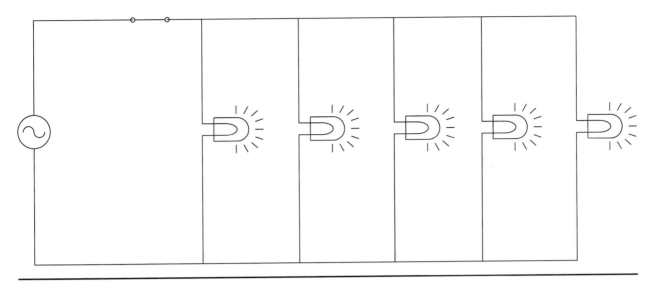

**FIGURE 5-10** A parallel circuit consisting of five branches.

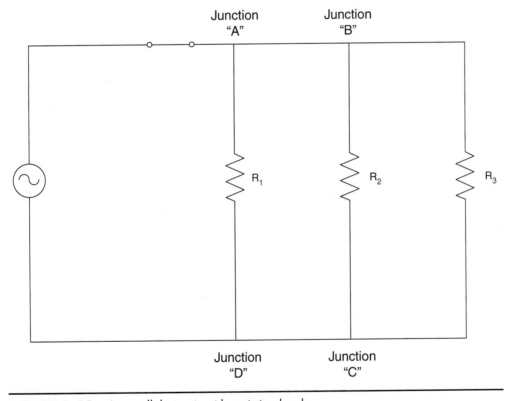

**FIGURE 5-11** A parallel circuit with resistive loads.

to current flow must be less. *In a parallel circuit, the total circuit resistance will be smaller than the smallest branch resistance.*

To determine the opposition to current flow in a parallel circuit, we will use the following formula:

$$R_T = \cfrac{1}{\cfrac{1}{R_1} + \cfrac{1}{R_2} + \cfrac{1}{R_3} + \cdots} \qquad \text{(Eq. 5.3)}$$

*The total resistance of a parallel circuit is equal to the reciprocal of the sum of the reciprocals of the individual resistances.* What is the total resistance of the circuit shown in Figure 5–12?

What do we know?

$$R_1 = 100 \ \Omega$$
$$R_2 = 330 \ \Omega$$
$$R_3 = 470 \ \Omega$$

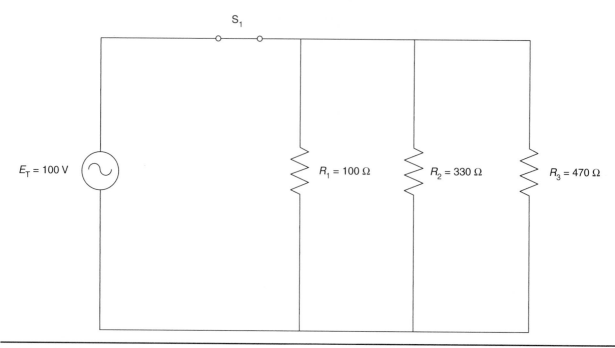

**FIGURE 5–12** A parallel circuit with values added.

What do we not know?

$$R_T = ?$$

What formula could we use?

$$R_T = \cfrac{1}{\cfrac{1}{R_1} + \cfrac{1}{R_2} + \cfrac{1}{R_3} + \ldots}$$

Substitute the known values and solve for $R_T$:

$$R_T = \cfrac{1}{\cfrac{1}{R_1} + \cfrac{1}{R_2} + \cfrac{1}{R_3} + \ldots}$$

$$R_T = \cfrac{1}{\cfrac{1}{100\ \Omega} + \cfrac{1}{330\ \Omega} + \cfrac{1}{470\ \Omega}}$$

$$R_T = 65.97\ \Omega$$

Therefore, the total resistance of the circuit shown in Figure 5–12 is 65.97 Ω. Now that we know the total resistance of the circuit, and we know the total applied voltage to the circuit, we can determine the total circuit current.

What do we know?

$$E_T = 100\ V$$
$$R_T = 65.97\ \Omega$$

What do we not know?

$$I_T = ?$$

What formula could we use?

$$I_T = \frac{E_T}{R_T}$$

Substitute the known values and solve for $I_T$:

$$I_T = \frac{E_T}{R_T}$$

$$I_T = \frac{100\ V}{65.97\ \Omega}$$

$$I_T = 1.52\ A$$

Therefore, the total current flowing in the circuit in Figure 5–12 is 1.52 A.

If the total applied voltage is 100 V, how much voltage is dropped across $R_1$, $R_2$, and $R_3$? Would you believe 100 V? Look carefully at Figure 5–12. Imagine placing one probe of your voltmeter at the top of $R_1$. Now, imagine placing the other probe of your voltmeter at the bottom of $R_1$. This is how you would measure the voltage dropped across $R_1$. Look at what you are actually measuring. Follow the connection from the top of $R_1$ to the left and to the top of the power source. Now, follow the connection from the bottom of $R_1$ to the left and to the bottom of the power source. You have actually placed your voltmeter across the power source. This means that you will measure the voltage of the power source, 100 V. Therefore, the voltage drop across $R_1$ is the same as the voltage of the power source, 100 V. The voltage drop across $R_2$ will also be 100 V. Likewise, the voltage dropped across $R_3$ will be 100 V. *In a parallel circuit, the voltage across the individual branch circuits will be the same.* Expressed as a formula:

$$E_T = E_1 = E_2 = E_3 = \ldots \qquad \text{(Eq. 5.4)}$$

If the total current flowing in the circuit of Figure 5–12 is 1.52 A, how much current is flowing through $R_1$? Since we know the voltage drop across $R_1$ is 100 V, and we know the resistive value of $R_1$ is 100 $\Omega$, we can determine the current flowing through $R_1$.

What do we know?

$$E_1 = 100 \text{ V}$$
$$R_1 = 100 \ \Omega$$

What do we not know?

$$I_1 = \ ?$$

What formula could we use?

$$I_1 = \frac{E_1}{R_1}$$

Substitute the known values and solve for $I_1$.

$$I_1 = \frac{E_1}{R_1}$$
$$I_1 = \frac{100 \text{ V}}{100 \ \Omega}$$
$$I_1 = 1 \text{ A}$$

Therefore, there is 1 A of current flowing through $R_1$. How much current is flowing through $R_2$?

What do we know?

$$E_2 = 100 \text{ V}$$
$$R_2 = 330 \ \Omega$$

What do we not know?

$$I_2 = \ ?$$

What formula could we use?

$$I_2 = \frac{E_2}{R_2}$$

Substitute the known values and solve for $I_2$.

$$I_2 = \frac{E_2}{R_2}$$
$$I_2 = \frac{100 \text{ V}}{330 \ \Omega}$$
$$I_2 = 303.03 \text{ mA}$$

Therefore, there is 303.03 mA of current flowing through resistor $R_2$. How much current is flowing through $R_3$?

What do we know?

$$E_3 = 100 \text{ V}$$
$$R_3 = 470 \ \Omega$$

What do we not know?

$$I_3 = \ ?$$

What formula could we use?

$$I_3 = \frac{E_3}{R_3}$$

Substitute the known values and solve for $I_3$.

$$I_3 = \frac{E_3}{R_3}$$
$$I_3 = \frac{100 \text{ V}}{470 \ \Omega}$$
$$I_3 = 212.77 \text{ mA}$$

Recall from our tracing of the current flow in our parallel circuit, that the current divided at junctions A and B and recombined at junctions C and D. This means that the total circuit current of 1.52 A divided into three branch currents of 1 A, 303.03 mA, and 212.77 mA. These three branch currents must recombine into the total circuit current. We can express this as a formula:

$$I_T = I_1 + I_2 + I_3 + \ldots \qquad \text{(Eq. 5.5)}$$

Let us see if this is true.

What do we know?

$$I_1 = 1 \text{ A}$$
$$I_2 = 303.03 \text{ mA}$$
$$I_3 = 212.77 \text{ mA}$$

What do we not know?

$$I_T = \ ?$$

What formula could we use?

$$I_T = I_1 + I_2 + I_3 + \ldots$$

Substitute the known values and solve for $I_T$:

$$I_T = I_1 + I_2 + I_3$$
$$I_T = 1 \text{ A} + 303.03 \text{ mA} + 212.77 \text{ mA}$$
$$I_T = 1.52 \text{ A}$$

As you can see, this is the same value of current that we calculated earlier. The current leaving the power source divides through the individual branch circuits. The branch circuit currents then recombine and return to the power source. Therefore, the current that leaves the power source must equal the current that returns to the power source. Or, expressed another way, *in a parallel circuit, the sum of the individual branch currents will equal the total circuit current.*

There is one other parameter of a circuit with which we must be concerned, that is, the amount of power consumed by the circuit and the individual components of the circuit. We will again refer to Figure 5–12. We now know the total circuit voltage, the total circuit current, and the total circuit resistance. We do not know the total circuit power.

We also know the voltage drop across $R_1$, $R_2$, and $R_3$, as well as the current flowing through each resistor, and the resistive value of each resistor. We do not know how much power each resistor must dissipate. We will begin by finding the total power, $P_T$. Recall again the three formulas for finding power:

$$P = I \times E$$

$$P = \frac{E^2}{R}$$

$$P = I^2 \times R$$

We can use any of these formulas to determine the total circuit power. We know $I_T$ and $E_T$; therefore, we could use the first formula. We know $E_T$ and $R_T$; therefore, we could use the second formula. Finally, we know $I_T$ and $R_T$; therefore, we could use the third formula. For this example, we will use the second formula. You may wish to try to solve the example using either or both of the remaining formulas. Your results should be the same. Now let us find $P_T$.

What do we know?

$$E_T = 100 \text{ V}$$
$$R_T = 65.97 \text{ }\Omega$$

What do we not know?

$$P_T = ?$$

What formula could we use?

$$P_T = \frac{E^2}{R}$$

Substitute the known values and solve for $P_T$:

$$P_T = \frac{E^2}{R}$$

$$P_T = \frac{100^2 \text{ V}}{65.97 \text{ }\Omega}$$
$$P_T = 151.58 \text{ W}$$

Therefore, this circuit will consume 151.58 W of power. Now, let us turn our attention to the individual resistors, $R_1$, $R_2$, and $R_3$. How much power must each resistor dissipate? We will begin with $R_1$.

What do we know?

$$E_1 = 100 \text{ V}$$
$$I_1 = 1 \text{ A}$$
$$R_1 = 100 \text{ }\Omega$$

What do we not know?

$$P_1 = ?$$

What formula could we use? (We could use any one of the following.)

$$P_1 = I_1 \times E_1$$

$$P_1 = \frac{E_1^2}{R_1}$$

$$P_1 = I_1^2 \times R_1$$

Substitute the known values and solve for $P_1$. (We will use the first equation.)

$$P_1 = I_1 \times E_1$$
$$P_1 = 1 \text{ A} \times 100 \text{ V}$$
$$P_1 = 100 \text{ W}$$

Therefore, resistor $R_1$ must dissipate 100 W of heat. Recall that a good rule of thumb is that the resistor wattage rating should be twice the actual wattage that must be handled. Therefore, $R_1$ should be rated at 200 W. Next, we will determine how much power resistor $R_2$ must dissipate.

What do we know?

$$E_2 = 100 \text{ V}$$
$$I_2 = 303.03 \text{ mA}$$
$$R_2 = 330 \text{ }\Omega$$

What do we not know?

$$P_2 = ?$$

What formula could we use? (We could use any one of the following.)

$$P_2 = I_2 \times E_2$$

$$P_2 = \frac{E_2^2}{R_2}$$

$$P_2 = I_2^2 \times R_2$$

Substitute the known values and solve for $P_2$. (We will use the third equation.)

$$P_2 = I_2^2 \times R_2$$
$$P_2 = 303.03^2 \text{ mA} \times 330 \text{ }\Omega$$
$$P_2 = 30.30 \text{ W}$$

Therefore, resistor $R_2$ must dissipate 30.30 W of heat. In the actual circuit, resistor $R_2$ should be rated at a minimum of 60.60 W. Now, let us determine how much power resistor $R_3$ must dissipate.

What do we know?

$$E_3 = 100 \text{ V}$$
$$I_3 = 212.77 \text{ mA}$$
$$R_3 = 470 \text{ }\Omega$$

What do we not know?

$$P_3 = ?$$

What formula could we use? (We could use any one of the following.)

$$P_3 = I_3 \times E_3$$

$$P_3 = \frac{E_3^2}{R_3}$$

$$P_3 = I_3^2 \times R_3$$

Substitute the known values and solve for $P_3$. (We will use the second equation.)

$$P_3 = \frac{E_3^2}{R_3}$$

$$P_3 = \frac{100^2 \text{ V}}{470 \text{ }\Omega}$$

$$P_3 = 21.28 \text{ W}$$

Therefore, resistor $R_3$ must dissipate 21.28 W of heat. In the actual circuit, resistor $R_3$ should be rated at a minimum of 42.56 W.

It is interesting that if we take the wattage dissipated by $R_1$ and add it to the wattage dissipated by $R_2$ and $R_3$, we will find the total wattage consumed by the circuit.

$$P_1 = 100 \text{ W}$$
$$P_2 = 30.30 \text{ W}$$
$$P_3 = 21.28 \text{ W}$$
$$P_T = P_1 + P_2 + P_3$$
$$P_T = 100 \text{ W} + 30.30 \text{ W} + 21.28 \text{ W}$$
$$P_T = 151.58 \text{ W}$$

As we learned earlier in our discussion of series circuits, the same is true for a parallel circuit: *The sum of all individual wattages will equal the total circuit wattage.*

## COMBINATION CIRCUITS

The third type of circuit that we will study is the **combination circuit**. As the name implies, a combination circuit consists of components that are connected in both series and parallel. Figures 5–13 and 5–14 show two types of combination circuits. In Figure 5–13, resistor $R_1$ is connected in series with the parallel combination of $R_2$ and $R_3$. In Figure 5–14, resistors $R_1$ and $R_2$ are connected in series. This series combination is then paralleled by resistor $R_3$. By looking at each of these circuits in more detail, we may better understand how to analyze them.

The circuit in Figure 5–13 consists of a power source (indicated by the AC sine wave), conductors, three loads ($R_1$, $R_2$, and $R_3$), and a control device (the switch). When you look at this circuit, imagine yourself as the current flowing from the power source. As you leave the power source, you flow through the conductor to the switch. In Figure 5–13, the switch is closed. The current can now flow through the closed switch, through the conductor to resistor $R_1$, and through resistor $R_1$ to the junction at A. At junction A, the current can take one of two paths; it can flow through resistor $R_2$ or continue toward resistor $R_3$. Actually, the current will do both. The current will *divide*, with some of the current flowing through $R_2$ and some of the current flowing through $R_3$. The current flowing through $R_2$ and $R_3$ must return to the power source. Therefore, after the current flows through $R_3$, it will combine at junction B with the current flowing through $R_2$. The total recombined current will then flow through the conductor back to the power source. Notice that the current could also flow in the opposite direction, that is, from the power source, through the conductor to junction B where the current will divide. Some of the current will flow through $R_2$, and some of the current will flow through $R_3$. The current from $R_3$ will combine with the current from $R_2$ at junction A. This current will then flow through resistor $R_1$, through the conductor to the closed switch, through the switch, and back to the power source. In this circuit, there is a portion of the circuit where all of the circuit current must flow. This is between the power source and junction A and between the power source and junction B. The current will be the same value at either of these two points. Since this is the only path that current can follow, these portions of the circuit are series connected. At junctions A and B the current can follow two paths. This tells us that this portion of the circuit is parallel connected.

Figure 5–14 is also a combination circuit. However, it looks a little different. Let us trace the current path through this circuit to help us understand better.

This circuit also consists of a power source (indicated by the AC sine wave), conductors, three loads ($R_1$, $R_2$, and $R_3$), and a control device (the switch). Again, imagine yourself as the current flowing from the power source. As you leave the power source, you flow through the conductor to the switch. In Figure 5–14, the switch is closed; the current can now flow through the closed switch and through the conductor to junction A. At junction A, the current can take one of two paths. Some of the current will flow through resistor $R_1$, through resistor $R_2$ to the junction at B. Some of the current will flow from junction A through resistor $R_3$ to junction "B." The current flowing through resistor $R_1$ and resistor $R_2$ has only one path to follow. This tells us that resistors $R_1$ and $R_2$ must be connected in series. This also tells us that the current flowing through $R_1$ must be the same value as the current flowing through $R_2$. Since the current divided at junction A, we know that resistor $R_3$ is connected in parallel with the series combination of $R_1$ and $R_2$.

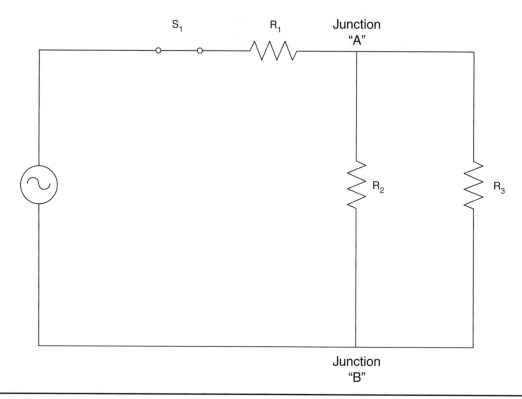

**FIGURE 5–13** One type of combination circuit.

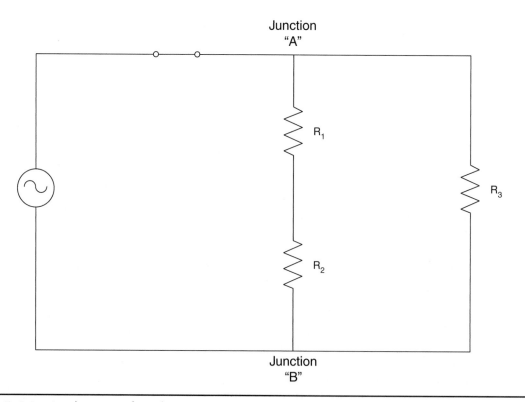

**FIGURE 5–14** Another type of combination circuit.

The current flowing through the $R_1$, $R_2$ branch and the current flowing through the $R_3$ branch must return to the power source. These two branch currents will combine at junction B. The total combined current will then flow through the conductor back to the power source. Notice that the current could also flow in the opposite direction, that is, from the power source and through the conductor to junction B, where the current will divide. Some of the current will flow through the $R_1$, $R_2$ series branch, and some of the current will flow through the $R_3$ branch. The current from $R_3$ will combine with the current from the $R_1$, $R_2$ series branch at junction A. This current will then flow through the conductor to the closed switch, through the switch, and back to the power source.

Once we have identified the circuit as being a combination circuit, we can determine other useful information about it. Figure 5–15 is the same circuit as seen in Figure 5–13, except we have added values for the amount of voltage present as well as the values of $R_1$, $R_2$, and $R_3$. We will use the same values as in our previous examples so that you can see the effects of connecting components in a combination circuit. Since we know the amount of voltage present, can we determine the amount of current flowing in the circuit?

If we were to try to determine the amount of current flowing in this circuit, we would have some difficulty. What do we use for the value of $R_T$? Do we use the value of $R_1$, or do we use the value of $R_2$, $R_3$, or a combination? We must first look at the circuit

and determine which components are connected in series and which are connected in parallel. We cannot combine resistor $R_1$ with $R_2$ because resistor $R_3$ is connected in parallel with $R_2$. Likewise, we cannot combine resistor $R_1$ with $R_3$ because resistor $R_2$ is connected in parallel with $R_3$. We must first find the resistance of the parallel combination of resistors $R_2$ and $R_3$. This resistance can then be combined with resistor $R_1$ to give us the total circuit resistance.

To determine the resistance of the parallel combination of resistors $R_2$ and $R_3$, we will use the following formula:

$$R_A = \cfrac{1}{\cfrac{1}{R_2} + \cfrac{1}{R_3}}$$

(We will use $R_A$ to represent the parallel combination of $R_2$ and $R_3$.)

What do we know?

$$R_2 = 330 \ \Omega$$
$$R_3 = 470 \ \Omega$$

What do we not know?

$$R_A = \ ?$$

What formula could we use?

$$R_A = \cfrac{1}{\cfrac{1}{R_2} + \cfrac{1}{R_3}}$$

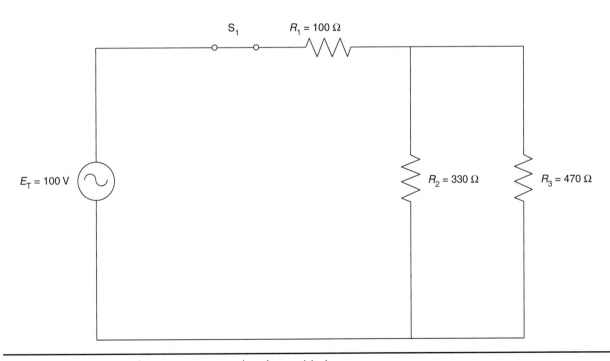

**FIGURE 5–15**   A combination circuit with values added.

Substitute the known values and solve for $R_A$:

$$R_A = \frac{1}{\dfrac{1}{R_2} + \dfrac{1}{R_3}}$$

$$R_A = \frac{1}{\dfrac{1}{330\ \Omega} + \dfrac{1}{470\ \Omega}}$$

$$R_A = 193.88\ \Omega$$

Therefore, the resistance of the parallel combination of $R_2$ and $R_3$ is 193.88 $\Omega$. We will now redraw the circuit shown in Figure 5–15 to reflect the combination of $R_2$ and $R_3$. The redrawn circuit appears in Figure 5–16. Notice that resistor $R_1$ is connected in series with $R_A$ (the parallel combination of $R_2$ and $R_3$). We can simply add the value of $R_1$ to the value of $R_A$ to find the total circuit resistance, $R_T$.

What do we know?

$$R_1 = 100\ W$$
$$R_A = 193.88\ \Omega$$

What do we not know?

$$R_T = ?$$

What formula could we use?

$$R_T = R_1 + R_A$$

Substitute the known values and solve for $R_T$:

$$R_T = R_1 + R_A$$
$$R_T = 100\ \Omega + 193.88\ \Omega$$
$$R_T = 293.88\ \Omega$$

Therefore, the total circuit resistance in Figure 5–15 is 293.88 $\Omega$. We will now redraw the circuit shown in Figure 5–16 to reflect the combination of $R_1$ and $R_A$. The redrawn circuit appears in Figure 5–17. Notice that this circuit is very simple in appearance. It now consists of a power source, a switch, conductors, and one resistor, $R_T$.

With the circuit simplified in this manner, we can now determine the total circuit current.

What do we know?

$$E_T = 100\ V$$
$$R_T = 293.88\ \Omega$$

What do we not know?

$$I_T = ?$$

What formula could we use?

$$I_T = \frac{E_T}{R_T}$$

Substitute the known values and solve for $I_T$:

$$I_T = \frac{E_T}{R_T}$$

$$I_T = \frac{100\ V}{293.88\ \Omega}$$

$$I_T = 340.27\ mA$$

Therefore, the total current flowing in the circuit in Figure 5–17 is 340.27 mA.

While we have the circuit simplified, we can determine the total power consumed by the circuit. We will begin by finding the total power, $P_T$. Recall again the three formulas for finding power:

$$P = I \times E$$
$$P = \frac{E^2}{R}$$
$$P = I^2 \times R$$

We can use any of these formulas to determine the total circuit power. We know $I_T$ and $E_T$; therefore, we could use the first formula. We know $E_T$ and $R_T$; therefore, we could use the second formula. Finally, we know $I_T$ and $R_T$; therefore, we could use the third formula. For this example, we will use the second formula. You may wish to try to solve the example using either or both of the remaining formulas. Your results should be the same. Now, let us find $P_T$.

What do we know?

$$E_T = 100\ V$$
$$R_T = 293.88\ \Omega$$

What do we not know?

$$P_T = ?$$

What formula could we use?

$$P_T = \frac{E^2}{R}$$

Substitute the known values and solve for $P_T$:

$$P_T = \frac{E^2}{R}$$

$$P_T = \frac{100^2\ V}{293.88\ \Omega}$$

$$P_T = 34.03\ W$$

Therefore, this circuit will consume 34.03 W of power. However, at this point, we do not know how much voltage is dropped across $R_1$, $R_2$, or $R_3$. We also do not know how much current is flowing through the three resistors or how much power each resistor must dissipate. In order to solve these unknowns, we must rebuild our circuit back to the original circuit one step at a time.

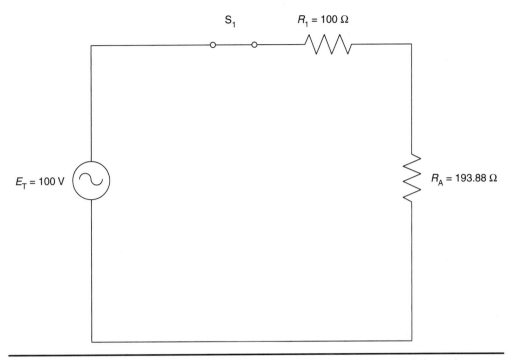

**FIGURE 5-16** The redrawn circuit of Figure 5–15 showing the parallel combination, $R_A$, of resistors $R_2$ and $R_3$.

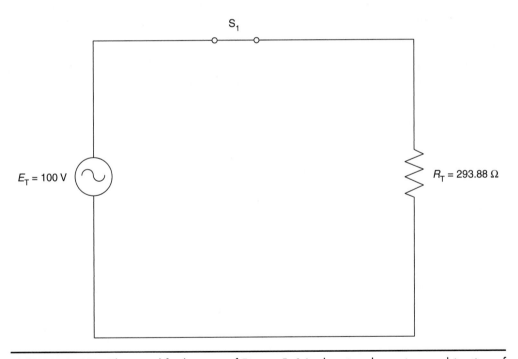

**FIGURE 5-17** The simplified circuit of Figure 5–16, showing the series combination of $R_1$ and $R_A$.

Compare Figure 5–17 to Figure 5–16. Recall that Figure 5–17 was created by combining the series resistances of $R_1$ and $R_A$. Since these resistances are connected in series, the current flowing through each resistance must be the same. This means there must be 340.27 mA of current flowing through $R_1$ and 340.27 mA of current flowing through $R_A$. This is shown in Figure 5–18.

When we look at Figure 5–18, we can see that we now know two things about each resistor. We know the value of the resistance and we know the amount of current flowing through each resistance. This means that we can now determine the amount of voltage dropped across each resistance.

What do we know?

$$I_1 = 340.27 \text{ mA}$$
$$R_1 = 100 \ \Omega$$

What do we not know?

$$E_1 = ?$$

What formula could we use?

$$E_1 = I_1 \times R_1$$

Substitute the known values and solve for $E_1$:

$$E_1 = I_1 \times R_1$$
$$E_1 = 340.27 \text{ mA} \times 100 \ \Omega$$
$$E_1 = 34.03 \text{ V}$$

Therefore, the voltage dropped across $R_1$ is 34.03 V. How much voltage is dropped across resistor $R_A$? We will use the same value for current, but we must use the resistance value for $R_A$.

What do we know?

$$I_A = 340.27 \text{ mA}$$
$$R_A = 193.88 \ \Omega$$

What do we not know?

$$E_A = ?$$

What formula could we use?

$$E_A = I_A \times R_A$$

Substitute the known values and solve for $E_A$:

$$E_A = I_A \times R_A$$
$$E_A = 340.27 \text{ mA} \times 193.88 \ \Omega$$
$$E_A = 65.97 \text{ V}$$

Therefore, the voltage dropped across $R_A$ is 65.97 V. Now let us check our work. Recall that the sum of the individual voltage drops in a series circuit must total the applied voltage. Is this true in this example?

$$E_1 = 34.03 \text{ V}$$
$$E_A = 65.97 \text{ V}$$
$$E_T = E_1 + E_A$$
$$E_T = 34.03 \text{ V} + 65.97 \text{ V}$$
$$E_T = 100 \text{ V}$$

The sum of the voltage drops is equal to the applied voltage. This means that we are on the right track. Now, let us turn our attention to resistor $R_A$. Recall that the value of this resistor was a result of the parallel combination of resistors $R_2$ and $R_3$. Let us now rebuild our circuit back to the original circuit by separating resistor $R_A$ back into resistors $R_2$ and $R_3$.

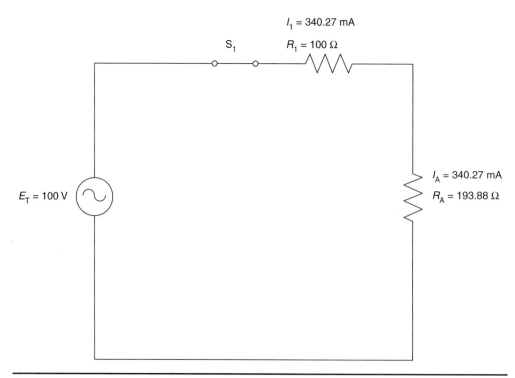

**FIGURE 5–18** The current is the same value throughout a series circuit.

Since resistors $R_2$ and $R_3$ are connected in parallel, the voltage drop across $R_2$ and $R_3$ must be the same value. This is the value of the voltage that is dropped across resistor $R_A$. In other words, there will be 65.97 V dropped across resistor $R_2$ and 65.97 V dropped across resistor $R_3$.

Figure 5–19 shows the circuit as it originally appeared, only now all of the known values have been added. If we turn our attention to resistors $R_2$ and $R_3$, we will see that we know two things about each of these resistors. We know the voltage drop across each resistor, and we know the amount of resistance of each resistor. We are, therefore, able to determine the amount of current flowing through each resistor. We will begin with resistor $R_2$.

What do we know?

$$E_2 = 65.97 \text{ V}$$
$$R_2 = 330 \text{ }\Omega$$

What do we not know?

$$I_2 = ?$$

What formula could we use?

$$I_2 = \frac{E_2}{R_2}$$

Substitute the known values and solve for $I_2$:

$$I_2 = \frac{E_2}{R_2}$$
$$I_2 = \frac{65.97 \text{ V}}{330 \text{ }\Omega}$$
$$I_2 = 199.91 \text{ mA}$$

Therefore, the current flowing through resistor $R_2$ in the circuit in Figure 5–19 is 199.91 mA. Now let us determine the amount of current flowing through resistor $R_3$.

What do we know?

$$E_3 = 65.97 \text{ V}$$
$$R_3 = 470 \text{ }\Omega$$

What do we not know?

$$I_3 = ?$$

What formula could we use?

$$I_3 = \frac{E_3}{R_3}$$

Substitute the known values and solve for $I_3$:

$$I_3 = \frac{E_3}{R_3}$$

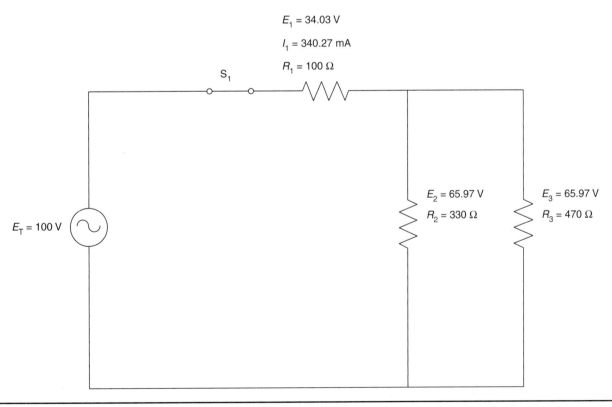

$E_1 = 34.03$ V

$I_1 = 340.27$ mA

$R_1 = 100 \text{ }\Omega$

$S_1$

$E_2 = 65.97$ V

$R_2 = 330 \text{ }\Omega$

$E_3 = 65.97$ V

$R_3 = 470 \text{ }\Omega$

$E_T = 100$ V

**FIGURE 5–19** The voltage across each parallel branch is the same value. The branch currents in a parallel circuit will add up to equal the total current.

$$I_3 = \frac{65.97 \text{ V}}{470 \text{ }\Omega}$$
$$I_3 = 140.36 \text{ mA}$$

Therefore, the current flowing through resistor $R_3$ in the circuit in Figure 5–19 is 140.36 mA.

Follow the current from the power source, through the switch, the resistors, and back to the power source. Refer to Figure 5–20 as we discover something fascinating.

When the switch is closed, 340.27 mA of current flows from the power source, through the switch, and through resistor $R_1$ to junction A. At junction A, this current divides, and 199.91 mA of current will flow through resistor $R_2$ toward junction B. Then, 140.36 mA of current will flow from junction A, through resistor $R_3$, toward junction B. At junction B, the 199.91 mA of current from $R_2$ combines with the 140.36 mA of current from $R_3$. This produces a total of 340.27 mA of current flow back to the power source. *The amount of current that left the power source is the amount of current that returned to the power source. The amount of current that flowed into junction A is the amount of current that flowed out of junction B.*

We have three final calculations to make. Earlier, we calculated the total circuit power. We now need to determine the amount of power that each resistor must dissipate. We will begin with $R_1$.

What do we know?

$$E_1 = 34.03 \text{ V}$$

$$I_1 = 340.27 \text{ mA}$$
$$R_1 = 100 \text{ }\Omega$$

What do we not know?

$$P_1 = ?$$

What formula could we use? (We could use any one of the following.)

$$P_1 = I_1 \times E_1$$
$$P_1 = \frac{E_1^2}{R_1}$$
$$P_1 = I_1^2 \times R_1$$

Substitute the known values and solve for $P_1$. (We will use the first equation.)

$$P_1 = I_1 \times E_1$$
$$P_1 = 340.27 \text{ mA} \times 34.03 \text{ V}$$
$$P_1 = 11.58 \text{ W}$$

Therefore, resistor $R_1$ must dissipate 11.58 W of heat. Recall that a good rule of thumb is that the resistor wattage rating should be twice the actual wattage that must be handled. Therefore, $R_1$ should be rated at 23.16 W. Now let us determine how much power resistor $R_2$ must dissipate.

What do we know?

$$E_2 = 65.97 \text{ V}$$
$$I_2 = 199.91 \text{ mA}$$
$$R_2 = 330 \text{ }\Omega$$

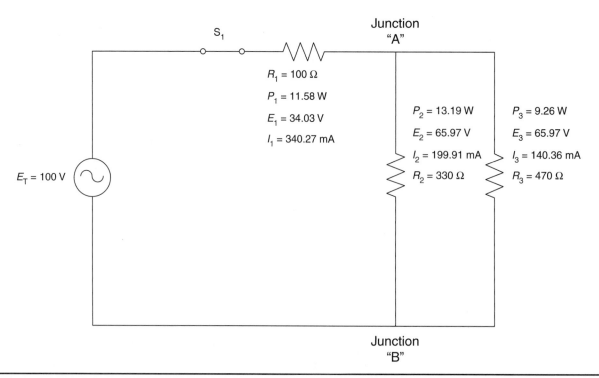

**FIGURE 5–20**   A combination circuit showing all computed and known values.

What do we not know?

$$P_2 = ?$$

What formula could we use? (We could use any one of the following.)

$$P_2 = I_2 \times E_2$$
$$P_2 = \frac{E_2^2}{R_2}$$
$$P_2 = I_2^2 \times R_2$$

Substitute the known values and solve for $P_2$. (We will use the third equation.)

$$P_2 = I_2^2 \times R_2$$
$$P_2 = 199.91^2 \text{ mA} \times 330 \text{ }\Omega$$
$$P_2 = 13.19 \text{ W}$$

Therefore, resistor $R_2$ must dissipate 13.19 W of heat. In the actual circuit, resistor $R_2$ should be rated at a minimum of 26.38 W. Now, let us determine how much power resistor $R_3$ must dissipate.

What do we know?

$$E_3 = 65.97 \text{ V}$$
$$I_3 = 140.36 \text{ mA}$$
$$R_3 = 470 \text{ }\Omega$$

What do we not know?

$$P_3 = ?$$

What formula could we use? (We could use any one of the following.)

$$P_3 = I_3 \times E_3$$
$$P_3 = \frac{E_3^2}{R_3}$$
$$P_3 = I_3^2 \times R_3$$

Substitute the known values and solve for $P_3$. (We will use the second equation.)

$$P_3 = \frac{E_3^2}{R_3}$$
$$P_3 = \frac{65.97^2 \text{ V}}{470 \text{ }\Omega}$$
$$P_3 = 9.26 \text{ W}$$

Therefore, resistor $R_3$ must dissipate 9.26 W of heat. In the actual circuit, resistor $R_3$ should be rated at a minimum of 18.52 W.

It is interesting that if we take the wattage dissipated by $R_1$ and add it to the wattage dissipated by $R_2$ and $R_3$, we will find the total wattage consumed by the circuit.

$$P_1 = 11.58 \text{ W}$$
$$P_2 = 13.19 \text{ W}$$

$$P_3 = 9.26 \text{ W}$$
$$P_T = P_1 + P_2 + P_3$$
$$P_T = 11.58 \text{ W} + 13.19 \text{ W} + 9.26 \text{ W}$$
$$P_T = 34.03 \text{ W}$$

Compare the total power that was calculated earlier (34.03 W) to the total power found by adding the individual powers (34.03 W). This proves that in a combination circuit, the sum of all individual wattages will equal the total circuit wattage.

Figure 5–21 is the same circuit as seen earlier in Figure 5–14. We have already determined that this circuit is a combination circuit. Resistors $R_1$ and $R_2$ are connected in series. Resistor $R_3$ is connected in parallel with the series combination of $R_1$ and $R_2$. We have now added values for the amount of voltage present as well as the values of $R_1$, $R_2$, and $R_3$. We will use the same values found in our previous examples so that you can see the effects of connecting components in a combination circuit of different designs. Since we know the amount of voltage present, can we determine the amount of current flowing in the circuit?

If we were to try to determine the amount of current flowing in this circuit, we would have some difficulty. What do we use for the value of $R_T$? Do we use the value of $R_1$, or do we use the value of $R_2$ or $R_3$ or a combination? We must first look at the circuit and determine which components are connected in series and which are connected in parallel. We can combine resistor $R_1$ with $R_2$ because these resistors are connected in series. We cannot combine resistor $R_1$ with $R_3$ because resistor $R_2$ is connected in series with $R_1$. Likewise, we cannot combine resistor $R_2$ with $R_3$ because resistor $R_1$ is connected in series with $R_2$. We must first find the resistance of the series combination of resistors $R_1$ and $R_2$. This resistance can then be combined with resistor $R_3$ to give us the total circuit resistance.

To determine the resistance of the series combination of resistors $R_1$ and $R_2$, we will use the following formula:

$$R_A = R_1 + R_2$$

(We will use $R_A$ to represent the series combination of $R_1$ and $R_2$.)

What do we know?

$$R_1 = 100 \text{ }\Omega$$
$$R_2 = 330 \text{ }\Omega$$

What do we not know?

$$R_A = ?$$

What formula could we use?

$$R_A = R_1 + R_2$$

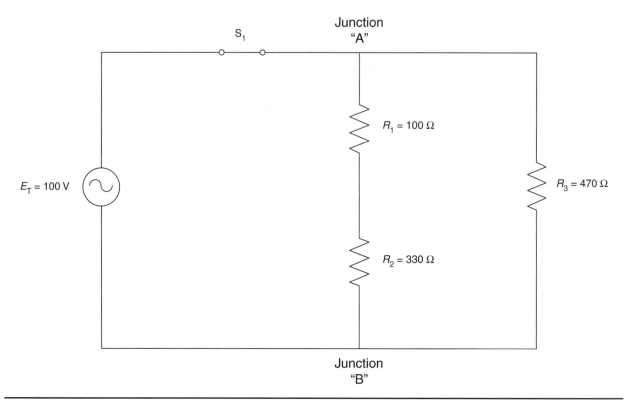

**FIGURE 5–21**   The combination circuit shown in Figure 5–14 with values added.

Substitute the known values and solve for $R_A$:

$$R_A = R_1 + R_2$$
$$R_A = 100\ \Omega + 330\ \Omega$$
$$R_A = 430\ \Omega$$

Therefore, the resistance of the series combination of $R_1$ and $R_2$ is 430 Ω. We will now redraw the circuit shown in Figure 5–21 to reflect the combination of $R_1$ and $R_2$. The redrawn circuit appears in Figure 5–22. Notice that resistor $R_3$ is connected in parallel with $R_A$ (the series combination of $R_1$ and $R_2$). We must now determine the total circuit resistance, $R_T$.

What do we know?

$$R_A = 430\ \Omega$$
$$R_3 = 470\ \Omega$$

What do we not know?

$$R_T = ?$$

What formula could we use?

$$R_T = \cfrac{1}{\dfrac{1}{R_A} + \dfrac{1}{R_3}}$$

Substitute the known values and solve for $R_T$:

$$R_T = \cfrac{1}{\dfrac{1}{R_A} + \dfrac{1}{R_3}}$$

$$R_T = \cfrac{1}{\dfrac{1}{430\ \Omega} + \dfrac{1}{470\ \Omega}}$$
$$R_T = 224.56\ \Omega$$

Therefore, the total circuit resistance in Figure 5–22 is 224.56 Ω. We will now redraw the circuit shown in Figure 5–22 to reflect the combination of $R_A$ and $R_3$. The redrawn circuit appears in Figure 5–23. Notice that this circuit is very simple in appearance. The circuit now consists of a power source, a switch, conductors, and one resistor, $R_T$.

With the circuit simplified in this manner, we can now determine the total circuit current.

What do we know?

$$E_T = 100\ V$$
$$R_T = 224.56\ \Omega$$

What do we not know?

$$I_T = ?$$

What formula could we use?

$$I_T = \frac{E_T}{R_T}$$

Substitute the known values and solve for $I_T$:

$$I_T = \frac{E_T}{R_T}$$

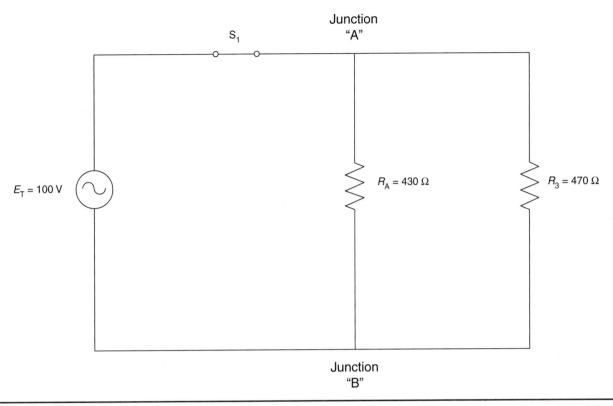

**FIGURE 5-22**  The redrawn circuit of Figure 5–21 showing the series combination, $R_A$, of resistors $R_1$ and $R_2$.

$$I_T = \frac{100 \text{ V}}{224.56 \text{ } \Omega}$$
$$I_T = 445.32 \text{ mA}$$

Therefore, the total current flowing in the circuit in Figure 5–23 is 445.32 mA.

While we have the circuit simplified, we can determine the total power consumed by the circuit. We will begin by finding the total power, $P_T$. Recall again the three formulas for finding power:

$$P = I \times E$$
$$P = \frac{E^2}{R}$$
$$P = I^2 \times R$$

We can use any of these formulas to determine the total circuit power. We know $I_T$ and $E_T$; therefore, we could use the first formula. We know $E_T$ and $R_T$; therefore, we could use the second formula. Finally, we know $I_T$ and $R_T$; therefore, we could use the third formula. For this example, we will use the second formula. You may wish to try to solve the example using either or both of the remaining formulas. Your results should be the same. Now let us find $P_T$.

What do we know?

$$E_T = 100 \text{ V}$$
$$R_T = 224.56 \text{ } \Omega$$

What do we not know?

$$P_T = ?$$

What formula could we use?

$$P_T = \frac{E^2}{R}$$

Substitute the known values and solve for $P_T$:

$$P_T = \frac{E^2}{R}$$
$$P_T = \frac{100^2 \text{ V}}{224.56 \text{ } \Omega}$$
$$P_T = 44.53 \text{ W}$$

Therefore, this circuit will consume 44.53 W of power. However, at this point, we do not know how much voltage is dropped across $R_1$, $R_2$, or $R_3$. We also do not know how much current is flowing through the three resistors or how much power each resistor must dissipate. In order to solve these unknowns, we must rebuild our circuit back to the original circuit one step at a time.

Compare Figure 5–23 to Figure 5–22. Recall that Figure 5–23 was created by combining the parallel resistances of $R_A$ and $R_3$. Since these resistances are connected in parallel, the voltage dropped across each resistance must be the same. That is, there

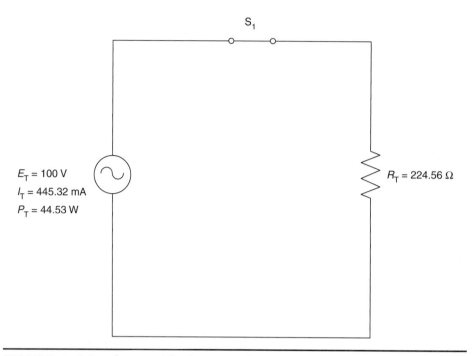

**FIGURE 5–23** The simplified circuit of Figure 5–22, showing the parallel combination of $R_A$ and $R_3$.

must be 100 V across $R_A$, and 100 V across $R_3$. This is shown in Figure 5–24.

When we look at Figure 5–24, we can see that we now know two things about each resistor. We know the value of the resistance and we know the amount of voltage dropped across each resistance. This means that we can now determine the amount of current flowing through each resistance.

What do we know?

$$E_A = 100 \text{ V}$$
$$R_A = 430 \text{ } \Omega$$

What do we not know?

$$I_A = \text{ ?}$$

What formula could we use?

$$I_A = \frac{E_A}{R_A}$$

Substitute the known values and solve for $I_A$:

$$I_A = \frac{E_A}{R_A}$$
$$I_A = \frac{100 \text{ V}}{430 \text{ } \Omega}$$
$$I_A = 232.56 \text{ mA}$$

Therefore, the current flowing through $R_A$ is 232.56 mA. How much current is flowing through resistor $R_3$? We will use the same value for voltage,

but we must use the resistance value for $R_3$. What do we know?

$$E_3 = 100 \text{ V}$$
$$R_3 = 470 \text{ } \Omega$$

What do we not know?

$$I_3 = \text{ ?}$$

What formula could we use?

$$I_3 = \frac{E_3}{R_3}$$

Substitute the known values and solve for $I_3$:

$$I_3 = \frac{E_3}{R_3}$$
$$I_3 = \frac{100 \text{ V}}{470 \text{ } \Omega}$$
$$I_3 = 212.77 \text{ mA}$$

Therefore, the current flowing through $R_3$ is 212.77 mA. Now, let us check our work. Recall that the sum of the individual branch currents in a parallel circuit must equal the total circuit current. Is this true in this example?

$$I_A = 232.56 \text{ mA}$$
$$I_3 = 212.77 \text{ mA}$$
$$I_T = I_A + I_3$$
$$I_T = 232.56 \text{ mA} + 212.77 \text{ mA}$$
$$I_T = 445.33 \text{ mA}$$

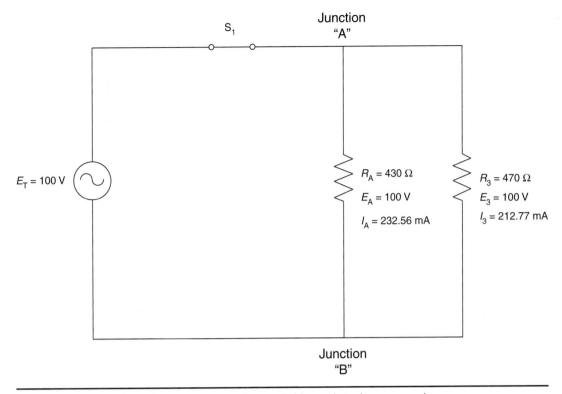

S$_1$

Junction "A"

$E_T = 100$ V

$R_A = 430\ \Omega$

$E_A = 100$ V

$I_A = 232.56$ mA

$R_3 = 470\ \Omega$

$E_3 = 100$ V

$I_3 = 212.77$ mA

Junction "B"

**FIGURE 5–24**  The voltage across each parallel branch is the same value.

The sum of the individual branch circuit currents is equal to the total circuit current. This means that we are on the right track. (Do not be concerned with the 0.01 mA difference between the two answers. This is a result of the rounding of numbers to two decimal places during calculations.)

We must turn our attention to resistor $R_A$. Recall that the value of this resistor was a result of the series combination of resistors $R_1$ and $R_2$. Let us now rebuild our circuit back to the original circuit by separating resistor $R_A$ back into resistors $R_1$ and $R_2$.

Since resistors $R_1$ and $R_2$ are connected in series, the current flowing through $R_1$ and $R_2$ must be the same value. This is the value of the current flowing through resistor $R_A$. In other words, there will be 232.56 mA of current flowing through resistor $R_1$ and 232.56 mA of current flowing through resistor $R_2$.

Figure 5–25 shows the circuit as it originally appeared, only now all of the known values have been added. If we turn our attention to resistors $R_1$ and $R_2$, we will see that we know two things about each of these resistors. We know the amount of current flowing through each resistor, and we know the amount of resistance of each resistor. We are, therefore, able to determine the amount of voltage dropped across each resistor. We will begin with resistor $R_1$.

What do we know?

$$I_1 = 232.56 \text{ mA}$$
$$R_1 = 100\ \Omega$$

What do we not know?

$$E_1 = \text{?}$$

What formula could we use?

$$E_1 = I_1 \times R_1$$

Substitute the known values and solve for $E_1$:

$$E_1 = I_1 \times R_1$$
$$E_1 = 232.56 \text{ mA} \times 100\ \Omega$$
$$E_1 = 23.26 \text{ V}$$

Therefore, the voltage dropped across resistor $R_1$ in the circuit in Figure 5–25 is 23.26 V. Now, let us determine the amount of voltage dropped across resistor $R_2$.

What do we know?

$$I_2 = 232.56 \text{ mA}$$
$$R_2 = 330\ \Omega$$

What do we not know?

$$E_2 = \text{?}$$

What formula could we use?

$$E_2 = I_2 \times R_2$$

Substitute the known values and solve for $E_2$:

$$E_2 = I_2 \times R_2$$
$$E_2 = 232.56 \text{ mA} \times 330\ \Omega$$
$$E_2 = 76.74 \text{ V}$$

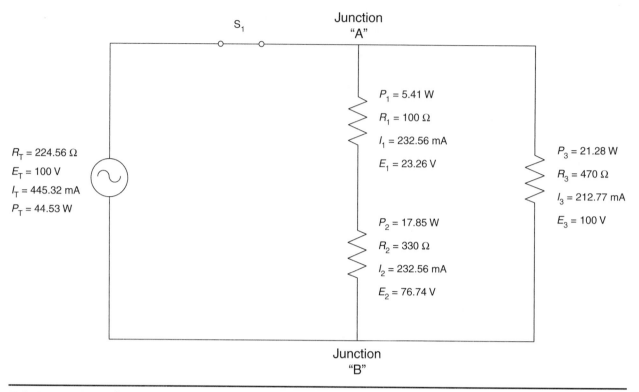

**FIGURE 5–25** The current is the same value throughout a series circuit. The branch currents in a parallel circuit will add to equal the total current.

Therefore, the voltage dropped across resistor $R_2$ in the circuit in Figure 5–25 is 76.74 V.

It is interesting to note that this circuit is supplied with 100 V. Imagine placing one probe of your voltmeter at the top of $R_1$. Now, imagine placing the other probe of your voltmeter at the bottom of $R_2$. Look at what you are actually measuring. Follow the connection from the top of $R_1$ to the left, to the top of the power source. Now, follow the connection from the bottom of $R_2$ to the left, to the bottom of the power source. You have placed your voltmeter across the power source. This means that you will actually measure the voltage of the power source, 100 V. Notice the voltage drops across $R_1$ and $R_2$. $R_1$ has a voltage drop of 23.26 V. $R_2$ has a voltage drop of 76.74 V. Your meter is measuring the total voltage across resistors $R_1$ and $R_2$, or 23.26 V + 76.74 V, which equals 100 V.

There are three final calculations to make. Earlier, we calculated the total circuit power, and now we need to determine the amount of power that each resistor must dissipate. We will begin with $R_1$.

What do we know?

$$E_1 = 23.26 \text{ V}$$
$$I_1 = 232.56 \text{ mA}$$
$$R_1 = 100 \ \Omega$$

What do we not know?

$$P_1 = ?$$

What formula could we use? (We could use any one of the following.)

$$P_1 = I_1 \times E_1$$
$$P_1 = \frac{E_1{}^2}{R_1}$$
$$P_1 = I_1{}^2 \times R_1$$

Substitute the known values and solve for $P_1$. (We will use the first equation.)

$$P_1 = I_1 \times E_1$$
$$P_1 = 232.56 \text{ mA} \times 23.26 \text{ V}$$
$$P_1 = 5.41 \text{ W}$$

Therefore, resistor $R_1$ must dissipate 5.41 W of heat. Recall that a good rule of thumb is that the resistor wattage rating should be twice the actual wattage that must be handled. Therefore, $R_1$ should be rated at 10.82 W. Now, we can determine how much power resistor $R_2$ must dissipate.

What do we know?

$$E_2 = 76.74 \text{ V}$$
$$I_2 = 232.56 \text{ mA}$$
$$R_2 = 330 \ \Omega$$

What do we not know?

$$P_2 = \text{?}$$

What formula could we use? (We could use any one of the following.)

$$P_2 = I_2 \times E_2$$
$$P_2 = \frac{E_2^2}{R_2}$$
$$P_2 = I_2^2 \times R_2$$

Substitute the known values and solve for $P_2$. (We will use the third equation.)

$$P_2 = I_2^2 \times R_2$$
$$P_2 = 232.56^2 \text{ mA} \times 330 \text{ }\Omega$$
$$P_2 = 17.85 \text{ W}$$

Therefore, resistor $R_2$ must dissipate 17.85 W of heat. In the actual circuit, resistor $R_2$ should be rated at a minimum of 35.70 W. Now, let us determine how much power resistor $R_3$ must dissipate.

What do we know?

$$E_3 = 100 \text{ V}$$
$$I_3 = 212.77 \text{ mA}$$
$$R_3 = 470 \text{ }\Omega$$

What do we not know?

$$P_3 = \text{?}$$

What formula could we use? (We could use any one of the following.)

$$P_3 = I_3 \times E_3$$
$$P_3 = \frac{E_3^2}{R_3}$$
$$P_3 = I_3^2 \times R_3$$

Substitute the known values and solve for $P_3$. (We will use the second equation.)

$$P_3 = \frac{E_3^2}{R_3}$$
$$P_3 = \frac{100^2 \text{ V}}{470 \text{ }\Omega}$$
$$P_3 = 21.28 \text{ W}$$

Therefore, resistor $R_3$ must dissipate 21.28 W of heat. In the actual circuit, resistor $R_3$ should be rated at a minimum of 42.56 W.

Now, notice something interesting. If we take the wattage dissipated by $R_1$ and add it to the wattage dissipated by $R_2$ and $R_3$, we will find the total wattage consumed by the circuit.

$$P_1 = 5.41 \text{ W}$$
$$P_2 = 17.85 \text{ W}$$

$$P_3 = 21.28 \text{ W}$$
$$P_T = P_1 + P_2 + P_3$$
$$P_T = 5.41 \text{ W} + 17.85 \text{ W} + 21.28 \text{ W}$$
$$P_T = 44.54 \text{ W}$$

Compare the total power that was calculated earlier (44.53 W) to the total power found by adding the individual powers (44.54 W). This proves that, in a combination circuit, the sum of all individual wattages will equal the total circuit wattage. (Again, the difference is due to the rounding of numbers during calculations.)

## REVIEW QUESTIONS

*Multiple Choice*

1. An electric circuit is the
   a. flow of electrons.
   b. path over which electrons flow.
   c. force that causes electrons to flow.
   d. opposition to electron flow.

2. A series circuit is a circuit that
   a. has only one path over which the current flows.
   b. has more than one path over which the current flows.
   c. has three current paths.
   d. contains many devices.

3. The value of current flowing in a series circuit is
   a. equal to the sum of the currents through the various branches.
   b. the same value through all parts of the circuit.
   c. equal to the resistance divided by the voltage.
   d. equal to the applied voltage.

4. The total resistance in a series circuit is
   a. equal to the sum of the resistances of each part of the circuit.
   b. inversely proportional to the voltage.
   c. always less than any resistance in the circuit.
   d. equal to the current divided by the voltage.

5. A parallel circuit is a circuit that
   a. has only one path over which electrons can flow.
   b. has more than one path through which electrons can flow.
   c. contains many devices.
   d. has very low resistance.

6. The value of current flowing in a parallel circuit is
   a. equal to the sum of the currents through the various branches.
   b. the same value through all parts of the circuit.
   c. equal to the resistance divided by the voltage.
   d. equal to the applied voltage.

7. The voltage across any branch of a parallel circuit is
   a. determined by the resistance of the branch.
   b. greater if the current is greater.
   c. equal to the supply voltage.
   d. equal to the current divided by the resistance.

*Give Complete Answers*

1. State a rule about current, voltage, resistance, and power as it applies to series circuits.
   a. Current: _____
   b. Voltage: _____
   c. Resistance: _____
   d. Power: _____

2. State a rule about current, voltage, resistance, and power as it applies to parallel circuits.
   a. Current: _____
   b. Voltage: _____
   c. Resistance: _____
   d. Power: _____

*Solve each problem, showing the method used to arrive at the solution.*

1. Given the circuit in the following figure, find the unknown values.

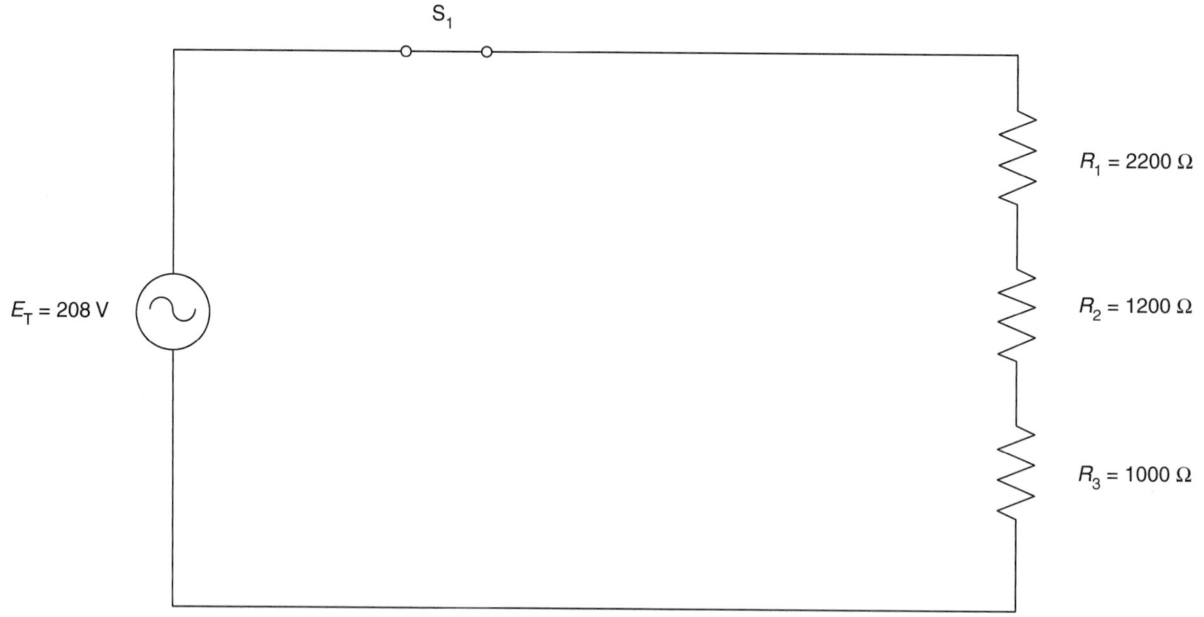

**FIGURE 5–26**

2. Given the circuit in the following figure, find the unknown values.

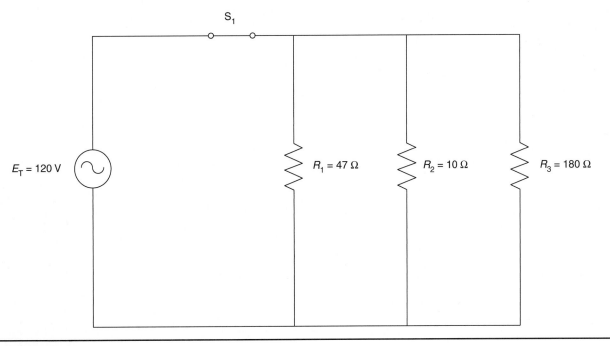

**FIGURE 5–27**

3. Given the circuit in the following figure, find the unknown values.

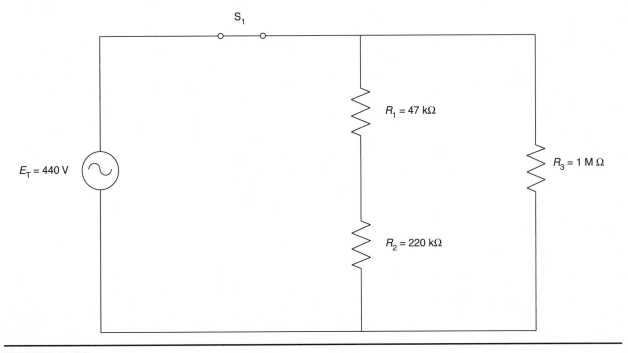

**FIGURE 5–28**

# Magnets and Magnetism

**OBJECTIVES**

After studying this chapter, the student will be able to:

- Describe various types of magnets.
- Describe the nature of magnetic fields and forces.
- Explain the theories of magnetism.
- List the various uses for magnetism and methods of controlling magnetic forces.
- Explain the relationship between magnetism and electricity.

A good understanding of magnetism is of major importance in the study of electricity. Magnetism is involved in the operation of most electrical apparatuses. Some examples are motors, generators, transformers, controllers, relays, meters, and lifting magnets.

# MAGNETS

A magnet is a material that attracts other metals. Magnets are divided into three classes: *natural*, *artificial*, and *electromagnets*. The first known magnets were stones found in Asia. These stones, composed of an iron ore called magnetite, had the unusual property of being able to attract small pieces of iron, steel, and other metals. It was later discovered that the stone, in an elongated form, would align itself in a nearly north and south direction if freely suspended. This property of magnetite made it useful as a compass, which was the first practical use of a magnet. For this reason, the stone was called a leading stone, or lodestone.

In 1600, Dr. William Gilbert showed that the Earth itself acts like a magnet, attracting the ends of a needle of lodestone. He called the end of the needle that pointed north the north-seeking pole, or north (N) pole. The end that pointed south he called the south-seeking pole, or south (S) pole.

Gilbert found that when two north poles of two magnets are brought close to each other and then released, they move apart. When two south poles are brought together, they also repel each other. Figure 6–1 illustrates this phenomenon. He also learned that if a north pole is placed near a south pole, they attract each other, as shown in Figure 6–2. From these observations, the following laws for magnetic attraction and repulsion were formulated:

1. Like magnetic poles repel each other.
2. Unlike magnetic poles attract each other.
3. The nearer the magnets are to each other, the greater the attraction or repulsion.

These rules are very important in the study of motors, generators, controllers, and solenoids.

Continued research led to the development of artificial magnets, which are stronger and more useful than natural magnets. Some materials used in the manufacture of artificial magnets are aluminum, nickel, cobalt, iron, and vanadium. When mixed in the proper proportions, these metals make strong permanent magnets.

# MAGNETIC FIELDS AND FORCES

If a piece of paper is placed over a bar magnet and iron powder is sprinkled on the paper, the powder forms a pattern similar to that shown in Figure 6–3. The arrangement of the powdered iron is more pronounced if the paper is tapped gently while sprinkling the powder.

The space around the magnet where the iron powder forms a pattern shows where the force exists. The pattern of the powder appears as lines drawn from one end of the bar to the other (Figure 6–4). Because of this pattern, the force is referred to as **magnetic lines of force**. The space in which this force exists is called the **magnetic field**.

Another experiment to identify the field of force can be performed with a magnetic compass. If the compass is placed in the magnetic field, the needle

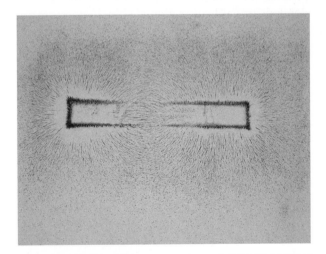

**FIGURE 6–3**  Powdered iron used to outline the area of magnetic force.

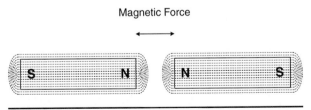

**FIGURE 6–1**  Like magnetic poles repel.

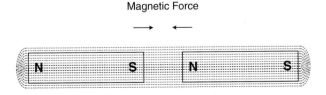

**FIGURE 6–2**  Unlike magnetic poles attract.

**FIGURE 6–4**  Pattern of magnetic lines of force.

will line up with the lines of force. Moving the compass to different locations within the field shows that the lines have a definite direction. The lines appear to leave the magnet by the north pole and reenter at the south pole. Within the magnet, the lines of force continue from the south pole to the north pole.

## MAGNETIC THEORIES

If a single bar magnet is cut in half, each half will be a magnet having north and south poles. If one should continue to separate each half into smaller and smaller particles, each particle will be a magnet. This separation of a single magnet into many smaller magnets has led to the belief that a magnetic substance is composed of molecular magnets (but not necessarily that each molecule is a magnet). This theory is called the *molecular theory of magnetism.*

When a material is magnetized, the molecules are aligned as shown in Figure 6–5. When the material is not magnetized, the molecules are arranged as in Figure 6–6.

Another theory that is widely accepted is called the *electron theory of magnetism.* The electrons of each atom rotate in orbits about the nucleus and also spin, just as the Earth rotates on its axis. According to this theory, as the electrons spin, they produce a magnetic field. The direction of this field depends upon the direction of spin. In most atoms, there are an equal number of electrons spinning in

opposite directions. These atoms are magnetically neutral because the magnetic fields of the electrons are equal and opposite.

In atoms of magnetic materials, however, there are more electrons spinning in one direction than in the other direction. These atoms produce weak magnetic fields. When a large number of these magnetized atoms group together with their magnetic fields aligned, they form a *domain.* Each domain produces a magnetic field in a specific direction. In nonmagnetized materials, the arrangement of these domains is such that their magnetic effect is neutralized. When such material is placed under the influence of another magnetic force, the domains arrange themselves so that their north poles are in one direction and their south poles are in the opposite direction. The material thus becomes a magnet.

## MAGNETIC MATERIALS

Artificial magnets can be classified as permanent or temporary. Their classification depends upon the materials from which they are made. **Permanent magnets** retain their magnetic properties for many years, possibly 100 years or more. *Temporary magnets* lose their magnetism almost as soon as they are removed from the magnetizing influence. Steel and alloys of steel are materials that make good permanent magnets. One of the most common types of permanent magnets is made of a material called *alnico.* Alnico is a mixture of aluminum, nickel, cobalt, and iron. Soft iron and iron alloys are used for temporary magnets.

### Magnetic Shields

Because all materials have some ability to conduct magnetic lines of force, it is not practical to make a magnetic insulator. Strong magnets placed near electrical instruments can cause permanent damage. Therefore, it is necessary to prevent the magnetic flux (lines of force) from passing through the instrument.

Magnetic shields can be constructed because magnetism passes more readily through some materials than others. When a piece of soft iron is placed near a magnet, the field is distorted and the lines tend to pass through the iron instead of the air (Figure 6–7). Iron is a better conductor of magnetism than air. This fact is utilized in constructing magnetic shields. Many expensive instruments that may be subjected to magnetic influences are surrounded by iron or iron alloys. The iron provides a magnetic path around the sensitive parts (Figure 6–8).

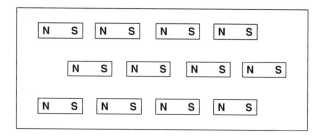

**FIGURE 6–5**  Magnetized material.

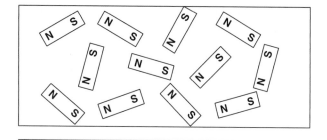

**FIGURE 6–6**  Unmagnetized material.

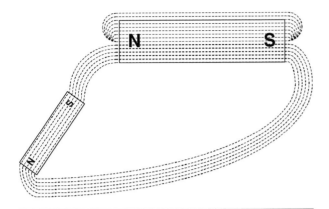

**FIGURE 6–7**  Soft iron provides a path for magnetic lines of force.

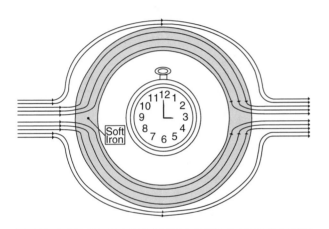

**FIGURE 6–8**  Soft iron provides a magnetic shield for expensive instruments.

## ELECTROMAGNETISM

Magnetism is produced by electron spin. Therefore, it follows that electricity and magnetism are closely related. An electric current (the movement of electrons along a conductor) produces a magnetic field. The field strength varies with the amount of current. The magnetic field produced by current flowing in a single conductor is illustrated in Figure 6–9.

The magnetic effect of current flow in a conductor can be demonstrated by arranging a conductor and a piece of cardboard, as shown in Figure 6–10. Using direct current, allow the current to flow while sprinkling powdered iron on the cardboard. The powder will take the shape of the magnetic field. An examination of the pattern shows that the field is circular, with the conductor in the center. Another method used to see the magnetic reaction of current flow is to place compasses at

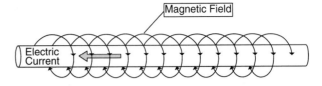

**FIGURE 6–9**  Magnetic field produced by current flow.

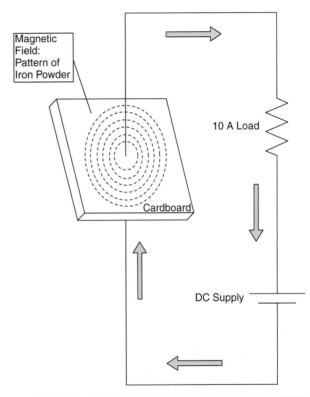

**FIGURE 6–10**  Iron powder takes the form of the magnetic field around a current-carrying conductor.

various positions on the cardboard (Figure 6–11). When the current flows through the wire, the compass needles line up with the magnetic field. The needles indicate the direction of the lines of force.

### Left-Hand Rule for a Single Conductor

The direction of current flow through a conductor and the direction of the lines of force of the magnetic field can be determined by the *left-hand rule*. Grasp the conductor in the left hand with the thumb extended in the direction of current flow (Figure 6–12). The fingers encircling the conductor will point in the direction of the lines of force. If the direction of the lines of force is known, the direction of current flow can be determined in the same manner.

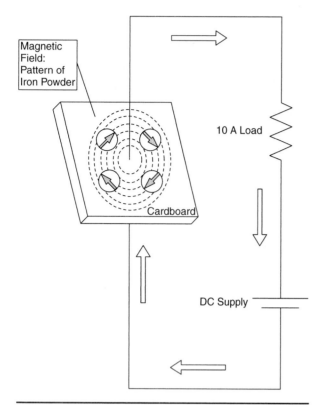

**FIGURE 6–11** Compass needles line up with the magnetic field.

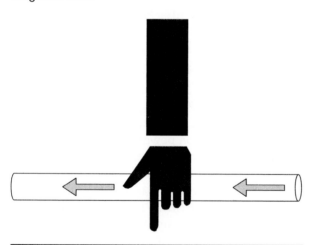

**FIGURE 6–12** Left-hand rule. The thumb points in the direction of current flow. The fingers encircle the conductor in the direction of the lines of force.

### Rules

1. Grasp the conductor in the left hand with the thumb pointing in the direction of current flow. The fingers will point in the direction of the magnetic force.

2. Grasp the conductor in the left hand with the fingers pointing in the direction of the magnetic force. The extended thumb will point in the direction of current flow.

## Magnetic Forces

If two magnetic forces are within magnetic reach of each other, their fields will react according to the laws of attraction and repulsion. Figure 6–13 shows a loop of wire carrying a current. The direction of the current is indicated by the arrows.

The lines of force are illustrated by the circles. Figure 6–14 shows only the end views of the same loop. The dot indicates that the current is coming out of the wire toward the viewer. The "X" indicates that the current is flowing into the wire away from the viewer. Notice in Figure 6–14 that the two magnetic fields are within magnetic reach of each other and tend to force the loop apart.

If two conductors carrying current in the same direction are placed near each other (Figure 6–15), the magnetic forces tend to pull them together.

If several loops of wire are placed loosely together to form a coil, most of the lines of force will thread through the whole coil (Figure 6–16). If these loops are wound very closely together, nearly all the flux will thread through the coil (Figure 6–17). This produces a magnetic field with the same shape as a bar magnet.

## Left-Hand Rule for a Coil

The magnetic polarity of a coil can be determined by grasping the coil in the left hand with the fingers pointing in the direction of current flow. The extended thumb will point in the direction of the lines of force, which is toward the north pole of the magnet (Figure 6–18).

### Rules

1. Grasp the coil with the left hand so that the fingers are pointing in the direction of current flow through the loops. The extended thumb will point in the direction of flux toward the north pole.

2. Grasp the coil with the left hand so that the thumb points to the north pole of the magnet. The fingers will point in the direction of current flow through the loops.

## Electromagnets

A coil, as described previously, is known as an **electromagnet**. It has the same properties as a bar magnet. The advantage is that the magnetic force can be turned on and off with the current. For this reason, it is classified as a temporary magnet. Another advantage of an electromagnet is the ability to vary the field strength. An increase in current through the coil causes an increase in the strength

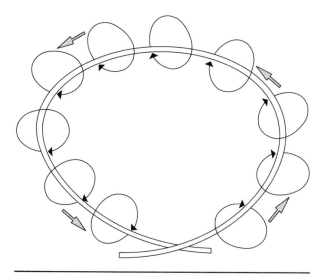

**FIGURE 6-13**  Loop of wire carrying current.

Magnetic Force

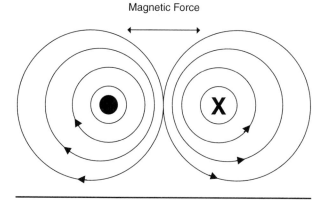

**FIGURE 6-14**  Conductors carrying current in opposite directions. The magnetic field forces the conductors apart.

Magnetic Force

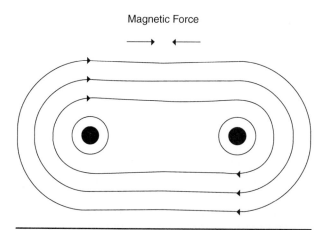

**FIGURE 6-15**  Two conductors carrying current in the same direction. Magnetic forces pull the conductors together.

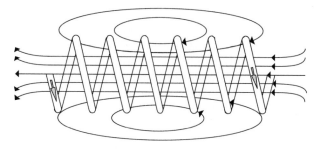

**FIGURE 6-16**  Loosely formed coil.

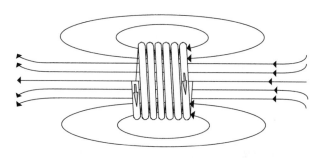

**FIGURE 6-17**  Lines of force through the entire coil.

**FIGURE 6-18**  Left-hand rule for a coil.

of the magnetic field. Decreasing the current decreases the strength of the magnetic field.

Another way to increase the strength of an electromagnet is to insert an iron core into the coil. Iron is a better conductor of magnetism than air is and, therefore, increases the field strength.

In the construction of an electromagnet, there are several factors to consider:

- The number of loops of wire to be placed on the coil
- The amount of current the coil can carry without overheating
- The ability of the core to conduct magnetic lines of force

The first two items involve the electrical conductor used to form the coil. The material and size of the conductor determine the maximum amount of current it can carry. Most conductors for electromagnets are made of soft-drawn copper, which has relatively low resistance and is easy to form into a coil. This type of wire is called *magnet wire.*

Magnet wire is formed into close-fitting loops. It, therefore, cannot dissipate heat as rapidly as can conductors in cables or conduit. The ampacity is much lower than the values listed in *NEC* Article 310 for the same size of wire. Data on ampacity of magnetic wire are available from magnet wire manufacturers.

The core for an electromagnet is generally made of soft iron or iron alloys. Iron is a good conductor of magnetism but it loses its magnetic properties almost instantaneously when removed from the magnetic influence. For most electromagnets, it is desirable that they lose their magnetic properties when the current is turned off.

An electromagnet is sometimes referred to as a **solenoid**. Although the names are used interchangeably, the term *electromagnet* usually refers to coils with stationary iron cores. The term *solenoid* usually refers to coils with movable iron cores.

## Application of Electromagnets

The lifting magnet is one of the common uses of electromagnets. Lifting magnets are used to move large amounts of iron and steel. These magnets can lift as much as 200 pounds per square inch (14 kilograms per square centimeter) of magnet surface. Figures 6–19A and 6–19B show a typical lifting magnet.

Many industrial machines use magnetic clutches to connect and disconnect the load from the driving source or to provide speed control. The amount of slip is varied by adjusting the distance between the driving magnet and the driven component. Another method is to vary the strength of the magnetic field.

A solenoid can be used to open and close valves. Relays, circuit breakers, and door chimes are other examples of electromagnetism. Motors, generators, and transformers also depend upon electromagnetism for their operation.

## MAGNETIC CIRCUITS AND MEASUREMENTS

The strength of an electromagnet depends upon its ability to conduct magnetism. In this respect, one may compare the path of the lines of force with the path of an electric current. They both form a complete path (Figures 6–20A and 6–20B).

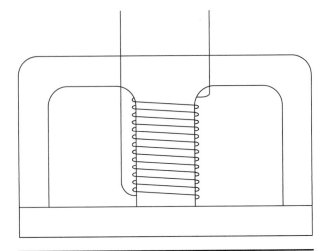

**FIGURE 6–19A**  Internal construction of a lifting magnet.

**FIGURE 6–19B**  Common lifting magnet used to move heavy metal objects.

The amount of flux in a magnetic circuit is determined by the number of lines of force and is measured in *maxwells* (Mx). One **maxwell** is equal to one magnetic line of force.

In an electromagnet, the amount of flux produced depends upon the *magnetomotive force* (mmf). The mmf is the product of the coil current ($I$) and the number of turns ($T$) of wire on the coil. **Magnetomotive force** is measured in *gilberts* (Gb). The unit **gilbert** is frequently used when working with permanent magnets. One gilbert is the mmf that will establish a flux of 1 maxwell in a magnetic

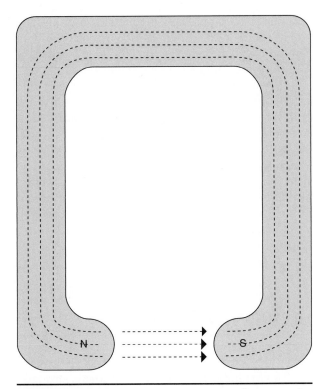

**FIGURE 6–20A** Horseshoe magnet illustrates the magnetic circuit (from the north pole, through the air, to the south pole, through the magnet, and back to the north pole).

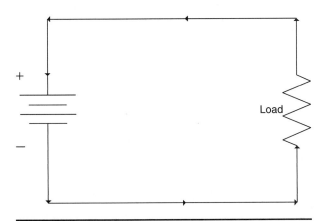

**FIGURE 6–20B** Electric circuit (current flows from the negative terminal of the battery and through the load to the positive terminal).

circuit having a reluctance (rel) of 1 unit. Magnetomotive force is also measured in *ampere-turns;* 1 ampere-turn is equal to 1.257 gilberts.

**Reluctance** is the opposition to the magnetic flux and can be compared to resistance in an electric circuit. No unit has been established for the measurement of reluctance. However, it can be said that 1 unit of reluctance is the opposition produced by a portion of the magnetic circuit 1 inch (2.54 centimeters)

long and 1 square inch (6.451 square centimeters) in cross-section, having unit permeability.

**Permeability** is the ability of a material to conduct lines of force. Permeability varies with the type of material used. The more permeable a material is, the greater the number of lines of force the material can conduct per square inch.

The strength of an electromagnet and/or solenoid can be varied by varying the mmf (ampereturns). This is generally accomplished in the field by increasing or decreasing the current through the coil. The mmf can also be increased by increasing the flux density of the core. Figures 6–21A and 6–21B illustrate this condition. When an iron core is inserted into the coil (Figure 6–21B), the flux density is increased considerably.

Flux density is expressed in the number of lines of force (maxwells) per square inch. To convert to the metric value, 1 maxwell per square centimeter (0.155 square inch) is equal to 1 *gauss* (G). A flux density of 10,000 gauss or more is common for electrical machinery.

Other factors that are important to consider when designing magnets are:

- Retentivity, the ability of a material to retain its magnetism after being removed from the magnetizing influence.
- Residual magnetism, the magnetism that remains in a material after being removed from the magnetizing influence.
- Magnetic saturation, the point at which a material being magnetized by an electric current reaches saturation. Beyond this point, a large increase in current results in only a small increase in the magnetic strength.

These three factors depend upon the type, size, and length of the core. The greater the cross-sectional area of the core, the more permeable is the magnet. The longer the core, the more reluctance there is to the magnetic lines of force.

The residual magnetism and the retentivity vary with the material of the core. Soft iron is very permeable but has low retentivity. Hardened steel is not as permeable as soft iron, but it has high retentivity.

In the design of an electromagnet/solenoid, the most important factor to consider is the type of core. The selection of materials depends upon the use of the core. For example, a lifting magnet should have a core with good permeability and low retentivity. The field cores of an electric generator should have good permeability and be able to retain magnetism for an indefinitely long period of time.

The magnetomotive force and the core reluctance are the qualities that determine the amount

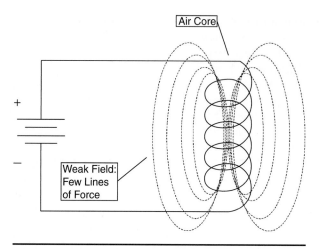

**FIGURE 6–21A**   Air core coil—weak electromagnet.

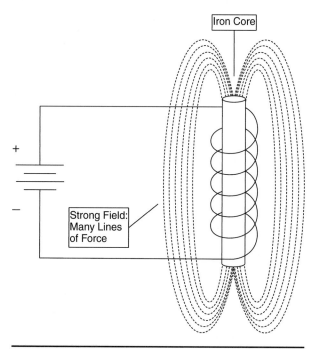

**FIGURE 6–21B**   Iron core coil—strong electromagnet.

of flux for any electromagnet. These values can be calculated by the following formula:

$$\Phi = \frac{f}{\Re}$$   (Eq. 6.1)

where   $\Phi$ = flux, in maxwells (Mx)
        $f$ = mmf, in gilberts (Gb)
        $\Re$ = units of reluctance

## SOLENOIDS

A solenoid is a type of electromagnet with a movable core. It is constructed by winding magnet wire on a hollow fiber or plastic form. The core is made

of a material that is easily magnetized but does not retain its magnetism when the current no longer flows through the coil. The core, commonly called the **armature**, is arranged to move in and out of the coil.

When the coil is energized, the magnetic field established by the current pulls the core into the coil. When the coil is de-energized, the core returns to its original position.

Solenoids are commonly used in door chimes. When the coil is energized, the armature is pulled into the coil in such a way that it strikes a brass chime bar; as it returns to its original position, it strikes another chime bar.

Industrial uses for solenoids include such applications as opening and/or closing valves, operating clutches, and operating relays.

## REVIEW QUESTIONS

*Multiple Choice*

1. The magnetic force on a bar magnet is strongest
   a. in the center of the bar.
   b. at the north magnetic pole.
   c. at both ends.
   d. at the south magnetic pole.

2. Two unlike poles placed within magnetic reach of each other
   a. neutralize each other.
   b. attract each other.
   c. repel each other.
   d. none of the above.

3. Artificial magnets are
   a. stronger than natural magnets.
   b. less effective than natural magnets.
   c. weaker than natural magnets.
   d. the same as natural magnets.

4. The space around a magnet in which the magnetic force exists is called the magnetic
   a. range.
   b. field.
   c. area.
   d. path.

5. If a bar magnet is cut in half,
   a. one half will be the north pole and the other half will be the south pole.
   b. it will lose its magnetism.
   c. each half will be a magnet having a north pole and a south pole.
   d. each half will have better retentivity.

6. A nonmagnetized atom has
   a. most of the electrons spinning in one direction.
   b. half of the electrons spinning in one direction and half spinning in the opposite direction.
   c. none of the electrons spinning.
   d. no pattern to the spin.

7. An electric current produces
   a. a square magnetic field.
   b. an elongated magnetic field.
   c. a circular magnetic field.
   d. a rectangular magnetic field.

8. Solenoids are used in
   a. transformers.
   b. relays.
   c. motors.
   d. generators.

9. Retentivity is the
   a. ability of a material to retain its magnetism after being removed from the magnetizing influence.
   b. magnetism that remains in a material after being removed from the magnetizing influence.
   c. state of a magnetic material being saturated.
   d. the ability of a material to be magnetized.

10. Residual magnetism is the magnetism
    a. produced by an electric current.
    b. that remains in a material after being removed from the magnetizing influence.
    c. produced by another magnet.
    d. that is lost when the magnet is removed from the magnetizing force.

*Give Complete Answers*

1. Define the term magnet.
2. Name the three basic kinds of magnets.
3. What was the first practical use of a magnet?
4. Write the rules for magnetic attraction and repulsion.
5. List four materials used in making permanent magnets.
6. What are magnetic lines of force?
7. Define magnetic flux.
8. Define magnetic field.
9. Describe the experiment in which powdered iron and a bar magnet are used. What does the experiment prove?
10. How can it be proven that the lines of force leave from the north pole and return by the south pole?
11. Explain the molecular theory of magnetism.
12. Explain the electron theory of magnetism.
13. Do the electron and molecular theories of magnetism conflict with each other? Explain.
14. Name at least two types of alloys used to make permanent magnets.
15. Explain the purpose of a magnetic shield, and describe how it works.
16. Describe how electromagnetism is produced.
17. Write the rule for determining the direction of the magnetic field around a current-carrying conductor.
18. Two wires carrying an electric current are placed next to each other. They are forced apart. Explain why this phenomenon takes place.
19. Why does a coil of wire carrying current produce a magnetic field similar to a bar magnet?
20. Write the rule for determining the polarity of an electromagnet.
21. What is the advantage of an electromagnet compared to a permanent magnet?

# Alternating Current

## OBJECTIVES

After studying this chapter, the student will be able to:

■ Explain the difference between alternating current and direct current.

■ Describe the basic principles of alternating current.

■ List the advantages and disadvantages of alternating current.

■ Describe the characteristics of alternating current with regard to resistance, inductance, and capacitance.

■ Define *power in AC circuits* and *power factor*.

# BASIC AC THEORY

Alternating current is the primary source of electrical energy today. It is less expensive to generate and transmit than direct current. AC equipment is generally more economical to maintain and requires less space per unit of power than DC equipment. While direct current is unidirectional and of a constant value, alternating current not only reverses direction periodically but also varies in strength.

The standard AC generator (alternator) produces a voltage that, if plotted on a graph, forms a curve similar to that in Figure 7–1. This curve shows that the voltage increases from zero to a maximum value in one direction, drops to zero, increases to a maximum value in the opposite direction, and drops to zero again. When such a change in values has taken place, a cycle has been completed. From then on, the cycle is merely repeated.

The number of cycles completed in 1 second is called the **frequency**. The modern unit of measurement of frequency is the **hertz** (Hz). "Hertz" refers to cycles per second. The most common frequency in the United States is 60 hertz. Many European and Asian countries operate on a frequency of 50 hertz.

## Hydraulic Analogy of Alternating Current

The flow of alternating current can be compared to the flow of water in a closed system (Figure 7–2). When the piston, P, is moved toward A, the water is forced out at A and drawn in at C, flowing through the pipe in the direction ABC. When the piston reaches the end of the stroke, it moves back toward C. The water is then forced out at C and drawn in at A. The flow is reversed, flowing through the pipe in the direction CBA. As the piston is moved back and forth, it sets up an alternating pressure, causing the current of water to change direction at the end of each stroke. Thus, the current of water alternates in direction, and the rate of flow varies widely.

## Generation of a Voltage Curve

The amount of voltage produced by a coil rotating in a magnetic field depends upon the following factors:

- The strength of the magnetic field.
- The speed at which the coil moves through the field.
- The number of turns of wire in series on the coil.
- The angle at which the coil moves through the field.

The direction of the induced voltage depends upon the direction of motion of the coil. Figure 7–3 portrays various positions of the rotor in an elementary alternator. Assume that the armature is rotating in a clockwise direction. In position A, the coil is moving parallel to the lines of force and no voltage is produced. As the coil continues to rotate, it begins to cut across the flux, and a voltage is produced. When the coil is moving directly across the lines of force (position B), the maximum voltage is generated. Further rotation of the coil causes the angle to decrease. The induced voltage also decreases. When the coil is again moving parallel with the lines of force (position C), the voltage is zero. As the coil rotates further, side X is moving up through the field and Y is moving down. This causes the induced voltage to reverse direction. When the coil is again moving at right angles to the field, the maximum voltage is produced. Notice that the polarity of the voltage has reversed. Position E shows the coil back in the original position (one cycle of alternating voltage has been completed).

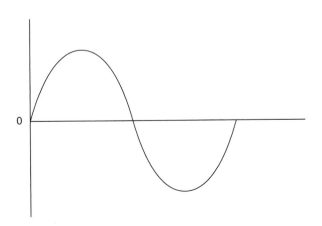

**FIGURE 7–1** Graph of an AC voltage.

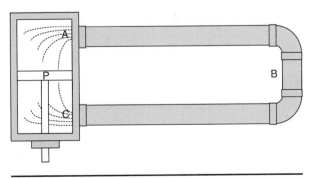

**FIGURE 7–2** Closed system filled with water. The piston forces water through the pipe, producing an alternating water flow similar to AC flow in an electric circuit.

If the maximum voltage of the alternator is known, the generated voltage can be plotted to form a curve. Draw a circle with the radius representing the maximum value of voltage. Any convenient scale may be used. Divide the circle into equal parts (Figure 7–4). Draw a horizontal line to scale, along which one voltage cycle will be plotted. Divide the line into the same number of equal parts as the circle. Draw horizontal and vertical lines, as illustrated by the dashed lines in Figure 7–4. The intersection of the lines represents the value of voltage at that instant. For example, a horizontal and vertical line intersect at point X. Using the same scale as that used for the radius of the circle, one can measure the value of voltage. This value is the emf produced when the coil is cutting the lines of force at a 30-degree angle.

## Use of Vector Diagrams

The change that occurs in the value of an alternating voltage and/or current during a cycle can also be shown by using vector diagrams. A **vector** is a line segment that has a definite length and direction. A **vector diagram** is two or more vectors joined together to convey information. Vector diagrams drawn to scale can be used to determine instantaneous values of current and/or voltage.

Figure 7–4 can be analyzed by means of vector diagrams, according to the following procedure. Draw a horizontal line as a reference line (Figure 7–5). Starting at point O, 30 degrees from the reference line, draw OA to scale to represent a maximum voltage ($V_m$) of 100 volts. From the end of vector OA, draw a vertical dashed line. This line should form a 90-degree angle with the reference line. The vertical dashed line is labeled AB and represents the instantaneous value of voltage ($E_i$) when the coil is cutting the lines of force at a 30-degree angle. Measure vector AB. It should scale to 50 volts.

The same procedure can be followed for any degree of rotation. The vector diagram shown in

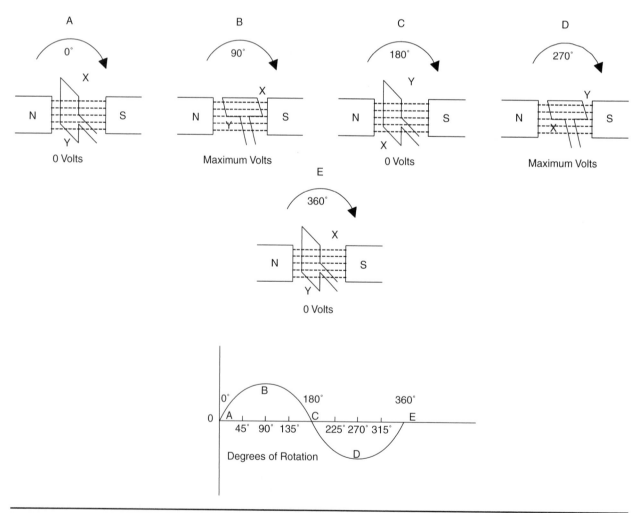

**FIGURE 7–3** Generation of an alternating voltage. As the loop rotates through the magnetic field, the amount and polarity of the voltage change with the angle and direction of motion.

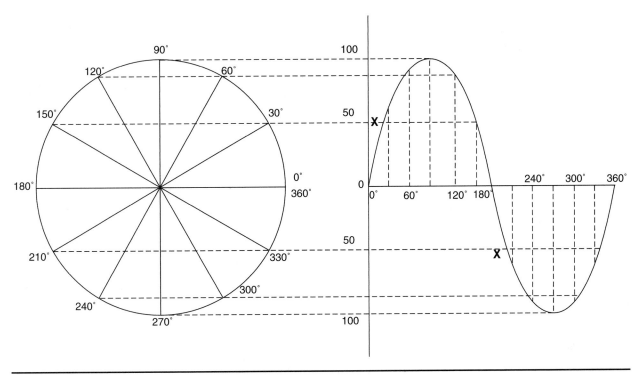

**FIGURE 7–4** Plotting a curve of alternating voltage.

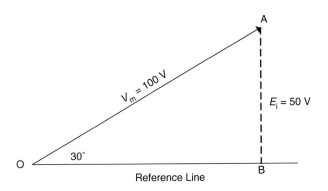

**FIGURE 7–5** Vector diagram drawn to scale to determine the voltage when the coil is moving at a 30-degree angle to the lines of force.

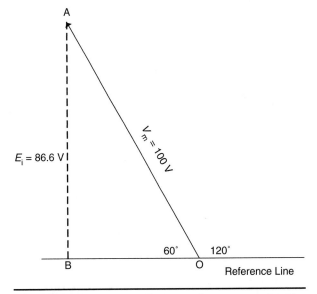

**FIGURE 7–6** Vector diagram drawn to scale. The coil has rotated 120 degrees and is cutting the flux at 60 degrees.

Figure 7–6 is used to determine the value of voltage when the coil has rotated 120 degrees. Although the coil has rotated 120 degrees, the angle it is making with the lines of force is only 60 degrees. It is this angle that determines the value of the instantaneous voltage. For example, if the coil rotates 210 degrees, it cuts the lines of force at an angle of 30 degrees (Figure 7–7).

Referring back to Figure 7–4, it can be seen that each division of the circle can represent vector OA. Vector AB can be represented by points along the voltage curve. The angle between the horizontal diameter of the circle and the radius $V_m$ is the angle at which the coil is cutting the flux. Although vector diagrams are seldom used alone, they are a simple way of presenting a visual illustration of a problem. Vector diagrams are usually used with trigonometric functions.

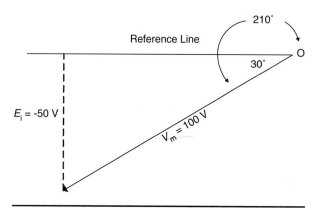

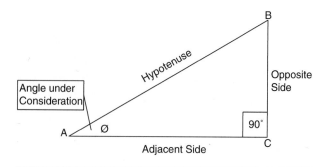

**FIGURE 7-7** Vector diagram drawn to scale. The coil has rotated 210 degrees and is cutting the flux at 30 degrees. The emf has reversed direction, indicated by the "−" sign.

**FIGURE 7-9** Right triangle.

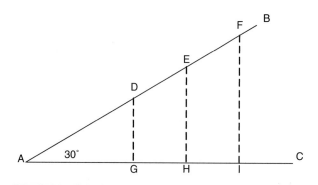

**FIGURE 7-8** Vector diagram forming three 30-degree–60-degree–90-degree triangles.

## Trigonometry

Many electrical problems are solved through the use of trigonometry. The vector diagrams used with *trigonometric functions* are generally in the form of triangles and/or parallelograms. Consider Figure 7–8. Two vectors are drawn to form a vector diagram. Vector AB joins vector AC, forming a 30-degree angle. If three lines are drawn perpendicular to AC, three right triangles are formed. If these triangles are drawn to scale, the ratio of two sides of one triangle is equal to the ratio of the corresponding sides of either of the other two triangles. This is because the corresponding angles of the three triangles are equal. This ratio can be stated mathematically as follows:

$$DG/DA = EH/EA = FI/FA = 0.5$$
$$GA/DA = HA/EA = IA/FA = 0.866$$
$$DG/GA = EH/HA = FI/IA = 0.577$$

These values illustrate that for a given angle A (∠A), the ratio of the two sides of a triangle has the same value regardless of the length of the sides. For any acute angle (an angle less than 90 degrees) of a right triangle, there are six possible ratios. These ratios are called trigonometric functions of the angle.

In the right triangle ABC in Figure 7–9, AB is the hypotenuse. The **hypotenuse** is always the side opposite to the 90-degree angle (the right angle) and is the longest side of the triangle. BC is opposite angle A (the angle under consideration), and it is called the *opposite side*. AC is touching angle A; therefore, AC is called the *adjacent side*. The ratios are stated as follows:

$$\frac{BC}{AB} = \frac{\text{side opposite } \angle A}{\text{hypotenuse}}$$
$$= \text{sine of } \angle A, \text{ abbreviated sin A}$$

$$\frac{AC}{AB} = \frac{\text{side adjacent } \angle A}{\text{hypotenuse}}$$
$$= \text{cosine of } \angle A, \text{ abbreviated cos A}$$

$$\frac{BC}{AC} = \frac{\text{side opposite } \angle A}{\text{side adjacent } \angle A}$$
$$= \text{tangent of } \angle A, \text{ abbreviated tan A}$$

$$\frac{AC}{BC} = \frac{\text{side adjacent } \angle A}{\text{side opposite } \angle A}$$
$$= \text{cotangent of } \angle A, \text{ abbreviated cot A}$$

$$\frac{AB}{AC} = \frac{\text{hypotenuse}}{\text{side adjacent } \angle A}$$
$$= \text{secant of } \angle A, \text{ abbreviated sec A}$$

$$\frac{AB}{BC} = \frac{\text{hypotenuse}}{\text{side opposite } \angle A}$$
$$= \text{cosecant of } \angle A, \text{ abbreviated csc A}$$

These ratios can be used for either of the acute angles of a right triangle. The ratios are generally written in formula form as follows:

$$\sin \phi = \frac{\text{opp}}{\text{hypt}} \quad \cos \phi = \frac{\text{adj}}{\text{hypt}} \quad \tan \phi = \frac{\text{opp}}{\text{adj}}$$
$$\cot \phi = \frac{\text{adj}}{\text{opp}} \quad \sec \phi = \frac{\text{hypt}}{\text{adj}} \quad \csc \phi = \frac{\text{hypt}}{\text{opp}}$$
(Eq. 7.1)

Because the ratio of the sides of a right triangle is a constant for any given angle, the values for any angle from 0 degrees through 90 degrees can be listed.

# ALTERNATING CURRENT AND VOLTAGE VALUES

The voltage and current in ac circuits are always changing in value. This poses the questions of how to measure these values and what is the overall effect of these changing values on the circuit.

## Effective Values

The **effective value** of an alternating current is that value that will produce the same heating effect as a specific value of a steady direct current. In other words, an alternating current has an effective value of 1 ampere if it produces heat at the same rate as the heat produced by 1 ampere of direct current, both flowing in the same value of resistance.

Another name for the effective value of an alternating current or voltage is the *root-mean-square* (rms) *value*. This term was derived from one method used to compute the value. The rms is calculated as follows: The instantaneous values for one cycle are selected for equal periods of time. Each value is squared, and the average of the squares is calculated. (Values are squared because the heating effect varies as the square of the current or voltage.) The square root of this answer is the rms value.

A more accurate method requires the use of calculus. Either method shows that the effective or rms value is 0.707 times the maximum value. A simple equation for calculating the effective value is shown here:

$$\text{For voltage, } E = 0.707E_m$$
$$\text{For current, } I = 0.707I_m \qquad \text{(Eq. 7.2)}$$

where   subscript "m" refers to the maximum value

When an alternating current or voltage is specified, it is always the effective value that is meant, unless otherwise stated. Standard AC meters indicate effective values.

## Average Values

It is sometimes useful to know the **average value** for one-half cycle. If the current changed at the same rate over the entire half cycle, the average value would be one-half of the maximum value. Because current does not change at the same rate, another method must be used. This value can also be found by using

calculus. However, it has been determined that the average value is equal to 0.637 times the maximum value. The equations are as follows:

$$\text{For voltage, } E_{av} = 0.637E_m$$
$$\text{For current, } I_{av} = 0.637I_m \qquad \text{(Eq. 7.3)}$$

where   subscript "av" refers to the average value and subscript "m" refers to the maximum value

## Instantaneous Values

The **instantaneous values** of an alternating current or voltage can be determined by drawing a vector diagram to scale and measuring the resultant vector. However, because most alternators produce a voltage that, when plotted on a graph, forms a curve that coincides with the sine table, the following formulas may be used:

$$\text{For voltage, } e = E_m \sin \phi$$
$$\text{For current, } i = I_m \sin \phi \qquad \text{(Eq. 7.4)}$$

where   $e$ = instantaneous value of voltage
   $i$ = instantaneous value of current
   $E_m$ = maximum value of voltage
   $I_m$ = maximum value of current
   $\sin \phi$ = trigonometric function for the angle at which the flux is being cut

In a plotting of instantaneous values, it should be noted that the distance along the horizontal line represents not only the angle that the coil makes with the lines of force but also the passage of time. Therefore, the values along the horizontal line are called *electrical time degrees*. One cycle represents 360 electrical time degrees.

# ADVANTAGES AND DISADVANTAGES OF AC

Alternating current can be generated at higher voltages than DC with fewer problems of heating and arcing. Some standard values of voltage are 2300, 4600, 6900, 13,800, and 33,100 volts. These values are frequently increased to 100,000, 200,000, and 800,000 volts for transmission over long distances. At the load area, the voltage is decreased to working values of 120, 208, 240, 277, 440, 480, and 550 volts. Figure 7–10 shows a transmission line.

The ease with which the voltage can be raised and lowered makes AC ideal for transmission purposes. Large amounts of power can be transmitted at high voltages and low currents with minimum line loss. Because $P = I^2 \times R$ (Equation 3.4), the lower the value of transmission current, the less will be

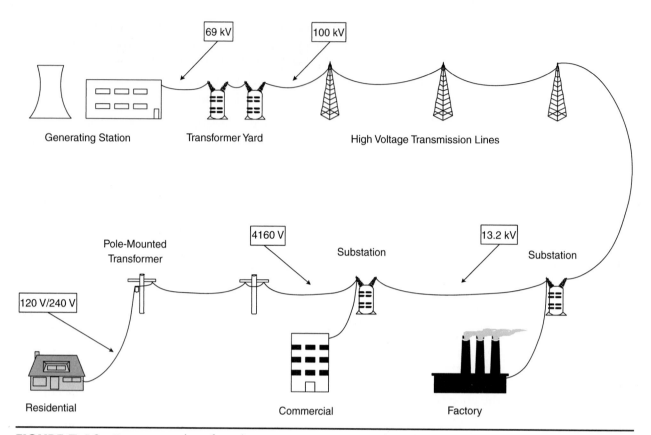

**FIGURE 7–10**  Transmission lines from the generating station to industrial, commercial, and residential customers.

the line loss. Because the current is low, smaller transmission wires can be used to reduce the installation and maintenance costs.

Direct current generators, because of their construction, limit their output voltage to 2000 volts or less. The voltage cannot be raised or lowered through the use of transformers. Long-distance transmission requires heavier cables and generally results in greater power loss.

Alternating current generators can be driven at high speeds and constructed in large sizes. Because they generally have rotating fields and stationary armatures, their rotor windings are small and light in weight, thus reducing the centrifugal force. Modern alternators are built with capacities of up to 500,000 kilowatts.

Because of the need for commutators, DC generators are limited in capacity. The maximum power available from any one unit is generally 10,000 kilowatts. If DC were the main source of supply, many more generating stations would be needed. Each station would require a source of power to drive the generators. With the supply of fossil fuels dwindling, this would be a gross waste of energy. AC, on the other hand, can be produced in large central stations and distributed over greater distances with maximum efficiency.

An advantage of AC is that it produces a varying magnetic field. This changing field is used in

the distribution transformer for raising and lowering the voltage. Lighting units that use transformers produce better and more efficient light. Induction motors utilize the transformer principle for their operation. These motors are less expensive to build, install, and maintain than are DC motors. They also require less space than that required by DC motors of the same horsepower.

DC motors, however, have one distinct advantage over AC motors: they have better speed control. In general, DC motors are used when wide variations of speed and accurate speed adjustments are required. However, with the advent of electronic variable speed drives (inverters), the speed control capabilities of AC motors rivals and, in many cases exceeds, that of DC motors.

# ELECTROMAGNETIC INDUCTION

**Electromagnetic induction** is the process by which a voltage is produced in a coil as the result of a moving magnetic field's passing across the coil, or the coil's moving through the magnetic field. In a transformer, the AC flowing in the primary produces an alternating magnetic field that moves across the secondary. This action induces an emf in the secondary winding. This is an example of **mutual induction**. Current flowing in one coil induces a voltage into another coil. There is no electrical connection between the two coils.

Another type of induction that takes place in an ac circuit is called **self-induction**, or *self-inductance*. Induction takes place whenever the amount of current flowing in a circuit changes in value. The amount of inductance is generally negligible unless there is a coil in the circuit.

If DC is applied to a circuit containing pure resistance (no coils), the current rises rapidly, and then levels off at a steady value (Figure 7–11). If the circuit contains a coil, it takes a longer period of time for the current to reach its maximum value. As the current begins to increase in the coil, the magnetic field begins to expand, moving across the turns of wire. The moving magnetic field induces

another voltage into the coil. This voltage of self-induction is in the direction opposite to the applied voltage, and it tends to retard the increasing current. The final result is that it takes the current a longer period of time to reach the maximum value (Figure 7–12).

When the applied voltage is removed from the circuit, the current begins to decrease. The magnetic field contracts, moving across the turns on the coil in the opposite direction. The induced voltage is, therefore, in a direction that tends to retard the decreasing current (Figure 7–13).

In an AC circuit, the current value is changing continuously. Therefore, the voltage of self-induction is always present. If it were possible to have a pure inductive circuit, then the voltage/current relationship would be as shown in Figure 7–14.

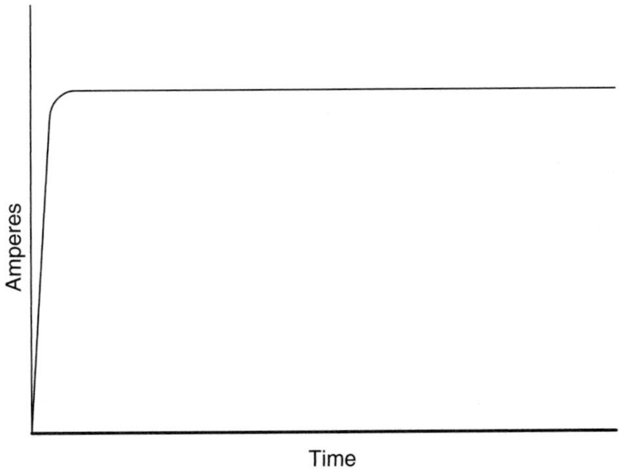

**FIGURE 7–11** Graph of DC applied to a pure resistive circuit.

**FIGURE 7–13** Graph of DC when the applied voltage is removed from an inductive circuit.

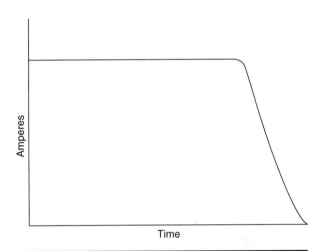

**FIGURE 7–12** Graph of DC applied to an inductive circuit.

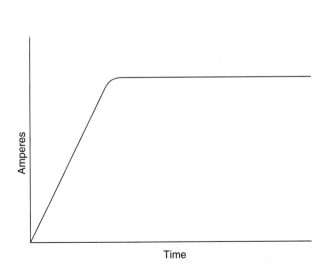

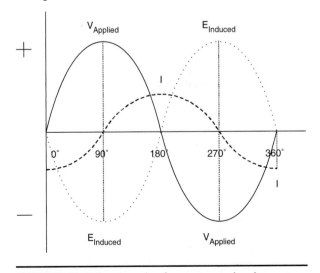

**FIGURE 7–14** Graph of current and voltage in a pure inductive circuit. The current lags the voltage by 90 electrical time degrees. E represents the voltage of self-induction.

In this circuit, the current reaches its maximum value 90 electrical time degrees after the applied voltage. Thus, in a pure inductive circuit, the current lags the applied voltage by 90 degrees. The voltage of self-induction lags the current by 90 degrees, and also lags the applied voltage by 180 degrees. In other words, the current is 90 degrees out of phase with the applied voltage, and the voltage of self-induction is 180 degrees out of phase with the applied voltage.

A pure inductive circuit cannot be obtained, however, since all circuits have some resistance. Therefore, the phase relationship will differ, depending on the value resistance. A more typical circuit is illustrated in Figure 7–15. In this circuit, the current lags the applied voltage by 60 degrees, and the voltage of self-induction lags the current by 90 degrees. The voltage of self-induction lags the applied voltage by 60 degrees + 90 degrees = 150 degrees.

A vector diagram of the two voltages is shown in Figure 7–16. Vector $c$ is the resultant of $a$ (applied voltage) and $b$ (induced voltage). The resultant

voltage can also be determined by the following formula:

$$c = \sqrt{a^2 + b^2 \pm 2ab \cos \phi} \qquad \text{(Eq. 7.5)}$$

where   $c$ = resultant value of voltage
$a$ = applied voltage
$b$ = voltage of self-induction
$\cos \phi$ = 0.866 (from the cosine table)

### Example 1

If the applied voltage equals 10 V in the circuit in Figure 7–16, what is the resultant voltage if the voltage of self-induction lags the applied voltage by 150 degrees? The voltage of self-induction is equal to the applied voltage (10 V).

$$c = \sqrt{a^2 + b^2 \pm 2ab \cos \phi}$$
$$c = \sqrt{100 + 100 - (2 \times 10 \times 10 \times 0.866)}$$
$$c = \sqrt{200 - 173.2}$$
$$c = \sqrt{26.8}$$
$$c = 5.177 \text{ V}$$

This method of calculation is based on trigonometric functions and the parallelogram method for solving vector diagrams. Figure 7–17 shows how the parallelogram is constructed and the resultant voltage is determined. Lay out vector $a$ on the horizontal from point O. Draw vector $b$, beginning at point O, 150 degrees from vector $a$. To form a parallelogram, begin at the end of vector $a$, and draw a dashed line parallel and equal to vector $b$. Beginning at the end of vector $b$, draw a dashed line parallel and equal to vector $a$. The point where the two dashed lines intersect is called point Y. Draw vector $c$ from point O to point Y. Vector $c$ represents the resultant voltage.

To determine where to use the plus (+) or minus (−) sign in the formula $c = \sqrt{a^2 + b^2 \pm (2bc \cos \phi)}$, use the following procedure. If the angle between the applied voltage and the voltage of self-induction is from 0 degrees to 90 degrees, the sign is +. If the angle is greater than 90 degrees to 270 degrees, the sign is −. From 270 degrees to 360 degrees, the

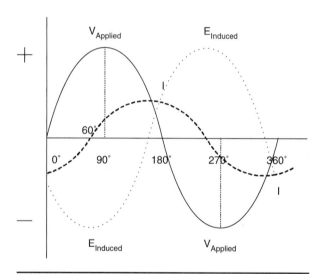

**FIGURE 7–15** Graph of current and voltage in a circuit containing resistance and inductance. The current lags the voltage by 60 electrical time degrees. E represents the voltage of self-induction.

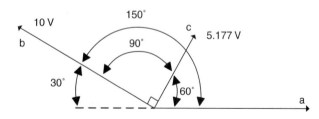

**FIGURE 7–16** Vector diagram of voltage in a circuit containing resistance and inductance.

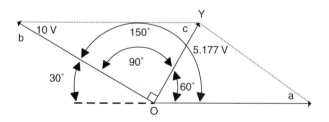

**FIGURE 7–17** Constructing a parallelogram to determine the resultant voltage of a circuit containing resistance and inductance.

sign is +. In Example 1, the angle is 150 degrees; therefore, the sign is −.

## Inductance

*Inductance* is a measurement of the induced voltage caused by the changing current. To avoid confusion between the induced voltage and the applied voltage, the unit **henry** (H) has been established. A coil has an inductance of 1 henry if a current changing at the rate of 1 ampere per second produces an induced voltage of 1 volt.

The factors that affect the inductance of a coil are the physical and geometric characteristics of the core and the coil, as well as the frequency of the circuit. In other words, the type, length, and area of the core, the number and proximity of the turns of wire, and the rate at which the current is changing all affect the value of inductance.

A formula for determining the inductance is:

$$L = \frac{0.4\pi N^2 \mu A}{10^8 \ell} \qquad \text{(Eq. 7.6)}$$

where $L$ = inductance, in henrys (H)
$0.4\pi = 1.25664$ (a constant value for this formula)
$N$ = number of turns of wire on the coil
$\mu$ = permeability of the core (determined by the material used)
$A$ = area of the core, in square inches (in.²)
$\ell$ = length of the core, in inches (in.)

## Example 2

A coil containing 100 turns of wire is wound on a metal core 3 in. long with a surface area of 6.28 in.² The permeability of the core is $5 \times 10^3$. Calculate the inductance of the coil.

$$L = \frac{0.4\pi N^2 \mu A}{10^8 \ell}$$
$$L = \frac{1.25664 \times 100^2 \times 5 \times 10^3 \times 6.28}{10^8 \times 3}$$
$$L = \frac{1.25664 \times 5 \times 6.28}{30}$$
$$L = \frac{39.4585}{30}$$
$$L = 1.315 \text{ H}$$

If the inductance of the coil is known, the voltage of self-inductance can be calculated by using the following formula:

$$E = -L\left(\frac{\Delta I}{\Delta t}\right) \qquad \text{(Eq. 7.7)}$$

where $E$ = voltage of self-inductance
$L$ = inductance, in henrys (H)
$\Delta I$ = change in the value of current, in the time $\Delta t$
$\Delta t$ = change, in time

## Example 3

Calculate the induced voltage for the coil in Example 2 if the current changes from 5 A to 50 A in 1 s.

$$E = -L\left(\frac{\Delta I}{\Delta t}\right)$$
$$E = -1.3153\left(\frac{45}{1}\right)$$
$$E = -1.3153 \times 45$$
$$E = -59.1885 \text{ V}$$

The negative sign is used to indicate that it is the voltage of self-inductance.

## Inductive Time Constant

Inductive coils are often used in DC time-delay circuits. The time that it takes a direct current to reach its maximum steady value is determined by the ratio of the inductance to the resistance. One time constant is required for the current to increase from zero to 63.2 percent of its maximum value. Five time constants are required for the current to rise from zero to its maximum value. The formula to calculate the inductive time constant is:

$$t = \frac{L}{R} \qquad \text{(Eq. 7.8)}$$

where $t$ = inductive time constant, in seconds (s)
$L$ = inductance, in henrys (H)
$R$ = resistance, in ohms (Ω)

## Example 4

Calculate the amount of time it will take direct current to reach its maximum value, if the coil in Example 2 has a resistance of 2 Ω.

$$t = \frac{L}{R}$$
$$t = \frac{1.3153}{2}$$
$$t = 0.65765 \text{ s} = \text{one time constant}$$
$$0.65765 \times 5 = 3.28825 \text{ s} = \text{the time to reach the maximum value}$$

Many time delay circuits are used in controlling industrial machinery.

## Inductive Reactance

Equation 7.5 indicates that the resultant voltage is less than the applied voltage. If equal values of alternating voltage and direct voltage are applied to the same circuit, less current would flow for AC than for DC. With DC, the current is opposed by the induced voltage only when it is rising to its maximum value. Once it has reached a steady value, there is no inductive effect. With alternating voltage, the current is continually changing. Therefore, inductance is present at all times.

This opposition to the flow of AC is called *inductive reactance*, and its unit of measurement is the ohm. The amount of inductive reactance produced by a coil depends upon the frequency of the current and the amount of inductance. The inductive reactance of a circuit or circuit component can be calculated by the following formula:

$$X_L = 2\pi f L \qquad \text{(Eq. 7.9)}$$

where   $X_L$ = inductive reactance, in ohms ($\Omega$)
    $2\pi$ = 6.2832 (a constant for this formula)
    $f$ = frequency, in hertz (Hz)
    $L$ = inductance, in henrys (H)

### Example 5

Calculate the inductive reactance of the coil in Example 2 if it is connected to a 60-Hz circuit.

$$X_L = 2\pi f L$$
$$X_L = 6.2832 \times 60 \times 1.3153$$
$$X_L = 496 \ \Omega$$

## Noninductive Coil

It is sometimes necessary to wind a coil so that it will have a high resistance and no inductance. This can be accomplished by winding the coil as shown in Figure 7–18. The current through such a coil is in the opposite direction through adjacent turns. Consequently, the magnetizing action of one-half of the turns neutralizes the magnetizing action of the other half. No flux is produced, and the coil is noninductive.

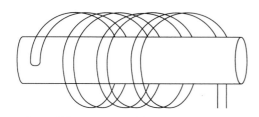

**FIGURE 7–18**   Noninductive coil.

## Impedance in Inductive Circuits

Because inductive circuits contain both inductive reactance and resistance, it is apparent that there are two factors that oppose current flow. The combined effect of these two factors determines the amount of current flow. The total opposition to the flow of current in an AC inductive circuit is called **impedance** and is measured in ohms. The Ohm's law formula for calculating impedance is:

$$Z = \frac{E}{I} \qquad \text{(Eq. 7.10)}$$

where   $Z$ = total opposition to current flow, in ohms ($\Omega$)
    $E$ = applied voltage
    $I$ = amount of current flowing in the circuit

Another method for calculating the impedance is used when only the resistance and the inductive reactance are known (Figure 7–19). This method uses vector diagrams and trigonometric functions. Draw vectors at right angles to one another to represent the resistance and the inductive reactance. Scale the resistance vector along the horizontal line and the inductive reactance vector along the vertical line. Connect the ends of the two vectors by a dashed line. The dashed line represents the impedance. Measuring $Z$ will give the value of impedance. A formula for calculating the impedance is:

$$Z = \sqrt{R^2 + X_L^2} \qquad \text{(Eq. 7.11)}$$

where   $Z$ = impedance, in ohms ($\Omega$)
    $R$ = resistance, in ohms ($\Omega$)
    $X_L$ = inductive reactance, in ohms ($\Omega$)

## Phase Relationship in Inductive Circuits

In an AC inductive circuit, the current always lags the applied voltage. That is, the current reaches its maximum and zero values some time after the

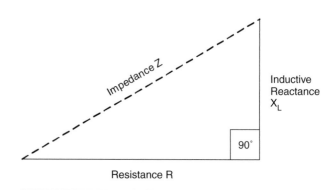

**FIGURE 7–19**   Impedance vector diagram.

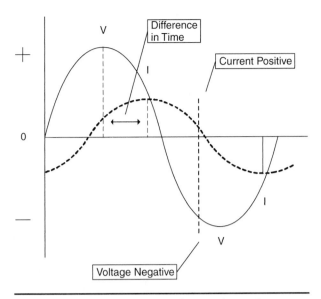

**FIGURE 7–20** Phase relationship of current and voltage in a circuit containing resistance and inductance.

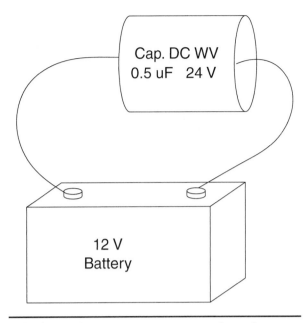

**FIGURE 7–21** Capacitor connected to a battery.

voltage does. Figure 7–20 illustrates this phenomenon. To help you remember when the current reaches its peak, memorize the phrase *ELI the ICE man*. The letters *E, L,* and *I* in the word *ELI* are used to represent *voltage, inductance,* and *current,* respectively. Notice that the letter *I* appears after the letter *E*. This should help you remember that *current lags* (or *follows*) *the voltage* in an inductor. We will discuss the word *ICE* later.

## Summary of Inductance

Inductance in AC circuits causes the current to lag the voltage. The amount of lag depends upon the ratio of the inductance to the resistance. Inductance also reduces the amount of AC that will flow in the circuit. The greater the value of inductance, the less will be the current. Inductance affects DC only when it is changing in value. The amount of inductance in a coil depends upon the physical and geometric characteristics of the core and the coil and the frequency of the circuit.

## | CAPACITANCE

A **capacitor,** also called a **condenser,** consists of two conductors separated by an insulating material. The conductors are called the *plates,* and the insulating material is called the *dielectric.* When a direct voltage is placed across a capacitor, electrons are forced from one plate and caused to accumulate on the other plate. This flow continues until the charge built up

across the plate is equal to that of the DC source. If the DC is removed, the capacitor remains charged until a conductor is placed across the plates. Placing a conductor across the plates allows the electrons to return to their original plate, and both plates become neutral. Figure 7–21 illustrates a capacitor connected to a battery.

Note that current does not flow through the dielectric. The electrons simply leave one plate and build up on the other plate. The flow of electrons stops when the charge on the plates is equal to the emf of the battery.

If an alternating voltage is placed across a capacitor, a somewhat different phenomenon takes place. Because alternating voltage is continually changing in strength and direction, current continually flows back and forth in the circuit. The current does not flow through the capacitor but flows alternately in one direction and then in the other direction as each plate charges and discharges.

The flow of electrons in a circuit containing a capacitor can be compared to a hydraulic system. The capacitor reaction is similar to a device installed in a closed-loop system that contains a rubber diaphragm stretched across its center. Figure 7–22A illustrates the device, and Figure 7–22B shows the closed-loop system. A centrifugal pump is installed, and the entire system is filled with water.

When the pump is started, there is a flow of water in the pipe while the diaphragm is being stretched. When the pressure of the stretched diaphragm is equal to that of the pump, the flow stops. If the valve is closed while the diaphragm is

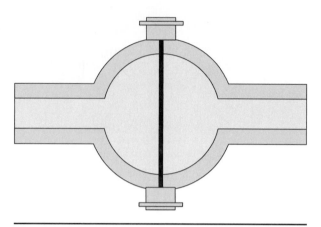

**FIGURE 7–22A** Device containing a rubber diaphragm.

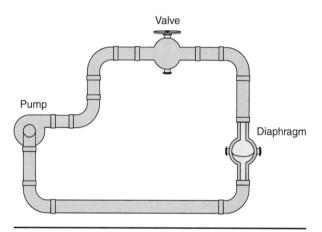

**FIGURE 7–22B** Closed-loop system containing a rubber diagram.

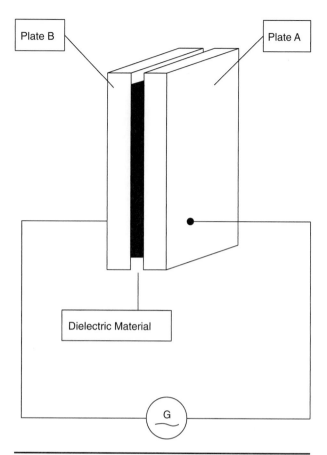

**FIGURE 7–23** Alternator connected to a capacitor.

stretched, the pump may be stopped and no water will flow. Opening the valve allows the water to flow back through the pump until the strain is relieved from the diaphragm.

If the valve is left open and the pump is arranged to reverse direction periodically, the water flows first in one direction and then reverses and flows in the opposite direction. Figure 7–23 shows an alternator connected to a capacitor. When the voltage is in one direction, the electrons are pulled off plate A and forced to build up on plate B. When the voltage reverses direction, the electrons are pulled off plate B and forced to build up on plate A. This reversal of electron flow continues as long as the alternator is in operation. Note that the current does not flow through the capacitor.

## Capacity

The size of the capacitor is determined by the size of the plates, the type of dielectric, and the distance between the plates. A capacitor is said to have a capacity of 1 **farad** (F) when a charge of 1 volt per second across its plates produces an average current of 1 ampere. A farad is an extremely large unit of measurement. Therefore, for practical purposes, the *microfarad* ($\mu$F) is generally used. One microfarad is equal to one-millionth of a farad.

It can also be said that a capacitor has a capacity of 1 farad when 1 volt DC applied to its plates causes it to charge to 1 coulomb. The formula for calculating capacitance is:

$$C = \frac{Q}{E} \qquad \text{(Eq. 7.12)}$$

where   $C$ = capacitance, in farads (F)
   $Q$ = charge on one plate, in coulombs (C)
   $E$ = voltage applied to the plates

## Capacitive Time Constant

As the electrons build up on one plate of the capacitor, they begin to repel any additional electrons. The larger the number of electrons on the plate, the greater the repelling force. The accumulation of electrons on one plate and the removal of electrons from the other plate cause a potential difference across the plates. The potential difference is in the direction opposite to the applied voltage.

The length of time it takes to charge a capacitor to its rated value is a factor of the resistance of the plate circuit and the capacity of the capacitor. The formula for calculating the actual charging rate is:

$$t = RC \qquad \text{(Eq. 7.13)}$$

where  $t$ = one time constant, in microseconds ($\mu$s)
$R$ = resistance of the plate circuit, in ohms ($\Omega$)
$C$ = capacitance in microfarads ($\mu$F)

Five time constants are required to fully charge or discharge a capacitor. Figures 7–24A and 7–24B show typical curves of the voltage buildup and discharge.

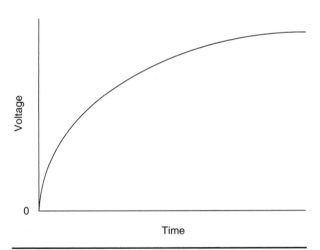

**FIGURE 7–24A**  Graph of voltage building up in a DC circuit containing capacitance.

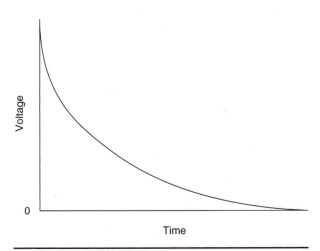

**FIGURE 7–24B**  Graph of voltage discharge in a DC circuit containing capacitance.

## Effect of Capacitance on Current Flow

In a DC circuit, a capacitor allows the current to flow only until the charge on the capacitor is equal to that of the applied voltage (Figure 7–25). Once the capacitor has met this requirement, current flow ceases. It can be said, then, that capacitors block the flow of DC.

When both AC and DC are applied to a circuit, but certain parts of the circuit are restricted to AC only, capacitors can be used to block out the DC. Another use for capacitors in DC circuits is to smooth out the ripple. If a pulsating DC is applied to a circuit, a capacitor connected across the circuit discharges when the applied voltage drops below the charge on the capacitor (Figure 7–26).

In AC circuits, capacitors cause the current to lead the voltage. In other words, the current reaches its maximum and zero values ahead of the voltage. The amount of lead depends upon the ratio of the resistance to the capacitance. If it were possible to have a pure capacitive circuit, the current would lead the voltage by 90 degrees.

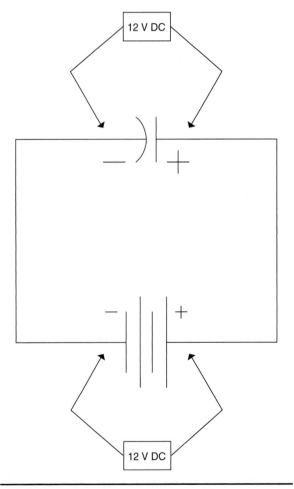

**FIGURE 7–25**  The charge on the capacitor equals the charge on the battery. These forces are equal and opposite, and no current flows.

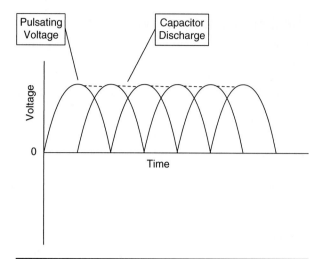

**FIGURE 7–26** Graph of a pulsating voltage with a capacitor connected to smooth out the ripple.

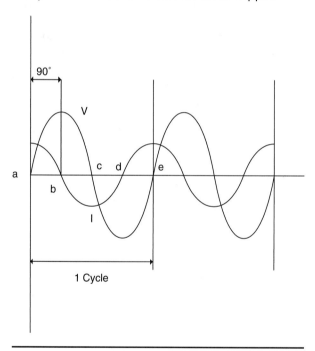

**FIGURE 7–27** Graph showing current/voltage relationship in a pure capacitive circuit.

When the voltage begins to build up in the positive direction, maximum current flows because there is no charge on the capacitor. As the electrons begin to accumulate on plate A (Figure 7–23), they oppose any further buildup, and the current flow decreases. When the voltage reaches its maximum value, the capacitor is fully charged and the current stops flowing. As the applied voltage decreases, the charge on the capacitor is greater than the applied voltage, and the current begins to flow from plate A to plate B. Therefore, while the applied voltage is still in the positive direction but decreasing, the current is flowing in the negative direction (Figure 7–27).

When the applied voltage decreases to zero, the charge on the capacitor causes the current to increase to the maximum value. At this point, the applied voltage reverses direction and tries to keep the current flowing in the negative direction (point c, Figure 7–27). However, the electrons have now built up on plate B and oppose any further buildup, causing the current flow to decrease (point d, Figure 7–27). When the applied voltage reaches the maximum value in the negative direction, the current has dropped to zero. This cycle of current and voltage changes continues as long as ac is applied to the circuit. To help you remember this relationship, recall the saying "ELI the ICE man." We learned that ELI represented the voltage (E) leading the current (I) in an inductor (L). Now, focus on the word "ICE." The letters I, C, and E in the word ICE are used to represent *current, capacitance,* and *voltage,* respectively. Notice that the letter I appears before the letter E. This should help you remember that *current leads the voltage in a capacitor.* Notice that in a circuit containing only capacitance (purely capacitive), the current and voltage are 90 degrees out of phase. We can therefore state, *in a purely capacitive circuit, the current will lead the voltage by 90 degrees.*

## Capacitive Reactance

The opposition to the flow of an alternating current, offered by a capacitor, is called *capacitive reactance.* Its unit of measurement is the ohm, because it opposes current flow. The amount of opposition to current flow produced by a capacitor is determined by its capacitance and the frequency of the circuit. The formula for capacitive reactance is:

$$X_C = \frac{1}{2\pi fC} \qquad \text{(Eq. 7.14)}$$

where  $X_C$ = capacitive reactance, in ohms ($\Omega$)
$2\pi$ = 6.2832 (a constant for this formula)
$f$ = frequency of the applied voltage, in hertz (Hz)
$C$ = capacitance, in farads (F)

## Impedance in Capacitive Circuits

Because capacitive circuits contain both capacitive reactance and resistance, it is apparent that there are two factors that oppose current flow. The combined effect of these factors determines the amount of current flow. Figure 7–28 shows capacitive reactance and resistance combined vectorially. The resultant of the two factors is the impedance of the circuit.

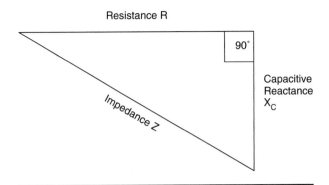

**FIGURE 7–28** Vector diagram of resistance, capacitive reactance, and impedance.

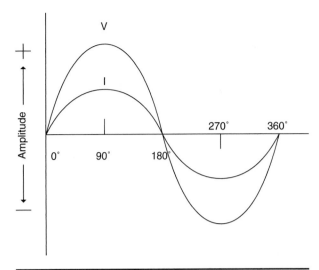

**FIGURE 7–29** Current and voltage curves in a pure resistive circuit.

The formula for vector diagrams that form right triangles may be used to solve for impedance. The formula is:

$$Z = \sqrt{R^2 + X_C^2} \qquad \text{(Eq. 7.15)}$$

where   $Z$ = impedance, in ohms ($\Omega$)
          $R$ = resistance, in ohms ($\Omega$)
          $X_C$ = capacitive reactance, in ohms ($\Omega$)

## Summary of Capacitance

Capacitance in AC circuits causes the current to lead the voltage. The amount of lead depends upon the ratio of the capacitance to the resistance. Capacitance also has a reactive component, which opposes current flow. The greater the capacitive reactance, the lesser the current flow. Capacitance affects DC only when it is changing in value. A capacitor blocks out the flow of DC. The capacity of a capacitor depends upon the size and type of the plates, the type of dielectric, the distance between the plates, and the frequency of the circuit.

## POWER IN AC CIRCUITS

The power expended by a DC circuit is equal to the product of the current flowing in the circuit and the voltage impressed across the circuit. In an AC circuit, both the current and voltage are changing in value. Therefore, it can be said that the power at any instant is equal to the product of the current and voltage at that instant. Figure 7–29 shows a current and voltage wave for a pure resistive circuit. In a pure resistive circuit, the current and voltage are in phase. "In phase" means that the current and voltage start at the same instant, reach their maximum values at the same instant and in the same direction, and then drop back to zero and reverse

direction at the same instant. To calculate the power at any instant, multiply the instantaneous values of current and voltage together (Equation 3.2). (Lowercase letters are used to indicate instantaneous values.) With these products a new curve, P, is plotted. Curve P is the power curve.

Assume that at instant *a* in Figure 7–30, the current is 2 amperes (indicated by *ab*). At the same instant, the pressure is 3 volts (indicated by *ac*). The power at that instant is found by multiplying the instantaneous current by the instantaneous voltage (Equation 3.2), or $p = iv$. Therefore, 2 amperes $\times$ 3 volts = 6 watts. Other points are found in a similar manner, and the curve is plotted. The power curve is positive during both alternations (half cycles) because, when multiplying like signs together, the product is always positive. Notice that the power curve forms a sine wave having twice the frequency of the current or voltage.

The maximum power in a pure resistive circuit is equal to the maximum current times the maximum voltage ($P_m = I_m V_m$). Because the power curve is all positive, the average power is equal to the maximum power divided by 2, and the average power is equal to the effective power. Therefore, for a pure resistive circuit, the effective current ($I$) times the effective voltage ($V$) equals the effective power ($P$), or $P = IV$.

## Positive and Negative Power

In Figure 7–31A, a battery is supplying power to a resistance (R). The emf of the battery, which is 6 volts, forces a current through the circuit from the negative terminal, through the resistance, and back to the positive terminal. In this circuit, both

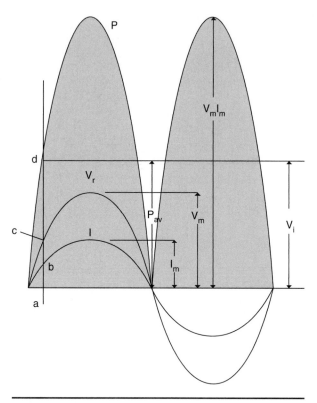

**FIGURE 7–30** Graph of current, voltage, and power in a pure resistive circuit.

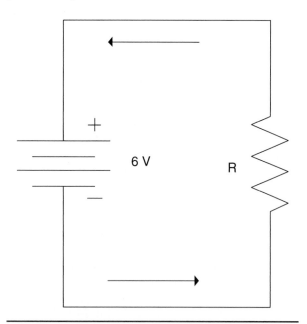

**FIGURE 7–31A** Battery supplying power to a resistance. The arrows indicate the direction of current flow.

the voltage and the current are in the same direction. If signed numbers are applied to the values of current and voltage, it can be said that both have positive values and that the power delivered to the resistance is also positive.

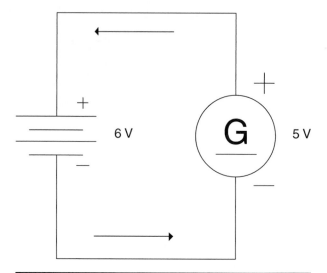

**FIGURE 7–31B** Generator connected across a 6-V battery. The arrows indicate the direction of current flow.

Figure 7–31B shows a 6-volt battery with a generator connected across its terminals. The positive terminal of the generator is connected to the positive terminal of the battery. The negative terminal of the generator is connected to the negative terminal of the battery. If the emf of the generator is equal to that of the battery, no current will flow.

If the generator emf is lower than that of the battery, current will flow from the negative terminal of the battery to the negative terminal of the generator. In this case, as in the resistive circuit, the current is in the same direction as the battery voltage, and the power is positive. The battery is delivering power to the generator and tends to drive it like a motor. Relative to the current, voltage, and power of the battery, the circuit conditions are the same as in Figure 7–31A.

If the voltage of the generator is adjusted to exceed that of the battery, the current flow will be reversed. The flow will now be from the negative terminal of the generator to the negative terminal of the battery (Figure 7–31C). The emf of the battery remains in the same direction, but the current is reversed. If the battery emf is assigned a positive value, then the current must be negative. The battery is being charged and is receiving power from the circuit. Because the battery is considered to be an energy source, under these conditions, it is delivering negative power to the circuit. Viewing it another way, the current and voltage of the battery are in opposite directions. Therefore, the power is negative.

This example shows that any device that is capable of delivering power and receiving power may have positive and negative power components. The

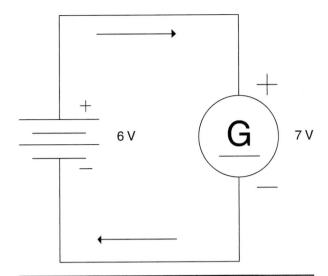

**FIGURE 7-31C** Generator charging a battery. The current flow is in the direction opposite to the battery voltage.

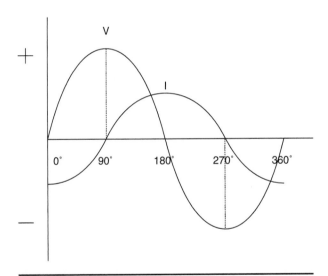

**FIGURE 7-32** Graph of voltage/current relationships in a pure inductive circuit. The current lags the voltage by 90 electrical time degrees.

power is positive when the current and voltage are in the same direction. The power is negative when the current and voltage are in opposite directions.

## Power in an Inductive Circuit

In a theoretical AC circuit of pure inductance (no resistance), the current lags the applied voltage by 90 degrees. This condition is illustrated in Figure 7–32. To determine the power curve, multiply the instantaneous values of current and voltage together. In Figure 7–33, when either the current or the voltage is zero, the power at those instances is zero (points a,

b, c, d, and e). Between points a and b, the voltage is positive and the current is negative; thus, the power is negative. This means that the magnetic energy stored in the coil is delivering power to the source, just as the generator is doing in Figure 7–31C. Between points b and c, the current and voltage are both positive because they are acting in the same direction. Thus, the power during this period of time is positive. The source is supplying power to the inductor. Between c and d, the voltage is negative and the current is positive. As a result, the power is negative, again. Between d and e, the voltage and current are both negative. Because they are both in the same direction, the power is positive. For one cycle of current, the positive area of the power curve equals

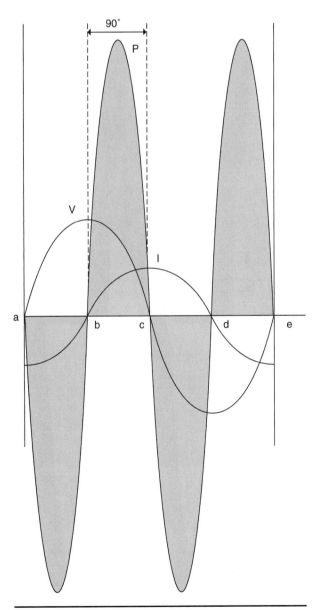

**FIGURE 7-33** Current, voltage, and power curves for a pure inductive circuit.

the negative area. Therefore, the power received by the inductor from the source is equal to the power returned to the source. The average power input to the inductor is zero, and a wattmeter connected in the circuit indicates zero.

## Power in a Capacitive Circuit

Figure 7–34 shows a theoretical circuit of pure capacitance in which the current leads the applied voltage by 90 degrees. At points a, b, c, d, and e, either the voltage or the current is zero, resulting in zero power. Between points a and b, the voltage and current are positive, and the power is also positive. Between b and c, the voltage is positive and the

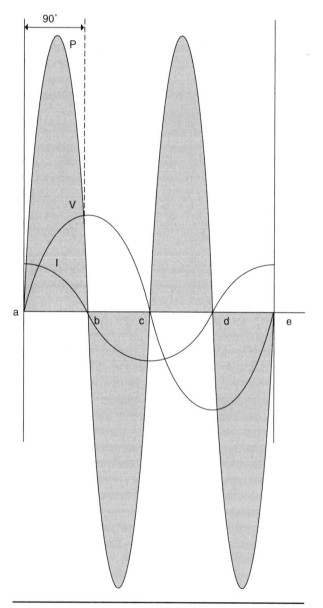

**FIGURE 7–34** Current, voltage, and power curves for a pure capacitive circuit.

current is negative. The resulting power is negative. Between c and d, the voltage and current are both negative, resulting in a positive power. Between d and e, the voltage is negative and the current is positive; therefore, a negative power exists. For one cycle of current, the positive area of power is equal to the negative area of power, and the result is zero power.

## Power in Circuits Containing Resistance and Inductance

In AC circuits containing pure resistance, the current and voltage are in phase, and all the power delivered to the circuit by the source is utilized by the circuit.

In an AC circuit of pure inductance, the current lags the voltage by 90 degrees, and all the power delivered to the circuit is returned to the source. The utilized power is zero.

Because all circuits contain some resistance, a circuit containing a coil is a resistive/inductive circuit (RL circuit). In an RL circuit, the current lags the voltage by a certain number of electrical time degrees. The number of degrees is between zero and 90. The angle of lag depends upon the ratio of the resistance (R) to the inductance (L).

In the RL circuit in Figure 7–35, it can be seen that most of the power is positive and is utilized by the circuit. The small amount of negative power is returned to the source by the magnetic energy from the coil. The negative power does not register on a wattmeter. The wattmeter indicates the **true power**, which is the power utilized by the circuit.

When an ammeter and a voltmeter are connected in this circuit, they indicate the effective values of current and voltage. The product of the effective values of current and voltage in an RL circuit is equal to the apparent power of the circuit. The apparent power is measured in volt-amperes and is always equal to or greater than the true power. In an RL circuit, the apparent power is always greater than the true power.

The ratio of the true power to the apparent power is called the **power factor** of the circuit. The power factor is usually stated as a decimal value or as a percentage. It is sometimes referred to as the percentage of the total power that is utilized by the circuit. The formula for power factor is:

$$Pf = \frac{P}{P_{app}} \times 100 \qquad \text{(Eq. 7.16)}$$

where   Pf = power factor, in percent (%)
  $P$ = true power
  $P_{app}$ = apparent power

Vector diagrams can be used to solve power problems. Referring to Figure 7–36, vector AC

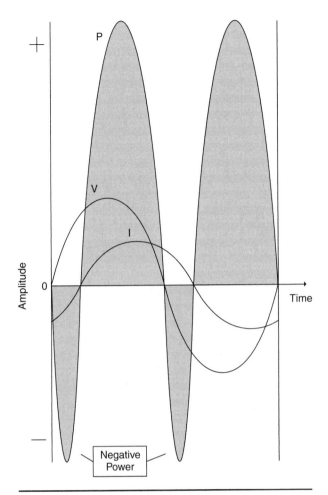

**FIGURE 7–35** Graph of power in an RL AC circuit.

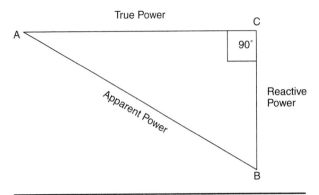

**FIGURE 7–36** Vector diagram of power in an RL AC circuit.

Most industrial loads are highly inductive and cause the current to lag the voltage. One way to overcome this lagging power factor is to install a capacitor or a group of capacitors at strategic points throughout the system. Capacitors cause the current to lead the voltage. The effect is opposite to that of an inductor.

## Power Formulas and Units

For calculations of the true power of an AC circuit, the following formulas may be used:

$$P = IE \cos \phi \qquad \text{(Eq. 7.17)}$$

$$P = I^2R \qquad \text{(Eq. 7.18)}$$

$$P = \sqrt{P_{app}^2 - P_{XL}^2} \qquad \text{(Eq. 7.19)}$$

The units of measurement for power in AC systems are as follows:

- True power (P) is measured in watts (W).
- Apparent power (Papp) is measured in volt-amperes (VA).
- Reactive power ($P_{XL}$) is measured in volt-amperes reactive (VAR).

# THREE-PHASE SYSTEMS

For most purposes, alternating current is generated and transmitted in the form of three-phase. Three-phase generators produce three separate voltages that are 120 electrical time degrees apart. Figure 7–37 shows the voltage curves for a three-phase system.

Three-phase systems provide smoother power than single-phase systems. Also, most three-phase equipment requires less space than single-phase equipment of the same rating. Three-phase

represents the true power. Vector AB represents the apparent power. The power factor is the ratio of the true power AC to the apparent power AB. In other words, the power factor is the cosine of angle A. Because the reactive power (negative power) is 90 degrees out of phase with the true power, angle A represents the number of degrees that the current lags the voltage. Vector BC represents the average power that is returned to the source by the magnetic energy stored in the coil.

The more inductive a circuit becomes, the greater is the angle between the true power and the apparent power. As angle A increases, the power factor decreases. This means that the reactive power becomes greater, and the power utilized by the load becomes smaller.

A low power factor is undesirable. With a low power factor, a large amount of power is circulating through the circuit, but only a small amount is transferred to the workload. A power factor lower than 80 percent is considered below industrial standards; steps should be taken to remedy this situation. Many utility companies increase a customer's rates if the power factor drops below a specified value.

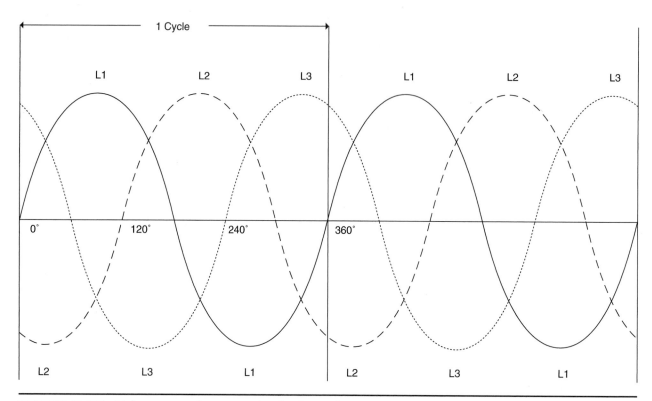

**FIGURE 7–37**  Voltage curve for one cycle of a three-phase system.

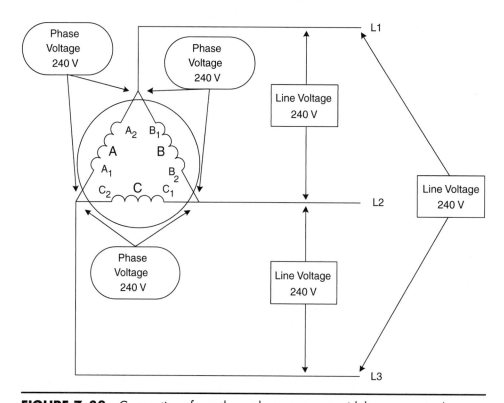

**FIGURE 7–38**  Connections for a three-phase generator (delta connection).

equipment is more efficient and less expensive. Transmission of three-phase power requires less conductor material than single-phase power, and single-phase circuits can be tapped from three-phase systems.

Figure 7–38 illustrates the coil connections for a three-phase generator. This arrangement is called a *delta connection*. Another frequently used arrangement is the four-wire wye connection (Figure 7–39).

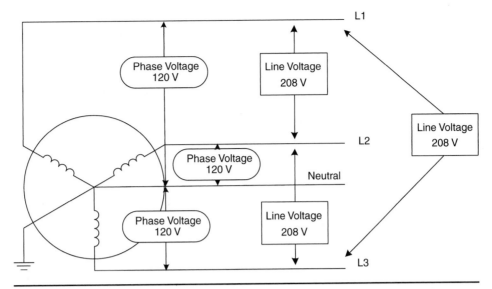

**FIGURE 7–39**   Connections for a three-phase generator (wye connection).

With the three-phase delta system, the phase voltage ($E_p$) is the voltage between any two of the phase conductors. Single-phase equipment can be supplied by using any two of the three conductors. For three-phase equipment, all three conductors are used. The delta system, therefore, supplies the same value of voltage for single-phase and three-phase circuits.

With the three-phase wye system, the phase voltage is the voltage between any one of the phase conductors and the neutral conductor. The voltage between any two-phase conductors is called the *line voltage* ($E_L$). Two values of voltage are available for single conductors. For a wye connection, the line voltage is always 1.73 times as much as the phase voltage. All three-phase conductors are used to supply three-phase circuits.

When a three-phase system is supplying single-phase circuits, the single-phase loads should be balanced between all three phases. On a perfectly balanced system, the current through all three-phase conductors is equal. It is seldom possible to maintain a perfect balance; however, severely unbalanced systems can cause conductors and equipment to overheat and overcurrent devices to operate.

## REVIEW QUESTIONS

*Multiple Choice*

1. An alternating voltage is an emf that is
   a. continually changing in value.
   b. alternately reversing direction.
   c. both a and b.
   d. neither a nor b.

2. The number of cycles of an alternating voltage completed in 1 s is called the
   a. frequency.
   b. alternation.
   c. hertz.
   d. fluctuation.

3. The most common frequency used in the United States is
   a. 40 cycles per second.
   b. 50 cycles per second.
   c. 60 cycles per second.
   d. 25 cycles per second.

4. The modern unit of measurement for frequency, which means cycles per second, is the
   a. hertz.
   b. watt.
   c. henry.
   d. farad.

5. The standard AC generator produces a voltage that, when plotted on a graph, produces a
   a. tangent curve.
   b. frequency curve.
   c. sine curve.
   d. cosine curve.

6. A vector is
   a. an arc.
   b. a line segment that has a definite length and direction.
   c. curved lines.
   d. a triangle.

7. A vector diagram is
   a. a rectangular diagram.
   b. two or more vectors joined together to convey information.
   c. a triangle.
   d. a circle.

8. The hypotenuse of a right triangle is the
   a. longest side.
   b. shortest side.
   c. sum of all the sides.
   d. side opposite the acute angle.

9. The side of a right triangle that is opposite to the angle under consideration is called the
   a. tangent side.
   b. adjacent side.
   c. opposite side.
   d. hypotenuse.

10. The side of a right triangle that is opposite to the 90° angle is called the
    a. tangent side.
    b. adjacent side.
    c. opposite side.
    d. hypotenuse.

11. The effective value of an alternating current is the
    a. value at a particular instant.
    b. value that produces the same heating effect as a specific value of a steady direct current.
    c. average value of one cycle.
    d. maximum value produced.

12. The rms value of an alternating current is the same as the
    a. instantaneous value.
    b. average value.
    c. effective value.
    d. maximum value.

13. Standard AC ammeters and voltmeters indicate
    a. instantaneous values.
    b. average values.
    c. effective values.
    d. maximum values.

14. The maximum value of an alternating voltage is
    a. equal to the effective value.
    b. greater than the effective value.
    c. smaller than the effective value.
    d. none of the above.

15. The primary source of electrical energy today is in the form of
    a. varying current.
    b. alternating current.
    c. pulsating direct current.
    d. steady direct current.

16. AC generators generally produce
    a. higher voltages than DC generators.
    b. lower voltages than DC generators.
    c. voltages equal to those produced by DC generators.
    d. different values of voltage.

17. AC is
    a. less expensive to transmit than DC.
    b. more expensive to transmit than DC.
    c. about the same cost to transmit as DC.
    d. less expensive to transmit than DC but more expensive to maintain.

18. Large alternators have
    a. rotating fields and stationary armatures.
    b. rotating armatures and stationary fields.
    c. rotating fields and rotating armatures.
    d. stationary fields and stationary armatures.

19. In an inductive circuit, the current
    a. lags the voltage.
    b. leads the voltage.
    c. is in phase with the voltage.
    d. is always greater than the voltage.

20. In a capacitive circuit, the current
    a. lags the voltage.
    b. leads the voltage.
    c. is in phase with the voltage.
    d. is always greater than the voltage.

21. The unit of measurement for impedance is the
    a. volt-ampere.
    b. henry.
    c. farad.
    d. ohm.

22. Power factor is the ratio of the
    a. true power to the reactive power.
    b. true power to the apparent power.
    c. reactive power to the apparent power.
    d. inactive power to reactive power.

23. For most purposes, alternating current is generated and transmitted in the form of
    a. single-phase.
    b. two-phase.

c. three-phase.

d. four-phase.

24. Three-phase systems are generally
    a. more efficient than single-phase systems.
    b. less efficient than single-phase systems.
    c. just as efficient as single-phase systems.
    d. less efficient than DC.

25. Two types of connections for three-phase generators are
    a. delta and wye.
    b. star and wye.
    c. delta and theta.
    d. wye and theta.

26. The three-phase delta system supplies
    a. one value of voltage.
    b. two values of voltage.
    c. three values of voltage.
    d. four values of voltage.

27. On a three-phase system, the voltages are
    a. 60° apart.
    b. 90° apart.
    c. 120° apart.
    d. 180° apart.

*Give Complete Answers*

1. Define *alternating current*.

2. Define the term *frequency*.

3. List four factors that determine the amount of voltage induced into a coil that is rotating in a uniform magnetic field.

4. Describe one method used to plot an alternating voltage curve if the maximum value of voltage is known.

5. Define a *vector*.

6. With either acute angle of a right triangle, there are six possible ratios. List these six ratios in formula form.

7. What is meant by the "effective value" of an alternating current?

8. What do the letters *rms* stand for?

9. Write the formula for calculating the effective value of an alternating voltage when the maximum value is known.

10. Write the formula for calculating the instantaneous value of an alternating current when the maximum value is known.

11. In a plot of a voltage curve, what unit of measurement is indicated along the horizontal line?

12. List at least three advantages of AC compared to DC.

13. List one advantage of DC compared to AC.

14. Define the term *mutual induction*.

15. What is the difference between mutual induction and self-induction?

16. Define *inductance*.

17. What is the unit of measurement of inductance?

18. List three factors that affect the amount of inductance of a coil.

19. What determines the length of time that it takes for a steady direct current to reach a steady value in an inductive circuit?

20. Explain why a specific inductive circuit allows more current to flow when 120 volts DC are applied than when 120 VAC are applied.

21. Define *inductive reactance*. What is its unit of measurement?

22. How can a coil be wound so that it does not produce an inductive effect?

23. Define *impedance*. What is its unit of measurement?

24. What is the phase relationship between the current and the voltage in an AC inductive circuit?

25. What is a capacitor?

26. Define the term *dielectric*.

27. Describe the effect of connecting a capacitor across a steady DC circuit.

28. Describe the effect of connecting a capacitor across an AC circuit.

29. Define *capacitance*. What is its unit of measurement?

30. List two factors that determine the length of time necessary to charge a capacitor to its rated value.

31. What is the phase relationship between the current and the voltage in an ac circuit containing capacitance?

32. Under what conditions is it possible to have the applied voltage in one direction while the current is flowing in the opposite direction?

33. Define *capacitive reactance*. What is its unit of measurement?

34. List three types of power present in AC circuits, and state their units of measurement.

35. Define the term *negative power*.

36. What power is indicated on an AC wattmeter?

37. If the current and the voltage of an AC circuit are measured and their values are multiplied together, the result indicates what type of power?

38. Define *power factor*.

39. Draw a power triangle, and label each side.

40. Why is a low power factor undesirable?

41. Are most industrial loads inductive or capacitive?

42. Name one method frequently used to improve the power factor of an industrial load.

43. Why is it sometimes said that the power factor is the cosine of the angle between the current and the voltage?

44. Give one practical use for a capacitor in a dc circuit.

45. List three formulas that may be used to calculate the true power of an ac circuit.

46. What form of alternating current is generally used for the transmission of electrical power?

47. List four advantages of three-phase power compared to single-phase power.

48. Draw a diagram of a three-phase delta connection.

49. Draw a diagram of a three-phase, four-wire connection.

50. In a three-phase delta system, what is meant by *phase voltage*? What is the symbol for phase voltage?

51. What is the difference between the phase voltage and the line voltage in a three-phase, four-wire wye connection?

52. How many values of voltage are available from a three-phase delta system?

53. How many values of voltage are available from a three-phase, four-wire wye system?

54. What is the symbol for the line voltage in a three-phase system?

55. Why is it necessary to balance a three-phase system?

*Solve each problem, showing the method used to arrive at the solution.*

1. Find the angle that coincides with the trigonometric functions listed below.
   a. cosine = 0.8660
   b. sine = 0.5000
   c. tangent = 0.0699
   d. cotangent = 57.29
   e. secant = 1.566
   f. cosecant = 1.701

2. The maximum value of an alternating voltage is 170 V. Calculate the effective value.

3. What is the average value of the voltage in Problem 2?

4. Calculate the instantaneous values of voltage in Problem 2 at the following angles:
   a. 30°
   b. 60°
   c. 90°
   d. 120°

5. A voltage has an effective value of 277 V. What is the maximum value?

6. An ac inductive circuit has an applied voltage of 240 V, and the current lags the applied voltage by 45°. Draw a vector diagram showing the applied voltage, the voltage of self-induction, and the resultant voltage. Calculate the resultant voltage.

7. A coil containing 500 turns is wound on a metal core 5-in. long with a cross-sectional area of 10 in.2 The permeability of the core is $6 \times 103$. Calculate the inductance of the coil.

8. Calculate the induced voltage for the coil in Problem 7 if the current changes from 2 A to 80 A in 1 s.

9. Calculate the amount of time it will take for the current in Problem 8 to change from 10 A to 40 A.

10. Calculate the amount of time required for a direct current to reach its maximum steady value if the coil in Problem 7 has a resistance of 5 W.

11. What is the inductive reactance of a 20-H coil connected across a 60-Hz supply?

12. What is the inductive reactance of the coil in Problem 11 if it is connected across a 50-Hz supply?

13. If the coil in Problem 11 has a resistance of 10 Ω, what is its impedance?

14. Determine the amount of current that will flow through the coil in Problem 13 if it is connected to a 120-V, 60-Hz supply.

15. Calculate the capacitance of a capacitor that will charge to 0.05 C when 600 V is applied to its terminals.

16. How long will it take to charge a capacitor if the resistance of its plates is 0.05 Ω and it has a capacitance of 100 $\mu$F?

17. Calculate the resistance of the plates of a capacitor that can be fully charged to 50 $\mu$F in 5 $\mu$s.

18. A 100-$\mu$F capacitor is connected to a 480-V, 60-Hz supply. What is its capacitive reactance?

19. A capacitor has a resistance of 0.5 $\Omega$ and a capacitive reactance of 10 $\Omega$. Calculate the impedance.

20. An ammeter in a circuit indicates 20 A, and the voltmeter indicates 100 V. If a wattmeter indicates 1500 W, what is the reactive power of the circuit?

21. What is the power factor of the circuit in Problem 20?

22. What is the phase angle between the current and the voltage in Problem 20?

23. What is the impedance of the circuit in Problem 20?

24. If the resistance of the circuit in Problem 20 is 0.15 $\Omega$, what is the reactance?

25. What is the power factor of the circuit in Problem 20 if the ammeter indicates 15 A instead of 20 A?

# AC Circuits

## OBJECTIVES

After studying this chapter, the student will be able to:

■ Discuss the characteristics of various types of alternating current circuits.

■ Describe the effects of inductance and capacitance on alternating current circuits.

■ Describe the effects of high- and low-power factors on alternating current circuits.

# PURE RESISTIVE CIRCUITS

The current and voltage are in phase in pure resistive circuits. Figure 8–1 illustrates this phase relationship. The capacitive and/or inductive effect is negligible, and the characteristics are much the same as for DC. Resistive heating units and incandescent lighting are considered to be pure resistive loads.

## Effective Resistance

In general, pure resistive circuits react much the same for AC or DC. There are, however, some differences that must be considered. These differences vary with the frequency and are generally negligible at low frequencies.

The following five factors affect the amount of current flowing in a pure resistive circuit:

- DC resistance
- Skin effect
- Eddy currents
- Hysteresis effect
- Dielectric stress

The *DC resistance* is the resistance measured with a very accurate ohmmeter. It is the total resistive effect to pure DC.

The fact that alternating current changes in value and direction tends to make it flow along the outer surface of the conductor. This phenomenon, known as **skin effect**, reduces the inner conductive effect of the conducting material and increases the circuit resistance.

Alternating current produces a magnetic flux that changes polarity with each reversal or current flow. The change in polarity causes the molecules in the metal parts near the circuit to be in motion, thus producing heat. The heat either radiates back into the circuit conductors or retards the dissipation of heat produced by current flowing in the conductors. This *hysteresis effect* increases the effective resistance of the circuit.

**Eddy currents** are caused by voltages induced into the conductors and other surrounding metal parts. They vary with and are directly proportional to the frequency of the supply. Heat produced by these currents tends to increase the effective resistance of the circuit.

As the alternating voltage varies in strength, the stress on the conductor insulation increases and decreases. This variation in *dielectric stress* also produces heat, which increases the circuit resistance.

The effective resistance of an AC circuit is equal to the DC resistance plus the effects of eddy currents, hysteresis, dielectric stress, and skin effect. The effective resistance may be written as follows:

$$R = R_o + R_i + R_m + R_d + R_s \qquad \text{(Eq. 8.1)}$$

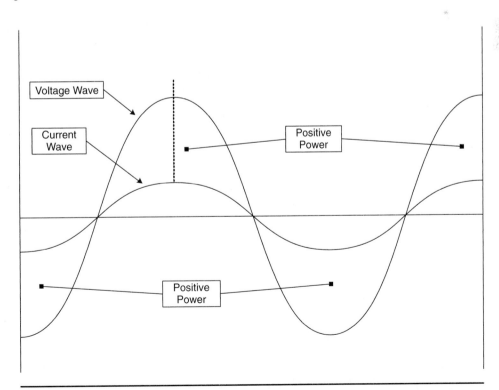

**FIGURE 8–1**  Graph of current and voltage in a pure resistive AC circuit.

where  $R$ = effective resistance
$R_o$ = pure DC resistance
$R_i$ = increase in resistance caused by eddy currents
$R_m$ = increase in resistance caused by the hysteresis effect
$R_d$ = increase in resistance caused by dielectric stress variations
$R_s$ = increase in resistance caused by the skin effect

# AC SERIES CIRCUITS

When a circuit contains inductive components, these components contain both resistance and inductive reactance. It is more convenient to consider each inductor as two components—one of pure resistance and one of pure inductance. A single coil in a circuit would then appear as two components in series—one of pure resistance and the other of pure inductance. Figure 8–2 illustrates this practice.

The rules for current, voltage, and resistance in series circuits hold true for AC circuits. However, there are slight variations. The rules for AC series circuits are as follows:

- The instantaneous value of current is the same through all parts of a series circuit.

- The applied voltage is equal to the vectorial (sometimes called phasor) sum of the individual voltages around the circuit.

- The combined resistance of an AC series circuit is equal to the sum of the individual resistances of the circuit.

- The combined reactance of an AC series circuit is equal to the vectorial sum of the individual reactances in the circuit.

- The combined impedance of an AC series circuit is equal to the vectorial sum of the individual impedances of the circuit.

Because the current at any instant has the same value in all parts of a series circuit, the current can be used as a phase reference. Actually, any circuit quantity can be used as a reference, but it is generally more convenient to use the current because it is common to all components of the circuit.

## Series RL Circuits

The phase relationship of the applied voltage and the voltage drops for the circuit in Figure 8–2 can be illustrated in a vector diagram, as shown in Figure 8–3.

The phase relationship of the voltages across the parts of the circuit can be expressed with reference to the current vector. Vector $E_R$ is equal to $I_R$ and represents the voltage across the resistance $R$. Because in a pure resistance the current and voltage are in phase, vectors $E_R$ and $I_R$ are laid out along the horizontal. Vector $E_L$ is equal to $IX_L$ and represents the voltage across the inductance L. Because the current lags the voltage by 90 degrees in a pure inductive circuit, vector $E_L$ is laid out 90 degrees out of phase with $E_R$. Note that $E_R$ is in phase with the current. Vector $E_L$ is drawn vertically upward because vector rotation is always considered to be counterclockwise.

The vector sum of $E_R$ and $E_L$ is equal to the applied voltage $E$ and is calculated by the following formula:

$$E = \sqrt{E_R^2 + E_L^2} \qquad \text{(Eq. 8.2)}$$

where  $E$ = applied voltage
$E_R$ = voltage across the resistance
$E_L$ = voltage across the inductance

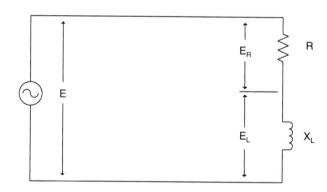

**FIGURE 8–2**  AC circuit containing a resistor and an inductor.

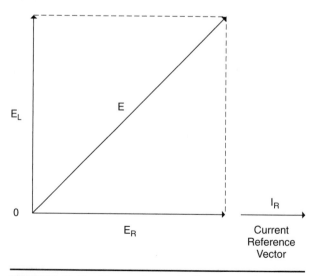

**FIGURE 8–3**  Vector diagram illustrating the phase relationships of the voltages for the circuit in Figure 8–2.

## Example 1

The inductor just described has a resistance of 10 Ω and an inductance of 0.04 H. It is connected to a 120-V, 60-Hz supply. What is the impedance of the coil? Calculate the current through the circuit, the voltage across the resistance ($E_R$), the voltage across the inductance ($E_L$), the power factor, and the power dissipated.

a. $X_L = 2\pi f L$ (*Equation 7.9*)

$X_L = 2 \times 3.1416 \times 60 \times 0.04$

$X_L = 15\ \Omega$ (inductive reactance of the coil)

b. $Z = \sqrt{R^2 + X_L^2}$ (*Equation 7.11*)

$Z = \sqrt{10^2 + 15^2}$

$Z = \sqrt{325}$

$Z = 18\ \Omega$ (coil impedance)

c. $I = \dfrac{E}{Z}$ (*Equation 7.10*)

$I = \dfrac{120}{18}$

$I = 6.667\ A$ (circuit current)

d. $E_R = IR$ (*from Ohm's law*)

$E_R = 6.667 \times 10$

$E_R = 66.67\ V$ (voltage across resistance $R$)

e. $E_L = IX_L$ (*from Ohm's law*)

$E_L = 6.667 \times 15$

$E_L = 100\ V$ (voltage across inductance L)

f. $P = I^2 R$ (*Equation 7.18*)

$P = (6.667)^2 \times 10$

$P = 44.5 \times 10$

$P = 445\ W$ (effective power or true power)

g. $P_{app} = IE$ (*from Equation 3.2*)

$P_{app} = 6.667 \times 120$

$P_{app} = 800\ VA$ (apparent power)

h. $Pf = \dfrac{P}{P_{app}} \times 100$ (*Equation 7.16*)

$Pf = \dfrac{445}{800}$

$Pf = 0.556 = 55.6\%$ (power factor)

Another formula for power factor in a series circuit is:

$$Pf = \frac{R}{Z} \qquad \text{(Eq. 8.3)}$$

where  Pf = power factor
R = resistance
Z = impedance

The same answer would result if this equation were used in place of Equation 7.16, used in calculation h of Example 1.

## Example 2

A resistance of 10 Ω is connected in series with a coil of negligible resistance with 0.05-H inductance. If the current through the coil is 10 A, determine the voltage across each component and the supply voltage from a 60-Hz source.

a. $X_L = 2\pi f L$ (*Equation 7.9*)

$X_L = 2 \times 3.1416 \times 60 \times 0.05$

$X_L = 18.85\ \Omega$ (inductive reactance of coil)

b. $E_L = IX_L$ (*from Ohm's law*)

$E_L = 10 \times 18.85$

$E_L = 188.5$ (voltage across coil)

c. $E_R = IR$ (*from Ohm's law*)

$E_R = 10 \times 10$

$E_R = 100\ V$ (voltage across the resistance)

d. $E = \sqrt{E_R^2 + E_L^2}$ (*Equation 8.2*)

$E = \sqrt{100^2 + 188.5^2}$

$E = \sqrt{45,532}$

$E = 213\ V$ (applied voltage)

## Example 3

Calculate the following for the circuit in example 2: (a) the impedance, (b) the true power, (c) the apparent power, (d) the reactive power, and (e) the power factor.

a. $Z = \dfrac{E}{I}$ (*Equation 7.10*)

$Z = \dfrac{213}{10}$

$Z = 21.3\ \Omega$ (impedance)

*or*

$Z = \sqrt{R^2 + X_L^2}$ (*Equation 7.11*)

$Z = \sqrt{10^2 + 18.85^2}$

$Z = 21.3\ \Omega$

b. $P = I^2 R$ (*Equation 7.18*)

$P = 10^2 \times 10$

$P = 1000\ W = 1\ kW$ (effective or true power)

c. $P_{app} = IE$ (*Equation 3.2*)

$P_{app} = 10 \times 213$

$P_{app} = 2130\ VA = 2.13\ kVA$ (apparent power)

d. $P_L = IE_L$ (*Equation 3.2*)

$P_L = 10 \times 188.5$

$P_L = 1885\ VAR$ (reactive power)

*or*

$P_L = \sqrt{P_{app}^2 - P_{XL}^2}$ (*Equation 7.19*)

$P_L = \sqrt{2130^2 - 1000^2}$

$P_L = 1881\ VAR$

The difference of 4 VAR is due to rounding off of numbers and is negligible.

e. $\text{Pf} = \dfrac{P}{P_{\text{app}}} \times 100$ (*Equation 7.16*)

$\text{Pf} = \dfrac{1000}{2130}$

$\text{Pf} = 0.469 = 46.9\%$ (power factor lagging)

*or*

$\text{Pf} = \dfrac{R}{Z}$ (*Equation 8.3*)

$\text{Pf} = \dfrac{10}{21.3}$

$\text{Pf} = 0.469 = 46.9\%$

When more than one resistance and/or inductance are connected in series, calculate the values for each component and then apply the rules for series circuits.

## Series RC Circuits

Figure 8–4 shows a resistance connected in series with a capacitance. The combination is connected to an AC supply. Because this is a series circuit, all the rules for series circuits apply. The voltage across the resistor ($E_R$) is in phase with the current. The current at any instant is the same value through all the components. Therefore, the vector representing $E_R$ is laid out along the current vector on the horizontal (Figure 8–5). The current leads the voltage in a capacitive circuit. Therefore, vector $E_C$ is laid out to indicate the current leading the voltage by 90 degrees. $E_C$ is drawn vertically downward from $E_R$. This also indicates that $E_C$ leads $E_R$ by 90 degrees. The applied voltage is the vector sum of these two values:

$$E = \sqrt{E_R^2 + E_C^2} \qquad \text{(Eq. 8.4)}$$

## Example 4

The capacitor in Figure 8–4 has a capacity of 900 μF. The resistance $R$ has a value of 5 Ω. Assume the circuit is connected to a 120-V, 60-Hz supply. Calculate the impedance, the current through the circuit, the voltage across the resistance, the voltage across the capacitance, the power dissipated by the circuit, and the power factor.

a. $X_C = \dfrac{1}{2\pi f C}$ (*Equation 7.14*)

$X_C = \dfrac{1}{2 \times 3.1416 \times 60 \times 0.009}$

$X_C = \dfrac{1}{0.3392928}$

$X_C = 2.947\ \Omega$ (capacitive reactance)

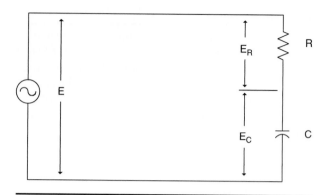

**FIGURE 8–4**　Resistance and capacitance connected in series across an AC supply (RC circuit).

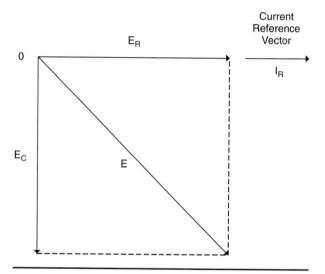

**FIGURE 8–5**　Vector diagram illustrating the phase relationships for the voltages in an RC series circuit. $E = \sqrt{E_R^2 + E_C^2}$.

b. $Z = \sqrt{R^2 + X_C^2}$ (*Equation 7.15*)

$Z = \sqrt{5^2 + 2.947^2}$

$Z = \sqrt{25 + 8.6848}$

$Z = \sqrt{33.6848}$

$Z = 5.8\ \Omega$ (impedance)

c. $I = \dfrac{E}{Z}$ (*from Equation 7.10*)

$I = \dfrac{120}{5.8}$

$I = 20.7\ A$ (current)

d. $E_R = IR$ (*from Ohm's law*)

$E_R = 20.7 \times 5$

$E_R = 103.5\ V$ (voltage across the resistance)

e. $E_C = IX_C$ (*from Ohm's law*)

$E_C = 20.7 \times 2.947$

$E_C = 61$ V (voltage across the capacitance)

f. $P = I^2R$ (*Equation 7.18*)

$P = 20.7^2 \times 5$

$P = 428.49 \times 5$

$P = 2142$ W (true power)

g. $P_{app} = IE$ (*Equation 3.2*)

$P_{app} = 20.7 \times 120$

$P_{app} = 2484$ VA (apparent power)

h. $Pf = \dfrac{P}{P_{app}} \times 100$ (*Equation 7.16*)

$Pf = \dfrac{2142}{2484}$

$Pf = 0.862 = 86.2\%$ (power factor)

## Series RLC Circuits

Figure 8–6A shows a resistance, an inductance, and a capacitance connected in series across an AC supply. By applying the rules for series circuits and vector analysis, we can determine all values.

**Example 5**

The components in Figure 8–6A have the following values: $R = 5$ Ω, $C = 900$ μF, and $L = 0.02$ H. Assume that the circuit is connected to a 120-V, 60-Hz supply. Calculate the impedance, the current through the circuit, the voltage across the inductor, the voltage across the capacitor, the power dissipated by the circuit, and the power factor.

a. $X_L = 2\pi fL$ (*Equation 7.9*)

$X_L = 2 \times 3.1416 \times 60 \times 0.02$

$X_L = 7.54$ Ω (inductive reactance)

b. Draw a vector diagram representing the resistance and the inductive reactance of the circuit (Figure 8–6B).

c. $X_C = \dfrac{1}{2\pi fC}$ (*Equation 7.14*)

$X_C = \dfrac{1}{2 \times 3.1416 \times 60 \times 0.0009}$

$X_C = \dfrac{1}{0.3392928}$

$X_C = 2.947$ Ω (capacitive reactance)

d. Draw a vector diagram representing the resistance and capacitive reactance of the circuit (Figure 8–6C).

e. Combine the vector diagrams in Steps (b) and (d). See Figure 8–6D.

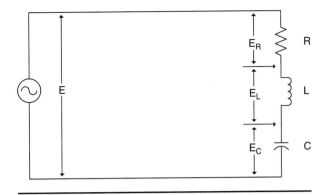

**FIGURE 8–6A** Resistance, capacitance, and inductance connected in series across an AC supply (RLC circuit).

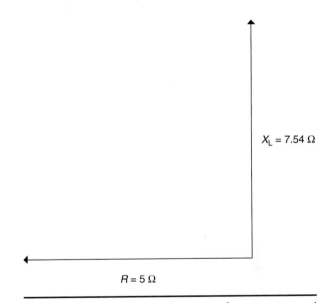

**FIGURE 8–6B** Vector diagram of resistance and inductive reactance.

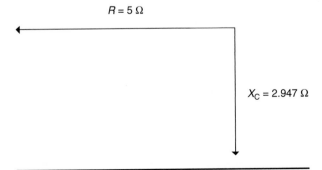

**FIGURE 8–6C** Vector diagram of resistance and capacitive reactance.

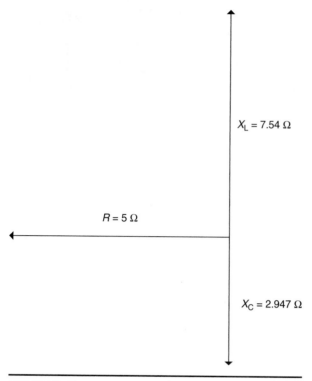

**FIGURE 8–6D** Vector diagram of resistance, inductive reactance, and capacitive reactance. Because $X_L$ and $X_C$ are drawn in opposite directions, they are given opposite signs (signs of opposition). It is customary to assign a positive sign to $X_L$ and a negative sign to $X_C$.

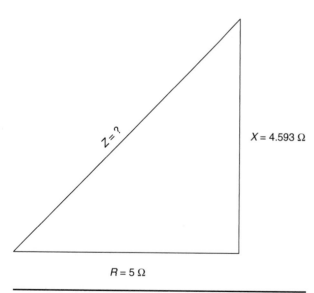

**FIGURE 8–6E** Impedance vector diagram. The reactance vector is drawn upward from the resistance vector because the circuit is primarily inductive.

f. Calculate the vector sum of $X_L$ and $X_C$.

$$X = X_L + (-X_C)$$
$$X = 7.54 + (-2.947)$$
$$X = 4.593 \ \Omega \ (\text{reactance})$$

g. Draw an impedance vector diagram of the combined reactance and the resistance (Figure 8–6E).

h. Calculate the impedance.

$$Z = \sqrt{R^2 + X^2} \ (\text{Equation 7.11})$$
$$Z = \sqrt{25 + 21}$$
$$Z = \sqrt{46}$$
$$Z = 6.78 \ \Omega \ (\text{impedance})$$

i. $I = \dfrac{E}{Z}$ (from Equation 7.10)

$$I = \frac{120}{6.78}$$
$$I = 17.7 \ \text{A} \ (\text{circuit current})$$

j. $E_L = IX_L$ (from Ohm's law)

$$E_L = 17.7 \times 7.54$$
$$E_L = 133.5 \ \text{V} \ (\text{voltage across the inductor})$$

k. $E_C = IX_C$ (from Ohm's law)

$$E_C = 17.7 \times 2.947$$
$$E_C = 52.2 \ \text{V} \ (\text{voltage across the capacitance})$$

l. $P = I^2R$ (Equation 7.18)

$$P = 17.7^2 \times 5$$
$$P = 1566 \ \text{W} \ (\text{true power})$$

m. $P_{app} = IE$ (Equation 3.2)

$$P_{app} = 17.7 \times 120$$
$$P_{app} = 2124 \ \text{VA} \ (\text{apparent power})$$

n. $\text{Pf} = \dfrac{P}{P_{app}} \times 100$ (Equation 7.16)

$$\text{Pf} = \frac{1566}{2124}$$
$$\text{Pf} = 0.737 = 73.7\% \ (\text{power factor})$$

In a series AC circuit, the supply voltage is equal to the vector sum of the voltages across the individual components.

**Example 6**

Perform vector addition of the voltages across the components in Figure 8–6A to determine the supply voltage.

a. Calculate the voltage drop across $R$.

$$E_R = IR \ (\text{from Ohm's law})$$
$$E_R = 17.7 \times 5$$
$$E_R = 88.5 \ \text{V} \ (\text{voltage across } R)$$

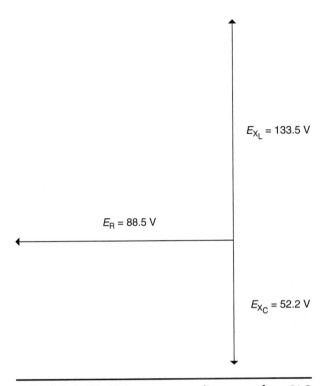

**FIGURE 8-7** Voltage vector diagram of an RLC circuit.

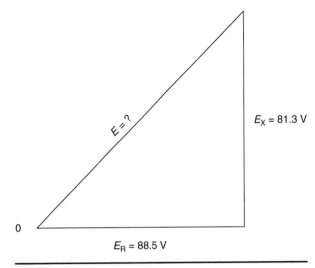

**FIGURE 8-8** Voltage vector diagram.

b. Draw a vector diagram of the voltages in Figure 8-6A. See Figure 8-7.

c. Calculate the vector sum of $E_{XL}$ and $E_{XC}$.

$E_X = E_{XL} + (-E_{XC})$

$E_X = 133.5 + (-52.2)$

$E_X = 81.3 \text{ V (reactive voltage)}$

d. Draw a vector diagram of $E_X$ and $E_R$ (Figure 8-8).

e. Calculate the supply voltage.

$E = \sqrt{E_R^2 + E_X^2} \quad (\text{Equation 7.11})$

$E = \sqrt{88.5^2 + 81.3^2}$

$E = \sqrt{14{,}442}$

$E = 120 \text{ V (supply voltage)}$

The preceding formulas can be combined into the following equation:

$$E = \sqrt{E_R^2 + (E_L - E_C)^2} \qquad (\text{Eq. 8.5})$$

$E = \sqrt{7832.25 + 6609.69}$

$E = \sqrt{14{,}442}$

$E = 120 \text{ V}$

In series AC circuits, the power factor, which is equal to the cosine of the angle between the current

and the voltage, may be determined by the following equations:

$$Pf = \frac{E_R}{E} \quad \text{or} \quad \cos\phi = \frac{E_R}{E} \qquad (\text{Eq. 8.6})$$

$$Pf = \frac{IR}{IZ} \quad \text{or} \quad \cos\phi = \frac{IR}{IZ}$$

$$Pf = \frac{R}{Z} \quad \text{or} \quad \cos\phi = \frac{R}{Z}$$

$$Pf = \frac{P}{P_{app}} \quad \text{or} \quad \cos\phi = \frac{P}{P_{app}}$$

A reexamination of example 5 shows how the inductance and capacitance affect the power factor. Taking various combinations of the components and connecting them across 120 volts, 60 hertz will bring about varying power factors.

**Example 7**

1. Connect the inductor and resistor from example 5 in series across the 120-V, 60-Hz supply. Compute the power factor of the circuit.

a. Find the impedance.

$Z = \sqrt{R^2 + X_L^2} \quad (\text{Equation 7.11})$

$Z = \sqrt{5^2 + 7.54^2}$

$Z = \sqrt{25 + 56.85}$

$Z = \sqrt{81.85}$

$Z = 9.05 \text{ }\Omega \text{ (impedance)}$

b. Calculate the power factor.

$Pf = \frac{R}{Z} \quad (\text{Equation 8.3})$

$Pf = \frac{5}{9.05}$

$Pf = 0.552 = 55.2\% \text{ (power factor)}$

2. Connect the capacitor and the resistor in series across the 120-V, 60-Hz supply.

   a. Calculate the impedance.

   $Z = \sqrt{R^2 + X_C^2}$ (Equation 7.11)

   $Z = \sqrt{5^2 + 2.947^2}$

   $Z = \sqrt{25 + 8.68}$

   $Z = \sqrt{33.68}$

   $Z = 5.8\ \Omega$ (impedance)

   b. Determine the power factor.

   $Pf = \dfrac{R}{Z}$ (Equation 8.3)

   $Pf = \dfrac{5}{5.8}$

   $Pf = 0.862 = 86.2\%$ (power factor)

3. Connect the capacitor and the inductor in series across the 120-V, 60-Hz supply.

   a. Calculate the impedance.

   $Z = X_L - X_C$

   $Z = 7.54 - 2.947$

   $Z = 4.59\ \Omega$ (impedance)

Because $Z$ is the result of inductive reactance only, the circuit becomes a pure inductive circuit, and the power factor is zero. This circuit can exist only in theory, because all circuits have some resistance.

It can be seen from the preceding calculations that the RL circuit has a poor power factor (55.2%). Adding the capacitance improves the power factor. The RLC circuit has a power factor of 73.7%.

## Series Resonance

A resonant circuit contains resistance, inductance, and capacitance. However, the current and voltage of a resonant circuit are in phase. In order to accomplish this, $X_L$ must be equal to $X_C$.

In other words, a series resonant circuit is an RLC circuit having a power factor of 100 percent.

**Example 8** _____

What size capacitor must be connected in series with the resistance and inductance in example 5 to bring the power factor to unity (100%)?

   a. In order to obtain a 100% power factor, $X_C$ must be equal to $X_L$.

   $X_C = X_L$

   $\dfrac{1}{2\pi fC} = 2\pi fL$ (Equations 7.9 and 7.14)

   $C = \dfrac{1}{(2\pi f)^2 L}$

   $C = \dfrac{1}{(2 \times 3.1416 \times 60)^2 \times 0.02}$

   $C = \dfrac{1}{2842.46}$

   $C = 351.8\ \mu F$

   or

   $C = \dfrac{1}{2\pi fX_C}$

   $C = \dfrac{1}{2 \times 3.1416 \times 60 \times 7.54}$

   $C = \dfrac{1}{2842.52}$

   $C = 351.8\ \mu F$

A 351.8-$\mu$F capacitor must be connected in series with the resistance and inductance.

Proof:

$X_L$ must equal $X_C$.

   $X_C = \dfrac{1}{2\pi fC}$

   $X_C = \dfrac{1}{2 \times 3.1416 \times 60 \times 0.0003518}$

   $X_C = 7.54\ \Omega$

   $X_L = 7.54\ \Omega$

Therefore, $X_C = X_L$.

   $Pf = \dfrac{R}{Z}$ (Equation 8.3)

   $Pf = \dfrac{5}{5}$

   $Pf = 1 = 100\% =$ unity power factor

**Example 9** _____

What size inductor must be connected in series with the resistance and capacitance in example 5 to bring the power factor to unity?

   $X_L = X_C$

   $X_C = 2.947$ and $X_L = 2.947$

   $L = \dfrac{X_L}{2\pi f}$ (Equation 7.9)

   $L = \dfrac{2.947}{2 \times 3.1416 \times 60}$

   $L = \dfrac{2.947}{377}$

   $L = 0.007817\ H$

Another way to obtain resonance in a circuit is to adjust the frequency to the proper value. When the frequency is low, a coil has low reactance and a capacitor has high reactance. Under these conditions, the circuit has high impedance and the current is low. When the frequency is high, a coil has high reactance and a capacitor has low reactance. Under these conditions, the circuit again has a high impedance and a low current.

At a certain frequency, the reactance of the coil is equal to the reactance of the capacitor, and the combined reactance is zero. Then, the only opposition to current flow is the resistance. When a coil and a capacitor having equal reactance are connected in series, the circuit is in resonance and the power factor is 100%.

## Example 10
A 5-$\Omega$ resistance is connected in series with a 200-$\mu$F capacitance and an inductance of 0.02 H. What is the resonant frequency of the circuit?

From the formulas $X_L = 2\pi fL$ and $X_C = \dfrac{1}{2\pi fC}$, and where $X_L$ must equal $X_C$, as in example 8, then

$$2\pi fL = \frac{1}{2\pi fC}$$

$$f = \sqrt{\frac{1}{(2\pi)^2 LC}}$$

$$f = \sqrt{\frac{1}{39.479 \times 0.02 \times 200}}$$

$$f = \sqrt{\frac{1}{157.9}}$$

$$f = \sqrt{0.006333}$$

$$f = 80 \text{ Hz (resonant frequency)}$$

In a series-resonant circuit with low resistance and high reactances, the current will be high and the voltages across the reactances will be equal and high. These voltages may be considerably higher than the supply voltage.

## Impedance Vector Diagrams of a Series Circuit

Impedance in a series circuit containing resistance, inductance, and capacitance is found in the following way. Combine, at right angles, the difference between the inductive reactance and the capacitive reactance with the resistance. To draw an impedance vector diagram for Figure 8–6A, first draw a resistance/reactance vector (Figure 8–6D). The resistance is drawn along the horizontal. The inductive reactance is drawn vertically upward from the resistance, and the capacitive reactance is drawn vertically down from the resistance. The impedance diagram is then drawn as shown in Figure 8–6E.

## AC PARALLEL CIRCUITS

The following rules apply to parallel AC circuits:

- The line current (total current) is equal to the vector sum of the currents through the individual branches.
- At any instant, the voltage is the same value across all branches of a parallel AC circuit, and is equal to the supply voltage at that instant.
- The combined (equivalent) resistance of a parallel circuit is smaller than the resistance of any of the branches.

Because the voltage at any instant must have the same value across all branches, the voltage is a convenient phase reference. The voltage vector is laid out on the horizontal. The current through each branch may be represented with reference to the common voltage vector. The total current is the vector sum of the individual currents.

The impedance and power factor formulas for parallel circuits are as follows:
The formula for impedance is

$$Z = \frac{RX_L}{\sqrt{R^2 + X_L^2}} \qquad \text{(Eq. 8.7)}$$

The formula for power factor is

$$Pf = \frac{Z}{R} \qquad \text{(Eq. 8.8)}$$

## RL Parallel Circuits

An RL parallel circuit (Figure 8–9A) is a circuit with branches containing resistance and inductance.

## Example 11
In Figure 8–9A, determine the angle of lag between the current and voltage, the impedance and power factor of the circuit, and the value of the line current. The resistance of the inductor is negligible.

The vector diagram for the current is shown in Figure 8–9B.

a. $I_R = \dfrac{E}{R}$ (*from Ohm's law*)

$I_R = \dfrac{120}{30}$

$I_R = 4$ A (current through $R$)

b. $I_L = \dfrac{E}{X_L}$ (*from Ohm's law*)

$I_L = \dfrac{120}{20}$

$I_L = 6$ A (current through $X_L$)

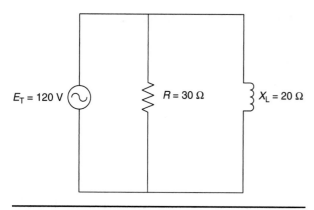

**FIGURE 8–9A**  Parallel RL circuit.

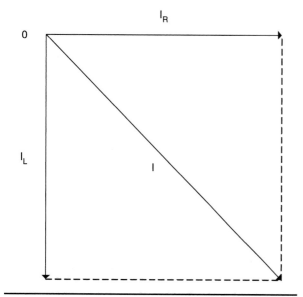

**FIGURE 8–9B**  Current vector diagram for an AC parallel RL circuit.

c.  $Z = \dfrac{RX_L}{\sqrt{R^2 + X_L^2}}$ (*Equation 8.7*)

$Z = \dfrac{30 \times 20}{\sqrt{30^2 + 20^2}}$

$Z = \dfrac{600}{\sqrt{900 + 400}}$

$Z = \dfrac{600}{\sqrt{1300}}$

$Z = \dfrac{600}{36}$

$Z = 16.667\ \Omega$ (circuit impedance)

d.  $I = \dfrac{E}{Z}$ (*Equation 7.10*)

$I = \dfrac{120}{16.667}$

$I = 7.2$ A (line current)

e.  Pf $= \dfrac{Z}{R}$ (*Equation 8.8*)

Pf $= \dfrac{16.667}{30}$

Pf $= 0.555 = 55.5\%$ (power factor of the circuit)

Pf $=$ cosine of the phase angle between the total current and the line voltage

The angle whose cosine is 0.555 falls between 56.0 degrees and 56.5 degrees. The phase angle between the current and the voltage is 56.3 degrees. This indicates that the line current lags the voltage by 56.3 degrees.

## Example 12

In the circuit in Figures 8–10A and 8–10B, solve for the following values: the current through each branch, the line current, and the power factor of the circuit.

a.  $I_a = \dfrac{E}{R}$ (*from Ohm's law*)

$I_a = \dfrac{250}{100}$

$I_a = 2.5$ A (current through $a$)

b.  $I_b = \dfrac{E}{X_L}$

$I_b = \dfrac{250}{250}$

$I_b = 1$ A (current through $b$)

c.  $I_c = \dfrac{E}{R}$

$I_c = \dfrac{250}{125}$

$I_c = 2$ A (current through $c$)

d.  $I_d = \dfrac{E}{X_L}$

$I_d = \dfrac{250}{50}$

$I_d = 5$ A (current through $d$)

e.  The currents through $a$ and $c$ are in phase with the voltage. Therefore, they can be laid out along the horizontal voltage vector end to end. The currents through $b$ and $d$ are lagging the voltage by 90 degrees, as indicated in Figure 8–10B.

The total current is equal to the square root of the sum of their squares.

$I = \sqrt{6^2 + 4.5^2}$

$I = \sqrt{36 + 20.25}$

$I = \sqrt{56.25}$

$I = 7.5$ A (line current)

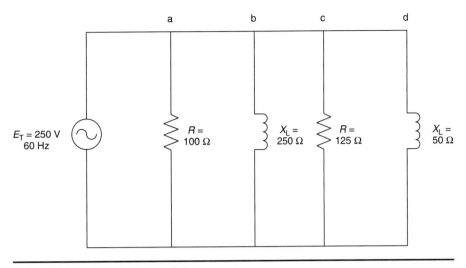

**FIGURE 8–10A** AC parallel RL circuit.

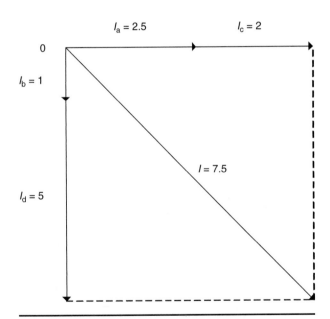

**FIGURE 8–10B** Vector addition of currents in a parallel AC RL circuit.

f. $Z = \dfrac{E}{I}$ *(Equation 7.10)*

$Z = \dfrac{250}{7.5}$

$Z = 33.333\ \Omega$ (circuit impedance)

A formula frequently used to calculate the combined resistance of only two resistances connected in parallel is

$$R_t = \frac{R_1 R_2}{R_1 + R_2} \qquad \text{(Eq. 8.9)}$$

where $R_t$ = combined resistance of the two resistances
$R_1$ = resistance of the first resistance
$R_2$ = resistance of the second resistance

Using Equation 8.9, the combined resistance of resistances in Figure 8–10A can be found as follows:

g. $R_t = \dfrac{R_a R_c}{R_a + R_c}$

$R_t = \dfrac{12{,}500}{225}$

$R_t = 55.56\ \Omega$ (resistance of the circuit)

h. $\mathrm{Pf} = \dfrac{Z}{R}$ *(Equation 8.8)*

$\mathrm{Pf} = \dfrac{33.33}{55.56}$

$\mathrm{Pf} = 0.6 = 60\%$ (power factor of the circuit)

## RC Parallel Circuits

An RC parallel circuit (Figure 8–11) contains capacitance and resistance connected in parallel.

### Example 13

In Figure 8–11, calculate the total resistance, the total capacitance, the capacitive reactance, the current flowing through each branch, the circuit impedance, and the power factor of the circuit.

a. $R = \dfrac{R_a R_b}{R_a + R_b}$

$R = \dfrac{24 \times 12}{24 \times 12}$

$R = \dfrac{288}{36}$

$R = 8\ \Omega$ (circuit resistance)

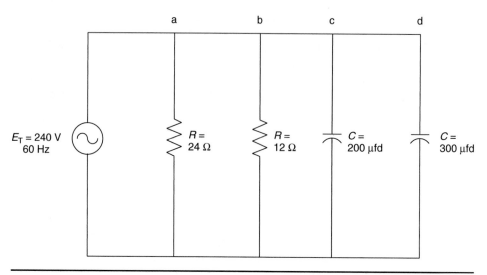

**FIGURE 8-11**　RC parallel AC circuit.

b.　$C = C_c + C_d$
　　$C = 200 + 300$
　　$C = 500 \; \mu F$ (circuit capacitance)

c.　$X_C = \dfrac{1}{2\pi f C}$ *(Equation 7.14)*

　　$X_C = \dfrac{1}{2 \times 3.1416 \times 60 \times 0.000500}$

　　$X_C = \dfrac{1}{0.188496}$ or $\dfrac{1}{0.188500}$

　　$X_C = 5.3 \; \Omega$ (circuit reactance)

d.　$X_C = \dfrac{1}{2\pi f C}$ *(Equation 7.14)*

　　$X_C = \dfrac{1}{2 \times 3.1416 \times 60 \times 0.000200}$

　　$X_C = \dfrac{1}{0.0753984}$ or $\dfrac{1}{0.075400}$

　　$X_C = 13.3 \; \Omega$ (reactance of $c$)

　　$X_C = \dfrac{1}{2\pi f C}$

　　$X_C = \dfrac{1}{2 \times 3.1416 \times 60 \times 0.000300}$

　　$X_C = \dfrac{1}{0.1130976}$ or $\dfrac{1}{0.113100}$

　　$X_C = 8.84 \; \Omega$ (reactance of $d$)

e.　$I_a = \dfrac{E}{R}$ *(from Ohm's law)*

　　$I_a = \dfrac{240}{24}$

　　$I_a = 10$ A (current through $a$)

　　$I_b = \dfrac{E}{R}$

　　$I_b = \dfrac{240}{12}$

$I_b = 20$ A (current through $b$)

$I_c = \dfrac{E}{X_C}$

$I_c = \dfrac{240}{13.3}$

$I_c = 18$ A (current through $c$)

$I_d = \dfrac{E}{X_C}$

$I_d = \dfrac{240}{8.84}$

$I_d = 27$ A (current through $d$)

$I_R = 10 + 20 = 30$ A through branches $a$ and $b$

$I_c = 18 + 27 = 45$ A through branches $c$ and $d$

Because currents $I_R$ and $I_c$ are 90 degrees out of phase with each other, their total is equal to the square root of the sum of their squares.

$$I = \sqrt{I_R^2 + I_c^2}$$

$$I = \sqrt{900 + 2025}$$

$$I = \sqrt{2925}$$

$$I = 54 \text{ A (line current)}$$

f.　$Z = \dfrac{E}{I}$ *(Equation 7.10)*

　　$Z = \dfrac{240}{54}$

　　$Z = 4.4 \; \Omega$ (circuit impedance)

g.　$Pf = \dfrac{Z}{R}$ *(Equation 8.8)*

　　$Pf = \dfrac{4.4}{8}$

　　$Pf = 0.55 = 55\%$ (circuit power factor)

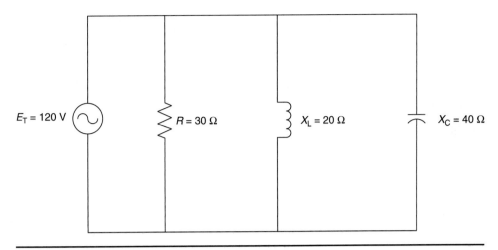

**FIGURE 8–12A**   RLC parallel AC circuit.

## RLC Parallel Circuits

RLC parallel circuits consist of branches containing capacitance, inductance, and resistance. To solve these circuits, the LC portion is usually solved first, and then the RL or RC portion is considered.

**Example 14** _____

Figure 8–12A illustrates an RLC parallel circuit connected to 120 V AC. Calculate the power factor, the line current, and the total power dissipated by the circuit.

a.   $I_R = \dfrac{E}{R}$ *(From Ohm's law)*

$I_R = \dfrac{120}{30}$

$I_R = 4$ A (current through $R$)

$I_L = \dfrac{E}{X_L}$

$I_L = \dfrac{120}{20}$

$I_L = 6$ A (current through $X_L$)

$I_c = \dfrac{E}{X_C}$

$I_c = \dfrac{120}{40}$

$I_c = 3$ A (current through $X_C$)

The vector diagram for the currents is shown in Figure 8–12B.

$I = \sqrt{I_R{}^2 + (I_L - I_c)^2}$ *(Equation 8.5)*

$I = \sqrt{16 + (6 - 3)^2}$

$I = \sqrt{16 + 9}$

$I = \sqrt{25}$

$I = 5$ A (line current)

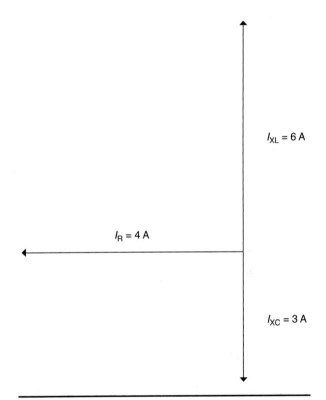

**FIGURE 8–12B**   Vector diagram of currents in an RLC circuit.

b.   $Z = \dfrac{E}{I}$ *(Equation 7.10)*

$Z = \dfrac{120}{5}$

$Z = 24\ \Omega$ (circuit impedance)

$Pf = \dfrac{Z}{R}$ *(Equation 8.8)*

$Pf = \dfrac{24}{30}$

$Pf = 0.8 = 80\%$ (circuit power factor)

c. $P = IE \cos \phi$ (*Equation 7.17*)

$P = 5 \times 120 \times 0.8$

$P = 480$ W (power utilized by the circuit)

## Parallel Resonance

Calculations for parallel resonant circuits are similar to those for series resonance. The inductive reactance must be equal to the capacitive reactance ($X_L = X_C$). The following equation can be applied to a parallel circuit as well as to a series circuit:

$$f = \sqrt{\frac{1}{(2\pi)^2 LC}}$$ (Eq. 8.10)

where $f$ = resonant frequency, in hertz (Hz)
$\pi$ = 3.1416 (a constant)
$L$ = inductance, in henrys (H)
$C$ = capacitance, in farads (F)

Parallel circuits, however, have quite different characteristics from series resonant circuits.

Figure 8–13A shows a circuit containing an inductor and a capacitor connected in parallel. To illustrate that the coil has resistance, a resistance is shown in series with the inductor.

Figure 8–13B shows the current/impedance curves at varying frequencies. When the frequency is low, the inductor takes a high lagging current and the capacitor takes a low leading current. The circuit current is high and lags the voltage; thus, the impedance is low. When the frequency is high, the capacitor takes a high leading current and the inductor takes a low lagging current. The circuit current is again high. In this case, however, the current leads the voltage.

If this circuit has zero resistance at resonant frequency, then the current taken by the inductor will equal the current taken by the capacitor, and the line current will be zero. The capacitor will discharge into the inductor, and the inductor will feed back into the capacitor. There will be a continuous exchange of energy.

To further analyze this condition, assume that the resistance is removed from the circuit in Figure 8–13A and a DC source is momentarily connected across the circuit. The DC supply will charge the capacitor. When the supply is removed, the capacitor will discharge through the inductor. Current flowing through the inductor will set up a magnetic field. As the capacitor discharges, the current will decrease. The decreasing current will produce a moving magnetic field that induces a voltage into the coil. The induced voltage causes the current to continue to flow, charging the capacitor with the opposite polarity. When the field around the

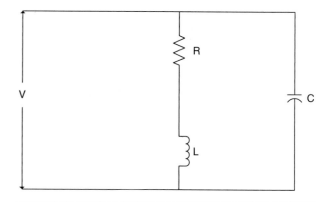

**FIGURE 8–13A**  Parallel LC circuit.

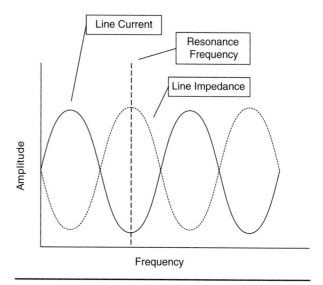

**FIGURE 8–13B**  Current/impedance curves at varying frequencies.

inductor has dissipated, the capacitor will again discharge, but in the direction opposite to the original discharge. This exchange of energy will continue indefinitely in a circuit of zero resistance.

In a practical, parallel LC resonant circuit, the inductor and capacitor have minimum resistance. Therefore, the exchange of energy and the current flow decrease with each exchange of energy. Figure 8–14 shows a current curve for this type of circuit.

If alternating current is applied to the circuit shown in Figure 8–13A, the line current will be just enough to compensate for the initial decrease in current caused by the resistance. Thus, the current between the capacitor and the inductor will be much greater than the line current.

In either series or parallel circuits, resonance may be obtained in any one of three ways: by adjusting the frequency of the supply, by adjusting the inductance, or by adjusting the capacitance. To obtain resonance, $X_L$ must be equal to $X_C$.

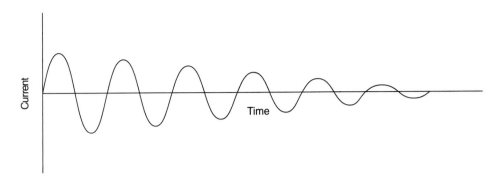

**FIGURE 8-14**  Current curve showing the exchange of energy in an LC circuit (tank circuit).

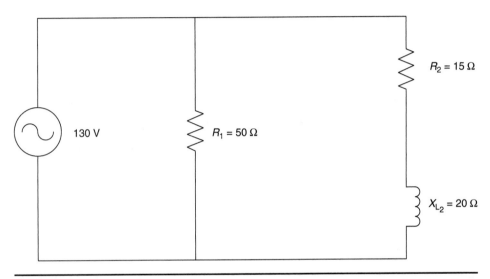

**FIGURE 8-15A**  Inductor and resistor connected in parallel across 130-V AC.

## AC COMBINATION CIRCUITS

Most circuits contain more than pure resistance, pure inductance, or pure capacitance; therefore, each path of a parallel circuit must be treated as a separate series circuit. The current for each branch must be determined and a vector diagram drawn. The individual currents are indicated with reference to a common voltage vector.

### Example 15

Figure 8–15A shows an inductor and a resistance connected in parallel. The inductor has a reactance of 20 Ω and a resistance of 15 Ω. The pure resistive branch contains 50 Ω of resistance. This resistance and the inductor are connected in parallel across a 130-V AC source. Calculate the current through each branch, the power factor of the circuit, and the total current.

a. $I_1 = \dfrac{E}{R_1}$ (*from Ohm's law*)

$I_1 = \dfrac{130}{50}$

$I_1 = 2.6$ A (current through branch 1)

b. $Z_2 = \sqrt{R_2^2 + X_{L_2}^2}$ (*from Equation 7.11*)

$Z_2 = \sqrt{15^2 + 20^2}$

$Z_2 = 25$ Ω (impedance of branch 2)

c. $I_2 = \dfrac{E}{Z_2}$ (*from Ohm's law*)

$I_2 = \dfrac{130}{25}$

$I_2 = 5.2$ A (current through branch 2)

Branch 2 contains both resistance and reactance. Therefore, $I_2$ is neither in phase nor 90 degrees out of phase with E. It is necessary to determine the power factor of branch 2.

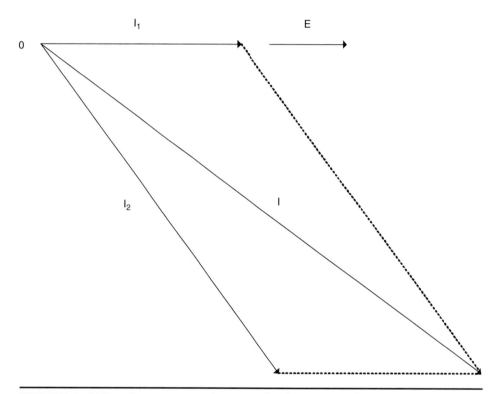

**FIGURE 8–15B** Current vector diagram of inductance and resistance in parallel.

d. $Pf = \dfrac{R}{Z}$ (*Equation 8.3*)

$Pf = \dfrac{15}{25}$

$Pf = 0.6$ (power factor for branch 2)

The angle for cosine 0.6 is 53 degrees lagging.

A current vector diagram may be drawn as shown in Figure 8–15B. Using the parallelogram method to solve vector diagrams enables the line current to be calculated.

e. $I = \sqrt{I_1^2 + I_2^2 + (2I_1 I_2 \cos\phi)}$ (*Equation 7.5*)

$I = \sqrt{2.6^2 + 5.2^2 + (2 \times 2.6 \times 5.2 \times 0.6)}$

$I = \sqrt{50.024}$

$I = 7.07$ A (line current)

## Example 16

Referring to the circuit in Figure 8–16A, calculate the current through each branch, the power factor of the circuit, the line current, the impedance, and the power utilized by the circuit.

a. $I_1 = \dfrac{E}{R_1}$ (*from Ohm's law*)

$I_1 = \dfrac{240}{10}$

$I_1 = 24$ A (current through branch 1)

b. $Z_2 = \sqrt{R_2^2 + X_{L_2}^2}$ (*Equation 7.11*)

$Z_2 = \sqrt{12^2 + 16^2}$

$Z_2 = \sqrt{400}$

$Z_2 = 20\ \Omega$ (impedance of branch 2)

$I_2 = \dfrac{E}{Z_2}$ (*from Ohm's law*)

$I_2 = \dfrac{240}{20}$

$I_2 = 12$ A (current through branch 2)

c. $Z_3 = \sqrt{R_3^2 + X_{C_3}^2}$ (*Equation 7.11*)

$Z_3 = \sqrt{(15)^2 + (30)^2}$

$Z_3 = \sqrt{225 + 900}$

$Z_3 = \sqrt{1125}$

$Z_3 = 33.54\ \Omega$ (impedance of branch 3)

$I_3 = \dfrac{E}{Z_3}$ (*from Ohm's law*)

$I_3 = \dfrac{240}{33.5}$

$I_3 = 7.16$ A (current through branch 3)

d. $Pf = \dfrac{R_2}{Z_2}$ (*Equation 8.3*)

$Pf = \dfrac{12}{20}$

$Pf = 0.6$ (power factor for branch 2)

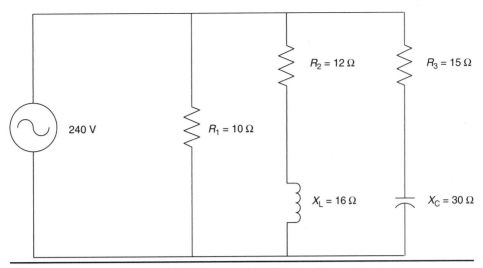

**FIGURE 8–16A**   RLC parallel AC circuit. Each branch contains resistance.

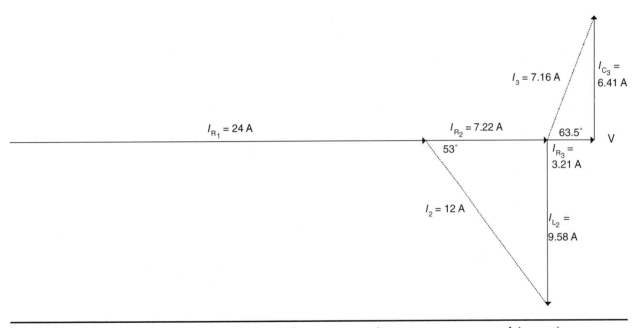

**FIGURE 8–16B**   Current vectors for calculating the resistive and reactive components of the total current.

The angle for cosine 0.6 is 53 degrees.

$$\text{Pf} = \frac{R_3}{Z_3} \text{ (Equation 8.3)}$$

$$\text{Pf} = \frac{15}{33.5}$$

Pf = 0.448 (power factor for branch 3)

The angle for cosine 0.448 is 63.5 degrees.

   With complex circuits such as this one, it is simpler to calculate the total in-phase current first. All three branches have resistance. Therefore, all three branches have some in-phase current.

1. Lay out the voltage vector along the horizontal (see Figure 8–16B).

2. Because branch 1 is pure resistance, $I_1$ is laid out along the voltage vector.

3. Branch 2 has a resistive component and a reactive component of current. The resistive component is in phase with the voltage and the reactive component is 90 degrees out of phase with the voltage. To construct the vector diagram, begin at the end of $I_1$, and lay out the in-phase current of $I_2$ along the voltage vector ($I_{R_2}$). At the end of $I_{R_2}$ and at a 90-degree angle to the voltage vector, lay out the reactive current $I_{L_2}$.

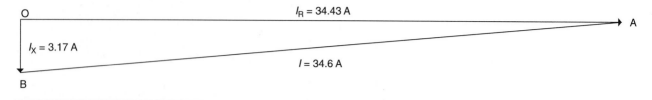

**FIGURE 8–16C**  Vector diagrams of line current in an RLC parallel circuit.

4. Draw vector $I_2$, representing the total current through branch 2, or 12 A.

5. The power factor for branch 2 is 0.6. The cosine table shows the angle for 0.6 is 53°. Therefore, the angle between $I_2$ and $I_{R_2}$ is 53 degrees.

6. The values of $I_{R_2}$ and $I_{L_2}$ can be found by using the trigonometric functions.

$$\sin \phi = \frac{\text{opp}}{\text{hypt}} \qquad \cos \phi = \frac{\text{adj}}{\text{hypt}}$$

$$\begin{aligned}
\text{opp} &= h(\sin \phi) & \text{adj} &= h(\cos \phi) \\
I_{L_2} &= I_2(\sin \phi) & I_{R_2} &= I_2(\cos \phi) \\
I_{L_2} &= 12 \times 0.7986 & I_{R_2} &= 12 \times 0.6018 \\
I_{L_2} &= 9.58 \text{ A} & I_{R_2} &= 7.22 \text{ A}
\end{aligned}$$

7. Referring to branch 3, lay out the in-phase current vector. Begin at the end of $I_{R_2}$, and draw $I_{R_3}$ along the voltage vector, as in Figure 8–16B.

8. From the end of $I_{R_3}$ and at a 90-degree angle to the voltage vector, draw vector $I_{C_3}$. Because the capacitive component of current is 180 degrees out of phase with the inductive component, vector $I_{C_2}$ must be drawn up.

9. Draw vector $I_3$, representing the total current through branch 3.

10. The power factor of branch 3 is 0.448. The cosine table shows the angle for 0.448 is 63.5 degrees. Therefore, the angle between $I_3$ and $I_{R_3}$ is 63.5 degrees.

11. Using the trigonometric formula, solve for $I_3$ and $I_{R_3}$.

$$\sin \phi = \frac{\text{opp}}{\text{hypt}} \qquad \cos \phi = \frac{\text{adj}}{\text{hypt}}$$

$$\begin{aligned}
\text{opp} &= h(\sin \phi) & \text{adj} &= h(\cos \phi) \\
I_{C_3} &= 7.16 \times 0.8949 & I_{R_3} &= 7.16 \times 0.448 \\
I_{C_3} &= 6.41 \text{ A} & I_{R_3} &= 3.21 \text{ A}
\end{aligned}$$

12. The total in-phase current is

$$\begin{aligned}
I_{R_t} &= I_{R_1} + I_{R_2} + I_{R_3} \\
I_{R_t} &= 24 + 7.22 + 3.21 \\
I_{R_t} &= 34.43 \text{ A}
\end{aligned}$$

13. The total reactive current is

$$\begin{aligned}
I_X &= I_L - I_C \\
I_X &= 9.58 - 6.41 \\
I_X &= 3.17 \text{ A}
\end{aligned}$$

14. The final current vector diagram is shown in Figure 8–16C. To solve for the total line current, use the formula for solving right triangles.

$$\begin{aligned}
I &= \sqrt{I_{R_t}^2 + I_X^2} \\
I &= \sqrt{1185.42 + 10} \\
I &= \sqrt{1195.42} \\
I &= 34.6 \text{ A (total line current)}
\end{aligned}$$

15. Figure 8–16C shows that OA represents the in-phase current and is laid out along the voltage vector. AB represents the total line current. Angle A then represents the phase angle between the line current and the supply voltage. The cosine of A is the power factor of the circuit.

$$\cos \phi = \frac{\text{adj}}{\text{hypt}} \quad (\textit{Equation 7.1})$$

$$\cos \phi = \frac{34.43}{34.6}$$

$$\cos \phi = 0.995 = 99.5\% \text{ (power factor of the circuit)}$$

16. $P = IE \cos \phi$ (*Equation 7.17*)

$$P = 34.6 \times 240 \times 0.995$$

$$P = 8262 \text{ W} = 8.262 \text{ kW (power dissipated by the circuit)}$$

17. $Z = \dfrac{E}{I}$ (*Equation 7.10*)

$$Z = \frac{240}{34.6}$$

$$Z = 6.94 \text{ V (circuit impedance)}$$

## Power Factor Correction

The vector diagram for power is shown in Figure 8–17. Vector $P$, drawn along the horizontal, designates the true power delivered to the load. The true power is also called effective power because it is the power that has a direct effect on the load. The product of the current and the voltage ($IE$) is

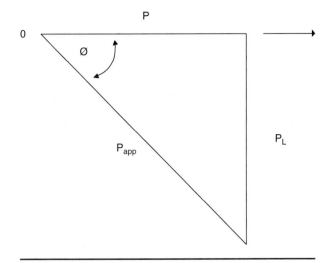

**FIGURE 8-17** Vector diagram of power in an AC circuit.

represented by the hypotenuse of the triangle and is $\phi$ degrees out of phase with the effective power. This power vector represents the apparent power.

The side opposite angle $\phi$ represents the reactive power. This power circulates between the generator at the power station and the loads. It causes the conductors through which it passes to produce heat, and it performs no useful work. In order to maintain an efficient operation, the reactive power must be kept to a minimum.

Because the power factor of a system varies according to the type of load, certain types of electrical equipment, such as alternators and transformers, are rated in volt-amperes (VA) or kilovolt-amperes (kVA). This type of equipment has a fixed value for the maximum current and voltage ratings. The value depends upon the type of insulation and the size and type of conductors used. The number of watts they can safely deliver depends upon the power factor of the load.

In most industrial establishments, a large percentage of the loads consists of induction motors or other equipment containing coils. Induction motors are simple and rugged and can be manufactured to meet most industrial requirements. These machines, as well as welding machines and induction furnaces, operate with a lagging power factor. In other words, they cause the current to lag the voltage. Although many establishments operate on a power factor as low as 60 percent, anything below 80 percent is considered to be below industrial standards.

A power factor below unity (100 percent) has the following detrimental effects:

1. More current is required in order to deliver a given amount of power to the load. An increase in current causes a greater line loss ($I^2R$), a greater line drop ($IR$), and more power loss and voltage drop within the system. If an excessive amount of current is required, it may be necessary to increase the size of the system and circuit conductors.

2. Large out-of-phase currents can cause transformers and alternators to operate with poor voltage regulation.

3. Increased currents caused by a low power factor may cause overcurrent devices to operate.

In an effort to conserve fuel and provide an efficient operation, utility companies strive to sell electrical energy at a high power factor. A consumer whose load has a low power factor may be charged more per kilowatt-hour than if the power factor were high. This induces the consumer to employ various means to improve the power factor.

Customers who have highly inductive loads can install capacitors and/or synchronous motors to improve the power factor. This equipment should be connected on the load side of the supply transformer to keep the transformer losses to a minimum and allow for better voltage regulation. The ideal location for capacitors is as near as practical to the load they are servicing to help reduce line losses in the feeder and branch-circuit conductors. A large industrial establishment may have several capacitors or capacitor banks located throughout the plant.

A capacitor causes the current to lead the voltage, thereby compensating for the lagging current caused by inductive loads. Individual capacitors may be connected directly across the terminals of the inductive load. The action that takes place is the same as that described in the section on parallel resonance in this chapter. The reactive current will circulate between the inductance and capacitance, and the supply current will be kept to a minimum. Frequently, in order to conserve space and reduce installation costs, consumers connect capacitors to the feeder conductors.

Capacitors used for power factor correction are generally rated in vars (VAR) or kilovars (kVAR) instead of farads (F). This is more convenient because when the reactive power of the capacitor is equal to the reactive power of the load, unity power factor has been established.

The reactive power of a synchronous motor depends upon two factors: the DC field excitation and the mechanical load driven by the machine. Maximum leading power factor is developed under maximum field excitation with zero load. Synchronous motors are often more practical for power factor improvement than capacitors are because they can drive a mechanical load as well as develop a leading power factor. When they are serving this dual purpose, it is advisable that they be used to drive a constant load at a constant speed.

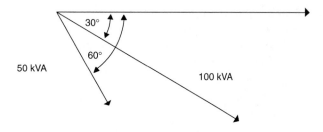

**FIGURE 8–18A** Vector diagram of the apparent power in two branches of a parallel circuit.

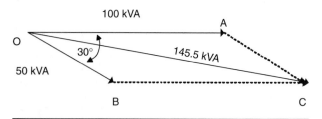

**FIGURE 8–18B** Vector addition of apparent power.

Motors and other electrical equipment are provided with information on the nameplate to aid the electrical worker in installing the equipment, analyzing circuits, and solving problems. The nameplate information includes the rating in kilovolt-amperes and/or kilowatts, voltage, amperes, and frequency. The total apparent power required by a machine, or the power required from the alternator, may be calculated through vector addition.

## Example 17

A load in a factory requires 50 kVA and has a 50% lagging power factor. Another load connected to the same power lines requires 100 kVA and has a lagging power factor of 86.6%. Calculate the effective power, the reactive power, the power factor, and the apparent power of the line.

a. The apparent power taken by each load, expressed with reference to the voltage, is represented in Figure 8–18A. The angle corresponding to a power factor of 50% (cos 0.5000) is 60 degrees. The angle for 86.6% (cos 0.8660) is 30 degrees. To determine the resultant power, draw the vector diagram as shown in Figure 8–18B.

b. Draw the 100-kVA value along the horizontal and label it OA.

c. Figure 8–18A shows that the angle between 100 kVA and 50 kVA is 60 degrees − 30 degrees = 30 degrees.

d. Thirty degrees from OA, draw the 50-kVA vector OB (Figure 8–18B).

$$OC = \sqrt{(OA)^2 + (OB)^2 + [2(OA)(OB)(\cos\phi)]}$$

$$(Equation\ 7.5)$$

$$OC = \sqrt{10^4 + (25 \times 100) + (2 \times 100 \times 50 \times 0.866)}$$

$$OC = \sqrt{10^4 + (25 \times 100) + 8660}$$

$$OC = \sqrt{21,160}$$

$$OC = 145.5\ kVA\ (total\ apparent\ power)$$

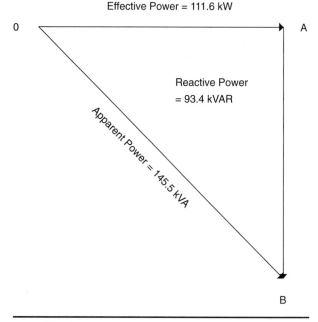

**FIGURE 8–18C** Power vector diagram of the total apparent, effective, and reactive power.

e. Calculate the effective power for each load.

$P = IE \cos\phi$ (*Equation 7.17*) and $IE = 50$ kVA

$P = 50 \times 0.5$

$P = 25$ kW

$P = IE \cos\phi$ and $IE = 100$ kVA

$P = 100 \times 0.866$

$P = 86.6$ kW

f. The effective power for each load is in phase with the voltage, and the total effective power is 25 + 86.6 = 111.6 kW

g. Draw a power vector diagram for the total apparent, effective, and reactive power (Figure 8–18C). Draw the effective power on the horizontal. Draw the reactive power vertically downward from the effective power. The apparent power is the hypotenuse of the right triangle.

h. Solve for the reactive power.

$P_L = \sqrt{P_{app}^2 - P^2}$ (*Equation 7.19*)

$P_L = \sqrt{21,170 - 12,454}$

$P_L = \sqrt{8716}$

$P_L = 93.4$ kVAR (total reactive power)

i. $Pf = \dfrac{P}{P_{app}} \times 100$ (*Equation 7.16*)

$Pf = \dfrac{111.6}{145.5}$

$Pf = 0.767 = 76.7\%$ (power factor of the circuit)

## Example 18

A group of induction motors require 100 kVA and operate at a power factor of 84% lagging. If a synchronous motor is installed that requires 60 kVA and operates at a power factor of 70.7% leading, calculate the total effective power, apparent power, reactive power, and power factor of the load.

a. Figure 8–19A shows a vector diagram of the apparent power of the two loads with reference to the voltage supply. The lagging power factor load is represented below the reference (voltage) vector, and the leading power factor load is represented above the reference vector. The angles indicated correspond to the power factors of the respective loads.

b. To determine the resultant apparent power, draw a vector diagram indicating the phase relationship with reference to each other (Figure 8–19B). 45 degrees + 33 degrees = 78 degrees.

$C = \sqrt{A^2 + B^2 + 2AB \cos \phi}$ (*Equation 7.5*)

$C = \sqrt{100^2 + 60^2 + (2 \times 100 \times 60 \times 0.2079)}$

$C = \sqrt{10,000 + 3600 + 2495}$

$C = \sqrt{16,095}$

$C = 126.87$ kVA (apparent power for the circuit)

c. Calculate the effective power for each load and the total effective power.

$P = IE \cos \phi$ (*Equation 7.17*)

$P = 100 \times 0.84$

$P = 84$ kW (effective power of the induction motors)

$P = IE \cos \phi$

$P = 60 \times 0.707$

$P = 42.42$ kW (effective power of the synchronous motor)

$P_t = P_1 + P_2$

$P_t = 84 + 42.42$

$P_t = 126.42$ kW (effective power for the total load)

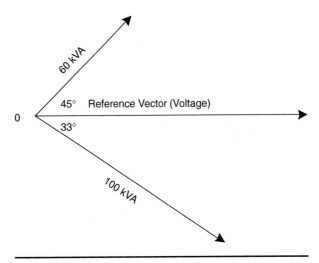

**FIGURE 8–19A**  Vector diagram of the apparent power with reference to the voltage supply.

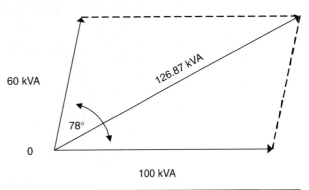

**FIGURE 8–19B**  Vector diagram indicating the phase relationship of the apparent power in two branches of a parallel circuit.

d. Draw a power triangle (Figure 8–19C).

e. Solve for the reactive power.

$P_L = \sqrt{P_{app}^2 - P^2}$ (*Equation 7.19*)

$P_L = \sqrt{126.87^2 - 126.42^2}$

$P_L = \sqrt{16,096 - 15,982}$

$P_L = \sqrt{114}$

$P_L = 10.68$ kVAR – (reactive power for the total load)

f. Solve for the power factor.

$Pf = \dfrac{P}{P_{app}} \times 100$ (*Equation 7.16*)

$Pf = \dfrac{126.42}{126.87}$

$Pf = 0.996 = 99.6\%$ (power factor of total load)

An analysis of the circuit in Example 18 shows that the induction motors require 100 kVA and operate at a power factor of 84 percent *lagging*. The synchronous motor requires 60 kVA and operates at a power factor of 70.7 percent *leading*.

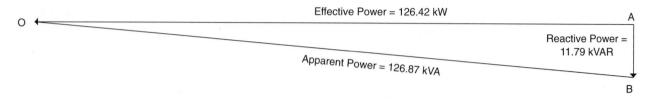

**FIGURE 8–19C**   Vector diagram for power in a high power factor circuit.

Because the induction motors have a lagging power factor and the synchronous motor has a leading power factor, the synchronous motor is returning power to the line during most of the cycle when the induction motors are utilizing it. The induction motors return power to the line during most of the cycle when the synchronous motor is utilizing it. The end result is a more efficient operation than if all the motors were of the induction type. Synchronous motors are explained in Chapter 19.

### Example 19

Replace the synchronous motor in example 18 with an induction motor of the same size and operating at the same power factor as the synchronous motor. Calculate the total effective power, apparent power, and reactive power. Determine the power factor of the total load.

a.  The power taken by each load, expressed with reference to the supply voltage, is represented in Figure 8–20A. The angle corresponding to the power factor of 70.7% (cos 0.7071) is 45 degrees. To determine the resultant power, draw a vector diagram as shown in Figure 8–20B. Remember that all induction motors have a lagging power factor.

b.  Draw the 100-kVA value along the horizontal, and label it OA.

c.  Figure 8–20A shows that the angle between the 100-kVA load and the 60-kVA load is 45 degrees − 33 degrees = 12 degrees.

d.  Twelve degrees from OA, draw the 60-kVA vector OB (Figure 8–20B).

e.  Solve for the total apparent power.
$$OC = \sqrt{(OA)^2 + (OB)^2 + 2(OA)(OB)(\cos \phi)}$$
$$OC = \sqrt{100^2 + 60^2 + (2 \times 100 \times 60 \times 0.9781)}$$
$$OC = \sqrt{10,000^2 + 3600 + 11,737.2}$$
$$OC = \sqrt{25,337.2}$$
$$OC = 159.177 \text{ kVA (total apparent power)}$$

f.  Calculate the effective power of each load.
$$P = IE \cos \phi \qquad P = IE \cos \phi$$
$$P = 60 \times 0.707 \qquad P = 100 \times 0.84$$
$$P = 42.42 \text{ kVA} \qquad P = 84 \text{ kVA}$$

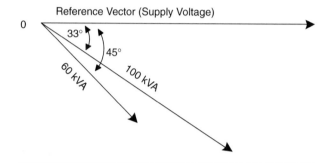

**FIGURE 8–20A**   Vector diagram of the apparent power of two inductive loads.

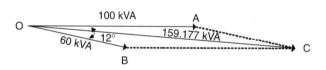

**FIGURE 8–20B**   Vector addition of the apparent power of two inductive loads.

g.  Calculate the total effective power.
$$P_t = P_1 + P_2$$
$$P_t = 42.42 + 84$$
$$P_t = 126.42 \text{ kW (total effective power)}$$

h.  Draw a power vector diagram for the total apparent power, effective power, and reactive power (Figure 8–20C). Draw the effective power on the horizontal. Draw the reactive power vertically downward from the effective power. The apparent power is the hypotenuse of the right triangle.

i.  Solve for the reactive power.
$$P_L = \sqrt{P_{app}^2 - P^2}$$
$$P_L = \sqrt{159.177^2 - 126.42^2}$$
$$P_L = \sqrt{25,337.317 - 15,982.016}$$
$$P_L = \sqrt{9355.301}$$
$$P_L = 96.732 \text{ kVAR (total reactive power)}$$

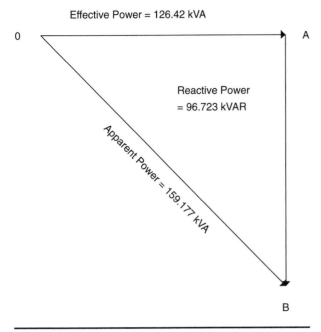

**FIGURE 8–20C** Power vector diagram of the total apparent, effective, and reactive power.

j. Calculate the power factor.

$$Pf = \frac{P}{P_{app}}$$

$$Pf = \frac{126.42}{159.177}$$

$$Pf = 0.794 = 79.4\% \text{ (power factor of the total load)}$$

From the calculations in examples 18 and 19, it can be seen that the power factor is improved considerably by using a synchronous motor in place of an induction motor. Maintaining a high power factor provides a more efficient operation.

## REVIEW QUESTIONS

*Multiple Choice*

1. The effective resistance of an AC circuit varies with the
   a. applied voltage.
   b. frequency.
   c. inductive reactance.
   d. impedance.

2. The fact that alternating current tends to flow along the outer surface of the conductor is called
   a. skin effect.
   b. eddy currents.
   c. hysteresis.
   d. inductance.

3. The phase relationship of the voltages across various parts of a series AC circuit can be expressed with reference to the
   a. resistance.
   b. effective power.
   c. current.
   d. apparent power.

4. The voltage across a pure resistive load in an AC series circuit is
   a. out of phase with the current (lagging).
   b. in phase with the current.
   c. out of phase with the current (leading).
   d. none of the above.

5. An RLC circuit is one that contains
   a. resistance and capacitance.
   b. inductance and capacitance.
   c. resistance, inductance, and capacitance.
   d. resistance, reactance, and capacitance.

6. In a resonant circuit, the current and the voltage are
   a. out-of-phase leading.
   b. out-of-phase lagging.
   c. in phase.
   d. out-of-phase with the inductance.

7. A circuit can be brought into resonance by adjusting the
   a. voltage.
   b. current.
   c. frequency.
   d. power.

8. In a parallel RLC circuit, when the frequency is low, the inductor takes a
   a. high lagging current.
   b. low lagging current.
   c. high leading current.
   d. low leading current.

9. Another name for true power is
   a. effective power.
   b. reactive power.
   c. apparent power.
   d. inductive power.

10. Most transformers and alternators are rated in
   a. kilowatts.
   b. kilovolt-amperes.
   c. kilovars.
   d. kilograms.

*Give Complete Answers*

1. Identify the factors that affect the resistance of an AC circuit but that do not affect a DC circuit.

2. Define *skin effect*.

3. Write the rule for voltage in an AC series circuit.

4. Why is the current vector usually used as a reference vector for AC series circuits?

5. What is the phase relationship of the current and the voltage in a pure resistive AC series circuit?

6. What factors affect the inductive reactance of a coil in an AC circuit?

7. What is the phase relationship of the current and the voltage in a pure inductive AC series circuit?

8. Does the current lead or lag the voltage in a series AC circuit containing a large value of capacitance?

9. List three components that make up impedance.

10. What is a series RC circuit?

11. How does inductance affect the power factor of an AC circuit?

12. Define *resonant circuit*.

13. List three factors that affect the resonance of a circuit.

14. What is the rule for resistance in a parallel AC circuit?

15. Why is the voltage vector usually used as a reference vector for an AC parallel circuit?

16. What two quantities must be equal in order for a circuit to be in resonance?

17. Describe the exchange of energy between a capacitor and an inductor in an AC resonant parallel circuit.

18. What effect does a low power factor have on a circuit or a system?

19. Do most industrial establishments have a leading or a lagging power factor?

20. Identify two methods frequently used to improve the power factor in an industrial plant.

*Solve each problem, showing the method used to arrive at the solution.*

1. A capacitor with a capacitance of 410 μF is connected to a 120-V, 60-Hz line. Calculate the capacitor current.

2. A resistance of 50 Ω and a capacitance with a reactance of 35 Ω are connected in series across a 120-V, 60-Hz supply. Determine:
   a. The circuit current
   b. The voltage across the resistance
   c. The voltage across the capacitance

d. The phase angle between the line current and the voltage
   e. The power taken by each component
   f. The effective power of the circuit
   g. The power factor of the circuit

3. A coil with an inductance of 0.1 H and negligible resistance is connected in series with a 500-μF capacitor. The combination is connected to a 230-V, 60-Hz line. Calculate the reactance of the combination and the line current.

4. A 60-μF capacitor and a 0.22-H coil of negligible resistance are connected in series across 120 V, 60 Hz. Determine the reactance of the combination and the line current.

5. A coil with a resistance of 15 Ω and an inductance of 0.1 H is connected in series with a 500-μF capacitor. The combination is connected to a 60-Hz supply. Calculate:
   a. The circuit voltage if 3 A flow in the line
   b. The voltage across each component
   c. The impedance of the circuit
   d. The power factor of the circuit
   e. The power factor of the coil

6. A circuit contains a coil of 0.1 H inductance and 50-Ω resistance. What current will flow through it when it is connected to a 120-V, 60-Hz supply?

7. At what frequency will resonance occur in a series circuit containing an inductor of negligible resistance and 0.1 H inductance and a 2-μF capacitor?

8. What size capacitor must be connected in series with a 5-H coil that has 300-Ω resistance in order to produce resonance at 60 Hz?

9. If the coil in Problem 8 is connected to a 120-V, 60-Hz AC supply, determine:
   a. The line current
   b. The voltage across each component
   c. The power factor of the circuit
   d. The effective power of the circuit

10. A resistance of 10 Ω and an inductive reactance of 20 Ω are connected in parallel across 120-V AC. Calculate:
   a. The power factor of the circuit
   b. The line current
   c. The circuit impedance
   d. The effective power
   e. The apparent power
   f. The reactive power

11. An inductor of negligible resistance and 30-$\Omega$ reactance is connected in parallel with a capacitor of 50-$\Omega$ reactance. The combination is connected across 150-V AC. What is the line current?

12. A resistance of 20 $\Omega$, an inductive reactance of 15 $\Omega$, and a capacitive reactance of 35 $\Omega$ are connected in parallel across a 120-V supply. Determine:
    a. The phase angle between the line current and the voltage
    b. The total current taken by the circuit
    c. The current through each component
    d. The circuit impedance
    e. The effective power of the circuit

13. Determine the impedance of a circuit containing a resistance of 32 $\Omega$ and an inductive reactance of 10 $\Omega$ connected in parallel.

14. At what frequency will resonance occur in a parallel circuit containing a coil of 20-$\Omega$ resistance and 0.5-H inductance that is connected in parallel with a 0.1-$\mu$F capacitor?

15. What size capacitor must be connected in parallel with a coil that has 100-$\Omega$ resistance and 0.2-H inductance to produce resonance at 180 Hz?

16. A circuit has two paths connected across 110 V. One path has a resistance of 25 $\Omega$; the other path has a resistance of 10 $\Omega$ and an inductive reactance of 20 $\Omega$. Calculate:
    a. The circuit power factor
    b. The total current of the circuit
    c. The effective power of the circuit
    d. The impedance of the circuit

17. A 5-hp motor connected to a 120-V, 60-Hz line has an efficiency of 80% at full load and a power factor of 60%.
    a. What size capacitor must be connected across the motor terminals to increase the power factor to unity?
    b. Calculate the line current at unity power factor.
    c. What size capacitor is necessary to increase the power factor to 85% lagging?
    *Note:* Power factor correcting capacitors are rated in kilovars (kVAR).

18. A circuit has two paths. One path has a coil with a resistance of 8 $\Omega$ and an inductive reactance of 20 $\Omega$. The other path has a resistance of 8 $\Omega$ in series with a capacitive reactance of 16 $\Omega$. Calculate the impedance of the circuit.

19. An alternator delivers 50 kW and 83.3 kVAR to a load. Determine the apparent power and the power factor of the load.

20. A 40-kW load operating at a power factor of 70% is supplied through 10 miles of No. 6 AWG copper wire. The wire has a resistance of 0.410 $\Omega$ per 1000 ft. (The load is 5 miles from the supply.) If the voltage at the load is 13,200 V, calculate:
    a. The reactive power of the load
    b. The apparent power of the load
    c. The line current
    d. The line loss (watts)
    e. The voltage drop

21. Repeat the calculations for Problem 20 using a power factor of 90%.

22. A synchronous motor taking 60 kVA at 10% leading power factor is connected in parallel with an induction motor taking 100 kVA at 70% lagging power factor. Calculate:
    a. The total kW of the circuit
    b. The power factor of the combination
    c. The total kVA
    d. The kVAR of the circuit

23. An induction motor of 30 kW at 75% power factor is in parallel with a synchronous motor taking 20 kVAR at 82% power factor (leading). Determine:
    a. The power factor of the circuit
    b. The total apparent power
    c. The total effective power

24. A load connected to a 60-Hz, 440-V line requires 10 kW. The power factor is 60% (lagging). Calculate the kVAR rating of the capacitor that will increase the power factor to unity.

25. An induction motor is rated at 10 kW. If an ammeter indicates 60 A on a 240-V line, what is the power factor of the motor?

# Conductor Types and Sizes

## OBJECTIVES

After studying this chapter, the student will be able to:

- Define the units of measurement of electrical conductors and calculate their area.
- Describe the tools and methods used to determine the size of electrical conductors.
- Define the term *resistivity* and calculate the resistance of electrical conductors.
- Explain the effect of temperature on the resistance of conductors and calculate any change in resistance.
- Determine the ampacity of conductors under various conditions of use.
- Describe a variety of types of conductors and insulations used in the electrical industry.
- Define and calculate *line drop* and *line loss*.
- Describe and know how to make various types of splices and connections used in the electrical industry.

A *conductor* is any material that offers little opposition to the flow of an electric current. Although a conductor may be in the form of a solid, liquid, or gas, most conductors used in the electrical industry are solids. Copper and aluminum are the most common materials used. Other good conductors are silver, zinc, brass, platinum, iron, nickel, tin, and lead. Electrical conductors used for general transmission and distribution are manufactured in circular shapes known as *wires* and/or *cables* and in rectangular shapes called *bus bars*. Wires, cables, and bus bars are used extensively in the electrical industry.

# UNITS OF MEASUREMENT

The size of electrical conductors varies widely according to their use and the amount of current they carry. Therefore, special units of measurement have been established for conductors.

## Circular Mils

Many applications require very small wires. A unit of measurement much smaller than the inch is needed. The unit **mil** has been selected as the basis for measuring electrical conductors. One mil is equal to one-thousandth of an inch (1/1000 inch or 0.001 inch). Rather than stating that a conductor's diameter is 0.050 inch, it is stated that it measures 50 mils.

In the metric system, there are 1,000,000 *micrometers* ($\mu$m) in 1 meter (m). Because there is 0.0254 meter in 1 inch, and 1 mil is equal to 0.001 inch, it can be calculated that 1 mil is equal to 25.4 micrometers. *Note:* Micrometers are sometimes called *microns*.

Conductor sizes are usually expressed as unit area rather than diameter. To measure circular conductors (wires and cables), the unit *circular mil* is used. To measure rectangular conductors (bus bars), the unit *square mil* is used.

One circular mil is the area of a circle 1 mil in diameter. One square mil is the area of a square in which each side measures 1 mil. The linear measurement in mils is denoted by the abbreviation "m." The area in circular mils is abbreviated "cm," and square mils is abbreviated "sq m."

*Note:* In the metric system of measurement, the symbol "m" denotes meter, and the symbol "cm" denotes centimeter. When dealing with the size of conductors, the reader is cautioned to recognize the abbreviations "cm" and "sq m" as circular mil and square mil.

To determine the circular mil area of a wire, the following formula may be used:

$$A = d^2 \qquad \text{(Eq. 9.1)}$$

where  $A$ = area, in circular mils (cm)
$d$ = diameter, in mils (m)

### Example 1

Calculate the circular mil area of a wire that measures 0.025 in. in diameter.

*Customary System*

$$0.025 \times 1000 = 25 \text{ mils (m)}$$
$$A = d^2$$
$$A = 25^2$$
$$A = 625 \text{ cm}$$

The circular mil (English units) area of a conductor may be changed to circular millimeters (metric units, abbreviated "c mm") by this formula:

$$c \text{ mm} = \frac{cm}{1550} \qquad \text{(Eq. 9.2)}$$

### Example 2

Calculate the circular millimeter value of the conductor in Example 1.

$$c \text{ mm} = \frac{cm}{1550}$$
$$c \text{ mm} = \frac{625}{1550}$$
$$c \text{ mm} = 0.403$$

### Example 3

What is the diameter of a wire in inches if it has an area of 5184 cm?

$$d = \sqrt{cm}$$
$$d = \sqrt{5184}$$
$$d = 72 \text{ mils}$$
$$72 \div 1000 = 0.072 \text{ in.}$$

### Example 4

What is the diameter of the wire in Example 3, in millimeters (mm)?

$$1 \text{ in.} = 25.4 \text{ mm}$$
$$0.072 \times 25.4 = 1.829 \text{ mm}$$

## Square Mils

Areas of bus bars are more conveniently expressed in square mils. The square mil area of a conductor may be determined by multiplying the thickness in mils by the width in mils. Use the formula:

$$A = TW \qquad \text{(Eq. 9.3)}$$

where  $A$ = area, in square mils (sq m)
$T$ = thickness, in mils (m)
$W$ = width, in mils (m)

### Example 5

Calculate the square mil area of a bus bar 1/4-in. thick and 2-in. wide.

(1) Change inches to mils
$$\frac{1}{4} \text{ in.} = 0.25 \text{ in.}$$
$$0.25 \times 1000 = 250 \text{ m}$$
$$2 \times 1000 = m$$

(2) $A = TW$
$$A = 250 \times 2000$$
$$A = 500,000 \text{ sq m}$$

## Example 6

Determine the number of square mils in a bus bar that measures 1/2 in. by 1 in.

$$\frac{1}{2} \text{ in.} = 0.5 \text{ in.} \qquad A = TW$$
$$0.5 \times 1000 = 500 \text{ m} \qquad A = 500 \times 1000$$
$$1 \times 1000 = 1000 \text{ m} \qquad A = 500,000 \text{ sq m}$$

The square mil area of a conductor may be changed to circular mils by the formula:

$$cm = \frac{sq \text{ m}}{0.7854} \qquad \text{(Eq. 9.4)}$$

## Example 7

What is the circular mil area of the conductor in Example 6?

$$cm = \frac{sq \text{ m}}{0.7854}$$
$$cm = \frac{500,000}{0.7854}$$
$$cm = 636,617$$

The square mil area of a conductor may be changed to square millimeters (sq mm) by the formula:

$$sq \text{ mm} = \frac{sq \text{ m}}{1550} \qquad \text{(Eq. 9.5)}$$

## Example 8

Calculate the square millimeter area of the conductor in Example 6.

$$sq \text{ mm} = \frac{sq \text{ m}}{1550}$$
$$sq \text{ mm} = \frac{500,000}{1550}$$
$$sq \text{ mm} = 322.58$$

## Example 9

What is the circular mil area of a bus bar that is 1/8-in. (3.175-mm) thick and 1-in. (25.4-mm) wide?

$$1/8 \text{ in.} = 0.125 \text{ in.}$$
$$0.125 \text{ in} \times 1000 = 125 \text{ m}$$
$$1 \times 1000 = 1000 \text{ m}$$
$$A = TW$$
$$A = 125 \times 1000$$
$$A = 125,000 \text{ sq m}$$
$$cm = \frac{sq \text{ m}}{0.7854}$$
$$cm = \frac{125,000}{0.7854}$$
$$cm = 159,154$$

## American Wire Gauge

A special scale has been established for the more common wires used in the electrical industry. The common name for this scale is the *American wire gauge* (AWG). The wire sizes range from No. 50 AWG, which is the smallest standard size, to No. 4/0 AWG, which is the largest standard size for this scale. Figure 9–1 illustrates the instrument used to measure the gauge of a wire. All wire sizes larger and smaller than the AWG scale are expressed in circular mils.

The *NEC* lists standard wire sizes from No. 18 AWG to 4/0 (0000, or four aught) AWG. Also listed are the standard sizes from 250,000 circular mils (250 kcmil) to 2,000,000 circular mils (2000 kcmil). (kcmil means thousands of circular mils.) Number 18 wire has an area of 1620 circular mils and 4/0 has an area of 211,600 circular mils. Table A–1 in the appendix of this text is taken from Chapter 9, Table 8 of the 2008 *NEC*.

Rectangular conductors are measured with a micrometer (Figure 9–2). Micrometers are very accurate measuring instruments and are calibrated in 1/1000-inch increments. Determining conductor size is relatively easy because 1 mil is equal to 1/1000 inch. Therefore, each 1/1000 inch on the micro-meter is equal to 1 mil. (*Note:* Do not confuse the instrument "micrometer" with the unit of measurement "micrometer.")

To measure a rectangular conductor, measure the width and thickness and then multiply the values to obtain the square mil area. Micrometers can

**FIGURE 9–1**   Gauge for measuring wire sizes.

**FIGURE 9–2**  Micrometer.

also be used to measure circular conductors. Measure the diameter and then square the number to obtain the circular mil area.

All electrical personnel should be able to use a micrometer and an American wire gauge.

## MIL-FOOT WIRE

A wire 1 mil in diameter and 1 foot long is called a **mil-foot** of wire (Figure 9–3). One foot of any wire may be considered to be composed of a number of mil-foot wires in parallel. In other words, a foot of wire with a cross-sectional area of 5 circular mils equals five 1 mil-foot wires.

The resistance of any number of identical wires in parallel is equal to the resistance of 1 wire divided by the total number of wires. For example, the resistance of 1 mil-foot of copper building wire is 10.4 ohms. A wire of the same material and length, but 10 circular mils in cross-section, is 10.4 ohms ÷ 10 circular mils, or 1.04 ohms. The resistance of a wire is directly proportional to its length and inversely proportional to its cross-sectional area. If a 10-circular-mils wire is 1000 feet long, its resistance is 1.04 ohms × 1000 feet, or 1040 ohms (See equation 9.6.)

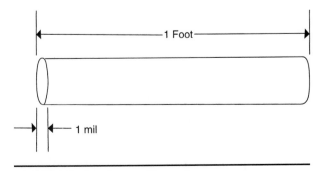

**FIGURE 9–3**  One mil-foot of wire.

## RESISTIVITY

The **resistivity** (K) of a conducting material is the resistance per mil-foot at a temperature of 68 degrees Fahrenheit (F), or 20 degrees Celsius (C). The resistance of the conductor depends upon the material from which it is made, its cross-sectional area, its length, and the operating temperature. (The operating temperature is the ambient temperature plus the increase in temperature caused by the current flow.)

Table 9–1 lists the resistivity of various conducting materials. To calculate the resistance of any conductor at 68 degrees Fahrenheit (20 degrees Celsius), divide the resistance of 1 mil-foot by the cross-sectional area (in circular mils) and multiply by its length in feet. From this information, the following equation can be constructed:

$$R = \frac{K\ell}{A} \qquad \text{(Eq. 9.6)}$$

where   $R$ = resistance of the conductor, in ohms ($\Omega$)
   $K$ = resistivity
   $\ell$ = length, in feet (ft)
   $A$ = area, in circular mils (cm)

**Example 10**
Determine the resistance of a copper wire 0.15 in. in diameter and 1 mile long.

$$0.15 \times 1000 = 150 \text{ m}$$
$$150 \times 150 = 22{,}500 \text{ cm}$$
$$R = \frac{K\ell}{A}$$
$$R = \frac{10.4 \times 5280}{22{,}500}$$
$$R = 2.44 \ \Omega$$

**Example 11**
Calculate the length of a copper wire 0.025 in. in diameter, if it has a resistance of 5 $\Omega$.

$$0.025 \times 1000 = 25 \text{ m}$$
$$25 \times 25 = 625 \text{ cm}$$
$$R = \frac{K\ell}{A}$$
$$R = \frac{10.4\ell}{625}$$
$$5 = \frac{10.4\ell}{625}$$
$$10.4\ell = 3125$$
$$\ell = 300 \text{ ft}$$

**TABLE 9–1** Resistivity and temperature coefficients of conducting materials.

| Materials | Resistivity (K) at 20°C or 68°F | Temperature Coefficient (a) per °C* |
|---|---|---|
| Aluminum | 17.7 | 0.0043 |
| Carbon | 20,000 | −0.0005 |
| Constantan | 296 | 0 |
| Copper | 10.4 | 0.0043 |
| German Silver Wire | 200 | 0.0004 |
| Iron wire | 60 | 0.006 |
| Iron (cast) | 500 | 0.0008 |
| Manganin | 266 | 0.00002 |
| Nichrome | 660 | 0.0002 |
| Nickel | 60 | 0.006 |
| Silver | 9.5 | 0.004 |
| Steel (soft) | 90 | 0.0044 |
| Steel (hard) | 275 | 0.0016 |
| Tungsten (annealed) | 26 | 0.005 |
| Tungsten (hard drawn) | 33 | 0.005 |

*Average of values between 0° and 100°C

## Example 12

What is the resistance of a feeder consisting of two copper bus bars, each 250 ft long? Each bar is ¼ in. thick and 2 in. wide.

$$0.25 \times 1000 = 250 \text{ m}$$
$$2 \times 1000 = 2000 \text{ m}$$
$$250 \times 2000 = 500,000 \text{ sq m}$$
$$cm = \frac{sq\ m}{0.7854}$$
$$cm = \frac{500,000}{0.7854}$$
$$cm = 636,618$$
$$R = \frac{K\ell}{A}$$
$$R = \frac{10.4 \times (2 \times 250)}{638,618}$$
$$R = 0.008\ \Omega$$

## THERMAL EFFECT

The average temperature coefficients of various materials between 0 degrees Celsius and 100 degrees Celsius (32 degrees Fahrenheit and 212 degrees Fahrenheit) are listed in Table 9–1. The temperature coefficient is the amount by which the resistance of a material changes for each degree of change in temperature, for each ohm of resistance.

The resistance of pure metals, such as silver, copper, and aluminum, increases as the temperature increases. The resistance of some metal alloys, such as constantan and manganin, remains almost unaffected by the temperature. Other materials, such as carbon and most electrolytes, react in an opposite way to pure metals in that their resistance decreases as the temperature increases.

To calculate the resistance of a material at a specific temperature, use the following formula:

$$R_o = R + Rat_o \text{ or } R_o = R(1 + at_o) \quad \text{(Eq. 9.7)}$$

where $R_o$ = resistance at the operating temperature

$R$ = resistance at zero degrees Celsius (0°C)

$a$ = temperature coefficient at zero degrees Celsius (°C)

$t_o$ = operating temperature, in degrees Celsius (°C)

## Example 13

The resistance of a nichrome heating element is 50 Ω at 0°C. What is its resistance at 1000°C?

$$R_o = R(1 + at_o)$$
$$R_o = 50[1 + (0.0002 \times 1000)]$$
$$R_o = 50(1 + 0.2)$$
$$R_o = 50 \times 1.2$$
$$R_o = 60\ \Omega$$

## Example 14

A carbon electrode has a resistance of 0.02 Ω at 0°C. What is its resistance at 1500°C?

$$R_o = R(1 + at_o)$$
$$R_o = 0.02[1 + (−0.0005 \times 1500)]$$
$$R_o = 0.02[1 + (−0.75)]$$
$$R_o = 0.02 \times 0.25$$
$$R_o = 0.005\ \Omega$$

## Example 15

The resistance of a coil wound with copper wire is 100 Ω at 0°C. What is the resistance at 60°C?

$$R_o = R(1 + at_o)$$
$$R_o = 100[1 + (0.0043 \times 60)]$$
$$R_o = 100(1 + 0.258)$$
$$R_o = 100 \times 1.258$$
$$R_o = 125.8\ \Omega$$

If the operating temperature is expressed in degrees Fahrenheit, the temperature can be converted to degrees Celsius by using a temperature conversion chart or the formula:

$$°C = \frac{5}{9}(°F - 32) \qquad \text{(Eq. 9.8)}$$

It is frequently necessary to determine the operating resistance when a change in temperature occurs from a value other than 0 degrees Celsius. This may be accomplished as follows:

$$R = \frac{R_i}{1 + at_i} \qquad \text{(Eq. 9.9)}$$

where   $R$ = resistance at zero degrees Celsius (0°C)
$R_i$ = resistance at the initial temperature
$a$ = temperature coefficient at zero degrees Celsius (0°C)
$t_i$ = initial temperature, in degrees Celsius (°C)

## Example 16

The resistance of a coil wound with iron wire is 200 Ω at 20°C. What is the resistance at 80°C?

$$R = \frac{R_i}{1 + at_i}$$
$$R = \frac{200}{1 + (0.006 \times 20)}$$
$$R = \frac{200}{1.12}$$
$$R = 178.571 \text{ Ω at 0°C}$$
$$R_o = R(1 + at_o)$$
$$R_o = 178.571[1 + (0.006 \times 80)]$$
$$R_o = 178.571[1 + 0.48]$$
$$R_o = 178.571 \times 1.48$$
$$R_o = 264.3 \text{ Ω at 80°C}$$

It is not necessary to first find the resistance at 0 degrees Celsius if the two formulas are combined into one. The new formula becomes:

$$R_o = \frac{R_i(1 + at_o)}{1 + at_i} \qquad \text{(Eq. 9.10)}$$

## Example 17

Solve the problem in Example 16, using equation 9.10.

$$R_o = \frac{R_i(1 + at_o)}{1 + at_i}$$
$$R_o = \frac{R_i(1 + 0.006t_o)}{(1 + 0.006t_i)}$$
$$R_o = \frac{200(1 + 0.006 \times 80)}{1 + 0.006 \times 20}$$

$$R_o = \frac{200 \times 1.48}{1.12}$$
$$R_o = 264.3 \text{ Ω}$$

## Example 18

The copper winding of a motor has a resistance of 10 Ω when the machine is started. The resistance increases to 12 Ω after it has run several hours at full load. If the room temperature is 20°C, what is the temperature rise of the winding?

$$R_o = \frac{R_i(1 + at_o)}{1 + at_i}$$
$$R_o = \frac{R_i(1 + 0.0043t_o)}{(1 + 0.0043t_i)}$$
$$12 = \frac{10(1 + 0.0043t_o)}{(1 + 0.0043 \times 20)}$$
$$10(1 + 0.0043t_o) = 12(1 + 0.0043 \times 20)$$
$$1 + 0.0043t_o = \frac{12(1 + 0.0043 \times 20)}{10}$$
$$1 + 0.0043t_o = \frac{13.032}{10}$$
$$1 + 0.0043t_o = 1.3032$$
$$0.0043t_o = 0.3032$$
$$t_o = 70.5°C$$

The temperature rise is 70.5°C − 20°C = 50.5°C.

Chapter 9, Table 8, of the 2008 *NEC* indicates the resistance of copper and aluminum wires at 75 degrees Celsius. From this information, the resistance at any other temperature can be calculated by using equation 9.10.

# INSULATION AND AMPACITY OF CONDUCTORS

Heat is generated whenever current flows through a conductor. The amount of heat depends upon the value of current and the resistance of the conductor. As the heat is generated, the operating temperature of the conductor increases. This rise in temperature continues until the rate at which heat leaves the conductor equals the rate at which heat is produced. When the two are equal, the temperature remains constant.

*The current, in amperes, that a conductor can carry continuously under the conditions of use without exceeding its temperature rating* is the definition of ampacity as defined by the *NEC. National Electrical Code* and *NEC* are registered trademarks of the National Fire Protection Association, Inc., Quincy, MA 02269.

*Reprinted with permission from NFPA 70-2008, *National Electrical Code*® Copyright© 2007, National Fire Protection Association, Quincy, MA 02269. This reprinted material is not the referenced subject which is represented only by the standard in its entirety.

The ampacity of a conductor is limited by the temperature that its insulation can tolerate without deteriorating and/or losing its insulating quality. When considering the ampacity of a conductor, it is also necessary to consider its operating temperature. Table 310.17 of the *NEC* lists the allowable ampacity of insulated conductors having an insulation voltage rating from 0 volts to 2000 volts and installed as a single conductor in free air (not enclosed). It is based on an ambient temperature of 30 degrees Celsius. The maximum allowable operating temperature is listed at the top of each column. This value is determined by the type of insulation used. The operating temperatures range from 60 degrees Celsius to 90 degrees Celsius.

Tables B–1 and C–1 are taken from *NEC* Tables 310.16 and 310.17 and are found in the Appendix. These tables list the allowable ampacities of the same types of conductors. Appendix Table B–1 is for conductors installed in a raceway, a cable, or earth. Appendix Table C–1 applies to single insulated conductors in free air. The values in Appendix Table B–1 apply only when there are not more than three conductors in the raceway, cable, or earth. Ambient and operating temperatures are the same for both tables. The ampacity ratings in Appendix Table B–1 are lower because heat cannot be dissipated as rapidly when the conductors are grouped and/or placed in an enclosure.

## Correction Factors

The operating temperature of a conductor depends upon the ambient temperature and the amount of current flowing in the conductor. For example, Appendix Table B–1 indicates that insulation types TW and UF have a maximum operating temperature of 60 degrees Celsius (140 degrees Fahrenheit). Insulations that are subjected to temperatures higher than their maximum operating temperatures will deteriorate.

The values indicated in Appendix Tables B–1 and C–1 are for use where the ambient temperature does not exceed 30 degrees Celsius (86 degrees Fahrenheit). For conductors installed in areas where the ambient temperature is above 30 degrees Celsius (86 degrees Fahrenheit), the ampacity is determined by applying the correction factor indicated at the bottom of the table.

### Example 19

A No. 12 AWG copper conductor is covered with RHW insulation. If this conductor is installed in an area where the ambient temperature is 122°F (50°C), what is the ampacity of the conductor?

Appendix Table B–1 indicates that the load current rating of No. 12 RHW copper wire is 20 A

(see footnote of Appendix Table B–1), provided that not more than three conductors are installed in the same raceway and the maximum ambient temperature does not exceed 86 degrees Fahrenheit (30 degrees Celsius). The correction factor for 122 degrees Fahrenheit (50 degrees Celsius) is 0.75. Therefore,

$$20 \times 0.75 = 15 \text{ A}$$

The safe ampacity for the conductor in this example is 15 A.

A correction factor must also be applied when there are more than three conductors in the raceway.

### Example 20

If the raceway in Example 19 contains six conductors, what is the ampacity of the conductors?

The correction factor for more than three conductors is listed in *NEC* Table 310.15(B)(2)(a). The factor indicated for six conductors is 80% or 0.8, so that

$$20 \times 0.75 \times 0.8 = 12 \text{ A}$$

The safe ampacity for this conductor is 12 A.

## Types of Insulations

Plastic is, by far, the most common insulating material used on conductors. For general-purpose wiring in buildings, types TW, THW, and THWN are used more than any other insulations. Tables 9–2A, 9–2B, 9–3A and 9–3B list specifications for various types of wires.

Insulations such as TFN and MTW are designed for special purposes. Type TFN is designed for wiring lighting and similar fixtures. Type MTW is used for wiring machine tools and other applications exposed to harsh liquids. Tables 9–4A, 9–4B, 9–5A, and 9–5B are specification tables for types TFN, TFFN, and MTW.

*NEC* Table 310.13(A) lists the various insulations that meet the *Code* standards. This table provides information as to the maximum operating temperatures, types of installations (dry or wet), and maximum voltages. For installing electrical conductors, it is necessary to consider all factors that may affect the size and material of the conductors and the type of insulation used.

## FLEXIBLE CORDS AND CABLES

Some of the most common types of cords and cables used in the electrical industry are lamp cord, hard service cord, range and dryer cable, data processing cable, and elevator cable. The conductors in these cords and cables are made up of fine strands in order to obtain maximum flexibility.

**SPEC 2020**

January 1, 2003

# ROME THWN or THHN
## PVC Insulation, Nylon Jacket, 600 Volts

APPLICATION: General purpose wiring in accordance with the National Electrical Code, maximum conductor temperature of 90°C in dry locations and 75°C in wet locations, 600 volts, for installation in conduit or other recognized raceway. Also used for wiring of machine tools (stranded), appliances, and control circuits not exceeding 600 volts.

STANDARDS:
1. Listed by UL as Type THHN or THWN per Standard 83, and as Type MTW per Standard 1063 (stranded items).
2. Listed by UL as Gasoline and Oil Resistant II.
3. Listed by UL as Sunlight Resistant (1/0 AWG and larger, black only).
4. 1/0 AWG and larger pass UL and IEEE-383 ribbon burner flame test and are listed for CT Use.
5. Listed by UL as 105°C Appliance Wiring Material, 60°C where exposed to oil (stranded items only).
6. C(UL) listed as Type T90 Nylon or TWN75, FT1. (Solid Colors Only)

CONSTRUCTION: Annealed uncoated copper conductor, PVC insulation, nylon jacket, surface printed.

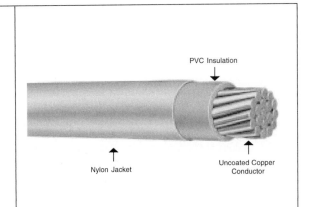

PVC Insulation

Nylon Jacket

Uncoated Copper Conductor

| Size AWG or kcmil | No. of Strands | Thickness in Mils PVC Insulation | Thickness in Mils Nylon Jacket | Nominal Diam. Inches | NEC Ampacity* 75°C THWN | NEC Ampacity* 90°C THHN | Approx. Wt. lb./1000 Ft. Net | Approx. Wt. lb./1000 Ft. Shipping | Standard Package Length | Standard Package Put-up | 1 | 2 | 3 | 4 | 5 | 6 | 7 | 8 | 9 | 10 | 11 | 12 |
|---|---|---|---|---|---|---|---|---|---|---|---|---|---|---|---|---|---|---|---|---|---|---|
| **Solid (THWN or THHN)** |||||||||||||||||||||||
| 14 | Solid | 15 | 4 | .105 | 20 t | 25 t | 16 | 17 | 500' spls. | 4 per ctn. | S | S | S | S | S | S | S | S | S | | S | |
| | | | | | | | | 17 | 2500' | NR reel | | | | | | | | | | | | |
| 12 | Solid | 15 | 4 | .122 | 25 t | 30 t | 24 | 25 | 500' spls. | 4 per ctn. | S | S | S | S | S | S | S | S | S | S | S | |
| | | | | | | | | 26 | 2500' | NR reel | | | | | | | | | | | | |
| 10 | Solid | 20 | 4 | .153 | 35 t | 40 t | 38 | 39 | 500' spls. | 2 per ctn. | S | S | S | S | S | S | S | S | | | S | |
| | | | | | | | | 40 | 2500' | NR reel | | | | | | | | | | | | |
| **Stranded (MTW or THWN or THHN)** |||||||||||||||||||||||
| 14 | 19 | 15 | 4 | .112 | 20 t | 25 t | 16 | 18 | 500' spls. | 4 per ctn. | S | S | S | S | S | S | S | S | S | S | S | S |
| | | | | | | | | 17 | 2500' | NR reel | | | | | | | | | | | | |
| 12 | 19 | 15 | 4 | .130 | 25 t | 30 t | 24 | 25 | 500' spls. | 4 per ctn. | S | S | S | S | S | S | S | S | S | S | S | |
| | | | | | | | | 26 | 2500' | NR reel | | | | | | | | | | | | |
| 10 | 19 | 20 | 4 | .164 | 35 t | 40 t | 38 | 40 | 500' spls. | 2 per ctn. | S | S | S | S | S | S | S | S | S | S | S | |
| | | | | | | | | 40 | 2500' | NR reel | | | | | | | | | | | S | |
| 8 | 19 | 30 | 5 | .220 | 50 | 55 | 64 | 71 | 500' | NR reel | S | S | S | S | S | S | | | | | | |
| | | | | | | | | | 1000', NS | NR reel | S | | | | | | | | | | | |
| 6 | 19 | 30 | 5 | .256 | 65 | 75 | 98 | 99 | 500' | NR reel | S | S | S | S | S | S | S | S | | | | |
| | | | | | | | | | 1000', NS | NR reel | S | | | | | | | | | | | |
| 4 | 19 | 40 | 6 | .325 | 85 | 95 | 155 | 168 | 500' | NR reel | S | | | | | | | | | | | |
| | | | | | | | | | 1000' | NR reel | S | S | | | | | | | | | | |
| | | | | | | | | | NS | NR reel | S | | S | | S | | | | | | | |
| 3 | 19 | 40 | 6 | .353 | 100 | 110 | 190 | 204 | 500', 1000', NS | NR reel | S | | | | | | | | | | | |
| 2 | 19 | 40 | 6 | .386 | 115 | 130 | 236 | 254 | 500', 1000' | NR reel | S | | | | | | | | | | | |
| | | | | | | | | | NS | NR reel | S | S | S | | S | | | | | | | |
| 1 | 19 | 50 | 7 | .443 | 130 | 150 | 300 | 319 | 1000', NS | NR reel | S | | | | | | | | | | | |
| 1/0 | 19 | 50 | 7 | .484 | 150 | 170 | 372 | 395 | 1000', NS | NR reel | S | | | | S | | | | | | | |
| 2/0 | 19 | 50 | 7 | .529 | 175 | 195 | 462 | 485 | 1000', NS | NR reel | S | | | | S | | | | | | | |
| 3/0 | 19 | 50 | 7 | .579 | 200 | 225 | 575 | 600 | 1000', NS | NR reel | S | | | | | | | | | | | |
| 4/0 | 19 | 50 | 7 | .635 | 230 | 260 | 716 | 745 | 1000', NS | NR reel | S | | | | S | | | | | | | |
| 250 | 37 | 60 | 8 | .703 | 255 | 290 | 846 | 905 | 1000', NS | NR reel | S | | | | | | | | | | | |
| 300 | 37 | 60 | 8 | .756 | 285 | 320 | 1005 | 1060 | NS | NR reel | S | | | | | | | | | | | |
| 350 | 37 | 60 | 8 | .806 | 310 | 350 | 1165 | 1225 | 1000', NS | NR reel | S | | | | | | | | | | | |
| 400 | 37 | 60 | 8 | .851 | 335 | 380 | 1325 | 1380 | NS | NR reel | S | | | | | | | | | | | |
| 500 | 37 | 60 | 8 | .934 | 380 | 430 | 1640 | 1725 | 1000', NS | NR reel | S | | | | | | | | | | | |
| 600 | 61 | 70 | 9 | 1.03 | 420 | 475 | 1995 | 2090 | NS | NR reel | S | | | | | | | | | | | |
| 750 | 61 | 70 | 9 | 1.14 | 475 | 535 | 2480 | 2580 | NS | NR reel | S | | | | | | | | | | | |
| 1000 | 61 | 70 | 9 | 1.32 | 545 | 615 | 3300 | | | | | | | | | | | | | | | |

*Ampacity in accordance with NEC for not more than three conductors in raceway. As THHN: 90°C conductor temperature and 30°C ambient in dry locations. As THWN: 75°C conductor temperature and 30°C ambient in wet or dry locations.

t The over current protection shall not exceed 15 amperes for 14 AWG, 20 AWG amperes for 12 AWG and 30 amperes for 10 AWG copper.

(1) Color Code: 1 black, 2 white, 3 red, 4 blue, 5 green, 6 yellow, 7 orange, 8 brown, 9 purple, 10 pink, 11 gray, 12 tan.

**Information on this sheet subject to change without notice.**

**TABLE 9–2A** Specifications for Type THWN or THHN wire (Courtesy of Rome Cable Corporation).

Specification

ROME THWN or THHN

PVC - Nylon, 600 Volts

## 1. SCOPE

1.1 This specification describes single conductor Rome THWN or THHN, a general purpose building wire insulated with polyvinyl chloride (PVC) and covered with a tough protective sheath of nylon intended for lighting and power circuits at 600 Volts or less, in residential, commercial and industrial buildings. The wire may be operated at 90°C maximum continuous temperature in dry locations and 75°C in wet locations and is listed by Underwriters Laboratories for use in accordance with Article 310 of the National Electrical Code. The wire shall also be C(UL) listed as Types T90 Nylon or TWN75, FT1 indicating suitability for use in accordance with the Canadian Code.

## 2. APPLICABLE STANDARDS

2.1 The following standards form a part of this specification to the extent specified herein:

2.1.1 Underwriters Laboratories Standard 83 for Thermoplastic Insulated Wires.

2.1.2 Underwriters Laboratories Standard 1063 for Machine-Tool Wires and Cables (Stranded items only).

2.1.3 Underwriters Laboratories Standard 758 for 105°C Appliance Wiring Materials (Stranded items only).

2.1.4 CSA Standard C22.2 No. 75 and Electrical Bulletin No. 1451 for Type T90 Nylon or TWN75.

## 3. CONDUCTORS

3.1 Conductors shall be solid, Class B or Class C stranded, annealed uncoated copper per UL Standards 83 or 1063.

## 4. INSULATION

4.1 Each conductor shall be insulated with PVC and sheathed with nylon complying with requirements of UL Standard 83 for Types THHN or THWN, UL Standard 1063 for Type MTW and CSA C22.2 No. 75 for T90 Nylon. In addition, Types THWN or THHN shall comply with the optional Gasoline and Oil Resistant II rating of UL Standard 83. The insulation on stranded sizes shall also comply with UL requirements for 105°C Appliance Wiring Material.

4.2 The average thickness of PVC insulation, for a given conductor size, shall be as specified in UL Standard 83 for Types THWN or THHN. The minimum thickness at any point, of the PVC insulation, shall be not less than 90% of the specified average thickness. The minimum thickness at any point of the nylon sheath, shall be as specified in UL Standard 83 for Types THWN or THHN. The PVC insulation shall be applied tightly to the conductor and shall be free-stripping.

## 5. IDENTIFICATION

5.1 The wire shall be identified by surface marking indicating manufacturer's identification, conductor size and metal, voltage rating, UL Symbol, type designations and optional ratings. The wire shall also be identified as C(UL) Type T90 Nylon or TWN75, FT1.

## 6. TESTS

6.1 Wire shall be tested in accordance with the requirements of UL Standard 83 for Types THWN or THHN wire and for the optional Gasoline and Oil Resistant II listings; as Type MTW to UL Standard 1063 (stranded items); as AWM to UL Standard 758 (stranded items); and as C(UL) Type T90 Nylon or TWN75.

## 7. LABELS

7.1 The wire shall bear the Underwriters Laboratories labels for Types THWN or THHN (solid conductors) and Type MTW (stranded conductors) and the C(UL) label for Types T90 Nylon or TWN75, FT1.

---

**TABLE 9–2B**  Specifications for Type THWN or THHN wire *(continued from Table 9–2A) (Courtesy of Rome Cable Corporation).*

**SPEC 2150**

January 1, 2003

## ROME USE-2 or RHW-2 or RHH
Rome-XLPE Insulation, 600 Volts

APPLICATION:
For lighting and power applications in accordance with the National Electrical Code and for other general purpose wiring applications. Suitable for use in circuits not exceeding 600 volts at conductor temperatures not exceeding 90°C in wet or dry locations. May be installed in raceway, duct, direct burial and aerial installations.

STANDARDS:
1. Listed by UL as Type USE-2 (90°C wet or dry) per Standard 854 for Service Entrance Cables.
2. Listed by UL as Types RHW-2 (90°C wet or dry) or RHH (90°C dry) per Standard 44.
3. Conforms to ICEA S-95-658/NEMA WC70, utilizing Column A insulation thicknesses.
4. Conforms to Federal Specification CID A-A-59544.

CONSTRUCTION: Annealed copper conductor, Rome-XLPE thermosetting chemically crosslinked polyethylene insulation, surface printed.

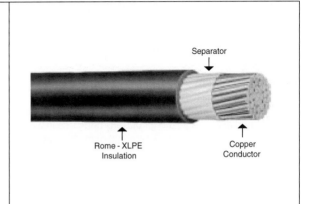

Separator

Rome - XLPE Insulation

Copper Conductor

| Size AWG or kcmil | No. of Strands | Insulation Thickness Mils | Nom. Diam. Inches | Copper | | Approx. Wt. lb./1000 Ft. | | Stock Items |
|---|---|---|---|---|---|---|---|---|
| | | | | Ampacity | | | | |
| | | | | 90°C * USE-2 RHW-2 RHH | 75°C ** USE RHW | Net | Shipping | |
| **Solid** | | | | | | | | |
| 14 | Solid | 45 | .16 | 25 [t] | 20 [t] | 21 | 24 | - |
| 12 | Solid | 45 | .18 | 30 [t] | 25 [t] | 30 | 33 | S |
| 10 | Solid | 45 | .20 | 40 [t] | 35 [t] | 43 | 46 | S |
| 8 | Solid | 60 | .25 | 55 | 50 | 68 | 73 | - |
| **Stranded** | | | | | | | | |
| 14 | 7 | 45 | .17 | 25 [t] | 20 [t] | 23 | 25 | - |
| 12 | 7 | 45 | .19 | 30 [t] | 25 [t] | 32 | 33 | S |
| 10 | 7 | 45 | .21 | 40 [t] | 35 [t] | 45 | 49 | S |
| 8 | 7 | 60 | .27 | 55 | 50 | 72 | 79 | S |
| 6 | 7 | 60 | .31 | 75 | 65 | 105 | 115 | S |
| 4 | 7 | 60 | .36 | 95 | 85 | 155 | 175 | S |
| 2 | 7 | 60 | .41 | 130 | 115 | 234 | 255 | S |
| 1 | 19 | 80 | .49 | 150 | 130 | 305 | 330 | S |
| 1/0 | 19 | 80 | .53 | 170 | 150 | 380 | 405 | S |
| 2/0 | 19 | 80 | .58 | 195 | 175 | 470 | 495 | S |
| 3/0 | 19 | 80 | .63 | 225 | 200 | 580 | 615 | S |
| 4/0 | 19 | 80 | .68 | 260 | 230 | 725 | 765 | S |
| 250 | 37 | 95 | .76 | 290 | 255 | 865 | 925 | S |
| 300 | 37 | 95 | .81 | 320 | 285 | 1025 | 1090 | - |
| 350 | 37 | 95 | .86 | 350 | 310 | 1190 | 1250 | S |
| 400 | 37 | 95 | .91 | 380 | 335 | 1345 | 1410 | - |
| 500 | 37 | 95 | .99 | 430 | 380 | 1665 | 1760 | S |
| 600 | 61 | 110 | 1.10 | 475 | 420 | 2020 | 2110 | - |
| 750 | 61 | 110 | 1.20 | 535 | 475 | 2500 | 2600 | S |
| 1000 | 61 | 110 | 1.35 | 615 | 545 | 3225 | 3420 | - |
| 1250 | 91 | 125 | 1.51 | 665 | 590 | 4130 | 4310 | - |
| 1500 | 91 | 125 | 1.63 | 705 | 625 | 4930 | 5110 | - |
| 1750 | 127 | 125 | 1.74 | 735 | 650 | 5720 | 5990 | - |
| 2000 | 127 | 125 | 1.85 | 750 | 665 | 6510 | 6910 | - |

*AMPACITY in accordance with NEC for not more than three conductors. As RHW-2: in raceway, 90°C conductor temperature and 30°C ambient in wet or dry locations. As RHH: in raceway, 90°C conductor temperature and 30°C ambient in dry locations. As USE-2: direct burial, 90°C conductor temperature and 30°C ambient in wet locations.

**AMPACITY in accordance with NEC for not more than three conductors. As RHW: in raceway, 75°C conductor temperature and 30°C ambient in wet or dry locations. As USE: direct burial, 75°C conductor temperature and 30°C ambient in wet locations.

[t] The overcurrent protection shall not exceed 15 amperes for 14 AWG, 20 amperes for 12 AWG and 30 amperes for 10 AWG.

**Information on this sheet subject to change without notice.**

**TABLE 9–3A** Specifications for Type USE-2, RHW-2, or RHH wire *(Courtesy of Rome Cable Corporation).*

Specification

ROME USE-2 or RHW-2 or RHH

Rome-XLPE Insulation, 600 Volts

## 1. SCOPE

1.1 This specification describes single conductor Rome-XLPE, Type USE-2 or RHW-2 or RHH, crosslinked polyethylene insulated cables for use in circuits not exceeding 600 volts. Cables are listed by UL as Type USE-2 and are recognized for underground use in wet locations at a maximum continuous conductor temperature of 90°C in accordance with Article 338 of the National Electrical Code. The cables are also listed by UL as Type RHH or RHW-2 for general purpose wiring applications at maximum continuous conductor temperature of 90°C in dry locations (RHH) or 90°C in wet or dry locations (RHW-2) and may be installed in air, conduit or other recognized raceways in accordance with Article 310 of the National Electrical Code.

## 2. APPLICABLE STANDARDS

2.1 The following standards form a part of this specification to the extent specified herein:

    2.1.1    Underwriters Laboratories Standard 854 for Service Entrance Cables.

    2.1.2    Underwriters Laboratories Standard 44 for Rubber-Insulated Wires and Cables.

    2.1.3    ICEA Pub. No. S-95-658, NEMA Pub. No. WC70 for Nonshielded Power Cables Rated 2000 Volts or Less.

    2.1.4    Federal Specification CID A-A-59544.

## 3. CONDUCTORS

3.1 Conductors shall be solid or Class B stranded annealed uncoated copper per UL Standard 854 and 44.

## 4. SEPARATOR

4.1 A suitable separator over the conductor may be used at the option of the manufacturer.

## 5. INSULATION

5.1 Each conductor shall be insulated with Rome-XLPE, a crosslinked polyethylene complying with the physical and electrical requirements of UL Standard 854 for Type USE-2 and UL Standard 44 for Types RHW-2 or RHH and Table 3-7, Class X-2 of ICEA.

5.2 The average thickness of insulation shall be as specified in UL Standard 44 for Types RHH and RHW-2 and Table 3-4, Column A of ICEA. The minimum thickness at any point shall be not less than 90% of the specified average thickness. The insulation shall be applied tightly to the conductor and shall be free-stripping.

## 6. IDENTIFICATION

6.1 The wire shall be identified by surface marking indicating manufacturer's identification, conductor size and metal, voltage rating, UL Symbol and type designations.

## 7. TESTS

7.1 Wire shall be tested in accordance with the requirements of UL Standard 854 for Type USE-2, UL Standard 44 for Types RHW-2 or RHH and ICEA S-95-658.

## 8. LABELS

8.1 The wire shall bear the Underwriters Laboratories label for Type USE-2.

---

**TABLE 9–3B** Specifications for Type USE-2, RHW-2, or RHH wire (continued from Table 9–3A) (Courtesy of Rome Cable Corporation).

# Rome Cable
## C O R P O R A T I O N

SPEC 5115

January 1, 2003

## ROME TFFN FIXTURE WIRE

PVC - Nylon, 600 Volts
Multi-Rated Wire

APPLICATION: For wiring of lighting and similar fixtures in accordance with the National Electrical Code, 90 °C maximum conductor temperature, 600 volts. Also used for wiring of machine tools, appliances and control circuits not exceeding 600 volts.

STANDARDS: Listed by Underwriters Laboratories as follows:
1. Type TFFN - 16 AWG, 90 °C
2. Gasoline and Oil Resistant II
3. Type MTW - 90°C Machine Tool Wire
4. AWM Style 1316 - 105°C, 80°C where exposed to oil
5. AWM Style 1452 - 90°C, 1000 volts.

CONSTRUCTION: Annealed uncoated copper conductor, PVC insulation, nylon jacket.

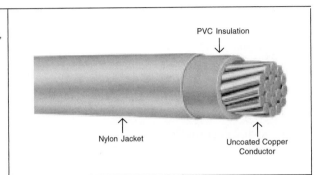

### Type MTW or TFFN

| Size AWG | No. of Strands | Thickness in Mils | | Nominal Diam. Inches | NEC Ampacity* | Approx. Net Wt. lb./1000 Ft. | Standard Package | |
|---|---|---|---|---|---|---|---|---|
| | | PVC Insula-tion | Nylon Jacket | | | | Length | Put-up |
| 16 | 19 | 15 | 4 | .099 | 8 | 12 | 500' spls. | 4 per ctn. |

`Ampacity in accordance with Article 402 of the NEC.

NOTES: 1. Available in following colors; black, white, red, blue, green, yellow, orange, brown, purple, pink and gray.

**Information on this sheet subject to change without notice.**

**TABLE 9–4A** Specifications for Type TFFN fixture wire *(Courtesy of Rome Cable Corporation)*.

Flexible cords and cables are permitted for use as pendants, fixture wiring, connections to portable lamps and/or appliances, elevators, cranes and hoists, and some stationary equipment.

They may not be used for fixed wiring of a building. *NEC* Article 400 covers the use of flexible cords and cables. Various types of cords are shown in Figures 9–4A, 9–4B, and 9–4C.

## | ELECTRICAL DISTRIBUTION

Conductors for distributing electricity to various parts of a building are classified as feeders and branch circuits. A *feeder circuit* consists of the conductors between the service equipment and the final branch-circuit overcurrent device. A *branch circuit* consists of all the conductors between the final circuit overcurrent protective device and the outlets.

Transmitting electricity in large buildings frequently requires that conductors be installed over long distances. It is necessary to consider the distance between the supply and the load because the resistance of a conductor increases with its length.

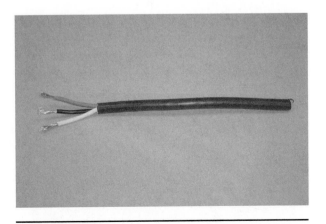

**FIGURE 9–4A** Power cord.

SPEC 5115

1-1-03

Specification

## ROME TFFN FIXTURE WIRE

PVC-Nylon, 600 Volts
Multi-Rated Wire

**1. SCOPE**

1.1    This specification describes multi-rated single conductor, 600 volt Rome PVC-Nylon Type TFFN Fixture Wire suitable for wiring of lighting fixtures and in similar equipment where enclosed or protected or for connecting lighting fixtures to the branch-circuit conductors supplying the fixtures where exposed to temperatures not exceeding 90°C, or where exposed to oil at a temperature not exceeding 80°C. Also suitable for internal wiring of appliances at 90°C, 1000 volts.

**2. APPLICABLE STANDARDS**

2.1    The following standards form a part of this specification to the extent specified herein:

    2.1.1   Underwriters Laboratories Standard 83 for Thermoplastic-Insulated Wire.

    2.1.2   Underwriters Laboratories Standard 66 Fixture Wire.

    2.1.3   Underwriters Laboratories Standard 758 for Appliance Wiring Material.

    2.1.4   Underwriters Laboratories Standard 1063 for Machine Tool Wire.

**3. CONDUCTOR**

3.1    The conductor shall be stranded, annealed uncoated copper per UL Standards 83, 62 and 1063.

**4. INSULATION**

4.1    The conductor shall be insulated with PVC and sheathed with nylon. The PVC shall comply with the physical and electrical requirements of UL Standard 83, Class 12 and the nylon sheath to the requirements of UL Standard 62 and UL Standard 758.

4.2    The insulation and sheath thicknesses shall be as specified by UL Standards 62 and 1063.

**5. IDENTIFICATION**

5.1    The wire shall be identified by surface marking indicating manufacturer's identification, conductor size, voltage rating, UL symbol, and type designations.

**6. TESTS**

6.1    The wire shall be tested in accordance with the requirements of UL Standard 62.

**7. LABELS**

7.1    The wire shall bear the Underwriters Laboratories Machine Tool Wire label.

---

**TABLE 9–4B**   Specifications for Type TFFN fixture wire *(continued from Table 9–4A)* *(Courtesy of Rome Cable Corporation)*.

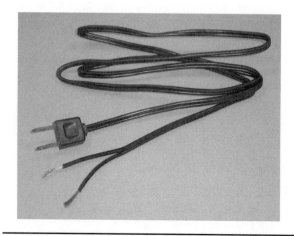

**FIGURE 9–4B**   Lamp (zip) cord.

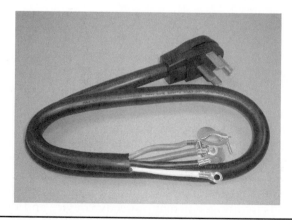

**FIGURE 9–4C**   Four-conductor range cable, type SRDT, 2-6 AWG, and 2-8 AWG.

**ROME MACHINE TOOL WIRE**

SPEC 5160

January 1, 2001

PVC Insulation, 600 Volts
Multi-Rated MTW, THW, AWM, TEW105°C

APPLICATIONS: For wiring of machine tools, appliances and internal wiring in electrical equipment in control and power circuits not exceeding 600 volts. Suitable for use as Machine Tool Wire at 90°C and lower temperatures in dry locations and at 60°C where exposed to moisture, oil, or coolants - such as cutting oils. Also suitable for general purpose wiring as Type THW in accordance with the NEC.

STANDARDS:
1. Conforms to Underwriters Laboratories Standard 1063 for Machine Tool Wires and Cables.
2. Conforms to Underwriters Laboratories Standard 83 for Thermoplastic Insulated Wires.
3. Also multi-listed by Underwriters Laboratories as Appliance Wiring Material as follows:
   (a) AWM Styles 1344, 1345, 1346-105°C in air, 75°C moisture resistant, 60°C oil resistant.
   (b) AWM Styles 1230, 1231, 1232-105°C in air, 60°C moisture resistant, 60°C oil resistant.
   (c) AWM Styles 1015, 1028, 1283-105°C in air, 60°C oil resistant.
4. Meets VW-1 flame test requirement of UL and FT-1 flame test requirement of CSA.
5. Approved by Canadian Standards Association as Thermoplastic Equipment Wire Type TEW rated 600 volts, 105°C in air, 60°C in oil.
6. Conforms to National Fire Protection Association Electrical Standard for Metal Working Machine Tools (ANSI/NFPA79).

CONSTRUCTION: Annealed uncoated copper conductor, PVC insulation, surface printed.

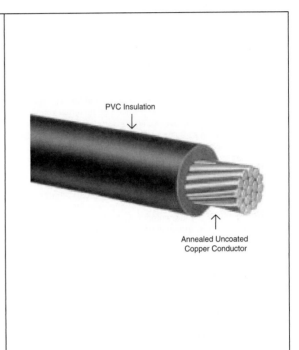

PVC Insulation

Annealed Uncoated
Copper Conductor

| Size AWG | No. of Strands | Insulation Thickness Mils | Nominal Diam. Inches | Approx. NetWt. Lb./1000 | UL AWM StyleNumbers |
|---|---|---|---|---|---|
| **Multi-Listed as: MTW or THW or 105°C AWM or TEW 105°C** | | | | | |
| 14 | 19 | 30 | .14 | 18 | 1345,1230, 1015 |
| 12 | 19 | 30 | .16 | 27 | 1345,1230, 1015 |
| 10 | 19 | 30 | .18 | 40 | 1345,1230, 1015 |
| 8 | 19 | 45 | .24 | 67 | 1344,1231, 1028 |
| **Multi-Listed as: MTW or THW or 105°C AWM or TEW 105°C** | | | | | |
| 14 | 19 | 45 | .17 | 22 | 1344,1231, 1028 |
| 12 | 19 | 45 | .19 | 31 | 1344,1231, 1028 |
| 10 | 19 | 45 | .21 | 45 | 1344,1231, 1028 |
| 8 | 19 | 60 | .27 | 75 | 1346,1232, 1283 |
| 6 | 19 | 60 | .31 | 108 | 1346,1232, 1283 |
| 4 | 19 | 60 | .36 | 162 | 1346,1232, 1283 |
| 2 | 19 | 60 | .42 | 245 | 1346,1232, 1283 |

**Information on this sheet subject to change without notice.**

**TABLE 9–5A**  Specifications for Type Machine Tool Wire *(Courtesy of Rome Cable Corporation).*

SPEC 5160

1-1-01

Specification

# ROME MACHINE TOOL WIRE

PVC Insulated, 600 Volts
Multi-Rated MTW, THW, AWM, TEW105°C

## 1. SCOPE

1.1   This specification describes multi-rated single conductor, 600 volt Rome PVC Machine Tool Wire suitable for wiring of machine tools, appliances and internal wiring in electrical equipment in control and power circuits where exposed to temperatures not exceeding 90°C in dry locations and at 60°C and lower temperatures where exposed to moisture, oil and coolants-such as cutting oils. Also suitable for use as Type THW for general purpose wiring in accordance with the National Electrical Code.

## 2. APPLICABLE STANDARDS

2.1   The following standards form a part of this specification to the extent specified herein:
2.1.1 Underwriters Laboratories Standard 1063 for Machine Tool Wire.
2.1.2 Underwriters Laboratories Standard 83 for Thermoplastic Insulated Wires.
2.1.3 Underwriters Laboratories Standard 758 for Appliance Wiring Material.
2.1.4 National Fire Protection Association Electrical Standard No. 79 for Metal Working Machine Tools.
2.1.5 Canadian Standards Association Standard C22.2 No. 127 for Equipment Wires.

## 3. CONDUCTORS

3.1   The conductor shall be stranded, annealed, uncoated copper per UL Standards 1063, 83, 758 and National Fire Protection Association Electrical Standard No. 79.

## 4. INSULATION

4.1   The conductor shall be insulated with PVC complying with the physical and electrical requirements of UL Standard 1063, UL Standard 83 and CSA Standard C22.2 No.127 for Type TEW 105°C Thermoplastic Equipment Wire.
4.2   The insulation shall comply with the requirements of UL's VW-1 (Vertical Wire) flame test and CSA's FT-1 flame test.

## 5. IDENTIFICATION

5.1   The wire shall be identified by surface marking indicating manufacturer's identification, conductor size, voltage rating, UL Symbol, VW-1, UL and CSA designations.

## 6. TESTS

6.1   The wire shall be tested in accordance with the requirements of UL Standard 1063 for Machine Tool Wire, UL Standard 83 for Type THW and CSA Standard C22.2 No.127 for Type TEW 105°C Equipment Wire.

## 7. LABELS

7.1   The wire shall bear the UL label for Machine Tool Wire and the CSA label for Equipment Wire.

---

**TABLE 9–5B**   Specifications for Type Machine Tool Wire *(continued from Table 9–5A)* *(Courtesy of Rome Cable Corporation).*

## Line Drop

The resistance of the conducting wires can become so great that it affects the operating characteristics of the equipment. An electrical pressure (voltage) is required to force the current through the resistance of the wires. This extra voltage must be supplied in addition to the rated voltage of the load. This extra pressure is called **line drop**, the voltage drop caused by current flowing though the resistance of the wires.

The *NEC* allows a maximum voltage (line) drop of 3 percent for branch circuits and 5 percent for combined feeder and branch-circuit conductors. To calculate the percentage of voltage drop, the voltage must be measured at the source and at the point farthest from the source while the circuit is fully loaded. The following equation is used:

$$\% \text{ voltage drop} = \frac{E_1 - E_2}{E_1} \times 100 \quad \text{(Eq. 9.11)}$$

where $E_1$ = source voltage

$E_2$ = voltage at the load

### Example 21

The source voltage of a circuit is 120 V, and at the point farthest from the source, a voltmeter indicates 118 V at full load. Calculate the percent voltage drop caused by the circuit conductors.

$$\% \text{ voltage drop} = \frac{E_1 - E_2}{E_1} \times 100$$

$$\% \text{ voltage drop} = \frac{120 - 118}{120} \times 100$$

$$\% \text{ voltage drop} = 1.67\%$$

### Example 22

A lighting load is located 500 ft (152.4 m) from the supply. If the circuit conductors are No. 12 AWG solid copper wire, how much voltage is necessary to force 10 A through the conductors? (See Figure 9–5.)

Five hundred feet (152.4 m) of No. 12 solid copper wire have a resistance of 0.965 Ω at 167°F (75°C). (See Table 9–1.)

Figure 9–6 illustrates an equivalent circuit to Figure 9–5. The following procedure may be used to calculate the voltage (line) drop.

| | | |
|---|---|---|
| $E_{L1} = IR_{L1}$ | $E_{L2} = IR_{L2}$ | $E_{Lt} = E_{L1} + E_{L2}$ |
| $E_{L1} = 10 \times 0.965$ | $E_{L2} = 10 \times 0.965$ | $E_{Lt} = 9.65 + 9.65$ |
| $E_{L1} = 9.65$ V | $E_{L2} = 9.65$ V | $E_{Lt} = 19.3$ V |

To force 10 A through the conductors, 19.3 V are necessary. Therefore, the voltage line drop is 19.3 V.

### Example 23

If the supply voltage for the circuit in Example 22 is 120 V, what is the voltage across the load?

$$120 - 19.3 = 100.7 \text{ V}$$

### Example 24

In Example 22, is the percent voltage drop within the maximum limits permitted by the *NEC*?

$$\% \text{ voltage drop} = \frac{120 - 100.7}{120}$$

$$\% \text{ voltage drop} = 16\%$$

The maximum percent voltage drop permitted by the *NEC* is 3%. Therefore, Example 22 does not conform to the standards.

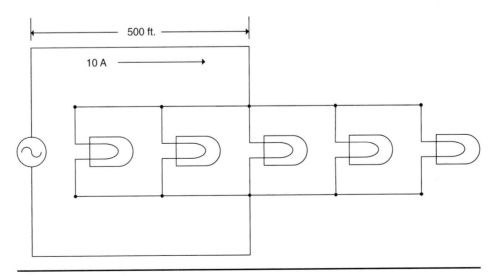

**FIGURE 9–5**  Lighting circuit.

## Example 25

The circuit in Figure 9–7 is wired with No. 10 AWG solid copper wire. Each lamp requires 1 A and is rated at 115 V. What is the voltage drop from the source to group B? Are the lamps in group B delivering full light?

According to Chapter 9, Table 8, of the *NEC*, No. 10 AWG solid copper wire has a resistance of 1.21 Ω per 1000 feet (304.8 m) at a temperature of 167°F (75°C).

The resistance of the circuit conductors between the source and group A is 1.21 ÷ 2 = 0.605 Ω. Ten amperes are flowing through these conductors. The voltage drop between the source and group A is:

$$E_{L1} = IR_{L1}$$
$$E_{L1} = 10 \times 0.605$$
$$E_{L1} = 6.05 \text{ V}$$

The resistance of the circuit conductors between group A and group B may be calculated as follows:

$$1.21 \div 1000 = 0.00121 \ \Omega$$

The resistance of 1 ft of No. 10 AWG copper wire is 0.00121 Ω.

$$0.00121 \times 300 = 0.363 \ \Omega$$

The resistance of the circuit conductors between group A and group B is 0.363 Ω. Five amperes flow in the conductors between group A and group B.

$$E_{L2} = IR_{L2}$$
$$E_{L2} = 5 \times 0.363$$
$$E_{L2} = 1.815 \text{ V}$$

The voltage drop between group A and group B is 1.815 V.

The total voltage (line) drop is:

$$6.05 + 1.815 = 7.865 \text{ V}$$

The voltage across group B is:

$$120 - 7.865 = 112.135 \text{ V}$$

Group B is glowing at less than full light because 115 V are required for the lamps to deliver their full light.

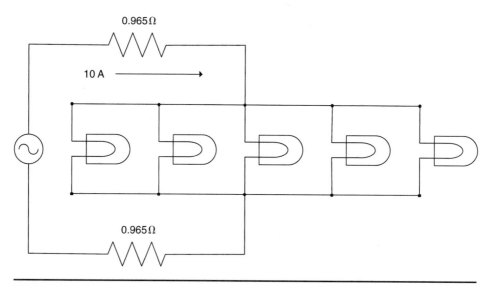

**FIGURE 9–6**    Equivalent circuit for Figure 9–5.

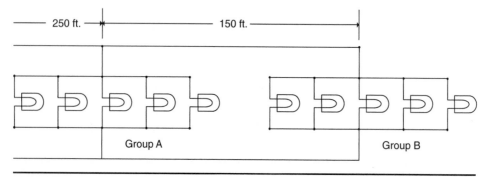

**FIGURE 9–7**    Branch circuit for lighting load.

## Line Loss

**Line loss** is the power dissipated as heat. It is caused by the current flowing through the feeder and branch circuit conductors. Large line losses result in very inefficient distribution systems and greater operating costs for the customer. Line losses should always be kept to a minimum.

With reference to Example 22, the line loss can be calculated in any one of the following ways:

$$P_L = IE_L \qquad \text{(Eq. 9.12)}$$
$$P_L = 10 \times 19.3$$
$$P_L = 193 \text{ W}$$

$$P_L = I^2R_L \qquad \text{(Eq. 9.13)}$$
$$P_L = 100 \times 1.93$$
$$P_L = 193 \text{ W}$$

$$P_L = \frac{E_L^2}{R_L} \qquad \text{(Eq. 9.14)}$$
$$P_L = \frac{372.49}{1.93}$$
$$P_L = 193 \text{ W}$$

The power dissipated as heat is 193 W caused by current flowing through the circuit conductors. This loss of power can be reduced by increasing the size of the circuit conductors.

## Example 26

If the conductors in Example 22 are increased in size to No. 4 AWG, what are the voltage drop and power loss in the circuit conductors?

Number 4 AWG copper wire has a resistance of 0.308 Ω per 1000 ft (1000 ft of wire are used for the circuit).

$$E_L = IR_L \qquad\qquad P_L = I^2R_L$$
$$E_L = 10 \times 0.308 \qquad P_L = 100 \times 0.308$$
$$E_L = 3.08 \text{ V} \qquad\quad P_L = 30.8 \text{ W}$$

The voltage drop is 3.08 V, and the line loss is 30.8 W. This results in a considerable savings of power over the original circuit. The savings in operating costs over a short time will offset the increase in installation costs for the heavier wire.

Using No. 4 copper wire keeps the voltage drop within the NEC standards.

When planning an electrical installation, it is advisable to locate the distribution panel close to the center of the area. This will minimize the distance between the panel and the loads.

The following two requirements must be met when selecting conductor sizes:

- The conductor must be large enough to carry the current without overheating.

- The conductor must be sized to keep the voltage drop within the maximum limits of the NEC.

Electrical equipment is designed to operate best at its rated voltage. Efficiency and operating characteristics are adversely affected when equipment is operated at voltages other than rated values. With a lighting load, a small decrease in voltage causes a great decrease in light output. Motors subject to reduced voltage have poor starting torque and poor speed regulation. They also draw more current than they would at their rated voltage, and they are inefficient and tend to overheat.

# TERMINAL CONNECTIONS AND SPLICES

It is frequently necessary to connect two or more wires together when installing electrical circuits. These connections are called *taps* or **splices**. There are several methods used for connecting wires together. The method selected depends upon the type of installation.

Before conductors are connected together, the insulation must be removed and the conductor thoroughly cleaned. During removal of the insulation, care must be taken to not nick the conductor. A nick in the conductor causes a weak point, which may result in a break when subject to vibration or tension. Insulation may be removed from the conductor with a knife or a wire stripper. When a knife is used, the insulation should be cut at about a 30-degree angle (Figure 9–8). Never cut through the insulation at right angles to the conductor. When using a wire stripper (Figure 9–9), always follow the manufacturer's instructions. The conductors may be cleaned by scraping with a knife or using very fine emery cloth.

## Splices

A splice is a point in the wiring system where two or more wires are joined together. Splices should be avoided whenever possible because they are the weakest point in the wiring system. When it is necessary to make a splice, it must be mechanically and electrically secure.

New insulation must be applied after the splice is completed. This insulation must be equivalent to that which was removed. In other words, the new insulation must contain all the physical properties of the original insulation. If the original insulation was Type T, it means it was made out of a flame-retardant thermoplastic compound

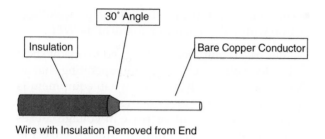

FIGURE 9–8  Wire with insulation removed from end.

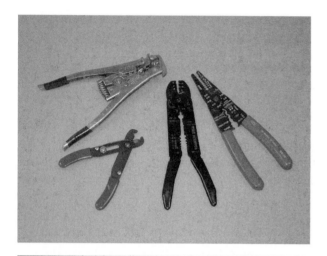

FIGURE 9–9  Four types of wire strippers.

that can safely withstand operating temperatures up to 140 degrees Fahrenheit (60 degrees Celsius) in dry locations. The replacement insulation must meet these standards. If the original insulation is rated at 600 volts, the replacement insulation must be thick enough to prevent current leakage at 600 volts.

Type MTW insulation is listed in the *NEC* as being flame-retardant and moisture-, heat-, and oil-resistant thermoplastic.

Its maximum operating temperature is 140 degrees Fahrenheit (60 degrees Celsius) for wet locations and 194 degrees Fahrenheit (90 degrees Celsius) for dry locations. If a splice is made in a conductor having Type MTW insulation, the replacement insulation must have the previously stated characteristics. To determine the insulation characteristics, refer to the *NEC* or the manufacturer's specifications.

Replacement insulation is generally supplied on a roll and is referred to as *tape*. The most common types are made of a thermoplastic compound.

> ⚠ **CAUTION** Although various types may look alike, their characteristics may differ considerably. It is important to check the manufacturer's specifications to determine if the compound meets the requirements for the particular type of insulation.

When applying tape to a splice, allow each turn to overlap slightly, and also overlap the original insulation. Always keep the tape taut while it is being applied, and apply enough layers to provide electrical and mechanical protection equal to the original insulation.

Other types of replacement insulation are rubber, silicone rubber, varnish cambric, and silk.

## Splicing Devices

Some of the most common types of splicing devices are the *wire nut, sleeve connector, bolt connector,* and *clamp connector*. For any of these splices, the insulation is removed and the conductor is cleaned.

For small wires (No. 14 AWG to No. 10 AWG), the wire nut or sleeve-type connectors are generally used. The wire nut (Figures 9–10A and 9–10B) threads onto the wires. Sleeve connectors are slid over the stripped ends of the conductors and then crimped as shown in Figure 9–11.

The bolt connector is usually used for wires larger than No. 10 AWG (Figure 9–12). The clamp type (Figure 9–13) is generally used for ground connections.

When connecting electrical conductors together, always strive to use conductors of the same material. A chemical reaction between different materials can cause conductors to deteriorate rapidly, producing heat or other hazardous conditions.

Aluminum has become a popular conducting material because of its light weight and low cost. As a result, it is sometimes necessary to connect copper and aluminum conductors together. In order to make such a connection, special precautions must be taken. Special chemical compounds have been developed that must be applied to the conducting materials.

> ⚠ **CAUTION** Follow the manufacturer's directions precisely when applying the compound. Only by adhering to recommended procedures will the installation be safe.

**FIGURE 9–10A** Standard wire nuts.

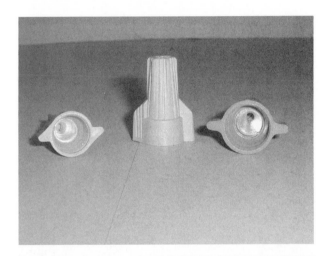

**FIGURE 9–10B** Wire nuts with built-in wrench (wings).

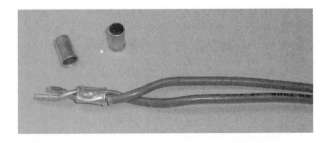

**FIGURE 9–11** Crimp-type sleeve connector.

Devices such as switches, receptacles, and wire-connecting devices are rated for connections to copper, aluminum, or both. These devices are identified as shown in Figure 9–10B. ALR indicates aluminum conductors, and CO indicates copper. The wire nut in Figure 9–10B is manufactured for connections of aluminum to aluminum (ALR/ALR), copper to

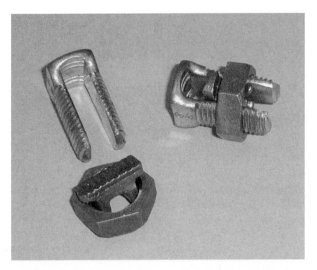

**FIGURE 9–12** Bolt-type (split bolt) wire connector for splicing together two or more conductors.

**FIGURE 9–13** Grounding conductor to grounding electrode clamp.

aluminum (CO/ALR), and copper to copper (CO/CO). (See Article 110 of the *NEC*.)

## Terminal Connections

In most installations, the ends of conductors are connected to terminals of switchboards, panel boards, or devices. The most common method used to connect conductors to devices is shown in Figure 9–14. The device has an upturned lug and setscrew. The insulation is removed from the conductor, the conducting material is cleaned, and a loop is formed in the conductor (Figure 9–15). After the screw is tightened, the excess wire is removed and the screw is tightened again to ensure a good connection (Figure 9–16). Wires should always be wrapped

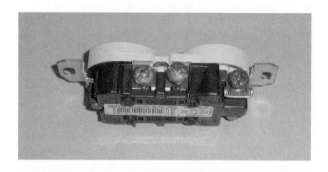

**FIGURE 9-14** Connecting lugs (upturned) on a device (receptacle).

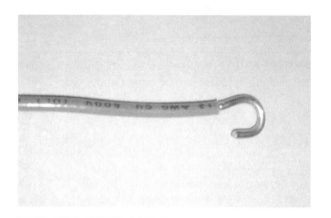

**FIGURE 9-15** Conductor formed into loop for connecting to lug on the device in Figure 9-14. The loop is positioned so as to close around the screw in a clockwise fashion.

around the screw in the direction the screw is tightened.

Conductor ends connected to terminal lugs are illustrated in Figure 9-17. All connections should be made tight and then rechecked to ensure mechanical security and electrical continuity.

> ⚠ **CAUTION** Loose connections can cause overheating of conductors.

## ▌ REVIEW QUESTIONS

*Multiple Choice*

1. A conductor is
   a. any material that offers very little opposition to current flow.
   b. a material with an excess of electrons.
   c. a material with an excess of protons.
   d. a material with an excess of neutrons.

**FIGURE 9-16** Connecting wires to the device in Figure 9-14.

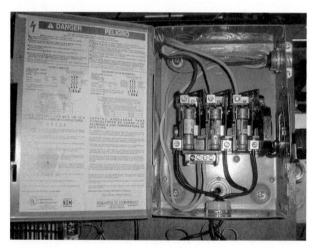

**FIGURE 9-17** Proper dressing (squaring) of conductors within an equipment enclosure.

2. Most conductors used in the electrical industry are
   a. solids.
   b. liquids.
   c. gases.
   d. none of the above.

3. One mil is equal to
   a. 1/10 in.
   b. 1/100 in.
   c. 1/1000 in.
   d. 1/10,000 in.

4. Circular conductors are usually measured in
   a. mils.
   b. circular mils.
   c. square mils.
   d. circular inches.

5. Areas of bus bars are usually expressed in
   a. circular mils.
   b. square mils.
   c. inches.
   d. mils.

6. The standard scale for measuring wires is called the
   a. American wire gauge.
   b. circular mil gauge.
   c. Universal wire gauge.
   d. International wire gauge.

7. An instrument used to measure rectangular conductors is the
   a. millimeter.
   b. Rectangular wire gauge.
   c. micrometer.
   d. American wire meter.

8. A wire 1 mil in diameter and 1 ft long is called a
   a. foot-mil of wire.
   b. mil-foot of wire.
   c. circular mil-foot of wire.
   d. square mil-foot of wire.

9. The resistivity of copper building wire is
   a. 8.5 Ω.
   b. 6.2 Ω.
   c. 10.4 Ω.
   d. 12.3 Ω.

10. The resistance of a conductor depends upon the
    a. material from which it is made.
    b. operating temperature.
    c. room temperature.
    d. both a and b.

11. The temperature coefficient of a conductor is the
    a. increase in temperature from the ambient temperature to the operating temperature.
    b. amount by which the resistance of a material changes per degree change in temperature for each ohm of resistance.
    c. percentage of temperature increase.
    d. percentage of resistance change.

12. The amount of heat generated by current flowing in a conductor depends upon the
    a. amount of current flowing.
    b. resistance of the conductor.
    c. conductor insulation.
    d. both a and b.

13. The amount of current a conductor can safely carry is called its
    a. volume.
    b. ampacity.
    c. current ability.
    d. safety factor.

14. The ambient temperature of a conductor is
    a. the same as the operating temperature.
    b. lower than the operating temperature.
    c. higher than the operating temperature.
    d. none of the above.

15. The most common insulating material used on an electrical conductor is
    a. plastic.
    b. rubber.
    c. varnish.
    d. cotton.

16. A feeder is
    a. the conductors between the service equipment and the final branch-circuit overcurrent device.
    b. all the conductors between the final circuit overcurrent protective device and the load.
    c. the conductors entering the building.
    d. both a and b.

17. Voltage (line) drop is the
    a. current loss in the conductors.
    b. voltage loss in the conductors.
    c. power loss in the conductors.
    d. voltage applied to the circuit.

18. The *National Electrical Code* allows a maximum of a
    a. 3% voltage drop for branch circuits.
    b. 3% voltage drop for feeders.
    c. 6% voltage drop for combined feeder and branch circuits.
    d. all of the above.

19. Line loss is the
    a. current loss in the conductors.
    b. voltage loss in the conductors.
    c. power loss in the conductors.
    d. none of the above.

20. The most common method used to connect conductors to devices is the use of the
    a. wire nut.
    b. upturned lug.
    c. screw connector.
    d. device connector.

*Give Complete Answers*

1. Define the unit of measurement for electrical wires.

2. In what unit are bus bars commonly measured?

3. Describe an American wire gauge.

4. Describe the instrument called a micrometer.

5. Define the term *resistivity*.

6. Describe the effect of temperature change on the resistance of aluminum wire.

7. Does an increase in temperature affect the resistance of constantan?

8. How does an increase in temperature affect the resistance of carbon?

9. Name two of the most common materials used in building wiring.

10. Define the terms *voltage drop* and *line loss*.

11. What is an upturned lug?

12. What is the most common effect of loose connections?

13. List four types of connectors.

14. What is a splice?

15. Describe one method used to remove insulation from a wire.

*Solve each problem, showing the method used to arrive at the solution.*

1. Calculate the circular mil area of a wire that measures 1/4 in. in diameter.

2. What is the area of the wire in Problem 1, in circular millimeters?

3. A wire has an area of 250,000 circular mils. What is its diameter, in mils?

4. What is the diameter of the wire in Problem 3, in millimeters?

5. Calculate the square mil area of a bus bar 1/4-in. thick and 3/4-in. wide.

6. What is the circular mil area of the bus bar in Problem 5?

7. Calculate the square millimeter area of the conductor in Problem 5.

8. Calculate the length of a copper building wire 1/4 in. in diameter, if it has a resistance of 0.25 $\Omega$.

9. What is the resistance of 500 ft (152.4 m) of No. 12 AWG aluminum wire at a temperature of 20°C (68°F)?

10. What is the resistance of 500 ft (152.4 m) of No. 12 AWG copper wire at a temperature of 20°C (68°F)?

11. The resistance of a tungsten (annealed) filament is 10 $\Omega$ at 0°C. What is its resistance at 1500°C?

12. The resistance of a coil of aluminum wire is 20 W at 20°C. What is its resistance at 70°C?

13. If the coil in Problem 12 is wound with copper wire instead of aluminum, and all other facts remain the same, what is its resistance at 70°C?

14. The copper windings of a motor have a resistance of 10 $\Omega$ when the motor is operating at no load. The resistance is increased to 13 $\Omega$ when the motor is operating at full load. If the no-load temperature of the windings is 30°C, what is the temperature rise of windings?

15. Three No. 10 AWG aluminum wires covered with TW insulation are installed in a 1/2-in. conduit. If the ambient temperature is 100°F (37.8°C), what is the ampacity of the conductors?

16. If three more No. 10 AWG aluminum conductors are installed in the raceway in Problem 15, what is the ampacity of all the conductors in the raceway?

17. Table 310.16 of the *NEC* lists the ampacity of No. 6 AWG, type TW aluminum wire as 40 A. What is the voltage drop for 500 ft (152.4 m) of this wire if the wires are conducting at full capacity?

18. For the circuit in Problem 17, if the source voltage is 120 V, what is the percent voltage drop?

19. What is the line loss for the conductors in Problem 17?

20. The resistance of a coil wound with aluminum wire is 2 $\Omega$ at 20°C. What is its resistance at 80°C?

# Wiring Methods

**OBJECTIVES**

After studying this chapter, the student will be able to:

■ Explain the purpose of electrical codes and of the National Fire Protection Association.

■ List the type of installations covered by the **National Electrical Code**.

■ List and describe some common wiring methods used in the electrical industry.

■ Describe the conditions of use of various wiring methods.

## ELECTRICAL CODES

Electrical codes are rules and regulations governing the installation of electrical systems. In general, the purpose of an electrical code is to ensure safe installations.

There have been many electrical codes developed on local, state, and national levels. Most states have enacted laws setting forth minimum requirements for electrical safety. By far, the most accepted standards are those developed by the National Fire Protection Association (NFPA). The primary goal of the NFPA is "to promote the science and improve the methods of fire protection."

The NFPA publishes materials pertaining to safe practices in the electrical industry. All electrical personnel should become familiar with this literature. The *National Electrical Code* is the most frequently used publication. These are some other popular publications:

- *Life Safety Code Handbook*
- *National Fire Alarm Code*
- *NFPA 70E Handbook For Electrical Safety in the Workplace*
- *NFPA 79 Electrical Standard for Industrial Machinery*
- *NFPA 5000 Building Construction and Safety Code*
- *NFPA 780 Standard for the Installation of Lightning Protection Systems*

The wiring methods and installations discussed in this text are covered by the *National Electrical Code*. Article 90 of the *NEC* lists the types of installations covered by the *Code*. The types of installations include electrical equipment within or on public or private buildings and some other types of premises. Installations that are not covered by the *NEC*, such as the wiring in ships, aircraft, underground mines, and automobiles, are covered by other codes designed specifically for each purpose.

In the design and installation of electrical systems, safety must be the top priority. Reference should be made to the *NEC* and other safety regulations to determine the methods and materials to be used.

## ARMORED AND METAL-CLAD CABLE

Armored cable is also known as Type AC cable. According to the *National Electrical Code*, AC cable is *a*

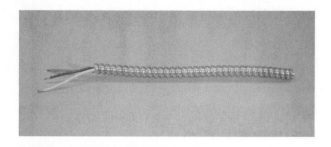

**FIGURE 10–1**  Metal-clad (type MC) cable.

*fabricated assembly of insulated conductors in a flexible metallic enclosure,\* as stated in Article 320.2. In addition, the *NEC* further defines AC cable, according to Article 320.100:

> *Type AC cable shall have an armor of flexible metal tape and shall have an internal bonding strip of copper or aluminum in intimate contact with the armor for its entire length.\**

Article 320.104 sets the specifications for the conductors as:

> *Insulated conductors shall be of a type listed in Table 310.13(A) or those identified for use in this cable. In addition, the conductors shall have an overall moisture-resistant and fire-retardant fibrous covering. For Type ACT, a moisture-resistant fibrous covering shall be required only on the individual conductors.\**

Article 320 of the *NEC* contains additional information relating to AC cable.

Similar to armored cable is metal-clad cable, or MC cable. According to *NEC* Article 330.2, MC cable is:

> *A factory assembly of one or more insulated circuit conductors with or without optical fiber members enclosed in an armor of interlocking metal tape, or a smooth or corrugated metallic sheath.\**

Article 330.104 gives the specifications for the conductors as:

> *The conductors shall be of copper, aluminum, or copper-clad aluminum, solid or stranded. The minimum conductor size shall be 18 AWG copper and 12 AWG aluminum or copper-clad aluminum.\**

For additional information on MC cable, look in Article 330 of the *NEC*. Figure 10–1 shows a photograph of Type MC cable.

---

\*Reprinted with permission from NFPA 70-2008, *National Electrical Code®* Copyright© 2007, National Fire Protection Association, Quincy, MA 02269. This reprinted material is not the referenced subject which is represented only by the standard in its entirety.

# NONMETALLIC-SHEATHED CABLE

Perhaps the most common name for nonmetallic-sheathed cable is Romex. The *NEC* defines nonmetallic-sheathed cable in *Article 334.2* and states:

*A factory assembly of two or more insulated conductors enclosed within an overall nonmetallic jacket.\**

Article 334.104 gives the specifications for the conductors:

*The 600 V insulated conductors shall be sizes 14 AWG through 2 AWG copper conductors or sizes 12 AWG through 2 AWG aluminum or copper-clad aluminum conductors. The communications conductors shall comply with Part V of Article 800.\**

Article 334.112 of the 2008 edition of the *NEC* states that the insulated power conductors contained within the nonmetallic-sheathed cable to have insulation rated at 90 degrees Celsius (194 degrees Fahrenheit). In the fine print note (FPN), types NM, NMC, and NMS cable that meet the 90 degrees Celsius requirement will be identified by the markings NM-B, NMC-B, and NMS-B.

In addition, the NM cable is now color coded for easy conductor size identification. NM cable containing 14 AWG conductors will have an outer sheath that is white in color. If the conductors are 12 AWG, the outer sheath will be yellow in color. An orange-colored outer sheath denotes 10 AWG conductors. Finally, a black outer sheath indicates that the conductors are either 8 AWG or 6 AWG.

Additional information on nonmetallic-sheathed cable can be found in Article 334 of the *NEC*. Figure 10–2 shows several types of NM-B cable. Figure 10–3 shows a tool for stripping the outer sheath from the conductors without damage. Figure 10–4 shows different types of staples that are used to secure the NM cable to a surface.

# SERVICE-ENTRANCE CABLE

Service-entrance cable is defined in *Article 338.2* of the 2008 *NEC* as:

*A single conductor or multiconductor assembly provided with or without an overall covering, primarily used for services, and of the following types:*

*Type SE. Service-entrance cable having a flame-retardant, moisture-resistant covering.*

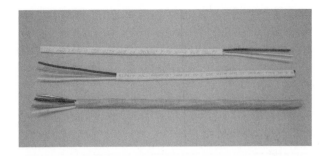

**FIGURE 10–2**  14 AWG, 12 AWG, and 10 AWG type NM-B cable.

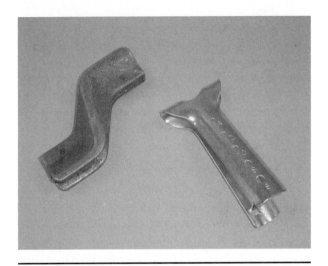

**FIGURE 10–3**  Two styles of NM cable strippers.

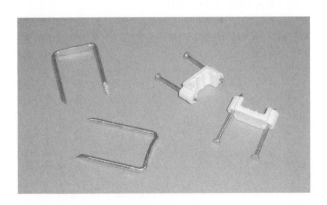

**FIGURE 10–4**  NM cable staples.

*Type USE. Service-entrance cable, identified for underground use, having a moisture-resistant covering, but not required to have a flame-retardant covering.\**

Additional information on service-entrance cable can be found in Article 338 of the *NEC*. A photograph of Type SE service-entrance cable appears in Figure 10–5.

---

\*Reprinted with permission from NFPA 70-2008, *National Electrical Code®* Copyright© 2007, National Fire Protection Association, Quincy, MA 02269. This reprinted material is not the referenced subject which is represented only by the standard in its entirety.

**FIGURE 10–5**   Type SE cable.

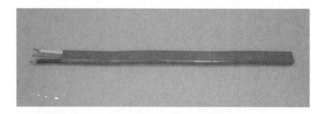

**FIGURE 10–6**   Type UF cable.

# UNDERGROUND FEEDER AND BRANCH-CIRCUIT CABLE

Underground feeder and branch-circuit cable is known as Type UF cable. The specifications for Type UF cable can be found in Article 340.2 of the *NEC*. The *NEC* definition is:

> *A factory assembly of one or more insulated conductors with an integral or an overall covering of nonmetallic material suitable for direct burial in the earth.**

Other information pertaining to type UF cable can be found in Article 340 of the *NEC*. Figure 10–6 shows a photograph of type UF cable.

# RACEWAYS

The *National Electrical Code* defines a raceway as:

> *An enclosed channel of metal or nonmetallic materials designed expressly for holding wires, cables, or busbars, with additional functions as permitted in this* Code. *Raceways include, but are not limited to, rigid metal conduit, rigid nonmetallic conduit, intermediate metal conduit, liquidtight flexible conduit, flexible metallic tubing, flexible metal conduit, electrical nonmetallic tubing, electrical metallic tubing, underfloor raceways, cellular concrete floor raceways, cellular metal floor raceways, surface raceways, wireways, and busways.**

For the purpose of this text, we will focus our attention on the type of raceways more commonly referred to as conduit. This will include electrical nonmetallic tubing (ENT), intermediate metal conduit (IMC), rigid metal conduit (RMC), liquidtight flexible conduit (LFMC), flexible metallic tubing (FMT), flexible metal conduit (FMC), and electrical metallic tubing (EMT).

Article 362 of the *NEC* addresses electrical nonmetallic tubing, otherwise known as ENT. ENT is described as a nonmetallic, pliable, corrugated raceway of circular cross-section with integral or associated couplings, connectors, and fittings for the installation of electrical conductors. ENT is composed of a material that is resistant to moisture and chemical atmospheres and is flame retardant. A pliable raceway is a raceway that can be bent by hand with a reasonable force but without other assistance. Article 362.10 of the *NEC* explains the permitted uses of ENT, while Article 362.12 explains the uses or areas for which ENT is not approved. Intermediate metal conduit (IMC) is covered in *NEC* Article 342. IMC is described as a steel threadable raceway of circular cross-section designed for the physical protection and routing of conductors and cables and for use as an equipment grounding conductor when installed with its integral or associated coupling and appropriate fittings. IMC has thicker wall construction than EMT, but thinner wall construction than RMC. IMC is permitted to be threaded. Article 342.10 of the *NEC* lists the permitted uses of IMC.

Article 344 of the *NEC* addresses rigid metal conduit, otherwise known as RMC. RMC is described as a threadable raceway of circular cross-section designed for the physical protection and routing of conductors and cables and for use as an equipment grounding conductor when installed with its integral or associated coupling and appropriate fittings. RMC has thicker wall construction than either EMT or IMC. RMC is permitted to be threaded and may be constructed of galvanized steel or aluminum. Article 344.10 explains the permitted uses of RMC.

Article 358 of the *NEC* addresses electrical metallic tubing, otherwise known as EMT. EMT is described as a listed, metallic tubing of circular cross-section approved for the installation of electrical conductors when joined together with listed fittings. EMT has the thinnest wall construction when compared to IMC and RMC. Article 358.10 explains the permitted uses of EMT, while Article 358.12 identifies the uses not permitted.

Flexible metallic tubing is covered in *NEC* Article 360. FMT is described as a raceway that is circular in cross-section, flexible, metallic, and liquidtight without a nonmetallic jacket. Article 360.10 of the *NEC* lists the permitted uses of FMT, and Article 360.12 lists the uses where FMT is not permitted.

Article 348 of the *NEC* addresses flexible metal conduit, otherwise known as FMC. FMC is described as a raceway of circular cross-section made of helically wound, formed, interlocked metal strip. Article 348.10 explains the permitted uses of FMC, while Article 348.12 identifies the uses not permitted.

Liquidtight flexible metal conduit (LFMC) and liquidtight flexible nonmetallic conduit (LFNC) are covered in *NEC* Articles 350 and 356 respectively. Article 350.2 defines LFMC as *a raceway of circular cross section having an outer liquidtight, nonmetallic, sunlight-resistant jacket over an inner flexible metal core with associated couplings, connectors, and fittings for the installation of electric conductors.* Article 356.2 defines LFNC as a raceway of circular cross section of various types as follows:

(1) *A smooth seamless inner core and cover bonded together and having one or more reinforcement layers between the core and covers, designated as Type LFNC-A.*

(2) *A smooth inner surface with integral reinforcement within the conduit wall, designated as Type LFNC-B.*

(3) *A corrugated internal and external surface without integral reinforcement within the conduit wall, designated as Type LFNC-C.*

*LFNC is flame resistant and with fittings and is approved for the installation of electrical conductors.*

*FPN: FNMC is an alternative designation for LFNC.*

The permitted and not-permitted uses for LFMC are covered in Articles 350.10 and 350.12 respectively. Articles 356.10 and 356.12 respectively cover the permitted and not-permitted uses for LFNC.

# FITTINGS, CONDUIT BODIES, AND BOXES

Cable and conduits make up only part of the complete wiring methods employed in most facilities.

To contain the cables and conductors when splicing or terminating to a device, fittings, conduit bodies, and boxes are used. In this section, we will focus on some of the more common fittings, conduit bodies, and boxes. These are by no means the only types available. You will need to familiarize yourself with the many vendors who manufacture and sell these items. There are many types to choose from and will quite often satisfy a unique need that you may encounter.

## Fittings

The *NEC* defines a fitting as:

> *An accessory such as a locknut, bushing, or other part of a wiring system that is intended primarily to perform a mechanical rather than an electrical function.**

The type of cable, conductor, and/or raceway being used will determine the type of fitting needed for the installation at hand. Figures 10–7, 10–8, 10–9, and 10–10 show some of the types of fittings used when working with an armored cable (AC) and a flexible metal conduit (FMC).

Some metal-clad (MC) cables have an outer jacket fully enclosing the corrugated metal tube. This is called jacketed metal-clad cable. The outer jacket provides protection against corrosion. When using this type of cable, fittings different from those used on AC or MC cables must be used. Some of these fittings appear in Figure 10–11.

Nonmetallic-sheathed cable requires yet another type of fitting. These fittings may also be used with Type SE cable. Figure 10–12 shows some of these types of fittings.

It should be noted, however, that when using service-entrance cable, additional fittings are used. These include entrance caps, sill plates, cable straps, and some type of sealing compound to keep moisture from entering. Examples of these items are shown in Figures 10–13, 10–14, and 10–15.

As mentioned previously, different types of raceways will also determine the type of fitting needed. For example, intermediate metal conduit and rigid metal conduit use threaded fittings and hubs for securing the conduits to bodies and boxes. In addition, there are specific nipples, enlargers, reducers, couplings, and bushings that must be used. Figures 10–16 through 10–25 show examples of these types of fittings.

Electrical metallic tubing (EMT) does not use threaded fittings. Therefore, when working with

---

*Reprinted with permission from NFPA 70-2008, *National Electrical Code*® Copyright© 2007, National Fire Protection Association, Quincy, MA 02269. This reprinted material is not the referenced subject which is represented only by the standard in its entirety.

# BC

## Speedlock™

### Steel Box Connectors

### Type AMC

**Use:**
For connecting MC, AC, and Flex Cable

**Features:**
• No locknut required.
• Single-unit or duplex construction with captive clamp.
• Connects various sizes of MC, AC and Flex RW cables with just four fittings.
• Easy to install. Just tilt, insert and snap down.
• Integral red plastic insulated throat bushing provides maximum protection for wire installation.

**Materials/Finish:**
Steel/Zinc Plated

**Applicable Third Party Standards:**
UL Standard: 514B
CSA Standard: C22.2 No. 18
Fed. Spec: W-F-406E

Exceeds UL specifications for electrical continuity and is suitable as a grounding means under the NEC 350-5 and for hazardous locations under NEC 501-4(b), Class 1, Div. 2.

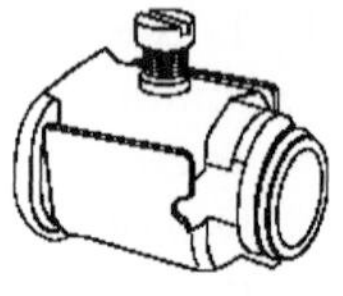

**AMC-50** 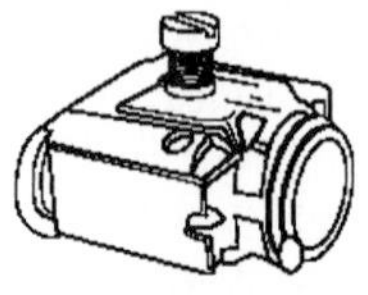          **AMC-52**

| Trade Size (inches) | Catalog Number | Dimensions in Inches | | | Wt. Lbs. Per 100 | Ctn. Qty. | Std. Pkg. |
|---|---|---|---|---|---|---|---|
| | | A | B | C | | | |
| 1/2 Single | **AMC-50** | 1-3/8 | 1 | 13/32 | 6.5 | 50 | 500 |
| 3/4 Single | **AMC-75** | 1-5/8 | 1-5/16 | 9/16 | 7.5 | 25 | 250 |
| 1/2 KO; 3/4 capacity | **AMC-5075** | 1-5/8 | 1-5/16 | 9/16 | 7.5 | 25 | 250 |
| 1/2 Duplex | **AMC-52** | 1-9/16 | 1-23/32 | 9/16 | 12.0 | 25 | 250 |

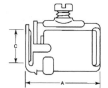

| AMC SELECTION CHART | | | | |
|---|---|---|---|---|
| Cable Type | AMC-50 1/2" KO | AMC-5075 1/2" KO 3/4" Capacity | AMC-75 3/4" KO | AMC-52** 1/2" KO Duplex |
| AC Steel | 14/2 14/3 14/4 12/2 12/3 12/4 10/2 10/3 | 10/4 10/5 8/3 8/4 6/3 | 10/4 10/5 8/3 8/4 6/3 | 14/2 14/3 14/4 12/2 12/3 12/4 |
| HFC Steel Health Care Facilities | 14/2 14/3 12/2 12/3 10/2 | | | 14/2 14/3 12/2 12/3 10/2 |
| AC Aluminum Interlocked | 14/2 14/3 14/4 12/2 12/3 12/4 | 10/4 8/3 8/4 | 10/4 8/3 8/4 | 14/2 14/3 14/4 12/2 12/3 12/4 |
| MC Steel | *14/2 *14/3 *14/4 *12/2 *12/3 *12/4 *10/2 | *10/3 *10/4 *8/2 *8/3 *6/2 | *10/3 *10/4 *8/2 *8/3 *6/2 | 14/2 14/3 14/4 12/2 12/3 12/4 10/2 |
| MC Aluminum Interlocked | 14/2 14/3 14/4 12/2 12/3 12/4 10/2 | 10/3 10/4 8/2 | 10/3 10/4 8/2 | 14/2 14/3 14/4 12/2 12/3 12/4 |
| MC Aluminum | 14/3 14/4 *12/2 *12/3 12/4 10/2 10/3 | | | |
| MC Aluminum Flat | 12/3 | | | |
| Flex RW | 3/8 Steel & Al. | 1/2 Steel | 1/2 Steel | |

*Concrete tight when taped and used with PVC jacketed cord.
**Accepts 2 cables of each size listed.

Effective October, 2001
Copyright 2001  Printed in U.S.A.

**PAGE 2**

7770 N. Frontage Road
Skokie, Illinois 60077

**FIGURE 10–7**  AC and FMC cable connectors *(Courtesy of Appleton Electric).*

# BC

LISTED
File #
E14813

BC

## Straight Armored Cable
## and Flexible Metal Conduit Connectors

| Catalog No. | Flexible Conduit Size (In.) | Cable Opening (In.) Max. | Cable Opening (In.) Min. | Bushed Hole Dia. (In.) | Dimensions in Inches A | B | C | Wt. Lbs. Per 100 | Ctn. Qty. | Std. Pkg. |
|---|---|---|---|---|---|---|---|---|---|---|
| **Pressure Cast – Screw In – Flexible Metal Conduit** | | | | | | | | | | |
| SGC-38DC | 3/8 (1/2" KO) | | | | 15/16 | 3/8 | 1-1/32 | 5.0 | 50 | 500 |
| SGC-50DC | 1/2 | | | | 1 | 3/8 | 1-1/16 | 6.0 | 50 | 500 |
| SGC-75DC | 3/4 | | | | 1-1/4 | 3/8 | 1-9/16 | 9.0 | 25 | 250 |
| **Pressure Cast – Armored Cable\* and Flexible Metal Conduit** | | | | | | | | | | |
| 7226V†† | 3/8 (1/2" KO) | .625 | .421 | | 7/16 | 3/8 | 1-1/16 | 5.0 | 50 | 500 |
| **Pressure Cast – Combination – 3/8" Flexible Flexible Metal Conduit to 1/2" EMT – Couplings** | | | | | | | | | | |
| TW38-50DC† | 3/8 Flex | .625 | .218 | | 7/8 | 13/32 | 1-1/4 | 7.5 | 25 | 100 |
| **Pressure Cast – Armored Ground Wire and Nonmetallic Sheathed Cable\* Connector** | | | | | | | | | | |
| 7287V | 1/2 | .937 | .406 | 19/32 | 1-3/32 | 13/32 | 1-7/8 | 17.0 | 25 | 100 |
| 7288V | 1/2 (3/4" KO) | .937 | .593 | 25/32 | 1-3/32 | 7/16 | 1-7/8 | 19.0 | 25 | 100 |
| 7289V | 3/4 | 1.125 | .718 | 25/32 | 1-5/32 | 1/2 | 2-1/16 | 19.0 | 25 | 100 |
| 7230V | 3/8 (1/2" KO) | .625 | .343 | | 5/8 | 7/16 | 1 | 7.0 | 50 | 500 |
| **Duplex – Malleable – Armored Cable\* and Flexible Metal Conduit** | | | | | | | | | | |
| 7240V | 3/8 (1/2" KO) | .625 | .343 | 13/32 | 31/32 | 15/32 | 1-3/4 | 21.0 | 25 | 100 |

† UL Listed and CSA Certification not applicable.
†† Not CSA Certified
\* Select Connectors by Physical Dimensions of Cable as Dimensions May Vary Between Manufacturers.

Effective October, 2001
Copyright 2001  Printed in U.S.A.

PAGE 4

7770 N. Frontage Road
Skokie, Illinois 60077

**FIGURE 10–8**  AC and FMC cable connectors *(Courtesy of Appleton Electric).*

CHAPTER 10

BC

## Straight Armored Cable and Flexible Metal Conduit Connectors

| Catalog No. | Flexible Conduit Size (In.) | Cable Opening (In.) Max. | Cable Opening (In.) Min. | Bushed Hole Dia. (In.) | Dimensions in Inches A | B | C | Wt. Lbs. Per 100 | Ctn. Qty. | Std. Pkg. |
|---|---|---|---|---|---|---|---|---|---|---|
| **Malleable – Armored Cable\* and Flexible Metal Conduit** |
| 7265V | 3/8† | .625 | .468 | | 17/32 | 7/16 | 1-5/32 | 7.0 | 50 | 500 |
| 7480V | 3/8† | .687 | .531 | 5/8 | 27/32 | 7/16 | 1-5/32 | 11.4 | 25 | 100 |
| 7481V | 1/2 | .937 | .750 | 5/8 | 7/8 | 7/16 | 1-1/2 | 12.0 | 25 | 100 |
| 7482V | 3/4 | 1.125 | .906 | 13/16 | 1-1/16 | 7/16 | 1-21/32 | 17.5 | 25 | 100 |
| **Malleable – One Screw – Flexible Metal Conduit** |
| 7483 | 1 | 1.380 | 1.031 | 1 | 1-1/8 | 1/2 | 1-31/32 | 32.0 | 5 | 25 |
| 7484 | 1-1/4 | 1.630 | 1.375 | 1-3/8 | 1-1/8 | 9/16 | 2-1/4 | 43.0 | 5 | 25 |
| **Malleable – Two Screw – Flexible Metal Conduit** |
| 7485 | 1-1/2 | 1.950 | 1.625 | 1-19/32 | 1-27/32 | 9/16 | 2-11/16 | 67.0 | 5 | 25 |
| 7486 | 2 | 2.450 | 2.156 | 2 | 2 | 3/4 | 3-5/16 | 138.0 | 5 | 10 |
| 7487 | 2-1/2 | 3.062 | 2.718 | 2-7/16 | 2 | 13/16 | 4-3/32 | 169.0 | 1 | 10 |
| 7488 | 3 | 3.563 | 3.031 | 3-1/16 | 2-3/16 | 15/16 | 4-9/16 | 230.0 | 1 | 5 |
| 7489 | 3-1/2 | 4.125 | 3.750 | 3-1/2 | 2-7/16 | 1 | 5-1/8 | 300.0 | – | 1 |
| 7490 | 4 | 4.665 | 4.187 | 4 | 2-11/16 | 1 | 5-11/16 | 400.0 | – | 1 |
| **Insulated** |
| **Malleable – Armored Cable\* and Flexible Metal Conduit††** |
| 7480VI | 3/8† | .687 | .531 | 5/8 | 27/32 | 7/16 | 1-5/32 | 11.4 | 25 | 100 |
| 7481VI | 1/2 | .937 | .750 | 5/8 | 7/8 | 7/16 | 1-1/2 | 12.0 | 25 | 100 |
| 7482VI | 3/4 | 1.125 | .906 | 13/16 | 1-1/16 | 7/16 | 1-21/32 | 17.5 | 25 | 100 |
| **Malleable – One Screw – Flexible Metal Conduit††** |
| 7483I | 1 | 1.380 | 1.031 | 1 | 1-1/8 | 1/2 | 1-31/32 | 32.0 | 5 | 25 |
| 7484I | 1-1/4 | 1.630 | 1.375 | 1-3/8 | 1-1/8 | 9/16 | 2-1/4 | 43.0 | 5 | 25 |
| **Malleable – One Screw – Flexible Metal Conduit††** |
| 7485I | 1-1/2 | 1.950 | 1.625 | 1-19/32 | 1-27/32 | 9/16 | 2-11/32 | 67.0 | 5 | 25 |
| 7486I | 2 | 2.450 | 2.156 | 2 | 2 | 3/4 | 3-5/16 | 138.0 | 5 | 10 |
| 7487I | 2-1/2 | 3.062 | 2.718 | 2-7/16 | 2 | 13/16 | 4-3/32 | 169.0 | 1 | 10 |
| 7488I | 3 | 3.563 | 3.031 | 3-1/16 | 2-3/16 | 15/16 | 4-9/16 | 230.0 | 1 | 5 |
| 7489I | 3-1/2 | 4.125 | 3.750 | 3-1/2 | 2-7/16 | 1 | 5-1/8 | 300.0 | – | 1 |
| 7490I | 4 | 4.665 | 4.187 | 4 | 2-11/16 | 1 | 5-11/16 | 400.0 | – | 1 |

† 1/2" KO
†† Not CSA Certified
\* Select Connectors by Physical Dimensions of Cable as Dimensions May Vary Between Manufacturers.

Effective October, 2001
Copyright 2001  Printed in U.S.A.

7770 N. Frontage Road
Skokie, Illinois 60077

**PAGE 5**

**FIGURE 10–9**  AC and FMC cable connectors *(Courtesy of Appleton Electric).*

# BC

LISTED
File #
E14813
E14814

## 45° and 90° Armored Cable and Flexible Metal Conduit Connectors

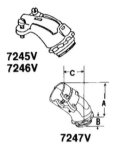

**7245V**
**7246V**

**7247V**

**7380V**

**7381V-7384**

**7385-7388**

**7380VI**

**7381VI-7384I**

**7385I-7388I**

| Catalog No. | Flexible Conduit Size (In.) | Cable Opening (In.) Max. | Cable Opening (In.) Min. | Bushed Hole Dia. (In.) | Dimensions in Inches A | B | C | Wt. Lbs. Per 100 | Ctn. Qty. | Std. Pkg. |
|---|---|---|---|---|---|---|---|---|---|---|
| **45°** | | | | | | | | | | |
| **Steel/Malleable – Armored Cable\* and Flexible Metal Conduit** | | | | | | | | | | |
| 7245V | 3/8 (1/2" KO) | .625 | .375 | 17/32 | 1-5/8 | 7/16 | 1 | 16.0 | 25 | 100 |
| 7246V† | 1/2 | .937 | .812 | 17/32 | 1-11/16 | 7/16 | 1-1/4 | 22.5 | 25 | 100 |
| 7247V† | 3/4 | 1.125 | .953 | 13/16 | 1- 3/4 | 7/16 | 1-1/2 | 28.0 | 10 | 50 |
| **90°** | | | | | | | | | | |
| **Steel/Malleable – Armored Cable\* and Flexible Metal Conduit†** | | | | | | | | | | |
| 7380V | 3/8 (1/2" KO) | .625 | .312 | – | 3/4 | 7/16 | 1-1/4 | 14.6 | 25 | 100 |
| **Steel – Armored Cable\* and Flexible Metal Conduit** | | | | | | | | | | |
| 7381V | 1/2 | .937 | .593 | – | 1-1/2 | 7/16 | 2 | 22.6 | 25 | 100 |
| **Steel/Malleable – Armored Cable\* and Flexible Metal Conduit** | | | | | | | | | | |
| 7382V | 3/4 | 1.125 | .812 | – | 1-11/16 | 7/16 | 2 | 33.0 | 10 | 50 |
| **Malleable – Flexible Metal Conduit** | | | | | | | | | | |
| 7383 | 1 | 1.380 | 1.187 | – | 2-5/16 | 1/2 | 2-1/8 | 56.0 | 5 | 50 |
| 7384 | 1-1/4 | 1.781 | 1.468 | – | 2-13/16 | 9/16 | 2-3/4 | 118.0 | 5 | 25 |
| 7385 | 1-1/2 | 2.000 | 1.718 | – | 2-31/32 | 9/16 | 3-1/2 | 176.0 | 5 | 25 |
| 7386 | 2 | 2.437 | 2.000 | – | 4-1/8 | 3/4 | 4-3/4 | 290.0 | 1 | 5 |
| 7387 | 2-1/2 | 3.000 | 2.656 | – | 4-7/8 | 1-1/16 | 5-7/8 | 450.0 | – | 1 |
| 7388 | 3 | 3.500 | 3.156 | – | 5-1/2 | 1-1/16 | 6-5/8 | 612.0 | – | 1 |
| **90° Insulated** | | | | | | | | | | |
| **Steel/Malleable – Armored Cable\* and Flexible Metal Conduit†** | | | | | | | | | | |
| 7380VI | 3/8 (1/2" KO) | .625 | .312 | – | 3/4 | 7/16 | 1-1/4 | 14.6 | 25 | 100 |
| **Steel – Armored Cable\* and Flexible Metal Conduit†** | | | | | | | | | | |
| 7381VI | 1/2 | .937 | .593 | – | 1-1/2 | 7/16 | 2 | 22.6 | 25 | 100 |
| **Steel/Malleable – Armored Cable\* and Flexible Metal Conduit†** | | | | | | | | | | |
| 7382VI | 3/4 | 1.125 | .812 | – | 1-11/16 | 7/16 | 2 | 33.0 | 10 | 50 |
| **Malleable – Flexible Metal Conduit†** | | | | | | | | | | |
| 7383I | 1 | 1.380 | 1.187 | – | 2-5/16 | 1/2 | 2-1/8 | 56.0 | 5 | 50 |
| 7384I | 1-1/4 | 1.781 | 1.468 | – | 2-13/16 | 9/16 | 2-3/4 | 118.0 | 5 | 25 |
| 7385I | 1-1/2 | 2.000 | 1.718 | – | 2-31/32 | 9/16 | 3-1/2 | 176.0 | 5 | 25 |
| 7386I | 2 | 2.437 | 2.000 | – | 4-1/8 | 3/4 | 4-3/4 | 290.0 | 1 | 5 |
| 7387I | 2-1/2 | 3.000 | 2.656 | – | 4-7/8 | 1-1/16 | 5-7/8 | 450.0 | – | 1 |
| 7388I | 3 | 3.500 | 3.156 | – | 5-1/2 | 1-1/16 | 6-5/8 | 612.0 | – | 1 |

† Not CSA Certified * Select Connectors by Physical Dimensions of Cable as Dimensions May Vary Between Manufacturers.

Effective October, 2001
Copyright 2001  Printed in U.S.A.

**PAGE 6**

7770 N. Frontage Road
Skokie, Illinois 60077

**FIGURE 10–10**  AC and FMC cable connectors *(Courtesy of Appleton Electric).*

# CC

## TMC CableReady® Connectors For Jacketed Metal Clad Cable

TMC For Non-Hazardous Locations

### Applications
• CableReady® connectors are designed for use with the following cables: Type MC – jacketed corrugated aluminum, interlocked aluminum and interlocked steel and Teck cable.

• CableReady® connectors provide a means for terminating jacketed type MC cable, forming a mechanical watertight connection and providing ground continuity for cable armor.

• For use on horizontal or vertical runs.

• For use in non-hazardous locations.

• For Class I and Class II hazardous locations – See TMCX.

• Watertight-NEMA Type 4.

### Features
• Simplified construction reduces installation time. No disassembly is required for installation.

• 14 connectors cover a cable armor range of 0.400" to 4.020" in 1/2" to 4" NPT thread sizes.

• Stainless steel grounding spring provides positive ground continuity as well as excellent pull-out resistance.

• Integral "O" ring ensures watertight connection (1/2" thru 2").

• EPDM/Neoprene bushing provides a watertight seal.

• Hex design for easy wrenching.

• 1/2" thru 2" are provided with integral hub sealing ring.

• Watertight-NEMA Type 4

### Standard Materials
• Bodies, Sleeve and gland nut are copper-free aluminum or steel.

• Grounding spring is stainless steel.

• Bushing is neoprene/EPDM

• Stop and slip washers are nylon.

• O-ring is Buna N

### Size Ranges
• Trade sizes 1/2" thru 4".

• Cable Jacket Diameter Range: 0.500" thru 4.220".

• Cable Armor Diameter Range: 0.400" thru 4.020".

### Compliances
• UL Listed

• CSA Certified

• UL Standard 514B

• CSA Standard C22.2 No. 18-98

• NEMA Type 4

• Also CSA Certified for Class II, Division 1, Groups E,F and G; Class III.

• NEMA FB1-1977

• Fed. Spec. A-A-50552 (formerly W-F-406)

| Catalog Number Aluminum | Steel | Trade Size | Cable Jacket Diameter Range | Cable Armor Diameter Range | Turning Diameter | Max Core Diameter |
|---|---|---|---|---|---|---|
| TMC-5076 | TMC-5076S | 1/2 | 0.500 - 0.760 | 0.400 - 0.660 | 1.34 | .054 |
| TMC-5099 | TMC-5099S | 1/2 | 0.725 - 0.990 | 0.625 - 0.890 | 1.63 | .061 |
| TMC-75121 | TMC-75121S | 3/4 | 0.880 - 1.210 | 0.800 - 1.110 | 2.00 | 0.82 |
| TMC-100138 | TMC-100138S | 1 | 1.150 - 1.375 | 1.050 - 1.275 | 2.23 | 1.04 |
| TMC-125163 | TMC-125163S | 1-1/4 | 1.350 - 1.625 | 1.250 - 1.525 | 2.68 | 1.27 |
| TMC-125188 | TMC-125188S | 1-1/4 | 1.575 - 1.875 | 1.475 - 1.750 | 2.81 | 1.37 |
| TMC-150200 | TMC-150200S | 1-1/2 | 1.700 - 2.000 | 1.580 - 1.880 | 3.06 | 1.59 |
| TMC-150220 | TMC-150220S | 1-1/2 | 1.900 - 2.200 | 1.780 - 2.080 | 3.19 | 1.61 |
| TMC-200238 | TMC-200238S | 2 | 1.900 - 2.375 | 1.780 - 2.250 | 3.62 | 2.00 |
| TMC-200275 | TMC-200275S | 2 | 2.300 - 2.750 | 2.180 - 2.630 | 4.01 | 2.05 |
| TMC-251250 | — | 2-1/2 | 2.100 - 2.660 | 1.930 - 2.510 | 4.22 | 2.25 |
| TMC-304300 | — | 3 | 2.600 - 3.320 | 2.450 - 3.040 | 4.98 | 2.75 |
| TMC-359350 | — | 3-1/2 | 3.110 - 3.780 | 2.950 - 3.590 | 5.50 | 3.34 |
| TMC-402400 | — | 4 | 3.700 - 4.220 | 3.500 - 4.020 | 5.84 | 3.78 |

All dimensions are in inches.

Effective October, 2001
Copyright 2001  Printed in U.S.A.

PAGE 8

7770 N. Frontage Road
Skokie, Illinois 60077

**FIGURE 10–11**  MC cable connectors *(Courtesy of Appleton Electric).*

**BC**

# Nonmetallic Sheathed and Service Entrance Cable Connectors

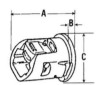

| Catalog No. | Size (In.) | Cable Opening (In.) Max. | Cable Opening (In.) Min. | Dimensions in Inches A | B | C | Wt. Lbs. Per 100 | Ctn. Qty. | Std. Pkg. |
|---|---|---|---|---|---|---|---|---|---|
| **Thermoplastic – Positive-Stop Locking Action – Nonmetallic Sheathed Cable\* Connector – with Cam Lock** | | | | | | | | | |
| A-1† | 1/2 | | | 1-3/32 | 1/8 | 1 | 1.6 | 100 | 500 |
| For 14-2, 12-2 and 10-2 (With or Without Ground) | | | | | | | | | |

| Catalog No. | Size (In.) | Max. | Min. | A | B | C | Wt. Lbs. Per 100 | Ctn. Qty. | Std. Pkg. |
|---|---|---|---|---|---|---|---|---|---|
| **Malleable – Service Entrance and Nonmetallic Sheathed Cable\* Connector** | | | | | | | | | |
| 15233 | 3/4 | .735 | .437 | 21/32 | 7/16 | 1-13/16 | 14.4 | 25 | 100 |
| 15234 | 1 | .968 | .500 | 21/32 | 9/16 | 2-1/32 | 21.0 | 10 | 100 |
| 15235 | 1-1/4 | 1.312 | .812 | 21/32 | 9/16 | 2-3/8 | 31.0 | 5 | 50 |
| 15236 | 1-1/2 | 1.593 | .812 | 13/16 | 3/4 | 2-5/8 | 53.0 | -- | 10 |
| 15237 | 2 | 2.000 | 1.000 | 15/16 | 3/4 | 3-1/4 | 80.0 | -- | 10 |

| Catalog No. | Size (In.) | Max. | Min. | A | B | C | Wt. Lbs. Per 100 | Ctn. Qty. | Std. Pkg. |
|---|---|---|---|---|---|---|---|---|---|
| **Pressure Cast – Service Entrance and Nonmetallic Sheathed Cable\* Connector** | | | | | | | | | |
| 15233-DC | 3/4 | .812 | .250 | 3/8 | 9/16 | 1-11/16 | 11.3 | 50 | 200 |
| 15234-DC | 1 | 1.031 | .375 | 9/16 | 1/2 | 1-15/16 | 17.5 | 25 | 100 |
| 15235-DC | 1-1/4 | 1.031 | .375 | 5/8 | 5/8 | 2-3/16 | 23.0 | 25 | 100 |

| Catalog No. | Size (In.) | Max. | Min. | A | B | C | Wt. Lbs. Per 100 | Ctn. Qty. | Std. Pkg. |
|---|---|---|---|---|---|---|---|---|---|
| **Pressure Cast – Armored Ground Wire and Nonmetallic Sheathed Cable\* Connector** | | | | | | | | | |
| 7286 | 1/2 | 1/2 | 1/4 | 1/2 | 3/8 | 1-3/16 | 5.6 | 50 | 500 |
| 1/4" to 7/16" Dia. Armored Ground Wire and 14-2, 14-3, 12-2 and 12-3 (With or Without Ground) | | | | | | | | | |

† Not CSA Certified
\* Select Connectors by Physical Dimensions of Cable as Dimensions May Vary Between Manufacturers.

Effective October, 2001
Copyright 2001  Printed in U.S.A.

7770 N. Frontage Road
Skokie, Illinois 60077

**PAGE 3**

**FIGURE 10–12**  Nonmetallic-sheathed and SE cable connectors *(Courtesy of Appleton Electric).*

# EF

LISTED
File #          File #
E14817         65178

## Entrance Caps

| Cat. No. | Size (In.) | Number of and Dia. Of Holes | Dimensions in Inches | | | Wt. Lbs. Per 100 | Ctn. Qty. | Std. Pkg. |
|---|---|---|---|---|---|---|---|---|
| | | | A | B | C | | | |
| **Set Screw Entrance Caps—Exclusive "Snap-On" Cover** | | | | | | | | |
| EF-125 | 1-1/4 | (3) 5/8" and (2) 13/32" | 4-5/16 | 2-7/8 | 3-5/8 | 64.0 | — | 10 |
| EF-150 | 1-1/2 | (3) 25/32", (2) 9/16" and (1) 3/8" | 5 | 3-5/16 | 4-11/32 | 93.0 | — | 10 |
| EF-200 | 2 | (3) 1", (2) 3/4" and (1) 17/32" | 6-11/32 | 4-1/2 | 5-9/16 | 183.0 | — | 5 |
| EF-250 | 2-1/2 | (3) 7/8", (1) 1" and (3) 1-5/16" | 9 | 6-1/4 | 7-3/4 | 350.0 | — | 1 |
| **Set Screw Entrance Caps—Aluminum—Exclusive "Snap-On" Cover** | | | | | | | | |
| EFM-125† | 2 | (3) 5/8" and (2) 13/32" | 4-5/16 | 2-7/8 | 3-5/8 | 62.0 | — | 10 |
| **Threaded Entrance Caps—Malleable Iron/Aluminum** | | | | | | | | |
| F50 | 1/2 | (4) 5/16" | 2-3/4 | 2-5/32 | 2-9/32 | 59.0 | 10 | 100 |
| F75 | 3/4 | (3) 13/32" and (2) 3/8" | 3-5/32 | 2-9/32 | 2-15/32 | 40.0 | 10 | 50 |
| F100 | 1 | (3) 1/2" and (2) 7/16" | 3-13/16 | 2-21/32 | 2-29/32 | 1135.0 | 5 | 25 |
| **Threaded Entrance Caps—Aluminum—With Exclusive "Snap-On" Cover** | | | | | | | | |
| F125 | 1-1/4 | (3) 5/8" and (2) 7/16" | 4-5/16 | 2-7/8 | 3-5/8 | 63.0 | — | 10 |
| F150 | 1-1/2 | (3) 3/4", (2) 19/32" and (1) 7/16" | 5 | 3-5/16 | 4-11/32 | 95.0 | — | 10 |
| F200 | 2 | (3) 1", (2) 3/4" and (1) 17/32" | 6-11/32 | 4-1/2 | 5-9/16 | 187.0 | — | 5 |
| F250 | 2-1/2 | (3) 1-5/16", (3) 7/8" and (1) 1" | 9 | 6-1/4 | 7-3/4 | 381.0 | — | 1 |
| **Threaded Entrance Caps—Malleable Iron** | | | | | | | | |
| F500-3 | 5 | (3) 1-31/32" | 13 | 10-1/4 | 14-7/8 | 4175.0 | — | 1 |
| F500-4† | 5 | (4) 1-21/32" | 13 | 10-1/4 | 14-7/8 | 4175.0 | — | 1 |
| F500-6† | 5 | (6) 1-15/32" | 13 | 10-1/4 | 14-7/8 | 4175.0 | — | 1 |
| F600-3 | 6 | (3) 2-3/16" | 15-3/8 | 13-1/8 | 18-1/4 | 8075.0 | — | 1 |
| F600-4† | 6 | (4) 2-3/32" | 15-3/8 | 13-1/8 | 18-1/4 | 8075.0 | — | 1 |
| F600-6† | 6 | (6) 1-21/32" | 15-3/8 | 13-1/8 | 18-1/4 | 8075.0 | — | 1 |

†Not UL Listed

Effective July, 2002
Copyright 2001  Printed in U.S.A.

**PAGE 2**

7770 N. Frontage Road
Skokie, Illinois 60077

**FIGURE 10–13**   Service-entrance caps *(Courtesy of Appleton Electric)*.

File #
E10627
E14814

**EF**

# Entrance Cable Connectors and Elbows

| Cat. No. | Size (In.) | Cable Gromm. Min. | Open. Max. | Nom. Size Cable* | A | B | C | Wt. Lbs. Per 100 | Ctn. Qty. | Std. Pkg. |
|---|---|---|---|---|---|---|---|---|---|---|
| **Oval Service Entrance Cable Connectors** | | | | | | | | | | |
| SEO-15 | 1/2 | .150 X .325 | .250 X .425 | 2 #14 2 #12 | 19/32 | 1-1/32 | 1 | 29.0 | 25 | 100 |
| SEO-16 | 3/4 | .165 X .575 | .290 X .710 | 3 #14 3 #12 3 #10 | 19/32 | 1-1/8 | 1-5/16 | 29.0 | 25 | 100 |
| SEO-22 | 3/4 | .359 X .672 | .520 X .812 | 3 #6 3 #4 3 #3 | 7/16 | 1-3/16 | 1-5/8 | 29.5 | 10 | 100 |
| SEO-32 | 1 | .359 X .672 | .520 X .812 | 3 #6 3 #4 3 #3 | 7/16 | 1-3/16 | 1-5/8 | 29.0 | 10 | 100 |
| SEO-33 | 1 | .469 X .844 | .656 X .968 | 3 #4 3 #3 3 #2 | 7/16 | 1-3/16 | 1-5/8 | 29.0 | 10 | 100 |
| SEO-41 | 1-1/4 | .547 X .890 | .687 X 1.031 | 3 #2 3 #1 | 17/32 | 1-11/32 | 2-1/32 | 52.0 | 5 | 50 |
| SEO-61† | 2 | .790 X 1.390 | .968 X 1.500 | 3 #3/0 3 #4/0 | 11/16 | 1 | 2-25/32 | 101.0 | 5 | 10 |

| Cat. No. | Size (In.) | Description | Dimensions in Inches A | B | C | Wt. Lbs. Per 100 | Ctn. Qty. | Std. Pkg. |
|---|---|---|---|---|---|---|---|---|
| **Threaded Entrance Elbows Type "LAY"—Malleable Iron** | | | | | | | | |
| LAY-250 | 2-1/2 | With large | 15-1/8 | 3-7/32 | 6-9/16 | 1675.0 | — | 1 |
| LAY-300 | 3 | bolt-on | 14-7/8 | 3-7/16 | 7-5/32 | 1938.0 | — | 1 |
| LAY-350 | 3-1/2 | gasketed | 19-7/16 | 4-7/16 | 9-19/32 | 4237.0 | — | 1 |
| LAY-400 | 4 | dome cover | 19-7/16 | 4-7/16 | 9-19/32 | 4037.0 | — | 1 |
| **Threaded Entrance Elbows Type "SLAY" and "LAY"—Aluminum** | | | | | | | | |
| SLAY-50A | 1/2 | | 2-17/32 | 9/16 | 2-15/16 | 21.0 | 10 | 50 |
| SLAY-75A | 3/4 | | 2-29/32 | 11/16 | 3-7/16 | 27.0 | 10 | 50 |
| SLAY-100A | 1 | With | 3-1/4 | 7/8 | 3-27/32 | 45.0 | 5 | 25 |
| SLAY-125A | 1-1/4 | gasket and | 4-11/32 | 1-1/16 | 5-7/32 | 73.0 | 2 | 10 |
| SLAY-150A | 1-1/2 | cover | 5-1/2 | 1-3/16 | 6-1/2 | 111.0 | 1 | 5 |
| SLAY-200A | 2 | | 6-3/16 | 1-15/32 | 7-15/32 | 178.0 | 1 | 5 |
| LAY-250A | 2-1/2 | | 15-1/8 | 3-7/32 | 6-9/16 | 285.0 | — | 1 |
| LAY-300A | 3 | | 14-7/8 | 3-7/16 | 7-5/32 | 456.0 | — | 1 |
| **Threaded Entrance Elbows Type "AY"—Malleable Iron** | | | | | | | | |
| AY-50 | 1/2 | | 1-15/16 | 2-3/32 | 3-3/16 | 70.0 | 10 | 50 |
| AY-75 | 3/4 | Threaded | 2-13/32 | 2-3/32 | 3 | 80.0 | 5 | 50 |
| AY-100 | 1 | removable | 2-7/16 | 2-5/8 | 3-5/8 | 120.0 | 5 | 25 |
| AY-125 | 1-1/4 | cover | 2-7/8 | 4-3/16 | 5-1/16 | 236.0 | 1 | 20 |
| AY-150 | 1-1/2 | | 3-1/8 | 4-3/16 | 5-1/8 | 285.0 | 1 | 10 |

*Physical dimensions and types of jackets of service entrance cable vary between manufacturers.
Select connectors by physical dimension of cable.
†Not UL Listed

Effective October, 2001
Copyright 2001 Printed in U.S.A.

7770 N. Frontage Road
Skokie, Illinois 60077

**PAGE 3**

**FIGURE 10–14**   SE cable connectors and elbows *(Courtesy of Appleton Electric).*

# EF

## Entrance Sill Plate and Clamps

EF

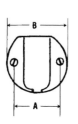

| Cat. No. | Description | Dimensions in Inches | | | Wt. Lbs. Per 100 | Ctn. Qty. | Std. Pkg. |
|---|---|---|---|---|---|---|---|
| | | A | B | C | | | |
| **Sill Plate—Aluminum—With Ductseal** | | | | | | | |
| SP-1/SP-2 | Screws furnished | 2-1/8 | 2-3/4 | — | 8.0 | 10 | 10 |
| **Service Entrance Cable Straps** | | | | | | | |
| SECL-1 | 2—#10, 3—#12 to 3—#10 | 1-1/4 | 7/16 | 7/32 | 1.3 | 100 | 1000 |
| SECL-1U | 3—#10 to 3—#6 | 1-3/8 | 9/16 | 9/32 | 2.5 | 100 | 1000 |
| SECL-2U | 3—#4 TO 3—#1 | 1-13/16 | 11/16 | 9/32 | 6.0 | 50 | 100 |
| SECL-4U | 3—1/0 or 3—2/0 | 2 | 1 | 9/32 | 7.4 | 25 | 100 |
| SECL-5U | 3—3/0 or 3—4/0 | 2-13/16 | 1-3/16 | 9/32 | 6.0 | 10 | 50 |
| **Ductseal Weather Proof Compound†** | | | | | | | |
| DUC-1 | 1-lb. Pkg. | — | — | — | 120.0 | 1 | 25 |
| DUC-5 | 5-lb. Pkg. | — | — | — | 542.0 | 1 | 10 |

†UL Listing Not Applicable

Effective October, 2001
Copyright 2001  Printed in U.S.A.

**PAGE 4**

7770 N. Frontage Road
Skokie, Illinois 60077

**FIGURE 10–15**  Entrance sill plates and clamps *(Courtesy of Appleton Electric).*

UL LISTED
File #
E14814

CSA
File #
65178

**CF-2**

# Threaded Rigid and IMC Conduit Hubs

Appleton's Uni-Seal Rigid Conduit Hubs eliminate the need for welded hubs. Efficiency of installation is built into the superior design; single wrench installation. Patented hex-hub wedge adapter fits nearly flush against the interior side walls of enclosures; provides maximum wiring room. Simple two piece construction. Protective insulated throats, positive grounding and water-tight sealing action. Flame resistant insulated throat eliminates need for end bushings. Locking edge of body bites into enclosure wall, makes hub self-locking, eliminates the need for locknuts, provides continuous 360° pressure on both sides of enclosure wall, forms positive grounding and vibration-resistant connection. Built-in recessed neoprene gasket.

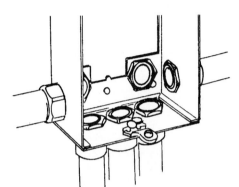

CF

½"—1"
Steel

1¼"–4"
Malleable Iron

| Catalog No. | Size (In.) | Hole Dia. (In.) Max. | Min. | Panel Thickness (In.) Max. | Min. | Dimensions (In.) A | B | Wt. Lbs. Per 100 | Ctn. Qty. | Std. Pkg. |
|---|---|---|---|---|---|---|---|---|---|---|
| **Straight** | | | | | | | | | | |
| HUB-50 | 1/2 | 7/8 | 15/16 | 1/16 | 5/64 | 1-5/64 | 1-7/64 | 16.0 | 25 | 100 |
| | | | 31/32 | 3/32 | 1/4 | | | | | |
| HUB-75 | 3/4 | 1-3/32 | 1-5/32 | 1/16 | 5/64 | 1-7/64 | 1-3/8 | 23.0 | 25 | 100 |
| | | | 1-7/32 | 3/32 | 1/4 | | | | | |
| HUB-100 | 1 | 1-11/32 | 1-13/32 | 1/16 | 5/64 | 1-1/4 | 1-23/32 | 39.0 | 10 | 50 |
| | | | 1-15/32 | 3/32 | 1/4 | | | | | |
| HUB-125 | 1-1/4 | 1-11/16 | 1-25/32 | 1/16 | 5/64 | 1-51/64 | 5-5/16 | 77.0 | 5 | 25 |
| | | | 1-27/32 | 3/32 | 5/16 | | | | | |
| HUB-150 | 1-1/2 | 1-15/16 | 2-1/32 | 1/16 | 5/64 | 1-53/64 | 2-5/8 | 92.0 | 5 | 10 |
| | | | 2-3/32 | 3/32 | 5/16 | | | | | |
| HUB-200 | 2 | 2-25/64 | 2-17/32 | 1/16 | 5/64 | 1-7/8 | 3-1/8 | 134.0 | – | 5 |
| | | | 2-19/32 | 3/32 | 5/16 | | | | | |
| HUB-250 | 2-1/2 | 2-57/64 | 3-1/64 | 3/32 | 5/16 | 2-25/64 | 3-5/8 | 236.0 | 1 | 5 |
| HUB-300 | 3 | 3-33/64 | 3-41/64 | 3/32 | 5/16 | 2-31/64 | 4-5//16 | 310.0 | 1 | 2 |
| HUB-350 | 3-1/2 | 4-1/64 | 4-1/8 | 3/32 | 5/16 | 2-19/16 | 4-13/16 | 400.0 | – | 1 |
| HUB-400 | 4 | 4-33/64 | 4-5/8 | 3/32 | 5/16 | 2-5/8 | 5-7/16 | 475.0 | – | 1 |
| **90° Malleable Iron** | | | | | | | | | | |
| HUB-9050 | 1/2 | 7/8 | 15/16 | 1/16 | 5/64 | 1-9/32 | 7/8 | 36.0 | 25 | 50 |
| | | | 31/32 | 3/32 | 1/4 | | | | | |
| HUB-9075 | 3/4 | 1-3/32 | 1-5/32 | 1/16 | 5/64 | 1-7/16 | 15/16 | 50.0 | 10 | 50 |
| | | | 1-7/32 | 3/32 | 1/4 | | | | | |
| HUB-90100 | 1 | 1-11/32 | 1-13/32 | 1/16 | 5/64 | 1-5/8 | 1-1/8 | 75.0 | 5 | 25 |
| | | | 1-15/32 | 3/32 | 1/4 | | | | | |

Effective October, 2001
Copyright 2001 Printed in U.S.A.

EGS Electrical Group   Appleton

7770 N. Frontage Road
Skokie, Illinois 60077

**PAGE 3**

**FIGURE 10–16**  RMC and IMC conduit hubs *(Courtesy of Appleton Electric).*

# CF-2

**Suitable for Hazardous Location**
**Class I, Div. 2**
**Class II, Div. 1 and 2**
**Class III, Div. 1 and 2**
**NEC 501-4(b), 502-4(a), 503-3(a)**

## Threaded Rigid and IMC Conduit Hubs With Bonding Screws

**Applications**
- Provides threaded termination in sheet metal enclosures.
- Suitable for service entrance.
- General purpose fitting.

**Features**
- Bonding locknut.
- Insulated throat.
- Recessed neoprene O-Ring to assure water and dust tight connections.

**Standard Materials and Finishes**
- Body:   1/2" - 1": Steel, zinc electroplate.
         1-1/4" - 6": Malleable iron, zinc electroplate, dichromate, epoxy powder coat.
- Wedge:  1-1/4" - 4": Malleable iron, zinc electroplate, dichromate, epoxy powder coat.
- Locknut: 1/2" - 1", 5" - 6": Steel, zinc electroplate.
- Screw:  1/2" - 6": Steel, zinc electroplate.
- Seals:  1/2" - 6": Neoprene.
- Liners: 1/2" - 6": Nylon..

**Size Range**
- Available in 1/2" through 6" conduit size.

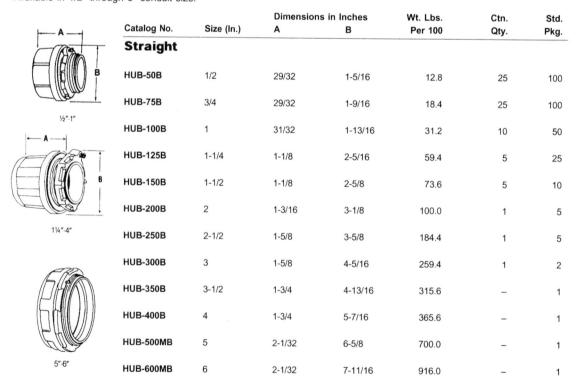

| Catalog No. | Size (In.) | Dimensions in Inches | | Wt. Lbs. | Ctn. | Std. |
| | | A | B | Per 100 | Qty. | Pkg. |
|---|---|---|---|---|---|---|
| **Straight** | | | | | | |
| HUB-50B | 1/2 | 29/32 | 1-5/16 | 12.8 | 25 | 100 |
| HUB-75B | 3/4 | 29/32 | 1-9/16 | 18.4 | 25 | 100 |
| HUB-100B | 1 | 31/32 | 1-13/16 | 31.2 | 10 | 50 |
| HUB-125B | 1-1/4 | 1-1/8 | 2-5/16 | 59.4 | 5 | 25 |
| HUB-150B | 1-1/2 | 1-1/8 | 2-5/8 | 73.6 | 5 | 10 |
| HUB-200B | 2 | 1-3/16 | 3-1/8 | 100.0 | 1 | 5 |
| HUB-250B | 2-1/2 | 1-5/8 | 3-5/8 | 184.4 | 1 | 5 |
| HUB-300B | 3 | 1-5/8 | 4-5/16 | 259.4 | 1 | 2 |
| HUB-350B | 3-1/2 | 1-3/4 | 4-13/16 | 315.6 | – | 1 |
| HUB-400B | 4 | 1-3/4 | 5-7/16 | 365.6 | – | 1 |
| HUB-500MB | 5 | 2-1/32 | 6-5/8 | 700.0 | – | 1 |
| HUB-600MB | 6 | 2-1/32 | 7-11/16 | 916.0 | – | 1 |

Effective October, 2001
Copyright 2001  Printed in U.S.A.

**PAGE 4**

7770 N. Frontage Road
Skokie, Illinois 60077

**FIGURE 10–17**   RMC and IMC conduit hubs with bonding screws *(Courtesy of Appleton Electric).*

File #
E14814

File #
LR65178

# CF-2

## Watertight Hubs
### Die Cast, Insulated and Gasketed

### Applications
• Used in applications to connect rigid metal conduit or IMC to a threadless opening in an enclosure.

• May be used in wet or dry locations, indoors or outdoors.

• Suitable for hazardous locations: Class I, Division 2; Class II, Divisions 1 & 2; Class III, Per NEC 501-4(b), 502-4(a), 503-3(a)

### Features
• Rugged metallic construction ensures mechanical protection.

• Die cast design provides clean lines and aesthetic appeal.

• Insulated throat, rated up to 105°C, prevents wire abrasion.

• Integral O-ring provides watertight seal.

• Grounding version provides positive ground for external conductor.

• Unique profile permits easy wrenching.

• Smooth, accurately tapped threads facilitate installation.

• Watertight and corrosion resistant: NEMA 4X.

### Standard Materials
• Bodies and nuts: Die Cast Zinc
• O-ring: BUNA-N
• Insulator: Polycarbonate
• Ground Screw: Mild Steel
• Size Ranges: 1/2" through 4"

### Optional Finishes
• Chrome plating optional add suffix **C**; consult factory.

### Compliances
• UL Listed
• CSA Certified
  (incl. Class II, Groups E, F, G)
• UL Standard 514B
• CSA Standard C22.2-18-98
• NEMA Type 2, 3, 3R, 4, 4X, 5, 12, 13
• NEMA Standard FB-1
• Appleton UL File No. E14814
• Appleton CSA File No. LR65178

HUB

HUBG

### Die Cast Hubs– Insulated  All dimensions are in inches

| Catalog Number | Trade Size | Hole Dia. Min – Max | Panel Thickness Min – Max | Dimensions A | B | C | Wt. Per 100 (lbs) | Carton Qty | Package Qty |
|---|---|---|---|---|---|---|---|---|---|
| HUB-50D | 1/2" | 0.86 – 0.91 | 0.026 – 0.25 | 0.84 | 0.64 | 1.44 | 20 | 25 | 125 |
| HUB-75D | 3/4" | 1.08 – 1.14 | 0.026 – 0.25 | 0.94 | 0.64 | 1.69 | 32 | 25 | 125 |
| HUB-100D | 1" | 1.33 – 1.41 | 0.026 – 0.25 | 1.03 | 0.78 | 2.00 | 40 | 25 | 125 |
| HUB-125D | 1-1/4" | 1.67 – 1.77 | 0.026 – 0.25 | 1.12 | 0.78 | 2.38 | 60 | 10 | 50 |
| HUB-150D | 1-1/2" | 1.92 – 2.02 | 0.062 – 0.25 | 1.12 | 0.78 | 2.75 | 70 | 10 | 50 |
| HUB-200D | 2" | 2.36 – 2.50 | 0.062 – 0.25 | 1.12 | 0.78 | 3.25 | 90 | 10 | 50 |
| HUB-250D | 2-1/2" | 2.86 – 3.00 | 0.062 – 0.31 | 1.56 | 1.06 | 3.72 | 200 | 5 | 25 |
| HUB-300D | 3" | 3.48 – 3.63 | 0.062 – 0.31 | 1.56 | 1.06 | 4.38 | 250 | 2 | 10 |
| HUB-350D | 3-1/2" | 3.98 – 4.16 | 0.062 – 0.31 | 1.56 | 1.06 | 4.94 | 300 | 2 | 10 |
| HUB-400D | 4" | 4.48 – 4.67 | 0.062 – 0.31 | 1.56 | 1.06 | 5.47 | 350 | 2 | 10 |

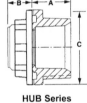

HUB Series

HUBG Series

### Grounding Style Hubs– Insulated  All dimensions are in inches

| Catalog Number | Trade Size | Hole Dia. Min – Max | Panel Thickness Min – Max | D | E | Dimensions F | G | H | Wt. Per 100 (lbs) | Carton Qty | Package Qty |
|---|---|---|---|---|---|---|---|---|---|---|---|
| HUBG-50D | 1/2" | 0.86 – 0.91 | 0.026 – 0.25 | 0.84 | 0.64 | 1.44 | 10–32x1/4 | 0.69 | 20 | 25 | 125 |
| HUBG-75D | 3/4" | 1.08 – 1.14 | 0.026 – 0.25 | 0.94 | 0.64 | 1.69 | 10–32x1/4 | 0.78 | 32 | 25 | 125 |
| HUBG-100D | 1" | 1.33 – 1.41 | 0.026 – 0.25 | 1.03 | 0.78 | 2.00 | 10–32x1/4 | 0.78 | 40 | 25 | 125 |
| HUBG-125D | 1-1/4" | 1.67 – 1.77 | 0.026 – 0.25 | 1.12 | 0.78 | 2.38 | 1/4–20x3/8 | 0.81 | 60 | 10 | 50 |
| HUBG-150D | 1-1/2" | 1.92 – 2.02 | 0.062 – 0.25 | 1.12 | 0.78 | 2.75 | 1/4–20x3/8 | 1.00 | 70 | 10 | 50 |
| HUBG-200D | 2" | 2.36 – 2.50 | 0.062 – 0.25 | 1.12 | 0.78 | 3.25 | 1/4–20x1/2 | 1.00 | 90 | 10 | 50 |
| HUBG-250D | 2-1/2" | 2.86 – 3.00 | 0.062 – 0.31 | 1.56 | 1.06 | 3.72 | 1/4–20x1/2 | 1.25 | 200 | 5 | 25 |
| HUBG-300D | 3" | 3.48 – 3.63 | 0.062 – 0.31 | 1.56 | 1.06 | 4.38 | 1/4–20x1/2 | 1.25 | 250 | 2 | 10 |
| HUBG-350D | 3-1/2" | 3.98 – 4.16 | 0.062 – 0.31 | 1.56 | 1.06 | 4.94 | 1/4–20x1/2 | 1.31 | 300 | 2 | 10 |
| HUBG-400D | 4" | 4.48 – 4.67 | 0.062 – 0.31 | 1.56 | 1.06 | 5.47 | 1/4–20x1/2 | 1.31 | 350 | 2 | 10 |

Effective October, 2001
Copyright 2001  Printed in U.S.A.

7770 N. Frontage Road
Skokie, Illinois 60077

**PAGE 5**

EGS Electrical Group  Appleton

**FIGURE 10–18**  RMC and IMC watertight hubs *(Courtesy of Appleton Electric).*

UL
LISTED
File #
E14814

# CF-2

# Threaded Rigid Conduit and IMC Bolt-On™ Couplings

Series "SCC" Threaded Bolt-On™ Coupling for use with thread-ed rigid conduit and IMC.

Concrete tight.
Neoprene gasketed steel construction.

Provides a fast inexpensive method of joining two sections of threaded rigid conduit in areas of close conduit spacing, tight clearance locations and irregular conduit bends which may nor-mally require an expensive union.

**Fast Installation**
• Slip Bolt-On™ Coupling over threaded end of one conduit section.
• Butt the end of the second threaded conduit section to the first.
• Slip Bolt-On™ Coupling to the center of the two conduit threads and tighten bolt(s).

**Standard Materials and Finishes**
• 1/2" - 2": Steel coupling and hardware, zinc electroplate.
• 2-1/2" - 4": Malleable iron coupling, zinc electroplate, dichromate, epoxy powder coat. Hardware, zinc electroplate.
• 1/2" - 4": Neoprene gasket.

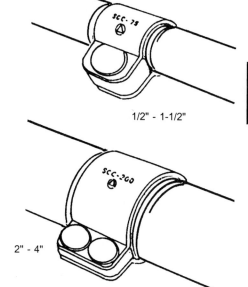

1/2" - 1-1/2"

2" - 4"

CF

| Catalog No. | Size (In.) | Dimensions in Inches | | | Wt. Lbs. | Ctn. | Std. |
| | | A | B | C | Per 100 | Qty. | Pkg. |
|---|---|---|---|---|---|---|---|
| SCC-50 | 1/2 | 1-3/4 | 1-1/8 | 1-1/4 | 20.0 | 10 | 50 |
| SCC-75 | 3/4 | 2 | 1-11/32 | 1-1/4 | 24.0 | 10 | 50 |
| SCC-100 | 1 | 2-9/32 | 1-5/8 | 1-1/2 | 35.0 | 10 | 50 |
| SCC-125 | 1-1/4 | 2-5/8 | 1-31/32 | 1-1/2 | 50.0 | 5 | 25 |
| SCC-150 | 1-1/2 | 2-7/8 | 2-7/32 | 1-1/2 | 65.0 | 5 | 25 |
| SCC-200 | 2 | 3-3/8 | 2-11/16 | 1-3/4 | 103.0 | 1 | 5 |
| SCC-250 | 2-1/2 | 4-19/32 | 3-7/16 | 3 | 136.0 | 1 | 5 |
| SCC-300 | 3 | 5-7/32 | 4-1/16 | 3 | 174.0 | 1 | 5 |
| SCC-350 | 3-1/2 | 5-25/32 | 4-9/16 | 3-1/4 | 228.0 | 1 | 5 |
| SCC-400 | 4 | 6-9/32 | 5-1/16 | 3-1/4 | 289.0 | 1 | 5 |

1/2" — 1-1/2"

2" — 4"

**FIGURE 10–19** RMC and IMC Bolt-On™ couplings *(Courtesy of Appleton Electric).*

# CF-2

## Threaded Rigid Conduit and IMC Couplings and Connectors

½"-¾"—Steel

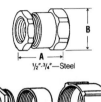

1"-6"—Malleable Iron

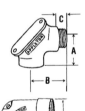

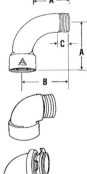

**CF**

| Catalog No. | Size (In.) | A | B | C | Wt. Lbs. Per 100 | Ctn. Qty. | Std. Pkg. |
|---|---|---|---|---|---|---|---|
| **Three Piece Union– Concrete Tight** | | | | | | | |
| EC-50 | 1/2 | 1-7/16 | 1-7/16 | – | 16.5 | 10 | 100 |
| EC-75 | 3/4 | 1-9/16 | 1-9/16 | – | 24.0 | 10 | 100 |
| EC-100 | 1 | 1-11/16 | 2 | – | 58.0 | 5 | 25 |
| EC-125 | 1-1/4 | 2-1/8 | 2-7/16 | – | 98.0 | 5 | 25 |
| EC-150 | 1-1/2 | 2-3/16 | 2-11/16 | – | 134.0 | 5 | 25 |
| EC-200 | 2 | 2-7/16 | 3-1/4 | – | 198.0 | 5 | 25 |
| EC-250 | 2-1/2 | 2-5/8 | 3-1/16 | – | 255.0 | 2 | 10 |
| EC-300 | 3 | 2-5/8 | 4-5/8 | – | 394.0 | 1 | 10 |
| EC-350 | 3-1/2 | 2-5/8 | 5-3/16 | – | 380.0 | 1 | 5 |
| EC-400 | 4 | 3-7/16 | 5-11/16 | – | 700.0 | 1 | 5 |
| EC-500 | 5 | 3-1/2 | 6-15/16 | – | 788.0 | 1 | 2 |
| EC-600 | 6 | 3-1/2 | 8-1/8 | – | 825.0 | – | 1 |
| **90° Female Gasketed Pulling Elbows– Watertight – Malleable Iron** | | | | | | | |
| FFL-50 | 1/2 | 1-3/16 | – | – | 31.0 | 10 | 100 |
| FFL-75 | 3/4 | 1-11/32 | – | – | 46.0 | 10 | 50 |
| FFL-100 | 1 | 1-19/32 | – | – | 76.0 | 5 | 25 |
| FFL-125 | 1-1/4 | 3-7/8 | – | – | 120.0 | 5 | 25 |
| FFL-150 | 1-1/2 | 4-1/2 | – | – | 160.0 | 5 | 10 |
| FFL-200 | 2 | 5-5/8 | – | – | 293.0 | 5 | 10 |
| **90° Male to Female Gasketed Pulling Elbows– Watertight– Malleable Iron** | | | | | | | |
| MFL-50 | 1/2 | 1-3/16 | 1-7/16 | 15/32 | 36.0 | 10 | 100 |
| MFL-75 | 3/4 | 1-11/32 | 1-17/32 | 1/2 | 55.0 | 10 | 50 |
| MFL-100 | 1 | 1-19/32 | 1-3/4 | 9/16 | 90.0 | 5 | 25 |
| MFL-125* | 1-1/4 | 3-7/8 | 2-5/8 | 15/16 | 141.0 | 5 | 25 |
| MFL-150* | 1-1/2 | 4-1/2 | 3 | 1 | 192.0 | 5 | 10 |
| MFL-200* | 2 | 5-5/8 | 3-11/16 | 1-1/4 | 335.0 | 5 | 10 |
| **90° Female Pulling Elbows†† – Malleable Iron** | | | | | | | |
| PFFL-50 | 1/2 | 2-7/16 | 1-1/8 | – | 49.0 | 10 | 100 |
| PFFL-75 | 3/4 | 2-23/32 | 1-3/8 | – | 76.0 | 5 | 50 |
| PFFL-100 | 1 | 3-1/4 | 1-23/32 | – | 121.0 | 5 | 25 |
| **90° Female Elbows†† – Malleable Iron** | | | | | | | |
| LF90-50 | 1/2 | 1-1/2 | – | – | 39.0 | 10 | 100 |
| LF90-75 | 3/4 | 1-11/16 | – | – | 53.0 | 10 | 100 |
| **90° Male to Female Long Bushed Elbows†† – Malleable Iron** | | | | | | | |
| LMFL90-50 | 1/2 | 1-3/4 | 1-25/32 | 19/32 | 32.5 | 25 | 100 |
| LMFL90-75 | 3/4 | 2-1/4 | 2-9/32 | 11/16 | 50.0 | 5 | 25 |
| LMFL90-100 | 1 | 2-21/32 | 2-11/16 | 3/4 | 87.0 | 5 | 25 |
| LMFL90-125 | 1-1/4 | 3-5/16 | 3-11/32 | 1 | 126.0 | 5 | 20 |
| LMFL90-150 | 1-1/2 | 3-15/16 | 3-31/32 | 1-1/16 | 170.0 | 5 | 10 |
| LMFL90-200 | 2 | 4 | 4 | 1-1/16 | 410.0 | 1 | 10 |
| **90° Male to Female Short Bushed Elbows†† – Malleable Iron** | | | | | | | |
| LMF90-50 | 1/2 | 1-1/4 | 1-13/32 | 1/2 | 23.0 | 25 | 100 |
| LMF90-75 | 3/4 | 1-7/16 | 1-11/16 | 13/16 | 36.0 | 25 | 100 |
| LMF90-100 | 1 | 1-21/32 | 1-15/16 | 1 | 58.0 | 10 | 100 |
| **90° Male to Female Short Box Connectors†† – Malleable Iron** | | | | | | | |
| LMF90-50L | 1/2 | 1-1/4 | 1-3/32 | 7/16 | 26.0 | 25 | 50 |
| LMF90-75L | 3/4 | 1-7/16 | 1-1/4 | 7/16 | 34.0 | 25 | 50 |

†† Not CSA Certified.          * Furnished with Removable Nipple

Effective October, 2001
Copyright 2001  Printed in U.S.A.

 EGS Electrical Group  Appleton

7770 N. Frontage Road
Skokie, Illinois 60077

**PAGE 9**

**FIGURE 10–20**  Threaded RMC and IMC couplings and connectors *(Courtesy of Appleton Electric).*

# CF-2

LISTED
File #        File #
E14814       65178

## Threaded Rigid Conduit and IMC Nipples and Enlargers

**CF**

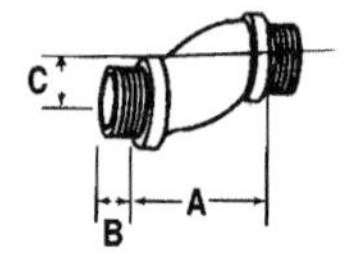

1/2"– 1"
Pressure Cast
1 1/4"–6"
Malleable Iron

1/2"– 1"
Pressure Cast
1 1/4"–6"
Malleable Iron

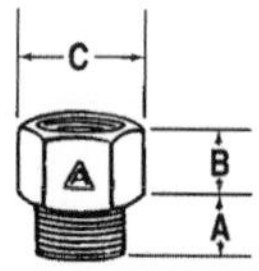

1/2"– 3/4" Steel
1" Malleable Iron/Steel

| Catalog No. | Size (In.) | Dimensions in Inches | | | Wt. Lbs. Per 100 | Ctn. Qty. | Std. Pkg. |
|---|---|---|---|---|---|---|---|
| | | A | B | C | | | |
| **Offset Nipples– Malleable Iron** | | | | | | | |
| OFN-50 | 1/2 | 1-5/8 | 7/16 | 3/4 | 23.0 | 25 | 100 |
| OFN-75 | 3/4 | 1-5/8 | 7/16 | 3/4 | 31.0 | 25 | 100 |
| OFN-100 | 1 | 1-5/8 | 1/2 | 3/4 | 44.0 | 10 | 50 |
| OFN-125 | 1-1/4 | 1-5/8 | 9/16 | 3/4 | 60.0 | 5 | 25 |
| OFN-150 | 1-1/2 | 1-5/8 | 5/8 | 3/4 | 82.0 | 5 | 25 |
| OFN-200 | 2 | 1-5/8 | 3/4 | 3/4 | 112.0 | 2 | 10 |
| **Bushed Nipples** | | | | | | | |
| CN-38 | 3/8 | 3/8 | 1/8 | 15/16 | 3.0 | 50 | 100 |
| CN-50 | 1/2 | 7/16 | 1/8 | 1 | 3.4 | 50 | 500 |
| CN-75 | 3/4 | 1/2 | 1/8 | 1-1/4 | 4.4 | 50 | 500 |
| CN-100 | 1 | 5/8 | 1/8 | 1-1/2 | 8.5 | 25 | 100 |
| CN-125 | 1-1/4 | 11/16 | 1/4 | 2-1/8 | 22.0 | 25 | 200 |
| CN-150 | 1-1/2 | 7/8 | 1/4 | 2-3/8 | 28.5 | 25 | 200 |
| CN-200 | 2 | 15/16 | 3/8 | 2-7/8 | 48.0 | 10 | 100 |
| CN-250 | 2-1/2 | 1-1/8 | 3/8 | 3-9/16 | 77.0 | 5 | 25 |
| CN-300 | 3 | 1-3/16 | 3/8 | 4-5/16 | 114.0 | 5 | 25 |
| CN-350 | 3-1/2 | 1-1/4 | 3/8 | 4-9/16 | 136.0 | 5 | 25 |
| CN-400 | 4 | 1-1/4 | 3/8 | 5-1/8 | 169.0 | 1 | 5 |
| CN-500 | 5 | 1-1/4 | 7/16 | 6-5/16 | 314.0 | 1 | 5 |
| CN-600 | 6 | 1-5/16 | 7/16 | 7-1/2 | 320.0 | 1 | 5 |
| **Bushed Nipples– Insulated** | | | | | | | |
| CN-50I | 1/2 | 7/16 | 1/8 | 1 | 3.3 | 50 | 500 |
| CN-75I | 3/4 | 1/2 | 1/8 | 1-1/4 | 3.8 | 50 | 500 |
| CN-100I | 1 | 5/8 | 1/8 | 1-1/2 | 7.2 | 25 | 100 |
| CN-125I | 1-1/4 | 11/16 | 1/4 | 2-1/8 | 20.0 | 25 | 200 |
| CN-150I | 1-1/2 | 7/8 | 1/4 | 2-3/8 | 25.0 | 25 | 200 |
| CN-200I | 2 | 15/16 | 3/8 | 2-7/8 | 44.0 | 10 | 100 |
| CN-250I | 2-1/2 | 1-1/8 | 3/8 | 3-9/16 | 74.0 | 5 | 25 |
| CN-300I | 3 | 1-3/16 | 3/8 | 4-5/16 | 112.0 | 5 | 25 |
| CN-350I | 3-1/2 | 1-1/4 | 3/8 | 4-9/16 | 124.0 | 5 | 25 |
| CN-400I | 4 | 1-1/4 | 3/8 | 5-1/8 | 160.0 | 1 | 5 |
| CN-500I | 5 | 1-1/4 | 7/16 | 6-5/16 | 350.0 | 1 | 5 |
| CN-600I | 6 | 1-5/16 | 7/16 | 7-1/2 | 360.0 | 1 | 5 |
| **Male Enlargers††** | | | | | | | |
| ME50-75 | 3/4F x 1/2M | 9/16 | 25/32 | 1-3/8 | 16.0 | 25 | 250 |
| ME75-100 | 1F x 3/4M | 1/2 | 15/16 | 1-3/4 | 26.0 | 25 | 250 |
| ME100-125 | 1-1/4F x 1M | 9/16 | 1-1/16 | 2-1/4 | 47.0 | 25 | 100 |
| ME50-75EP* | 3/4F x 1/2M | 5/8 | 15/16 | 1-3/16 | 16.0 | 25 | 250 |
| ME75-100EP* | 1F x 3/4M | 11/16 | 1-1/16 | 1-7/16 | 26.0 | 25 | 250 |
| ME100-125EP* | 1-1/4F x 1M | 13/16 | 1-1/8 | 1-13/16 | 47.0 | 25 | 250 |

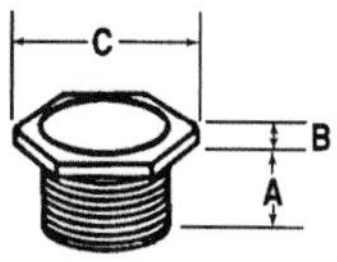

*Listed for Class I, Div. 1 and 2, Groups A, B, C, D; Class II, Div. 1 and 2, Groups E, F, G.

**Threaded Bell Reducing Couplings–** See Section I for an expanded offering.

†† Not CSA Certified.

Effective October, 2001
Copyright 2001  Printed in U.S.A.
**PAGE 10**

EGS
Electrical Group
**Appleton**

7770 N. Frontage Road
Skokie, Illinois 60077

**FIGURE 10–21**   Threaded RMC and IMC nipples and enlargers (*Courtesy of Appleton Electric*).

Class I, Groups A♦,B,C,D
Class II, Groups E,F,G
Class III

UL
LISTED
File #
E10444

CSA
File #
65181

# CF-2

# Threaded Rigid Conduit and IMC Reducers

Steel
RB50-13 – RB200-150

Malleable Iron
RB250-150 – RB600-500

| Catalog No. | Size (In.) | Wt. Lbs. Per 100 | Ctn. Qty. | Std. Pkg. |
|---|---|---|---|---|
| **Reducing Bushings– Steel** | | | | |
| RB50-13*♦ | 1/2 – 1/8 | 1.5 | 50 | 500 |
| RB50-25*♦ | 1/2 – 1/4 | 2.0 | 50 | 500 |
| RB50-38*♦ | 1/2 – 3/8 | 2.7 | 100 | 500 |
| RB75-50♦ | 3/4 – 1/2 | 6.0 | 50 | 500 |
| RB100-50♦ | 1 – 1/2 | 13.8 | 25 | 250 |
| RB100-75♦ | 1 – 3/4 | 8.0 | 25 | 100 |
| RB125-50 | 1-1/4 – 1/2 | 29.0 | 25 | 100 |
| RB125-75 | 1-1/4 – 3/4 | 23.0 | 25 | 100 |
| RB125-100 | 1-1/4 – 1 | 17.0 | 25 | 100 |
| RB150-50 | 1-1/2 – 1/2 | 39.0 | 25 | 50 |
| RB150-75 | 1-1/2 – 3/4 | 35.0 | 25 | 50 |
| RB150-100 | 1-1/2 – 1 | 31.0 | 25 | 50 |
| RB150-125 | 1-1/2 – 1-1/4 | 16.5 | 25 | 100 |
| RB200-50 | 2 – 1/2 | 68.0 | 5 | 50 |
| RB200-75 | 2 – 3/4 | 64.0 | 5 | 25 |
| RB200-100 | 2 – 1 | 57.0 | 5 | 25 |
| RB200-125 | 2 – 1-1/4 | 45.0 | 5 | 50 |
| RB200-150 | 2 – 1-1/2 | 38.0 | 5 | 50 |
| **Reducing Bushings– Malleable Iron** | | | | |
| RB250-100 | 2-1/2 – 1 | 75.0 | 5 | 25 |
| RB250-125 | 2-1/2 – 1-1/2 | 78.0 | 5 | 25 |
| RB250-150 | 2-1/2 – 1-1/2 | 82.0 | 5 | 25 |
| RB250-200 | 2-1/2 – 2 | 56.0 | 5 | 25 |
| RB300-100 | 3 – 1 | 130.0 | 5 | 25 |
| RB300-125 | 3 – 1-1/2 | 135.0 | 5 | 25 |
| RB300-150 | 3 – 1-1/2 | 142.0 | 5 | 25 |
| RB300-200 | 3 – 2 | 126.0 | 5 | 25 |
| RB300-250 | 3 – 2-1/2 | 83.0 | 5 | 25 |
| RB350-200 | 3-1/2 – 2 | 155.0 | 5 | 25 |
| RB350-250 | 3-1/2 – 2-1/2 | 164.0 | 5 | 25 |
| RB350-300 | 3-1/2 – 3 | 92.0 | 5 | 25 |
| RB400-200 | 4 – 2 | 165.0 | 2 | 10 |
| RB400-250 | 4 – 2-1/2 | 245.0 | 2 | 10 |
| RB400-300 | 4 – 3 | 174.0 | 2 | 10 |
| RB400-350 | 4 – 3-1/2 | 106.0 | 2 | 10 |
| RB500-350† | 5 – 3-1/2 | 180.0 | 1 | 5 |
| RB500-400† | 5 – 4 | 190.0 | 1 | 5 |
| RB600-400† | 6 – 4 | 192.0 | 1 | 5 |
| RB600-500† | 6 – 5 | 195.0 | 1 | 5 |

CF

* UL Recognized.
♦ Suitable for Class I, Group A in addition to Class I Groups B, C, D; Class II, Groups E, F, G; and Class III.
† Class I, Groups C,D; Class II, Groups E,F,G; Class III.

Effective October, 2001
Copyright 2001 Printed in U.S.A.

EGS
Electrical Group
Appleton®

7770 N. Frontage Road
Skokie, Illinois 60077

PAGE 11

**FIGURE 10–22** Threaded RMC and IMC reducers *(Courtesy of Appleton Electric).*

# CF-2

UL LISTED
File #
E10444

CSA
File #
65181

Class I, Groups A♦,B,C,D
Class II, Groups E,F,G
Class III

## Threaded Rigid Conduit and IMC Reducers

**Aluminum
RB50-13A – RB600-500A**

| Catalog No. | Size (In.) | Wt. Lbs. Per 100 | Ctn. Qty. | Std. Pkg. |
|---|---|---|---|---|
| **Reducing Bushings– Aluminum** | | | | |
| RB50-13A*♦ | 1/2 — 1/8 | 0.5 | 50 | 500 |
| RB50-25A*♦ | 1/2 — 1/4 | 0.7 | 50 | 500 |
| RB50-38A*♦ | 1/2 — 3/8 | 0.9 | 100 | 500 |
| RB75-50A♦ | 3/4 — 1/2 | 2.3 | 50 | 500 |
| RB100-50A♦ | 1 — 1/2 | 6.8 | 25 | 250 |
| RB100-75A♦ | 1 — 3/4 | 3.5 | 25 | 100 |
| RB125-50A | 1-1/4 — 1/2 | 9.2 | 25 | 100 |
| RB125-75A | 1-1/4 — 3/4 | 7.3 | 25 | 100 |
| RB125-100A | 1-1/4 — 1 | 6.0 | 25 | 100 |
| RB150-50A | 1-1/2 — 1/2 | 14.2 | 25 | 50 |
| RB150-75A | 1-1/2 — 3/4 | 12.8 | 25 | 50 |
| RB150-100A | 1-1/2 — 1 | 12.0 | 25 | 50 |
| RB150-125A | 1-1/2 — 1-1/4 | 6.5 | 25 | 100 |
| RB200-50A | 2 — 1/2 | 29.0 | 5 | 50 |
| RB200-75A | 2 — 3/4 | 26.5 | 5 | 25 |
| RB200-100A | 2 — 1 | 23.6 | 5 | 25 |
| RB200-125A | 2 — 1/4 | 18.7 | 5 | 50 |
| RB200-150A | 2 — 1-1/2 | 16.0 | 5 | 50 |
| RB250-100A | 2-1/2 — 1 | 27.0 | 5 | 25 |
| RB250-125A | 2-1/2 — 1-1/4 | 28.1 | 5 | 25 |
| RB250-150A | 2-1/2 — 1-1/2 | 82.0 | 5 | 25 |
| RB250-200A | 2-1/2 — 2 | 20.0 | 5 | 25 |
| RB300-100A | 3 — 1 | 46.8 | 5 | 25 |
| RB300-125A | 3 — 1-1/4 | 48.6 | 5 | 25 |
| RB300-150A | 3 — 1-1/2 | 51.1 | 5 | 25 |
| RB300-200A | 3 — 2 | 51.0 | 5 | 25 |
| RB300-250A | 3 — 2-1/2 | 35.0 | 5 | 25 |
| RB350-200A | 3 — 2 | 55.8 | 5 | 25 |
| RB350-250A | 3-1/2 — 2-1/2 | 57.4 | 5 | 25 |
| RB350-300A | 3-1/2 — 3 | 42.0 | 5 | 25 |
| RB400-200A | 4 — 2 | 59.4 | 2 | 10 |
| RB400-250A | 4 — 2-1/2 | 60.9 | 2 | 10 |
| RB400-300A | 4 — 3 | 77.5 | 2 | 10 |
| RB400-350A | 4 — 3-1/2 | 50.0 | 2 | 10 |
| RB500-350A† | 5 — 3-1/2 | 64.8 | 1 | 5 |
| RB500-400A† | 5 — 4 | 68.4 | 1 | 5 |
| RB600-400A† | 6 — 4 | 69.1 | 1 | 5 |
| RB600-500A† | 6 — 5 | 70.2 | 1 | 5 |

* UL Recognized.
♦ Suitable for Class I, Group A in addition to Class I Groups B, C, D; Class II, Groups E, F, G; and Class III.
† Class I, Groups C,D; Class II, Groups E,F,G; Class III.

Effective July, 2002
Copyright 2002  Printed in U.S.A.
**PAGE 12**

7770 N. Frontage Road
Skokie, Illinois 60077

**FIGURE 10–23**  Threaded RMC and IMC reducers *(Courtesy of Appleton Electric).*

# CF-2

## Threadless Rigid Conduit Connectors and Couplings

| Catalog No. | Size (In.) | A | B | C | Wt. Lbs. Per 100 | Ctn. Qty. | Std. Pkg. |
|---|---|---|---|---|---|---|---|
| **Compression Connectors– Concrete Tight** | | | | | | | |
| NTC-50 | 1/2 | 7/8 | 3/8 | 1-5/16 | 15.0 | 50 | 200 |
| NTC-75 | 3/4 | 7/8 | 3/8 | 1-5/8 | 23.0 | 25 | 200 |
| NTC-100 | 1 | 1 | 5/8 | 1-7/8 | 40.0 | 5 | 25 |
| NTC-125 | 1-1/4 | 1-5/16 | 5/8 | 2-5/16 | 64.0 | 5 | 50 |
| NTC-150 | 1-1/2 | 1-3/8 | 5/8 | 2-11/16 | 90.0 | 2 | 10 |
| NTC-200 | 2 | 1-7/16 | 5/8 | 3-1/4 | 153.0 | 1 | 5 |
| NTC-250 | 2-1/2 | 2-3/8 | 1 | 3-5/8 | 258.0 | 1 | 10 |
| NTC-300 | 3 | 2-7/8 | 1 | 4-1/4 | 350.0 | 1 | 10 |
| NTC-350 | 3-1/2 | 2-11/16 | 1 | 4-7/8 | 450.0 | 1 | 10 |
| NTC-400 | 4 | 2-11/16 | 1 | 5-3/8 | 510.0 | 1 | 5 |
| **Compression Connectors– Insulated– Concrete Tight** | | | | | | | |
| RNTC-50 | 1/2 | 7/8 | 3/8 | 1-5/16 | 14.0 | 50 | 200 |
| RNTC-75 | 3/4 | 7/8 | 3/8 | 1-5/8 | 22.0 | 25 | 200 |
| RNTC-100 | 1 | 1 | 5/8 | 1-7/8 | 40.0 | 5 | 25 |
| RNTC-125 | 1-1/4 | 1-5/16 | 5/8 | 2-5/16 | 64.0 | 5 | 50 |
| RNTC-150 | 1-1/2 | 1-3/8 | 5/8 | 2-11/16 | 90.0 | 2 | 10 |
| RNTC-200 | 2 | 1-7/16 | 5/8 | 3-1/4 | 145.0 | 2 | 10 |
| **Compression Couplings– Concrete Tight** | | | | | | | |
| NTCC-50 | 1/2 | 2-1/8 | 1-5/16 | – | 25.0 | 25 | 100 |
| NTCC-75 | 3/4 | 2-5/16 | 1-5/8 | – | 35.0 | 25 | 100 |
| NTCC-100 | 1 | 2-1/2 | 1-7/8 | – | 57.0 | 5 | 25 |
| NTCC-125 | 1-1/4 | 2-7/8 | 2-5/16 | – | 91.0 | 5 | 25 |
| NTCC-150 | 1-1/2 | 3-5/16 | 2-11/16 | – | 129.0 | 5 | 25 |
| NTCC-200 | 2 | 4-1/16 | 3-1/4 | – | 273.0 | 1 | 10 |
| NTCC-250 | 2-1/2 | 5 | 3-5/8 | – | 375.0 | 1 | 10 |
| NTCC-300 | 3 | 5-1/16 | 4-1/4 | – | 465.0 | 1 | 10 |
| NTCC-350 | 3-1/2 | 5-1/4 | 4-7/8 | – | 605.0 | 1 | 5 |
| NTCC-400 | 4 | 5-3/8 | 5-3/8 | – | 680.0 | 1 | 5 |
| **90° Compression Box Connectors– Concrete Tight** | | | | | | | |
| NL90-50L | 1/2 | | | | 25.0 | 25 | 100 |
| NL90-75L | 3/4 | | | | 40.0 | 10 | 50 |

½"—¾" Steel
1"—1¼" Steel & Malleable

1½"—4" Malleable

½"-¾"—Steel
1". -1¼" Steel & Malleable

2" Malleable

½"—¾" Steel
1"—2" Steel & Malleable

2½"—4" Malleable

Effective October, 2001
Copyright 2001 Printed in U.S.A.

7770 N. Frontage Road
Skokie, Illinois 60077

PAGE 13

**FIGURE 10–24** Threadless RMC connectors and couplings *(Courtesy of Appleton Electric).*

# CF-2

## Rigid Conduit and IMC Bushings

| Catalog No. | Size (In.) | A | Dimensions in Inches B | C | Wt. Lbs. Per 100 | Ctn. Qty. | Std. Pkg. |
|---|---|---|---|---|---|---|---|
| **Bushings– Malleable Iron** | | | | | | | |
| BU50 | 1/2 | 1-1/16 | 11/32 | 19/32 | 2.0 | 100 | 1000 |
| BU75 | 3/4 | 1-1/4 | 3/8 | 13/16 | 3.2 | 100 | 1000 |
| BU100 | 1 | 1-5/8 | 1/2 | 1 | 8.2 | 50 | 500 |
| BU125 | 1-1/4 | 2 | 1/2 | 1-11/32 | 11.0 | 50 | 200 |
| BU150 | 1-1/2 | 2-5/16 | 1/2 | 1-9/16 | 13.0 | 50 | 200 |
| BU200 | 2 | 2-29/32 | 9/16 | 2 | 24.0 | 25 | 100 |
| BU250 | 2-1/2 | 3-1/4 | 3/4 | 2-1/2 | 36.0 | 10 | 50 |
| BU300 | 3 | 3-7/8 | 13/16 | 3 | 45.0 | 10 | 50 |
| BU350 | 3-1/2 | 4-9/16 | 13/16 | 3-17/32 | 85.0 | 5 | 20 |
| BU400 | 4 | 5-1/16 | 13/16 | 4 | 100.0 | 5 | 20 |
| BU500 | 5 | 6-5/16 | 1 | 4-7/8 | 155.0 | 1 | 10 |
| BU600 | 6 | 7-7/16 | 1 | 5-7/8 | 265.0 | 1 | 10 |
| **Bushings– Insulated– Malleable Iron– 150°C Temperature Rating** | | | | | | | |
| BU50I | 1/2 | 1-1/16 | 13/32 | 19/32 | 2.3 | 100 | 1000 |
| BU75I | 3/4 | 1-1/4 | 7/16 | 13/16 | 2.7 | 100 | 1000 |
| BU100I | 1 | 1-5/8 | 19/32 | 1 | 7.0 | 50 | 500 |
| BU125I | 1-1/4 | 2 | 19/32 | 1-11/32 | 12.0 | 50 | 200 |
| BU150I | 1-1/2 | 2-5/16 | 19/32 | 1-9/16 | 13.0 | 50 | 200 |
| BU200I | 2 | 2-29/32 | 21/32 | 2 | 23.0 | 25 | 100 |
| BU250I | 2-1/2 | 3-1/4 | 7/8 | 2-1/2 | 37.0 | 10 | 50 |
| BU300I | 3 | 3-7/8 | 29/32 | 3 | 41.0 | 10 | 50 |
| BU350I | 3-1/2 | 4-9/16 | 15/16 | 3-17/32 | 87.5 | 5 | 20 |
| BU400I | 4 | 5-1/16 | 15/16 | 4 | 102.5 | 5 | 20 |
| BU500I | 5 | 6-5/16 | 1-3/16 | 4-7/8 | 160.0 | 1 | 10 |
| BU600I | 6 | 7-7/16 | 1-3/16 | 5-7/8 | 271.0 | 1 | 10 |
| **Capped Bushings– Malleable Iron** | | | | | | | |
| BUC50 | 1/2 | 1-1/16 | 11/32 | – | 2.7 | 100 | 1000 |
| BUC75 | 3/4 | 1-1/4 | 3/8 | – | 4.0 | 100 | 1000 |
| BUC100 | 1 | 1-5/8 | 1/2 | – | 8.0 | 50 | 500 |
| BUC125 | 1-1/4 | 2 | 1/2 | – | 12.8 | 50 | 200 |
| BUC150 | 1-1/2 | 2-5/16 | 1/2 | – | 16.0 | 10 | 100 |
| BUC200 | 2 | 2-29/32 | 9/16 | – | 26.0 | 10 | 100 |
| BUC250 | 2-1/2 | 3-1/4 | 3/4 | – | 44.0 | 5 | 50 |
| BUC300 | 3 | 3-7/8 | 13/16 | – | 51.0 | 5 | 25 |
| BUC350 | 3-1/2 | 4-9/16 | 13/16 | – | 96.0 | 5 | 25 |
| BUC400 | 4 | 5-1/16 | 13/16 | – | 110.0 | 5 | 20 |
| **Impact Resistant Plastic Bushing– 105°C Temperature Rating** | | | | | | | |
| BBU50 | 1/2 | 1-1/16 | 13/32 | 19/32 | 0.6 | 100 | 400 |
| BBU75 | 3/4 | 1-5/16 | 13/32 | 25/32 | 0.8 | 100 | 400 |
| BBU100 | 1 | 1-9/16 | 9/16 | 1 | 1.5 | 50 | 200 |
| BBU125 | 1-1/4 | 1-29/32 | 9/16 | 1-5/16 | 2.3 | 25 | 100 |
| BBU150 | 1-1/2 | 2-3/16 | 9/16 | 1/9/16 | 3.0 | 25 | 100 |
| BBU200 | 2 | 2-11/16 | 5/8 | 2 | 4.0 | 25 | 50 |
| BBU250 | 2-1/2 | 3-3/16 | 23/32 | 2-13/32 | 7.8 | 10 | 20 |
| BBU300 | 3 | 3-27/32 | 3/4 | 3 | 10.0 | 10 | 20 |
| BBU350 | 3-1/2 | 4-11/32 | 3/4 | 3-13/32 | 13.0 | 5 | 10 |
| BBU400 | 4 | 4-27/32 | 25/32 | 3-29/32 | 11.0 | 5 | 10 |

Effective October, 2001
Copyright 2001  Printed in U.S.A.

**PAGE 16**

EGS Electrical Group  ⓐ Appleton

7770 N. Frontage Road
Skokie, Illinois 60077

**FIGURE 10–25**  RMC and IMC bushings *(Courtesy of Appleton Electric).*

EMT, you must be certain that you use the correct fitting. The walls of EMT are too thin to be threaded. Therefore, the fittings use either a setscrew or compression to fasten the fitting to the raceway. Figures 10–26 through 10–30 show the various types of connectors, couplings, caps/heads, adapters, clamps, and straps that are used with EMT.

## Conduit Bodies

The *NEC* defines a conduit body as:

> *A separate portion of a conduit or tubing system that provides access through a removable cover(s) to the interior of the system at a junction of two or more sections of the system or at a terminal point of the system.*

> *Boxes such as FS and FD or larger cast or sheet metal boxes are not classified as conduit bodies.*

There are many different types of conduit bodies, and they are made from many different materials. Figure 10–31 shows some aluminum conduit bodies that may be used with rigid metal conduit (RMC) and intermediate metal conduit (IMC), provided the conduit bodies are threaded. When working with electrical metal tubing (EMT), the conduit bodies will have the same physical appearance except that setscrews are used instead of threads. Figure 10–32 shows some of the different covers and gaskets that are available.

## Boxes

Boxes are used in a conduit system to allow access for the pulling of wires, contain splices, and allow for the mounting of devices (switches, receptacles, etc.). Boxes may be metallic or nonmetallic. They are available in a variety of shapes and sizes. In Figure 10–33, you will see various types of clamps and brackets that are used with boxes. Round boxes are shown in Figure 10–34. Figures 10–35, 10–36, and 10–37 show octagon boxes. Figure 10–35 shows octagon boxes with conduit knockouts. Figure 10–36 shows octagon boxes with nonmetallic-sheathed cable clamps. Finally, Figure 10–37 shows octagon boxes containing armored cable clamps. Covers for round and octagon boxes are shown in Figure 10–38.

Square boxes are shown in Figures 10–39 through 10–42. Figure 10–39 shows square boxes with conduit knockouts. Square boxes with nonmetallic-sheathed cable clamps are shown in Figure 10–40, while Figure 10–41 shows square boxes with armored cable/metal-clad cable clamps. Finally, Figure 10–42 shows square boxes that are used when mounting to metal studs. Figure 10–43 shows extension rings with conduit knockouts that can be used with square boxes, while Figure 10–44 shows some of the covers that are available for square boxes.

Handy boxes and covers are shown in Figures 10–45 through 10–47. Figures 10–48 through 10–50 show various types of switch boxes, and finally Figure 10–51 shows some accessories that are available for outlet and switch boxes.

## CONDUCTOR INSTALLATION

The most common method of installing conductors in raceways is to pull the conductors through the raceway with a tempered steel tape, called a fish tape, or fish. A hook is formed on one end of the fish tape, as seen in Figure 10–52. The hook serves two purposes: it permits the fish tape to slide through the raceway with ease, and it provides a means for attaching the conductors, as seen in Figures 10–53A and 10–53B.

# TW-1

## Steel EMT Set-Screw Connectors and Couplings

### Set Screw Connectors Type – 4000 Series

**Features**
- Heavy steel walls.
- Concretetight when taped.
- Pre-set/Pre-staked set screws.
- Male hub threads NPS.
- Available with insulated throat.
- Three-way combination screws.

**Standard Materials**
- Steel.

**Standard Finish**
- Zinc Plated.

**Third Party Certification**
- UL and cUL Listed: E-14815.

**Applicable Third Party Standards**
- UL 514B
- CSA Standard C22.2 No. 18
- NEMA FB-1
- Federal Spec. W-F-408E

Type 4050S — Type 4100S    Type 4125S — Type 4400S

| Trade Size (inches) | Catalog Number | Insulated Throat Catalog Number | Dimensions in Inches Body Length | Max Dia. | Thread Length | Wt. Lbs. Per 100 | Ctn. Qty. | Std. Pkg. |
|---|---|---|---|---|---|---|---|---|
| 1/2 | 4050S | 4050ST | 1-1/32 | 27/32 | 7/16 | 10 | 25 | 250 |
| 3/4 | 4075S | 4075ST | 1-3/32 | 1-5/64 | 7/16 | 15 | 20 | 200 |
| 1 | 4100S | 4100ST | 1-3/16 | 1-5/16 | 7/16 | 23 | 20 | 100 |
| 1-1/4 | 4125S | 4125ST | 1-3/4 | 1-11/16 | 9/16 | 42 | 10 | 40 |
| 1-1/2 | 4150S | 4150ST | 1-15/16 | 1-15/16 | 9/16 | 53 | 10 | 40 |
| 2 | 4200S | 4200ST | 2-3/16 | 2-3/8 | 5/8 | 75 | 5 | 20 |
| 2-1/2 | 4250S | 4250ST | 2-3/8 | 3-1/8 | 7/8 | 130 | — | 10 |
| 3 | 4300S | 4300ST | 2-7/8 | 3-3/4 | 7/8 | 230 | — | 5 |
| 3-1/2 | 4350S | 4350ST | 3 | 4-1/4 | 1 | 250 | — | 5 |
| 4 | 4400S | 4400ST | 3-3/16 | 4-3/4 | 1 | 250 | — | 5 |

### Set Screw Couplings Type – 5000 Series

**Features**
- Heavy steel walls.
- Concretetight when taped.
- Pre-set/Pre-staked set screws.
- Three-way combination screws.

**Standard Materials**
- Steel.

**Standard Finish**
- Zinc Plated.

**Third Party Certification**
- UL and cUL Listed: E-14815.

**Applicable Third Party Standards**
- UL 514B
- CSA Standard C22.2 No. 18
- NEMA FB-1
- Federal Spec. W-F-408E

All dimensions are approximate and subject to change.

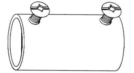

Type 5050S — Type 5100S    Type 5125S — Type 5400S

| Trade Size (Inches) | Catalog Number | Dimension in Inches Body Length | Max. Dia. | Wt. Lbs. Per 100 | Ctn. Qty. | Std. Pkg. |
|---|---|---|---|---|---|---|
| 1/2 | 5050S | 1-3/4 | 27/32 | 10 | 25 | 250 |
| 3/4 | 5075S | 2-1/8 | 1-5/64 | 16 | 20 | 200 |
| 1 | 5100S | 2-1/2 | 1-5/16 | 25 | 20 | 100 |
| 1-1/4 | 5125S | 3-1/4 | 1-11/16 | 45 | 10 | 40 |
| 1-1/2 | 5150S | 3-3/4 | 1-15/16 | 60 | 10 | 40 |
| 2 | 5200S | 4 | 2-3/8 | 82 | 5 | 20 |
| 2-1/2 | 5250S | 4-1/4 | 3-1/8 | 166 | — | 10 |
| 3 | 5300S | 4-1/2 | 3-3/4 | 213 | — | 5 |
| 3-1/2 | 5350S | 4-3/4 | 4-1/4 | 260 | — | 5 |
| 4 | 5400S | 5 | 4-3/4 | 300 | — | 5 |

**TW**

Effective October, 2001
Copyright 2001  Printed in U.S.A.
**PAGE 2**

EGS Electrical Group · Appleton

7770 N. Frontage Road
Skokie, Illinois 60077

**FIGURE 10–26**  Steel EMT set-screw connectors and couplings *(Courtesy of Appleton Electric).*

**TW-1**

# Steel EMT Compression Connectors and Couplings

## Gland Compression Couplings Type – 6000 Series

**Features**
- Concretetight.

**Standard Materials**
- Steel Body and Nut.

**Standard Finish**
- Zinc Plated

**Third Party Certification**
- UL Listed E14815
- Listed Concretetight 1/2" through 4".

**Applicable Third Party Standards**
- UL 514B
- CSA Standard C22.2 No. 18
- NEMA FB-1
- Federal Spec. W-F-408E

Type 6050S — Type 6400S

| Trade Size (Inches) | Catalog Number | Dimension in Inches Body Length | Max. Dia. | Wt. Lbs. Per 100 | Ctn. Qty. | Std. Pkg. |
|---|---|---|---|---|---|---|
| 1/2 | 6050S | 1-3/8 | 27/32 | 19 | 25 | 250 |
| 3/4 | 6075S | 1-1/2 | 1-3/8 | 25 | 20 | 200 |
| 1 | 6100S | 1-21/32 | 1-5/16 | 39 | 20 | 100 |
| 1-1/4 | 6125S | 2-1/2 | 1-11/16 | 88 | 5 | 25 |
| 1-1/2 | 6150S | 2-1/2 | 1-15/16 | 104 | 5 | 25 |
| 2 | 6200S | 2-1/2 | 2-3/8 | 149 | 2 | 10 |
| 2-1/2 | 6250S | 4-1/4 | 3-1/8 | 240 | — | 10 |
| 3 | 6300S | 4-1/2 | 3-3/4 | 350 | — | 5 |
| 3-1/2 | 6350S | 4-3/4 | 4-1/4 | 475 | — | 5 |
| 4 | 6400S | 5 | 4-3/4 | 600 | — | 5 |

## Gland Compression Connectors Type – 7000 Series

**Features**
- Male hub threads NPS.
- Available with insulated throat.
- Concretetight.

**Standard Materials**
- Steel Body and Nut.

**Standard Finish**
- Zinc Plated

**Third Party Certification**
- UL Listed E14815
- Listed Concretetight 1/2" through 4".

**Applicable Third Party Standards**
- UL 514B
- CSA Standard C22.2 No. 18
- NEMA FB-1
- Federal Spec. W-F-408E

All dimensions are approximate and subject to change.

Type 7050S — Type 7400S

| Trade Size (inches) | Catalog Number | Insulated Throat Catalog Number | Dimensions in Inches Body Length | Max Dia. | Thread Length | Wt. Lbs. Per 100 | Ctn. Qty. | Std. Pkg. |
|---|---|---|---|---|---|---|---|---|
| 1/2 | 7050S | 7050ST | 1-1/8 | 1-1/16 | 3/8 | 15 | 25 | 250 |
| 3/4 | 7075S | 7075ST | 1-1/32 | 1-1/4 | 7/16 | 21 | 25 | 250 |
| 1 | 7100S | 7100ST | 1-1/4 | 1-9/16 | 7/16 | 32 | 20 | 200 |
| 1-1/4 | 7125S | 7125ST | 1-13/16 | 2 | 9/16 | 72 | 5 | 25 |
| 1-1/2 | 7150S | 7150ST | 2 | 2-3/16 | 9/16 | 85 | 5 | 25 |
| 2 | 7200S | 7200ST | 2-1/4 | 2-11/16 | 11/16 | 124 | 2 | 10 |
| 2-1/2 | 7250S | 7250ST | 2-3/8 | 3-11/32 | 7/8 | 160 | — | 10 |
| 3 | 7300S | 7300ST | 2-7/8 | 4 | 7/8 | 265 | — | 5 |
| 3-1/2 | 7350S | 7350ST | 3 | 4-1/2 | 1 | 425 | — | 5 |
| 4 | 7400S | 7400ST | 3-3/16 | 5-1/16 | 1 | 600 | — | 5 |

**TW**

Effective October, 2001
Copyright 2001  Printed in U.S.A.

EGS
Electrical Group
▲ Appleton

7770 N. Frontage Road
Skokie, Illinois 60077

**PAGE 3**

**FIGURE 10–27**  EMT compression connectors and couplings *(Courtesy of Appleton Electric)*.

# TW-1

## EMT Connectors and Couplings

| Catalog No. | Trade Size (In.) | Dimensions in Inches A | B | C | Wt. Lbs. Per 100 | Ctn. Qty. | Std. Pkg. |
|---|---|---|---|---|---|---|---|
| **Gland Compression Connectors – Concrete Tight – Zinc Die Cast** | | | | | | | |
| TC-601 | 1/2 | 1.27 | 0.42 | 1.95 | 9.0 | 50 | 500 |
| TC-602 | 3/4 | 1.25 | 0.42 | 1.23 | 14.0 | 25 | 250 |
| TC-603 | 1 | 1.59 | 0.54 | 1.49 | 24.0 | 10 | 100 |
| TC-604 | 1-1/4 | 1.63 | 0.58 | 1.93 | 40.0 | – | 25 |
| TC-605 | 1-1/2 | 2.30 | 0.74 | 2.25 | 60.0 | – | 20 |
| TC-606 | 2 | 2.43 | 0.74 | 2.75 | 100.0 | – | 10 |
| TC-607 | 2-1/2 | 2.86 | 0.99 | 3.29 | 116.0 | – | 12 |
| TC-608 | 3 | 2.90 | 1.01 | 3.89 | 142.0 | – | 12 |
| TC-609 | 3-1/2 | 2.95 | 1.10 | 4.39 | 216.0 | – | 6 |
| TC-610 | 4 | 2.73 | 1.00 | 4.95 | 216.0 | – | 6 |
| **Gland Compression Connectors – Insulated – Concrete Tight – Zinc Die Cast** | | | | | | | |
| TCI-601 | 1/2 | 1.27 | 0.42 | 1.95 | 9.0 | 50 | 500 |
| TCI-602 | 3/4 | 1.25 | 0.42 | 1.23 | 14.0 | 25 | 250 |
| TCI-603 | 1 | 1.59 | 0.54 | 1.49 | 24.0 | 10 | 100 |
| TCI-604 | 1-1/4 | 1.63 | 0.58 | 1.93 | 40.0 | – | 25 |
| TCI-605 | 1-1/2 | 2.30 | 0.74 | 2.25 | 60.0 | – | 20 |
| TCI-606 | 2 | 2.43 | 0.74 | 2.75 | 100.0 | – | 10 |
| TCI-607 | 2-1/2 | 2.86 | 0.99 | 3.29 | 116.0 | – | 12 |
| TCI-608 | 3 | 2.90 | 1.01 | 3.89 | 142.0 | – | 12 |
| TCI-609 | 3-1/2 | 2.95 | 1.10 | 4.39 | 216.0 | – | 6 |
| TCI-610 | 4 | 2.73 | 1.00 | 4.95 | 216.0 | – | 6 |
| **Gland Compression Couplings – Concrete Tight – Zinc Die Cast** | | | | | | | |
| TC-611 | 1/2 | 1.52 | 0.95 | – | 11.0 | 50 | 500 |
| TC-612 | 3/4 | 1.66 | 1.26 | – | 18.0 | 25 | 250 |
| TC-613 | 1 | 2.02 | 1.50 | – | 31.0 | 10 | 100 |
| TC-614 | 1-1/4 | 2.18 | 1.92 | – | 60.0 | – | 25 |
| TC-615 | 1-1/2 | 3.10 | 2.25 | – | 76.0 | – | 20 |
| TC-616 | 2 | 3.06 | 2.75 | – | 65.0 | – | 10 |
| TC-617 | 2-1/2 | 3.63 | 3.27 | – | 142.0 | – | 12 |
| TC-618 | 3 | 3.52 | 3.89 | – | 192.0 | – | 12 |
| TC-619 | 3-1/2 | 3.77 | 4.40 | – | 233.0 | – | 6 |
| TC-620 | 4 | 4.06 | 4.89 | – | 267.0 | – | 6 |
| **Combination Couplings – EMT to Rigid Metal Conduit – Steel** | | | | | | | |
| TWR-50 | 1/2 | 11/16 | 1/2 | 1 | 10.0 | 50 | 200 |
| TWR-75 | 3/4 | 11/16 | 17/32 | 1-9/32 | 15.0 | 25 | 100 |
| TWR-100 | 1 | 7/8 | 17/32 | 1-19/32 | 17.0 | 25 | 100 |
| TWR-125 | 1-1/4 | 31/32 | 27/32 | 1-15/16 | 40.0 | 5 | 25 |
| **Connectors – Two Piece – Concrete Tight – Steel** | | | | | | | |
| 92T050 | 1/2 | 5/16 | 11/32 | 1-1/16 | 6.0 | 50 | 200 |
| 92T075 | 3/4 | 5/16 | 7/16 | 1-9/32 | 8.8 | 25 | 250 |
| 92T100 | 1 | 11/32 | 13/32 | 1-5/8 | 13.0 | 25 | 200 |

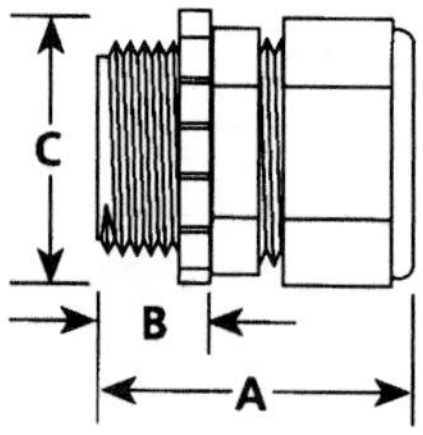

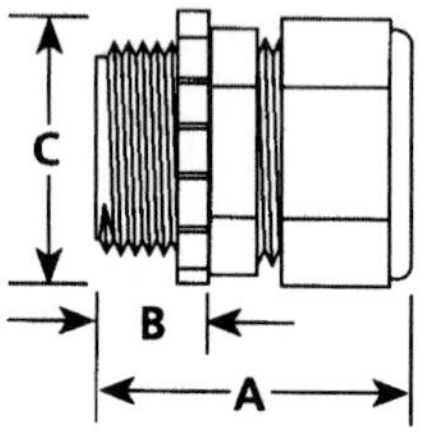

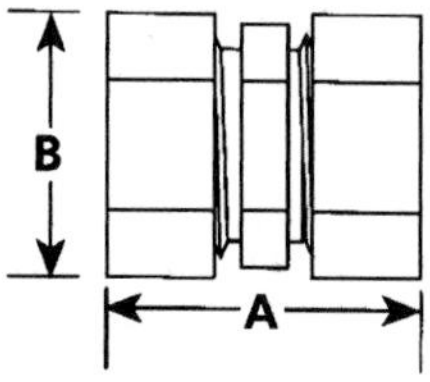

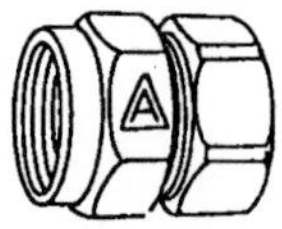

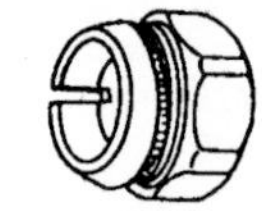

Effective October, 2001
Copyright 2001  Printed in U.S.A.
**PAGE 4**

7770 N. Frontage Road
Skokie, Illinois 60077

**FIGURE 10–28**  EMT connectors and couplings *(Courtesy of Appleton Electric).*

# TW-2

## EMT Couplings, 90° Connectors and Caps/Heads

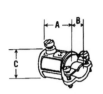

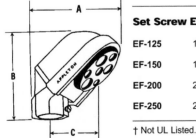

| Catalog No. | Trade Size (In.) | | Dimensions in Inches A | B | C | Wt. Lbs. Per 100 | Ctn. Qty. | Std. Pkg. |
|---|---|---|---|---|---|---|---|---|
| **Combination Coupling – EMT to Flexible Metal Conduit Pressure Cast** | | | | | | | | |
| TW38-50DC† | 1/2 | | 1-9/32 | 1-1/4 | 15/16 | 7.5 | 25 | 100 |
| **Malleable Body/Steel Nut †** | | | | | | | | |
| TWCC38-50 | 1/2 | | 27/32 | 5/8 | 1-1/16 | 12.0 | 10 | 100 |
| TWCC-50 | 1/2 | | 7/8 | 5/8 | 1-1/16 | 14.5 | 10 | 100 |
| TWCC-75 | 3/4 | | 1-1/16 | 21/32 | 1-5/16 | 25.0 | 10 | 100 |
| TWCC-100 | 1 | | 1-3/16 | 27/32 | 1-7/8 | 61.0 | 5 | 50 |
| **Gland Compression 90° Long Connectors – Concrete Tight Malleable – Steel Nut** | | | | | | | | |
| TWLL-50L | 1/2 | | 1-7/8 | 2-1/16 | 19/32 | 32.0 | 25 | 100 |
| TWLL-75L | 3/4 | Male End Has Locknut | 2-3/16 | 2-7/16 | 7/16 | 50.0 | 10 | 50 |
| TWLL-100L | 1 | | 2-19/32 | 3 | 7/16 | 105.0 | 5 | 25 |
| TWLL-125L | 1-1/4 | | 2-13/16 | 3-5/16 | 9/16 | 136.0 | 5 | 25 |
| **Gland Compression 90° Short Connectors – Concrete Tight Malleable – Steel Nut** | | | | | | | | |
| TWL-50L | 1/2 | Male End Has Locknut | 1-7/32 | 1-5/16 | 13/32 | 20.0 | 25 | 100 |
| TWL-75L | 3/4 | | 1-5/16 | 1-7/16 | 7/16 | 30.0 | 10 | 100 |
| TWL-100L | 1 | | 1-5/8 | 1-9/16 | 7/16 | 53.0 | 5 | 50 |
| **Set Screw Entrance Caps – Malleable Iron – Exclusive "Snap-On" Cover** | | | | | | | | |
| EF-125 | 1-1/4 | (3) 5/8" and (2) 13/32" | 4-5/16 | 2-7/8 | 3-5/8 | 64.0 | – | 10 |
| EF-150 | 1-1/2 | (3) 25/32", (2) 9/16" and (1) 3/8" | 5 | 3-5/16 | 4-11/32 | 93.0 | – | 10 |
| EF-200 | 2 | (3) 1", (2) 3/4 and (3) 17/32" | 6-11/32 | 4-1/2 | 5-9/16 | 183.0 | – | 5 |
| EF-250 | 2-1/2 | (3) 7/8", (1) 1" and (3) 1-5/16" | 9 | 6-1/4 | 7-3/4 | 350.0 | – | 1 |

† Not UL Listed.

**TW**

Effective October, 2001
Copyright 2001 Printed in U.S.A.

7770 N. Frontage Road
Skokie, Illinois 60077

**PAGE 1**

**FIGURE 10–29** EMT couplings, 90-degree connectors, and caps/heads *(Courtesy of Appleton Electric).*

# TW-2

## EMT Adapters, Clamps and Straps

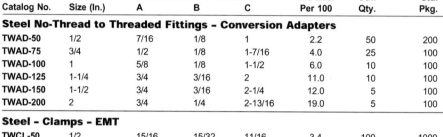

| Catalog No. | Trade Size (In.) | Dimensions in Inches | | | Wt. Lbs. Per 100 | Ctn. Qty. | Std. Pkg. |
|---|---|---|---|---|---|---|---|
| | | A | B | C | | | |
| **Steel No-Thread to Threaded Fittings – Conversion Adapters** | | | | | | | |
| TWAD-50 | 1/2 | 7/16 | 1/8 | 1 | 2.2 | 50 | 200 |
| TWAD-75 | 3/4 | 1/2 | 1/8 | 1-7/16 | 4.0 | 25 | 100 |
| TWAD-100 | 1 | 5/8 | 1/8 | 1-1/2 | 6.0 | 10 | 100 |
| TWAD-125 | 1-1/4 | 3/4 | 3/16 | 2 | 11.0 | 10 | 100 |
| TWAD-150 | 1-1/2 | 3/4 | 3/16 | 2-1/4 | 12.0 | 5 | 100 |
| TWAD-200 | 2 | 3/4 | 1/4 | 2-13/16 | 19.0 | 5 | 100 |
| **Steel – Clamps – EMT** | | | | | | | |
| TWCL-50 | 1/2 | 15/16 | 15/32 | 11/16 | 3.4 | 100 | 1000 |
| TWCL-75 | 3/4 | 1-1/32 | 17/32 | 29/32 | 4.6 | 100 | 1000 |
| TWCL-100 | 1 | 1-5/32 | 7/16 | 1-1/8 | 6.5 | 50 | 100 |
| TWCL-125 | 1-1/4 | 1-11/32 | 1/2 | 1-15/32 | 10.0 | 50 | 100 |
| TWCL-150 | 1-1/2 | 1-11/16 | 19/32 | 1-11/16 | 19.0 | 50 | 100 |
| TWCL-200 | 2 | 1-27/32 | 5/8 | 2-5/32 | 25.0 | 5 | 25 |
| CL-250 | 2-1/2 | 2-13/16 | 1-1/8 | 2-3/4 | 76.0 | 5 | 25 |
| CL-300 | 3 | 3-5/16 | 1-5/16 | 3-3/8 | 89.0 | – | 10 |
| CL-350 | 3-1/2 | 3-7/16 | 1-5/8 | 3-7/8 | 112.0 | – | 10 |
| CL-400 | 4 | 3-25/32 | 1-5/8 | 4-3/8 | 128.0 | – | 10 |

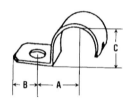

| Catalog No. | Trade Size (In.) | Dimensions in Inches | | | | Wt. Lbs. Per 100 | Ctn. Qty. | Std. Pkg. |
|---|---|---|---|---|---|---|---|---|
| | | A | B | C | D | | | |
| **Malleable – Clamps – EMT** | | | | | | | | |
| TWCL-50MN | 1/2 | 1-1/16 | 7/8 | 5/8 | 1/4 | 5.0 | 100 | 1000 |
| TWCL-75MN | 3/4 | 1-3/16 | 1-1/16 | 3/4 | 1/4 | 6.0 | 50 | 500 |
| TWCL-100MN | 1 | 1-3/8 | 1-3/8 | 13/16 | 1/4 | 10.0 | 50 | 250 |
| TWCL-125MN | 1-1/4 | 1-5/8 | 1-3/4 | 7/8 | 3/8 | 13.0 | 25 | 250 |
| TWCL-150MN | 1-1/2 | 1-13/16 | 1-15/16 | 15/16 | 3/8 | 15.0 | 25 | 100 |
| TWCL-200MN | 2 | 2-5/16 | 2-1/2 | 1-1/16 | 1/2 | 20.0 | 25 | 100 |
| CL-250MN | 2-1/2 | 2-9/16 | 3 | 1-5/16 | 1/2 | 71.0 | 5 | 25 |
| CL-300MN | 3 | 3-3/16 | 3-13/16 | 1-1/2 | 1/2 | 107.0 | 5 | 25 |
| CL-350MN | 3-1/2 | 3-3/8 | 4-7/16 | 1-11/16 | 5/8 | 172.0 | 2 | 10 |
| CL-400MN | 4 | 3-3/4 | 5 | 1-15/16 | 5/8 | 266.0 | 2 | 10 |

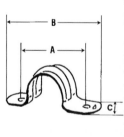

| Catalog No. | Trade Size (In.) | Dimensions in Inches | | | Wt. Lbs. Per 100 | Ctn. Qty. | Std. Pkg. |
|---|---|---|---|---|---|---|---|
| | | A | B | C | | | |
| **Steel Pipe Straps – Two Hole – EMT** | | | | | | | |
| TW2500 | 1/2 | 1-1/2 | 2-1/32 | 5/8 | 1.0 | – | 250 |
| TW2750 | 3/4 | 1-3/4 | 2-13/32 | 5/8 | 1.6 | – | 250 |
| TW2100 | 1 | 2-3/32 | 2-13/16 | 5/8 | 2.0 | – | 100 |
| TW2125 | 1-1/4 | 2-19/32 | 3-5/16 | 3/4 | 4.0 | – | 100 |
| TW2150 | 1-1/2 | 2-11/16 | 3-7/16 | 3/4 | 4.0 | – | 100 |
| TW2200 | 2 | 3-1/2 | 4-3/4 | 15/16 | 4.7 | 25 | 100 |
| CF250-H | 2-1/2 | 4-3/8 | 5-5/8 | 3-1/32 | 5.7 | 25 | 100 |
| CF300-H | 3 | 5 | 6-1/4 | 1 | 8.0 | 25 | 100 |
| CF350-H | 3-1/2 | 6-1/4 | 7-1/4 | 1 | 8.7 | 25 | 50 |
| CF400-H | 4 | 7 | 8 | 1-1/4 | 9.4 | 10 | 50 |

Effective October, 2001
Copyright 2001  Printed in U.S.A.

**PAGE 2**

7770 N. Frontage Road
Skokie, Illinois 60077

**FIGURE 10–30**   EMT adapters, clamps, and straps *(Courtesy of Appleton Electric).*

# A-12

## Form 85™ Aluminum Unilet®
## Conduit Outlet Bodies

Threaded Type for use with Rigid Metal Conduit and IMC.
Setscrew Type for use with Electrical Metal Tubing (EMT).

**Appleton Form 85™ Conduit Bodies: Threaded/SetScrew Type**   NOTE: Refer to page A-16 for Wiring Capacity Tables

| Hub Size (in.) | C Threaded Type | C Setscrew Type | E Threaded Type | E Setscrew Type | LB Threaded Type | LB Setscrew Type | LL Threaded Type | LL Setscrew Type | LR Threaded Type | LR Setscrew Type |
|---|---|---|---|---|---|---|---|---|---|---|
| 1/2 | C50-A | C50T-A | E50-A | E50T-A | LB50-A | LB50T-A | LL50-A | LL50T-A | LR50-A | LR50T-A |
| 3/4 | C75-A | C75T-A | E75-A | — | LB75-A | LB75T-A | LL75-A | LL75T-A | LR75-A | LR75T-A |
| 1 | C100-A | C100T-A | E100-A | — | LB100-A | LB100T-A | LL100-A | LL100T-A | LR100-A | LR100T-A |
| 1-1/4 | C125-A | C125T-A | — | — | LB125-A | LB125T-A | LL125-A | LL125T-A | LR125-A | LR125T-A |
| 1-1/2 | C150-A | C150T-A | — | — | LB150-A | LB150T-A | LL150-A | LL150T-A | LR150-A | LR150T-A |
| 2 | C200-A | — | — | — | LB200-A | LB200T-A | LL200-A | LL200T-A | LR200-A | LR200T-A |
| 2-1/2 | C250-A | — | — | — | LB250-A | — | LL250-A | — | LR250-A | — |
| 3 | C300-A | — | — | — | LB300-A | — | LL300-A | — | LR300-A | — |
| 3-1/2 | C350-A | — | — | — | LB350-A | — | LL350-A | — | LR350-A | — |
| 4 | C400-A | — | — | — | LB400-A | — | LL400-A | — | LR400-A | — |

| Hub Size (in.) | T Threaded Type | T Setscrew Type | TB Threaded Type | TB Setscrew Type | X Threaded Type | X Setscrew Type |
|---|---|---|---|---|---|---|
| 1/2 | T50-A | T50T-A | TB50-A | — | X50-A | — |
| 3/4 | T75-A | T75T-A | TB75-A | — | X75-A | — |
| 1 | T100-A | T100T-A | TB100-A | — | X100-A | — |
| 1-1/4 | T125-A | — | TB125-A | — | — | — |
| 1-1/2 | T150-A | — | TB150-A | — | — | — |
| 2 | T200-A | — | TB200-A | — | — | — |
| 2-1/2 | T250-A | — | — | — | — | — |
| 3 | T300-A | — | — | — | — | — |
| 3-1/2 | T350-A | — | — | — | — | — |
| 4 | T400-A | — | — | — | — | — |

Typical Form 85 Conduit Bodies with Setscrews. For use with Electrical Metal Tubing (EMT).

### Back Style for Form 85®

| Unilet Body | Flat Back | Round Back |
|---|---|---|
| C, LB, LL, LR, T | 1/2" – 2" | 2-1/2" – 4" |
| TB | 1-1/4", 1-1/2" | 1/2", 3/4", 1", 2" |
| E, X | 1/2" – 1" | |

All setscrew types are flatback design.

Effective October, 2001
Copyright 2001  Printed in U.S.A.

**PAGE 12**

7770 N. Frontage Road
Skokie, Illinois 60077

**EGS** Electrical Group  ⚠ **Appleton**®

**A**

---

**FIGURE 10–31**   RMC, IMC, and EMT conduit bodies *(Courtesy of Appleton Electric).*

# A-13

## Covers and Gaskets for Form 85™ Unilet® Conduit Outlet Bodies

Covers Furnished with Stainless Steel Fastening Screws.

**Appleton Form 85™ Covers and Gaskets** NOTE: Refer to page A-16 for Wiring Capacity Tables.

| Size | Blank Stamped Aluminum | Blank Cast Aluminum | Neoprene | Composition Fiber |
|------|------------------------|---------------------|----------|-------------------|
| | Domed: 1/2" - 3"  Flat: 3-1/2" - 4" | Flat: 1/2" - 2"  Domed: 2-1/2" - 4" | Tear out inner perforated section to convert to "open type" gasket. | |
| Form 85 Body Size (in.) | | | | |
| 1/2 | K50-A | K50-CA | GK50-N | GK50-V |
| 3/4 | K75-A | K75-CA | GK75-N | GK75-V |
| 1 | K100-A | K100-CA | GK100-N | GK100-V |
| 1-1/4 | K125 & 150-A | K125 & 150-CA | GK125-150-N | GK125-150-V |
| 1-1/2 | K125 & 150-A | K125 & 150-CA | GK125-150-N | GK125-150-V |
| 2 | K200-A | K200-CA | GK200-N | GK200-V |
| 2-1/2 | K250 & 300-A | K250 & 300-CA | GK250-300-N | GK-250-300-V |
| 3 | K250 & 300-A | K250 & 300-CA | GK250-300-N | GK-250-300-V |
| 3-1/2 | K350 & 400-A | K350 & 400-CA | GK350-400-N | GK-350-400-V |
| 4 | K350 & 400-A | K350 & 400-CA | GK350-400-N | GK-350-400-V |

**Wiring Device Covers**

| Lamp Receptacle with 1-1/2" Shade Holder Groove. Porcelain, 600 Watt, 600 Volt Rating | | Device Cover for Interchangeable (Despard) Wiring Devices, Cast Aluminum with Mounting Strap | | Switch Cover for Interchangeable (Despard) Devices, Cast Aluminum with Gasket and Steel Mounting Strap | |
|---|---|---|---|---|---|
| Form 85 Body Size (in.) | Cat. No. | Form 85 Body Size (in.) | Cat. No. | Form 85 Body Size (in.) | Cat. No. |
| 1/2 | KLR50 | 1/2 | KWD50-A | 3/4 | KVS75-A |
| 3/4 | KLR75 | | | | |

Effective October, 2001
Copyright 2001 Printed in U.S.A.

EGS Electrical Group  ⓐ Appleton®

7770 N. Frontage Road
Skokie, Illinois 60077

**PAGE 13**

**FIGURE 10–32** Covers and gaskets for conduit bodies *(Courtesy of Appleton Electric).*

# OB/SB-1

## Appleton Clamps

CL-9 Clamp
For nonmetallic cable

CL-20 Clamp
For nonmetallic cable

CL-24 Clamp
For nonmetallic cable

CL-26 Clamp
For nonmetallic cable

CL-13 Clamp
For armored cable and
flexible metal conduit

CL-25 Clamp
For armored cable and
flexible metal conduit

## Appleton Brackets

**OB/SB**

JB
Bracket

Plain
NL Bracket

Stud
NL Bracket

Snap-in Metal
Stud Bracket

MSC-OB
Outlet Box Metal
Stud Clip

892
LOX BOX® Switch
Box Supports

     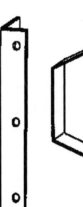 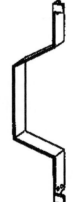 

Plain
Offset Side
Combination
Brackets

Barbed-Hook
Vertical

Plain
Vertical
Bracket

Vertical
Bracket
with Stud

Plain
Vertical
Bracket

Plain
Vertical
Angle
Bracket

Offset
Vertical
Bracket

Plain
Vertical
Bracket

Effective October, 2001
Copyright 2001  Printed in U.S.A.

PAGE 2

7770 N. Frontage Road
Skokie, Illinois 60077

**FIGURE 10–33**  Box clamps and brackets *(Courtesy of Appleton Electric).*

# OB/SB-1

LISTED
File #
E2527

## 3-1/4" Round Boxes

### With Nonmetallic Sheathed Cable Clamps

| Catalog Number (Universal No.) | Depth (inches) | Wiring Cubic Inch Capacity | Type Clamp | Knockout Description | | | Wt. Lbs. Per 100 | Std. Pkg. |
|---|---|---|---|---|---|---|---|---|
| | | | | Sides | Ends | Bottom | | |
| 510LC (36115) | 1/2 | 4.5 | CL-9 | -- | -- | (3)1/2" (4)21/32" | 29.0 | 50 |
| 510L | 1/2 | 4.5 | -- | -- | -- | (3)1/2" (4)21/32" | 48.0 | 50 |

510LC
1/2" Deep

510L
1/2" Deep

## 3-1/4" Round Covers

### For 3-1/4" Octagon and Round Boxes

| Catalog Number (Universal No.) | Depth (inches) | Wiring Cubic In. Cap. | Description | Wt. Lbs. Per 100 | Std. Pkg. |
|---|---|---|---|---|---|
| 8301A (24C1) | Flat | 0.0 | Blank | 17 | 50 |
| 8320 (24C6) | Flat | 0.0 | 1/2"KO | 17 | 50 |

8301A

8320

## 4" Round Box

### With Conduit Knockouts in Bottom

| Catalog Number (Universal No.) | Depth (inches) | Wiring Cubic Inch Capacity | Knockout Description | | | Wt. Lbs. Per 100 | Std. Pkg. |
|---|---|---|---|---|---|---|---|
| | | | Sides | Ends | Bottom | | |
| 4CL (56111) | 1/2 | 6.0 | -- | -- | (5)1/2" | 36.0 | 50 |

4CL
1/2" Deep

EGS Electrical Group · Appleton

7770 N. Frontage Road
Skokie, Illinois 60077

**FIGURE 10–34**  Round boxes and covers *(Courtesy of Appleton Electric).*

209

# OB/SB-1

## 4" Octagon Boxes
With Conduit Knockouts

4O-1/2
1-1/2" Deep

4O-3/4
1-1/2" Deep

4O-SPL
1-1/2" Deep

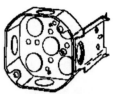
4OJB-1/2
1-1/2" Deep
Bracket set flush,
gauging notches 3/8" & 1/2"

4OD-1/2
2-1/8" Deep

4OD-3/4
2-1/8" Deep

4OD-SPL
2-1/8" Deep

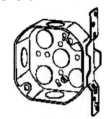
4OVB-1/2
1-1/2" Deep
Vertical bracket set back 5/8"

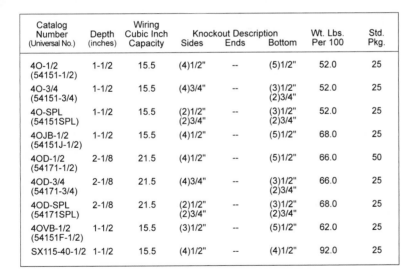

| Catalog Number (Universal No.) | Depth (inches) | Wiring Cubic Inch Capacity | Knockout Description | | | Wt. Lbs. Per 100 | Std. Pkg. |
|---|---|---|---|---|---|---|---|
| | | | Sides | Ends | Bottom | | |
| 4O-1/2 (54151-1/2) | 1-1/2 | 15.5 | (4)1/2" | -- | (5)1/2" | 52.0 | 25 |
| 4O-3/4 (54151-3/4) | 1-1/2 | 15.5 | (4)3/4" | -- | (3)1/2" (2)3/4" | 52.0 | 25 |
| 4O-SPL (54151SPL) | 1-1/2 | 15.5 | (2)1/2" (2)3/4" | -- | (3)1/2" (2)3/4" | 52.0 | 25 |
| 4OJB-1/2 (54151J-1/2) | 1-1/2 | 15.5 | (4)1/2" | -- | (5)1/2" | 68.0 | 25 |
| 4OD-1/2 (54171-1/2) | 2-1/8 | 21.5 | (4)1/2" | -- | (5)1/2" | 66.0 | 50 |
| 4OD-3/4 (54171-3/4) | 2-1/8 | 21.5 | (4)3/4" | -- | (3)1/2" (2)3/4" | 66.0 | 25 |
| 4OD-SPL (54171SPL) | 2-1/8 | 21.5 | (2)1/2" (2)3/4" | -- | (3)1/2" (2)3/4" | 68.0 | 25 |
| 4OVB-1/2 (54151F-1/2) | 1-1/2 | 15.5 | (3)1/2" | -- | (5)1/2" | 62.0 | 25 |
| SX115-40-1/2 | 1-1/2 | 15.5 | (4)1/2" | -- | (4)1/2" | 92.0 | 25 |

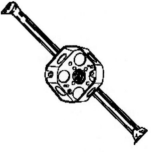
SX115-40-1/2
1-1/2" Deep
4O-1/2 box on SX115 stud bar hanger
adjusts from 11-1/2" to 18-1/2"

OB/SB

Effective October, 2001
Copyright 2001  Printed in U.S.A.

PAGE 6

EGS Electrical Group ⚡Appleton

7770 N. Frontage Road
Skokie, Illinois 60077

**FIGURE 10–35** Octagon boxes *(Courtesy of Appleton Electric).*

LISTED
File #
E2527

# OB/SB-1

## 4" Octagon Boxes

With Nonmetallic Sheathed Cable Clamps

**561L**
1-1/2" Deep

**561N**
1-1/2" Deep
Four nail holes each end

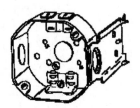

**561LJB**
1-1/2" Deep
Bracket set flush,
Gauging notches 3/8" & 1/2"

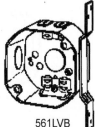

**561LVB**
1-1/2" Deep
Bracket set back 5/8"

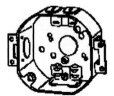

**561LXE**
1-1/2" Deep
Plaster ears

**561LD**
2-1/8" Deep

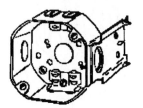

**561LDJB**
2-1/8" Deep
Bracket set back 1"
Gauging notches set back 1-3/8" & 1-1/2"

**SX18-561L**
1-1/2" Deep
561L box on SX18 clip bar
Hanger adjusts from 11-1/2" to 18-1/2"

**SX115-561L**
1-1/2" Deep
561L box on SX115 stud bar
Hanger adjusts from 11-1/2" to 18-1/2"

| Catalog Number Std. (Universal No.) | Depth (inches) | Wiring Cubic Inch Capacity | Type Clamp | Knockout Description | | | Per 100 | Wt. Lbs. Pkg. |
|---|---|---|---|---|---|---|---|---|
| | | | | Sides | Ends | Bottom | | |
| 561L (54151) | 1-1/2 | 15.5 | CL-26 | (2)1/2" | (4)21/32" | (1)1/2" | 57.0 | 50 |
| 561N (54151N) | 1-1/2 | 15.5 | CL-26 | (2)1/2" | (4)21/32" | (1)1/2" | 57.0 | 50 |
| 561LJB (54151JN) | 1-1/2 | 15.5 | CL-26 | (2)1/2" | (4)21/32" | (1)1/2" | 74.0 | 25 |
| 561LVB (54151FN) | 1-1/2 | 15.5 | CL-26 | (1)1/2" | (4)21/32" | (1)1/2" | 66.0 | 25 |
| 561LXE (54151NE) | 1-1/2 | 15.5 | CL-26 | (2)1/2" | (4)21/32" | (1)1/2" | 62.0 | 50 |
| 561LD (54171N) | 2-1/8 | 21.5 | CL-26 | (2)1/2" | (4)21/32" | (1)1/2" | 70.0 | 25 |
| 561LDJB (54171JN) | 2-1/8 | 21.5 | CL-26 | (2)1/2" | (4)21/32" | (1)1/2" | 84.0 | 25 |
| SX18-561L | 1-1/2 | 15.5 | CL-26 | (2)1/2" | (4)21/32" | -- | 96.0 | 25 |

Effective October, 2001
Copyright 2001  Printed in U.S.A.

**EGS** Electrical Group  ⚠ **Appleton**

7770 N. Frontage Road
Skokie, Illinois 60077

PAGE 7

**FIGURE 10–36**  Octagon boxes with nonmetallic-sheathed cable clamps *(Courtesy of Appleton Electric).*

# OB/SB-1

## 4" Octagon Boxes

### With Armored Cable Clamps

**551L**
1-1/2" Deep
Two nail holes each end

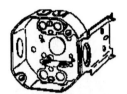

**551LJB**
1-1/2" Deep
Bracket set flush,
Gauging Notches 3/8" & 1/2"

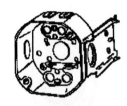

**551LDJB**
2-1/8" Deep
Bracket set flush,
Gauging Notches 1-3/8" & 1-1/2"

**SX115-551L**
1-1/2" Deep
551L box on SX115 stud bar
Hanger adjusts from 11-1/2" to 18-1/2"

| Catalog Number Lbs. Std. (Universal No.) | Depth (inches) | Wiring Cubic Inch Capacity | Type Clamp | Knockout Description | | | Per 100 | Wt Pkg. |
|---|---|---|---|---|---|---|---|---|
| | | | | Sides | Ends | Bottom | | |
| 551L (54151X) | 1-1/2 | 15.5 | CL-25 | (2)1/2" | (4)21/32" (4)21/32" | (1)1/2" | 61.0 | 50 |
| 551LJB | 1-1/2 | 15.5 | CL-25 | (2)1/2" | (4)21/32" | (1)1/2" (4)21/32" | 74.0 | 25 |
| 551LDJB | 2-1/8 | 21.5 | CL-25 | (2)1/2" | (4)21/32" | (1)1/2" (4)21/32" | 84.0 | 25 |

## 4" Octagon Box Extension Rings

### With Conduit Knockouts

**4OE-1/2**
1-1/2" Deep

**4OE-3/4**
1-1/2" Deep

**4OE-SPL**
1-1/2" Deep

| Catalog Number Lbs. Std. (Universal No.) | Depth (inches) | Wiring Cubic Inch Capacity | Type Clamp | Knockout Description | | | Per 100 | Wt Pkg. |
|---|---|---|---|---|---|---|---|---|
| | | | | Sides | Ends | Bottom | | |
| 4OE-1/2 (55151-1/2) | 1-1/2 | 15.5 | -- | (4)1/2" | -- | -- | 37.0 | 50 |
| 4OE-3/4 (55151-3/4) | 1-1/2 | 15.5 | -- | (4)3/4" | -- | -- | 39.0 | 50 |
| 4OE-SPL | 1-1/2 | 15.5 | -- | (2)1/2" | -- | -- | 39.0 | 50 |

Effective October, 2001
Copyright 2001  Printed in U.S.A.
PAGE 8

7770 N. Frontage Road
Skokie, Illinois 60077

**FIGURE 10–37**  Octagon boxes with armored cable clamps *(Courtesy of Appleton Electric).*

# OB/SB-1

## 4" Octagon and Round Covers

For 4" Octagon and Round Boxes

8409A
1/2" Raised

8409
5/8" Raised

8409B
3/4" Raised

8409C
1" Raised

8403

8413

8419LR

8420LR

| Catalog Number (Universal No.) | Depth (inches) | Wiring Cubic In. Cap. | Description | Wt. Lbs. Per 100 | Std. Pkg. |
|---|---|---|---|---|---|
| 8409A (54C3-1/2) | 1/2" Raised | 3.5 | Ears 2-3/4" C to C | 18.0 | 50 |
| 8409 (54C3-3/4) | 5/8" Raised | 4.5 | Ears 2-3/4" C to C | 20.0 | 50 |
| 8409B (54C-3/4) | 3/4" Raised | 5.5 | Ears 2-3/4" C to C | 24.0 | 25 |
| 8409C (54C3-1) | 1" Raised | 7.0 | Ears 2-3/4" C to C | 32.0 | 25 |
| 8403 (5401) | Flat | 0.0 | Blank | 26.0 | 50 |
| 8413 (54C6) | Flat | 0.0 | 1/2" Knockout | 26.0 | 50 |
| 8419LR (54-S) | Flat | 0.0 | One Single Flush Receptacle 1-13/32" D | 21.0 | 50 |
| 8420LR (54-D) | Flat | 0.0 | Duplex Receptacle | 19.0 | 50 |

Effective October, 2001
Copyright 2001  Printed in U.S.A.

7770 N. Frontage Road
Skokie, Illinois 60077

**FIGURE 10–38**  Octagon covers *(Courtesy of Appleton Electric).*

# OB/SB-1

## 4" Square Boxes
### With Conduit Knockouts

**4SL-1/2**
1-1/4" Deep, Drawn

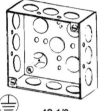

**4S-1/2**
1-1/2" Deep, Welded

**4S-1/2-DR**
1-1/2" Deep, Drawn

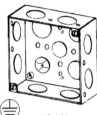

**4S-3/4**
1-1/2" Deep, Welded

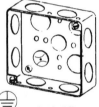

**4S-3/4-DR**
1-1/2" Deep, Drawn

**4SEK**
1-1/2" Deep, Welded

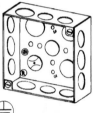

**4S-SPL**
1-1/2" Deep, Welded

**4S-SPL-DR**
1-1/2" Deep, Drawn

**4SP**
Plenum Box 1-1/2" Deep, Drawn
Without any holes
and with reset knockouts

| Catalog Number (Universal No.) | Depth (inches) | Wiring Cubic Inch Capacity | Knockout Description Sides | Ends | Bottom | Wt. Lbs. Per 100 | Std. Pkg. |
|---|---|---|---|---|---|---|---|
| **4SL-1/2** (52141-1/2) | 1-1/4 | 18.0 | (10)1/2" | -- | (5)1/2" | 61.0 | 50 |
| **4S-1/2** (52151-1/2) | 1-1/2 | 21.0 | (12)1/2" | -- | (4)1/2" | 73.0 | 50 |
| **4S-1/2-DR** (52151-1/2) | 1-1/2 | 21.0 | (12)1/2" | -- | (5)1/2" | 73.0 | 50 |
| **4S-3/4** (52151-3/4) | 1-1/2 | 21.0 | (8)3/4" | -- | (2)1/2" (2)3/4" | 67.0 | 50 |
| **4S-3/4-DR** (52151-3/4) | 1-1/2 | 21.0 | (8)3/4" | -- | (2)1/2" (2)3/4" | 67.0 | 50 |
| **4SEK** | 1-1/2 | 21.0 | (8)3/4" 4 Eccentric KO | -- | (2)1/2" 2 Eccentric KO | 72.0 | 50 |
| **4S-SPL** (52151-SPL) | 1-1/2 | 21.0 | (8)1/2" (4)3/4" | -- | (2)1/2" (2)3/4" | 73.0 | 50 |
| **4S-SPL-DR** (52151-SPL) | 1-1/2 | 21.0 | (8)1/2" (4)3/4" | -- | (2)1/2" (2)3/4" | 73.0 | 50 |
| **4SP** | 1-1/2 | 21.0 | (8)1/2" (4)3/4" | -- | (3)1/2" (2)3/4" | 70.0 | 50 |

**Groundskeeper™**
Raised ground feature on boss inside box provides one #10-32 tapped hole, allowing a 3/8" long green ground screw to be threaded into box without contacting mounting surface. Other boxes provided with flat ground tapped hole, except Plenum Boxes. See OB/SB-1 page 49 for ground screws, pigtails and accessories.

Effective October, 2002
Copyright 2002 Printed in U.S.A.

**PAGE 10**

EGS Electrical Group ▲ Appleton

800-621-1506
www.appletonelec.com

---

**FIGURE 10–39** Square boxes with conduit knockouts *(Courtesy of Appleton Electric).*

UL LISTED
File #
E2527

# OB/SB-1

## 4" Square Boxes

With Nonmetallic Sheathed Cable Clamps

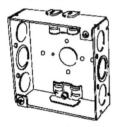

**4SR-EK**
1-1/2" Deep, Welded
Nail holes

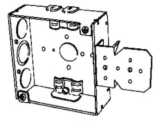

**4SRB-EK**
1-1/2" Deep, Welded
Bracket set flush
with front of box

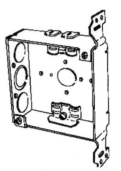

**4SRVB-EK-PL**
1-1/2" Deep, Welded
Plain vertical bracket

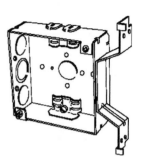

**4SROB-EK**
1-1/2" Deep, Welded
1-3/8" offset bracket set flush
with front of box

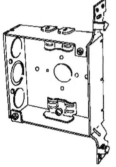

**4SRAB-EK**
1-1/8" Deep, Welded
Vertical bracket
with barbed hooks

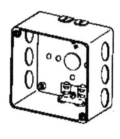

**4SRD**
2-1/8" Deep, Drawn
Nail holes

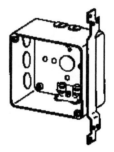

**4SRDVB-PL**
2-1/8" Deep, Drawn
Plain vertical bracket flush
with front of box

| Catalog Number | Depth (inches) | Wiring Cubic Inch Capacity | Type Clamp | Knockout Description Sides | Ends | Bottom | Wt. Lbs. Per 100 | Std. Pkg. |
|---|---|---|---|---|---|---|---|---|
| 4SR-EK | 1-1/2 | 21.0 | CL-26 | (4)1/2" (2)Eccentric KO | (4)21/32" | (1)1/2" | 74.0 | 25 |
| 4SRB-EK | 1-1/2 | 21.0 | CL-26 | (2)1/2" (1)Eccentric KO | (4)21/32" | (1)1/2" | 88.0 | 25 |
| 4SRVB-EK-PL | 1-1/2 | 21.0 | CL-26 | (2)1/2" (1)Eccentric KO | (4)21/32" | (1)1/2" | 84.0 | 25 |
| 4SROB-EK | 1-1/2 | 21.0 | CL-26 | (2)1/2" (1)Eccentric KO | (4)21/32" | (1)1/2" | 94.0 | 25 |
| 4SRAB-EK | 1-1/2 | 21.0 | CL-26 | (2)1/2" (1)Eccentric KO | (4)21/32" | (1)1/2" | 88.0 | 25 |
| 4SRD | 2-1/8 | 30.3 | CL-26 | (6)1/2" | (4)21/32" | (1)1/2" | 90.0 | 25 |
| 4SRDVB-PL | 2-1/8 | 30.3 | CL-26 | (3)1/2" | (4)21/32" | (1)1/2" | 100.0 | 25 |

Effective October, 2001
Copyright 2001  Printed in U.S.A.

EGS Electrical Group  ⚡ Appleton

7770 N. Frontage Road
Skokie, Illinois 60077

PAGE 13

**FIGURE 10–40**  Square boxes with nonmetallic-sheathed cable clamps *(Courtesy of Appleton Electric).*

# OB/SB-1

UL LISTED
File #
E2527

## 4" Square Boxes

With Armored Cable / Metal Clad Cable Clamps

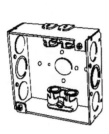

4SX-EK
1-1/2" Deep, Welded

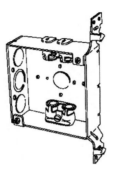

4SXAB-EK
1-1/2" Deep, Welded
Vertical bracket
with barbed hooks

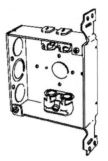

4SXVB-EK-PL
1-1/2" Deep, Welded
Plain vertical bracket

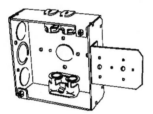

4SXB-EK-PL
1-1/2" Deep, Welded
Bracket set flush
with front of box

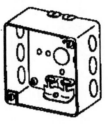

4SXD
2-1/8" Deep, Drawn

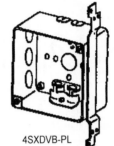

4SXDVB-PL
2-1/8" Deep, Drawn
Bracket set flush
with front of box

| Catalog Number (Universal No.) | Depth (inches) | Wiring Cubic Inch Capacity | Type Clamp | Knockout Description | | | Wt. Lbs. Per 100 | Std. Pkg. |
| --- | --- | --- | --- | --- | --- | --- | --- | --- |
| | | | | Sides | Ends | Bottom | | |
| 4SX-EK (52151VB) | 1-1/2 | 21.0 | CL-13 | (4)1/2" (2)Eccentric KO | (4)21/32" | (1)1/2" | 100.0 | 25 |
| 4SXAB-EK (52151VB) | 1-1/2 | 21.0 | CL-13 | (2)1/2" (1)Eccentric KO | (4)21/32" | (1)1/2" | 100.0 | 25 |
| 4SXVB-EK-PL | 1-1/2 | 21.0 | CL-13 | (2)1/2" (1)Eccentric KO | (4)21/32" | (1)1/2" | 114.0 | 25 |
| 4SXB-EK-PL | 1-1/2 | 21.0 | CL-13 | (2)1/2" (1)Eccentric KO | (4)21/32" | (1)1/2" | 116.0 | 25 |
| 4SXD | 2-1/8 | 30.3 | CL-13 | (6)1/2" | (4)21/32" | (1)1/2" | 100.0 | 25 |
| 4SXDVB-PL | 2-1/8 | 30.3 | CL-13 | (3)1/2" | (4)21/32" | (1)1/2" | 114.0 | 25 |

Effective October, 2001
Copyright 2001  Printed in U.S.A.
PAGE 14

7770 N. Frontage Road
Skokie, Illinois 60077

**FIGURE 10–41**  Square boxes with armored cable/metal-clad cable clamps *(Courtesy of Appleton Electric)*.

# OB/SB-1

## 4" Square Boxes

For 2-1/2" and 3-5/8" Metal Studs

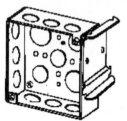

4S-1/2-SB2-1/2
1-1/2" Deep, Welded
For 2-1/2" metal stud

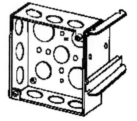

4S-1/2-SB3-5/8
1-1/2" Deep, Welded
For 3-5/8" metal stud

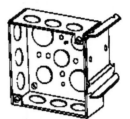

4S-SPL-SB2-1/2
1-1/2" Deep, Welded
For 2-1/2" metal stud

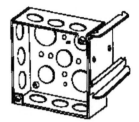

4S-SPL-SB3-5/8
1-1/2" Deep, Welded
For 3-5/8" metal stud

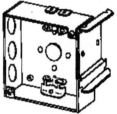

4SR-SB2-1/2
1-1/2" Deep, Welded
For 2-1/2" metal stud

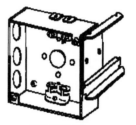

4SR-SB3-5/8
1-1/2" Deep, Welded
For 3-5/8" metal stud

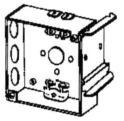

4SX-SB2-1/2
1-1/2" Deep, Welded
For 2-1/2" metal stud

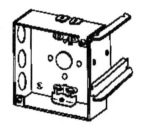

4SX-SB3-5/8
1-1/2" Deep, Welded
For 3-5/8" metal stud

| Catalog Number | Depth (inches) | Stud Size (inches) | Wiring Cubic Inch Capacity | Type Clamp | Knockout Description | | | Wt. Lbs. Per 100 | Std. Pkg. |
|---|---|---|---|---|---|---|---|---|---|
| | | | | | Sides | Ends | Bottom | | |
| **With Conduit Knockouts** | | | | | | | | | |
| 4S-1/2-SB2-1/2 | 1-1/2 | 2-1/2 | 21.0 | -- | (9)1/2" | -- | (5)1/2" | 92.0 | 25 |
| 4S-1/2-SB3-5/8 | 1-1/2 | 3-5/8 | 21.0 | -- | (9)1/2" | -- | (5)1/2" | 102.0 | 25 |
| 4S-SPL-SB2-1/2 | 1-1/2 | 2-1/2 | 21.0 | -- | (6)1/2" (3)3/4" | -- | (3)1/2" (2)3/4" | 92.0 | 25 |
| 4S-SPL-SB3-5/8 | 1-1/2 | 3-5/8 | 21.0 | -- | (6)1/2" (3)3/4" | -- | (3)1/2" (2)3/4" | 102.0 | 25 |
| **With Nonmetallic Sheathed Cable Clamps** | | | | | | | | | |
| 4SR-SB2-1/2 | 1-1/2 | 2-1/2 | 21.0 | CL-26 | (3)1/2" | (4)21/32" | (1)1/2" | 93.0 | 25 |
| 4SR-SB3-5/8 | 1-1/2 | 3-5/8 | 21.0 | CL-26 | (3)1/2" | (4)21/32" | (1)1/2" | 103.0 | 25 |
| **With Armored Cable/Metal Clad Cable Clamps** | | | | | | | | | |
| 4SX-SB2-1/2 | 1-1/2 | 2-1/2 | 21.0 | CL-13 | (3)1/2" | (4)21/32" | (1)1/2" | 119.0 | 25 |
| 4SX-SB3-5/8 | 1-1/2 | 3-5/8 | 21.0 | CL-13 | (3)1/2" | (4)21/32" | (1)1/2" | 129.0 | 25 |

Effective October, 2001
Copyright 2001  Printed in U.S.A.

EGS
Electrical Group  Appleton

7770 N. Frontage Road
Skokie, Illinois 60077

PAGE 15

**FIGURE 10–42**  Square boxes for mounting to metal studs *(Courtesy of Appleton Electric).*

# OB/SB-1

UL LISTED
File #
E2527

## 4" Square Box Extension Rings

### With Conduit Knockouts

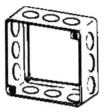

**4SE-1/2**
1-1/2" Deep, Drawn
Two 8-32 tapped holes
and two slide-on screw
slots in bottom

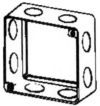

**4SE-3/4**
1-1/2" Deep, Drawn
Two 8-32 tapped holes
and two slide-on screw
slots in bottom

**4SES**
1-1/2" Deep, Drawn
Two 8-32 tapped holes
and two slide-on screw
slots in bottom

**4SSBE-SPL**
1-1/2" Deep, Welded
Bottom cutout to gang
to switch and handy box

| Catalog Number (Universal No.) | Depth (inches) | Wiring Cubic Inch Capacity | Type Clamp | Knockout Description | | | Wt. Lbs. Per 100 | Std. Pkg. |
|---|---|---|---|---|---|---|---|---|
| | | | | Sides | Ends | Bottom | | |
| 4SE-1/2 | 1-1/2 | 21.0 | -- | (12)1/2" | -- | -- | 44.0 | 50 |
| 4SE-3/4 (53151-3/4) | 1-1/2 | 21.0 | -- | (8)3/4" | -- | -- | 44.0 | 50 |
| 4SES (53151-1/2) | 1-1/2 | 21.0 | -- | (6)1/2" (4)3/4" | -- | -- | 47.0 | 50 |
| 4SSBE-SPL | 1-1/2 | 21.0 | -- | (8)1/2" (4)3/4" | -- | -- | 64.0 | 50 |

## 4" Square Covers

### Raised Surface Covers

| Catalog Number (Universal No.) | Depth (inches) | Wiring Cubic Inch Capacity | Description | WT. Lbs. Per 100 | Std. Pkg. |
|---|---|---|---|---|---|
| 8360 | 1/2 | 8.0 | Blank | 45.0 | 25 |
| 8361 (52-1T) | 1/2 | 8.0 | One Toggle Switch | 44.0 | 25 |
| 8362 | 1/2 | 8.0 | One GFCI Device | 35.0 | 25 |
| 8363 | 1/2 | 8.0 | One Single Flush Receptacle 1-13/32"D | 40.0 | 25 |
| 8364 | 1/2 | 8.0 | Two Single Flush Receptacles 1-13/32" D | 38.0 | 25 |

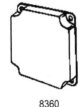

**8360**
1/2" Raised

**8361**
1/2" Raised

**8362**
1/2" Raised

**8363**
1/2" Raised
1-13/32" Opening

**8364**
1/2" Raised
1-13/32" Openings

Effective October, 2001
Copyright 2001  Printed in U.S.A.

PAGE 16

EGS Electrical Group  Appleton

7770 N. Frontage Road
Skokie, Illinois 60077

**FIGURE 10–43**  Square box extension rings with conduit knockouts *(Courtesy of Appleton Electric).*

LISTED
File #
E2527

# OB/SB-1

## 4" Square Covers

Raised Surface Covers

8365N
1/2" Raised

8367
1/2" Raised

8368
1/2" Raised
1-13/32" Opening

8371N
1/2" Raised

8373N
1/2" Raised

8374
1/2" Raised

8375N
1/2" Raised

8377
1/2" Raised
2-9/64" Opening

8378
1/2" Raised
2-1/2" Opening

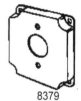

8379
1/2" Raised
1-19/32" Opening

| Catalog Number (Universal No.) | Depth (inches) | Wiring Cubic Inch Capacity | Description | Wt. Lbs. Per 100 | Std. Pkg. |
|---|---|---|---|---|---|
| 8365N | 1/2 | 8.0 | One Duplex Flush Receptacle* | 36.0 | 25 |
| 8367 (52-2T) | 1/2 | 8.0 | Two Toggle Switches | 42.0 | 25 |
| 8368 | 1/2 | 8.0 | One Toggle Switch, One Single Flush Receptacle 1-13/32"D | 39.0 | 25 |
| 8371N | 1/2 | 8.0 | Two Duplex Flush Receptacles* | 33.0 | 25 |
| 8373N | 1/2 | 8.0 | One Duplex Flush, One GFCI Device* | 32.0 | 25 |
| 8374 | 1/2 | 8.0 | One Toggle Switch, One GFCI Device | 38.0 | 25 |
| 8375N | 1/2 | 8.0 | One Toggle Switch, One Duplex Flush* | 38.0 | 25 |
| 8377 | 1/2 | 8.0 | 30-50A Single Flush Receptacle 2-9/64"D | 36.0 | 25 |
| 8378 | 1/2 | 8.0 | One Single Flush Receptacle 2-1/2" D | 35.0 | 25 |
| 8379 (52R54) | 1/2 | 8.0 | 20A Twist Lock Receptacle 1-19/32" D | 38.0 | 25 |

*Extra mounting screws and nuts provided.

Effective October, 2001
Copyright 2001 Printed in U.S.A.

7770 N. Frontage Road
Skokie, Illinois 60077

PAGE 17

**FIGURE 10–44**   Square covers *(Courtesy of Appleton Electric)*.

# OB/SB-1

## Handy Boxes

3-3/4" x 1-1/2" and 4" x 2-1/8" with Conduit Knockouts

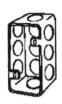

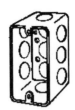

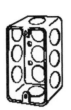

**4SS**
1-1/2" Deep, Drawn

**4CSL-1/2**
1-1/4" Deep, Drawn

**4SSLS-1/2**
1-1/2" Deep, Drawn

**4CS-1/2**
1-7/8" Deep, Drawn

**4CS-3/4**
1-7/8" Deep, Drawn

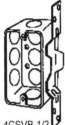
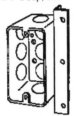
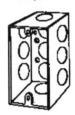
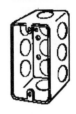
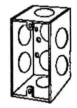

**4CSVB-1/2**
1-7/8" Deep, Drawn
Bracket set back 5/8"

**4CSX-1/2PL**
1-7/8" Deep, Drawn
Bracket set back 5/8"

**4SSL-1/2**
2-1/8" Deep, Welded

**4SSL-1/2DR**
2-1/8" Deep, Drawn

**4SSL-3/4**
2-1/8" Deep, Welded

**4SSL3/4-DR**
2-1/8" Deep, Drawn

| Catalog Number (Universal No.) | Depth (inches) | Wiring Cubic Inch Capacity | Knockout Description | | | Wt. Lbs. Per 100 | Std. Pkg. |
|---|---|---|---|---|---|---|---|
| | | | Sides | Ends | Bottom | | |
| 3-3/4" x 1-1/2" With Conduit Knockouts | | | | | | | |
| 4SS | 1-1/2 | 7.0 | (6)1/2" | (2)1/2" | (3)1/2" | 34.0 | 50 |
| 4" x 2-1/8" With Conduit Knockouts | | | | | | | |
| 4CSL-1/2 (58361-1/2) | 1-1/4 | 9.5 | (6)1/2" | (2)1/2" | (3)1/2" | 39.0 | 50 |
| 4SSLS-1/2 (58351-1/2) | 1-1/2 | 10.3 | (6)1/2" | (2)1/2" | (3)1/2" | 46.0 | 50 |
| 4CS-1/2 (58361-1/2) | 1-7/8 | 13.0 | (6)1/2" | (2)1/2" | (3)1/2" | 52.0 | 50 |
| 4CS-3/4 (58361-3/4) | 1-7/8 | 13.0 | (4)3/4" | (2)3/4" | (2)3/4" | 49.0 | 50 |
| 4CSVB-1/2 (58361-F1/2) | 1-7/8 | 13.0 | (3)1/2" | (2)1/2" | (3)1/2" | 65.0 | 50 |
| 4CSX-1/2PL (58361-A1/2) | 1-7/8 | 13.0 | (3)1/2" | (2)1/2" | (3)1/2" | 66.0 | 50 |
| 4SSL-1/2 (58371-1/2) | 2-1/8 | 14.5 | (6)1/2" | (2)1/2" | (3)1/2" | 54.0 | 50 |
| 4SSL1/2-DR (58371-1/2) | 2-1/8 | 14.5 | (6)1/2" | (2)1/2" | (3)1/2" | 54.0 | 50 |
| 4SSL-3/4 (58371-3/4) | 2-1/8 | 14.5 | (4)3/4" | (2)3/4" | (2)3/4" | 54.0 | 50 |
| 4SSL3/4-DR (58371-3/4) | 2-1/8 | 14.5 | (4)3/4" | (2)3/4" | (2)3/4" | 54.0 | 50 |

Effective October, 2001
Copyright 2001  Printed in U.S.A.
PAGE 30

EGS Electrical Group · Appleton

7770 N. Frontage Road
Skokie, Illinois 60077

**FIGURE 10–45**  Handy boxes with conduit knockouts *(Courtesy of Appleton Electric).*

# OB/SB-1

## Handy Boxes

### 4" x 2-1/8" Solid Gang with Conduit Knockouts

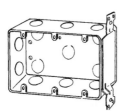

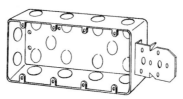

**133APFB-1/2**
2-1/8" Deep, Drawn
Three Gang – 5-3/4" Wide
Bracket set back 1/2"

**134APFB-1/2**
2-1/8" Deep, Drawn
Four Gang – 7-1/2" Wide
Bracket set back 1/2"

| Catalog Number | Depth (inches) | Wiring Cubic Inch Capacity | Knockout Description | | | Wt. Lbs. Per 100 | Std. Pkg. |
|---|---|---|---|---|---|---|---|
| | | | Sides | Ends | Bottom | | |
| 133APFB-1/2 | 2-1/8 | 44.5 | (2)1/2" | (6)1/2" | (5)1/2" | 129.0 | 20 |
| 134APFB-1/2 | 2-1/8 | 58.0 | (2)1/2" | (8)1/2" | (10)1/2" | 150.0 | 20 |

**OB/SB**

## Handy Boxes

### 4-1/8" x 2-3/8" with Conduit Knockouts

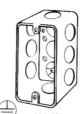

**180-1/2**
1-3/4" Deep, Drawn

**181-1/2**
2-1/4" Deep, Drawn

**184E**
1-1/2" Deep, Drawn

| Catalog Number (Universal No.) | Depth (inches) | Wiring Cubic Inch Capacity | Knockout Description | | | Wt. Lbs. Per 100 | Std. Pkg. |
|---|---|---|---|---|---|---|---|
| | | | Sides | Ends | Bottom | | |
| **With Conduit Knockouts** | | | | | | | |
| 180-1/2 (68361-1/2) | 1-3/4 | 15.0 | (6)1/2" | (2)1/2" | (2)1/2" | 54.0 | 50 |
| 181-1/2 (68371-1/2) | 2-1/4 | 18.3 | (6)1/2" | (2)1/2" | (2)1/2" | 60.0 | 50 |
| **Extension Ring for 4-1/8" x 2-3/8"** | **Handy Box with Conduit Knockouts** | | | | | | |
| 184E (69351) | 1-1/2 | 13.5 | (6)1/2" | (2)1/2" With Slide on | -- | 39.0 | 50 |

Groundskeeper™ Raised ground feature on boss inside box provides one #10-32 tapped hole, allowing a 3/8" long green ground screw to be threaded into box without contacting mounting surface. Other boxes provided with flat ground tapped hole, except Plenum Boxes. See OB/SB-1 page 49 for ground screws, pigtails and accessories.

Effective October, 2002
Copyright 2002 Printed in U.S.A.
**PAGE 32**

 EGS Electrical Group · Appleton

800-621-1506
www.appletonelec.com

**FIGURE 10–46** Ganged handy boxes *(Courtesy of Appleton Electric).*

UL
LISTED
File #
E2527

# OB/SB-1

## Handy Box Covers

For 3-3/4" x 1-1/2", 4" x 2-1/8" and 4-1/8" x 2-3/8" Boxes

| Catalog Number (Universal No.) | Description | Wt. Lbs. Per 100 | Ctn. Qty. | Std. Pkg. |
|---|---|---|---|---|
| For 3-3/4" x 1-1/2" Boxes | | | | |
| 2520 (48C1) | Blank | 12.0 | 50 | 100 |

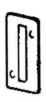

2520

| Catalog Number (Universal No.) | Description | Wt. Lbs. Per 100 | Ctn. Qty. | Std. Pkg. |
|---|---|---|---|---|
| For 4" x 2-1/8" Boxes | | | | |
| 2540 (58C1) | Blank | 11.0 | 25 | 100 |
| 2555 (58C6) | One 1/2" Knockout | 18.0 | 25 | 100 |
| 2538 (58C4) | Single Receptacle 1-19/32" D | 10.0 | 25 | 100 |
| 2539 (58C5) | Single Receptacle 1-13/32" D | 10.0 | 25 | 100 |
| 2510 (58C7) | Duplex Receptacle | 8.0 | 25 | 100 |
| 2594 (58C30) | One Toggle Switch | 10.0 | 25 | 100 |

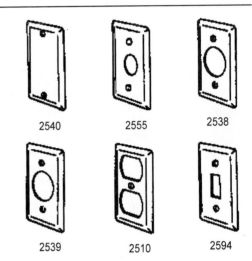

2540    2555    2538

2539    2510    2594

| Catalog Number (Universal No.) | Description | Wt. Lbs. Per 100 | Ctn. Qty. | Std. Pkg. |
|---|---|---|---|---|
| For 4-1/8" x 2-3/8" Boxes | | | | |
| 180A (68C-1) | Blank | 13.0 | 25 | 100 |
| 180T (68C-30) | One Toggle Switch | 12.0 | 25 | 100 |
| 180G | GFI Receptacle | 9.0 | 25 | 100 |
| 180W (68C-7) | Duplex Receptacle | 9.0 | 25 | 100 |
| 180X (68C-5) | Single Receptacle 1-13/32" D | 10.0 | 25 | 100 |

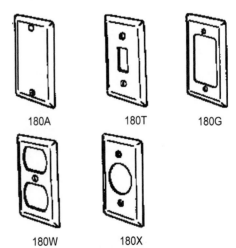

180A    180T    180G

180W    180X

Effective October, 2001
Copyright 2001 Printed in U.S.A.

EGS
Electrical Group    ⚠ Appleton

7770 N. Frontage Road
Skokie, Illinois 60077

PAGE 33

**FIGURE 10–47**   Handy box covers (*Courtesy of Appleton Electric*).

# OB/SB-1

## 3" x 2" Square Corner Switch Boxes
With Conduit Knockouts

OB/SB

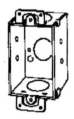 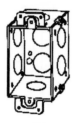 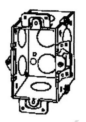 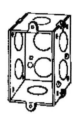 

**94**
1-1/2" Deep, Welded
Plaster ears*

**111**
2" Deep, Gangable
Plaster ears*

**111HB**
2" Deep, Gangable
"Loxbox" support on
each side
Plaster ears*

**111LE**
2" Deep, Gangable
Four nail holes
each side
side leveling bosses

**222**
2-1/2" Deep,
Gangable
Plaster ears*

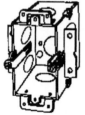 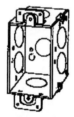 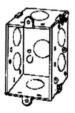 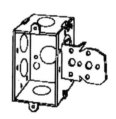 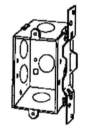

**222-OW**
2-1/2" Deep,
Gangable
Old work-support
clamps
Plaster ears*

**225**
2-1/2" Deep,
Gangable
Plaster ears*

**222LE**
2-1/2" Deep, Gangable
Four nail holes
each side
side leveling bosses

**222NL**
2-1/2" Deep, Gangable
Bracket set back 1/2"

**222VB-PL**
2-1/2" Deep,
Gangable
Bracket set back 3/4"

| Catalog Number | Depth (inches) | Wiring Cubic Inch Capacity | Knockout Description Sides | Ends | Bottom | Wt. Lbs. Per 100 | Std. Pkg. |
|---|---|---|---|---|---|---|---|
| 94* | 1-1/2 | Non-Gangable | 7.5 | — | (2)1/2" | (1)1/2" | 40.0 | 50 |
| 111* | 2 | Gangable | 10.0 | (4)1/2" | (2)1/2" | (2)1/2" | 51.0 | 50 |
| 111HB* | 2 | Gangable | 10.0 | (4)1/2" | (2)1/2" | (2)1/2" | 54.0 | 50 |
| 111LE | 2 | Gangable | 10.0 | (4)1/2" | (2)1/2" | (2)1/2" | 47.0 | 50 |
| 222* | 2-1/2 | Gangable | 12.5 | (4)1/2" | (2)1/2" | (1)1/2" | 58.0 | 50 |
| 222-OW* | 2-1/2 | Gangable | 12.5 | (4)1/2" | (2)1/2" | (1)1/2" | 68.0 | 50 |
| 225* | 2-1/2 | Gangable | 12.5 | (4)3/4" | (2)3/4" | (1)1/2" | 58.0 | 50 |
| 222LE | 2-1/2 | Gangable | 12.5 | (4)1/2" | (2)1/2" | (1)1/2" | 56.0 | 50 |
| 222NL | 2-1/2 | Gangable | 12.5 | (4)1/2" | (2)1/2" | (1)1/2" | 66.0 | 25 |
| 222VB-PL | 2-1/2 | Gangable | 12.5 | (2)1/2" | (2)1/2" | (1)1/2" | 60.0 | 25 |

* Plaster ears support GFCI and decorative cover plates.

Effective October, 2001
Copyright 2001 Printed in U.S.A.
PAGE 36

7770 N. Frontage Road
Skokie, Illinois 60077

**FIGURE 10–48**   Square corner switch boxes with conduit knockouts *(Courtesy of Appleton Electric).*

UL
LISTED
File #
E2527

# OB/SB-1

## 3" x 2" Bevel Corner Switch Boxes 2-1/4" Deep
### With Nonmetallic Sheathed Cable Clamps

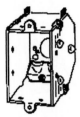

388LE-OR-44LE
2-1/4" Deep, Gangable
External nail brackets
leveling bosses

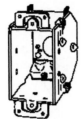

388LR-OR-44LR
2-1/4" Deep, Gangable
external nail brackets
leveling bosses
Plaster ears*

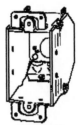

388-OR-44
2-1/4" Deep, Gangable
external nail brackets
Plaster ears*

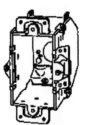

388HB
2-1/4" Deep, Gangable
"Loxbox" support
welded on each side
Plaster ears*

388VB-OR-44VB
2-1/4" Deep, Gangable
Bracket set back 3/4"

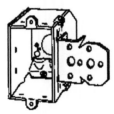

388DW
2-1/4" Deep, Gangable
Bracket set back 1/4"

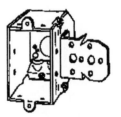

388NL-OR-44NL
2-1/4" Deep, Gangable
Bracket set back 1/2"

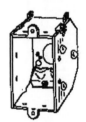

388LES
2-1/4" Deep, Welded
External nail brackets
leveling bosses

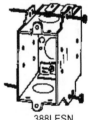

388LESN
2-1/4" Deep, Welded
Two 16 penny nails
in external brackets
leveling bosses

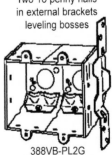

388VB-PL2G
2-1/4" Deep, Welded
Bracket set back 3/4"

| Catalog Number | Depth (inches) | Wiring Cubic Inch Capacity | Type Clamp | Knockout Description | | | Wt. Lbs. Per 100 | Std. Pkg. |
|---|---|---|---|---|---|---|---|---|
| | | | | Sides | Ends | Bottom | | |
| 388LE-OR-44LE Gangable | 2-1/4 | 10.5 | CL-20 | -- | (4)21/32" | (1)1/2" | 52.0 | 50 |
| 388LR-OR-44LR* Gangable | 2-1/4 | 10.5 | CL-20 | -- | (4)21/32" | (1)1/2" | 56.0 | 50 |
| 388-OR-44* Gangable | 2-1/4 | 10.5 | CL-20 | -- | (4)21/32" | (1)1/2" | 56.0 | 50 |
| 388HB* Gangable | 2-1/4 | 10.5 | CL-20 | -- | (4)21/32" | (1)1/2" | 60.0 | 50 |
| 388VB-OR-44VB Gangable | 2-1/4 | 10.5 | CL-20 | -- | (4)21/32" | (1)1/2" | 64.0 | 50 |
| 388DW Gangable | 2-1/4 | 10.5 | CL-20 | -- | (4)21/32" | (1)1/2" | 62.0 | 50 |
| 388NL-OR-44NL Gangable | 2-1/4 | 10.5 | CL-20 | -- | (4)21/32" | (1)1/2" | 62.0 | 50 |
| 388LES Non-Gangable | 2-1/4 | 10.5 | CL-20 | -- | (4)21/32" | (1)1/2" | 52.0 | 50 |
| 388LESN Non-Gangable | 2-1/4 | 10.5 | CL-20 | -- | (4)21/32" | (1)1/2" | 56.0 | 50 |
| 388VB-PL2G Welded Assembly | 2-1/4 | 21.0 | CL-20 | -- | (8)21/32" | (2)1/2" | 84.0 | 25 |

* Plaster ears support GFCI and decorative cover plates.

Effective October, 2001
Copyright 2001 Printed in U.S.A.

EGS
Electrical Group
Appleton

7770 N. Frontage Road
Skokie, Illinois 60077

PAGE 39

**FIGURE 10–49** Bevel corner switch boxes with nonmetallic-sheathed cable clamps *(Courtesy of Appleton Electric).*

UL LISTED
File #
E2527

# OB/SB-1

## 3" x 2" Square Corner Switch Boxes

### With Armored Cable/Metal Clad Cable Clamps

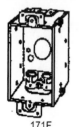

**171F**
2" Deep, Gangable
Plaster ears*

**173F**
2-1/2" Deep, Gangable
Plaster ears*

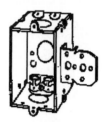

**173FNL**
2-1/2" Deep, Gangable
Bracket set back 1/2"

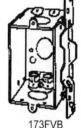

**173FVB**
2-1/2" Deep, Gangable
Bracket set back 3/4"

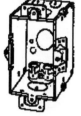

**173FHB**
2-1/2" Deep, Gangable
LOX BOX® supports
Plaster ears*

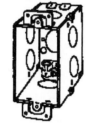

**174F**
2-3/4" Deep, Gangable
Plaster ears*

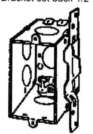

**174FVB-PL**
2-3/4" Deep, Gangable
Bracket set back 3/4"

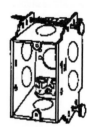

**18LESNX**
2-7/8" Deep, Welded
Two lb penny nails
in external brackets

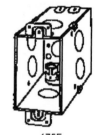

**175F**
3-1/2" Deep, Gangable
Plaster ears*

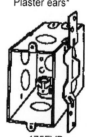

**175FVB**
3-1/2" Deep, Gangable
Bracket set back 3/4"

| Catalog Number | Depth (inches) | | Wiring Cubic Inch Capacity | Type Clamp | Knockout Description | | | Wt. Lbs. Per 100 | Std. Pkg. |
|---|---|---|---|---|---|---|---|---|---|
| | | | | | Sides | Ends | Bottom | | |
| 171F* | 2 | Gangable | 10.7 | CL-13 | -- | (4)21/32" | (1)1/2" | 55.0 | 50 |
| 173F* | 2-1/2 | Gangable | 12.5 | CL-13 | -- | (2)1/2" (4)21/32" | (1)1/2" | 63.0 | 50 |
| 173FNL | 2-1/2 | Gangable | 12.5 | CL-13 | -- | (2)1/2" (4)21/32" | (1)1/2" | 70.0 | 25 |
| 173FVB | 2-1/2 | Gangable | 12.5 | CL-13 | -- | (2)1/2" (4)21/32" | (1)1/2" | 72.0 | 50 |
| 173FHB* | 2-1/2 | Gangable | 12.5 | CL-13 | -- | (2)1/2" (4)21/32" | (1)1/2" | 63.0 | 50 |
| 174F* | 2-3/4 | Gangable | 14.0 | CL-13 | (4)1/2" | (2)1/2" (4)21/32" | (1)1/2" | 70.0 | 50 |
| 174FVB-PL | 2-3/4 | Gangable | 14.0 | CL-13 | (2)1/2" | (2)1/2" (4)21/32" | (1)1/2" | 72.0 | 50 |
| 18LESNX | 2-7/8 | Non-Gangable | 18.0 | CL-13 | (4)1/2" | (2)1/2" (4)21/32" | (1)1/2" | 115.0 | 25 |
| 175F* | 3-1/2 | Gangable | 18.0 | CL-13 | (4)1/2" | (2)1/2" (4)21/32" | (1)1/2" | 79.0 | 25 |
| 175FVB | 3-1/2 | Gangable | 18.0 | CL-13 | (4)1/2" | (2)1/2" (4)21/32" | (1)1/2" | 86.0 | 50 |

\* Plaster ears support GFCI and decorative cover plates.

Effective January, 2002
Copyright 2002 Printed in U.S.A.

EGS Electrical Group ⒶAppleton

7770 N. Frontage Road
Skokie, Illinois 60077

PAGE 43

**FIGURE 10-50** Square corner switch boxes with armored cable/metal-clad cable clamps *(Courtesy of Appleton Electric).*

# OB/SB-1

## Outlet Box/Switch Box Accessories

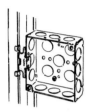

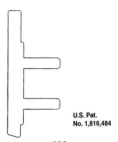

**MSC-OB**
**Outlet Box**
**Metal stud clip**

**1490**
**Switch Box**
**Extension**

U.S. Pat. No. 2,812,149
Pat. Canada 1954

**892**
**Lox Box®**
**Switch Box Supports**

U.S. Pat.
No. 1,816,484

**896**
**"E-Z-IN"**
**Switch box supports**

| Catalog Number | Description | Wt. Lbs. Per 100 | Ctn. Qty. | Std. Pkg. |
|---|---|---|---|---|
| **Outlet Box Metal Stud Clip** | | | | |
| MSC-OB | Mounts all types, sizes and depths of outlet boxes to any metal stud | 5.9 | -- | 50 |
| **Switch Box Extension** | | | | |
| 1490 | Fits all single-gang switch boxes with 1/8" screws – 7/8" Depth – 4.5 cubic inch wiring capacity | 11.0 | -- | 25 |
| **Switch Box Supports** | | | | |
| 892 | Lox Box® fits standard switch boxes – wall 1/4" minimum to 1-1/4" maximum | 2.6 | 100 | 500 |
| 896 | "E-Z-IN" wall 1/4" minimum to 1-1/4" maximum | 2.6 | 100 | 1000 |

**GR12**
**Grounding clip**

**SCR1032-PTL1**
**Insulated**
**Grounding screw**
**and pigtail**

**SCR1032-PTL**
**Bare Wire**
**Grounding screw**
**and pigtail**

**SCR1032**
**Grounding screw**

| Catalog Number | Description | Wt. Lbs. Per 100 | Std. Pkg. |
|---|---|---|---|
| **Grounding Clip** | | | |
| GR12 | For use with No. 14 or No. 12 non-metallic sheathed cable having ground wire | 0.4 | 1000 |
| **Washer Head Grounding Screws and Pigtails** | | | |
| SCR1032-PTL1 | 3/8" long – 10-32 thread with 6" insulated, No. 12 wire pigtail | 2.0 | 1000 |
| SCR1032-PTL | 3/8" long – 10-32 thread with 6" bare wire, No. 14 wire pigtail | 1.2 | 1000 |
| SCR1032 | 3/8" long – 10-32 thread – green | 0.5 | 1000 |

Effective October, 2001
Copyright 2001  Printed in U.S.A.

7770 N. Frontage Road
Skokie, Illinois 60077

**PAGE 49**

**FIGURE 10–51**  Outlet box/switch box accessories *(Courtesy of Appleton Electric).*

**FIGURE 10-52** Fish tape.

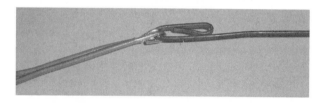

**FIGURE 10-53A** Conductors fastened to a fish tape before taping.

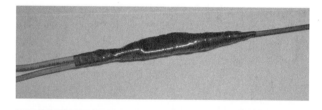

**FIGURE 10-53B** Conductors fastened to a fish tape, taped for pulling.

## | REVIEW QUESTIONS

*Multiple Choice*

1. The standard that is most frequently used to ensure safe electrical installation is the
   a. *Life Safety Code Handbook.*
   b. *NFPA 79 Electrical Standard for Industrial Machinery.*
   c. *National Electrical Code.*
   d. *NFPA 5000 Building Construction and Safety Code.*

2. The *National Electrical Code* is published by the
   a. U.S. Government Printing Office.
   b. Office of Safety and Health Administration.
   c. National Fire Protection Association.
   d. National Electrical Contractors Association.

3. Armored cable is a
   a. rigidly sheathed cable.
   b. flexible metal-covered cable.
   c. cable enclosed in copper armor.
   d. cable enclosed in rigid steel armor.

4. The outer jacket color of 12 AWG nonmetallic-sheathed cable is
   a. black.
   b. white.
   c. orange.
   d. yellow.

5. Which type of conduit has the thickest wall construction?
   a. RMC
   b. IMC
   c. EMT
   d. FMT

*Give Complete Answers*

1. Describe the general purpose of the *NEC*.
2. What is the primary goal of the National Fire Protection Association?
3. What are some of the types of electrical installations covered by the *NEC*?
4. List three types of electrical installations that are not covered by the *NEC*.
5. Describe armored cable.
6. What is nonmetallic-sheathed cable?
7. What is the difference between Type NM cable and Type NM-B cable?
8. Describe service-entrance cable.
9. What is underground feeder and branch-circuit cable?
10. Name three types of metallic raceway.

# Wiring Applications

**OBJECTIVES**

After studying this chapter, the student will be able to:

■ Apply the basic procedures used to calculate the size of various types of electrical services.

■ Explain how the number and size of feeders and branch circuits are determined for various installations.

■ Explain how to size the system grounding conductor and describe the various methods of grounding systems.

■ List the general safety requirements for various electrical installations.

■ Describe some common types of devices and equipment used in electrical installations and explain the reasons for using them.

There are certain basic procedures and methods that pertain to all electrical installations. When designing new installations or adding to existing ones, one must keep safety in mind as the prime consideration. Local and national codes must be observed. In a determination of the size of service, feeders, and branch circuits, allowances should be made for future needs. *National Electrical Code* Articles 215, 220, 225, and 230 provide some guidelines for calculating the ampacity of conductors and equipment.

The local electric utility company should be contacted to determine the type of power available. There are several different values of voltage and types of systems that are in common use. Some questions that must be considered are:

■ Does the system require AC, DC, or both?
■ If the supply is AC, what type of system is required?
■ In what part of the building will the service be located?
■ Is the supply overhead or underground?
■ What is the best method for grounding the system?

The requirements for obtaining a license vary for different areas. However, certain requirements are common to most areas. For a journeyman electrician's license, one must:

- Work a specific number of years under the direct supervision of a licensed journeyman electrician.
- Pass a written and practical test.

For a master's (contractor's) license, one must:

- Obtain a journeyman's license.

- Work a specific number of years as a licensed journeyman electrician.
- Pass a written test.

The tests required for the licenses are generally based on safety rules and regulations, usually as set forth in the *NEC* and local codes. They are also based on practical knowledge and knowledge of electrical theory. The contractor's test may require more knowledge of design and maintenance.

It is advisable to consult the local electrical inspector before the job is started. Most localities require that the electrical contractor file for a permit to do the work. Permits are obtained from the inspecting authority's office. Often, localities have special regulations that are peculiar to the area. The electrical inspector can make these regulations available beforehand.

Most states and/or municipalities require the licensing of electricians and/or electrical contractors. There are generally two types of licenses: *journeyman* and *master* (contractor). The journeyman's license is a certificate issued to electricians who work for contractors. These electricians are usually the workers who perform the actual installation. They are responsible for providing a complete installation according to all safety rules and regulations. The master's, or contractor's, license is a certificate issued to the person or company employing the electrical workers.

## RESIDENTIAL WIRING

The electrician or engineer should begin by referring to the basic floor plan. This plan provides the necessary information for calculating the general lighting load, the number and location of convenience outlets, and other electrical equipment. Other detailed information is provided in elevation and riser diagrams.

A simple and reliable method for determining the service size and the number of branch circuits is outlined in *NEC* Annex D. To determine the general lighting load, the "volt-amperes per square foot" method is used. The unit load for dwelling occupancies is 3 volt-amperes per square

foot ($VA/ft^2$). In a calculation of the area of the building, the outside dimensions should be used. Unoccupied areas, open porches, and similar areas need not be included. The area of each floor should be determined and the total used for the aforementioned calculations.

After determination of the general lighting load, it is necessary to add all other electrical loads. For each dwelling unit, a load of not less than 3000 volt-amperes must be included for small appliances. Each laundry must be wired with a 1500-volt-ampere circuit for appliances. Other loads, such as electric ranges, water heaters, air conditioners, dishwashers, electric heaters, and special lighting loads, must be included.

> *Note:* In a calculation of AC loads, the term volt-ampere is used to indicate the product of the voltage and the current.

The *NEC* does not require that the service be sized to carry the total connected load. In most residential occupancies, it is very unlikely that every light, appliance, and convenience outlet will be used at maximum capacity at the same time. For practical and economical purposes, the *NEC* allows the application of a **demand factor**. A demand factor is the ratio of the maximum load used at any one time to the total connected load.

Applying the demand factor reduces the size of the service conductors and equipment. Caution and common sense must be used in order to avoid the installation of an inadequate service. It is always advisable to allow for future increases in the load. (See *NEC* Article 220.)

The *NEC* provides the minimum requirements for safety. Adhering to these rules ensures an installation that is reasonably safe, but not necessarily convenient. In addition, the *Code* cannot cover every unique condition that may arise.

It is assumed that the electrical worker will want to make an installation of high quality. Some rules that will aid in performing a flexible, convenient, trouble-free installation are as follows:

1. Install a 100-ampere, three-wire or larger service for a single-family dwelling.

2. Install a 200-ampere, three-wire or larger service for two- and three-family dwellings.

3. Install at least one double convenience outlet over each counter space, 300 mm (12 inches) or wider, in the kitchen.

4. For wall counter space, install receptacle outlets so that no point along the wall is more than *600 mm (24 inches)*, measured horizontally from a receptacle outlet in that space.

5. Install one 20-ampere branch circuit for the laundry.

6. Install ground-fault interrupting devices as per *NEC* Article 210.8.

7. Do not load branch circuits to their maximum capacity; allow for future additions.

8. Always follow the standards prescribed by the *NEC*.

## Residential Service

A *residential service* consists of the service-entrance conductors, the kilowatt-hour meter, the main switch or circuit breaker, the overcurrent devices, and the common ground. Figure 11–1 shows a typical service for a single-family dwelling. Figure 11–2 shows one type of panel used for residential services.

*NEC* Article 230 prescribes the regulations for services. All electrical personnel should be familiar with these rules.

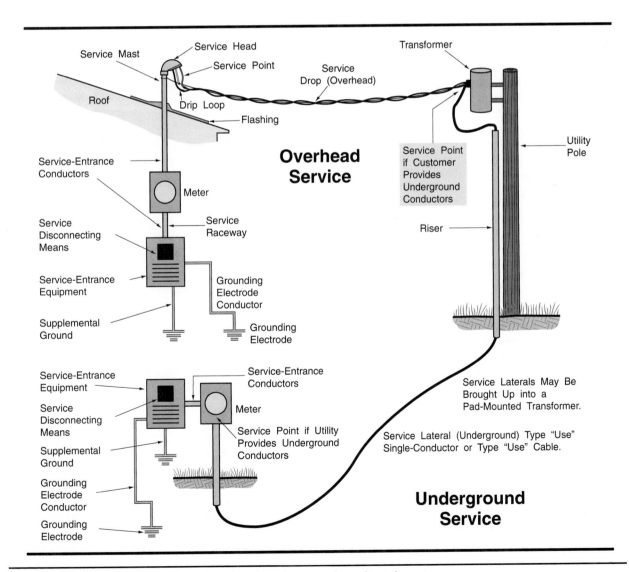

**FIGURE 11-1**   Examples of overhead and underground residential services.

**FIGURE 11–2** Panelboard for a single-family residence.

The electrical service is the heart of the wiring system. It contains the safety devices (fuses and/or circuit breakers) that protect the system from excessive current flow. The common ground provides a path for stray currents, reducing the possibility of fires.

The main electrical service terminates in branch circuits, using fuses or circuit breakers to protect the circuits. Each branch circuit supplies power to certain areas or equipment. These circuits should be divided equally throughout the house, when practical. Special circuits that may be required by the *NEC* and local codes must also be installed.

Fuses and circuit breakers are "safety valves." They disconnect the circuit when the current becomes too great. The operation of a fuse or circuit breaker indicates excessive current flow. The circuit should be checked and the fault eliminated before the fuse is replaced or the circuit breaker is reset.

> ⚠ **CAUTION** It is very important to install the proper size of fuse or circuit breaker. Oversized fuses or circuit breakers will not give adequate protection against excessive current flow.

All AC systems are grounded at the service. The grounding conductor provides a low-resistance path for current caused by lightning and leakage currents. It should not carry current under normal conditions.

## Calculating the Size of the Common Ground

The common grounding conductor must be large enough to conduct stray currents to ground without overheating. It is a very important part of the service. *NEC* Article 250 prescribes the rules for grounding. The common grounding conductor connects both the wiring system and the equipment to the grounding electrode.

The resistance of the **common ground** must be kept to a minimum. This can be done by ensuring tight connections and installing an adequately sized grounding conductor and electrode. *NEC* Table 250.66 lists the minimum size for the grounding conductor based on the size of service-entrance conductor. *NEC* Section 250-III sets the standards for the grounding electrode. The type of electrode to which the common grounding conductor is connected should be selected with care. A metal underground water-piping system generally provides an effective ground.

## Ground-Fault Circuit Interrupter

A **ground-fault circuit interrupter** is a device that senses a small imbalance of current between the ungrounded (hot) conductor and the grounded (neutral) conductor. The device opens the circuit when the current imbalance exceeds a predetermined value, usually around 6 milliamperes (6 mA).

Most circuits are protected against overcurrent by 15-ampere or larger fuses or circuit breakers. This protection is adequate against short circuits and overloads. Leakage currents to ground may be much less than 15 amperes and still be hazardous.

> ⚠ **CAUTION** The metal frame of electrical equipment will become energized if the ungrounded conductor (the hot wire) accidentally makes contact with it. A person touching the frame and the ground, or a grounded object, completes the circuit. Even a minute amount of current flowing through the heart, lungs, or brain can be fatal. Therefore, it is very important to take the necessary precautions to prevent leakage currents through the human body. The ground-fault circuit interrupter serves this purpose.

## Lighting Circuits

Most lighting fixtures are controlled by switches. *NEC* Article 404 prescribes methods to install switches. These switches are located in convenient places throughout the dwelling. They are usually installed in boxes set in the walls at a height of 42 inches to 48 inches.

There are three common styles of switches used to control lights. The single-pole switch controls the light from one specific location. Three-way switches are used to provide control from two locations. If control is desired from more than two locations, a combination of three-way and four-way switches is used. Two methods for connecting a single-pole switch to control a light are depicted in Figure 11–3. One method is seen in the upper

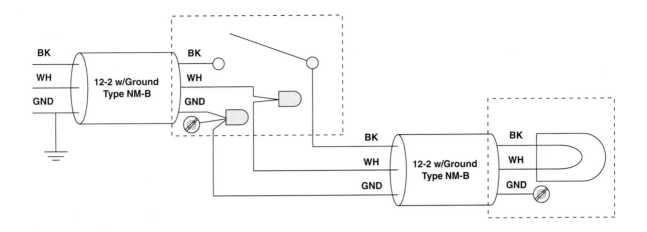

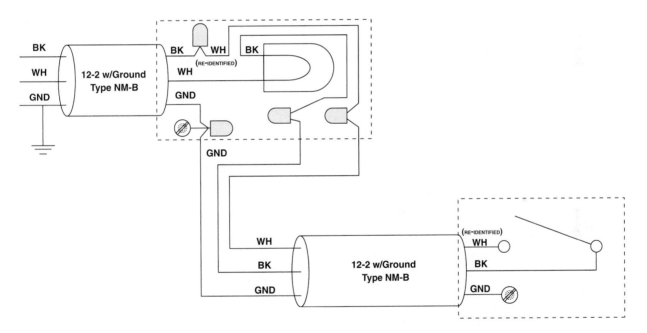

## NOTE: *NEC® Article 200.7(C) (2)* states:

*Where a cable assembly contains an insulated conductor for single-pole, 3-way, or 4-way switch loops and the conductor with white or gray insulation or a marking of three continuous white stripes is used for the supply to the switch but not as a return conductor from the switch to the switched outlet. In these applications, the conductor with white or gray insulation or with three continuous white stripes shall be permanently re-identified to indicate its use by painting or other effective means at its terminations and at each location where the conductor is visible and accessible.\**

**FIGURE 11-3**  Two methods for connecting a single-pole switch to control a light.

circuit. Notice the wire that is color coded black. This is the ungrounded (hot) conductor. The black conductor contained within the 12-2 w/ground Type NM-B cable in the upper left is the supply to the switch. Leaving the switch is a black conductor, which is part of the 12-2 w/ground Type NM-B cable that connects the switch to the lamp. This is called the switch leg. Notice the white conductor in the cable in the upper left. This is the grounded (neutral) conductor. Notice that the white conductor is spliced to another white conductor within the switch box. The second white conductor terminates at the lamp.

The lower circuit of Figure 11–3 shows a second method of connecting a single-pole switch to control a light. In this circuit, the power is fed into the light enclosure first. In the previous example, the power was fed into the switch enclosure first. Feeding the power into the light enclosure first results in some changes in the manner in which the circuit is wired. Notice the black conductor in the supply cable at the left. Notice also the 12-2 w/ground Type NM-B cable used between the light enclosure and the switch enclosure. This cable contains a black and a white conductor. The white conductor is spliced to the black supply conductor in the light enclosure. Therefore, the black supply conductor provides power to the switch by way of the white conductor. The return from the switch (switch loop) consists of the black conductor within the 12-2 w/ground Type NM-B cable between the switch enclosure and the light enclosure. This means that the white conductor supplies the power to the switch. In this situation, Article 200.7(C)(2) of the *NEC* requires the white conductor to be reidentified.

Figures 11–4A and 11–4B show four methods for connecting three-way switches to control a single light. Notice the color code for this arrangement. The black wire is the ungrounded (hot) conductor from the supply. The wires that connect the 2 three-way switches together are called the *travelers.* The white conductor is the conductor that connects the load to the grounded supply conductor.

The operation of this circuit can be better understood by tracing the path of current. Assume that the current is flowing out the black wire (Figure 11–4A, top) to the switch. If the switch contacts are made as shown, the circuit is complete, and the lamp will glow. Moving the left switch contacts breaks the circuit, and the lamp will be dark. Leaving the left switch contacts in this position and moving the right switch contacts will again complete the circuit, and the lamp will glow. In this way, the lamp can be controlled from both switches.

Figure 11–5 shows a combination of three-way and four-way switches. The four-way switch changes the connections between the traveler conductors, thus changing the current path each time it is switched.

To trace the current path, refer to Figure 11–5. The left and right switches are the three-way switches; the middle switch is a four-way switch. Assume that the current flows from the panel through the black wire to the left switch. With the left switch in the position shown, the current will continue to flow through the left switch contacts, through the black wire, and then to the middle switch. Now, the current leaves the middle switch by the red wire and flows to the right switch where the current path is open. At this point, the light will not light. Moving the position of the left switch causes the circuit from the red wire to be completed to the black wire, which is connected to the light. The current will then flow through the lamp filament into the white wire and back to the panel, causing the lamp to light.

While the lamp is lit, imagine changing the position of the middle switch. When the middle switch is moved, the black wires will be connected together, as will the red wires. Follow the current path to see the effect of changing the position of the four-way switch. The current will flow from the panel through the black wire to the left switch. With the left switch in the position shown, the current will continue to flow through the left switch contacts and through the black wire to the middle switch. As the position of the middle switch has been changed, the current enters and leaves the middle switch by way of the black wires. The current then continues to the right switch that was left in the "down" position. The circuit is open at this point, and the light will not light. As you can see by tracing the current paths with the various positions of the three switches, the lamp can be controlled by any of the three switches.

At times, lights are controlled by low-voltage switches and relays. This system is useful because it provides much more flexibility of control.

The individual switches of this system control a relay (an electromagnetic switch), which in turn controls the light. The relays may be located at the lighting fixtures or in one central location. The individual switches control the current to the low-voltage coil (usually 24 volts) in the relay. This eliminates the need to use three-way and four-way switches when control is desired from more than one location.

A master selector control may also be installed. This control permits all of the lights or specific groups of lights to be controlled from one location. The master control can also be wired with the burglar alarm system so that all the lights in the house will come on when the alarm is energized.

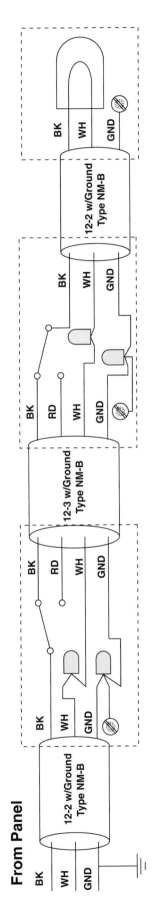

**Three-Way Switched Circuit with the Load at the End of the Run.**

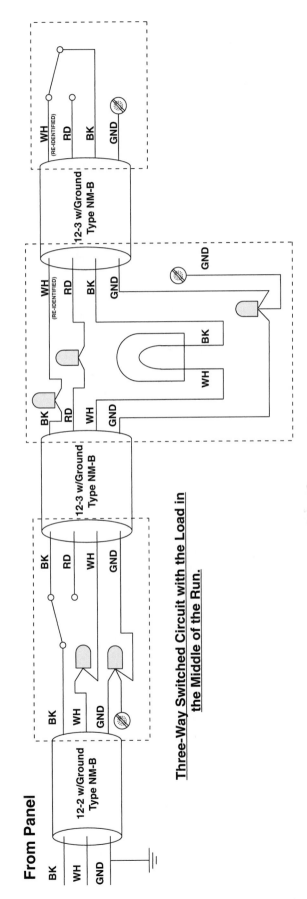

**Three-Way Switched Circuit with the Load in the Middle of the Run.**

**NOTE: NEC® Article 200.7(C)(2) states:**

*Where a cable assembly contains an insulated conductor for single-pole, 3-way, or 4-way switch loops and the conductor with white or gray insulation or a marking of three continuous white stripes is used for the supply to the switch but not as a return conductor from the switch to the switched outlet. In these applications, the conductor with white or gray insulation or with three continuous white stripes shall be permanently re-identified to indicate its use by painting or other effective means at its terminations and at each location where the conductor is visible and accessible.\**

**FIGURE 11–4A** Connections for three-way switches.

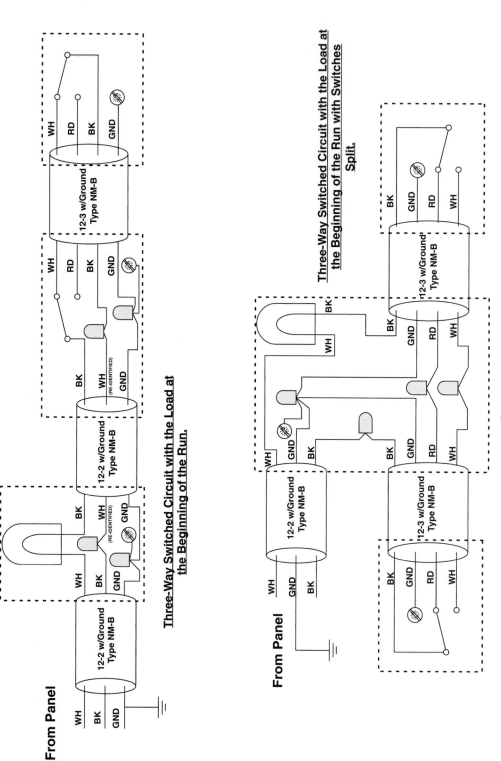

**From Panel**

12-2 w/Ground Type NM-B

WH
BK
GND

**Three-Way Switched Circuit with the Load at the Beginning of the Run.**

12-2 w/Ground Type NM-B

BK
WH (RE-IDENTIFIED)
GND

BK
WH (RE-IDENTIFIED)
GND

12-3 w/Ground Type NM-B

WH
RD
BK
GND

WH
RD
BK
GND

**Three-Way Switched Circuit with the Load at the Beginning of the Run with Switches Split.**

BK
GND
RD
WH

12-3 w/Ground Type NM-B

**From Panel**

WH
GND
BK

12-2 w/Ground Type NM-B

WH
GND
BK

BK
GND
RD
WH

12-3 w/Ground Type NM-B

BK
GND
RD
WH

**NOTE: *NEC*® *Article 200.7(C) (2)* states:**

*Where a cable assembly contains an insulated conductor for single-pole, 3-way, or 4-way switch loops and the conductor with white or gray insulation or a marking of three continuous white stripes is used for the supply to the switch but not as a return conductor from the switch to the switched outlet. In these applications, the conductor with white or gray insulation or with three continuous white stripes shall be permanently re-identified to indicate its use by painting or other effective means at its terminations and at each location where the conductor is visible and accessible.*

**FIGURE 11–4B** Connections for three-way switches (continued from Figure 11–4A).

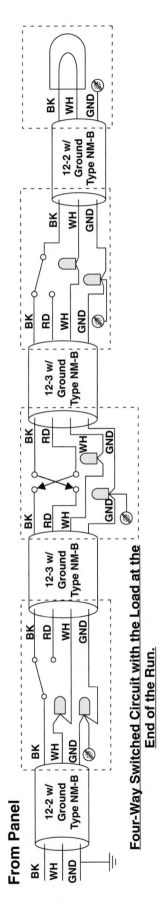

**FIGURE 11-5** Connections for control of light fixture using three-way and four-way switches.

## Alarm Systems

For the wiring of new houses or rewiring old ones, it is advisable to install burglar and fire alarm systems. There are many styles and types available. The selection of the type to use is a matter of personal preference. All systems should have a backup supply (usually batteries) in case of loss of power. When the main source of power is interrupted, a transfer switch automatically connects the batteries to supply power for the system. When the main power is restored, the transfer switch automatically disconnects the batteries and reconnects the main supply. The batteries should be checked periodically to ensure that they are adequately charged.

A fire alarm system may consist of smoke detectors, rate-of-rise heat detectors, and/or fixed-temperature heat detectors. A smoke detector reacts to smoke to start the alarm. A rate-of-rise heat detector energizes the alarm mechanism when the temperature increases very rapidly. A fixed-temperature heat detector starts the alarm when the temperature reaches a predetermined fixed value.

Burglar alarm systems fall into two major categories: those that signal when someone tries to enter a house (perimeter protection) and those that signal after someone enters the house (zone protection). Perimeter protection uses magnetic switches on doors and windows, which signal when a door or window is opened. They can be combined with other special devices that signal when the glass is broken. Zone protection, on the other hand, signals only after someone has entered the house. It operates by detection of changes in light, heat, or ultrasonic sound waves. With either system, contacts can be placed under carpets to signal when someone steps on them. Sometimes, carpet contacts are used alone. Very often, perimeter protection devices and zone protection devices are used together to provide better security.

## COMMERCIAL AND INDUSTRIAL WIRING

Commercial buildings are structures such as stores, offices, and warehouses. Commercial wiring is generally installed in rigid metal conduit, intermediate metal conduit, electrical metallic tubing, surface raceway, and cellular floor raceway. Circuit breakers and fuses are used for overcurrent protection.

Industrial installations consist of factories, foundries, machine shops, metal refineries, and similar structures. The wiring is generally installed in rigid metal conduit, wireways, and busways.

Industrial establishments also use fuses and circuit breakers for overcurrent protection.

Industrial and commercial installations frequently use large amounts of power, thus requiring the installation of transformers on the premises. The power is supplied at a high voltage, such as 2300 volts. The transformers lower the voltage to 277/480 volts and/or 120/208 volts. Other voltages sometimes used are 120/240 volts and 480 volts.

## Commercial and Industrial Services

The volt-amperes-per-square-foot method for calculating general lighting loads may be used for commercial and industrial buildings. (See Article 220 of the *NEC*.) All other loads connected to the installation must be added. The service may be as small as 100 amperes or as large as several thousand amperes. The service, feeders, and branch circuits should be of adequate size to allow for future additions. Figure 11–6 shows a typical commercial service.

## Calculating Loads

Determining the proper size of the service, feeders, and branch circuits requires accurate information and calculations. To make the calculations, it is common practice to start with the branch circuits and combine their loads at each panel.

To determine the feeder load, the combined branch-circuit loads are added to the load allowance for expansion, and the result is multiplied by the demand factor. (The demand factor varies with the type of installation.) After all the feeder loads have been calculated, they are combined to determine the size of the service. If the ampere rating of the service exceeds the maximum value for standard conductor sizes, multiple conductors must be installed in parallel. For example, if the ampacity of the service is 900 amperes, two Type THW, 2,000,000 circular-mil cables must be installed for each conductor of the service. For determining conductor ampacity and for multiple connections, see *NEC* Article 310.

## Overcurrent Protection

A very large amount of current is usually used in commercial and industrial areas. Therefore, it is important to use extreme care in selecting *overcurrent protection devices*. The most common types of overcurrent protection devices are fuses and circuit breakers. Fuses and circuit breakers have two current ratings. One rating is the current value at which they will disconnect the circuit from the supply. The second rating is the current value at which they can

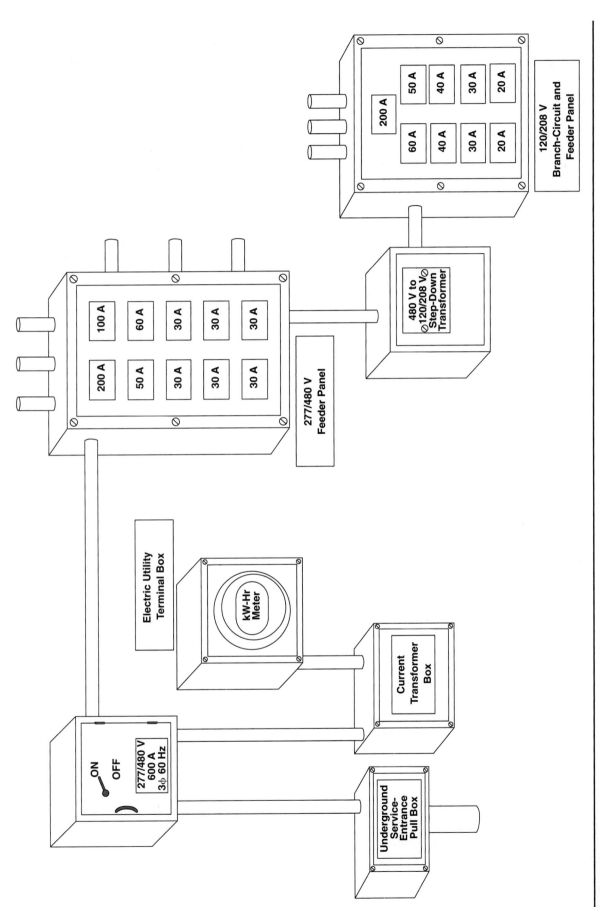

**FIGURE 11–6** Typical commercial service.

operate without damage to the equipment they are connected and in which they are housed. If a fuse is rated at 15 amperes, the link will begin to melt when more than 15 amperes pass through the fuse. The length of time it will take to open the circuit depends upon the amount of current flowing. For example, if 16 amperes are flowing in a circuit protected by a 15-ampere fuse, it may take several minutes for the link to melt. If 1000 amperes are flowing in the circuit, however, the link will melt in less than a second. This time factor is very important to consider when selecting the correct overcurrent device.

A **short circuit** is a place in the wiring system where conductors of opposite polarity make contact. In theory, one might say that a short circuit is a circuit with zero resistance, but this is not true in actual practice. The conductors and equipment between the generator and the point of contact have resistance. The resistance, along with other variables, places some restriction on the amount of current that will flow under short-circuit conditions.

If the opposition to current flow is very low, the short-circuit current will build up to several thousand amperes in less than a second. A fuse rated at 15 amperes will start to melt when the current exceeds 15 amperes, and it will eventually open the circuit. The ordinary zinc link fuse is rated at 10,000-amperes interrupting current.

> ⚠ **CAUTION** If, under short-circuit conditions, the current should build to a value in excess of 10,000 amperes before the link melts, the fuse could explode. Such an explosion may cause damage to equipment and injure personnel. It is the responsibility of the electrical worker to determine the available short-circuit current at the installation. This information can usually be obtained from the utility company. Fuses and circuit breakers are available for various current-interrupting capacities.

Fuses having a current-interrupting capacity of other than 10,000 amperes have the rating marked on the fuse. Circuit breakers having a current-interrupting rating of other than 5000 amperes have the rating indicated on the body of the breaker. Figure 11–7 illustrates some of the most common fuses and circuit breakers.

**Circuit breakers** are mechanical devices that respond to current changes. These devices interrupt current flow, within rated values, without injury to themselves. There are three general types of circuit breakers:

- Thermal
- Magnetic
- Thermal–magnetic

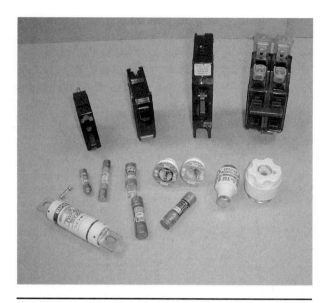

**FIGURE 11–7** Various types of overcurrent protective devices: circuit breakers and fuses.

The *thermal circuit breaker* responds to temperature. An increase in current produces an increase in the temperature of the sensing element. When the temperature reaches a predetermined value, the breaker opens the circuit.

The *magnetic circuit breaker* responds to a magnetic field produced by the current flowing through the breaker. As the value of current increases, the magnetic field becomes stronger. When the field strength reaches a predetermined value, the breaker disconnects the circuit from the supply.

The *thermal–magnetic circuit breaker* is a combination of both of the previous types. It responds to both temperature and magnetism.

In some cases, the short-circuit current may exceed the interrupting rating of the circuit breaker or fuse. It then becomes necessary to install overcurrent devices with a greater interrupting capacity or to use current-limiting overcurrent devices. A *current-limiting overcurrent device* is a fuse or circuit breaker that opens the circuit much more quickly than the ordinary device does. Fast-acting overcurrent devices open the circuit before the current can build to an excessive value. Current-limiting devices are used extensively in commercial and industrial establishments.

The main overcurrent device for a commercial building may be rated at several thousand amperes. Extending from this device are feeders that are used to distribute the power throughout the building. These feeders are large conductors that are protected by fuses or circuit breakers and terminate in branch-circuit panels. The branch circuits are the final branches of the wiring system, with conductors smaller than the feeder conductors. Branch circuits are also protected by overcurrent devices rated according to the conductor sizes. Figure 11–8 illustrates a typical commercial installation.

## Selection of Overcurrent Devices

The electrical system is the heart of a commercial or industrial establishment. The greatest operating expense is a shutdown. The major causes of shutdowns are short circuits, overloads, and burnouts. Careful selection of overcurrent devices can minimize shutdowns.

In the selection of overcurrent devices, the following factors should be considered:

- Maximum continuous current
- Maximum operating current
- Maximum interrupting current
- Frequency of the system
- Duty cycle
- Type of load

Because of their convenience, circuit breakers are often used in preference to fuses. Two of the more common types are the *molded-case circuit breaker* and the *air circuit breaker*. In the molded-case circuit breaker (Figure 11–9), the entire assembly is enclosed in a molded, nonmetallic case. Molded-case circuit breakers are available in continuous current ratings up to and including 800 amperes at 600 volts or less. In the air circuit breaker, the operating mechanism is enclosed in a container, but it is accessible for inspection and maintenance.

The circuit breaker must be able to carry the continuous current without nuisance tripping. A common practice is to allow a breaker to carry only 80 percent of its continuous current rating. It must be able to open the circuit under short-circuit conditions without injury to itself.

A circuit breaker should always be rated at a voltage equal to or greater than the system voltage. The frequency of the system can also have an effect on the operation of a circuit breaker. If the frequency is other than 60 hertz, consult the breaker manufacturer.

**Frequency** is the number of cycles of AC completed in 1 second. The unit of measurement of frequency is the **hertz** (Hz), which means cycles per second.

If a circuit breaker is protecting a load that is intermittent, such as frequent and periodic starting of motors or groups of flashing lights, a cumulative heating effect may occur. The ambient temperature and the temperature buildup caused by surges of current must always be considered when selecting circuit breakers. Time delay of operation is an important factor when considering motor-starting currents.

In summary, the circuit breaker should never be subject to currents in excess of its interrupting capacity. It should be installed so as to be protected against overheating during normal operation. The breaker should never be subject to voltages in excess of its rating. The frequency of the system should always be considered when selecting a circuit breaker.

The maintenance engineer should have equipment available to test circuit breakers that have been subjected to short-circuit currents. Circuit breakers should be periodically switched off and on to ensure that the contacts do not corrode together.

Air circuit breakers are available in continuous current ratings from 15 amperes to 4000 amperes, and at voltages up to 600 volts. These breakers are suitable for lighting and power circuits as well as motor starting and running service. They are equipped with heavy-duty contacts, arc quenchers, and an operating mechanism, which may be either manual or automatic. The air circuit breaker has the advantage of being easy to inspect, test, and maintain. In these respects, the air circuit breaker is preferred to the molded-case breaker. The disadvantages are initial cost, and in some instances, additional maintenance.

Air circuit breakers are adjustable for instantaneous, short-time, and/or time-delay operation. The standard time delays are 5 to 12 cycles, 12 to 20 cycles, and 20 to 30 cycles of time delay at 60 hertz. (In a 60-hertz circuit, a delay of 30 cycles is equivalent to a time delay of 1/2 second.)

**Fuses** are also used in industrial establishments and can provide maximum protection with minimum downtime. The two general types of fuses are the plug fuse and the cartridge fuse.

*Plug fuses* are available in ratings of up to 30 amperes, at 125 volts. There are several kinds of plug fuses: the standard Edison-base fuse, the Edison-base dual-element fuse, the type S, dual-element tamperproof fuse, and the DIN-style, type D, and neozed low-voltage fuse (also called "Bottle" fuse). The dual-element fuse provides time delay for motor-starting currents and instantaneous short-circuit protection. Figure 11–10 shows the type-S and DIN-style type-D fuses.

*Cartridge fuses* are the type most commonly used in industrial installations. They are available in single-element and dual-element styles as well as the renewable style: The renewable fuse has a replaceable link. The initial cost of the renewable fuse is higher than the onetime fuse. Using renewable links, however, is much less expensive than replacing the entire nonrenewable fuse.

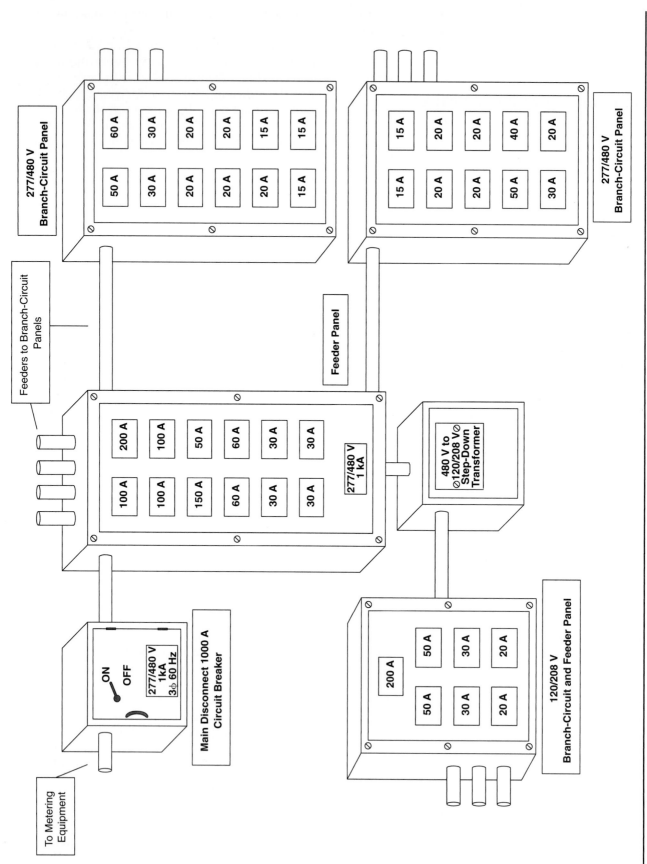

**FIGURE 11-8** Typical commercial installation.

**FIGURE 11-9** Molded-case circuit breakers.

**FIGURE 11-11** Various types of cartridge fuses.

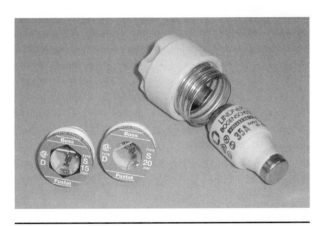

**FIGURE 11-10** Type-S fuses and DIN-style, type-D fuses.

Cartridge fuses are available in the ferrule and the knife-blade styles. The ferrule style is used for fuses rated at 60 amperes or less and up to 600 volts. The knife-blade style is used for fuses of more then 60 amperes and up to 600 volts. Figure 11–11 illustrates some types of cartridge fuses.

For circuits of over 600 volts, special overcurrent protective devices are used. The available short-circuit current in industrial areas frequently exceeds 10,000 amperes. The installation of standard fuses or circuit breakers backed up by a current-limiting overcurrent device maximizes protection. The current-limiting overcurrent device is designed both to function as a short-circuit protective device and to allow for overload.

Current-limiting overcurrent protective devices have high current-interrupting capabilities. This feature and their fast action under short circuits make them a valuable device in industrial installations.

## Voltage (Line) Drop

Voltage drop (line drop) is another very important factor to consider when wiring an industrial establishment. Because such installations cover large areas and generally require high values of current, voltage drop can become a serious problem. Services should be installed in a central location to minimize the distance to the loads. In calculations of feeder and branch-circuit sizes, the length of the conductors and the ambient temperature must always be considered. Conductor sizes must be increased to compensate for high temperatures and long distances.

## Grounding

Grounding of electrical systems and equipment is a very important safety factor. *Grounding* means to connect to the ground one wire of the electrical system and/or the metal enclosures of electrical conductors and equipment. Grounding of equipment is accomplished by connecting all the non-current-carrying metal parts together and then connecting them to the ground. All connections should be made wrench tight.

The main system ground for AC services is located on the line side of their service-disconnecting device. At this point, both the system ground and equipment ground are connected together to form a common ground. The grounding conductor connects to the common ground to the grounding electrode. All connections should be made tight to ensure good conductivity.

It may appear that it would be safer to keep all electrical systems ungrounded. If it were possible to have every installation perfect and to prevent any accidental grounds, it would be safer to have an ungrounded system. However, experience has shown that this is not possible. Oil damages insulation, and expansion of metal parts, causes insulation breakdown. Accumulation of moisture and other impurities causes leakage to ground.

> ⚠️ **CAUTION** A loose connection or break in the grounding conductor can cause a serious hazard. For example, if the grounding conductor becomes disconnected and a ground fault occurs in the system, the current caused by the fault will seek other paths to ground. The alternative paths may have high resistance, thus producing heat and possibly causing a fire. Another hazard could occur if the ungrounded (hot) wire accidentally makes contact with a metal enclosure and there is no path to ground. The metal enclosure becomes energized with a voltage equal to that of the system voltage to ground. A person making contact with the energized enclosure and ground completes the circuit, and the resulting shock could be fatal.
>
> A loose connection in the grounding conductor causes high resistance. Stray currents caused by lightning or by a ground fault flowing through the high resistance produce heat and can possibly cause a fire.

*NEC* Article 250 prescribes methods and standards for grounding.

## Preventive Maintenance

Once an installation has been completed, a good preventive maintenance program should be developed. All conductors and equipment should be tested periodically for deterioration, malfunction, and safe operation. A log should be maintained, listing the times and dates of inspection, the condition of the equipment and insulation, and any other pertinent information. A good maintenance program minimizes breakdowns.

## REVIEW QUESTIONS

*Multiple Choice*

1. Appliance circuits must be included in the wiring plan for dwelling occupancies. The total minimum volt-amperes for these circuits must be
   a. 1000 volt-amperes.
   b. 1500 volt-amperes.
   c. 2000 volt-amperes.
   d. 3000 volt-amperes.

2. The power included for the laundry circuit must be
   a. 1000 volt-amperes.
   b. 1500 volt-amperes.
   c. 2000 volt-amperes.
   d. 3000 volt-amperes.

3. By applying the demand factor when calculating the service size, one is allowed to
   a. increase the service size.
   b. decrease the service size.
   c. install a demand meter.
   d. install larger-size fuses.

4. The regulations for installing electrical services can be found in *NEC* Article
   a. 220.
   b. 230.
   c. 240.
   d. 250.

5. Fuses and circuit breakers are
   a. safety valves.
   b. testing devices.
   c. current increasers.
   d. current reducers.

6. The common grounding conductor provides a path to ground for
   a. leakage currents.
   b. overcurrents.
   c. high voltages.
   d. short circuits.

7. The size of the common ground is based upon
   a. the voltage of the system.
   b. the size of the largest service conductor.
   c. the volt-amperes per square foot.
   d. the lighting load.

8. A ground-fault circuit interrupter is a device that
   a. opens the common grounding conductor when excessive current flows.
   b. senses a small imbalance of current between the ungrounded (hot) conductor and the grounded (neutral) conductor and opens the circuit when a predetermined value is exceeded.
   c. opens the circuit when there is a fault on the grounding conductor.
   d. disconnects the system when lightning occurs.

9. A current-limiting overcurrent device
   a. limits the current to a specific value.
   b. interrupts the current flow quickly when a short circuit occurs.

c. limits the current to 100 A.

d. limits the current to 5000 A.

10. A circuit breaker is a

 a. mechanical device that interrupts current flow within its rated values, when short circuits or overloads occur, without injury to itself.

 b. device used instead of a fuse to prevent explosions.

 c. device that opens the circuit when the resistance increases.

 d. device used to open circuits from a remote area.

11. The demand factor for an industrial building is

 a. the same for all installations.

 b. different according to the type of installation.

 c. never used for industrial installations.

 d. limited to 20% of the total load.

12. A circuit breaker should always be rated at a voltage

 a. of 10,000 volts.

 b. equal to the system voltage.

 c. equal to or greater than the system voltage.

 d. equal to 150% of the system voltage.

13. Adjustable circuit breakers are

 a. air circuit breakers.

 b. molded-case circuit breakers.

 c. oil-immersed circuit breakers.

 d. all of the above.

14. One kind of plug fuse is the

 a. standard Edison-base type.

 b. prong type.

 c. ferrule type.

 d. renewable type.

15. One kind of cartridge fuse is the

 a. Edison-base type.

 b. knife-blade type.

 c. type S, dual-element tamperproof type.

 d. screw-shell type.

16. The main ground for an AC service is located

 a. as near as possible to the grounding electrode.

 b. anywhere convenient in the building.

 c. on the line side of the service-disconnecting device.

 d. on the load side of the service-disconnecting means.

17. The common grounding wire

 a. connects the system ground to the grounding electrode.

 b. connects the equipment ground to the grounding electrode.

 c. both a and b.

 d. none of the above.

18. The local utility company

 a. installs the service switch.

 b. determines the type of power and value of voltage available.

 c. develops the laws governing the wiring of the building.

 d. installs the main switch.

19. Demand factor is the ratio of the

 a. maximum demand of the system to the total connected load.

 b. input to the output.

 c. available current to the total current.

 d. applied voltage to the voltage drop.

20. The minimum-size service for a single-family house should be

 a. 50 A.

 b. 150 A.

 c. 100 A.

 d. 200 A.

*Give Complete Answers*

1. List six questions that should be answered by the utility company before an electrical service is installed.

2. What information is found on the electrical floor plan?

3. Describe how to determine the general lighting load of a residence using the volt-amperes per square foot method.

4. What is the minimum load that must be calculated for small appliances in a residence?

5. Define the term *demand factor.*

6. Why does the *NEC* allow a demand factor?

7. List eight rules for residential wiring that will aid in carrying out a flexible, convenient, and trouble-free installation.

8. List the components of a residential service.

9. What is the purpose of fuses and circuit breakers?

10. How can one ensure minimum resistance of the common ground?

11. Describe a ground-fault circuit interrupter.

12. List three common types of lighting switches.

13. Complete the diagrams A, B, and C in the following figure.

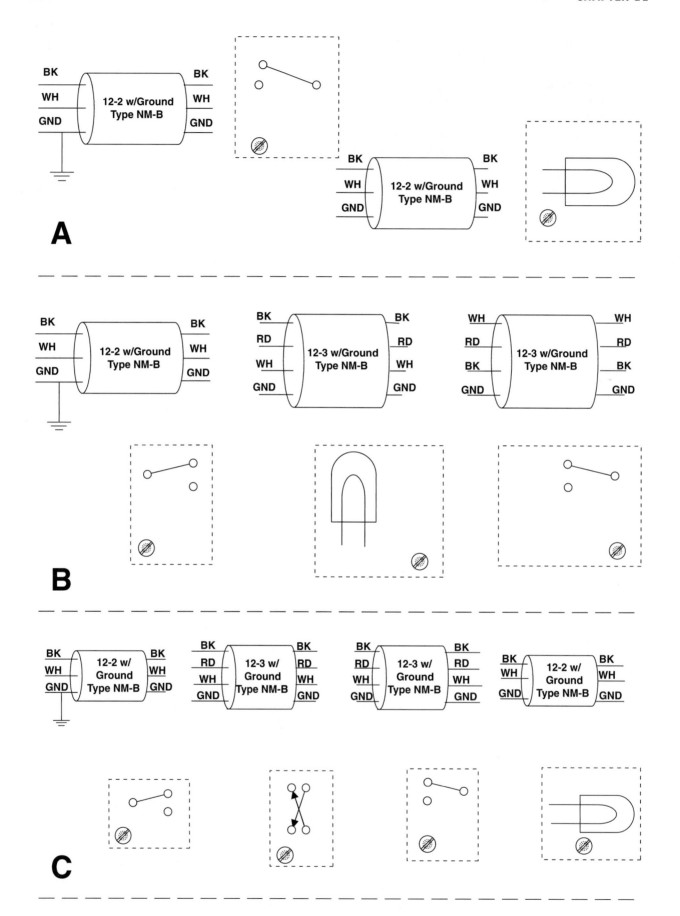

14. Explain the operation of low-voltage switching of lighting circuits.

15. List three types of detectors used with fire alarm systems.

16. What is the purpose of the backup supply used in conjunction with the fire and burglar alarm systems?

17. Describe a common type of residential burglar alarm system.

18. List five wiring methods used for commercial wiring.

19. What is the purpose of a power transformer in a commercial building?

20. Fuses and circuit breakers have two current ratings. What do these ratings indicate?

21. Using the *NEC*, describe one method for calculating the size of a service for a single-family dwelling.

22. Define the unit of measurement of frequency.

23. Name the two most common types of overcurrent protection devices.

24. List 10 general safety rules associated with electrical wiring.

25. Explain how to calculate the size of service, the size and number of feeders, and the size and number of branch circuits for an industrial establishment.

# Transformers

## OBJECTIVES

After studying this chapter, the student will be able to:

- Describe the construction and operating characteristics of transformers.
- Explain the theory of operation of various types of transformers.
- Describe the various methods used to prevent transformers from overheating.
- Illustrate various types of transformer connections and discuss the results of these connections.
- Describe the construction, use, and operating characteristics of special transformers.

# TRANSMISSION EFFICIENCY

Alternating current can be transmitted over great distances much more economically than direct current because of the transformer. The **transformer** is a device used to raise or lower voltage. It has no moving parts and is simple, rugged, and durable. The efficiency of a transformer may be as high as 99%.

Most electrical equipment is designed to be used on relatively low voltages. Common operating voltages are 120 volts, 208 volts, 240 volts, 277 volts, 440 volts, and 480 volts. Though it may be reasonably safe to do so, it would be very inefficient to transmit power over long distances at these voltages. The large quantities of power needed by cities, towns, and rural areas would require very high currents. The end result would be large power ($I^2R$) losses in the transmission lines, the use of very large conductors, or both.

Power-station alternators usually generate voltages in the vicinity of 13,000 volts to 100,000 volts. Power is transmitted over moderate distances at this range of voltage. At the load center, transformers are used to lower the voltage to the rated voltage of the equipment. If the power is being transmitted over a long distance, the voltage is increased at the generating station and then is decreased at the load center. When power is being transmitted 10 miles or more, the voltage is usually raised 1000 volts for each mile transmitted. In rural areas, transmission voltages can be as high as 1,000,000 volts.

# TRANSFORMER PRINCIPLE

The operation of a transformer depends upon electromagnetic induction. Most transformers consist of two or more coils. The coil receiving the energy is called the **transformer primary**, and the coil delivering the power to the load is called the **transformer secondary**. When an alternating voltage is applied to the primary, an alternating current flows. This current sets up an alternating magnetic field, which moves across the secondary coil (Figure 12–1). The moving magnetic lines of force induce a voltage into this coil.

In the transformer shown in Figure 12–1, there is no load connected to the secondary. Thus, no current is flowing in the secondary coil. Because the primary circuit has high inductance and low resistance, the current lags the voltage by nearly 90 degrees. The magnetic field produced by this current establishes a counter electromotive force (cemf) that is nearly 180 degrees out of phase with the applied voltage. Figure 12–2 illustrates this relationship.

The voltage induced into the secondary is a result of the current flowing in the primary and, therefore, is in phase with the cemf in the primary. In other words, at any instant, the secondary voltage of a transformer is in the direction opposite to the voltage applied to the primary.

The amount of voltage produced by electromagnetic induction depends upon the rate of speed at which the magnetic field moves across the loops and on the number of loops of wire on the coil. The

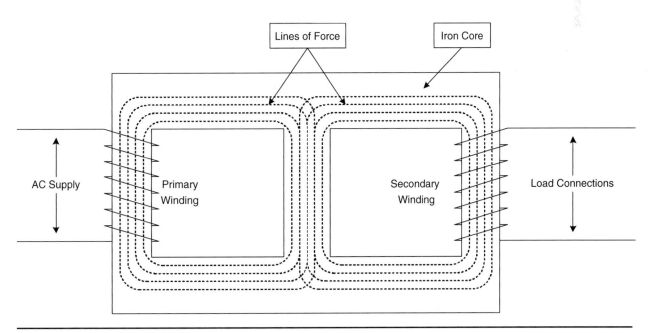

**FIGURE 12–1** Basic transformer: AC flowing in the primary winding establishes an alternating magnetic field, which induces a voltage into the secondary winding. No electrical connection is made between the primary winding and the secondary winding.

rate of speed depends upon the frequency of the primary circuit (which is a constant value). Therefore, the number of loops on the secondary coil must determine the value of the induced voltage. The more the turns of wire on the secondary coil, the greater will be the induced voltage.

Because the secondary voltage is a result of the primary current, their frequencies must be equal. Because the primary current is a result of the primary voltage, the frequency of the secondary voltage is equal to that of the primary voltage. Also, the number of volts per turn on the primary coil is equal to the number of volts per turn on the secondary coil. In other words, a transformer with 100 turns of wire on the primary and 200 turns on the secondary produces a voltage across the secondary that is double the voltage of the primary. If a pressure of 500 volts is applied to the primary, 1000 volts will appear across the secondary. Through basic mathematics it can be seen that there are 5 volts per turn on both the primary coil and the secondary coil.

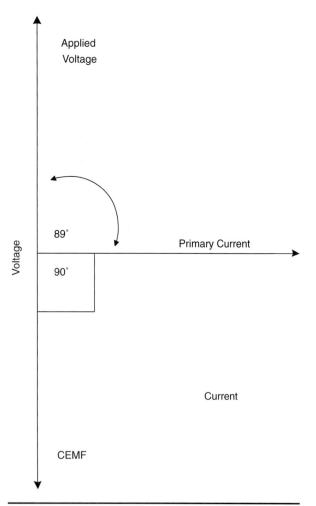

**FIGURE 12–2** Vector diagram of the relationship between the applied voltage, the primary current, and the cemf of a transformer at no load.

If a transformer increases the voltage, it is called a **step-up transformer**. If it decreases the voltage, it is called a **step-down transformer**. The same transformer can be used for either stepping up or stepping down the voltage. For example, if the transformer described previously is supplied with 1000 volts to the 200-turn coil, that coil becomes the primary coil. The 100-turn coil then becomes the secondary coil and will deliver 500 volts.

With no load on the secondary of a transformer, only a very small current flows in the primary winding. This is because the cemf is equal and almost opposite to the applied voltage. For example, if the applied voltage is 100 volts and the cemf lags the applied voltage by 179 degrees, then, from Equation 7.5,

$$E = \sqrt{E_a^2 + E_i^2 - (2E_aE_i \cos \phi)}$$
$$E = \sqrt{100^2 + 100^2 - (2 \times 100 \times 100 \times 0.9998)}$$
$$E = \sqrt{20,000 - 19,996}$$
$$E = \sqrt{4}$$
$$E = 2 \text{ V}$$

The value of current flowing in the primary winding is the result of 2 volts.

## Current Regulation under Load

When a load is connected across the secondary of a transformer, current flows in the secondary winding. The amount of current depends upon the impedance of the load. If the secondary load is primarily resistive, the secondary current, for all practical purposes, is 180 degrees out of phase with the primary current. As a result, this current produces a magnetic field that opposes the magnetic field produced by the primary current. This opposition weakens the primary magnetic field, resulting in less cemf. A reduction in the cemf allows the primary current to increase; the amount by which it increases is directly proportional to the value of current flowing in the secondary. Thus, the secondary current in a transformer regulates the amount of current flowing in the primary.

## Voltage Regulation under Load

The current flowing in the primary winding of a transformer produces a magnetic field that induces a voltage into the secondary winding. If the secondary current causes the primary flux to become weaker, how is the secondary voltage affected? To answer this question, refer to the statement about the primary current: "A reduction in the cemf allows the primary current to increase; the amount by which it increases is directly proportional to the value of current flowing in the secondary." This increase in current strengthens the primary field flux just enough to maintain the necessary voltage across the secondary windings.

## Capacitive and Inductive Effect

The power factor of the load determines the transformer's ability to maintain good secondary voltage regulation. As the load becomes more inductive, the secondary current begins to lag the voltage across the load. This changes the phase relationship between the primary current and the secondary current. The more inductive the load becomes, the greater is the change. As a result, highly inductive loads cause poor secondary voltage regulation even if the primary voltage is constant. Loads taking large lagging currents result in a high $IX_L$ drop in the secondary winding of a transformer. Capacitors or synchronous motors must be installed to overcome this problem. If enough capacitance is installed to bring the power factor back to 100%, the circuit is in resonance. Under this condition, the transformer will maintain a stable secondary voltage.

## Turns — Voltage — Current Relationships

The relationship between the voltage and the number of turns on each winding of a transformer is called the *turns ratio*.

When a transformer is supplying a load at or near 100% power factor, the voltage ratio between the primary and the secondary depends upon the number of turns of wire on each winding. If there are more turns on the secondary than on the primary, the secondary voltage will be greater than the voltage of the primary. If there are fewer turns on the secondary than on the primary, the secondary voltage will be less than the voltage of the primary. This ratio of voltage and turns can be expressed mathematically as follows:

$$\frac{E_p}{E_s} = \frac{N_p}{N_s} \quad \text{(Eq. 12.1)}$$

where   $E_p$ = primary voltage
   $E_s$ = secondary voltage
   $N_p$ = number of turns of wire on the primary winding
   $N_s$ = number of turns of wire on the secondary winding

## Example 1 _____

A transformer is being designed to decrease the voltage from 120 volts to 12 volts. If the primary requires 400 turns of wire, how many turns are required on the secondary?

$$\frac{E_p}{E_s} = \frac{N_p}{N_s}$$

$$\frac{120}{12} = \frac{400}{N_s}$$

$$N_s = \frac{4800}{120}$$

$$N_s = 40 \text{ turns (secondary turns)}$$

The amount of current flowing in the secondary depends upon the load connected to the transformer. This secondary current regulates the primary current. Therefore, there is a definite mathematical relationship between the two. Transformers are used to transfer power from one circuit to another with minimum loss. For most practical calculations, it can be assumed that transformers are 100% efficient. With this in mind, it can be observed that when the voltage increases, the current must decrease. Power is equal to current times voltage ($P = IE$). Thus, to maintain the same value of power, an increase in voltage will require a corresponding decrease in current. Likewise, a decrease in voltage will require an increase in current. This relationship between the voltage and the current in the two windings of a transformer may be expressed mathematically as follows:

$$\frac{E_p}{E_s} = \frac{I_s}{I_p} \quad \text{(Eq. 12.2)}$$

where   $E_p$ = primary voltage
   $E_s$ = secondary voltage
   $I_s$ = secondary current
   $I_p$ = primary current

## Example 2 _____

If the transformer in example 1 supplies a load requiring 5 amperes, how much current flows in the primary circuit?

$$\frac{E_p}{E_s} = \frac{I_s}{I_p}$$

$$\frac{120}{12} = \frac{5}{I_p}$$

$$I_p = \frac{60}{120}$$

$$I_p = 0.5 \text{ A (primary current)}$$

The combined relationship between voltage, current, and turns can be expressed mathematically as follows:

$$\frac{E_p}{E_s} = \frac{N_p}{N_s} = \frac{I_s}{I_p} \quad \text{(Eq. 12.3)}$$

When the term **transformer ratio** is used, it generally refers to the turns ratio. It is a common practice in industry to state the primary number first. For example, the transformer in example 1 has a ratio of 10 to 1 (written 10:1). This indicates that the transformer is being used as a step-down transformer. If the primary winding has 200 turns and the secondary winding has 600 turns, the ratio is 1:3, indicating a step-up transformer.

## LOSSES AND EFFICIENCY

The power losses in a transformer are made up of copper losses and magnetic losses. Manufacturers strive to keep losses to a minimum. The losses generally range from 1% to 15%, depending on the application. Transformers that are used to transfer large amounts of power are designed to be at least 98% efficient. Small transformers, such as those used for doorbell and signal circuits, are usually rather inefficient (even less than 85%). These transformers transfer small amounts of power for very brief periods of time; therefore, efficiency is not important.

### Copper Losses

The windings on transformers consist of copper wire, which has some resistance. Although the wire is sized to keep the resistance to a minimum, it cannot be completely eliminated. **Copper losses** are proportional to the square of the current ($P = I^2R$). Therefore, the proper sizing of the wire is very important.

When there is no load on a transformer, zero current flows in the secondary and only a very small current flows in the primary. Under this condition, the $I^2R$ loss is negligible.

### Magnetic Losses

The current flowing in the windings of a transformer is the source of the magnetic energy. The transformer is designed so that the magnetic lines of force will pass through the core. However, some of the magnetic lines leak out into the surrounding air. These serve no useful purpose and thus are wasted.

The core of a transformer is used to provide a low reluctance path for the magnetic lines of force. The type of material from which the core is made affects the efficiency of the transformer. Because an alternating current flows in the windings, the magnetic flux produced by the current is also alternating.

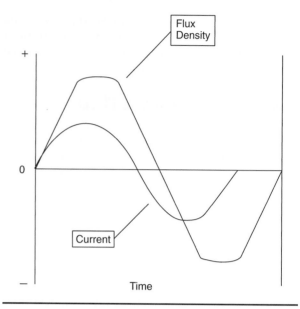

**FIGURE 12–3**  Magnetic flux lags the magnetizing force (current) that produces it.

This alternating flux tends to lag behind the current that produces it. Figure 12–3 illustrates the magnetizing current and its time relationship to the flux density. The shape of the flux curve and the amount of lag depend upon the material of the core.

When the **magnetizing current** begins to flow and increases in value, the flux tends to increase in a direction that is determined by the direction of the magnetizing current. The flux reaches a maximum value shortly after the current does. At this point, the flux tends to remain constant somewhat longer than the current. The amount of time depends upon the retentivity of the core. When the current reaches zero, the core still retains some magnetic flux. As the current reverses its direction of flow, it must produce a magnetic field that is strong enough to demagnetize the core before it can reverse the flux direction. The power required to demagnetize the core is considered a loss and, therefore, reduces the transformer efficiency.

Because the magnetizing and demagnetizing of the core cause molecular motion, the resulting friction produces heat. This heat loss plus the power loss caused by the magnetic lag is called the *hysteresis loss*.

Another magnetic loss caused by electromagnetic induction takes place in the core. The alternating magnetic field produced by the primary current induces a voltage into the core of the transformer. The voltage causes small currents to flow within the core. These eddy currents produce heat, which is dissipated into the surrounding air, resulting in a power loss.

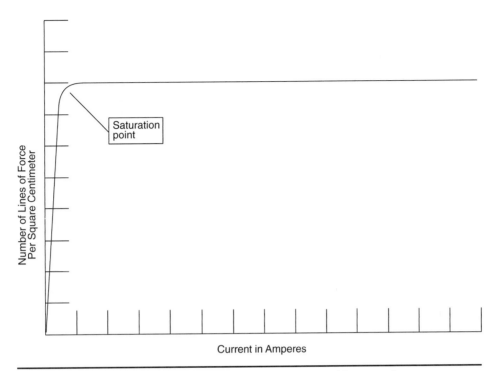

**FIGURE 12–4**  The relationship of the magnetic lines of force to the current.

A third magnetic loss is caused by a condition called *core saturation*. When the flux lines within the core become so dense that it is difficult to develop any more lines within it, the core is said to be magnetically saturated. At this point, a large increase in current will cause only a slight increase in the magnetic flux. After the saturation point is reached, a further increase in the primary current results in greater $I^2R$ loss, with very little increase in the magnetic effect. Figure 12–4 shows the relationship of the current and the lines of force through the saturation point.

Hysteresis and eddy current losses have a direct relationship to the frequency of the supply voltage. Therefore, in the design of transformers, the frequency should be of major concern in determining the type of core. When a metal core is used in a transformer, it should have good permeability and poor retentivity, thus reducing the hysteresis losses. If the core is constructed of thin laminations and the laminations are insulated from one another, the eddy currents can be kept to a minimum.

## TRANSFORMER CONSTRUCTION

The windings of commercial transformers are not placed on separate legs (as shown in Figure 12–1). A more efficient method, which reduces flux leakage, is to place the windings on top of one another (Figure 12–5). If a transformer is wound as shown in Figure 12–1, much of the flux produced by the primary current cannot reach the secondary winding. The leakage flux will induce a back voltage in the primary, causing a primary reactance drop. Similarly, much of the secondary flux will not reach the primary and, therefore, will not neutralize the primary flux but will produce a reactance drop in the secondary. The overall effect is similar to connecting a reactance in series with each winding.

In addition to reactance, each winding has resistance. The reactance and resistance of the windings may be represented as shown in Figure 12–6. In each winding of a transformer there is an $IX_L$ drop and an IR drop. The $IX_L$ drop is kept to a minimum by placing the primary and secondary windings on the same leg of the core. Frequently, they are made cylindrical in form and placed one inside the other. Another method is to build up thin, flat sections called *pancake coils*. These sections are sandwiched between layers of insulation. Figure 12–7 shows the cylindrical method; Figure 12–8 shows the pancake method. In the cylindrical method, the low-voltage winding is placed next to the core and the high-voltage winding is placed on the outside. This arrangement requires only one layer of high-voltage insulation placed between the two windings.

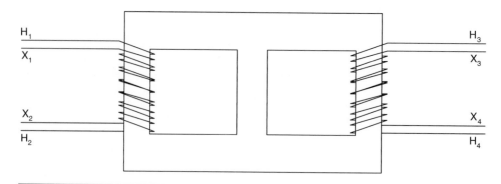

**FIGURE 12-5** Transformer with two high-voltage windings and two low-voltage windings. The windings may be connected either in series or in parallel, depending upon the voltage ratings and the desired results.

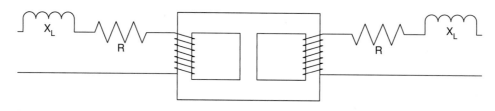

**FIGURE 12-6** Diagram showing the resistance and reactance of both windings of a transformer.

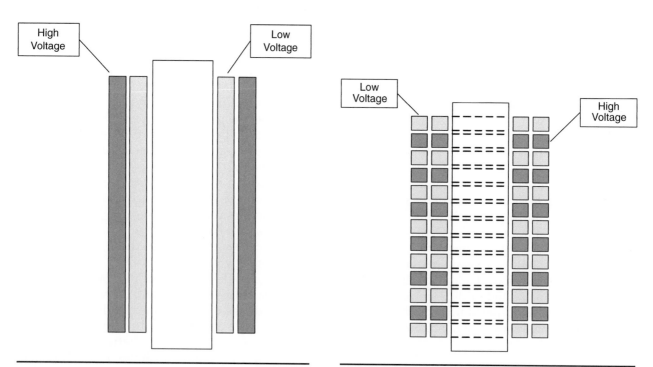

**FIGURE 12-7** Cutaway view of a transformer. The windings are made in the form of a cylinder and placed inside one another.

**FIGURE 12-8** Cutaway view of a transformer. The windings consist of thin, flat sections sandwiched between layers of insulation.

**FIGURE 12–9** The transformer core is made up of several layers of sheet steel, which are fastened together. This arrangement is known as laminated, core-type construction.

## Transformer Core Structure

Transformers used for power distribution are wound on cores made of iron or steel. The cores are built up in thin laminations, which are insulated from one another. Figure 12–9 illustrates one type of core construction. The purpose of the laminated core is to reduce eddy current losses.

## CONTROL TRANSFORMERS

Most transformers found in industrial applications fall into a category of transformers known as control transformers. A control transformer is a constant voltage transformer, most typically used to supply the proper operating voltage for control circuits involving switches, relays, contactors, motor starters, and others. Control transformers may also be used in signaling circuits.

The construction of control transformers is such that they are generally air cooled. The physical design of the control transformer enhances air flow around and through the windings, providing the necessary cooling. Some control transformers are of an open-frame design, while others may be enclosed in a metal housing.

Figures 12–10A, 12–10B, and 12–10C explain the differences in four types of control transformers as well as provide information in sizing control transformers. Figures 12–11A, 12–11B, 12–11C, and 12–11D provide additional information on the encapsulated type of control transformers. The open-style design

of control transformers is shown in Figures 12–12A and 12–12B. Control transformers that conform to international standards are shown in Figures 12–13A, 12–13B, and 12–13C. Finally, enclosed control transformers are shown in Figures 12–14A and 12–14B.

## Air Core Transformer

The coils of an air core transformer are wound on forms made of nonconducting materials. The magnetic circuit is through the air. Figure 12–15 illustrates an air core transformer. Air has much higher reluctance than iron or steel; therefore, this type of transformer is much less efficient than the iron/steel core.

A core having properties to form a perfect magnetic coupling between the primary and the secondary would have a coefficient of 1. Good-quality iron or steel cores used in distribution transformers have coefficients ranging from 0.975 to 0.988. The coefficients for most air core transformers range from 0.500 to 0.685.

The efficiency of an air core transformer is quite low. As a result, Equation 12.3, which was established for iron/steel core transformers ($E_p/E_s = N_p/N_s = I_s/I_p$), does not apply to air core transformers. These equations are based on 100% efficiency and an equal exchange of power.

Because air core transformers are inefficient, they are generally used for transferring small amounts of power when high frequencies make iron/steel core transformers impractical. (Hysteresis and eddy current losses increase rapidly as the frequency increases.)

## General Transformer Equation

The maximum flux density must be given careful consideration in the design of transformer cores. It has a major effect on the operating characteristics and efficiency of the transformer. The flux density is frequently measured by the number of lines of force per square centimeter or per square inch. The flux density depends upon many factors, but primarily upon the material of the core and the method of construction. The formula for flux density is

$$B_m = \frac{10^8 E_p}{4.44 f A N_p} \qquad \text{(Eq. 12.4)}$$

where   $B_m$ = flux density, in lines per square centimeter (lines/cm$^2$) or lines per square inch (lines/in.$^2$)

$E_p$ = voltage applied to the primary

$f$ = frequency, in hertz (Hz)

$A$ = area, in square centimeters (cm$^2$) or square inches (in.$^2$)

$N_p$ = number of turns on the primary

# Industrial Control Transformers

## Design Choices

Sola/Hevi-Duty offers a broad range of industrial control solutions to the most demanding industrial applications. Our products exceed NEMA ratings for inrush and regulation to ensure control systems are powered correctly. Electromagnetic control components demand inrush currents up to 10 times the transformer's nominal rating. While this inrush is occurring, the output side of the transformer must not fall below 85% of nominal as specified by NEMA ST-1, Part 4. Using a transformer that does not meet these ratings may cause erroneous shutdowns of downstream processes.

To meet your complete control needs, Sola/Hevi-Duty offers four series of control transformers, all of which exceed the NEMA standards. The Selection Chart can be used to identify the appropriate transformer for your application.

The **SBE series** is available from 50 - 5000 VA, 55°C Rise and features copper windings and encapsulation (through 1000 VA) for longer life and protection from the environment. This low temperature performance can mean smaller cabinet size or longer life for any electronic components that may be nearby.

The **SMT series** are 115°C rise, aluminum wound and for applications where good voltage regulation and higher power capacities (1000-5000 VA) are required.

The **International series** meets all IEC requirements including touchproof covers (IP20 ordered separately) for European applications.

The **HSZ series** rounds out Sola/Hevi-Duty's line with an enclosed series of control transformers from 1 - 10 KVA that feature either an UL-3R or NEMA 12 enclosure. This unique design, featuring copper windings and encapsulated construction, can help system designers meet harsher environmental standards or design for a safer installation outside of a control cabinet. The HSZ series is for applications where cost or heat issues make mounting the transformer outside the control panel necessary.

## Sizing an Industrial Control Transformer

For proper transformer selection, three characteristics of the load circuit must be determined in addition to the minimum voltage required to operate the circuit. These are total steady state (sealed) VA, total inrush VA, and inrush load power factor.

**A. Sealed VA** - Total steady state sealed VA is the volt-amperes that the transformer must deliver to the load circuit for an extended period of time.

**B. Inrush VA** - Total inrush VA is the volt-amperes that the transformer must deliver upon initial energization of the control circuit. Energization of electromagnetic devices takes 30-50 milliseconds. During this inrush period the electromagnetic control devices draw many times normal current – 3-10 times normal is typical.

**C. Inrush Load Power Factor** is difficult to determine without detailed vector analysis of all the load components. Generally such an analysis is not feasible, therefore, a safe assumption is 40% power factor. Until recently 20% PF was commonly used for transformer calculations, however, tests conducted on major brands of control devices indicate that 40% PF is a safer default assumption.

Visit our website at www.solaheviduty.com or
contact **Technical Services** at (800) 377-4384 with any questions.

**FIGURE 12–10A**   Control transformer types and sizing *(Courtesy of Sola/Hevi-Duty).*

**5** **Industrial Control Transformers**

## Selection Steps

1. Determine the supply and load voltages. The supply voltage is the available voltage to the control transformer. The load voltage is the operating voltage of the devices that will be connected to the transformer output.

2. Calculate the total sealed VA by adding the VA requirements of all components that will be energized together (timers, contactors, relays, solenoids, pilot lamps, etc.). Sealed VA data is available from the control device manufacturer.

3. Add the inrush VA of all components that will be energized together. Be sure to include the sealed VA of components that don't have an inrush, (lamps, timers, etc.) as they present a load to the transformer during maximum inrush.

4. Calculate selection inrush VA in one of the following two ways:

   A. Selection inrush VA =

   $$\sqrt{(VA\ sealed)^2 + (VA\ inrush)^2}$$

### Alternative Method

   B. VA sealed + VA inrush = Selection inrush

Method B will result in a slightly oversized transformer.

5. If your line voltage varies 10% or more, contact **Technical Services** for assistance.

6. From Chart A, select the transformer VA needed for your application from the "Transformer VA Rating" column. Check to be sure that the nameplate VA rating exceeds the sealed VA of the control circuit calculated in Step 1. If it does not, select a larger transformer VA that exceeds the circuit sealed VA.

By following the above procedure, the secondary voltage delivered by the transformer will be 90% of the nameplate secondary voltage under maximum inrush conditions at rated input voltage.

### Chart A: Regulation Data – Inrush VA at 20% and 40% Power Factor

| Selection Inrush VA* | | | | Transformer VA Rating |
|---|---|---|---|---|
| Type SBE | | Type SMT | | |
| 20% PF** | 40% PF** | 20% PF** | 40% PF** | |
| 294 | 207 | N/A | N/A | 50 |
| 515 | 363 | N/A | N/A | 75 |
| 696 | 490 | N/A | N/A | 100 |
| 1362 | 959 | N/A | N/A | 150 |
| 2131 | 1501 | N/A | N/A | 200 |
| 2883 | 2031 | N/A | N/A | 250 |
| 3608 | 2541 | N/A | N/A | 300 |
| 4777 | 3364 | N/A | N/A | 350 |
| 7601 | 5353 | N/A | N/A | 500 |
| 12939 | 9112 | N/A | N/A | 750 |
| 18703 | 13171 | 8277 | 5829 | 1000 |
| 23814 | 16066 | 17182 | 12100 | 1500 |
| 34586 | 24356 | 22834 | 16080 | 2000 |
| 45633 | 32770 | 34506 | 24300 | 3000 |
| 15800 | 111000 | 71284 | 50200 | 5000 |

*  Assuming the transformer is to deliver a minimum of 90% secondary voltage during inrush conditions.

Now refer to the Selection Tables on the following pages for the style you have chosen. Select your transformer according to your required voltage and VA capacity. By following the above procedure, the secondary voltage delivered by the transformer will be 90% of the nameplate secondary voltage under maximum inrush conditions at rated input voltage.

You can also use our online transformer product selector at www.solaheviduty.com/select. Enter your voltage requirements, hit the submit button and the models that meet your requirements will be listed.

**FIGURE 12–10B** Control transformer types and sizing (continued from Figure 12–10A) *(Courtesy of Sola/Hevi-Duty).*

# Industrial Control Transformers

**5**

## Choosing the Correct Series

The **SBE** series of industrial control transformers provide voltage regulation which exceeds NEMA standards. The SBE series are a 55°C rise and have copper windings and are 50/60 Hz rated. The SBE series can handle significant inrush with a minimal drop in output voltage.

The **SMT** series are 115°C rise, aluminum wound and are for applications where good voltage regulation and higher power capacities are required.

The **International** series have multiple voltage taps for easy application. These units also meet IEC 61558-1, 61558-2-2 and are CE marked for easy export to European countries.

The **HSZ** series is for applications where cost or heat issues make mounting the transformer outside the control panel necessary.

**Chart B: Voltage Code Chart**

| Voltage Code | Primary Voltage | Secondary Voltage |
|---|---|---|
| None | 240 x 480 | 120 |
| A | 230/460/575 | 115/95 |
| D | 240 x 480 | 24 |
| E | 120 x 240 | 24 |
| JL | 208/240/277 | 120/24 |
| JN | 208/240/480/600 | 120/24 |
| N | 240/415/480/600 | 130/120/99 |
| R | 480 | 240 |
| TC | 208/240/415 200/230/400 220/380 | 120/24 115/24 110/23 |
| TE | 208/240/415 277/480 200/230/400 220/380 | 24 24 24 23 |
| TF | 208/240/415/480/600 200/230/400/460/575 220/277/380 | 120 115 110 |
| TH | 240/415/480 230/400/460 220/380/440 | 120/240 115/230 110/220 |

## Selection Chart

| VA | SBE | | | | | | | SMT | | | International Series | | | | HSZ* | | |
|---|---|---|---|---|---|---|---|---|---|---|---|---|---|---|---|---|---|
| | -- | A | D | E | JL | JN | N | -- | A | N | TC | TE | TF | TH | -- | A | R |
| 50 | E050 | | E050D | E050E | E050JL | E050JN | | | | | E050TC | E050TE | E050TF | E050TH | | | |
| 75 | E075 | | | E075E | | | | | | | | | | | | | |
| 100 | E100 | | E100D | E100E | E100JL | E100JN | | | | | E100TC | E100TE | E100TF | E100TH | | | |
| 150 | E150 | | | E150E | | E150JN | | | | | E150TC | E150TE | E150TF | E150TH | | | |
| 200 | E200 | | | E200E | | | | | | | | | | | | | |
| 250 | E250 | | E250D | E250E | E250JL | E250JN | | | | | E250TC | E250TE | E250TF | E250TH | | | |
| 300 | E300 | | | E300E | | | | | | | | | | | | | |
| 350 | E350 | | | E350E | | | | | | | | | | | | | |
| 500 | E500 | | E500D | E500E | E500JL | E500JN | | | | | E500TC | E500TE | E500TF | E500TH | | | |
| 750 | E750 | | | E750E | | | | | | | | | E750TF | E750TH | | | |
| 1000 | E1000 | | | | | | Y1000N | | | T1000N | | | | | HZ1000 | HZ1000A | HZ1000R |
| 1500 | Y1500 | Y1500A | | | | | Y1500N | T1500 | T1500A | T1500N | | | | | HZ1500 | HZ1500A | HZ1500R |
| 2000 | Y2000 | Y2000A | | | | | Y2000N | T2000 | T2000A | T2000N | | | | | HZ2000 | HZ2000A | HZ2000R |
| 3000 | Y3000 | Y3000A | | | | | Y3000N | T3000 | T3000A | T3000N | | | | | HZ3000 | HZ3000A | HZ3000R |
| 5000 | Y5000 | Y5000A | | | | | Y5000N | T5000 | T5000A | T5000N | | | | | HZ5000 | HZ5000A | HZ5000R |

*\* Part Numbers listed are NEMA 3R, for NEMA 12 change HZ to HZ12*

Visit our website at www.solaheviduty.com or
contact **Technical Services** at **(800) 377-4384** with any questions.

**FIGURE 12–10C** Control transformer types and sizing (continued from Figure 12–10B) *(Courtesy of Sola/Hevi-Duty).*

# Industrial Control Transformers

## The SBE -- Encapsulated Series

The SBE Encapsulated industrial control transformers are epoxy encapsulated to seal the transformer windings against moisture, dirt and industrial contaminants. Extra deep, molded terminal barriers reduce the chance of electrical failure as the result of arcing or frayed lead wires. The rugged construction and proven reliability of the SBE design is uniquely suited for all industrial environments.

## Features

- 50 - 1000 VA, 50/60 Hz – suitable for world wide applications.
- Interleaved copper windings reduce $I^2R$ losses and maximize efficiency.
- 55°C Rise, 105°C insulation system to minimize heat.
- Epoxy encapsulated to protect cores and coils against moisture, dirt, and other contaminants.
- Meets or Exceeds NEMA Standard ST 1 and ANSI C89.1 for load inrush capability.
- Integrally molded, flame retardant (IEC 707 / ISO Class 1210) Terminal Blocks provide greater terminal contact area and improved conductivity.
- Heavy gauge steel mounting plate.
- Mounting dimensions are compatible with similar control transformers.
- *Secondary fuse holders (FB2X) included for 13/32 x 1- 1/2 cartridges (fuses not included).*
- *Factory-installed fuse holders are available (See WA & WB options).*
- 10 + 2 year warranty.

E77014

## Related Products

- Linear Power Supplies
- DIN Rail DC Power Supplies
- Constant Voltage Transformers
- Line Reactors

## SBE Mounting Profiles

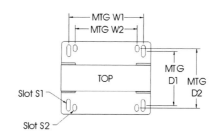

*Mounting Dimensions*

## Accessories

| Catalog Number | Description |
|---|---|
| FBP | Primary "CC" Rejection Type Fuse Holder (Finger Safe covers not available) |
| FB2 | Secondary Fuse Holder only (Glass or Ceramic, ¼" x 1¼" fuse). |
| FB2X | Secondary Fuse Holder only (Midget Cartridge Type, 13/32 x 1½" fuse). |
| FBPC1 | Primary "CC" Rejection Type Fuse Holder and Finger Safe Cover Kit |
| IP20 | IEC Touchproof Cover Kit |
| SBEDIN | IEC Fuse Holder Adaptor Kit |
| WA & WB | Factory installed Fuse Holders |

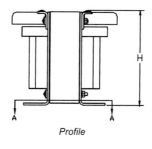

*Profile*

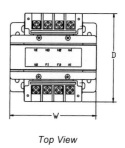

*Top View*

Visit our website at www.solaheviduty.com or contact **Technical Services** at **(800) 377-4384** with any questions.

**FIGURE 12–11A** Encapsulated control transformers *(Courtesy of Sola/Hevi-Duty).*

# Industrial Control Transformers

5

## SBE Encapsulated Series Selection Tables

### Group 1 – 120 x 240 Volt Primary, 24 Volt Secondary, 60 Hz

E77014

| VA | Catalog Number | Height (inch) | Width (inch) | Depth (inch) | Mtg Width W1 / W2 | Mtg Depth D1 / D2 | Slot Size S1 / S2 | Approx. Ship Weight (lbs) |
|---|---|---|---|---|---|---|---|---|
| 50 | E050E | 2.72 | 3.01 | 3.99 | 2.51 / NA | 2.02 / NA | .20 x .33 / .20 x .33 | 3 |
| 75 | E075E | 2.96 | 3.39 | 4.36 | 2.81 / 2.50 | 2.10 / NA | .20 x .50 / .20 x .50 | 3 |
| 100 | E100E | 2.96 | 3.39 | 4.61 | 2.81 / 2.50 | 2.37 / NA | .20 x .50 / .20 x .50 | 4 |
| 150 | E150E | 3.89 | 4.50 | 4.48 | 3.74 / 3.12 | 2.56 / 2.87 | .20 x .65 / .20 x .33 | 6 |
| 200 | E200E | 3.89 | 4.50 | 4.79 | 3.74 / 3.12 | 2.87 / 3.18 | .20 x .65 / .20 x .33 | 8 |
| 250 | E250E | 3.89 | 4.50 | 5.21 | 3.74 / 3.12 | 3.29 / 3.61 | .20 x .65 / .20 x .33 | 9 |
| 300 | E300E | 4.53 | 5.25 | 5.09 | 4.38 / 3.75 | 3.10 / NA | .31 x .71 / .31 x .71 | 10 |
| 350 | E350E | 4.53 | 5.25 | 5.53 | 4.38 / 3.75 | 3.54 / NA | .31 x .71 / .31 x .71 | 13 |
| 500 | E500E | 4.53 | 5.25 | 6.31 | 4.38 / 3.75 | 4.33 / NA | .31 x .85 / .31 x .85 | 17 |
| 750 | E750E | 5.56 | 6.38 | 6.93 | 5.32 / 4.37 | 4.25 / 5.75 | .31 x .85 / .31 x .85 | 25 |

**Note:** Includes FB2X Secondary fuse holder.

### Group 1A – Factory installed Primary Fuse Holder Class "CC" and Secondary Fuse Holder (Glass or Ceramic, ¼" x 1¼" fuse type).

| VA | Catalog Number | Height (inch) | Width (inch) | Depth (inch) | Mtg Width W1 / W2 | Mtg Depth D1 / D2 | Slot Size S1 / S2 | Approx. Ship Weight (lbs) |
|---|---|---|---|---|---|---|---|---|
| 50 | E050EWA | 4.18 | 3.01 | 3.99 | 2.51 / NA | 2.02 / NA | .20 x .33 / .20 x .33 | 3 |
| 75 | E075EWA | 4.41 | 3.39 | 4.36 | 2.81 / 2.50 | 2.10 / NA | .20 x .50 / .20 x .50 | 4 |
| 100 | E100EWA | 4.41 | 3.39 | 4.61 | 2.81 / 2.50 | 2.37 / NA | .20 x .50 / .20 x .50 | 5 |
| 150 | E150EWA | 5.36 | 4.50 | 4.48 | 3.74 / 3.12 | 2.56 / 2.87 | .20 x .65 / .20 x .33 | 8 |
| 200 | E200EWA | 5.36 | 4.50 | 4.79 | 3.74 / 3.12 | 2.87 / 3.18 | .20 x .65 / .20 x .33 | 10 |
| 250 | E250EWA | 5.36 | 4.50 | 5.21 | 3.74 / 3.12 | 3.29 / 3.61 | .20 x .65 / .20 x .33 | 11 |
| 300 | E300EWA | 5.99 | 5.25 | 5.09 | 4.38 / 3.75 | 3.10 / NA | .31 x .71 / .31 x .71 | 13 |
| 350 | E350EWA | 5.99 | 5.25 | 5.53 | 4.38 / 3.75 | 3.54 / NA | .31 x .71 / .31 x .71 | 15 |

**Note:** Includes Finger Safe covers.

### Group 1B – Factory installed Primary Fuse Holder Class "CC" and Secondary Fuse Holder (Midget Cartridge, 13/32" x 1½" fuse type).

| VA | Catalog Number | Height (inch) | Width (inch) | Depth (inch) | Mtg Width W1 / W2 | Mtg Depth D1 / D2 | Slot Size S1 / S2 | Approx. Ship Weight (lbs) |
|---|---|---|---|---|---|---|---|---|
| 50 | E050EWB | 4.18 | 3.01 | 3.99 | 2.51 / NA | 2.02 / NA | .20 x .33 / .20 x .33 | 3 |
| 75 | E075EWB | 4.41 | 3.39 | 4.36 | 2.81 / 2.50 | 2.10 / NA | .20 x .50 / .20 x .50 | 4 |
| 100 | E100EWB | 4.41 | 3.39 | 4.61 | 2.81 / 2.50 | 2.37 / NA | .20 x .50 / .20 x .50 | 5 |
| 150 | E150EWB | 5.36 | 4.50 | 4.48 | 3.74 / 3.12 | 2.56 / 2.87 | .20 x .65 / .20 x .33 | 8 |
| 200 | E200EWB | 5.36 | 4.50 | 4.79 | 3.74 / 3.12 | 2.87 / 3.18 | .20 x .65 / .20 x .33 | 10 |
| 250 | E250EWB | 5.36 | 4.50 | 5.21 | 3.74 / 3.12 | 3.29 / 3.61 | .20 x .65 / .20 x .33 | 11 |
| 300 | E300EWB | 5.99 | 5.25 | 5.09 | 4.38 / 3.75 | 3.10 / NA | .31 x .71 / .31 x .71 | 13 |
| 350 | E350EWB | 5.99 | 5.25 | 5.53 | 4.38 / 3.75 | 3.54 / NA | .31 x .71 / .31 x .71 | 15 |
| 500 | E500EWB | 5.99 | 5.25 | 6.31 | 4.38 / 3.75 | 4.33 / NA | .31 x .85 / .31 x .85 | 20 |
| 750 | E750EWB | 7.01 | 6.38 | 6.93 | 5.32 / 4.37 | 4.25 / 5.75 | .31 x .85 / .31 x .85 | 30 |

**Note:** Includes Finger Safe covers.

Visit our website at www.solaheviduty.com or
contact **Technical Services** at **(800) 377-4384** with any questions.

**FIGURE 12–11B** Encapsulated control transformers (continued from Figure 12–11A) *(Courtesy of Sola/Hevi-Duty).*

# 5     Industrial Control Transformers

## SBE Encapsulated Series Selection Tables

**Group 2 –**    220 x 440 Volt Primary, 110 Volt Secondary, 50/60 Hz
                 230 x 460 Volt Primary, 115 Volt Secondary, 50/60 Hz
                 240 x 480 Volt Primary, 120 Volt Secondary, 60 Hz

| VA | Catalog Number | Height (inch) | Width (inch) | Depth (inch) | Mtg Width W1 / W2 | Mtg Depth D1 / D2 | Slot Size S1 / S2 | Approx. Ship Weight (lbs) |
|---|---|---|---|---|---|---|---|---|
| 50 | E050 | 2.72 | 3.01 | 3.99 | 2.51 / NA | 2.02 / NA | .20 x .33 / .20 x .33 | 3 |
| 75 | E075 | 2.96 | 3.39 | 4.36 | 2.81 / 2.50 | 2.10 / NA | .20 x .50 / .20 x .50 | 3 |
| 100 | E100 | 2.96 | 3.39 | 4.61 | 2.81 / 2.50 | 2.37 / NA | .20 x .50 / .20 x .50 | 4 |
| 150 | E150 | 3.89 | 4.50 | 4.48 | 3.74 / 3.12 | 2.56 / 2.87 | .20 x .65 / .20 x .33 | 6 |
| 200 | E200 | 3.89 | 4.50 | 4.79 | 3.74 / 3.12 | 2.87 / 3.18 | .20 x .65 / .20 x .33 | 8 |
| 250 | E250 | 3.89 | 4.50 | 5.21 | 3.74 / 3.12 | 3.29 / 3.61 | .20 x .65 / .20 x .33 | 9 |
| 300 | E300 | 4.53 | 5.25 | 5.09 | 4.38 / 3.75 | 3.10 / NA | .31 x .71 / .31 x .71 | 10 |
| 350 | E350 | 4.53 | 5.25 | 5.53 | 4.38 / 3.75 | 3.54 / NA | .31 x .71 / .31 x .71 | 13 |
| 500 | E500 | 4.53 | 5.25 | 6.31 | 4.38 / 3.75 | 4.33 / NA | .31 x .85 / .31 x .85 | 17 |
| 750 | E750 | 5.56 | 6.38 | 6.93 | 5.32 / 4.37 | 4.25 / 5.75 | .31 x .85 / .31 x .85 | 25 |
| 1000 | E1000 | 5.56 | 6.38 | 7.36 | 5.32 / 4.37 | 4.68 / 6.18 | .31 x .85 / .31 x .85 | 32 |

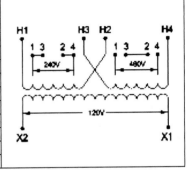

**Note:** *Includes FB2X Secondary fuse holder.*

**Group 2A –**    **Factory installed Primary Fuse Holder Class "CC" and Secondary Fuse Holder (Glass or Ceramic, ¼" x 1¼" fuse type).**

| VA | Catalog Number | Height (inch) | Width (inch) | Depth (inch) | Mtg Width W1 / W2 | Mtg Depth D1 / D2 | Slot Size S1 / S2 | Approx. Ship Weight (lbs) |
|---|---|---|---|---|---|---|---|---|
| 50 | E050WA | 4.18 | 3.01 | 3.99 | 2.51 / NA | 2.02 / NA | .20 x .33 / .20 x .33 | 3 |
| 75 | E075WA | 4.41 | 3.39 | 4.36 | 2.81 / 2.50 | 2.10 / NA | .20 x .50 / .20 x .50 | 4 |
| 100 | E100WA | 4.41 | 3.39 | 4.61 | 2.81 / 2.50 | 2.37 / NA | .20 x .50 / .20 x .50 | 8 |
| 150 | E150WA | 5.36 | 4.50 | 4.48 | 3.74 / 3.12 | 2.56 / 2.87 | .20 x .65 / .20 x .33 | 11 |
| 200 | E200WA | 5.36 | 4.50 | 4.79 | 3.74 / 3.12 | 2.87 / 3.18 | .20 x .65 / .20 x .33 | 10 |
| 250 | E250WA | 5.36 | 4.50 | 5.21 | 3.74 / 3.12 | 3.29 / 3.61 | .20 x .65 / .20 x .33 | 15 |
| 300 | E300WA | 5.99 | 5.25 | 5.09 | 4.38 / 3.75 | 3.10 / NA | .31 x .71 / .31 x .71 | 13 |
| 350 | E350WA | 5.99 | 5.25 | 5.53 | 4.38 / 3.75 | 3.54 / NA | .31 x .71 / .31 x .71 | 15 |
| 500 | E500WA | 5.99 | 5.25 | 6.31 | 4.38 / 3.75 | 4.33 / NA | .31 x .85 / .31 x .85 | 30 |
| 750 | E750WA | 7.01 | 6.38 | 6.93 | 5.32 / 4.37 | 4.25 / 5.75 | .31 x .85 / .31 x .85 | 30 |
| 1000 | E1000WA | 7.01 | 6.38 | 7.36 | 5.32 / 4.37 | 4.68 / 6.18 | .31 x .85 / .31 x .85 | 34 |

**Note:** *Includes Finger Safe covers.*

**Group 2B –**    **Factory installed Primary Fuse Holder Class "CC" and Secondary Fuse Holder (Midget Cartridge, 13/32" x 1½" fuse type).**

| VA | Catalog Number | Height (inch) | Width (inch) | Depth (inch) | Mtg Width W1 / W2 | Mtg Depth D1 / D2 | Slot Size S1 / S2 | Approx. Ship Weight (lbs) |
|---|---|---|---|---|---|---|---|---|
| 50 | E050WB | 4.18 | 3.01 | 3.99 | 2.51 / NA | 2.02 / NA | .20 x .33 / .20 x .33 | 3 |
| 75 | E075WB | 4.41 | 3.39 | 4.36 | 2.81 / 2.50 | 2.10 / NA | .20 x .50 / .20 x .50 | 4 |
| 100 | E100WB | 4.41 | 3.39 | 4.61 | 2.81 / 2.50 | 2.37 / NA | .20 x .50 / .20 x .50 | 8 |
| 150 | E150WB | 5.36 | 4.50 | 4.48 | 3.74 / 3.12 | 2.56 / 2.87 | .20 x .65 / .20 x .33 | 11 |
| 200 | E200WB | 5.36 | 4.50 | 4.79 | 3.74 / 3.12 | 2.87 / 3.18 | .20 x .65 / .20 x .33 | 10 |
| 250 | E250WB | 5.36 | 4.50 | 5.21 | 3.74 / 3.12 | 3.29 / 3.61 | .20 x .65 / .20 x .33 | 15 |
| 300 | E300WB | 5.99 | 5.25 | 5.09 | 4.38 / 3.75 | 3.10 / NA | .31 x .71 / .31 x .71 | 13 |
| 350 | E350WB | 5.99 | 5.25 | 5.53 | 4.38 / 3.75 | 3.54 / NA | .31 x .71 / .31 x .71 | 15 |
| 500 | E500WB | 5.99 | 5.25 | 6.31 | 4.38 / 3.75 | 4.33 / NA | .31 x .85 / .31 x .85 | 30 |
| 750 | E750WB | 7.01 | 6.38 | 6.93 | 5.32 / 4.37 | 4.25 / 5.75 | .31 x .85 / .31 x .85 | 30 |
| 1000 | E1000WB | 7.01 | 6.38 | 7.36 | 5.32 / 4.37 | 4.68 / 6.18 | .31 x .85 / .31 x .85 | 34 |

**Note:** *Includes Finger Safe covers.*

Visit our website at www.solaheviduty.com or
contact **Technical Services** at **(800) 377-4384** with any questions.

**FIGURE 12–11C**   Encapsulated control transformers (continued from Figure 12–11B) *(Courtesy of Sola/Hevi-Duty).*

## Industrial Control Transformers

**5**

### SBE Series Selection Tables - continued

#### Group 3 – 240 x 480 Volt Primary, 24 Volt Secondary, 60 Hz

| VA | Catalog Number | Height (inch) | Width (inch) | Depth (inch) | Mtg Width W1 / W2 | Mtg Depth D1 / D2 | Slot Size | Approx. Ship Weight (lbs) |
|----|----|----|----|----|----|----|----|----|
| 50 | E050D | 2.72 | 3.01 | 3.99 | 2.51 / NA | 2.02 / N/A | .20 x .33 | 3 |
| 100 | E100D | 2.96 | 3.39 | 4.61 | 2.81 / 2.50 | 2.37 / NA | .20 x .50 | 5 |
| 250 | E250D | 3.89 | 4.50 | 5.21 | 3.74 / 3.12 | 3.29 / 3.61 | .20 x .65 | 11 |
| 500 | E500D | 4.53 | 5.25 | 6.31 | 4.38 / 3.75 | 4.33 / NA | .31 x .71 | 20 |

**Note:** *Includes FB2X Secondary fuse holder.*

#### Group 4 – 208/240/277 Volt Primary, 120/24 Volt Secondary, 60 Hz

| VA | Catalog Number | Height (inch) | Width (inch) | Depth (inch) | Mtg Width W1 / W2 | Mtg Depth D1 / D2 | Slot Size | Approx. Ship Weight (lbs) |
|----|----|----|----|----|----|----|----|----|
| 50 | E050JL | 2.72 | 3.01 | 3.99 | 2.51 / NA | 2.02 / N/A | .20 x .33 | 3 |
| 100 | E100JL | 2.96 | 3.39 | 4.61 | 2.81 / 2.50 | 2.37 / NA | .20 x .50 | 5 |
| 250 | E250JL | 3.89 | 4.50 | 5.21 | 3.74 / 3.12 | 3.29 / 3.61 | .20 x .65 | 11 |
| 500 | E500JL | 4.53 | 5.25 | 6.31 | 4.38 / 3.75 | 4.33 / NA | .31 x .71 | 20 |

**Note:** *Will only accept one FB2 secondary fuse holder.*

#### Group 5 – 208/240/480/600 Volt Primary, 120/24 Volt Secondary, 60 Hz
#### 200/230/460/575 Volt Primary, 115/23 Volt Secondary, 60 Hz

| VA | Catalog Number | Height (inch) | Width (inch) | Depth (inch) | Mtg Width W1 / W2 | Mtg Depth D1 / D2 | Slot Size | Approx. Ship Weight (lbs) |
|----|----|----|----|----|----|----|----|----|
| 50 | E050JN | 2.96 | 3.39 | 4.36 | 2.81 / 2.50 | 2.10 / NA | .20 x .50 | 4 |
| 100 | E100JN | 3.89 | 4.50 | 4.48 | 3.74 / 3.12 | 2.56 / 2.87 | .20 x .65 | 8 |
| 150 | E150JN | 3.89 | 4.50 | 5.21 | 3.74 / 3.12 | 3.29 / 3.61 | .20 x .65 | 11 |
| 250 | E250JN | 4.53 | 5.25 | 5.53 | 4.38 / 3.75 | 3.54 / NA | .31 x .71 | 15 |
| 500 | E500JN | 5.56 | 6.38 | 6.93 | 5.32 / 4.37 | 4.25 / 5.75 | .31 x .85 | 30 |

**Note:** *Will only accept one FB2 secondary fuse holder.*

Visit our website at www.solaheviduty.com or
contact **Technical Services** at **(800) 377-4384** with any questions.

**FIGURE 12–11D** Encapsulated control transformers (continued from Figure 12–11C) *(Courtesy of Sola/Hevi-Duty).*

**5** **Industrial Control Transformers**

## SBE -- Copper Wound, Open Style Design

SBE performance in larger VA (1500 - 5000) sizes

The open style SBE Series provides voltage regulation in excess of NEMA recommendations without exceeding 55°C Rise. These higher power capacity transformers are the best choice when 80% or more of the load components are electromagnetic devices.

### Features

- Interleaved copper windings reduce I²R losses and maximize efficiency.
- Ratings 60 Hz unless noted 50/60 Hz.
- Meets or exceeds electrical requirements of NEMA, ANSI, NMTBA and JIC.
- UL listed (File #E77014).
- CSA certified (File #LR-14328-4).
- 55°C rise, 105°C insulation system.
- High quality silicon steel core.

### Related Products

- Linear Power Supplies
- DIN Rail DC Power Supplies
- Constant Voltage Transformers
- Line Reactors

### SBE Design Style

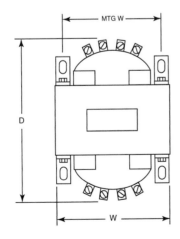

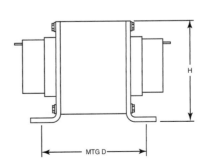

Visit our website at www.solaheviduty.com or
contact **Technical Services** at **(800) 377-4384** with any questions.

**FIGURE 12–12A**  Open style design control transformers *(Courtesy of Sola/Hevi-Duty).*

# Industrial Control Transformers

**5**

## SBE -- Open Style Design Selection Table

**Group 1 –** 240 X 480 Volt Primary, 120 Volt Secondary 60 Hz
230 X 460 Volt Primary, 115 Volt Secondary 50/60 Hz
220 X 440 Volt Primary, 110 Volt Secondary 50/60 Hz

| VA | Catalog Number | Height (inch) | Width (inch) | Depth (inch) | Mtg Width | Mtg Depth | Slot Size | Approx. Ship Weight (lbs) |
|---|---|---|---|---|---|---|---|---|
| 1500 | Y1500 | 6.25 | 6.75 | 8.75 | 5.75 | 6.38 | .44 x .69 | 43 |
| 2000 | Y2000 | 6.25 | 6.75 | 10.00 | 5.75 | 7.75 | .44 x .69 | 55 |
| 3000 | Y3000 | 8.00 | 9.00 | 9.63 | 8.00 | 6.00 | .44 x .69 | 74 |
| 5000 | Y5000 | 8.00 | 9.00 | 12.00 | 8.00 | 8.75 | .44 x .69 | 120 |

**Group 2 –** 240/480/600 Volt Primary, 120/99 Volt Secondary, 60 Hz
230/460/575 Volt Primary, 115/95 Volt Secondary, 60 Hz

| VA | Catalog Number | Height (inch) | Width (inch) | Depth (inch) | Mtg Width | Mtg Depth | Slot Size | Approx. Ship Weight (lbs) |
|---|---|---|---|---|---|---|---|---|
| 1500 | Y1500A | 6.25 | 6.75 | 8.95 | 5.75 | 6.38 | .44 x .69 | 54 |
| 2000 | Y2000A | 8.00 | 10.00 | 9.00 | 8.00 | 5.50 | .44 x .69 | 79 |
| 3000 | Y3000A | 8.00 | 9.00 | 10.00 | 8.00 | 6.38 | .44 x .69 | 115 |
| 5000 | Y5000A | 8.00 | 9.00 | 12.75 | 10.50 | 7.25 | .44 x .69 | 190 |

**Group 3 –** Universal Voltage Primary and Secondary, 50/60 Hz

| VA | Catalog Number | Height (inch) | Width (inch) | Depth (inch) | Mtg Width | Mtg Depth | Slot Size | Approx. Ship Weight (lbs) |
|---|---|---|---|---|---|---|---|---|
| 1000 | Y1000N | 6.25 | 6.75 | 9.13 | 5.75 | 6.75 | .44 x .69 | 43 |
| 1500 | Y1500N | 8.00 | 9.00 | 9.13 | 8.00 | 5.50 | .44 x .69 | 54 |
| 2000 | Y2000N | 8.00 | 9.00 | 10.00 | 8.00 | 6.38 | .44 x .69 | 79 |
| 3000 | Y3000N | 8.00 | 9.00 | 12.00 | 8.00 | 8.75 | .44 x .69 | 115 |
| 5000 | Y5000N | 11.00 | 12.00 | 12.13 | 10.50 | 8.25 | .44 x .69 | 190 |

## Universal Voltage Models – Primary and Secondary Voltage Chart

| Primary Voltage | | | | Secondary Voltage | | |
|---|---|---|---|---|---|---|
| H1-H2 | H1-H3 | H1-H4 | H1-H5 | X1-X2 | X1-X3 | X1-X4 |
| 240 | 416 | 480 | 600 | 99 | 120 | 130 |
| 230 | 400 | 460 | 575 | 95 | 115 | 125 |
| 220 | 380 | 440 | 550 | 91 | 110 | 120 |
| 208 | 346 | 400 | 500 | 85 | 100 | 110 |

*Notes:*

*Weights and dimensions may change and should not be used for construction purposes.*

*Fuse holders are not available for these voltage configurations.*

Visit our website at www.solaheviduty.com or
contact **Technical Services** at **(800) 377-4384** with any questions.

**FIGURE 12–12B** Open style design control transformers (continued from Figure 12–12A) *(Courtesy of Sola/Hevi-Duty).*

# 5  Industrial Control Transformers

## International Series Control Transformers

Electromagnetic control components demand inrush currents up to 10 times the transformers nominal rating without sacrificing secondary voltage stability beyond practical limits. The International series transformers fully comply with IEC and NEMA standards and are available with IEC touchproof covers (IP20).

### Features

- Epoxy encapsulated for cooler operation.
- Interleaved copper windings reduce $I^2R$ losses and maximize efficiency.
- 50/60 Hz
- 55°C Rise, 105°C insulation system for harsh, heavy duty applications.
- Exceeds IEC, NEMA, ANSI, NMTBA, JIC and automotive standards.

### Accessories

- IP20
    - Field installed Primary and Secondary IEC Touch Proof Cover Kit.
- SBEDIN
    - Field installed IEC Fuse Holder Adaptor Kit

### Related Products

- DIN Rail Power Supplies
- 63 Series Power Conditioners
- Surge Suppression Devices provide additional protection and longevity to any electronic equipment.

### International Certifications

| UL | CE |
|---|---|
| E77014 Vol. 1 | IEC 61558-1 61558-2-2 |

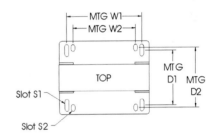

## Design Style

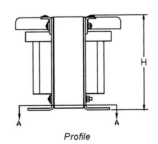

*Mounting Dimensions*

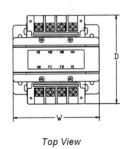

*Profile*                    *Top View*

Visit our website at www.solaheviduty.com or
contact **Technical Services** at **(800) 377-4384** with any questions.

**FIGURE 12–13A**  International standard control transformers *(Courtesy of Sola/Hevi-Duty).*

# Industrial Control Transformers

## Selection Tables: International Series

### Group 1 – 208/240/415 Volt Primary, 120/24 Secondary, 50/60 Hz
### 200/230/400 Volt Primary, 115/23 Secondary, 50/60 Hz

| Continuous VA | Instantaneous VA* | Catalog Number | Height (inch) | Width (inch) | Depth (inch) | Mtg Width W1 / W2 | Mtg Depth D1 / D2 | Slot Size | Approx. Ship Weight (lbs) |
|---|---|---|---|---|---|---|---|---|---|
| 50 | 105 | E050TC | 2.96 | 3.39 | 4.36 | 2.81 / 2.50 | 2.10 / NA | .20 x .50 / .20 x .50 | 4 |
| 100 | 230 | E100TC | 3.89 | 4.50 | 4.48 | 2.56 / 2.87 | 2.87 / 3.18 | .20 x .65 / .20 x .33 | 8 |
| 150 | 420 | E150TC | 3.89 | 4.50 | 5.21 | 3.74 / 3.12 | 3.29 / 3.61 | .20 x .65 / .20 x .33 | 11 |
| 250 | 675 | E250TC | 4.53 | 5.25 | 5.53 | 4.38 / 3.75 | 3.54 / NA | .31 x .71 / .31 x .71 | 15 |
| 500 | 1600 | E500TC | 5.56 | 6.38 | 6.93 | 5.32 / 4.37 | 4.25 / 5.75 | .31 x .85 / .31 x .85 | 30 |

*At 50% PF (Power Factor), 95% Nominal Secondary Voltage.*

### Group 2 – 208/240/415 Volt Primary, 24 Volt Secondary, 50/60 Hz
### 277/480 Volt Primary, 24 Volt Secondary, 60 Hz
### 200/230/400 Volt Primary, 24 Volt Secondary, 50/60 Hz
### 220/380 Volt Primary, 23 Volt Secondary, 50/60 Hz

| Continuous VA | Instantaneous VA* | Catalog Number | Height (inch) | Width (inch) | Depth (inch) | Mtg Width W1 / W2 | Mtg Depth D1 / D2 | Slot Size | Approx. Ship Weight (lbs) |
|---|---|---|---|---|---|---|---|---|---|
| 50 | 105 | E050TE | 2.96 | 3.39 | 4.36 | 2.81 / 2.50 | 2.10 / NA | .20 x .50 / .20 x .50 | 4 |
| 100 | 230 | E100TE | 3.89 | 4.50 | 4.48 | 2.56 / 2.87 | 2.87 / 3.18 | .20 x .65 / .20 x .33 | 8 |
| 150 | 420 | E150TE | 3.89 | 4.50 | 5.21 | 3.74 / 3.12 | 3.29 / 3.61 | .20 x .65 / .20 x .33 | 11 |
| 250 | 675 | E250TE | 4.53 | 5.25 | 5.53 | 4.38 / 3.75 | 3.54 / NA | .31 x .71 / .31 x .71 | 15 |
| 500 | 1600 | E500TE | 5.56 | 6.38 | 6.93 | 5.32 / 4.37 | 4.25 / 5.75 | .31 x .85 / .31 x .85 | 30 |

*At 50% PF (Power Factor), 95% Nominal Secondary Voltage.*

**Notes:**

*Weights and dimensions may change and should not be used for construction purposes.*

*Fuse holders are not available for these voltage configurations.*

Visit our website at www.solaheviduty.com or
contact **Technical Services** at **(800) 377-4384** with any questions.

**FIGURE 12–13B** International standard control transformers (continued from Figure 12–13A) *(Courtesy of Sola/Hevi-Duty).*

## Industrial Control Transformers

### Selection Tables: International Series

**Group 3 –** 208/240/415/480/600* Volt Primary, 120 Volt Secondary, 50/60 Hz
200/230/400/460/575* Volt Primary, 115 Volt Secondary, 50/60 Hz
220/277*/380 Volt Primary, 110 Volt Secondary, 50/60 Hz

| Continuous VA | Instantaneous VA** | Catalog Number | Height (inch) | Width (inch) | Depth (inch) | Mtg Width W1 / W2 | Mtg Depth D1 / D2 | Slot Size | Approx. Ship Weight (lbs) |
|---|---|---|---|---|---|---|---|---|---|
| 50 | 93 | E050TF | 2.96 | 3.39 | 4.36 | 2.81 / 2.50 | 2.10 / NA | .20 x .50 / .20 x .50 | 4 |
| 100 | 205 | E100TF | 3.89 | 4.50 | 4.48 | 3.74 / 3.12 | 2.56 / 2.87 | .20 x .65 / .20 x .33 | 8 |
| 150 | 390 | E150TF | 3.89 | 4.50 | 5.21 | 3.74 / 3.12 | 3.29 / 3.61 | .20 x .65 / .20 x .33 | 11 |
| 250 | 630 | E250TF | 4.53 | 5.25 | 5.53 | 4.38 / 3.75 | 3.54 / NA | .31 x .71 / .31 x .71 | 15 |
| 500 | 1200 | E500TF | 5.56 | 6.38 | 6.93 | 5.32 / 4.37 | 4.25 / 5.75 | .31 x .85 / .31 x .85 | 30 |
| 750 | 2290 | E750TF | 5.56 | 6.38 | 7.36 | 5.32 / 4.37 | 4.68 / 6.18 | .31 x .85 / .31 x .85 | 34 |

*\* 60 Hz Only*
*\*\* At 50% PF (Power Factor), 95% Nominal Secondary Voltage.*

**Group 4 –** 240/415/480 Volt Primary, 120/240 Volt Secondary, 50/60 Hz
230/400/460 Volt Primary, 115/230 Volt Secondary, 50/60 Hz
220/380/440 Volt Primary, 110/220 Volt Secondary, 50/60 Hz

| Continuous VA | Instantaneous VA* | Catalog Number | Height (inch) | Width (inch) | Depth (inch) | Mtg Width W1 / W2 | Mtg Depth D1 / D2 | Slot Size | Approx. Ship Weight (lbs) |
|---|---|---|---|---|---|---|---|---|---|
| 50 | 110 | E050TH | 2.96 | 3.39 | 4.36 | 2.81 / 2.50 | 2.10 / NA | .20 x .50 / .20 x .50 | 4 |
| 100 | 235 | E100TH | 3.89 | 4.50 | 4.48 | 3.74 / 3.12 | 2.56 / 2.87 | .20 x .65 / .20 x .33 | 8 |
| 150 | 470 | E150TH | 3.89 | 4.50 | 5.21 | 3.74 / 3.12 | 3.29 / 3.61 | .20 x .65 / .20 x .33 | 11 |
| 250 | 730 | E250TH | 4.53 | 5.25 | 5.53 | 4.38 / 3.75 | 3.54 / NA | .31 x .71 / .31 x .71 | 15 |
| 500 | 1670 | E500TH | 5.56 | 6.38 | 6.93 | 5.32 / 4.37 | 4.25 / 5.75 | .31 x .85 / .31 x .85 | 30 |
| 750 | 2250 | E750TH | 5.56 | 6.38 | 7.36 | 5.32 / 4.37 | 4.68 / 6.18 | .31 x .85 / .31 x .85 | 34 |

*\* At 50% PF (Power Factor), 95% Nominal Secondary Voltage.*

**Notes:**

*Weights and dimensions may change and should not be used for construction purposes.*

*Fuse holders are not available for these voltage configurations.*

### International Series - Fuse Recommendations

| VA | Maximum Current Rating of Fuse | | |
|---|---|---|---|
| | 24 VAC | 115 VAC | 230 VAC |
| 50 | 2 | 0.5 | 0.25 |
| 100 | 4 | 1 | 0.5 |
| 150 | 6 | 1.6 | 0.8 |
| 250 | 10 | 2 | 1 |
| 500 | 20 | 4 | 2 |
| 750 | -- | 6 | 4 |

**Primary Fusing:** *Consult local Electrical Code*
**Secondary Fusing:** *per IEC EN61588-2-2*

Visit our website at www.solaheviduty.com or
contact **Technical Services** at **(800) 377-4384** with any questions.

**FIGURE 12–13C** International standard control transformers (continued from Figure 12–13B) *(Courtesy of Sola/Hevi-Duty).*

## Industrial Control Transformers

5

### HSZ Series Industrial Control Transformers

The HSZ series of industrial control transformers are designed for applications requiring special mounting and are available in ratings from 1 through 10 KVA.

### Features

- Two case styles available, NEMA-3R and NEMA-12.
- NEMA 4 and 4X Enclosures are also available.
- UL Class 180°C insulation system, 80°C temperature rise under full load.
- Meets or exceeds NEMA regulation standards.
- Copper magnet wire windings.
- Encapsulated.

### HSZ Design Style

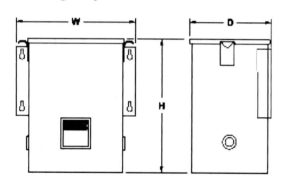

### Related Products

- Linear Power Supplies
- DIN Rail DC Power Supplies
- Constant Voltage Transformers
- Line Reactors

### HSZ Series Selection Tables

**Group 1 – 240/480, 230/460, 220/440 Volt Primary, 120/115/110 Volt Secondary, 50/60 Hz.**

| KVA | Catalog Number NEMA-3R | Catalog Number NEMA-12 | Height (inch) | Width (inch) | Depth (inch) | Approx. Ship Weight (lbs) |
|---|---|---|---|---|---|---|
| 1 | HZ1000 | HZ121000 | 12 | 10 | 7 | 43 |
| 1.5 | HZ1500 | HZ121500 | 12 | 10 | 7 | 55 |
| 2 | HZ2000 | HZ122000 | 12 | 10 | 7 | 68 |
| 3 | HZ3000 | HZ123000 | 17 | 14 | 9 | 108 |
| 5 | HZ5000 | HZ125000 | 17 | 14 | 9 | 138 |
| 7.5 | HZ7500 | HZ127500 | 17 | 14 | 9 | 173 |
| 10 | HZ10000 | HZ1210000 | 17 | 17 | 12 | 210 |

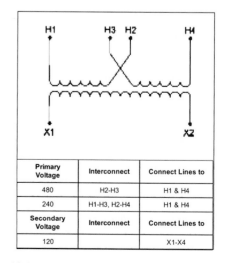

| Primary Voltage | Interconnect | Connect Lines to |
|---|---|---|
| 480 | H2-H3 | H1 & H4 |
| 240 | H1-H3, H2-H4 | H1 & H4 |
| Secondary Voltage | Interconnect | Connect Lines to |
| 120 | | X1-X4 |

Visit our website at www.solaheviduty.com or
contact **Technical Services** at **(800) 377-4384** with any questions.

**FIGURE 12–14A** Enclosed control transformers *(Courtesy of Sola/Hevi-Duty).*

# 5 Industrial Control Transformers

## HSZ Series Selection Tables and Electrical Connections

### Group 2 –  230/460/575 Volt Primary, 115/95 Volt Secondary, 50/60 Hz.

| KVA | Catalog Number (NEMA-3R) | Catalog Number (NEMA-12) | Height (inch) | Width (inch) | Depth (inch) | Approx. Ship Weight (lbs) |
|---|---|---|---|---|---|---|
| 1 | HZ1000R | HZ121000R | 12 | 10 | 7 | 43 |
| 1.5 | HZ1500R | HZ121500R | 12 | 10 | 7 | 55 |
| 2 | HZ2000R | HZ122000R | 12 | 10 | 7 | 68 |
| 3 | HZ3000R | HZ123000R | 17 | 14 | 9 | 108 |
| 5 | HZ5000R | HZ125000R | 17 | 14 | 9 | 138 |
| 7.5 | HZ7500R | HZ127500R | 17 | 14 | 9 | 173 |
| 10 | HZ10000R | HZ1210000R | 17 | 17 | 12 | 210 |

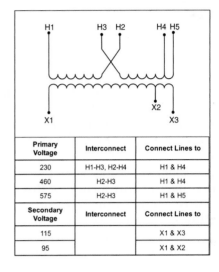

| Primary Voltage | Interconnect | Connect Lines to |
|---|---|---|
| 230 | H1-H3, H2-H4 | H1 & H4 |
| 460 | H2-H3 | H1 & H4 |
| 575 | H2-H3 | H1 & H5 |
| Secondary Voltage | Interconnect | Connect Lines to |
| 115 | | X1 & X3 |
| 95 | | X1 & X2 |

### Group 3 –  480 Volt Primary, 240 Volt Secondary, 50/60 Hz.

| KVA | Catalog Number (NEMA-3R) | Catalog Number (NEMA-12) | Height (inch) | Width (inch) | Depth (inch) | Approx. Ship Weight (lbs) |
|---|---|---|---|---|---|---|
| 1 | HZ1000R | HZ121000R | 12 | 10 | 7 | 43 |
| 1.5 | HZ1500R | HZ121500R | 12 | 10 | 7 | 55 |
| 2 | HZ2000R | HZ122000R | 12 | 10 | 7 | 68 |
| 3 | HZ3000R | HZ123000R | 17 | 14 | 9 | 108 |
| 5 | HZ5000R | HZ125000R | 17 | 14 | 9 | 138 |
| 7.5 | HZ7500R | HZ127500R | 17 | 14 | 9 | 173 |
| 10 | HZ10000R | HZ1210000R | 17 | 17 | 12 | 210 |

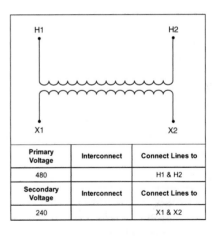

| Primary Voltage | Interconnect | Connect Lines to |
|---|---|---|
| 480 | | H1 & H2 |
| Secondary Voltage | Interconnect | Connect Lines to |
| 240 | | X1 & X2 |

**FIGURE 12–14B**  Enclosed control transformers (continued from Figure 12–14A) *(Courtesy of Sola/Hevi-Duty).*

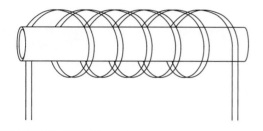

**FIGURE 12-15** Transformer wound on a nonconductive, nonmagnetic hollow tube (air core transformer).

From this equation, it can be seen that the primary voltage, primary turns, area of the core, and type of material affect the flux density. The equation is based on the standard iron/steel core material in general use.

The design engineer works from specifications that list the voltage and frequency for the design. The material is selected from a manufacturer's list of $B_m$ ratings. The area ($A$) and the number of turns ($N_p$) must be determined. If the area is made large and the number of turns small, the voltage regulation will be good, but the physical size of the transformer will be rather large for a given kilovolt-ampere (kVA) rating. If the area is made small, a large number of turns will be required. This arrangement provides a smaller transformer, but the voltage regulation is poor. No specific equation can be provided for calculating the best value for $A$ or $N_p$. Some standard curves have been developed that provide reasonably good reference data. Figure 12–16 shows curves for 25 hertz and 60 hertz at various kilovolt-ampere ratings. Using these curves and Equation 12.4, the design engineer can calculate the area of the core.

## Example 3

An engineer is requested to design a shell-type transformer, rated at 10 kVA, 440 volts to 110 volts, at 60 hertz. The core material has a flux density of 60,000 lines of force per square inch (9300 lines per square centimeter). (a) What will be the number of turns of wire required for the primary and secondary coils? (b) What cross-sectional area is required for the core?

a. For a transformer of this rating, Figure 12–16 recommends 3.6 V per turn. Therefore,

$$N_p = \frac{440}{3.6}$$

$$N_p = 122 \text{ turns on the primary}$$

$$\frac{N_p}{N_s} = \frac{E_p}{E_s}$$

$$\frac{122}{N_s} = \frac{440}{110}$$

$$N_s = \frac{13,420}{440}$$

$$N_s = 30.5 \text{ turns on the secondary}$$

b. $A = \dfrac{10^8 E_p}{4.44 f N_p B_m}$

$$A = \frac{10^8 \times 440}{4.44 \times 60 \times 122 \times 6 \times 10^4}$$

$$A = \frac{440}{19.5}$$

$$A = 22.56 \text{ in.}^2 \text{ cross-sectional area of center leg}$$

The center leg requires 22.56 square inches of sheet steel. The remainder of the magnetic circuit conducts only one half of the flux and, therefore, requires only one half of the cross-sectional area, or 11.28 square inches. These calculations do not allow for an $IR$ or $IX_L$ drop in the windings. Therefore, enough additional turns of wire are required on the secondary to compensate for these voltage drops.

## Noise Level

Because of their construction and their principle of operation, all transformers produce an audible "hum." The hum is a result of the laminations and windings vibrating as the magnetic field alternates. The amount of sound emitted depends upon the design and method of constructing the transformer. Wound core-type transformers are probably the least noisy. The noise level must be considered when recommending transformers for specific installations. The noise level should always be lower than the ambient noise level for the area in which the transformer is to be installed.

The hum of a transformer will be amplified if it is not mounted properly. The sound may also be transmitted through the conduit system. One method of mounting large transformers uses flexible mounts that are designed to absorb vibrations and are arranged to avoid metal-to-metal contact. To reduce the sound being transmitted through the wiring system, it is wise to enter the transformer through a flexible conduit or cable.

## Polarity

To ensure the correct connections of transformer leads, a standard marking system has been developed. The leads from the high-voltage winding are marked $H_1$, $H_2$, etc. The lead marked $H_1$ must always be located on the left when viewing the transformer from the side through which the low-voltage leads are brought out. The low-voltage leads are marked $X_1$, $X_2$, and so on.

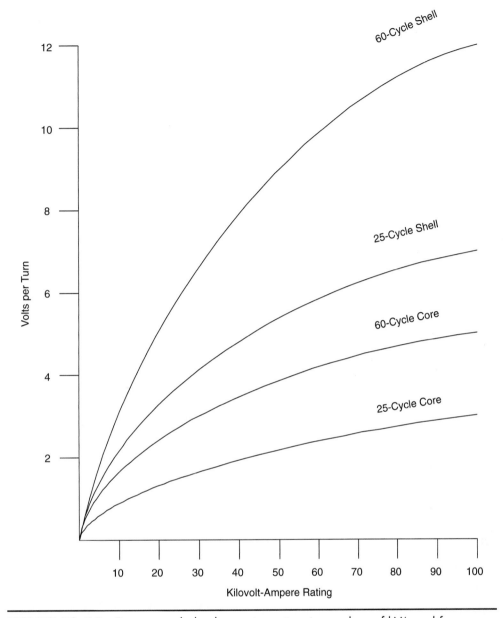

**FIGURE 12–16** Recommended volts per turn at various values of kVA and frequency.

A standard test procedure has been developed to ensure correct labeling. This procedure is as follows:

1. Connect one end of the high-voltage winding to one end of the low-voltage winding (Figure 12–17).
2. Connect a voltmeter between the two open ends.
3. Apply a voltage no greater than the rated voltage of the winding to either the high- or low-voltage winding.

Figure 12–18 illustrates this method.

⚠ **CAUTION** Notice that the high-voltage winding is rated at 2300 volts and the low-voltage winding is rated at 575 volts. Because these voltages are not always available and may present hazardous working conditions, lower voltages are used to perform the test.

It can be determined by basic mathematics that this transformer has a ratio of 4:1. Therefore, if 240 volts are impressed across the high-voltage winding, 60 volts will be induced into the low-voltage winding. The voltmeter selected for the test

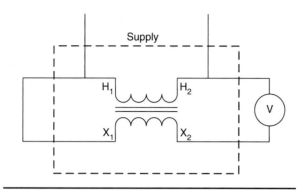

**FIGURE 12–17**  Connections for polarity test.

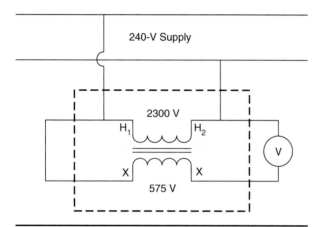

**FIGURE 12–18**  Using the voltmeter method to determine the polarity of a transformer.

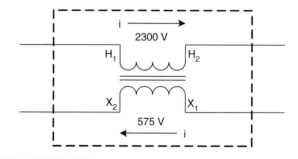

**FIGURE 12–19**  Distribution transformer arranged for additive polarity.

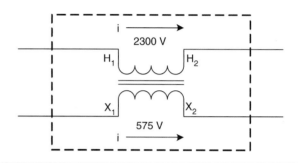

**FIGURE 12–20**  Distribution transformer arranged for subtractive polarity.

must be able to measure the sum of the two voltages (240 volts + 60 volts = 300 volts).

If the voltmeter indicates the sum of the two voltages, the transformer is additive and should be labeled as shown in Figure 12–19 ($H_1$ on the left and $X_1$ on the right). If the voltmeter indicates the difference between the two voltages (240 volts − 60 volts = 180 volts), the transformer is subtractive. Therefore, the leads should be marked as shown in Figure 12–20 ($H_1$ on the left and $X_1$ on the left).

The polarities of transformers should be such that at the instant when the current is entering $H_1$, it is leaving $X_1$ (see Figures 12–19 and 12–20). The polarity of the transformer can be changed by interchanging the low-voltage leads.

Some specific standards for transformer polarity are as follows:

1. Single-phase transformers with power ratings of up to 200 kilovolt-amperes and voltage ratings of not over 9000 volts shall be additive.

2. Single-phase transformers with power ratings of up to 200 kilovolt-amperes and voltage ratings greater than 9000 volts shall be subtractive.

3. Single-phase transformers with power ratings of over 200 kilovolt-amperes, regardless of the voltage rating, shall be subtractive.

If the transformer shown in Figure 12–19 is rated at 400 kilovolt-amperes, the leads should be arranged for subtractive polarity. This can be accomplished by interchanging $X_1$ and $X_2$.

Even though the leads are identified by the manufacturer, it is wise to check them in the field to eliminate any possibility of error.

## Methods of Cooling

Heat is generated within transformers as a result of (1) the current flowing in the windings, (2) eddy currents, and (3) hysteresis. The transformer must dissipate this heat as rapidly as it is generated or the transformer will overheat, causing the insulation to break down.

Transformers rated at 5 kilovolt-amperes or less are generally air cooled. The heat produced is dissipated into the surrounding air through natural radiation. In the installation of transformers, it is very important to consider the ambient temperature of the installation area.

Small- and medium-sized power and distribution transformers are cooled by being housed in tanks filled with oil or with a synthetic,

**FIGURE 12–21A** Oil-cooled distribution transformer (pole mounted).

**FIGURE 12–21B** Oil-cooled distribution transformer with tubes added to increase the ability to radiate heat.

**FIGURE 12–21C** Radiators may be added to achieve a further increase in the ability to radiate heat.

nonflammable insulating liquid. This type of coolant serves two purposes. First, it carries the heat from the windings to the surface of the tank, where it is dissipated into the air. Second, it serves as an insulation between the windings.

A moderately sized oil-cooled transformer is shown in Figure 12–21A. Figure 12–21B shows a larger transformer, which requires the use of tubes to increase the radiating surface. Some transformers require external radiators as shown in Figure 12–21C. With very large-capacity transformers (20,000 kilovolt-amperes and higher), the oil cannot carry away the heat fast enough. To overcome this problem, a coil of copper tubing is installed near the top of the tank, where the oil is the hottest. Water circulating through the coil absorbs the heat from the oil and dissipates it into the air outside of the tank.

Transformers are sometimes cooled with air forced up from the bottom. The air circulates around the core and windings and exhausts the excess heat out through the top. These types of transformers are frequently used when cost and space are major factors. They are generally lighter than water-cooled transformers, take up less space, and cost less. Their main disadvantage is the necessity for clean, dry air. Using forced air increases a dry transformer's capacity by about 33% over the self-cooled (natural circulation) type. A *forced-air-cooled transformer* may sometimes be called an *air-blast transformer*.

## SPECIAL TRANSFORMERS

Many installations require special types of transformers. For example, transformers installed a long distance from the supply may require a method to adjust the secondary voltage to compensate

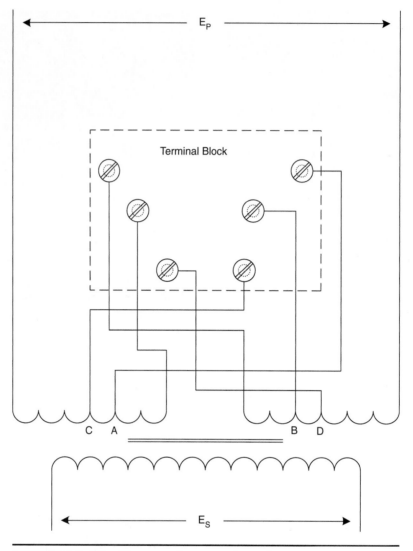

**FIGURE 12–22** Diagram of a tapped transformer.

for varying line drops as the load increases or decreases. Transformers installed where occasional heavy loads are used may require infrequent voltage adjustments. Some installations may require a constant current; others may require a constant voltage. Special apparatuses may require only a slightly different voltage from the supply voltage. Instruments, relays, lighting fixtures, and electronic equipment require various types of transformers.

## Tapped Transformers

A *tapped transformer* is used for installations where heavy loads are used infrequently or where it is apparent that future demand may require additional loading. For such installations, tap changes are required infrequently and are made by hand. Figure 12–22 shows a schematic diagram of a tapped transformer. To change the taps, the transformer

must be disconnected from the line, the change made, and then the transformer placed back in service. A switch installed in the primary lines permits this operation to be performed with ease and safety.

The taps from the secondary are connected to studs. On oil-filled transformers, these studs are inside the tank. To make the change, it is necessary to remove the cover and make the changes while the studs are immersed in oil. In addition to the inconvenience, certain hazards exist. Incorrect connections may result, the oil is exposed to contamination, and the worker may drop parts into the tank. Despite such disadvantages, however, this method is sometimes used because of the lower costs of initial installation.

A safer and more practical method for changing the connections is to use a tap-changing switch. With this method, the taps may be connected to either the high-voltage or low-voltage winding. It is not necessary to open the tank to operate the switch.

The transformer, however, must be disconnected from the supply before any changes are made.

The switch is mounted on a terminal block, usually located above the transformer core. It is operated from outside the tank by a shaft, which protrudes through the cover. A knob and an indicating plate are arranged to identify the positions in order to obtain the correct voltage.

A diagram showing the connections to the switch contacts is shown in Figure 12–23. When the movable contact is in the position shown, all the primary winding is connected into the circuit. In this position, the transformer is delivering the lowest voltage. When the contactor is moved to position 2, section 5 is removed from the circuit, causing the secondary voltage to increase. Further movement of the contactor to positions 3, 4, and 5 eliminates more of the primary turns, thereby increasing the secondary voltage. This increase can be verified by using the turns ratio equation ($E_p/E_s = N_p/N_s$).

Many installations require that adjustments in voltage and, therefore, tap changing, take place while the transformer is supplying a load. For this operation, the transformer usually has at least two high-voltage windings plus a reactor winding. The reactor shunts the switch during the tap-changing process. There is no open position on the switch.

This means that at least one set of contacts is made at all times, eliminating accidental opening of the high-voltage winding. Various combinations of switching arrangements increase or decrease the voltage across the load winding.

## Autotransformers

The **autotransformer** differs from the standard transformer in that it has only one winding. Figures 12–24A and 12–24B show how the windings are connected. One portion of the winding serves for both the primary and the secondary. If the transformer is being used to lower the voltage, the turns between $H_1$ and $H_2$ constitute the primary winding and those between $L_1$ and $L_2$ constitute the secondary winding. The ratio of the voltage, as in a two-winding transformer, is equal to the ratio of the primary turns to the secondary turns.

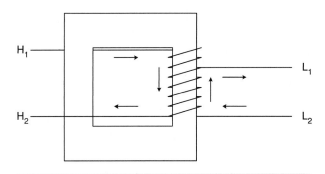

**FIGURE 12–24A**  Autotransformer.

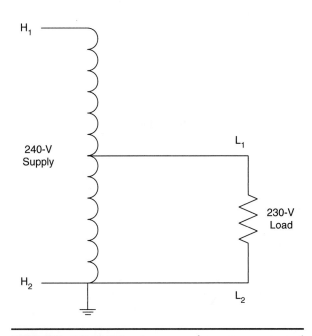

**FIGURE 12–24B**  Autotransformer connected to a load.

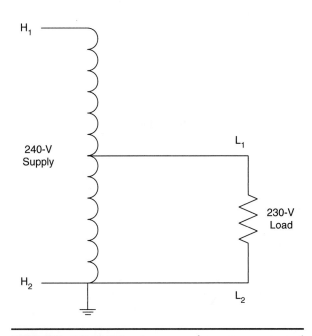

**FIGURE 12–23**  Transformer connections to a tap-changing switch.

If the exciting current is neglected, the primary and secondary ampere-turns in an autotransformer are equal. Therefore, the turns ratio equation ($E_p/E_s = N_p/N_s = I_s/I_p$) also applies to autotransformers.

At any instant, the currents in the primary and secondary of a transformer flow in opposite directions. In an autotransformer, the current in the portion of the winding that is common to both the primary and secondary is equal to the difference between the source current and the load current.

Autotransformers require less copper; thus, they are less expensive to manufacture than conventional transformers. They operate at higher efficiency and are smaller in physical size than a two-winding transformer of the same rating. In addition, most autotransformers have lower noise levels for the same load and frequency than do two-winding transformers.

One disadvantage of autotransformers is that they can present hazards if they are used to make large voltage changes. For example, if an autotransformer is used to lower the source voltage from 480 volts to 120 volts for lighting and small appliance loads, the difference between the two voltages is 480 volts − 120 volts = 360 volts. Figure 12–25A shows a diagram of this connection. If a break should occur in the winding that is common to both the primary and secondary, approximately 480 volts will appear across the load. Under

ground-fault conditions the potential to ground will be 480 volts, not 120 volts (Figure 12–25B).

Autotransformers are used to reduce the voltage to AC motors during the starting period and to increase the voltage in special situations. They are also used to compensate for voltage drops in transmission lines and to make minor changes in voltage to meet certain load requirements (Figure 12–25C).

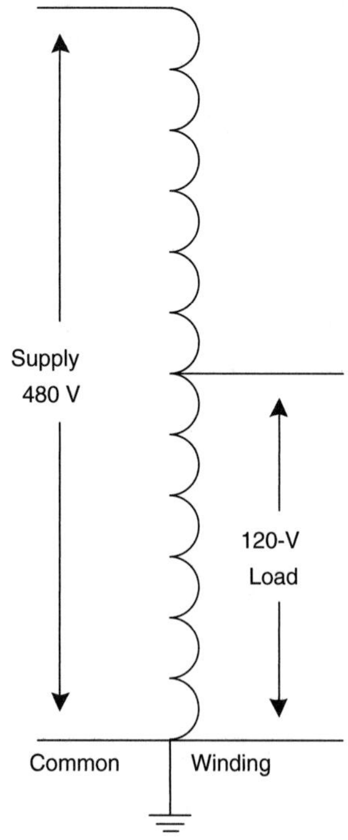

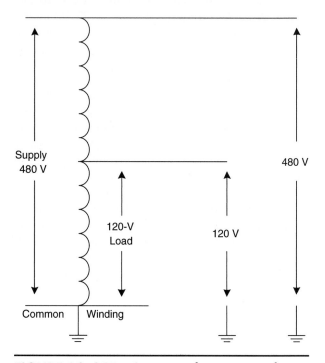

**FIGURE 12–25B**  Diagram of an autotransformer illustrating possible voltages to ground.

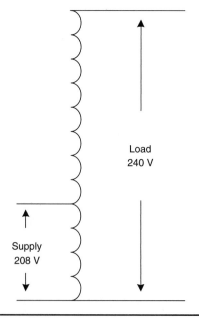

**FIGURE 12–25C**  Autotransformer used to increase the voltage from 208 V to 240 V for special equipment.

**FIGURE 12–25A**  Autotransformer used as a step-down transformer.

## Instrument Transformers

Certain measuring instruments and relays for AC systems should not be connected directly to high-voltage circuits. The *instrument transformer* is used to satisfy this condition. This type of transformer isolates the instrument or relay from the high voltage. It also makes it possible to standardize most electrical instruments for operation at lower values of current and voltage, such as 120 volts and 5 amperes.

*Voltage transformers*, also called *potential transformers*, are similar to the standard two-winding transformer. They are built in small sizes, usually not larger than 500 volt-amperes. They have a high-voltage winding, which is connected directly across the supply, and a low-voltage winding (generally wound to supply 120 volts), which is connected to the instruments. *Note:* The low-voltage winding should always be grounded for safety purposes and to eliminate static from the instruments.

Potential transformers, because of their use, must have very accurate voltage ratios. For most instrument transformers, the percentage of error is 0.5% or less. Because of the turns ratio on most potential transformers, a greater percentage of error could produce a major error in the voltage measurement or cause relays to malfunction. For example, if a transformer is used to reduce the voltage from 2400 volts to 120 volts, the ratio is 20:1. A voltmeter connected across the 120-volt winding should indicate 120 volts. This value, when multiplied by 20, is equal to 2400 volts. If there is a –4% error in the voltage ratio, the voltmeter will indicate 115.2 volts, and 20 = 115.2 volts = 2304 volts, resulting in an error of 96 volts.

To measure large direct currents, ammeters with separate shunts are used. Most of the current flows through the shunt; only a small percentage of the current flows through the meter. The use of shunts for AC is generally unsatisfactory because of the inductive effect. A *current transformer* is a much better device for this application.

The primary winding of a current transformer consists of one or more turns of heavy wire wound on an iron/steel core and connected in series with the supply line. Sometimes a cable or bus bar, passing through the center of the core, serves as the primary winding. Figure 12–26A shows a current transformer; Figure 12–26B illustrates the connections.

The secondary winding has many turns of fine wire. The current coils of instruments and/or relays are connected to this winding and act as the load on the transformer. Measuring instruments are usually designed for full-scale deflection at 5 amperes. Relay coils are generally rated at this value. Thus, the secondary winding of the current transformer is generally rated at 5 amperes.

**FIGURE 12–26A**   Current transformer.

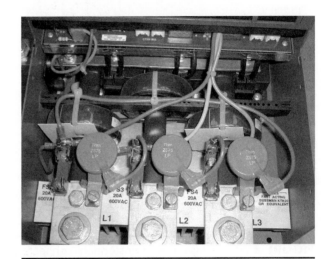

**FIGURE 12–26B**   Current transformers connected to monitor conductor current.

If the transformer has a current ratio of 20:1, for every 20 amperes flowing in the primary, 1 ampere flows in the secondary. This current ratio remains nearly constant as long as the load being served does not cause the transformer to exceed its volt-ampere (VA) rating. If the rating is exceeded, the core will become saturated and the secondary current will not increase in the same proportion as the primary current.

Current transformers differ from the transformers previously described in that the primary current depends upon the load being served and not upon the secondary load. If the secondary is opened while the primary is carrying a heavy load, the demagnetizing effect of the secondary current will no longer exist and the flux in the core will increase. This increase in flux induces a high voltage into the secondary winding, which may damage the insulation or cause a severe shock to anyone coming in contact with it.

> **⚠ CAUTION** The secondary circuit of a current transformer should never be opened while the primary is energized.

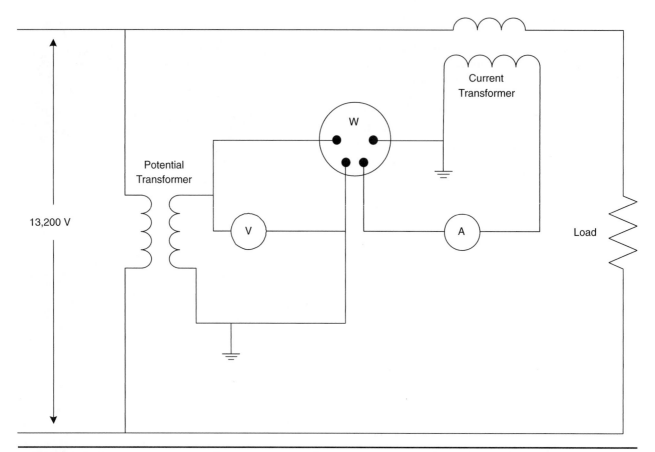

**FIGURE 12–27**  Instrument transformers connected to a 13,200-V supply, reducing the voltage and current for metering purposes.

Most current transformers have impedance that is high enough to limit the current to a safe value even when the secondary is short-circuited. A terminal block near the meters is arranged so that the secondary windings of the current transformers can be short-circuited in order to remove the meters without disconnecting the supply.

Figure 12–27 shows a diagram depicting the connections for a voltmeter, ammeter, and wattmeter to a 13,200-volt line. For permanent installations, the instruments are calibrated for direct reading.

In order to avoid confusion regarding transformer polarity, manufacturers have adopted a standard that requires that all instrument transformers have subtractive polarity.

## Buck-Boost Transformers

A *buck-boost transformer* is a two-winding transformer that is generally connected to operate as an autotransformer. Figure 12–28 shows a schematic diagram of one of the most common connections used.

The buck-boost transformer is used to increase or decrease the supply voltage by a small amount. For example, it may be used to increase the voltage

from 208 volts to 240 volts. Other common voltage changes are from 440 volts to 480 volts and vice versa and from 240 volts to 277 volts.

Because small voltage changes are involved, the buck-boost transformer is smaller in physical size and rating than the standard two-winding transformer for the same load. The buck-boost transformer is more efficient and produces less noise, and its purchase price is lower.

## Control Transformer

A control transformer is a two-winding transformer that provides a lower voltage supply for control circuits. It is frequently used in industrial establishments to provide reduced voltage for the control circuits of motor starters and controllers.

Common voltages supplied to industrial establishments are 208, 240, 277, 480, and 575 volts. The more common industrial controllers are manufactured with control circuits of 120 volts or less. These controls are less expensive to manufacture than those with higher voltage ratings.

The control transformer may have a primary winding with taps for various voltages (Figure 12–29),

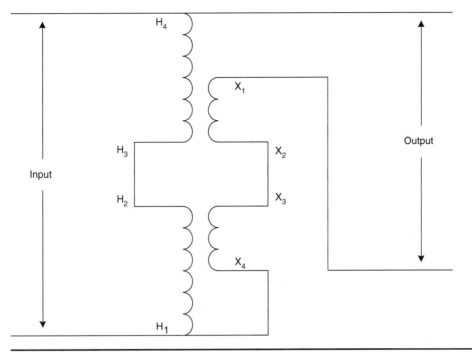

**FIGURE 12–28**   Connections for a buck-boost transformer.

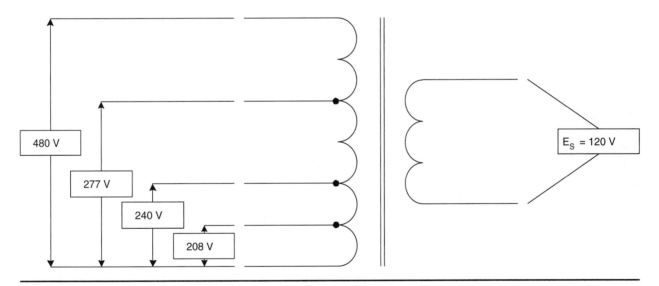

**FIGURE 12–29**   Control transformer with primary taps for connections to various voltages.

or the windings may be arranged for only one or two voltages. Figure 12–30 illustrates a transformer with two coils on the primary.

## TRANSFORMER CONNECTIONS

Transformers, like other electrical devices, may be connected into series, parallel, two-phase, or three-phase arrangements. When they are grouped together in any of these arrangements, the group is called a *transformer bank*.

In order to group transformers, it is necessary to comply with the following requirements:

1. Their voltage ratings must be equal.

2. Their impedance ratios (percent impedance) must be equal.

3. Their polarities must be determined and connections made accordingly.

4. Transformers are seldom connected in series. When they are, however, their current ratings must be large enough to carry the maximum current of the load. For the most efficient operation, their current ratings must be equal.

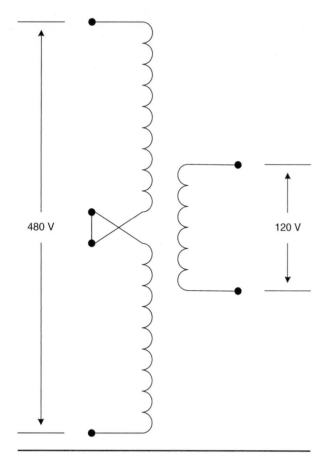

**FIGURE 12–30** Control transformer with two 240-V coils connected in series across 480 V.

## Series Connections

The main purpose of connecting transformers in series is to obtain higher voltage ratings. For example, if the supply voltage is 480 volts and the load requires two values of voltage (120 and 240), a transformer with two low-voltage windings is required (Figure 12–31). It is possible, however, to obtain the same result with two transformers connected in series. To perform this operation, the following procedure is used:

1. Check the voltage ratings of both transformers. They must be equal.

2. Check the percentage impedances of both transformers. They must be equal.

3. Check the current ratings of both transformers. They must be high enough to carry the maximum load.

4. Determine the polarity of each transformer.

5. Connect the high-voltage windings as shown in Figure 12–32A.

6. Connect the high-voltage windings to the supply voltage (Figure 12–32B).

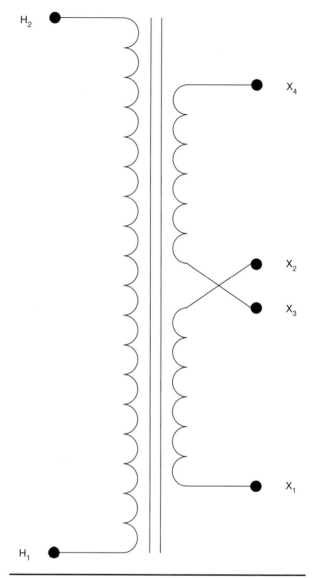

**FIGURE 12–31** A transformer with two low-voltage windings.

7. Measure the voltage across the secondary of each transformer. The voltage should be the same for each transformer.

8. Connect the secondaries as shown in Figure 12–32C. Be sure the polarities are correct.

9. Measure the voltage across the secondary of each transformer and across the two transformers (Figure 12–32D).

## Parallel Connections

Two transformers of equal ratings connected in parallel carry twice the kilovolt-ampere rating of either one. In other words, the kilovolt-ampere rating of transformers in parallel is equal to the sum of the individual ratings.

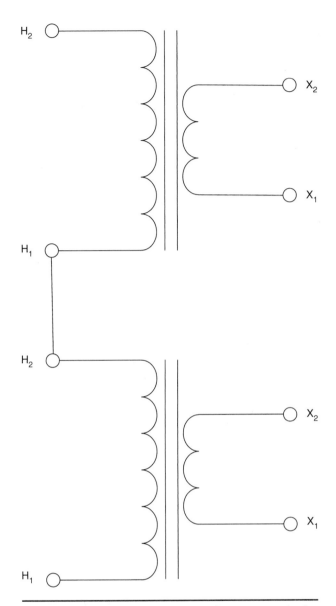

**FIGURE 12–32A** Two transformers with the high-voltage windings connected in series.

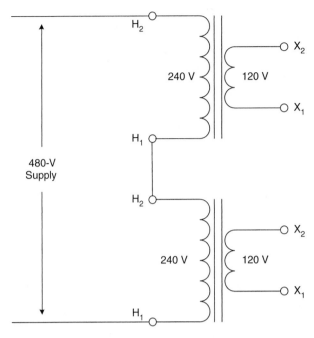

**FIGURE 12–32B** Two transformers with the high-voltage windings connected in series across a 480-volt supply.

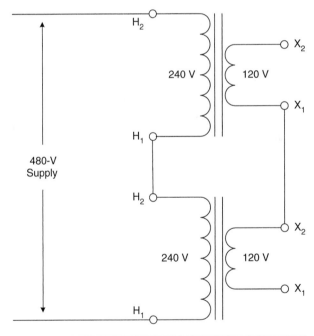

**FIGURE 12–32C** Two transformers with series connections on both the primary and secondary windings.

When an additional load is being installed in a plant, it is sometimes more practical to add another transformer rather than to change the existing one. In order to divide the load according to the rating of the transformers and to avoid circulating currents, it is necessary to meet the requirements previously stated for grouping transformers.

Before the final connections are made, a polarity test should be performed. This can be accomplished without disconnecting the original transformer. The second transformer is connected as shown in Figure 12–33A. Note that the secondary leads of the second transformer are not marked, but the voltmeter indicates 480 volts. The sum of the two secondary voltages is 480 volts. Therefore, the secondaries are connected in series rather than in parallel.

From the results of this test, it can be assumed that transformer A is subtractive and transformer B is additive. The secondary connections are changed as shown in Figure 12–33B. The voltmeter now should indicate zero volt. Remove the voltmeter and connect the transformers as shown in Figure 12–33C.

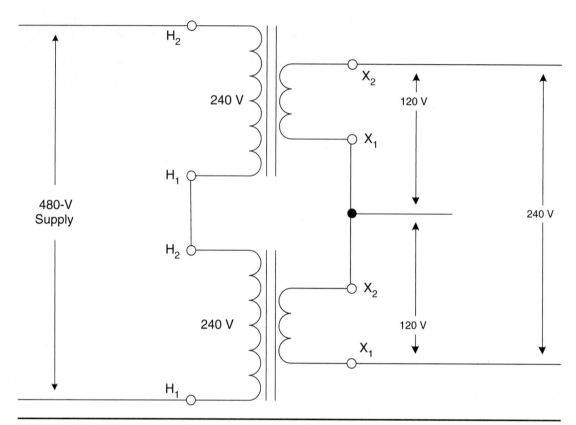

**FIGURE 12–32D**  Two transformers connected to form a three-wire system.

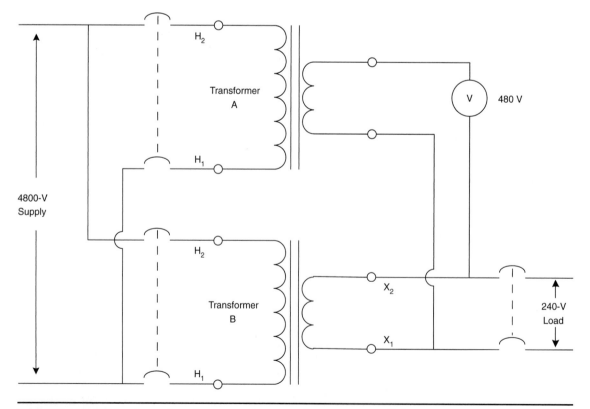

**FIGURE 12–33A**  Procedure for connecting transformers in parallel.

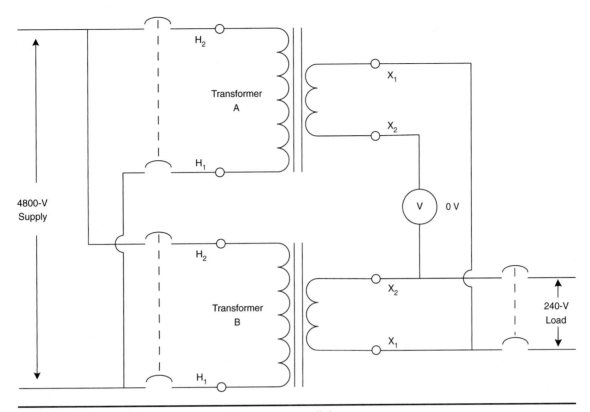

**FIGURE 12–33B**  Connecting transformers in parallel.

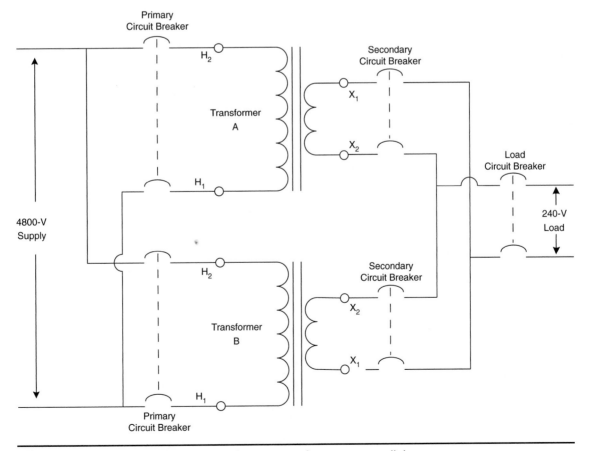

**FIGURE 12–33C**  Final connections for two transformers in parallel.

> **⚠ CAUTION** Never connect together leads that have a potential difference between them.

Parallel operation of transformers offers the advantage of being able to remove one transformer from service for replacement or maintenance without interrupting the entire operation.

> **⚠ CAUTION** When it is necessary to remove the transformer from the line, always be sure that both the primary and secondary switches are open before disconnecting the transformers.

If only the primary is open, the secondary winding will still be energized through the other transformers. Current flowing in the secondary will induce a voltage into the primary winding. It is also a safe practice to disconnect the low-voltage winding first, thereby ensuring that when the high-voltage winding is disconnected, there will be no induced emf across it.

On step-down transformers in parallel, a very hazardous condition can exist because of the possibility of this back feed. If the primary fuse or circuit breaker should open, the secondary takes on the characteristics of the primary. Transformers connected to a 13,200-volt supply could have this voltage induced back into the primary after it is disconnected from the supply.

## Two-Phase Connections

Two-phase installations are not common. However, there are still a few areas where such installations are in use. The two-phase, five-wire system is probably the most common arrangement. Again, in order to group transformers, they must meet the conditions previously stated. Connections for two-phase, three-wire installations; two-phase, four-wire installations; and two-phase, five-wire installations are shown in Figures 12–34A, 12–34B, and 12–34C.

## Three-Phase Connections

It is common practice to connect single-phase transformers into a three-phase bank. There are many different methods that may be utilized. One very common method is the *delta/delta connection* (Figures 12–35 and 12–36).

### Three-Phase Delta/Delta Connections

For connecting single-phase transformers into a three-phase bank, the following procedure is used. Assume that the transformers are rated for 4800 volts and 240 volts. They are to be installed as step-down transformers for a three-phase delta system.

1. Connect the high-voltage windings as shown in Figures 12–35 and 12–36.

2. Measure the voltage across each secondary winding. The meter should indicate 240 volts.

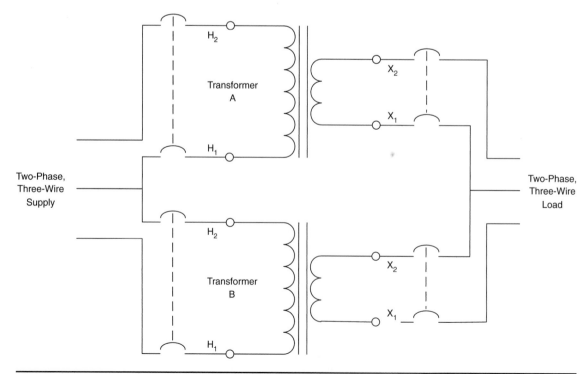

**FIGURE 12–34A** Two transformers connected to a two-phase, three-wire system.

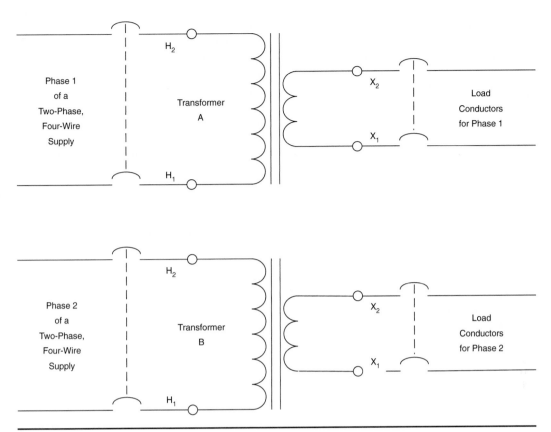

**FIGURE 12–34B** Two transformers connected to a two-phase, four-wire system.

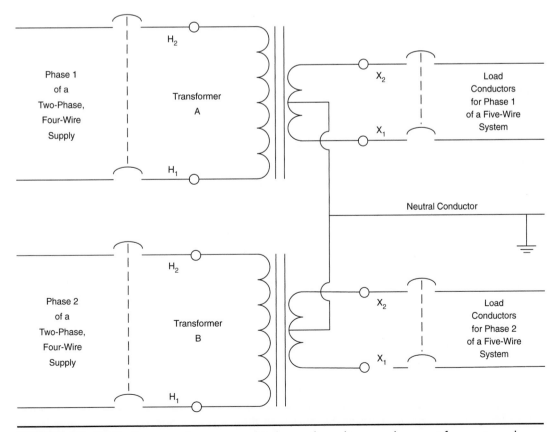

**FIGURE 12–34C** Two transformers used to lower the voltage and convert from a two-phase, four-wire system to a two-phase, five-wire system.

3. Connect $X_2$ of transformer A to $X_1$ of transformer B. Measure the voltage across the open ends (Figure 12–37A). The meter should indicate the vector sum of the voltage induced into each winding, which is 240 volts.

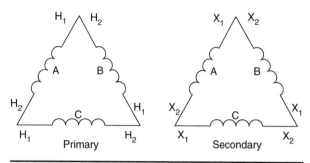

**FIGURE 12–35** Transformer windings connected to form a three-phase, delta/delta configuration.

Figure 12–37A shows the correct connection, and the arrows indicate the direction of the instantaneous voltage. Both voltages are in the same direction; therefore, vector addition is performed. Figure 12–37B shows a vector diagram of the two voltages, which are 120 degrees out of phase. Figure 12–38A illustrates incorrect connections. Note that the instantaneous voltages are in opposite directions. Therefore, in order to perform vector addition, one vector must be reversed. Figure 12–38B shows the vector diagram for this condition. The mathematical solutions for the circuits shown in Figures 12–37B and 12–38B are as follows:

For Figure 12–37B:

$$E = \sqrt{E_1^2 + E_2^2 - (2E_1E_2 \cos \phi)}$$

*(From Equation 7.5)*

$$E = \sqrt{240^2 + 240^2 - (2 \times 240 \times 240 \times 0.5)}$$
$$E = \sqrt{57,600}$$
$$E = 240 \text{ V}$$

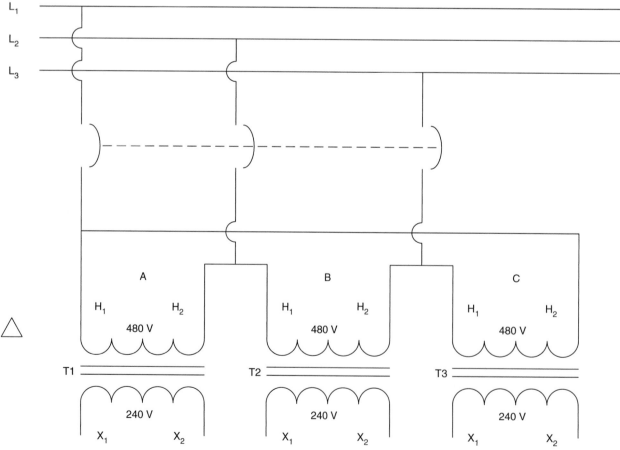

480-V, Three-Phase, Three-Wire Supply

**FIGURE 12–36** Three single-phase transformers with their high-voltage windings connected to form a three-phase, delta configuration.

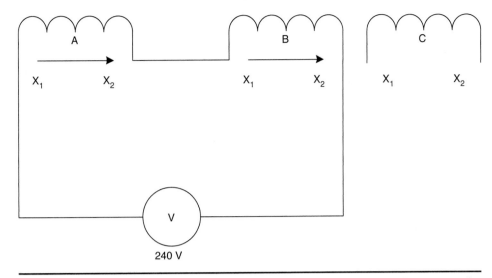

**FIGURE 12–37A** *Secondary windings of two transformers, illustrating the beginning of a three-phase, closed-delta connection.*

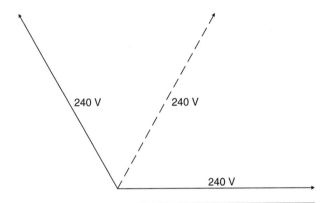

**FIGURE 12–37B** *Vector diagram of two voltages 120 electrical time degrees apart.*

For Figure 12–38B:

$$E = \sqrt{E_1^2 + E_2^2 + (2E_1E_2 \cos \phi)}$$

*(From Equation 7.5)*

$$E = \sqrt{240^2 + 240^2 + (2 \times 240 \times 240 \times 0.5)}$$
$$E = \sqrt{172,800}$$
$$E = 415.7 \text{ V}$$

4. With transformers A and B connected as shown in Figure 12–39A, connect $X_2$ of transformer B to $X_1$ of transformer C (Figure 12–39A). Measure the voltage between $X_1$ of transformer A and $X_2$ of transformer C. The meter should indicate zero volt. Figure 12–39B shows the vector diagram for determining their sum.

Figure 12–40A illustrates incorrect connections. Note that the instantaneous voltage of transformer C is opposite to that of transformers A and B. To perform vector addition, vector C must be reversed. This operation puts vector C in phase with the resultant voltage of transformers A and B, and adds arithmetically to equal 480 volts (Figure 12–40B).

> ⚠ **CAUTION** Never connect together two terminals if a potential difference exists between the two points.

5. With the transformer connected as shown in Figure 12–39A, zero potential difference between $X_2$ of transformer C and $X_1$ of transformer A, connect $X_2$ and $X_1$ together.

6. Connect the load conductors to the junction points of $X_1$ and $X_2$ as shown in Figures 12–41A and 12–41B.

The preceding connections are for a delta/delta, three-phase, three-wire system. *Delta/delta* means that both the primary and secondary windings are connected in a delta configuration.

### Three-Phase Wye/Wye Connections

Another method for connecting transformers is *wye/wye*. The following procedure is used. Assume that the transformers are rated at 4800 volts and 277 volts.

1. Connect $H_2$ of all three transformers together (Figures 12–42A and 12–42B).

2. Connect $H_1$ of each transformer to a supply conductor.

3. Measure the voltage across the secondary of each transformer. The meter should indicate 277 volts.

4. Connect $X_1$ of transformer A to $X_1$ of transformer B (Figure 12–43A). Measure the voltage between $X_2$ of transformer A and $X_2$

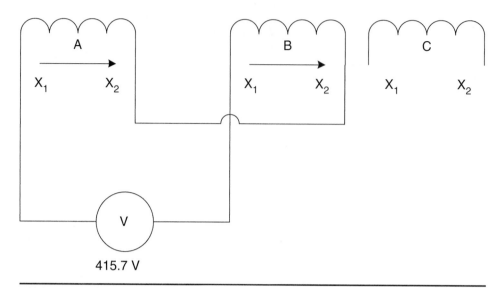

**FIGURE 12–38A** Incorrect connection for two transformers being connected into a closed-delta configuration.

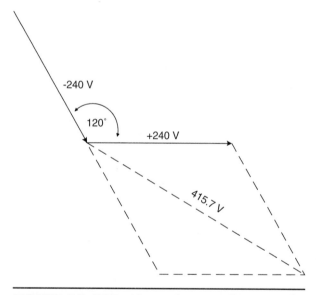

**FIGURE 12–38B** Vector diagram illustrating the resulting value of two voltages of opposite polarity 120 electrical time degrees apart.

of transformer B. The meter should indicate 480 volts. Figure 12–43B shows the vector diagram for this connection. Note that the instantaneous voltages are in the opposite directions. Therefore, to perform the vector addition, one vector must be reversed.

Figure 12–44A shows incorrect connections, which would result in a voltage of 277 volts. Figure 12–44B shows the vector diagram for this connection.

5. With transformers A and B connected as shown in Figure 12–43A, connect $X_2$ of transformer C

to $X_2$ of transformers A and B (Figure 12–45A). Measure the voltage between the various combinations of $X_1$. All combinations should indicate 480 volts. Figure 12–45B shows the vector diagram of this connection. Phase 1 is the vector sum of transformers A and B. Phase 2 is the vector sum of transformers B and C. Phase 3 is the vector sum of transformers C and A.

6. Connect the leads $X_1$ to the load conductors. A conductor connected to the midpoint ($X_1$) will provide a second voltage. Between conductor N (Figure 12–46) and any of the load conductors, the voltage will be 277 volts.

### Three-Phase Delta/Wye Connections

Another very common connection is the *delta/wye*. The primary is connected to form a delta configuration and the secondary is connected to form a wye. Figure 12–47 shows a diagram of this arrangement. The primary is supplied from a 13,850-volt, three-phase, three-wire line. The secondary is connected to a 277/480-volt, three-phase, four-wire load.

If the secondary is connected in delta instead of wye, the voltage supplied to the load will be 277 volts, three phase, three wire. The manner in which the secondary is connected will determine the value of the voltage supplied to the load. Note in the wye connection that two voltages are available: 277 volts, single phase, and 480 volts, three phase. The delta connection, however, supplies only one voltage, which is 277 volts, three phase.

### Three-Phase Wye/Delta Connections

Transformers may also be connected in *wye/delta*. Figure 12–48 illustrates this connection. Note that the secondary of each transformer has two separate

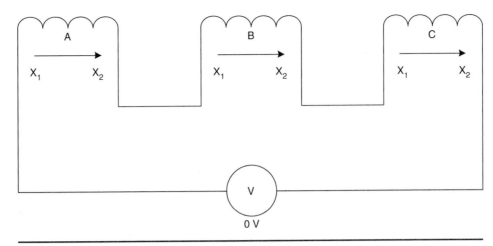

**FIGURE 12–39A**  Connections for a three-phase, closed-delta system. The voltmeter indicates 0 V.

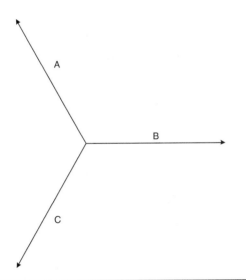

**FIGURE 12–39B**  Vector diagram of three equal voltages 120 electrical time degrees apart. The resultant voltage is zero.

windings. This is a common practice and permits a midpoint tap from one transformer.

The voltage supplied to the primary is 14,460 volts, three phase. The voltage across each primary winding is 14,460/1.73 = 8358 volts. If the ratio is 19:1, then the voltage across each secondary winding is 8358/19 = 440 volts. The secondary supplies 440 volts, three phase to the load. The voltage between the midpoint of transformer C and phase wires B and C is 220 volts, **single phase**. The voltage between the midpoint of transformer C and phase wire A is the vector sum of 440 and 220, which is 381 volts.

## Three-Phase, Open-Delta Connections

Another method sometimes used for three-phase transformer connections is called the *open delta*, or *V*, and requires only two transformers. In areas

where additional loading is anticipated, two transformers connected in open delta can be installed for the initial load. When the additional load is installed, a third transformer can be added to form a closed-delta system. The open-delta arrangement will carry 58% of the load that can be applied to a closed delta when using the same size of transformers.

The open-delta arrangement is also useful for maintenance purposes. If one transformer of a closed-delta system becomes defective, the transformer can be removed and the system reconnected in open delta. Under these conditions, the system can maintain 58% of its total load. Figure 12–49 shows an open-delta system.

## Three-Phase T Connection

In this arrangement, two transformers can be used with a three-phase system. The transformers must have a midpoint tap and an 86.6-percent tap. Figure 12–50 shows a schematic diagram of the T connection.

## Scott Connection

Most electrical equipment are constructed to operate on either single phase or three phase. In areas where the supply is two phase, transformers can be used to convert it to three phase. Figure 12–51 illustrates this arrangement, called a *Scott connection*.

## Isolation Transformers

Another common use of transformers is to isolate a system or part of a system from the supply. Frequently this requirement is fulfilled without a change in voltage. Transformers with a ratio of 1:1 are used. When used for this purpose, they are called **isolation transformers**.

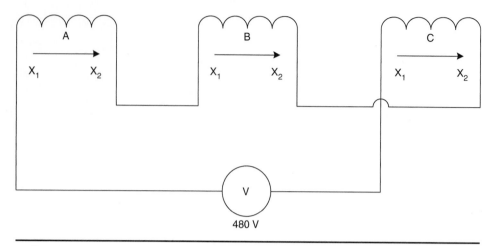

**FIGURE 12–40A**  Incorrect connections for a three-phase, closed-delta system.

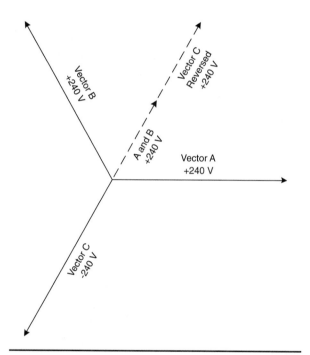

**FIGURE 12–40B**  Vector diagram of the voltages in Figure 12–40A.

## THREE-PHASE TRANSFORMER CALCULATIONS

Before beginning an analysis of the different three-phase transformer configurations, there are some new terms to understand. Since we are dealing with three-phase connections, there will be two different voltage and current measurements. One measurement will be called the *line* value and the other measurement will be called the *phase* value. Figure 12–52 shows three windings connected in a wye configuration. Notice that each winding is

connected to one of the three-phase power lines, $L_1$, $L_2$, and $L_3$. The voltage measured between $L_1$ and $L_2$, between $L_2$ and $L_3$, or between $L_1$ and $L_3$, is called the *line voltage* because this is the amount of voltage measured from *line to line*. On the other hand, if measuring the voltage from $L_1$ to the common point where all three windings are connected, the measurement would be the *phase voltage*. The phase voltage is the voltage measured across each phase winding. This would also be true if measuring the voltage from $L_2$ to common or $L_3$ to common. *In a wye-connected system, the line voltage will not be equal to the phase voltage.* In fact, *the line voltage will be 1.732 times greater than the phase voltage.*

If you imagine yourself as an electron flowing from $L_1$ to $L_2$, you would take the path that takes you through windings A and B. Essentially, you only have one path to follow. Hence, it appears as though windings A and B are connected in series. The same can be said for current flowing from $L_2$ to $L_3$. In this instance, the current can only flow through windings B and C. Likewise, current flowing from $L_1$ to $L_3$ can only flow through windings A and C. In each example, there is only one path for current flow; therefore, the line current must equal the phase current in a wye-connected circuit.

Looking next at a delta-connected circuit, Figure 12–53 shows three windings connected in a delta configuration. Notice that one end of winding A ($A_2$) is connected to one end of winding B ($B_1$). The other end of winding B ($B_2$) is connected to one end of winding C ($C_1$). The remaining end of winding C ($C_2$) is connected to the remaining end of winding A ($A_1$). The three-phase power connections are made at the junctions of two windings. As seen in the wye connection, the voltage measured between $L_1$ and $L_2$, between $L_2$ and $L_3$, or between $L_1$ and $L_3$ is called the line voltage because the

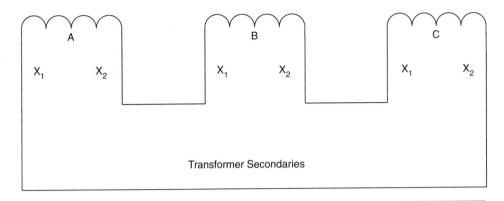

**FIGURE 12–41A** Three single-phase transformers connected to form a three-phase, closed-delta bank.

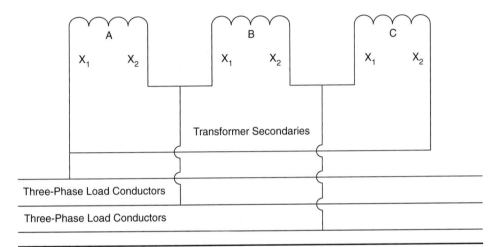

**FIGURE 12–41B** Three single-phase transformers connected into a closed-delta configuration with the load conductors connected.

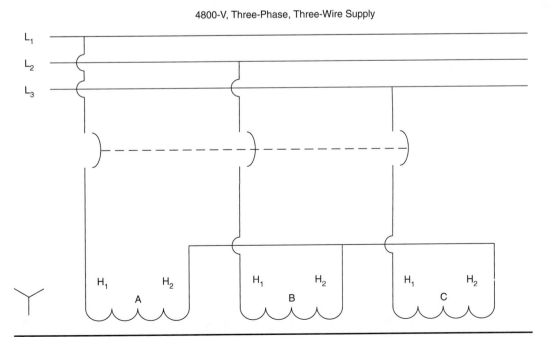

**FIGURE 12–42A** High-voltage windings of three single-phase transformers connected into a wye configuration and connected to a three-phase, three-wire supply.

4800-V, Three-Phase, Three-Wire Supply

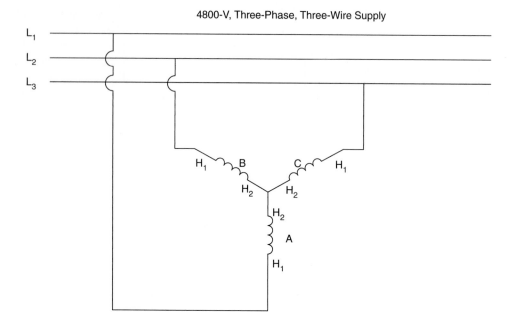

**FIGURE 12–42B** Primary windings of a three-phase transformer bank connected into a wye configuration.

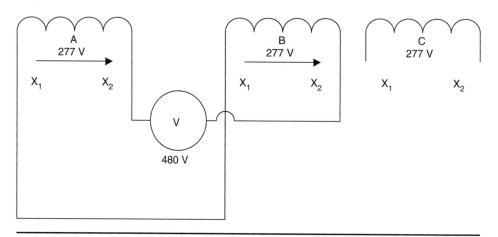

**FIGURE 12–43A** Connections for two of three single-phase transformers that are being connected into a three-phase wye configuration.

amount of voltage is measured from line to line. Likewise, if we were to measure the voltage across one winding, we would measure the phase voltage. *In a delta-connected system, the line voltage will equal the phase voltage.* Imagine placing the leads of your voltmeter across the line connections. Notice that your meter is also placed across one of the windings at the same time. This is why the phase voltage and the line voltage are equal in a delta-connected system.

An electron flowing from $L_1$ to $L_2$ would take the path through winding A. The electron could also take the path through windings B and C. Essentially, there are two paths to follow; therefore, it appears as though winding A is in parallel with the series combination of windings B and C. The same can be said for current flowing from $L_2$ to $L_3$. In this instance, the current can flow through winding C and through the series combination of windings A and B. Likewise, current flowing from $L_1$ to $L_3$ can flow through winding B and through the series combination of windings A and C. In each example, there are two paths for current flow. Therefore, the line current must be larger than the phase current in a delta-connected circuit because the line current must divide through the phase windings. The phase current then recombines to become the line current. Therefore, the line current must be larger than the phase current. In fact, *the line current will be 1.732 times greater than the phase current.*

We will use the following variables in our analysis of three-phase transformer circuits:

$E_{P(Phase)}$ = primary-phase voltage
$E_{P(Line)}$ = primary-line voltage
$E_{S(Phase)}$ = secondary-phase voltage
$E_{S(Line)}$ = secondary-line voltage
$E_{L(Phase)}$ = load-phase voltage
$E_{L(Line)}$ = load-line voltage
$I_{P(Phase)}$ = primary-phase current
$I_{P(Line)}$ = primary-line current
$I_{S(Phase)}$ = secondary-phase current
$I_{S(Line)}$ = secondary-line current
$I_{L(Phase)}$ = load-phase current
$I_{L(Line)}$ = load-line current

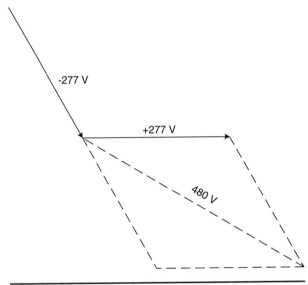

**FIGURE 12–43B**   Vector diagram of the voltages in Figure 12–43A.

Let us begin with the delta/delta configuration. Figure 12–54 shows a delta/delta connected transformer. The primary-line voltage, $E_{P(Line)}$ is 480 VAC. The load (a wye-connected motor) is designed to operate from 208-VAC power, $E_{L(Line)}$. Each motor winding has an impedance of 5 Ω. We need to determine the following:

What do we know?

$$E_{P(Line)} = 480 \text{ VAC}$$
$$E_{L(Line)} = 208 \text{ VAC}$$
$$k = 1.732$$
$$Z_{Load} = 5 \ \Omega$$

What do we not know?

$$E_{P(Phase)} = ?$$
$$E_{S(Phase)} = ?$$
$$E_{S(Line)} = ?$$
$$E_{L(Phase)} = ?$$
$$I_{P(Phase)} = ?$$
$$I_{P(Line)} = ?$$
$$I_{S(Phase)} = ?$$
$$I_{S(Line)} = ?$$
$$I_{L(Phase)} = ?$$
$$I_{L(Line)} = ?$$
$$\text{turns ratio} = ?$$

We begin by recognizing that the primary is connected in a delta configuration. As a result, the line voltage is equal to the phase voltage. Therefore:

$E_{P(Line)}$ = 480 VAC, and in a delta configuration
$E_{P(Line)} = E_{P(Phase)}$; therefore,                      (Eq. 12.5)
$E_{P(Phase)}$ = 480 VAC

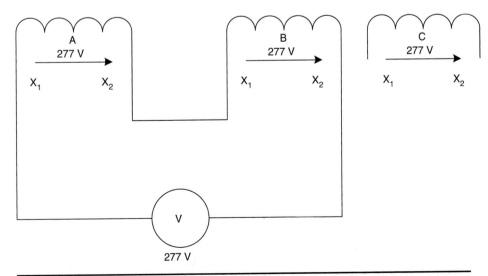

**FIGURE 12–44A**   Incorrect connections for two of three single-phase transformers being connected into a three-phase wye configuration.

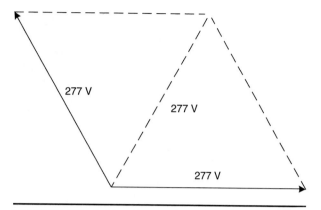

**FIGURE 12-44B** Vector diagram of the voltages in Figure 12-44A.

Notice that the secondary is also connected in a delta configuration and that the load operates from a 208-VAC line voltage. This means that the secondary-line voltage must be 208 VAC as well. If the secondary-line voltage is 208 VAC, and the secondary is connected in a delta configuration, the secondary-phase voltage must also be equal to 208 VAC. Therefore:

$$E_{L(Line)} = 208 \text{ VAC; therefore,}$$
$$E_{S(Line)} = 208 \text{ VAC, and in a delta configuration}$$
$$E_{S(Line)} = E_{S(Phase)}; \text{ therefore,} \qquad \text{(Eq. 12.6)}$$
$$E_{S(Phase)} = 208 \text{ VAC}$$

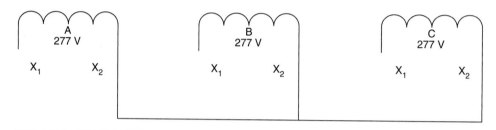

**FIGURE 12-45A** Low-voltage windings of three single-phase transformers connected to form a three-phase wye configuration.

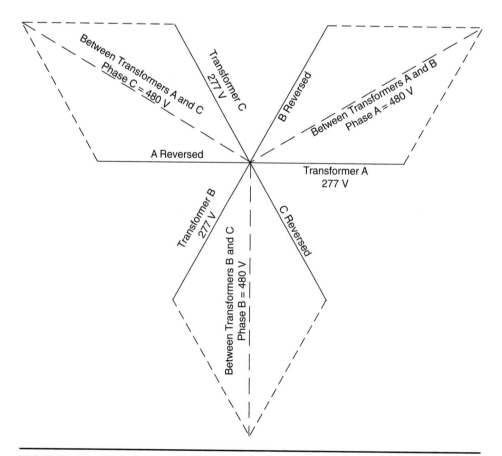

**FIGURE 12-45B** Vector diagram of three separate voltages obtained from a three-phase, three-wire wye connection.

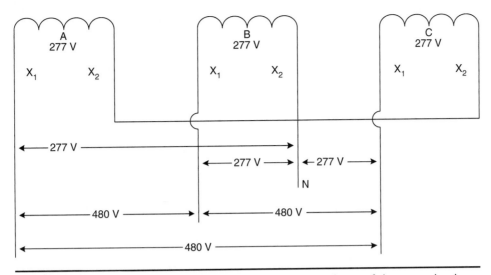

**FIGURE 12-46** Connections for the low-voltage windings of three single-phase transformers arranged to form a three-phase, four-wire wye system.

13,850-V, Three-Phase, Three-Wire Supply

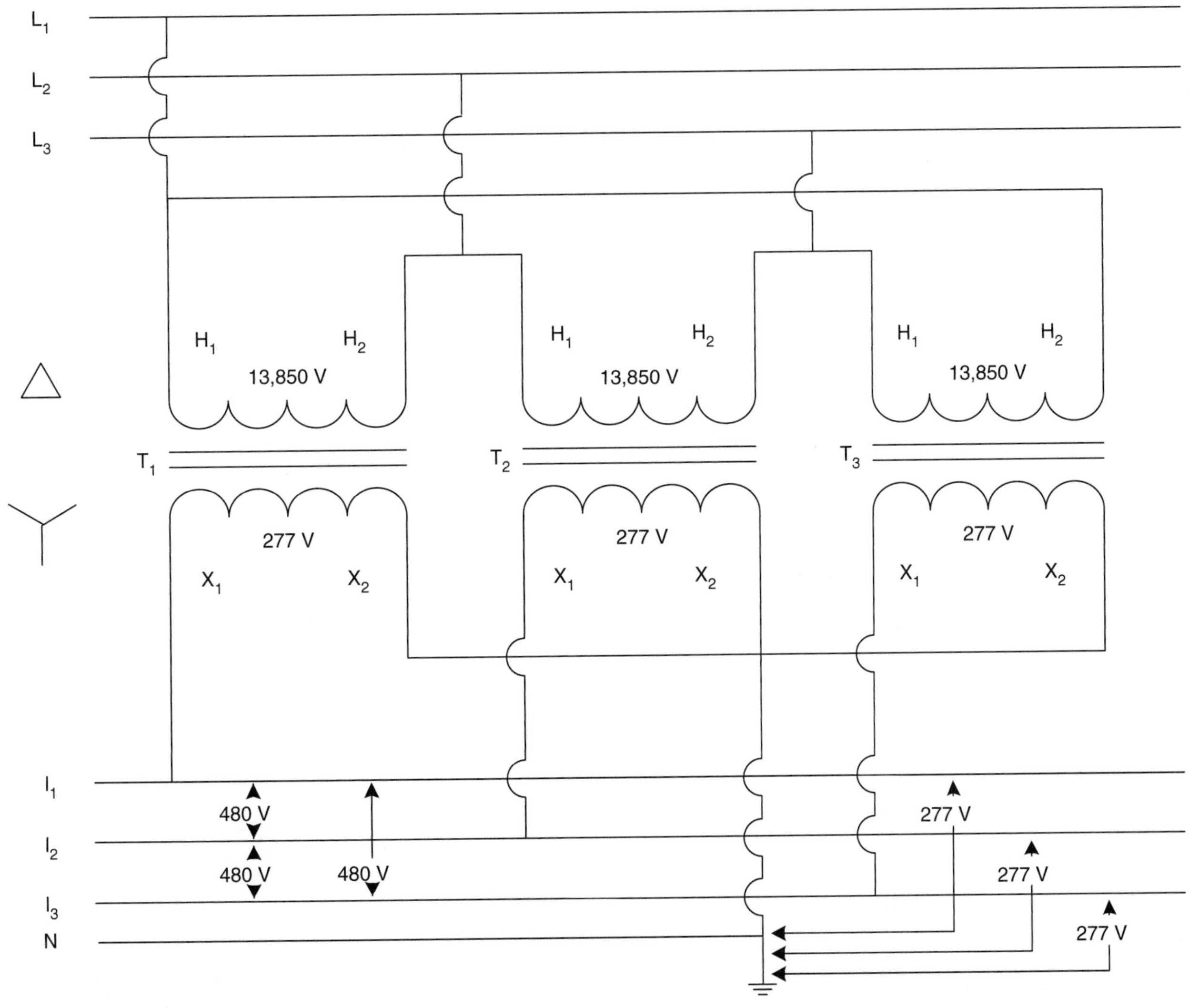

**FIGURE 12-47** Three single-phase transformers arranged to form a three-phase delta/wye bank.

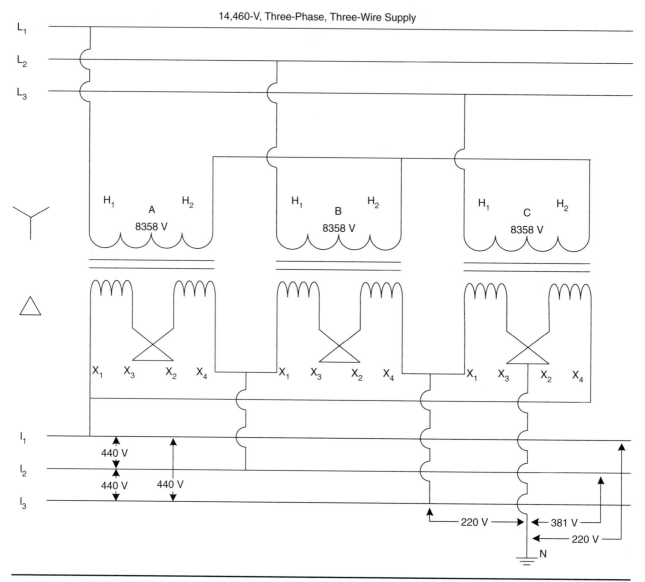

**FIGURE 12–48** Three single-phase transformers arranged into a three-phase wye/delta bank.

Before going further, there is an important rule to know when calculating transformer values. *Use only phase values of voltage and current when calculating transformer values.* Do not use line values of voltage and current because the transformation of voltage and current occurs in the windings of the transformer. The transformer windings produce the phase values of voltage and current. Now, we can determine the turns ratio of this transformer.

What do we know?

$$E_{P(Phase)} = 480 \text{ VAC}$$
$$E_{S(Phase)} = 208 \text{ VAC}$$

What do we not know?

$$\text{turns ratio} = \text{?}$$

What formula can we use?

$$\text{turns ratio} = \frac{E_{P(Phase)}}{E_{S(Phase)}}$$

Substitute the known values and solve for the turns ratio:

$$\text{turns ratio} = \frac{E_{P(Phase)}}{E_{S(Phase)}}$$
$$\text{turns ratio} = \frac{480 \text{ VAC}}{208 \text{ VAC}}$$
$$\text{turns ratio} = \frac{23.1}{1} \text{ or } 2.31{:}1$$

Therefore, the turns ratio of this transformer is 2.31:1. We will now turn our attention to the load. Recall that the load is connected in a wye configuration.

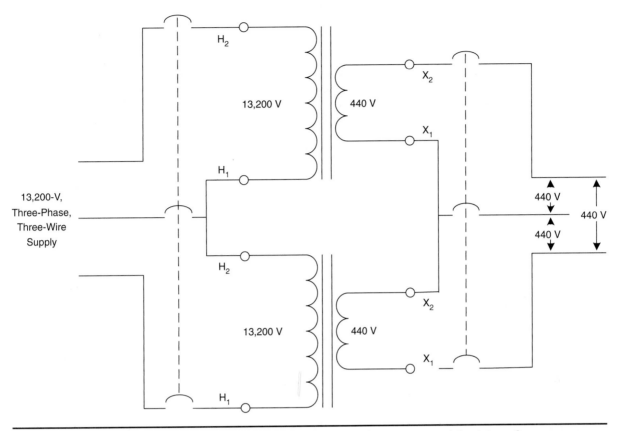

**FIGURE 12–49**  Two single-phase transformers connected to form an open-delta (V) configuration.

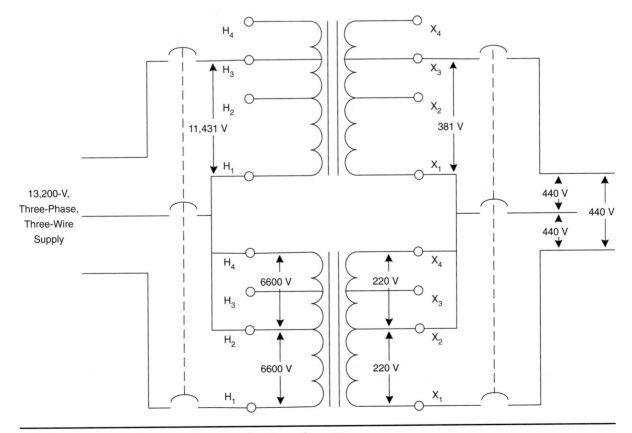

**FIGURE 12–50**  Schematic diagram of a three-phase T connection.

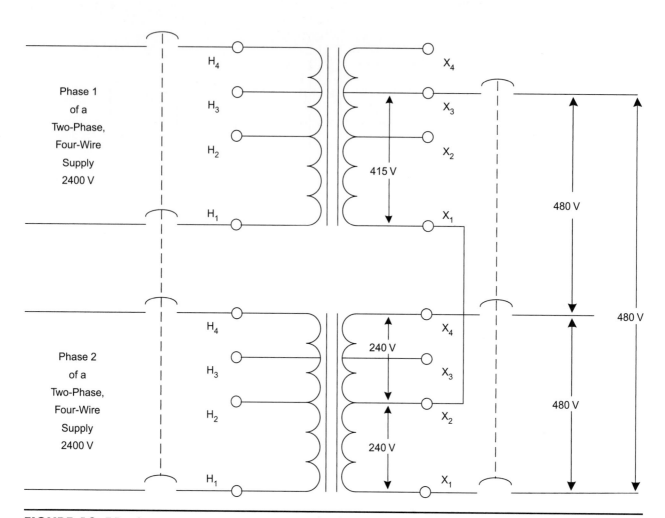

**FIGURE 12–51** Scott connection using two single-phase transformers to convert two-phase, four-wire to three-phase, three-wire.

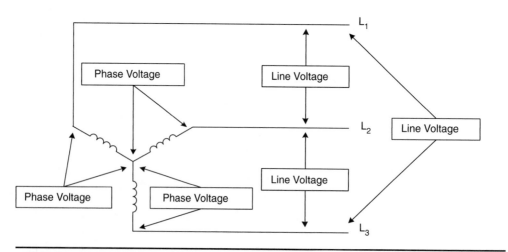

**FIGURE 12–52** Phase voltages and line voltages in a three-phase wye configuration.

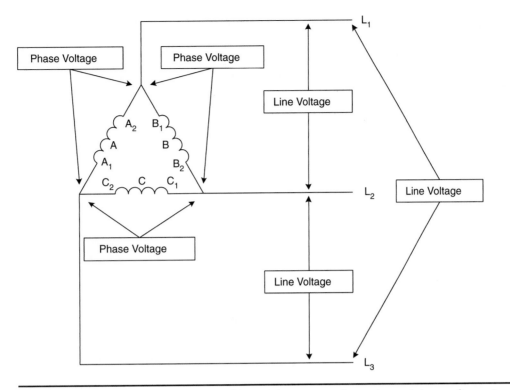

**FIGURE 12–53** Phase voltages and line voltages in a three-phase delta configuration.

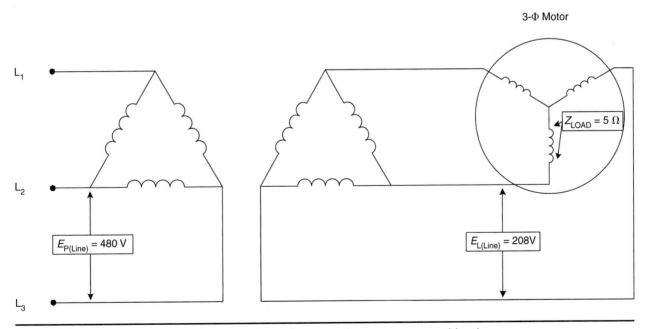

**FIGURE 12–54** A delta/delta-connected transformer with a wye-connected load.

As a result of the wye configuration, the load-line voltage, $E_{L(Line)}$, will be 1.732 times greater than the load-phase voltage $E_{L(Phase)}$. Recall that the load-line voltage, $E_{L(Line)}$, is equal to 208 VAC. We can now determine the load-phase voltage, $E_{L(Phase)}$.

What do we know?

$$E_{L(Line)} = 208 \text{ VAC}$$
$$k = 1.732$$

What do we not know?

$$E_{L(Phase)} = ?$$

What formula could we use?

$$E_{L(Phase)} = \frac{E_{L(Line)}}{k} \qquad \text{(Eq. 12.7)}$$

Now, substitute the known values and solve for $E_{L(Phase)}$:

$$E_{L(Phase)} = \frac{E_{L(Line)}}{k}$$
$$E_{L(Phase)} = \frac{208 \text{ VAC}}{1.732}$$
$$E_{L(Phase)} = 120.09 \text{ VAC}$$

Therefore, the load-line voltage is 120.09 VAC. Now that we know that each phase of the load consists of a 5-$\Omega$ impedance, and each load-phase voltage is equal to 120.09 VAC, we can determine the amount of load-phase current, $I_{L(Phase)}$, by using Ohm's law.

What do we know?

$$E_{L(Phase)} = 120.09 \text{ VAC}$$
$$Z_{Load} = 5 \text{ } \Omega$$

What do we not know?

$$I_{L(Phase)} = ?$$

What formula could we use?

$$I_{L(Phase)} = \frac{E_{L(Phase)}}{Z_{Load}}$$

Now, substitute the known values and solve for $I_{L(Phase)}$:

$$I_{L(Phase)} = \frac{E_{L(Phase)}}{Z_{Load}}$$
$$I_{L(Phase)} = \frac{120.09 \text{ VAC}}{5 \text{ } \Omega}$$
$$I_{L(Phase)} = 24.02 \text{ A}$$

Therefore, the load-phase current, $I_{L(Phase)}$, is equal to 24.02 amperes. Now we can determine the load-line current, $I_{L(Line)}$. Recall that the load is connected in a wye configuration. In a wye configuration,

the phase current and the line current are equal. Therefore:

$$I_{L(Phase)} = 24.02 \text{ A, and in a wye configuration}$$
$$I_{L(Phase)} = I_{L(Line)}; \text{ therefore,} \qquad \text{(Eq. 12.8)}$$
$$I_{L(Line)} = 24.02 \text{ A}$$

So, in our example, the load-line current, $I_{L(Line)}$, is equal to 24.02 amperes. The load-line current must be equal to the secondary-line current. After all, the secondary is the source of the line current that is supplied to the load. Therefore:

$$I_{L(Line)} = 24.02 \text{ A; therefore,}$$
$$I_{L(Line)} = I_{S(Line)}; \text{ therefore,} \qquad \text{(Eq. 12.9)}$$
$$I_{S(Line)} = 24.02 \text{ A}$$

The secondary-line current, $I_{S(Line)}$, is equal to 24.02 amperes. Now we can determine the secondary-phase current, $I_{S(Phase)}$. Recall that the secondary of our transformer is connected in a delta configuration. In a delta configuration, the line current will be 1.732 times greater than the phase current.

What do we know?

$$I_{S(Line)} = 24.02 \text{ A}$$
$$k = 1.732$$

What do we not know?

$$I_{S(Phase)} = ?$$

What formula could we use?

$$I_{S(Phase)} = \frac{I_{S(Line)}}{k} \qquad \text{(Eq. 12.10)}$$
$$I_{S(Phase)} = \frac{24.02 \text{ A}}{1.732}$$
$$I_{S(Phase)} = 13.87 \text{ A}$$

Therefore, the secondary-phase current, $I_{S(Phase)}$, will equal 13.87 amperes. Now that we know the secondary-phase current, we can determine the primary-phase current, $I_{P(Phase)}$. This is done by using the transformer ratio calculated earlier. Recall that when the primary voltage is stepped down, the current is stepped up by the same ratio. Likewise, when the primary voltage is stepped up, the current is stepped down by the same ratio. The transformer in our example is a step-down transformer. This tells us that if the primary voltage is stepped down, the current must be stepped up.

Since we are looking at this transformer from the secondary side, it is easy to get confused on this issue. We have determined that the secondary-phase current, $I_{S(Phase)}$, is equal to 13.87 amperes. We must now determine the primary-phase current, $I_{P(Phase)}$.

The primary-phase current must be lesser than the secondary-phase current. Also, we must remember to use the phase value of current when performing calculations. Let us calculate the primary-phase current, $I_{P(Phase)}$.

What do we know?

$$I_{S(Phase)} = 13.87 \text{ A}$$
$$\text{turns ratio} = 2.31:1$$

What do we not know?

$$I_{P(Phase)} = ?$$

What formula could we use?

$$I_{P(Phase)} = \frac{I_{S(Phase)}}{\text{turns ratio}} \qquad \text{(Eq. 12.11)}$$

Substitute the known values and solve for $I_{P(Phase)}$:

$$I_{P(Phase)} = \frac{I_{S(Phase)}}{\text{turns ratio}}$$
$$I_{P(Phase)} = \frac{13.87 \text{ A}}{2.31}$$
$$I_{P(Phase)} = 6.0 \text{ A}$$

Therefore, the primary-phase current of the transformer will be 6.0 amperes. Finally, we can determine the primary-line current, $I_{P(Line)}$. Recall that the primary of the transformer is connected in a delta configuration. In a delta configuration, the line current will be 1.732 times greater than the phase current.

What do we know?

$$I_{P(Phase)} = 6.0 \text{ A}$$
$$k = 1.732$$

What do we not know?

$$I_{P(Line)} = ?$$

What formula could we use?

$$I_{P(Line)} = I_{P(Phase)} \times k \qquad \text{(Eq. 12.12)}$$

Substitute the known values and solve for $I_{P(Line)}$:

$$I_{P(Line)} = I_{P(Phase)} \times k$$
$$I_{P(Line)} = 6.0 \text{ A} \times 1.732$$
$$I_{P(Line)} = 10.39 \text{ A}$$

Therefore, the primary-line current, $I_{P(Line)}$, for this example will be equal to 10.39 amperes. Figure 12–55 shows the same circuit with all of the values entered.

Figure 12–56 shows a wye/wye-configured transformer. In order to compare the effects of a wye/wye connection, we will use the same values

for $E_{P(Line)}$, $E_{L(Line)}$, and $Z_{Load}$. Therefore, the primary-line voltage, $E_{P(Line)}$, is 480 VAC, the load (a wye-connected motor) operates from 208-VAC power, $E_{L(Line)}$, and each motor winding has an impedance of 5 Ω. We need to determine the following:

What do we know?

$$E_{P(Line)} = 480 \text{ VAC}$$
$$E_{L(Line)} = 208 \text{ VAC}$$
$$k = 1.732$$
$$Z_{Load} = 5 \text{ Ω}$$

What do we not know?

$$E_{P(Phase)} = ?$$
$$E_{S(Phase)} = ?$$
$$E_{S(Line)} = ?$$
$$E_{L(Phase)} = ?$$
$$I_{P(Phase)} = ?$$
$$I_{P(Line)} = ?$$
$$I_{S(Phase)} = ?$$
$$I_{S(Line)} = ?$$
$$I_{L(Phase)} = ?$$
$$I_{L(Line)} = ?$$
$$\text{turns ratio} = ?$$

We begin by recognizing that the primary is connected in a wye configuration. As a result, the line voltage is 1.732 times greater than the phase voltage. Therefore:

$E_{P(Line)} = 480$ VAC, and in a wye configuration

$$E_{P(Phase)} = \frac{E_{P(Line)}}{1.732}; \text{ therefore,} \qquad \text{(Eq. 12.13)}$$
$$E_{P(Phase)} = \frac{480 \text{ VAC}}{1.732}$$
$$E_{P(Phase)} = 277.14 \text{ VAC}$$

Notice that the secondary is also connected in a wye configuration and that the load operates from a 208-VAC line voltage. This means that the secondary-line voltage must be 208 VAC as well. If the secondary-line voltage is 208 VAC, and the secondary is connected in a wye configuration, the secondary-phase voltage must be less than the secondary-line voltage by a factor of 1.732. Therefore:

$E_{L(Line)} = 208$ VAC; therefore,
$E_{S(Line)} = 208$ VAC, and in a wye configuration

$$E_{S(Phase)} = \frac{E_{S(Line)}}{1.732}; \text{ therefore,} \qquad \text{(Eq. 12.14)}$$
$$E_{S(Phase)} = \frac{208 \text{ VAC}}{1.732}$$
$$E_{S(Phase)} = 120.09 \text{ VAC}$$

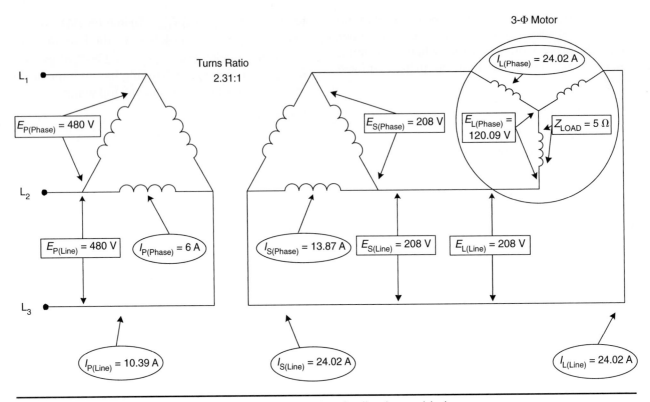

**FIGURE 12–55**  The same circuit as in Figure 12–54 with all values added.

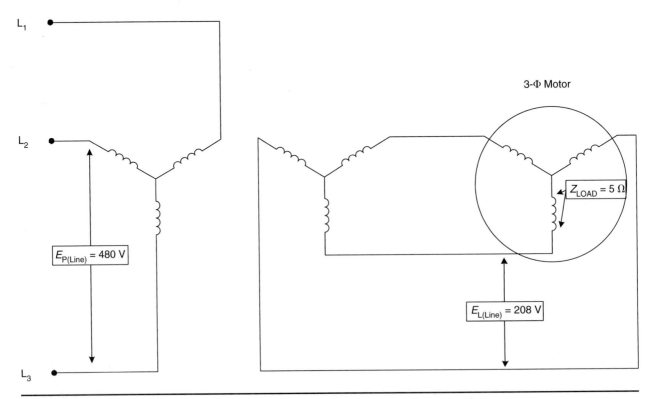

**FIGURE 12–56**  A wye/wye-connected transformer with a wye-connected load.

Recall the rule when calculating transformer values. *You must use only phase values of voltage and current when calculating transformer values.* Do not use line values of voltage and current. Now we can determine the turns ratio of this transformer.

What do we know?

$$E_{P(Phase)} = 277.14 \text{ VAC}$$
$$E_{S(Phase)} = 120.09 \text{ VAC}$$

What do we not know?

$$\text{turns ratio} = ?$$

What formula can we use?

$$\text{turns ratio} = \frac{E_{P(Phase)}}{E_{S(Phase)}}$$

Substitute the known values and solve for the turns ratio:

$$\text{turns ratio} = \frac{E_{P(Phase)}}{E_{S(Phase)}}$$
$$\text{turns ratio} = \frac{277.14 \text{ VAC}}{120.09 \text{ VAC}}$$
$$\text{turns ratio} = \frac{2.31}{1} \text{ or } 2.31{:}1$$

Therefore, the turns ratio of this transformer is 2.31:1. In turning our attention to the load, recall that the load is connected in a wye configuration. As a result of the wye configuration, the load-line voltage, $E_{L(Line)}$, will be 1.732 times greater than the load-phase voltage, $E_{L(Phase)}$. Recall that the load-line voltage, $E_{L(Line)}$, is equal to 208 VAC. We can now determine the load-phase voltage, $E_{L(Phase)}$.

What do we know?

$$E_{L(Line)} = 208 \text{ VAC}$$
$$k = 1.732$$

What do we not know?

$$E_{L(Phase)} = ?$$

What formula could we use?

$$E_{L(Phase)} = \frac{E_{L(Line)}}{k} \qquad \text{(Eq. 12.15)}$$

Now substitute the known values and solve for $E_{L(Phase)}$:

$$E_{L(Phase)} = \frac{E_{L(Line)}}{k}$$
$$E_{L(Phase)} = \frac{208 \text{ VAC}}{1.732}$$
$$E_{L(Phase)} = 120.09 \text{ VAC}$$

Therefore, the load-line voltage is 120.09 VAC. Now that we know that each phase of the load consists of a 5-$\Omega$ impedance, and each load-phase voltage is equal to 120.09 VAC, we can determine the amount of load-phase current, $I_{L(Phase)}$, by using Ohm's law.

What do we know?

$$E_{L(Phase)} = 120.09 \text{ VAC}$$
$$Z_{Load} = 5 \text{ }\Omega$$

What do we not know?

$$I_{L(Phase)} = ?$$

What formula could we use?

$$I_{L(Phase)} = \frac{E_{L(Phase)}}{Z_{Load}}$$

Now substitute the known values and solve for $I_{L(Phase)}$:

$$I_{L(Phase)} = \frac{E_{L(Phase)}}{Z_{Load}}$$
$$I_{L(Phase)} = \frac{120.09 \text{ VAC}}{5 \text{ }\Omega}$$
$$I_{L(Phase)} = 24.02 \text{ A}$$

Therefore, the load-phase current, $I_{L(Phase)}$, is equal to 24.02 amperes. Now we can determine the load-line current, $I_{L(Line)}$, which is connected in a wye configuration. In a wye configuration, the phase current and the line current are equal. Therefore:

$$I_{L(Phase)} = 24.02 \text{ A, and in a wye configuration}$$
$$I_{L(Phase)} = I_{L(Line)}; \text{ therefore,} \qquad \text{(Eq. 12.16)}$$
$$I_{L(Line)} = 24.02 \text{ A}$$

So, in this example, the load-line current, $I_{L(Line)}$, is equal to 24.02 amperes. The load-line current must be equal to the secondary-line current. After all, the secondary is the source of the line current that is supplied to the load. Therefore:

$$I_{L(Line)} = 24.02 \text{ A; therefore,}$$
$$I_{L(Line)} = I_{S(Line)}; \text{ therefore,} \qquad \text{(Eq. 12.17)}$$
$$I_{S(Line)} = 24.02 \text{ A}$$

The secondary line current, $I_{S(Line)}$, is equal to 24.02 amperes. Now we can determine the secondary-phase current, $I_{S(Phase)}$. The secondary of our transformer is connected in a wye configuration where the line current will be equal to the phase current. Therefore:

$$I_{S(Line)} = 24.02 \text{ A; therefore,}$$
$$I_{S(Phase)} = I_{S(Line)}; \text{ therefore,} \qquad \text{(Eq. 12.18)}$$
$$I_{S(Phase)} = 24.02 \text{ A}$$

The secondary-phase current, $I_{S(Phase)}$, will equal 24.02 amperes, and now we can determine the primary-phase current, $I_{P(Phase)}$, by using the transformer ratio calculated earlier. Recall that when the primary voltage is stepped down, the current is stepped up by the same ratio. Likewise, when the primary voltage is stepped up, the current is stepped down by the same ratio. The transformer in our example is a step-down transformer; therefore, if the primary voltage is stepped down, the current must be stepped up.

In looking at this transformer from the secondary side, it is easy to get confused on this issue. It has been determined that the secondary-phase current, $I_{S(Phase)}$, is equal to 24.02 amperes. Before determining the primary-phase current, $I_{P(Phase)}$, we must recognize that the primary-phase current must be smaller than the secondary-phase current. Also, remember to use the phase value of current when performing calculations. Let us calculate the primary-phase current, $I_{P(Phase)}$.

What do we know?

$$I_{S(Phase)} = 24.02 \text{ A}$$
$$\text{turns ratio} = 2.31{:}1$$

What do we not know?

$$I_{P(Phase)} = ?$$

What formula could we use?

$$I_{P(Phase)} = \frac{I_{S(Phase)}}{\text{turns ratio}} \qquad \text{(Eq. 12.19)}$$

Substitute the known values and solve for $I_{P(Phase)}$:

$$I_{P(Phase)} = \frac{I_{S(Phase)}}{\text{turns ratio}}$$
$$I_{P(Phase)} = \frac{24.02 \text{ A}}{2.31}$$
$$I_{P(Phase)} = 10.40 \text{ A}$$

Therefore, the primary-phase current of our transformer will be 10.40 amperes. Finally, we can determine the primary-line current, $I_{P(Line)}$. Recall that the primary of our transformer is connected in a wye configuration. In a wye configuration, the line current will be equal to the phase current. Therefore:

$$I_{P(Phase)} = 10.40 \text{ A; therefore,}$$
$$I_{P(Line)} = I_{P(Phase)}\text{; therefore,} \qquad \text{(Eq. 12.20)}$$
$$I_{P(Line)} = 10.40 \text{ A}$$

The primary line current, $I_{P(Line)}$, in this example will be equal to 10.40 amperes. Figure 12–57 shows the same circuit with all of the values entered.

Compare the values shown in Figure 12–55 with those shown in Figure 12–57. Notice that the primary-line-current values are essentially the same. However, the primary-phase current is larger in the wye-connected transformer of Figure 12–57. Now look at the secondary currents. The line currents of the wye- and delta-connected secondaries are essentially the same value. However, the secondary-phase current is larger in the wye-connected transformer of Figure 12–57.

The primary-phase voltage of the wye-connected transformer of Figure 12–57 is lower than the primary-phase voltage of the delta-connected transformer of Figure 12–55. Likewise, the secondary-phase voltage of the wye-connected transformer is lower than the delta-connected transformer.

The current and voltage values for the load did not change regardless of the type of transformer connection used. This means that should we decide to use a delta/delta configuration, the transformer could be constructed with smaller gauge wire for the primary- and secondary-phase windings. This could result in a smaller physical size and lower cost. On the other hand, should there be a need to develop other voltages than the 480 VAC and 208 VAC, we may opt to use the wye/wye configuration. This configuration will produce 480 VAC, 277 VAC, 208 VAC, and 120 VAC.

Figure 12–58 shows a delta/wye-connected transformer. The primary-line voltage, $E_{P(Line)}$, is 13.2 kVAC. The load (a delta-connected motor) is designed to operate from 480-VAC power, $E_{L(Line)}$. Each motor winding has an impedance of 10 Ω. We need to determine the following:

What do we know?

$$E_{P(Line)} = 13.2 \text{ kVAC}$$
$$E_{L(Line)} = 480 \text{ VAC}$$
$$k = 1.732$$
$$Z_{Load} = 10 \text{ Ω}$$

What do we not know?

$$E_{P(Phase)} = ?$$
$$E_{S(Phase)} = ?$$
$$E_{S(Line)} = ?$$
$$E_{L(Phase)} = ?$$
$$I_{P(Phase)} = ?$$
$$I_{P(Line)} = ?$$
$$I_{S(Phase)} = ?$$
$$I_{S(Line)} = ?$$
$$I_{L(Phase)} = ?$$
$$I_{L(Line)} = ?$$
$$\text{turns ratio} = ?$$

We begin by recognizing that the primary is connected in a delta configuration. As a result,

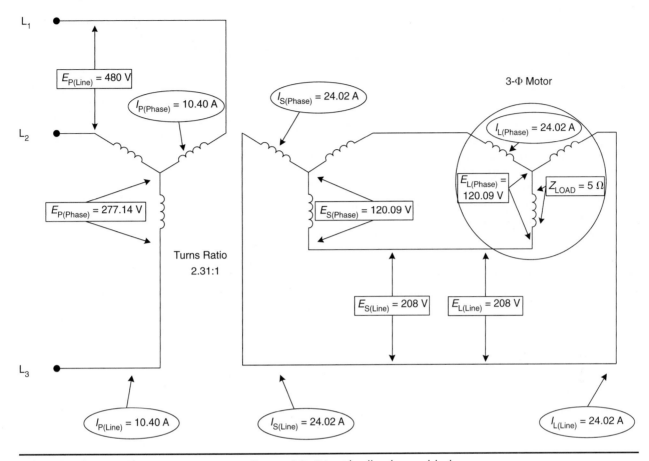

**FIGURE 12–57**   The same circuit as in Figure 12–56 with all values added.

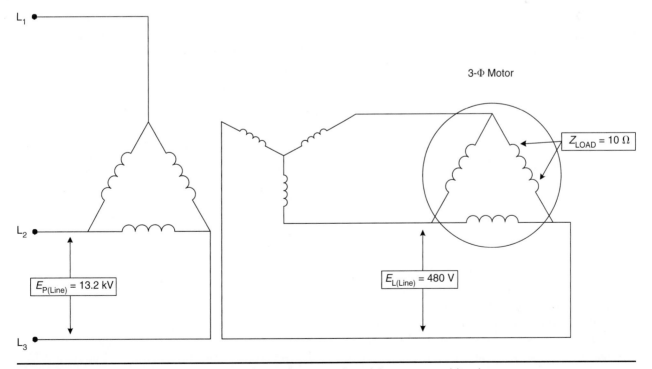

**FIGURE 12–58**   A delta/wye-connected transformer with a delta-connected load.

the line voltage is equal to the phase voltage. Therefore:

$$E_{P(Line)} = 13.2 \text{ kVAC, and in a delta configuration}$$
$$E_{P(Line)} = E_{P(Phase)}; \text{ therefore,}$$
$$E_{P(Phase)} = 13.2 \text{ kVAC}$$

Notice that the secondary is connected in a wye configuration and that the load operates from a 480-VAC line voltage. This means that the secondary-line voltage must be 480 VAC as well. If the secondary-line voltage is 480 VAC, and the secondary is connected in a wye configuration, the secondary-phase voltage must be less than the 480-VAC, secondary-line voltage by a factor of 1.732. Therefore:

$$E_{L(Line)} = 480 \text{ VAC; therefore,}$$
$$E_{S(Line)} = 480 \text{ VAC, and in a wye configuration}$$
$$E_{S(Phase)} = \frac{E_{S(Line)}}{1.732}; \text{ therefore,}$$
$$E_{S(Phase)} = \frac{480 \text{ VAC}}{1.732}$$
$$E_{S(Phase)} = 277.14 \text{ VAC}$$

Now we can determine the turns ratio of this transformer.

What do we know?

$$E_{P(Phase)} = 13.2 \text{ kVAC}$$
$$E_{S(Phase)} = 277.14 \text{ VAC}$$

What do we not know?

$$\text{turns ratio} = ?$$

What formula can we use?

$$\text{turns ratio} = \frac{E_{P(Phase)}}{E_{S(Phase)}}$$

Substitute the known values and solve for the turns ratio:

$$\text{turns ratio} = \frac{E_{P(Phase)}}{E_{S(Phase)}}$$
$$\text{turns ratio} = \frac{13.2 \text{ kVAC}}{277.14 \text{ VAC}}$$
$$\text{turns ratio} = \frac{47.63}{1} \text{ or } 47.63{:}1$$

Therefore, the turns ratio of this transformer is 47.63:1. In turning our attention to the load, recall that the load is connected in a delta configuration. As a result of the delta configuration, the load-line voltage, $E_{L(Line)}$, will be equal to the load-phase voltage, $E_{L(Phase)}$. Recall that the load-line voltage, $E_{L(Line)}$, is equal to 480 VAC. We can now determine the load-phase voltage, $E_{L(Phase)}$.

$$E_{L(Line)} = 480 \text{ VAC, and in a delta configuration}$$
$$E_{L(Phase)} = E_{L(Line)}; \text{ therefore,}$$
$$E_{L(Phase)} = 480 \text{ VAC}$$

The load-phase voltage is 480 VAC. Now that we know that each phase of the load consists of a 10-$\Omega$ impedance, and each load-phase voltage is equal to 480 VAC, we can determine the amount of load-phase current, $I_{L(Phase)}$, by using Ohm's law.

What do we know?

$$E_{L(Phase)} = 480 \text{ VAC}$$
$$Z_{Load} = 10 \text{ } \Omega$$

What do we not know?

$$I_{L(Phase)} = ?$$

What formula could we use?

$$I_{L(Phase)} = \frac{E_{L(Phase)}}{Z_{Load}}$$

Now substitute the known values and solve for $I_{L(Phase)}$:

$$I_{L(Phase)} = \frac{E_{L(Phase)}}{Z_{Load}}$$
$$I_{L(Phase)} = \frac{480 \text{ VAC}}{10 \text{ } \Omega}$$
$$I_{L(Phase)} = 48 \text{ A}$$

Therefore, the load-phase current, $I_{L(Phase)}$, is equal to 48 amperes. Now we can determine the load-line current, $I_{L(Line)}$. Recall that the load is connected in a delta configuration where the line current will be 1.732 times greater than the phase current. Therefore:

$$I_{L(Phase)} = 48 \text{ A, and in a delta configuration}$$
$$I_{L(Line)} = I_{L(Phase)} \times 1.732; \text{ therefore,}$$
$$I_{L(Line)} = 48 \text{ A} \times 1.732$$
$$I_{L(Line)} = 83.14 \text{ A}$$

In this example, the load-line current, $I_{L(Line)}$, is equal to 83.14 amperes. The load-line current must be equal to the secondary-line current. After all, the secondary is the source of the line current that is supplied to the load. Therefore:

$$I_{L(Line)} = 83.14 \text{ A; therefore,}$$
$$I_{S(Line)} = I_{L(Line)}; \text{ therefore,}$$
$$I_{S(Line)} = 83.14 \text{ A}$$

The secondary-line current, $I_{S(Line)}$, is equal to 83.14 amperes. Now we can determine the secondary-phase current, $I_{S(Phase)}$. Recall that the secondary of our transformer is connected in a wye configuration. In a wye configuration, the line current will be equal to the phase current. Therefore:

$$I_{S(Line)} = 83.14 \text{ A, and in a wye configuration}$$
$$I_{S(Phase)} = I_{S(Line)}; \text{ therefore,}$$
$$I_{S(Phase)} = 83.14 \text{ A}$$

Therefore, the secondary-phase current, $I_{S(Phase)}$, will equal 83.14 amperes. Now that we know the secondary-phase current, we can determine the primary-phase current, $I_{P(Phase)}$.

What do we know?

$$I_{S(Phase)} = 83.14 \text{ A}$$
$$\text{turns ratio} = 47.63:1$$

What do we not know?

$$I_{P(Phase)} = ?$$

What formula could we use?

$$I_{P(Phase)} = \frac{I_{S(Phase)}}{\text{turns ratio}}$$

Substitute the known values and solve for $I_{P(Phase)}$:

$$I_{P(Phase)} = \frac{I_{S(Phase)}}{\text{turns ratio}}$$
$$I_{P(Phase)} = \frac{83.14 \text{ A}}{47.63}$$
$$I_{P(Phase)} = 1.75 \text{ A}$$

Therefore, the primary-phase current of our transformer will be 1.75 amperes. Finally, we can determine the primary-line current, $I_{P(Line)}$. Recall that the primary of our transformer is connected in a delta configuration. In a delta configuration, the line current will be 1.732 times greater than the phase current. Therefore:

What do we know?

$$I_{P(Phase)} = 1.75 \text{ A}$$
$$k = 1.732$$

What do we not know?

$$I_{P(Line)} = ?$$

What formula could we use?

$$I_{P(Line)} = I_{P(Phase)} = k$$

Substitute the known values and solve for $I_{P(Line)}$:

$$I_{P(Line)} = I_{P(Phase)} \times k$$
$$I_{P(Line)} = 1.75 \text{ A} \times 1.732$$
$$I_{P(Line)} = 3.03 \text{ A}$$

Therefore, the primary-line current, $I_{P(Line)}$, for our example will be equal to 3.03 amperes. Figure 12–59 shows the same circuit with all of the values entered.

Figure 12–60 shows a wye/delta-connected transformer. The primary-line voltage, $E_{P(Line)}$, is 120 VAC. The load (a delta-connected motor) is designed to operate from 440-VAC power, $E_{L(Line)}$. Each motor winding has an impedance of 3 Ω. We need to determine the following:

What do we know?

$$E_{P(Line)} = 120 \text{ VAC}$$
$$E_{L(Line)} = 440 \text{ VAC}$$
$$k = 1.732$$
$$Z_{Load} = 3 \text{ Ω}$$

What do we not know?

$$E_{P(Phase)} = ?$$
$$E_{S(Phase)} = ?$$
$$E_{S(Line)} = ?$$
$$E_{L(Phase)} = ?$$
$$I_{P(Phase)} = ?$$
$$I_{P(Line)} = ?$$
$$I_{S(Phase)} = ?$$
$$I_{S(Line)} = ?$$
$$I_{L(Phase)} = ?$$
$$I_{L(Line)} = ?$$
$$\text{turns ratio} = ?$$

We begin by recognizing that the primary is connected in a wye configuration. As a result, the line voltage is 1.732 times greater than the phase voltage. Therefore:

$$E_{P(Line)} = 120 \text{ VAC, and in a wye configuration}$$
$$E_{P(Phase)} = \frac{E_{P(Line)}}{1.732}; \text{ therefore,}$$
$$E_{P(Phase)} = \frac{120 \text{ VAC}}{1.732}$$
$$E_{P(Phase)} = 69.28 \text{ VAC}$$

Looking at the secondary, notice that it is connected in a delta configuration and that the load operates from a 440-VAC line voltage. This means that the secondary-line voltage must be 440 VAC as well. If the secondary-line voltage is 440 VAC, and the secondary is connected in a delta configuration, the secondary-phase voltage must also be equal to 440 VAC. Therefore:

$$E_{L(Line)} = 440 \text{ VAC; therefore,}$$
$$E_{S(Line)} = 440 \text{ VAC, and in a delta configuration}$$
$$E_{S(Line)} = E_{S(Phase)}; \text{ therefore,}$$
$$E_{S(Phase)} = 440 \text{ VAC}$$

Now, we can determine the turns ratio of this transformer.

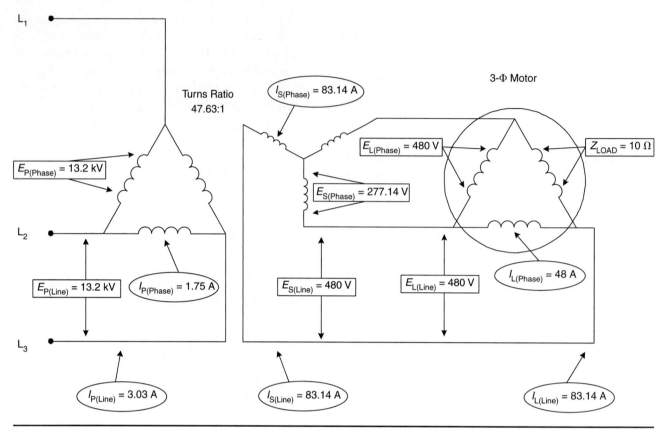

**FIGURE 12–59** The same circuit as in Figure 12–58 with all values added.

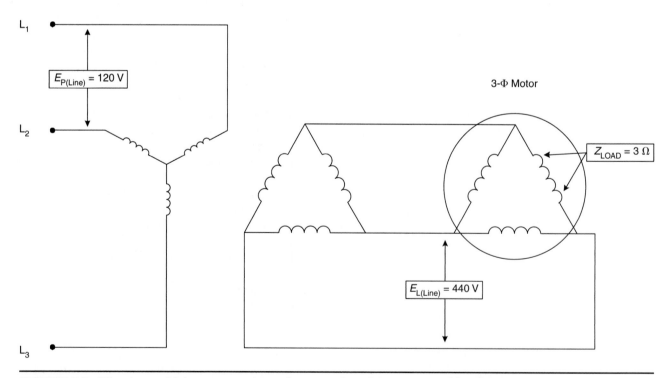

**FIGURE 12–60** A wye/delta-connected transformer with a delta-connected load.

What do we know?

$$E_{P(Phase)} = 69.28 \text{ VAC}$$
$$E_{S(Phase)} = 440 \text{ VAC}$$

What do we not know?

$$\text{turns ratio} = ?$$

What formula can we use?

$$\text{turns ratio} = \frac{E_{P(Phase)}}{E_{S(Phase)}}$$

Substitute the known values and solve for the turns ratio:

$$\text{turns ratio} = \frac{E_{P(Phase)}}{E_{S(Phase)}}$$
$$\text{turns ratio} = \frac{69.28 \text{ VAC}}{440 \text{ VAC}}$$
$$\text{turns ratio} = \frac{1}{6.34} \text{ or } 1{:}6.34$$

Therefore, the turns ratio of this transformer is 1:6.34. Notice that this is a step-up transformer. The primary-line voltage of 120 VAC is stepped-up to a secondary-line voltage of 440 VAC. We will now turn our attention to the load, which is connected in a delta configuration. As a result of the delta configuration, the load-line voltage, $E_{L(Line)}$, will be equal to the load-phase voltage, $E_{I(Phase)}$. Recall that the load-line voltage, $E_{L(Line)}$, is equal to 440 VAC. We can now determine the load-phase voltage, $E_{L(Phase)}$.

$$E_{L(Line)} = 440 \text{ VAC, and in a delta configuration}$$
$$E_{L(Phase)} = E_{L(Line)}; \text{ therefore,}$$
$$E_{L(Phase)} = 440 \text{ VAC}$$

Therefore, the load-phase voltage is 440 VAC. Now that we know that each phase of the load consists of a 3-Ω impedance, and each load-phase voltage is equal to 440 VAC, we can determine the amount of load-phase current, $I_{L(Phase)}$, by using Ohm's law.

What do we know?

$$E_{L(Phase)} = 440 \text{ VAC}$$
$$Z_{Load} = 3 \text{ Ω}$$

What do we not know?

$$I_{L(Phase)} = ?$$

What formula could we use?

$$I_{L(Phase)} = \frac{E_{L(Phase)}}{Z_{Load}}$$

Now substitute the known values and solve for $I_{L(Phase)}$:

$$I_{L(Phase)} = \frac{E_{L(Phase)}}{Z_{Load}}$$
$$I_{L(Phase)} = \frac{440 \text{ VAC}}{3 \text{ Ω}}$$
$$I_{L(Phase)} = 146.67 \text{ A}$$

The load-phase current, $I_{L(Phase)}$, is equal to 146.67 amperes. Now, we can determine the load-line current, $I_{L(Line)}$. Recall that the load is connected in a delta configuration, where the line current will be 1.732 times greater than the phase current. Therefore:

$$I_{L(Phase)} = 146.67 \text{ A, and in a delta configuration}$$
$$I_{L(Line)} = I_{L(Phase)} \times 1.732; \text{ therefore,}$$
$$I_{L(Line)} = 146.67 \text{ A} \times 1.732$$
$$I_{L(Line)} = 254.03 \text{ A}$$

So, in this example, the load-line current, $I_{L(Line)}$, is equal to 254.03 amperes. The load-line current must be equal to the secondary-line current. After all, the secondary is the source of the line current that is supplied to the load. Therefore:

$$I_{L(Line)} = 254.03 \text{ A; therefore,}$$
$$I_{S(Line)} = I_{L(Line)}; \text{ therefore,}$$
$$I_{S(Line)} = 254.03 \text{ A}$$

The secondary-line current, $I_{S(Line)}$, is equal to 254.03 amperes. Now, we can determine the secondary-phase current, $I_{S(Phase)}$, which is connected in a delta configuration. In a delta configuration, the line current will be 1.732 times greater than the phase current.

What do we know?

$$I_{S(Line)} = 254.03 \text{ A}$$
$$k = 1.732$$

What do we not know?

$$I_{S(Phase)} = ?$$

What formula could we use?

$$I_{S(Phase)} = \frac{I_{S(Line)}}{k}$$
$$I_{S(Phase)} = \frac{254.03 \text{ A}}{1.732}$$
$$I_{S(Phase)} = 146.67 \text{ A}$$

Therefore, the secondary-phase current, $I_{S(Phase)}$, will equal 146.67 amperes. Now that we know the

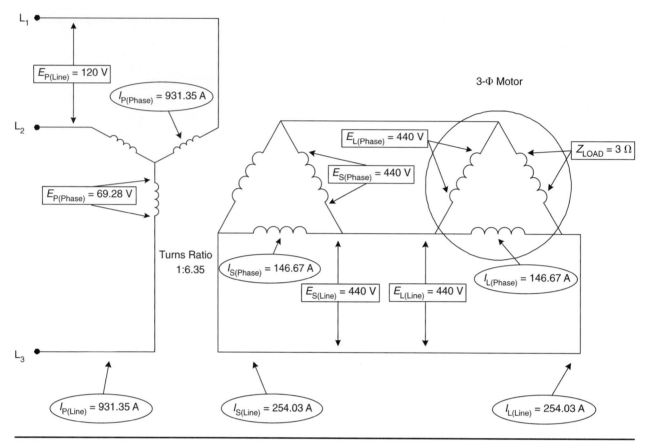

**FIGURE 12–61**  The same circuit as in Figure 12–60 with all values added.

secondary-phase current, we can determine the primary-phase current, $I_{P(Phase)}$.

What do we know?

$$I_{S(Phase)} = 146.67 \text{ A}$$
$$\text{turns ratio} = 1{:}6.35$$

What do we not know?

$$I_{P(Phase)} = ?$$

What formula could we use? (Remember, this is a step-up transformer!)

$$I_{P(Phase)} = I_{S(Phase)} \times \text{turns ratio}$$

Substitute the known values and solve for $I_{P(Phase)}$:

$$I_{P(Phase)} = I_{S(Phase)} \times \text{turns ratio}$$
$$I_{P(Phase)} = 146.67 \text{ A} \times 6.35$$
$$I_{P(Phase)} = 931.35 \text{ A}$$

The primary-phase current of our transformer will be 931.35 amperes. Finally, we can determine the primary-line current, $I_{P(Line)}$. Recall that the primary of our transformer is connected in a wye configuration. In a wye configuration, the line current will be equal to the phase current. Therefore:

$I_{P(Phase)} = 931.35$ A, and in a wye-connected primary
$I_{P(Line)} = I_{P(Phase)}$; therefore,
$I_{P(Line)} = 931.35$ A

Therefore, the primary-line current, $I_{P(Line)}$, for our example will be equal to 931.35 amperes. Figure 12–61 shows the same circuit with all of the values entered.

## REVIEW QUESTIONS

*Multiple Choice*

1. Alternating current can be transmitted over great distances much more economically than direct current because of a device called a
   a. transformer.
   b. transducer.
   c. motor–generator set.
   d. transmitter.

2. The operation of a transformer depends upon
   a. electrodynamics.
   b. electromagnetic induction.
   c. electromagnetic attraction.
   d. electromechanical induction.

3. The primary winding of a transformer is the coil that
   a. transfers the energy.
   b. receives the energy from the supply.
   c. changes magnetic energy into electrical energy.
   d. changes electrical energy into mechanical energy.

4. A transformer that is used to lower voltage is called a
   a. step-down transformer.
   b. step-up transformer.
   c. voltage-reducing transformer.
   d. voltage-lowering transformer.

5. The number of volts per turn on the primary winding of a transformer is
   a. equal to the number of volts per turn on the secondary winding.
   b. less than the number of volts per turn on the secondary winding.
   c. greater than the number of volts per turn on the secondary winding.
   d. equal to or greater than the number of volts per turn on the secondary winding.

6. When the secondary winding of a transformer is disconnected from the load,
   a. zero current flows in the primary.
   b. very little current flows in the primary.
   c. the primary current remains the same as under load.
   d. the primary current becomes very high.

7. The phase relationship between the primary voltage and the secondary voltage of a transformer is determined by the
   a. amount of load.
   b. demand factor of the load.
   c. power factor of the load.
   d. none of the above.

8. The iron/steel core of a transformer is
   a. solid.
   b. laminated.
   c. perforated.
   d. hollow.

9. Air core transformers are used when the frequency is
   a. high.
   b. low.
   c. medium.
   d. steady.

10. Air core transformers are generally used
    a. for distribution transformers.
    b. for signal transformers.
    c. in electronic equipment.
    d. in lighting transformers.

11. Transformer hum is a result of
    a. laminations and windings vibrating.
    b. the magnetic field alternating.
    c. high current.
    d. both a and b.

12. Heat generated by a transformer is a result of
    a. eddy currents and hysteresis.
    b. high voltages.
    c. high resistance.
    d. low frequency.

13. Air-cooled transformers are generally rated at
    a. 5 kilovolt-amperes or less.
    b. 5 kilovolt-amperes or more.
    c. 15 kilovolt-amperes or less.
    d. 15 kilovolt-amperes or more.

14. The purpose of oil in transformers is to
    a. lubricate.
    b. insulate and cool.
    c. lubricate and cool.
    d. insulate and lubricate.

15. A tapped transformer is one that has
    a. a tap for draining the oil.
    b. a T tap on the primary.
    c. several terminal posts to allow for voltage adjustments.
    d. taps for connecting the primary winding to the secondary winding.

16. Autotransformers are used
    a. to make small changes in voltage.
    b. in automobiles.
    c. to provide automatic voltage changes.
    d. to make large voltage changes.

17. Potential transformers are used
    a. with fluorescent lighting.
    b. with instruments and relays.
    c. to make small changes in voltage.
    d. to transfer large amounts of power.

18. Current transformers are used to
    a. increase the current to the load.
    b. decrease the current in special situations.
    c. maintain a constant current in the system.
    d. decrease large amounts of voltage.

19. A buck-boost transformer is a
    a. single-winding transformer.
    b. two-winding transformer.
    c. type of instrument transformer.
    d. heavy-duty transformer.

20. A control transformer is used to
    a. control the voltage to fluorescent lights.
    b. reduce the voltage for motor circuits.
    c. control the current to motors.
    d. reduce the voltage for motor control circuits.

*Give Complete Answers*

1. What is a transformer?

2. Describe the basic construction of a distribution transformer.

3. Explain the basic principle of the transformer operation.

4. What determines the amount of voltage produced in the secondary winding of a transformer?

5. Define the terms *primary winding* and *secondary winding*.

6. What determines the amount of current that will flow in the primary winding of a distribution transformer?

7. What determines the value of current that will flow in the secondary winding of a distribution transformer?

8. Define the term *current regulation* with regard to a distribution transformer.

9. How does the power factor of the load affect the operation of the transformer?

10. Define *turns ratio*.

11. What is the relationship of the volts per turn on the primary and secondary of standard transformers?

12. What causes power loss in a transformer?

13. Discuss hysteresis loss with regard to transformers.

14. Explain eddy currents and what causes them with regard to transformers.

15. How does core saturation affect the transformer operation?

16. List three types of core construction.

17. Describe a wound-core-type transformer.

18. List five advantages of the wound-core construction over the standard method of construction.

19. When is an air core transformer used?

20. What causes a transformer to hum?

21. Define the term *polarity* with regard to transformers.

22. Define the terms *additive* and *subtractive* with regard to transformers.

23. Transformers should have certain polarities according to their ratings. List the polarities and ratings.

24. List four methods used to cool transformers.

25. What is a tap-changing switch?

26. Where are tapped transformers generally used?

27. Describe an automatic tap-changing switch.

28. What is an autotransformer?

29. List two advantages of autotransformers.

30. What is the main disadvantage of an autotransformer?

31. For what purposes are autotransformers generally used?

32. What is the purpose of an instrument transformer?

33. Describe a potential transformer.

34. According to industrial standards, what is the maximum percentage of error for instrument transformers?

35. Describe a current transformer.

36. What determines the value of current flowing in the primary winding of a current transformer?

37. What determines the value of current flowing in the secondary winding of a current transformer?

38. List one important safety rule that must be followed when working with current transformers.

39. What is a buck-boost transformer?

40. What is a control transformer?

41. List four requirements necessary for connecting transformers into a transformer bank.

42. Describe the procedure to be followed when connecting two transformers in parallel.

43. Draw a diagram of two single-phase transformers connected to provide a two-phase, five-wire system.

44. Describe the procedure for connecting three single-phase transformers into a three-phase delta/delta bank.

45. Draw a diagram of three single-phase transformers connected into a delta/wye, three-phase, four-wire system.

46. Describe a three-phase, open-delta system.

47. Draw a diagram of two single-phase transformers used to convert two phase to three phase.

48. What is an isolation transformer?

49. Describe the T connection of two single-phase transformers used in a three-phase system.

*Solve each problem, showing the method used to arrive at the solution.*

1. A transformer is being designed to increase the voltage from 120 V to 480 V. If the primary winding requires 200 turns of wire, how many turns are required on the secondary?

2. If the load on the transformer in Problem 1 is 60 A, what is the primary current?

3. An engineer has been requested to design a core-type transformer. The core material has a flux density of 10,000 lines of force per square centimeter. The transformer is to be rated at 15 kVA, 480 V to 120 V, at 60 Hz. What will be the cross-sectional area of the core?

4. Calculate the number of turns required for the primary and secondary coils in Problem 3.

5. If the transformer in Problem 3 is supplying its rated load at 100% power factor, how much current is flowing in the primary and secondary coils?

6. A three-phase delta/delta transformer with a wye-connected load has a primary-line voltage of 277 VAC and a load-line voltage of 120 VAC. Find the following information about this circuit if the load impedance is 12 $\Omega$.

$$E_{P(Phase)} = ?$$
$$E_{S(Phase)} = ?$$
$$E_{S(Line)} = ?$$
$$E_{L(Phase)} = ?$$
$$I_{P(Phase)} = ?$$
$$I_{P(Line)} = ?$$
$$I_{S(Phase)} = ?$$
$$I_{S(Line)} = ?$$
$$I_{L(Phase)} = ?$$
$$I_{L(Line)} = ?$$
$$\text{turns ratio} = ?$$

7. A three-phase wye/wye transformer with a wye-connected load has a primary-line voltage of 460 VAC and a load-line voltage of 208 VAC. Find the following information about this circuit if the load impedance is 18 $\Omega$.

$$E_{P(Phase)} = ?$$
$$E_{S(Phase)} = ?$$
$$E_{S(Line)} = ?$$
$$E_{L(Phase)} = ?$$
$$I_{P(Phase)} = ?$$
$$I_{P(Line)} = ?$$
$$I_{S(Phase)} = ?$$
$$I_{S(Line)} = ?$$
$$I_{L(Phase)} = ?$$
$$I_{L(Line)} = ?$$
$$\text{turns ratio} = ?$$

8. A three-phase delta/wye transformer with a delta-connected load has a primary-line voltage of 560 VAC, and a load-line voltage of 277 VAC. Find the following information about this circuit if the load impedance is 7 $\Omega$.

$$E_{P(Phase)} = ?$$
$$E_{S(Phase)} = ?$$
$$E_{S(Line)} = ?$$
$$E_{L(Phase)} = ?$$
$$I_{P(Phase)} = ?$$
$$I_{P(Line)} = ?$$
$$I_{S(Phase)} = ?$$
$$I_{S(Line)} = ?$$
$$I_{L(Phase)} = ?$$
$$I_{L(Line)} = ?$$
$$\text{turns ratio} = ?$$

9. A three-phase wye/delta transformer with a delta-connected load has a primary-line voltage of 208 VAC, and a load-line voltage of 460 VAC. Find the following information about this circuit if the load impedance is 15 $\Omega$.

$$E_{P(Phase)} = ?$$
$$E_{S(Phase)} = ?$$
$$E_{S(Line)} = ?$$
$$E_{L(Phase)} = ?$$
$$I_{P(Phase)} = ?$$
$$I_{P(Line)} = ?$$
$$I_{S(Phase)} = ?$$
$$I_{S(Line)} = ?$$
$$I_{L(Phase)} = ?$$
$$I_{L(Line)} = ?$$
$$\text{turns ratio} = ?$$

# Electrical Distribution

**OBJECTIVES**

After studying this chapter, the student will be able to:

- Describe primary distribution systems.
- Describe residential, commercial, and industrial distribution systems.
- Explain the methods and purpose of grounding electrical systems.
- Explain the methods and purpose of grounding electrical equipment.
- Describe ground-fault protection.
- Describe the operating characteristics of unbalanced three-phase systems.
- Discuss the effect of harmonics on multi-wire systems.

# PRIMARY DISTRIBUTION SYSTEMS

The wiring between the generating station and the final distribution point is called the **primary distribution system**. There are several methods used for transmitting the power between these two points. The two most common methods are the **radial distribution system** and the loop distribution system.

## The Radial Distribution System

The term *radial* comes from the word *radiate*, which means to send out or emit from one central point. A *radial system* is an electrical transmission system that begins at a central station and supplies power to various substations.

In its simplest form, a radial system consists of a generating station that produces the electrical energy. This energy is transmitted from the generator(s) to the central station, which is generally part of, or adjacent to, the generating station. At the central station, the voltage is stepped up to a higher value for long-distance transmission.

From the central station, several lines carry the power to various substations. At the substations, the voltage is usually lowered to a value more suitable for distribution in populated areas. From the substations, lines carry the power to distribution transformers. These transformers lower the voltage to the value required by the consumer.

## The Loop Distribution System

The **loop distribution system** starts from the central station or a substation and makes a complete loop through the area to be served and back to the starting point. This results in the area being supplied from both ends, allowing sections to be isolated in case of a breakdown. An expanded version of the loop system consists of several central stations joined together to form a very large loop.

# CONSUMER DISTRIBUTION SYSTEMS

The type of distribution system the consumer uses to transmit power within the premises depends upon the requirements of the particular installation. Residential occupancies generally use the simplest type. Commercial and industrial systems vary widely with load requirements.

## Single-Phase Systems

Most single-phase systems are supplied from a three-phase primary. The primary of a single-phase transformer is connected to one phase of the three-phase system. The secondary contains two coils connected in series with a midpoint tap to provide a single-phase, three-wire system. This arrangement is generally used to supply power to residential occupancies and some commercial establishments. A schematic diagram is shown in Figure 13–1.

For residential occupancies, the service conductors are installed either overhead or underground. Single-family and small multifamily dwellings have kilowatt-hour meters that are installed on the outside of the building. From the kilowatt-hour meter, the conductors are connected to the main disconnect. Figures 13–2A and 13–2B show this arrangement.

From the main disconnect, the conductors supply power to the branch-circuit panels. For dwelling occupancies, there are three basic types of branch circuits: general lighting circuits, small appliance and laundry circuits, and individual branch circuits. The individual branch circuits are frequently used to supply central heating and/or air-conditioning systems, water heaters, and other special loads.

## Grounding Requirements

All AC services are required to be grounded on the supply side of the service-disconnecting means. This grounding conductor runs from the combination system and equipment ground to the grounding electrode.

For multifamily occupancies, it is permitted to use up to six service-disconnecting means. A single grounding conductor of adequate size should be used for the system ground (Figure 13–2B).

## Commercial and Industrial Installations

Commercial and industrial installations are more complex than small, residential installations. Large apartment complexes and condominiums, although classified as residential occupancies, often use commercial-style services. A single-phase, three-wire service or a three-phase, four-wire service may be brought into the building, generally from underground. The service-entrance conductors terminate in a main disconnect. From this point, the conductors are connected to the individual kilowatt-hour meters for each apartment and then to smaller disconnecting means and overcurrent protective devices. Branch-circuit panels are generally installed in each

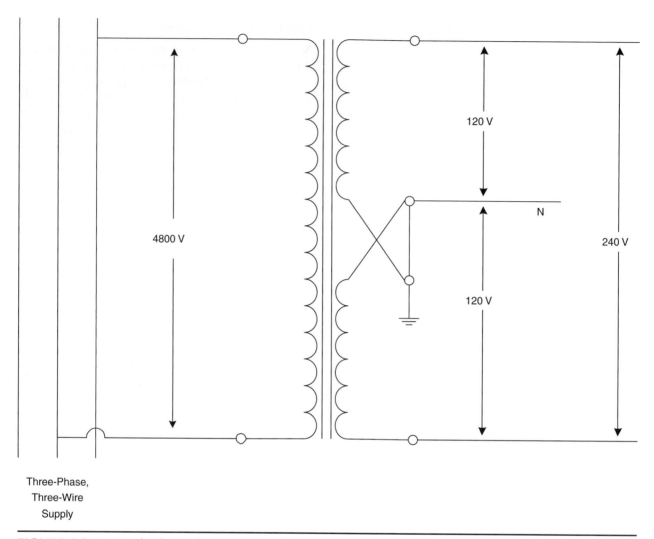

Three-Phase,
Three-Wire
Supply

**FIGURE 13–1** Single-phase, three-wire, 120/240-V system.

apartment. Feeder conductors connect the individual disconnecting means to the branch-circuit panels.

Commercial and/or industrial buildings may have more than one kilowatt-hour meter, depending upon the number of occupancies. The service sizes vary according to the demand. The service is usually a three-phase, four-wire system. The available voltages may be 120/208 volts or 277/480 volts. If the system provides 277/480 volts, a transformer must be installed in order to obtain 120 volts. If the building covers a large area, it is recommended that the service be installed near the center of the building. This arrangement minimizes line loss on feeder and branch-circuit conductors. Some utilities supply a three-phase, three-wire, and/or three-phase, four-wire delta system. The common voltages that may be obtained from the three-wire delta system are 240 volts, 440 volts, and 550 volts. With this arrangement, a transformer must be used to obtain 120 volts. The usual voltages supplied from the

four-wire delta system are 240 volts, three-phase, and 120 volts, single phase.

Many large consumers purchase electrical energy at the primary voltage, and transformers are installed on their premises. Three-phase voltages up to 15 kilovolts are often used.

The service for this type of installation generally consists of metal cubicles called a *substation unit*. The transformers are installed either within the cubicle or adjacent to it. Isolation switches of the drawer type are installed within the cubicle. These switches are used to isolate the main switch or circuit breaker from the supply during maintenance or repair.

## Consumer Loop Systems

Although the radial system of distribution is probably the most commonly used system of transmitting power on the consumer's property, the loop system is also employed. A block diagram of both systems

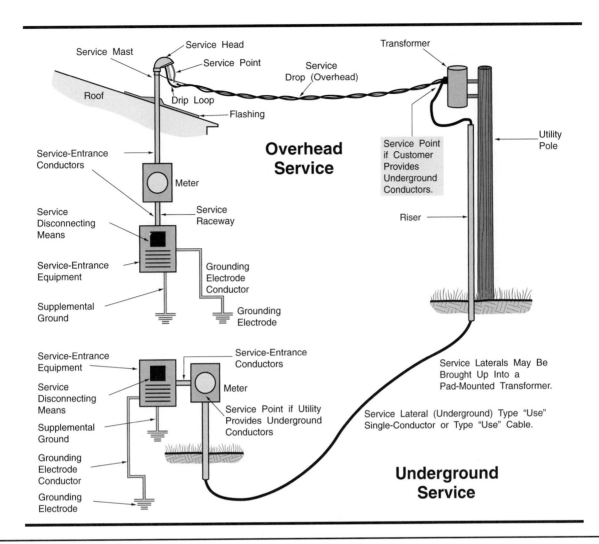

**FIGURE 13-2A** Examples of overhead and underground residential services.

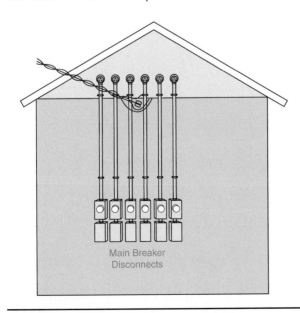

**FIGURE 13-2B** Three-wire, single-phase service for a multifamily dwelling. Six separate disconnecting means are used with one common ground.

is illustrated in Figures 13–3A and 13–3B. There are several variations of these systems in use in the industry, but the systems illustrated here show the basic structure.

In the installation of any system, overcurrent protection and grounding must be given primary consideration. Electrical personnel who design and install these systems must comply with the *NEC* and local requirements.

## Secondary High-Voltage Distribution

Large industrial establishments may find it more economical to distribute power at voltages higher than 600 volts. Depending upon the type of installation and the load requirements, voltages as high as 2300 volts may be used. Step-down transformers are installed in strategic locations to reduce the voltage to a practical working value. A diagram of a high-voltage radial system is shown in Figure 13–4A. Figure 13–4B illustrates a pad-mounted distribution transformer.

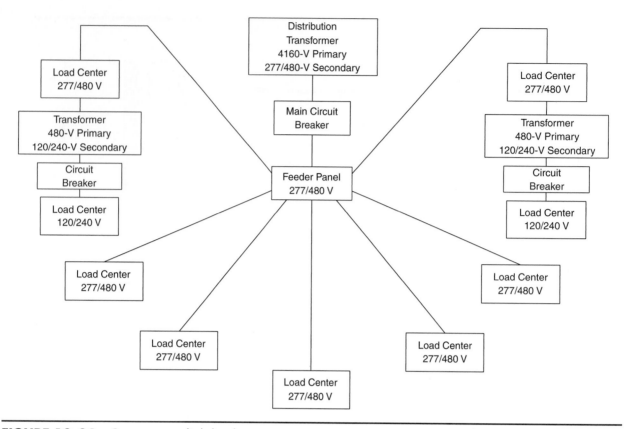

**FIGURE 13–3A** Consumer radial distribution system.

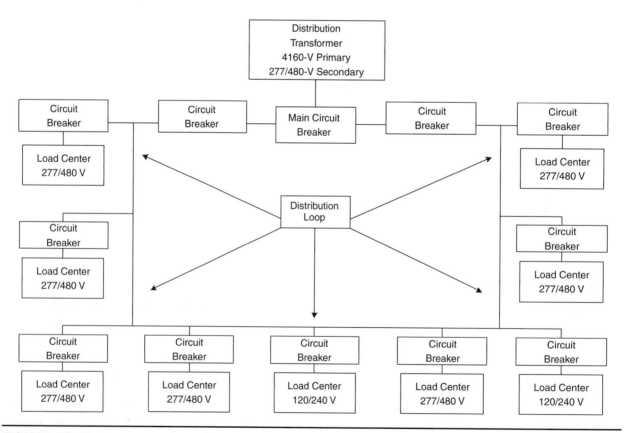

**FIGURE 13–3B** Consumer loop distribution system. Disconnecting means may be installed anywhere in the distribution loop to provide for isolating sections.

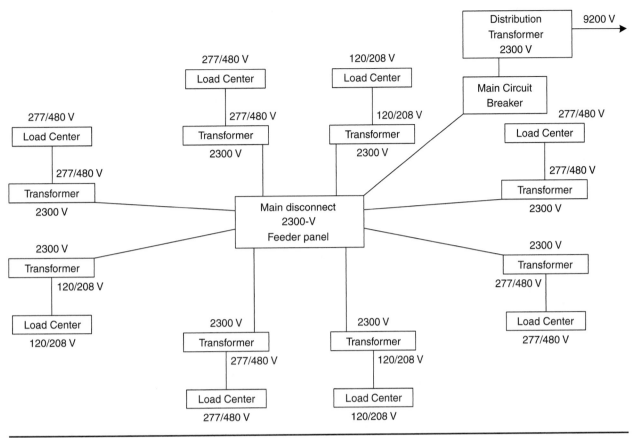

FIGURE 13-4A   Secondary high-voltage radial distribution system.

**FIGURE 13-4B**   Pad-mounted distribution transformer.

Sometimes the high-voltage (primary) system may be radial, and the low-voltage (secondary) system may be connected into a loop. Another method is to have both the primaries and secondaries connected to form a loop. Figures 13–5A and 13–5B show these methods.

## The Secondary Ties Loop System

It is frequently convenient to connect loads to the secondary conductors at points between transformers. These conductors are called **secondary ties**. Article 450 of the *NEC* gives specific requirements regarding the conductor sizes and overcurrent protection.

## GROUNDING OF ELECTRICAL SYSTEMS

In general, most electrical systems must be grounded. The purpose of grounding is to limit the magnitude of voltage surges caused by lightning, momentary surges, and accidental contact

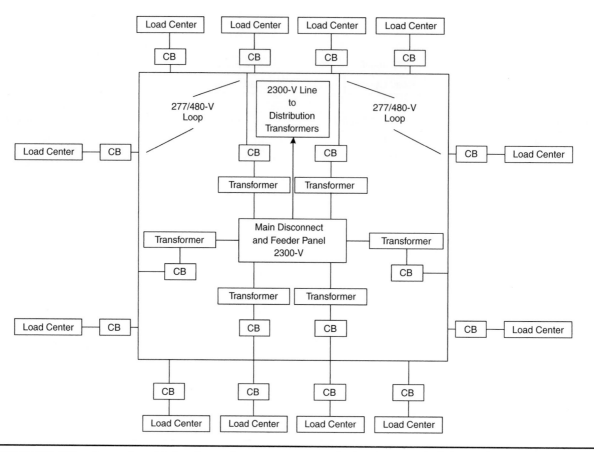

**FIGURE 13–5A** Secondary high-voltage distribution system: high-voltage radial, low-voltage loop.

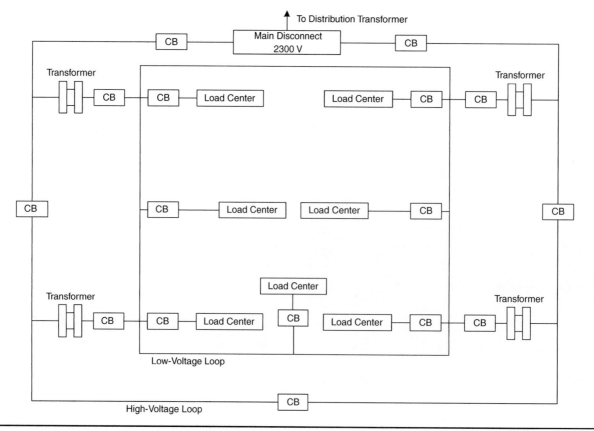

**FIGURE 13–5B** Consumer distribution system with high-voltage and low-voltage loops.

with higher voltages. System grounds must be arranged to provide a path of minimum impedance in order to ensure the operation of overcurrent devices when a ground fault occurs. Current should not flow through the grounding conductor during normal operation.

Direct current systems generally have the grounding conductor connected to the system at the supply station and not at the individual service. Alternating current systems, on the other hand, must be grounded on the supply side of the main disconnect at each individual service. For specific information on the location and method of grounding, refer to *NEC* Article 250.

# GROUNDING OF ELECTRICAL EQUIPMENT

Metal conduit and cases that enclose electrical conductors must be grounded. If the ungrounded (hot) conductor comes in contact with a metal enclosure that is not grounded, a voltage will be present between the enclosure and the ground. This presents a potential hazard. Persons coming in contact with the enclosure and ground will complete a circuit and be electrocuted.

All non–current-carrying metal parts of electrical installations should be tightly bonded together and connected to a grounding electrode. Good electrical continuity should be ensured through all metal enclosures. The current caused by accidental grounds will be conducted through the enclosures, the grounding conductor, and the grounding electrode to the earth. If the current is large enough, it will cause the overcurrent device to open.

# GROUND-FAULT PROTECTION

A **ground-fault protector** (GFP) is a device that senses ground faults and opens the circuit when the current to ground reaches a predetermined value. A *ground-fault circuit interrupter* (GFCI) is a device that opens the circuit when very small currents flow to ground.

There is no way to determine in advance the impedance of an accidental ground. Most circuits are protected by 15-ampere or larger overcurrent devices. If the impedance of a ground fault is low enough, such devices will open the circuit. What about currents less than 15 amperes? It has been proved that currents as small as 50 milliamperes through the heart, lungs, or brain can be fatal.

Electrical equipment exposed to moisture or vibration may develop high-impedance grounds. Arcing between a conductor and the frame of equipment may cause a fire, yet the current may be less than 1 ampere. Leakage current caused by dirt and/or moisture may take place between the conductor and the frame. Portable tools are frequently not properly grounded, and the only path to ground is through the body of the operator.

The GFCI was developed to provide protection against ground fault currents of less than 15 amperes. The GFCI is designed to operate on two-wire circuits in which one of the two wires is grounded. The standard circuit voltages are 120 volts and 277 volts. The time it takes to operate depends upon the value of the ground-fault current. Small currents of 10 milliamperes or less may flow for up to 5 seconds before the circuit is opened. A current of 20 milliamperes will cause the GFCI to operate in less than 0.04 seconds. This time/current element provides a sufficient margin of safety without nuisance tripping.

The GFCI operates on the principle that an equal amount of current is flowing through the two wires. When a ground fault occurs, some of the current flowing through the ungrounded (hot) wire does not flow through the grounded wire; it completes the circuit through the accidental ground. The GFCI senses the difference in the value of current between the two wires and opens the circuit. GFCIs may be incorporated into circuit breakers, installed in the line, or incorporated into a receptacle outlet or equipment.

GFPs are generally designed for use with commercial and/or industrial installations. They provide protection against ground fault currents from 2 amperes (special types go as low as 50 milliamperes) up to 2000 amperes. GFPs are generally installed on the main, submain, and/or feeder conductors. GFCIs are installed in the branch circuits. GFPs are generally used for three-wire, single-phase, and for three-phase installations; GFCIs are used for two-wire, single-phase circuits.

A GFP installed on supply conductors must enclose all the circuit conductors, including the neutral, if present. When operating under normal conditions, all the current to and from the load flows through the circuit conductors. The algebraic sum of the **flux** produced by these currents is zero. When a phase-to-ground fault occurs, the fault current returns through the grounding conductor. Under this condition, an alternating flux is produced within the sensing device. When the fault current reaches a predetermined value, the magnetic flux causes a relay to actuate a circuit breaker.

Sometimes the GFP is installed on the grounding conductor of the system. Under this condition, the unit senses the amount of phase-to-ground current flowing in the grounding conductor. When the current exceeds the setting of the GFP, it will cause the circuit breaker to open.

The GFP is actually a specially designed current transformer connected to a solid-state relay.

## THREE-PHASE SYSTEMS

The various three-phase systems in normal use were described in Chapter 12. Under ideal conditions, these systems operate in perfect balance, and if a neutral conductor is present, it carries zero current. In actual practice, perfectly balanced systems are seldom seen. The electrical worker, therefore, must be able to calculate values of current and voltage in unbalanced systems. Single-phase loads are frequently supplied from three-phase systems. The single-phase load requirements vary considerably, making it virtually impossible to maintain a perfect balance.

In a balanced three-phase system, the currents in the three lines are equal. The currents in the three phases are also equal. In other words, $I_{LX} = I_{LY} = I_{LZ}$ and $I_{pX} = I_{pY} = I_{pZ}$. If, however, $I_{LX} \neq I_{LY} \neq I_{LZ}$, then $I_{pX} \neq I_{pY} \neq I_{pZ}$, and the system is unbalanced. See Figure 13–6.

To calculate the line currents in an unbalanced three-phase system, the method in the following example may be used.

### Example 1

Three pure resistance, single-phase loads are connected in a delta configuration across a three-phase supply, as illustrated in Figure 13–6. Load X requires 30 A, load Y requires 50 A, and load Z requires 80 A. Calculate the current through each line wire.

1. Line $X = \sqrt{X^2 + Y^2 + (2XY \cos \phi)}$

   *(From Equation 7.5)*

   $X = \sqrt{30^2 + 50^2 + (2 \times 30 \times 50 \times 0.5)}$

   $X = \sqrt{900 + 2500 + 1500}$

   $X = \sqrt{4900}$

   $X = 70 \text{ A}$

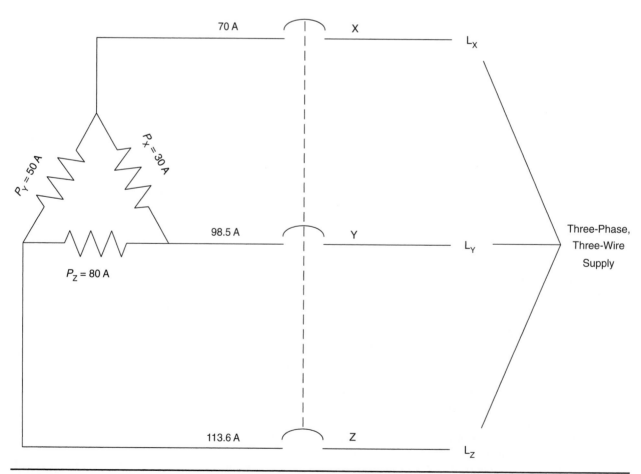

**FIGURE 13–6** Three-phase, unbalanced delta connection.

2. Line $Y = \sqrt{X^2 + Z^2 + (2XZ \cos \phi)}$
   $Y = \sqrt{30^2 + 80^2 + (2 \times 30 \times 80 \times 0.5)}$
   $Y = \sqrt{900 + 6400 + 2400}$
   $Y = \sqrt{9700}$
   $Y = 98.5$ A

3. Line $Z = \sqrt{Z^2 + Y^2 + (2XZ \cos \phi)}$
   $Z = \sqrt{80^2 + 50^2 + (2 \times 80 \times 50 \times 0.5)}$
   $Z = \sqrt{6400 + 2500 + 4000}$
   $Z = \sqrt{12,900}$
   $Z = 113.6$ A

Example 1 applies to loads of 100% power factor connected in delta. With loads of different power factors, the phase angle will vary from 120 degrees. For a wye connection, the line current is equal to the phase current.

Some connections may be a combination of single-phase and three-phase loads. Under these conditions, the phase angle between the three-phase load and the single-phase load must be considered.

# HARMONICS

Most distribution systems in the United States and Canada operate on a frequency of 60 hertz. Certain types of electronic equipment, such as computer systems and some **fluorescent** lighting systems, produce secondary frequencies, which are multiples of the supply frequency. These secondary frequencies are called **harmonics**. For example, the second harmonic of 60 hertz is 120 hertz, the third harmonic is 180 hertz, and so on.

The alternating flux developed by some transformers that are used in fluorescent lighting ballasts produces a voltage with a frequency of 180 hertz. This problem, however, has been reduced to a minimum in high-quality ballasts.

Equipment such as computers and programmable controls also produce voltages with frequencies that are harmonics of the supply frequency. These harmonics cause additional current to flow in the supply conductors. The additional current in the phase conductors is usually only a small percentage of the supply current. This harmonic current adds to the supply current, causing a greater heating effect in the conductors. The increase in heating effect is usually rather small, possibly in the vicinity of 3 percent to 5 percent.

The effect on the neutral conductor is quite different. The harmonic currents from the phase conductors add together, causing a large increase in the neutral current. The heating effect may be as much as 90 percent greater than if there were no harmonic effect.

---

> ⚠️ **CAUTION** In the installation of supply, feeder, and branch-circuit conductors for heavy fluorescent loads and computer loads, the size of the neutral conductor may have to be increased to allow for harmonic currents. See Sections 220.61(C) FPN No. 2 and 310.15(B)(4)(c) of the *NEC*.

# REVIEW QUESTIONS

*Multiple Choice*

1. A primary distribution system consists of the wiring between the
   a. generator and the substation.
   b. transformer and the final branch circuit.
   c. generating station and the final distribution point of the utility company.
   d. generating station and the transformer.

2. A radial distribution system is an arrangement of transmission lines that
   a. extend in many directions from the central station to various substations; at the substation, the lines divide again to carry the power to the distribution transformers.
   b. leave the central station, supplying many substations in succession, and return to the central station.
   c. distribute electrical power from one central station to another.
   d. extend in only one direction from the central station.

3. A loop-type distribution system is an arrangement of transmission lines that
   a. make a complete loop through each substation.
   b. start from the central station or a substation, make a complete loop through the area to be served, and return to the starting point.
   c. start from the central station, feed a substation, and loop back to the central station.
   d. make a complete loop to each transformer in the central system.

4. Most single-phase consumer distribution systems are supplied
   a. from a two-phase primary.
   b. directly from the generating station.
   c. from a three-phase primary.
   d. from a single-phase, three-wire primary.

5. Most single-family and small multifamily dwellings have the kilowatt-hour meters installed
   a. in the utility room.
   b. on the outside of the dwelling.
   c. in the basement.
   d. in the attic.

6. Central air-conditioning systems are generally connected to
   a. lighting circuits.
   b. small appliance circuits.
   c. individual branch circuits.
   d. any of the above.

7. All AC services are required to be grounded
   a. on the load side of the service disconnect.
   b. on the supply side of the service disconnect.
   c. within the housing for the kilowatt-hour meter.
   d. anywhere on the system.

8. Multifamily occupancies are permitted to use
   a. up to six service-disconnecting means.
   b. up to ten service-disconnecting means.
   c. only one main disconnecting means.
   d. not more than three service-disconnecting means.

9. Large apartment complexes and condominiums are classified as
   a. combination occupancies.
   b. residential occupancies.
   c. commercial occupancies.
   d. industrial occupancies.

10. The supply conductors for condominiums are generally installed
    a. overhead.
    b. in wireways.
    c. in busways.
    d. underground.

11. The most common type of service for commercial and industrial installations is the
    a. three-phase, four-wire wye system.
    b. three-phase, four-wire delta system.
    c. three-phase, three-wire, open-delta system.
    d. three-phase, three-wire delta system.

12. The service for a large industrial plant is made up of metal cubicles. When the primary voltage is brought to transformers located in or near these cubicles, the arrangement is called
    a. a distribution center.
    b. a substation.

c. a load center.
d. either a or b.

13. Consumer distribution systems are arranged to form
    a. a radial system.
    b. a loop system.
    c. a commercial system.
    d. either a or b.

14. Secondary high-voltage distribution systems are consumer distribution systems that use voltages as great as
    a. 300 V.
    b. 600 V.
    c. 1000 V.
    d. 2300 V.

15. Secondary ties are the conductors
    a. of a loop system that connect several transformer secondaries together.
    b. of a radial system that connect the transformer secondaries to the loads.
    c. between the kilowatt-hour meter and the main disconnect.
    d. that connect the primary of several transformers together.

16. Systems are grounded
    a. because the ground attracts electricity.
    b. to limit the magnitude of voltage caused by lightning.
    c. to eliminate inductance.
    d. to reduce the resistance of the supply conductors.

17. Alternating current systems are grounded
    a. at each individual service.
    b. at the supply station.
    c. only at the substation.
    d. only at the alternator.

18. Electrical conduits and other non–current-carrying metal equipment are grounded to
    a. eliminate possible hazards that may be caused by accidental grounds.
    b. conduct eddy currents to ground.
    c. reduce inductance.
    d. eliminate harmonics.

19. A ground-fault protector is a device that
    a. senses ground faults and opens the circuit when the current to ground reaches a predetermined value.
    b. prevents ground faults from occurring.

c. senses the location of the ground fault and eliminates it.

d. opens the grounding conductor when a fault occurs.

20. A ground-fault circuit interrupter is sometimes called a
    a. GFCI.
    b. GCI.
    c. GFC.
    d. ICFG.

21. When a ground-fault protector is being installed on the supply conductors, it must
    a. be installed on each conductor separately.
    b. enclose all the circuit conductors.
    c. enclose only the neutral conductor.
    d. be installed in series with the neutral conductor.

22. The ground-fault protector is a type of
    a. buck-boost transformer.
    b. potential transformer.
    c. current transformer.
    d. control transformer.

23. Harmonics produced by electronic equipment
    a. decrease the current in the neutral conductor of four-wire, three-phase systems.
    b. decrease the current in the supply conductors of four-wire, three-phase systems.
    c. cause current in the grounding conductor of four-wire, three-phase systems.
    d. increase the current in the neutral conductor of four-wire, three-phase systems.

*Give Complete Answers*

1. What is the difference between a primary distribution system and a secondary distribution system?

2. Where does one obtain information regarding the type of electrical service and systems available?

3. What is the function of an electric utility company?

4. Describe a radial transmission system.

5. Describe a loop transmission system.

6. What advantage does the loop system have over the radial system?

7. What type of service is usually supplied to a single-family residence?

8. What two points on an AC wiring system are joined by the grounding conductor?

9. What type of service is usually supplied to an industrial or commercial installation?

10. Why is it recommended to install the main service near the center of a large building?

11. Why would a consumer wish to distribute electrical power at a voltage greater than 600 volts?

12. What is a secondary tie?

13. Why are electrical systems grounded?

14. Why are metal parts of electrical equipment grounded?

15. What is a ground-fault protector?

16. What is a ground-fault circuit interrupter?

17. List two common causes of accidental grounds.

18. What type of hazards can occur because of accidental grounds?

19. Describe the operation of a GFCI.

20. Explain the principle operation of a GFP.

21. What causes unbalances in a three-phase system?

22. What are harmonics?

23. Explain the effect of harmonics in AC systems.

24. What type of equipment causes harmonic problems?

25. What does Section 220.61(C) FPN No. 2 and Section 310.15(B)(4)(c) of the *NEC* state regarding harmonics?

# Lighting

## OBJECTIVES

After studying this chapter, the student will be able to:

■ Describe the different units of measurement for light.

■ Determine the amount of light required for various areas and types of work.

■ Lay out and select correct lighting fixtures for various areas.

# LIGHTING MEASUREMENTS

When electric lighting was first developed, it was necessary to express in a specific unit the amount of light produced. Because the candle was in common use prior to electric lighting, it was selected as a standard reference. Candles differ in the amount of light they emit, and, therefore, a candle made to specific standards was used.

The unit for light intensity is based on the amount of light emitted from a standard candle in a horizontal direction. This unit is called the **candlepower**. The candlepower of a lamp does not indicate the total light emitted but does indicate the intensity in a given direction. The *mean spherical candlepower* is the average of the candlepowers in all directions from the lamp.

The unit for the total light emitted from a lighting source is the **lumen**. This unit can be better understood by using the spherical candlepower as a reference. A light source producing 1 candlepower in all directions is placed at the center of a sphere that has a radius of 1 foot. To prevent light from reflecting inside the sphere, the sphere is blackened on the inside. An opening of 1 square foot is cut into the sphere. The amount of light passing through the opening is equal to 1 lumen (1 lm.) (Figure 14–1).

*One lumen is the amount of light passing through an opening 1 square foot in area and located 1 foot from a light source that emits 1 candlepower in all directions.*

The area of a sphere that has a radius of 1 foot is 12.57 (rounded off from 12.5664) square feet. A light source that emits 1 candlepower in all directions, placed in the center of the sphere, emits 1 lumen for each square foot of sphere surface or a total of 12.57 lumens. This relationship between candlepower and lumens can be expressed mathematically as follows:

$$L = 12.57 C_m \qquad \text{(Eq. 14.1)}$$

where   $L$ = lumens
$C_m$ = means spherical candlepower

Candlepower and lumens are units of measurement of light emitted from a source. In the design of lighting systems, the amount of **illumination** on a given surface is of great importance. This unit of measurement is the **foot-candle**. One foot-candle is the intensity of illumination on a plane 1 foot away from a lighting source of 1 candlepower and at a right angle to the light rays from the source. Figure 14–2 illustrates a light source of 1 candlepower illuminating every point on surface A to an intensity of 1 foot-candle. In this case, 1 lumen of light is striking surface A. *One lumen of light covering an area of 1-square foot provides illumination at an intensity of foot-candle.*

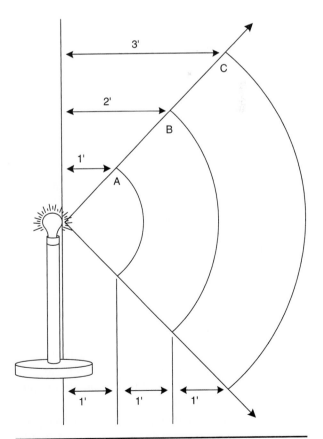

**FIGURE 14–2** Light source of 1 candlepower.

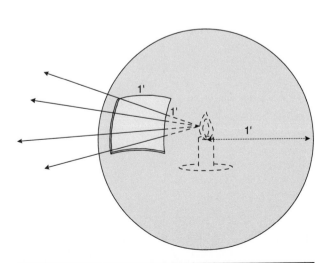

**FIGURE 14–1** Sphere containing a light source of 1 candlepower.

# INVERSE SQUARE LAW

In Figure 14–2, if plane A is removed, the light source now illuminates plane B, which is 2 feet away from the source. The same amount of light now covers four times the area, which is 4 square feet. The illumination per square foot on B is now 1/4 the amount of A, or 1/4 foot-candle. Removing B permits plane C to be illuminated. Plane C is 3 feet away from the light source, and thus the same amount of light will spread over an area nine times as large as A and illuminate plane C to 1/9 foot-candle. This illustration shows that *the intensity of illumination from a light source varies inversely with the square of the distance from the source.* Known as the *inverse square law,* it may be expressed mathematically as:

$$f_c = \frac{C}{d^2} \qquad \text{(Eq.14.2)}$$

where  $f_c$ = foot-candles
  $C$ = candlepower of the source
  $d$ = the distance in feet from the light source to a point on the plane that the light strikes

## Example 1

A ceiling fixture mounted 8 ft above a workbench emits light at an intensity of 100 candlepower downward. Calculate the illumination produced on the workbench at a point directly below the fixture:

$$f_c = \frac{C}{d^2}$$

$$f_c = \frac{100}{8^2}$$

$$f_c = 1.56 \text{ foot-candles}$$

Equation 14.2 applies only when the light strikes the surface at a right angle, as in Figure 14–3.

For many installations, the light source does not strike the surface at a right angle. For any other angle, Equation 14.2 must be modified as in Equation 14.3:

$$f_c = \frac{CH}{d^3} \qquad \text{(Eq. 14.3)}$$

where  $f_c$ = foot-candles
  $C$ = candlepower of the source
  $H$ = the distance in feet from the surface being illuminated to the light source
  $d$ = the distance in feet from the point being considered to the light source

## Example 2

What is the intensity of light at a point on a work surface 12 ft from a point directly under a lighting fixture? The fixture is 9 ft above the surface and emits a uniform illumination of 200 candlepower (see Figure 14–3).

Distance BC is equal to the square root of the sum of the squares of AB and AC.

$$d = \sqrt{12^2 + 9^2}$$
$$d = \sqrt{225}$$
$$d = 15 \text{ ft}$$
$$f_c = \frac{CH}{d^3}$$
$$f_c = \frac{200 \times 9}{15^3}$$
$$f_c = 0.53 \text{ foot-candles}$$

When the light is from long tubes like fluorescent lamps placed end to end, the rays strike at right angles to the surface, producing illumination that varies inversely with the distance from the source.

$$f_c = \frac{C}{d} \qquad \text{(Eq. 14.4)}$$

When the light rays from fluorescent lamps placed end to end strike the surface at any angle other than 90 degrees, the equation becomes

$$f_c = \frac{CH}{d^2} \qquad \text{(Eq. 14.5)}$$

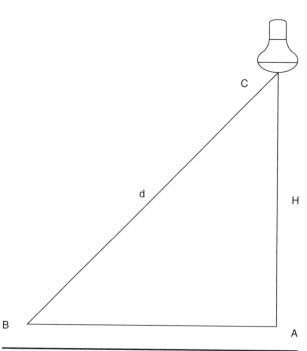

**FIGURE 14–3**  Lighting illuminating a work surface.

To determine the relationship between foot-candles and lumens, refer back to Figure 14–2. Surface A, 1 square foot in area, is illuminated to 1 foot-candle. After removing plane A, surface B, which is 4 sqr ft in area, is illuminated to 1/4 foot-candle. Removing A and B illuminates surface C, which is 9 sqr ft in area, to an intensity of 1/9 foot-candle. Each of these surfaces was illuminated by 1 lumen of light. The formula that expresses this relationship is:

$$L = f_c A \qquad \text{(Eq. 14.6)}$$

where   $L$ = total light in lumens
       $f_c$ = foot-candles of light intensity
       $A$ = the area of the surface in square feet

### Example 3

An area of 50 sqr. ft must be illuminated at an average intensity of 32 foot-candles. What size floodlight is required?

$$L = f_c A$$
$$L = 32 \times 50$$
$$L = 1600 \text{ lumens}$$

## LIGHT DISTRIBUTION

Light intensity in various directions around a lamp or fixture is shown by a graph. For the fixture in Figure 14–4, the variations are indicated by the intensity curve. The length of each radial line within the curve indicates the intensity of light in that direction.

| Angle | Candlepower |
|-------|-------------|
| 0° | 1450 |
| 15° | 1400 |
| 30° | 1300 |
| 45° | 1050 |
| 60° | 800 |
| 75° | 600 |
| 90° | 0 |

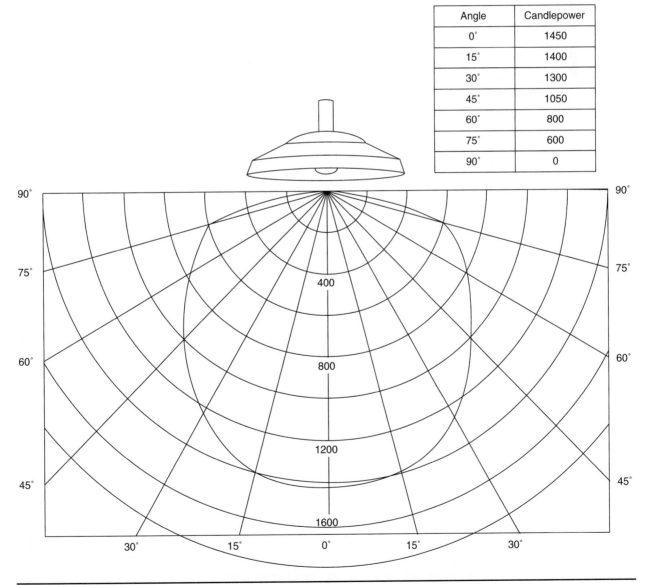

**FIGURE 14–4**   Intensity curve showing variation of light from a lighting source.

The direction of the rays is expressed by means of angles measured from a vertical line down from the source. In Figure 14–4, the intensity of light at 0 degrees is 1450 candlepower. The intensity of light at 15-degree increments is 1400, 1300, 1050, 800, and 600 candlepower.

For a given height, it is possible to determine the foot-candle intensity for various distances from the lighting source.

**Example 4**

In Figure 14–5, determine the foot-candle intensity of the surface at point A. The distribution curve indicates that the light intensity at point A is 1240 candlepower. The height to the light source is 8 ft. The horizontal distance from the source is 6 ft (see Figure 14–6).

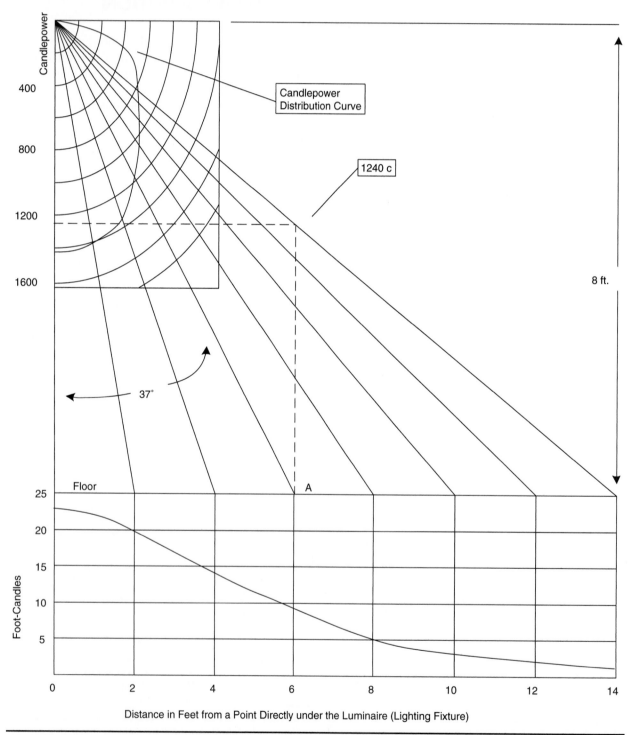

**FIGURE 14–5** Candlepower distribution curve.

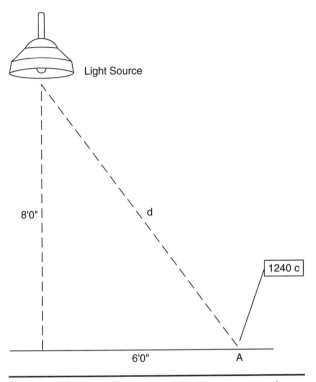

$$d = \sqrt{8^2 + 6^2}$$
$$d = \sqrt{64 + 36}$$
$$d = 10 \text{ feet}$$
$$f_c = \frac{1240 \times 8}{10^3}$$
$$f_c = \frac{9920}{10^3}$$
$$f_c = 9.92 \text{ foot-candles}$$

The light intensity at each of various points along the surface is found in the same manner.

When designing a lighting system, one must consider many lighting sources at specific distances. Figure 14–7 shows an example of more than one source illuminating a surface. The amount of illumination at point P is the sum of the light rays coming from all the units (Figure 14–7). The intensity from fixture B is 10 foot-candles computed by Equation 14.2. Equation 14.3 indicates that the intensity from A and C is 4.76 foot-candles each. The total amount of light at point P is 10 + 4.76 + 4.76 = 19.52 foot-candles.

**FIGURE 14–6** Light source illuminating a surface.

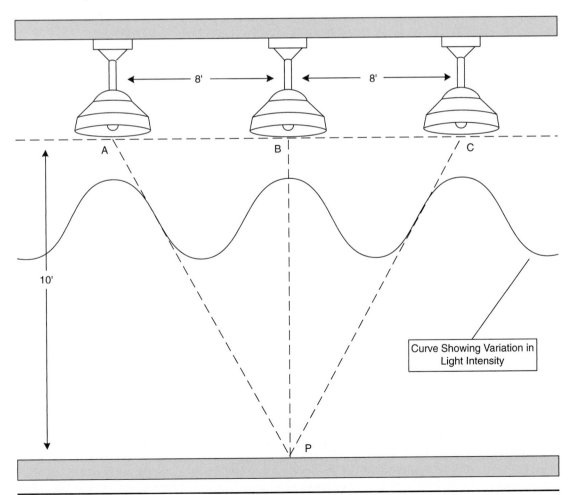

**FIGURE 14–7** More than one lighting source illuminating a surface.

## TYPES OF LUMINAIRES

The 2008 edition of the *National Electrical Code* defines a **luminaire** (lighting fixture) as

*A complete lighting unit consisting of a light source such as a lamp or lamps, together with the parts designed to position the light source and connect it to the power supply. It may also include parts to protect the light source or the **ballast** or to distribute the light. A lamp-holder itself is not a luminaire.\**

There are many different shapes and styles of luminaires available today. Figure 14–8 shows several different types and styles. The different styles of luminaires will use different types of lamps. Incandescent luminaires are designed to use incandescent lamps. Fluorescent lamps must be installed in fluorescent lamp luminaires. HID lamps must only

be installed in luminaires designed for HID lamps. Just as the different types of lamps produce different types and qualities of light, the different luminaires are designed to be mechanically mounted by different methods. Some luminaires use a pendant style of mounting, some are surface mounted, some recessed mounted, and some use other methods. The variations in luminaires also allow for different illumination patterns. In general, there are five different types of luminaires. They are classified as direct, semi-direct, general diffusing, semi-indirect, and indirect.

*Direct* lighting fixtures are usually equipped with reflectors or shades that direct at least 90 percent of the light downward. This type of fixture tends to produce **glare** and spotty lighting. Installing the fixtures as high as possible from the surface results in a more uniform distribution of light and less glare.

Fixtures that transmit 60 percent to 90 percent of the light downward are classified as *semi-direct*.

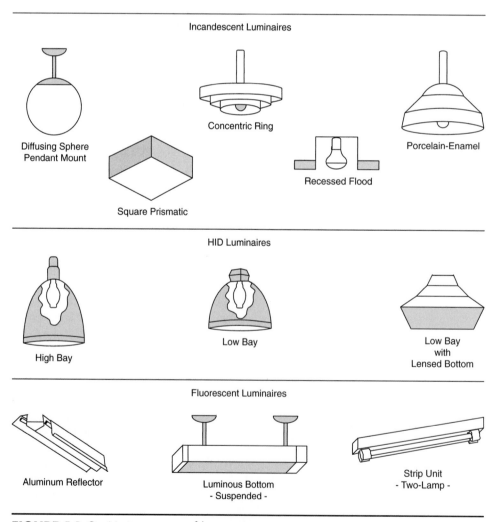

**FIGURE 14–8**  Various types of luminaires.

*Reprinted with permission from NFPA 70-2008, *National Electrical Code*® Copyright© 2007, National Fire Protection Association, Quincy, MA 02269. This reprinted material is not the referenced subject which is represented only by the standard in its entirety.

They usually consist of a lamp or lamps enclosed in a prismatic or opalescent glass. This arrangement conceals the light source, diffuses the light, and reduces glare.

Fixtures classified as *general diffusing* direct 40 percent to 60 percent of the light downward and the rest upward. They often consist of glass-diffusing globes or louvered units. Nearly all the light is transmitted up or down and practically none sideways. This arrangement is often used in office areas.

*Semi-indirect* lighting fixtures transmit 60 percent to 90 percent of the light upward. This arrangement produces very little glare and is more efficient than indirect types because more light is directed downward.

*Indirect* lighting fixtures transmit 90 percent or more of the light upward. The ceiling serves as a secondary light source by reflecting the light downward. The result is a diffused light of low intensity over a large area.

# REFLECTION

Materials from which lighting fixtures are made and the material, smoothness, and color of the surfaces within the area to be lighted all determine the lighting effect. Light passes through thin transparent material with very little scatter. Little or no light passes through opaque materials; it is either absorbed or reflected.

An opaque, highly polished, bright material reflects most of the light that strikes it. Even dark colors that are highly polished reflect much light. When a beam of light strikes a highly polished surface, it leaves at the same angle at which it struck the surface (Figure 14–9).

A person positioned so that the eyes are in the path of the reflected beam will be subject to glare, causing eye fatigue. Light striking a rough, opaque surface reflects in all directions (Figure 14–10).

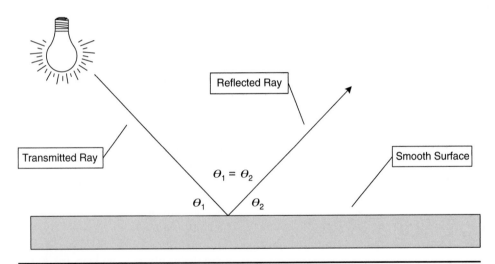

**FIGURE 14–9**   Light ray reflected from a smooth, polished surface.

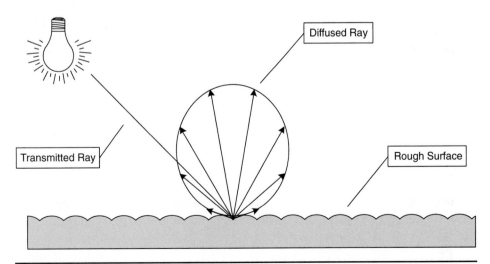

**FIGURE 14–10**   Light ray reflected from a rough, opaque surface.

The same intensity of **brightness** will be observed from the entire surface, thus reducing glare.

For uniformity of light and reduced glare, ceilings and walls should have a rough surface. The approximate percentages of light reflected by surfaces of different colors are given in Table 14–1.

## GLARE

When lighting is so bright that it causes eye discomfort, it is said to produce glare. There are many causes of glare, including a light source directly in line of vision, light reflected from highly polished surfaces, and high **contrast** between the light source and the background.

Brightness depends upon the intensity of the light from the source and the ability of a surface to reflect the light. Too much brightness causes discomfort, and too little causes eye strain.

A clear **incandescent lamp** emits a large amount of light from a small central point. It appears very bright and produces glare. Most incandescent lamps are frosted on the inside, thus scattering the rays and causing the light to appear distributed throughout the bulb.

Color also has an effect on brightness. If two sheets of paper, one white and one blue, are placed under a light, the white sheet appears brighter because it reflects more light.

## LIGHTING LAYOUT

In designing a lighting system, one must consider the purpose for which the area is to be used. Table 14–2 indicates recommended illumination for areas designed for various uses.

**TABLE 14–1**  Percentage of light reflected.

| Surface or Color | Percentage of Light Reflected |
|---|---|
| Polished silver | 92 |
| Mirrored glass | 85 |
| Polished chromium | 65 |
| Flat white | 80 |
| Bright yellow | 65 |
| Light green | 62 |
| Dark gray | 22 |
| Olive green | 14 |
| Black | 3 |

Foot-candle recommendations depend upon the fineness of details to be observed, the color, and the color contrast of the work. The ease with which a material can be seen depends upon the amount of light for the general area plus supplementary lighting in specific areas.

Other factors to consider are the coefficient of utilization, the maintenance factor, and color requirements. The lighting fixtures, ceilings, walls, floors, and other objects all absorb light. Thus the light striking the working plane is always less than the amount produced at the source.

The ratio between the light that reaches the working plane and the light produced by the luminaire provides a constant known as the **coefficient of utilization**. As a general rule, the coefficient of utilization for large rooms (20 feet by 20 feet or larger) may be considered to be 40 percent for indirect lighting, 50 percent for semi-indirect lighting, 60 percent for semi-direct lighting, and 70 percent for direct lighting.

Consideration must also be given to the decrease in illumination caused by depreciation of the lamp. In addition, as the reflecting surfaces get dirty, more light is absorbed and less reflected. This absorption of light by dirty reflecting surfaces must be considered. It is accomplished by applying the maintenance factor. The **maintenance factor** is the percentage of initial light to be expected with the usual cleaning and/or painting of the luminaires and reflecting surfaces.

Because some luminaires collect dust more readily than others, and some locations are dirtier than others, maintenance factors vary. In general, a maintenance factor of 70 percent can be used for normal conditions; for clean areas, a factor of 80 percent can be used; for dirty areas, 60 percent should be adequate. For more accurate values of both the coefficient of utilization and the maintenance factor, refer to the manufacturer's reference manual.

To determine the total lumens required for a certain area, one must consider both the coefficient of utilization and the maintenance factor. Equation 14.7 gives the total lumens without consideration to the coefficient of utilization or the maintenance factor. This equation can be revised as:

$$L = \frac{f_c A}{K_u M} \qquad \text{(Eq. 14.7)}$$

where   $L$ = total lumens
$f_c$ = foot-candles
$A$ = area in square feet
$K_u$ = coefficient of utilization
$M$ = maintenance factor

**TABLE 14–2**  Recommended illumination in foot-candles (fc).

| Area of Use | Purpose/Type of Work Area | Illumination | Area of Use | Purpose/Type of Work Area | Illumination |
|---|---|---|---|---|---|
| **Commercial** | | | **Industrial** | | |
| Drafting | Conventional | 150fc | Garages and service areas | Repair | 75fc |
|  | Computerized (with CRTs) | 75fc |  | Active traffic areas | 15fc |
| Libraries | Reading good print and typed originals | 30fc | Loading platform | | 20fc |
|  | Reading small print, handwriting, and photocopies | 75fc | Machine shops and assembly area | Rough bench/ machine work, and simple assembly | 50fc |
|  | Active stacks | 30fc |  | Medium bench and machine work with moderate difficulty | 100fc |
| Offices | Conference rooms | 30fc |  |  |  |
|  | Corridors, stairs, elevators | 15fc |  | Difficult machine work and assembly | 150fc |
|  | General | 100fc |  | Fine bench and machine work and assembly | 300fc |
|  | Lobbies and reception areas | 30fc |  |  |  |
|  | Private areas | 75fc | Receiving/ shipping | Active with large items | 15fc |
|  | Restrooms | 30fc | Warehouse/ Storage | Active with small items and labels | 30fc |
|  | Computer work stations | 75fc |  | Inactive | 5fc |
| School | Classrooms and laboratories | 75fc | **Outdoor Areas** | | |
|  | Shops | 100fc | Storage yards | Active | 20fc |
|  | Sight saving, hearing impaired, and exceptional children | 150fc |  | Inactive | 11fc |
| Store | Mass merchandising, high-activity areas | 100fc | Parking areas | Open with high activity | 2fc |
|  | Self-service/specialty merchandise areas medium activity | 75fc |  | Open with medium activity | 1fc |
|  | Clerk service, low activity areas | 30fc |  | Covered parking with pedestrian areas and entrances—daytime | 5fc |
|  | Feature displays, high activity | 5000fc |  |  |  |
|  | Feature displays, medium activity | 300fc |  | Covered parking with pedestrian areas and entrances—nighttime | 50fc |
|  | Feature displays, low activity | 150fc |  |  |  |

**Example 5**

A machine shop that is 1296 sqr ft in area has a fixture height of 15 ft and is used for medium work. Calculate the number of lumens needed for adequate lighting.

$$L = \frac{f_c A}{K_u M}$$

$$L = \frac{30 \times 1296}{0.70 \times 0.70}$$

$$L = \frac{38,880}{0.49}$$

$$L = 79,349 \text{ lumens}$$

In the selection of lighting fixtures, certain general factors must be considered. They are:

1. Type of work area to be lighted
2. Required number of lumens
3. Height of fixture from work surface

4. Spacing between fixtures and spacing between fixtures and wall

5. Size of lamps required

## COLOR

Color is a very important consideration in the design of lighting systems. Merchandise displayed in a salesroom may appear one color under artificial light and another color under natural light. In most areas, it is important to install a lighting system that will produce the same effect as natural daylight. Special installations may warrant colors that produce unnatural effects.

## TYPES OF LIGHTING FIXTURES

In general, there are two categories of lighting fixtures: **incandescent** and *electric discharge*. These two categories have many types.

Incandescent lamps are manufactured in many different styles and colors. These lamps produce light from a filament that is heated to incandescence (see Figure 14–11). The filament is made of a thin tungsten wire that has high resistance. Current flowing through the filament produces heat at a temperature high enough to cause the emission of white light. The white light from the lamp consists of the following colors: red, orange, yellow, green, blue, and violet. It also includes two invisible forms of light, known as *infrared* and *ultraviolet*. Ultraviolet light will not pass through ordinary glass.

White light striking a material is partly reflected, partly absorbed, and partly passed through the material. Colored light is usually produced by passing light through glass that absorbs light rays of all colors except one. If a light of a certain color, say blue, is desired, a source of white light is enclosed in a blue-colored glass that allows only blue rays to pass through. Some lamps are manufactured with a tint of blue in the bulb. As the light is transmitted through the bulb, some of the reds and yellows are absorbed, causing the appearance of a whiter light.

Incandescent lamps are manufactured in many different sizes and shapes. The four general types of screw-base lamps are shown in Figure 14–12. Figure 14–13 illustrates lamps of different shapes and bases. The letters and numbers beneath the lamps refer to the manufacturers' code for indicating the type of lamp.

In general, smaller-base lamps have a lower power rating than those with larger bases. The medium-base lamps are manufactured in ratings of 5 watts through 300 watts. Those with lower ratings are made with miniature or candelabra bases.

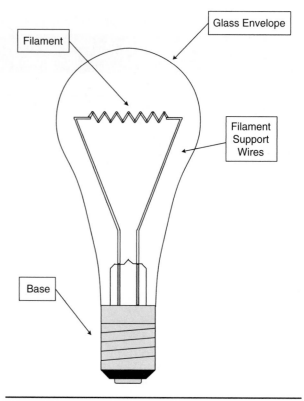

**FIGURE 14–11**  Construction of an incandescent lamp.

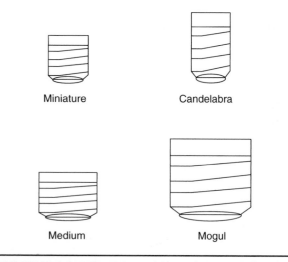

**FIGURE 14–12**  Screw bases for incandescent lamps.

The mogul base is used for lamps of greater than 300-watts rating. Lamps of very high power ratings, 1500 watts or more, are designed with bases to meet their requirements.

A very special type of incandescent lamp is the quartz lamp. Because the filament enclosure is made of quartz and filled with an iodine vapor, the filament does not deteriorate as rapidly as the one in the ordinary incandescent lamp.

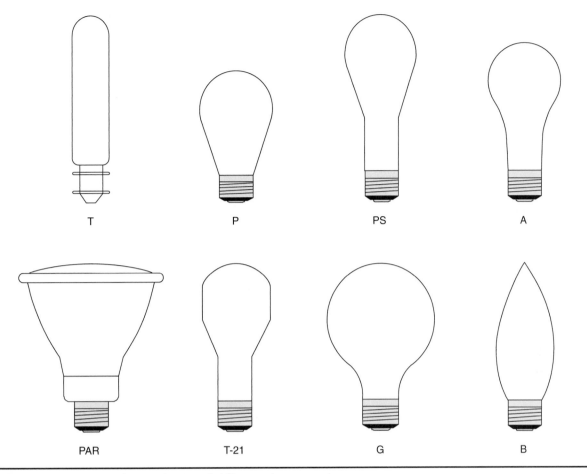

T     P     PS     A

PAR     T-21     G     B

**FIGURE 14–13** Incandescent lamps of different shapes and sizes.

The most common type of electric discharge lamp is the **fluorescent lamp**. This type of lighting is used extensively in commercial and industrial establishments. Its popularity is the result of its efficiency and diversity of colors. The average fluorescent lamp delivers about seven times the number of lumens per watt as the average incandescent.

Fluorescent lamps are manufactured in tubular form. Figure 14–14 illustrates some of the more common styles. The letters and numbers refer to the manufacturers' code. The inside of the glass tube is coated with phosphors. A small amount of argon or krypton gas and a bit of mercury are sealed within the tube. Each end contains an electrode, and under proper conditions, current flows from one electrode to the other through the ionized gas. Coating the inner surface of the glass with different phosphors makes it possible to obtain lamps that emit light of most any desired color.

Figure 14–15 illustrates the basic circuit for a fluorescent lamp. When the "on" switch is pressed, current flows through the circuit and heats the filaments, causing them to emit electrons into the space around them. When the switch is released, the

magnetic field around the induction coil collapses rapidly, inducing a high voltage across filaments A and B. This emf causes the emitted electrons to flow through the tube, ionizing the gas. The ionized gas becomes a good conductor, and current flow is established. The lamp may be turned off by pressing the "off" button, which breaks the circuit from the power source.

The operation of a fluorescent fixture depends upon the design. One type of construction is the *preheat* (Figure 14–16). This fixture has a **glow switch**. This switch, commonly called a starter, is similar to a neon light with a bimetal electrode (Figures 14–16 and 14–17). When a voltage is applied to the circuit, the starter is energized and glows, heating the electrodes. The bimetal electrode bends and makes contact with the other electrode. The circuit is now complete from $L_1$ through the induction coil, filament A, the starter, and filament B to $L_2$.

Current flowing through the circuit heats the filaments and causes them to emit electrons. Because the starter contacts are closed, heat is no longer generated by the neon light. The bimetal contacts cool and break the circuit. The magnetic field around the

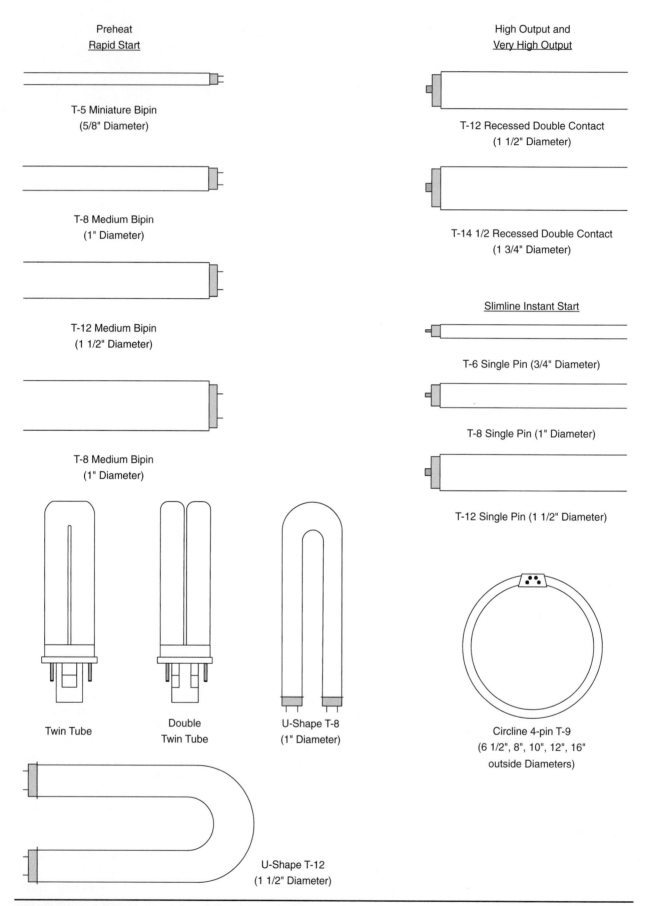

**FIGURE 14–14** Various styles of fluorescent lamps.

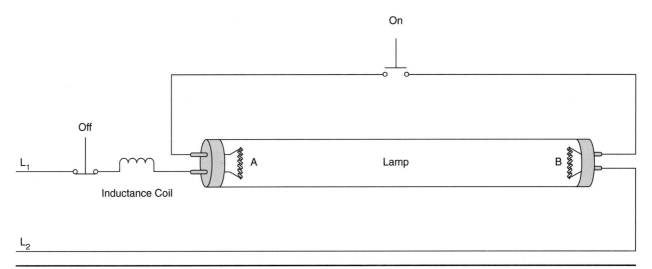

**FIGURE 14–15**   Basic circuit for a fluorescent lamp.

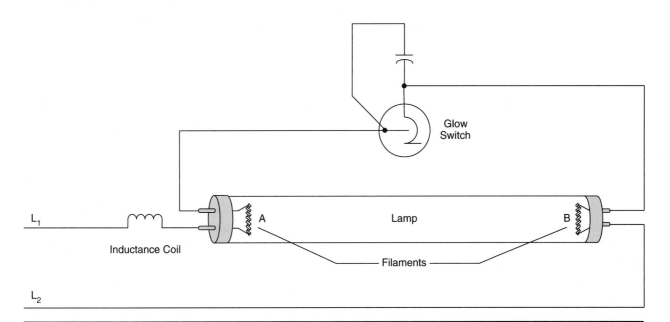

**FIGURE 14–16**   Schematic diagram for a preheat fluorescent fixture.

induction coil collapses rapidly, inducing a high voltage across filaments A and B. This high emf establishes current flow through the lamp.

Besides producing a high voltage across filaments A and B, the induction coil serves as a current-limiting device. The heat produced by the filaments causes the mercury in the tube to turn to a gas, changing the electrical characteristics within the tube. As current flows from A to B, the resistance within the tube begins to decrease, allowing more current to flow. The greater the current, the greater the decrease in resistance. If no means is provided to limit the value of current, this action will continue until the circuit breaker opens or the tube explodes.

The inductance coil, connected in series with the lamp, provides the impedance necessary to limit the current to a safe value. The major disadvantage of this type of fixture is the time required for preheating. The primary advantage is the initial cost of the fixture and the life span of the lamps compared to other types of fluorescent fixtures.

Two types of fixtures that require less starting time than the preheat are the instant-start and the rapid-start. The *hot-cathode, instant-start, bipin lamp* has two electrodes (Figure 14–18). These electrodes contain filaments that are short-circuited. The lamp is started without preheating the filaments, thus eliminating the need for a starter. After the lamp has started, the voltage supplied to it must be

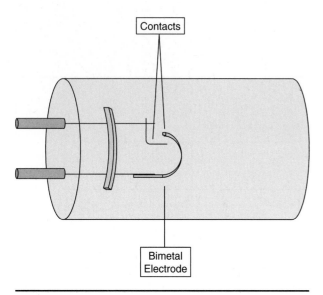

**FIGURE 14-17** Glow switch (starter) for a preheat fluorescent fixture.

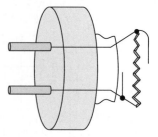

**FIGURE 14-18** Electrodes connected to a short-circuited filament in a hot-cathode, instant-start, bipin fluorescent lamp.

reduced to a value just large enough to maintain a filament temperature sufficient to emit electrons. This reduction in voltage is accomplished by a specially designed transformer. Because of the high voltage available from the transformer, the fixture is wired so that the primary circuit is open when the lamp is removed. Figure 14–19 illustrates the connections for the hot-cathode, instant-start fixture.

Because of the short-circuited filaments, these lamps cannot be used with ballasts for preheat starting. Conversely, a preheat lamp cannot be used with a ballast intended for an instant-start lamp. If this is attempted, the high-voltage surge would cause the filaments of the preheat lamps to reach a temperature high enough to cause rapid deterioration of the filaments.

Another type of hot-cathode, instant-start lamp has a single pin and is called *slimline*. It consists of a long, slender tube with only one prong at each end (Figure 14–20). This lamp is manufactured in specific lengths, from 42 inches to 96 inches and 3/4 inch or 1 inch in diameter. The slimline lamp operates at a higher voltage and lower current than the preheat lamp. In addition, this fixture is wired so that the primary circuit is open when the lamp is removed.

The *cold-cathode lamp* is also a type of instant start. In place of the filament, it has large, thimble-type iron cathodes. These cathodes are coated with an active material to provide large emitting areas (Figure 14–21). This design provides longer lamp life than the hot-cathode type and still has all the advantages of instant start. Other advantages are

simplicity of design and less flicker. The circuitry for this fixture is the same as for the slimline.

The *rapid-start, hot-cathode lamp* is one of the most popular types of fluorescent lamps in use today. It has longer life than the instant start, yet requires minimal starting time. This lamp requires a special ballast and a metal starting aid, which is at ground potential. One popular arrangement is to use the fixture reflector for this purpose.

Rapid-start ballasts are designed to provide smooth, quick starting. The lamp reaches full brightness in about 2 seconds. The ballast of this fixture must be grounded. This is usually accomplished by contact with the metal of the fixture, which is grounded through the equipment ground.

Proper polarity is very important when connecting a rapid-start fixture. The black wire from the ballast must be connected to the ungrounded (hot) wire from the supply. Figure 14–22 illustrates the connections for a rapid-start fixture.

Another type of ballast, known as *trigger-start*, is designed to operate with regular fluorescent lamps and requires no starter. It contains preheat windings that bring the filaments to the required temperature within 1 second. To provide quick starting, the trigger-start ballast must provide a higher open-circuit voltage than the preheat- or rapid-start-type.

The average fluorescent fixture operates best in ambient temperatures between 65 degrees Fahrenheit and 80 degrees Fahrenheit (18 degrees Celsius and 27 degrees Celsius). At other temperatures, the efficiency and lighting output are reduced.

For outdoor installations, the fixtures must be weatherproof, and the ballast must be designed for operation at ambient temperatures for the area. If temperatures drop below 50 degrees Fahrenheit (10 degrees Celsius) special ballasts must be used. Some fixtures are designed to confine the air around the lamps. When the lamp is in operation, it heats the air. Continuous operation can heat the air to a point of maximum light output.

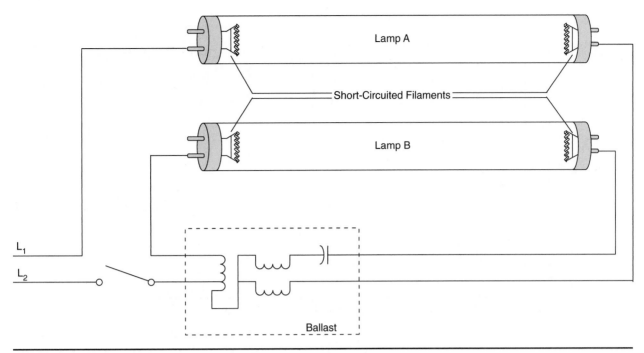

**FIGURE 14–19**  Connections for a hot-cathode, instant-start, fluorescent fixture.

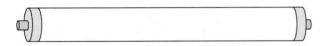

**FIGURE 14–20**  Tube for a slimline, fluorescent fixture.

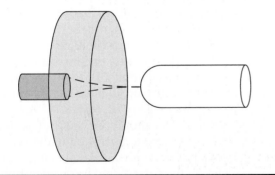

**FIGURE 14–21**  Thimble-type iron cathodes used in a cold-cathode fluorescent lamp.

Fluorescent fixtures are designed to operate on various source voltages. One must always check the input voltage rating to ensure proper operation. Some common voltage ratings are 120 volts and 277 volts. Some additional connection methods are shown in Figures 14–23A and 14–23B.

## High-Intensity Discharge (HID) Lamps

HID lamps consist of a glass envelope, a base, and an arc tube, as seen in Figure 14–24. The arc tube is filled with a gas, and inside there is a set of electrodes. When the HID lamp is energized, a ballast provides high voltage to the electrodes within the arc tube. An electrical arc is created between the electrodes, and a current passes through the gas filling the arc tube. The arc is the source of light produced by the HID lamp.

HID lamps require a warm-up period before they begin to emit light. Typically, several minutes of warm-up are required before full light output is realized. Many HID lamps produce a humming or buzzing sound when they are operating. This is normal and does not indicate a defect in the lamp or ballast. Figure 14–25 shows several different wiring methods for HID lamps. Always follow the wiring diagram that appears on the ballast you are installing.

There are several types of HID lamps. These are the *low-pressure sodium lamp*, the **mercury-vapor lamp**, the *metal-halide lamp*, and the *high-pressure sodium lamp*. As the names would imply, the basic differences between the various types of HID lamps is the gas used to fill the arc tube and the pressure of the gas.

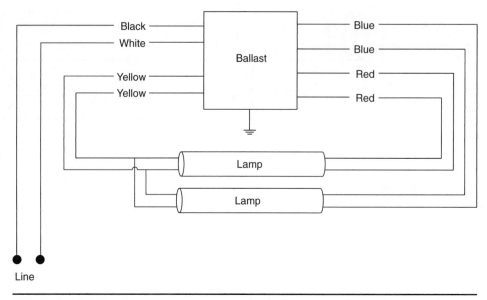

**FIGURE 14-22** Connections for a rapid-start fluorescent fixture.

## Comparison

Table 14-3 shows a comparison between the incandescent, fluorescent, and HID types of lamps. Notice that one lamp is not best suited for all applications. Due to the different methods used to produce light, alteration in color rendition is a concern. For example, low-pressure sodium lights, which produce a yellow-orange color, are not used in locations where the appearance of colors is important. It is unlikely that low-pressure sodium lamps would be used in a photography studio as the colors would not appear true. Likewise, mercury-vapor lights make it difficult to distinguish reds and blues. Fluorescent lamps tend to cause photographs to have a yellowish-green tint. Some additional items of concern when selecting lamps would be:

- Initial cost
- Ballast required
- Size and shape
- Warm-up required
- Efficiency
- Life expectancy
- Operating temperature
- Light output
- Ability to dim
- Color distortion
- Operating voltage
- Effects of ambient temperature

## LIGHTING MAINTENANCE

As previously stated, the lumen output of a lamp decreases with usage. The amount of depreciation varies with the type of lamp and the conditions of use. The average life of an incandescent lamp is 1000 hours of usage. The average life of a fluorescent lamp is 7500 hours of usage. As a lamp approaches its maximum life span, the lumen output decreases rapidly.

Another factor that affects the amount of illumination is the reflecting quality of the fixture, ceilings, walls, and other surfaces. As these surfaces become dirty, their ability to reflect the light rays decreases. The amount of dirt collecting on lamps and globes also reduces the amount of light passing through to the area to be illuminated. All these factors must be considered in a maintenance program.

The illumination should be checked periodically to ensure adequate lighting for the type of use. This can be accomplished by placing a footcandle meter on the surface under consideration. If the meter indicates a light measurement less than that recommended for the type of use, the condition must be evaluated.

Many manufacturers provide, free of charge, computerized analysis of a lighting system. A computer analysis will not only determine the necessary amount of illumination for a specific work area but will also recommend ways to improve the lighting efficiency.

Fluorescent Lamp Ballast

Wiring Diagrams

(Always Follow the Diagram which Appears on the Ballast that You Are Installing)

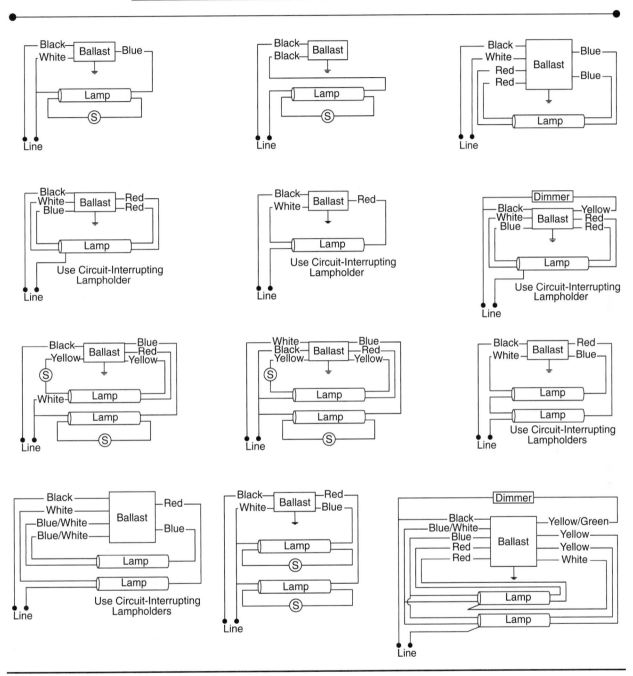

**FIGURE 14–23A** *Various wiring diagrams for different fluorescent ballasts.*

Computerized instruments measure the amount of illumination, determine if the light is evenly distributed, and make recommendations for improvement. Corrective measures may be to clean the fixtures and the lamps. If the lighting is still inadequate after performing these steps, it may be necessary to clean the ceilings, walls, and other surfaces.

A record should be maintained showing when new lamps have been installed and indicating the number of hours the fixtures are in use each day. As the lamps near the end of their life expectancy,

Fluorescent Lamp Ballast

Wiring Diagrams (Continued)

(Always Follow the Diagram which Appears on the Ballast that You Are Installing)

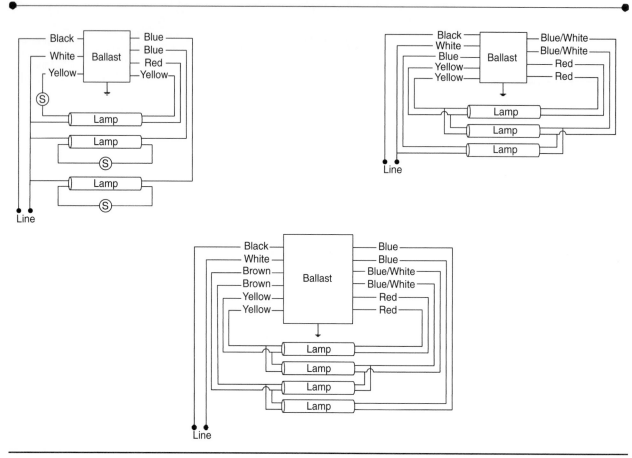

**FIGURE 14–23B** Various wiring diagrams for different fluorescent ballasts (continued from Figure 14–23A).

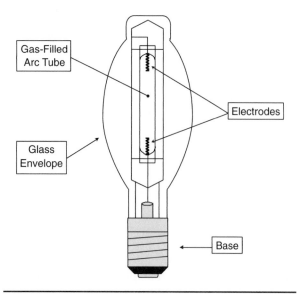

**FIGURE 14–24** Construction of a high-intensity discharge (HID) lamp.

they should be replaced before they deteriorate to a point of inadequate lighting.

Other than cleaning and replacing lamps, incandescent fixtures require very little maintenance. Socket contacts sometimes become loose, causing overheating. Replacement of the socket may be necessary if damage has occurred because of overheating.

After lamps have been replaced many times, the center contact of the socket may lose its spring tension. This causes it to lie flat in the bottom of the socket, providing a poor connection or no connection. If the lamp does not complete the connection to the center contact, an open circuit exists and the lamp will not glow. If the lamp makes a poor connection, current flowing through the lamp will cause the socket to overheat.

In the cleaning of fixtures, the sockets should be checked for loose connections. The center contact

High-Intensity Discharge (HID) Lamp Ballast

Wiring Diagrams (Continued)

(Always Follow the Diagram which Appears on the Ballast that You Are Installing)

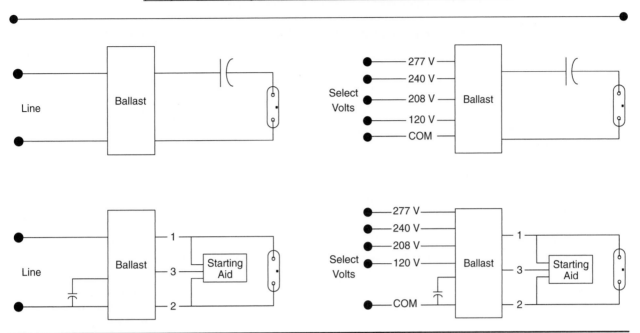

**FIGURE 14–25**  Various wiring diagrams for different HID ballasts.

should be checked and the socket inspected for signs of overheating.

Fluorescent fixtures also require periodic cleaning. Although the sockets on fluorescent fixtures are not of the screw-shell type, they also must be checked for loose connections and dirty, loose, or pitted contacts.

Other components of fluorescent fixtures that must be considered are the ballast and the starter. If only the ends of the lamp glow and the lamp does not blink, the problem is likely to be the starter. Replace it with a new one. A lamp that blinks while the ends glow generally indicates lamp failure. If, however, the lamp is known to be good, then the trouble may be a defective ballast. It is always good

practice to ensure that the lamp and starter are good before replacing the ballast.

An arrangement in the maintenance shop whereby lamps, starters, and ballasts can be tested will eliminate a lot of work in the area being illuminated.

Electric discharge lamps, especially the fluorescent type, can be hazardous if mishandled. They are under a partial vacuum and, if dropped or damaged, may explode. Also, the material contained in some lamps is poisonous if inhaled, ingested, or it enters the bloodstream through a cut.

Methods of disposing of lamps vary. One method is to place them into the original cartons to be disposed of with similar hazardous waste.

**TABLE 14-3**  Comparison between incandescent, fluorescent, and HID lamps.

|  | Incandescent | Fluorescent | HID |
|---|---|---|---|
| Cost | • Low initial cost | • Higher initial cost than incandescent | • Highest initial cost |
|  | • Low replacement cost | • Higher replacement cost than incandescent | • Highest replacement cost |
| Efficiency | • Low | • Better than incandescent | • Better than incandescent |
| Ballast required | • No | • Yes | • Yes |
| Warm-up or restart time required | • No | • Short | • Long |
| Dimmable | • Yes | • Yes—but expensive | • Expensive—or not possible |
| Operating temperature | • High | • Low | • High |
| Light output | • Low | • Higher than incandescent | • Highest |
| Operating voltage | • Low | • Low | • High |
| Life expectancy | • Short (900 hours) | • Long (16,000 hours) | • Low-pressure sodium (1900 hours)<br>• High-pressure sodium (10,000 hours)<br>• Mercury-vapor (20,000 hours)<br>• Metal-halide (11,500 hours) |
| Ambient temperature effects | • None | • Light output levels affected<br>• Color of light affected<br>• Difficulty starting in cold temperatures | • Difficulty starting in cold temperatures |

# REVIEW QUESTIONS

*Multiple Choice*

1. The unit for the total light emitted from a lighting source is the
   a. candlepower.
   b. lumen.
   c. foot-candle.
   d. watt.

2. The amount of illumination on a given surface is measured in units called
   a. candlepower.
   b. lumen.
   c. lunars.
   d. foot-candle.

3. The intensity of illumination from a lighting source
   a. varies inversely with the distance from the source.
   b. varies directly with the square of the distance from the source.
   c. varies inversely with the square of the distance from the source.
   d. varies directly with the distance from the source.

4. A direct lighting fixture directs at least
   a. 90% of the light downward.
   b. 85% of the light downward.
   c. 85% of the light upward.
   d. 80% of the light downward.

5. Fixtures classified as general diffusing direct
   a. 70% to 80% of the light downward and the rest upward.
   b. 60% to 70% of the light downward and the rest upward.
   c. 40% to 60% of the light downward and the rest upward.
   d. 30% to 40% of the light downward and the rest upward.

6. Thin, transparent material allows light to pass through with
   a. very little scatter.
   b. a great amount of scatter.
   c. no scatter.
   d. a moderate amount of scatter.

7. Opaque, highly polished, bright material
   a. absorbs most of the light that strikes it.
   b. passes most of the light that strikes it.
   c. reflects most of the light that strikes it.
   d. diffuses most of the light that strikes it.

8. Glare is caused by
   a. poor placement of the light source.
   b. insufficient illumination.
   c. rough surfaces.
   d. dull surfaces.

9. The intensity of illumination needed for a specific area depends upon
   a. the type of work to be performed in the area to be illuminated.
   b. the number of people working in the area.
   c. the type of fixtures to be used.
   d. the height of the lighting fixture.

10. The coefficient of utilization refers to
    a. the ratio of the power input to the power output.
    b. the ratio of the power input to the lighting output.
    c. the ratio between the light reaching the working plane and the light produced by the luminaires.
    d. the ratio of the life expectancy of the lamp to the actual life of the lamp.

11. In a determination of the amount of illumination provided by lighting fixtures, a maintenance factor must be considered because
    a. the lighting output deteriorates with the life of the lamp.
    b. the reflecting surfaces and the lighting fixtures become dirty over a period of time.
    c. the lighting fixtures must be disconnected from the supply when performing maintenance.
    d. the lighting output increases with time.

12. In general, there are two categories of lighting fixtures. They are
    a. resistance and fluorescent.
    b. electric discharge and resistance.
    c. incandescent and electric discharge.
    d. resistance and incandescent.

13. A lamp that produces light from a filament that is heated to a very high temperature is called
    a. fluorescent.
    b. incandescent.
    c. thermal.
    d. resistance.

14. Lamps that are rated greater than 300 W have a
    a. candelabra base.
    b. mogul base.
    c. medium base.
    d. miniature base.

15. The most common electric discharge lamp is the
    a. incandescent.
    b. thermal.
    c. fluorescent.
    d. resistance.

16. The glow switch is required with
    a. rapid-start fluorescent fixtures.
    b. preheat fluorescent fixtures.
    c. cold-cathode fluorescent fixtures.
    d. all of the above.

17. The cold-cathode fluorescent lamp is a type of
    a. preheat lamp.
    b. rapid-start lamp.
    c. instant-start lamp.
    d. resistance-start lamp.

18. The average fluorescent lamp operates best in ambient temperatures between
    a. 32°F and 65°F.
    b. 45°F and 75°F.
    c. 65°F and 80°F.
    d. 65°C and 80°C.

19. The average life of an incandescent lamp is
    a. 800 hours of usage.
    b. 1000 hours of usage.
    c. 5000 hours of usage.
    d. 9000 hours of usage.

20. The average life of a fluorescent lamp is
    a. 7500 hours of usage.
    b. 8000 hours of usage.
    c. 9500 hours of usage.
    d. 15,000 hours of usage.

21. It is good maintenance practice to check the illumination of work areas on a regularly scheduled basis. This can be accomplished with a
    a. lumen meter.
    b. foot-candle meter.
    c. candlepower meter.
    d. lunar meter.

22. A record should be maintained showing when new lamps have been installed. The purpose of this record is to determine
    a. if the manufacturer's ratings are correct.
    b. when the lamps should be replaced.
    c. how long the lamps last.
    d. the cost of lighting equipment.

23. When performing regular maintenance on lighting fixtures, one should always inspect the
    a. reflectors.
    b. circuit connections.
    c. lamp sockets.
    d. circuit breakers.

24. A fluorescent lamp that blinks while the ends glow usually indicates
    a. a defective starter.
    b. a lamp failure.
    c. a defective ballast.
    d. all of the above.

*Give Complete Answers*

1. What is the unit of measurement for the total light emitted from a lighting source?

2. Define the term *foot-candle*.

3. Write the inverse square law for light.

4. What does a light intensity curve indicate?

5. List five types of luminaires.

6. Describe a general diffusing fixture.

7. List four factors that affect light distribution.

8. Under what conditions do opaque materials reflect rather than absorb light?

9. Define *glare*.

10. What factors affect the brightness of illumination?

11. List three factors that must be considered when determining the amount of illumination required for a specific area.

12. Define the coefficient of utilization.

13. What is meant by the term *maintenance factor*?

14. What maintenance factor is usually applied under normal conditions of use?

15. List five factors that should be considered when selecting lighting fixtures.

16. List two general categories of lighting fixtures.

17. Describe how light is produced with an incandescent lamp.

18. List four general types of screw-shell bases.

19. What is the advantage of a quartz lamp over an ordinary incandescent lamp?

20. Describe the construction of a fluorescent lamp.

21. How is it possible to obtain different colors of light from fluorescent lamps?

22. Draw a diagram illustrating the basic circuit for a fluorescent fixture.

23. List two functions of the induction coil in a fluorescent fixture.

24. What is the main difference between a preheat fluorescent fixture and a rapid-start fluorescent fixture?

25. List two types of fluorescent fixtures that require less starting time than the preheat type.

26. Can fluorescent lamps designed for use with preheat ballast be used in fixtures designed for instant start?

27. Between what ambient temperatures do fluorescent fixtures provide their best performance?

28. Under what conditions can fluorescent fixtures be installed while exposed to the weather?

29. List two general types of electric discharge lamps.

30. What is the average life of an incandescent lamp? a fluorescent lamp?

31. List four factors that affect the amount of illumination in an area.

32. Describe a good maintenance program for lighting systems.

33. List four common causes of improper operation of fluorescent fixtures.

*Solve each problem, showing the method used to arrive at the solution.*

1. A fixture is to be positioned 6 ft directly above a secretary's desk. According to the illumination table (Table 14–2), what is the amount of illumination in foot-candles required on the desk surface? What must be the minimum candlepower of the fixture?

2. A fixture is secured 10 ft above a reading table in a library. If it emits light at an intensity of 1000 candlepower downward, does it provide adequate illumination on the table surface?

3. What is the intensity of light on a workbench if a fixture is 8 ft above the work surface and 6 ft horizontally from the work area and emits a uniform illumination of 1000 candlepower?

4. A room 20 ft long and 15 ft wide must be illuminated to an average intensity of 20 foot-candles. Calculate the number of lumens required.

5. Using Figure 14–7, calculate the light at point P if the candlepower of each fixture is 3000.

# CHAPTER 15

# Electric Heat

## OBJECTIVES

After studying this chapter, the student will be able to:

- Describe three methods used for heat transfer.
- Discuss the factors that must be considered when calculating heat losses.
- Describe various types of electric heating equipment.
- Explain the operation and function of controls used in conjunction with electric heating systems.

# SPACE HEATING

*Space heating* refers to the heating of specific areas of commercial and industrial buildings. Heat energy is generally transferred from one place or object to another by one of three methods: conduction, convection, or radiation. In most instances, all three methods are used in varying degrees. The system is named according to the primary method used.

**Conduction heating** is a method by which heat is transferred from one object to another. One might say the object conducts heat. An example of this method is cooking. A pan is placed on the range in contact with the heating element. Heat is transferred from the element to the pan and from the pan to the ingredients in the pan.

**Convection heating** is a method by which heat is transferred through a fluid or air. Hot-water, steam, and warm-air heating systems use the convection method.

**Heat radiation** is accomplished when the heat rays are transmitted through space. The rays are absorbed by the objects they contact, thereby warming the objects. They do not warm the air but warm the objects into which they are absorbed.

One common source of radiant heat is the sun. All objects radiate some heat; therefore, as heat rays strike objects, some are absorbed and some are radiated back into space.

All three methods of space heating can be employed when using electric energy as a heat source.

# CALCULATIONS

The unit of measurement for heat is the British thermal unit (Btu). A Btu is the amount of heat required to raise the temperature of 1 pound of water through 1 degree Fahrenheit.

To provide a comfortable area, it is recommended to maintain a temperature of 72 degrees Fahrenheit (22 degrees Celsius) with a **relative humidity** of 35 percent.

The greater the difference of temperature between two areas, the quicker heat is transferred from the warm area to the cold area. Other factors to consider when designing a heating system are the amount and type of materials separating the warm area from the cold area. The type, density, thickness, and color of the material all affect **heat transfer**. To determine the amount of heat required to maintain body comfort, all the aforementioned factors must be considered.

Because temperatures vary considerably in different parts of the world, it is necessary to refer to the engineer's recommendations for the specific area. For example, in northeastern United States, it is generally recommended to maintain a temperature of 72 degrees Fahrenheit (22 degrees Celsius) when the outside temperature is −10 degrees Fahrenheit (−23 degrees Celsius) with a wind of 10 miles per hour (16 kilometers per hour).

The heat that is transferred from the area being heated to the colder surrounding area is considered a heat loss. To calculate this loss, it is necessary to determine the square foot area of the walls, ceilings, and floors that are exposed to the colder area. Window and door areas must also be considered, in addition to the type and amount of insulation to be used.

In many areas, the local utility company provides charts and formulas for calculating heat loss. Some utility companies do the calculations as an incentive to install electric heat. And they often provide estimates of the annual heating cost.

Electric heat cost is based on the number of kilowatt hours of energy dissipated. Many utility companies give special rates for heating. In many areas, particularly in the colder regions, the price of electric heat may not be competitive with other fuels. It has, however, the advantage of providing quick, clean heat at a very even temperature. Electric heat requires little or no space for the heating plant and needs no area for storage of fuel. Another advantage is that electric heating systems are far more efficient than fossil fuel systems.

# HEATING EQUIPMENT

One system is *resistance baseboard heating*. The units consist of resistance heaters placed along the wall near the floor line. They are installed in areas of greatest heat loss (i.e., outside walls, under windows). Air circulates through the units, warming the walls and the surrounding air. This method of heating represents examples of both conduction and natural convection. A small amount of **radiation** also takes place. It is referred to as a convection system because the major portion of the heat is transmitted by convection. The heat is produced by current flowing through a resistance. The amount of heat produced is directly proportional to the resistance and the square of the current.

Perimeter heating provides a uniform heat throughout an area. It is extremely efficient because all of the heat produced is dissipated into the area to be heated. With this method of heating, each room can have its own thermostat.

*Electric heating cables,* often installed in ceilings, are used only when cost is not a factor. The heat from this system is transmitted primarily by radiation.

Another type of radiant heating consists of plaster-board panels containing a conductive film that produces heat when energized. Both of these systems produce the heat within the area being heated and provide uniform heat throughout the area. An individual thermostat is usually provided in each room.

An *electric furnace* is a central unit in which resistance heating elements are installed. A fan is placed near the elements. When the thermostat calls for heat, the circuit is completed to the heaters. When the air around the heaters reaches a predetermined temperature, the fan is energized. The fan pulls the cold air from the area to be heated and forces the warm air into the area. The cooler air entering the furnace is then heated by the elements. This cycle of operation continues until the thermostat is satisfied.

*Duct heaters* are similar to the electric furnace. Instead of having all the elements in one location, they are installed in the ducts for each specific area. This type of installation provides the flexibility of zone control.

The **heat pump** is a machine that can heat an area during cold weather and can cool the area during warm weather. It operates on a principle similar to a refrigerator. The pump circulates a liquid (refrigerant) through two sets of coils. A compressor is connected between the coils (Figure 15–1). The liquid is the type that boils at a very low temperature. When it boils, it vaporizes, just as water turns to steam when it boils. This cold vapor absorbs heat from the surrounding air. By compressing the vapor, the temperature is raised further. Before leaving the compressor, the vapor reaches a temperature of 100 degrees Fahrenheit (37.8 degrees Celsius) or higher. The heated vapor enters the other coil, called a condenser. The air surrounding the condenser is cooler than the vapor and absorbs the heat. As the vapor cools, it returns to liquid form, and the

process is repeated. For air conditioning, the process is reversed. The method used to transmit the heated air to a specific area is much the same as described for an electric furnace.

The heat pump is probably the most efficient type of electric heat. Depending on the type of installation, the operating cost can be competitive with, or even less than, systems using fossil fuels.

# SYSTEM CONTROLS

All automatic heating systems must have controls for their operation. There are two basic types: operating controls and safety controls. These two types can be further divided according to their particular function.

One of the more simple systems is used in conjunction with baseboard resistance heaters. Here, one type of operating control and one type of safety control are required.

The room thermostat functions as the operating control. When the temperature of the room drops below a predetermined value, the thermostat contacts close, completing the circuit to the heater(s). When the room temperature reaches the thermostat setting, the contacts open, disconnecting the heater(s) from the supply.

The accuracy of the thermostat depends upon its design and quality. The sensitivity of the heat-sensing device and the ability of the contacts to make and break quickly play important roles in maintaining a constant temperature.

Thermostats are manufactured for use with various values of voltage. Standard values are 24 volts, 120 volts, and 240 volts. Resistance baseboard heaters generally require 240 volts.

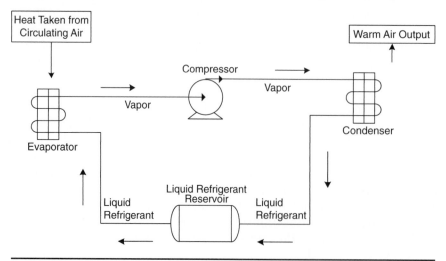

**FIGURE 15–1** Simplified sketch of a warm-air heating system using a heat pump.

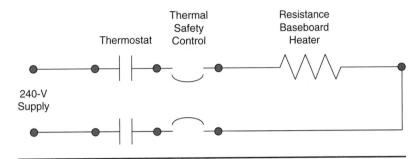

**FIGURE 15–2**  Circuit for a resistance baseboard heater.

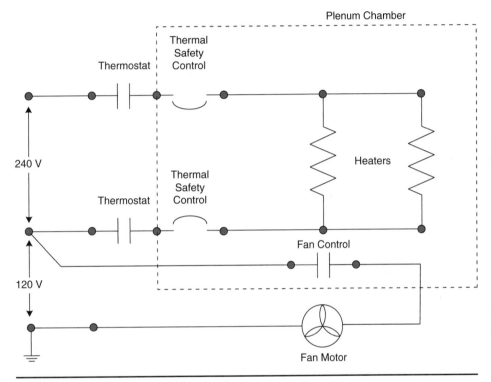

**FIGURE 15–3**  Basic circuit for an electric furnace.

Built into each baseboard heater is a safety control. It is a thermal device that opens the circuit if excessive heat is produced. The purpose of this control is twofold: it prevents damage to the unit caused by overheating, and it reduces the possibility of fire.

Baseboard resistance heating units, ceiling panels, and heating cables usually require only a thermostat and a thermal device. Figure 15–2 shows the circuit for a 240-volt baseboard heater.

The electric furnace requires additional controls. Beside the thermostat, an operating control is installed near the heaters in the plenum chamber. The purpose of this control is to stop and start the fan (Figure 15–3).

When the thermostat calls for heat, the circuit is completed to the heaters. When the air temperature

in the plenum reaches 85 degrees Fahrenheit (30 degrees Celsius), the fan control completes the circuit to the fan motor. The air is circulated through the area to be heated and through the system. If the temperature in the plenum drops below 82 degrees Fahrenheit (28 degrees Celsius), the fan control opens the circuit and stops the fan. With adequate design, the air temperature in the plenum will remain above 82 degrees Fahrenheit (28 degrees Celsius) until the thermostat is satisfied and for a short time thereafter.

When the thermostat is satisfied, the circuit to the heaters is opened, disconnecting them from the supply. The air in the plenum quickly cools, causing the fan control to open the circuit to the fan motor.

Another control that is placed in the plenum chamber is for safety. This device opens the circuit to the heaters if the temperature in the plenum exceeds a safe value.

Each installation is unique in its design. In addition, the type and number of controls vary according to the requirements. Manufacturers of various systems provide instructions and diagrams to meet their requirements. It is important to study this information carefully.

## HUMIDITY CONTROL

Humidity is a very important consideration in the design of space heating. Many heating systems tend to remove humidity from the air. The result is dry, unhealthy air and the need for higher temperatures to attain body comfort. To compensate for the moisture loss, **humidifiers** must be installed. Humidifiers replace lost moisture.

Other heating systems tend to cause excessive moisture. When this occurs, exhaust fans must be installed. Such fans pull the moist air to the outside. The moist air is then replaced by drier air.

Another method of removing moisture is to install a **dehumidifier**. This machine removes the moisture from the air and collects it in a reservoir or disposes of it into a drain.

For health and comfort, a relative humidity of 30 percent to 40 percent is desirable. A **humidistat** is a switch that makes and breaks contacts as the humidity changes. Humidistats should be installed to control the humidifier, dehumidifier, or exhaust fans, whichever is needed. The design of a building and the type of heating system are the determining factors as to what is needed. The amount of insulation and the type of vapor barriers are also factors. Some designs may include both exhaust fans and humidifiers. The controls are arranged to operate whichever is needed.

If exhaust fans are installed, they should have good dampers to prevent heat loss when the fans are not in operation. Nonmetallic ducts that are insulated are also recommended. Such an arrangement reduces heat loss and prevents condensation from forming inside the duct and returning to the heated air.

## DIELECTRIC HEATING

**Dielectric heating** is based on the principle of capacitance. An insulating material is placed between two conducting plates. The plates are connected to a high-frequency AC supply. The alternating polarity of the plates causes the electrons in the insulating material to strain first in one direction, then in the other. The continuous and rapid strain reversal produces heat.

This method of heating has many applications, such as cooking, softening of materials such as plastics, and rapid drying of materials. It is also used for medical purposes. Intense pain caused by strain or tense muscles can often be relieved by warming the affected part of the body. This treatment is called *diathermy*. The portion of the body to be treated is placed between the plates. A warm sensation is felt throughout the area being treated. The electron strain and the heat cause the muscles to relax, thus reducing pain.

The advantage of dielectric heating is its ability to instantly produce heat uniformly within a material. It is also easy to control by varying the frequency of the applied voltage.

## INDUCTION HEATING

**Induction heating** is used for metal treatment and in some electric furnaces. Induction heating, as the name implies, uses electromagnetic induction to produce heat. A conducting material is placed within a coil that is attached to a high-frequency AC supply. The increase and decrease of the magnetic field around the coil cause relative motion between the field and the conductor. The result is to produce eddy currents on the surface of the conducting material. The eddy currents produce heat.

The amount of heat produced depends upon the frequency of the supply voltage and the amount of current flowing in the coil. The depth to which the heat penetrates depends upon the frequency of the supply voltage. The higher the frequency, the lesser the depth. With lower frequencies, a longer heating time is required, allowing the heat to penetrate to a greater depth.

Other factors that affect the amount of heat and depth penetration are the resistivity and permeability of the material being heated.

## RESISTANCE HEATING

**Resistance heating** utilizes the fact that current flowing through a material produces heat. The amount of heat depends upon the value of current and the size and type of material through which the current is flowing. Some of the most common materials are aluminum, nickel, chromium, copper, iron, and cobalt in various proportions. Resistance heating elements are commonly used in cooking appliances, electric space heating, clothes dryers, and irons.

Some types of electric welders also utilize the theory of resistance heating. Two types of resistance welders are the spot welder and the seam welder.

With spot welding, thin sheets of metal are joined together in spots. The metal sheets are placed between two electrodes. The electrodes are moved together, clamping the sheets between them. A high current passes through the sheets from one electrode to the other. The current produces heat, causing the metal to weld. As soon as the spot reaches welding temperature, the circuit is opened. When the material cools, the electrodes are released.

Seam welding is similar to spot welding except the weld is continuous. The metal is joined together and placed between two electrodes in the form of rollers. As the metal is moved between the electrodes, a high current passes from one roller to the other. One continuous seam is formed by the welding heat produced.

## ARC HEATING

An electric arc is produced by ionizing the air between two electrodes. The heat from the arc is used for welding and melting metal.

With arc welding, the material to be welded forms one electrode, and the welding rod the other. To form the arc, the welding rod is touched to the material to be welded. The heat produced by the current flowing through the point of contact causes the metal to vaporize, causing the surrounding air to ionize. When the rod is pulled away from the material, the ionized air conducts the current, forming an arc. Once the arc is produced, the heat generated will cause the material and the rod to melt. The metal from the rod flows with that from the material to form a continuous seam.

## INFRARED HEAT

*Infrared heating* is a type of radiant heating. Electromagnetic waves transmitted through space are absorbed by the materials they strike.

These rays are usually produced by electric lamps containing a filament. Current flowing through the filament produces electromagnetic rays whose wavelengths are greater than those that produce visible light but shorter than microwaves. They have a frequency slightly below that of the visible light range. The lamps that are used to contain the filament are generally of the quartz type, in tubular shape or shaped like the lamp in Figure 14–13 (PAR).

Infrared heating is used for drying operations, food warming, and medical purposes.

## REVIEW QUESTIONS

*Multiple Choice*

1. Convection is a method of transferring heat
   a. by contact.
   b. through a fluid or air.
   c. by rays traveling through space.
   d. through a vacuum.

2. The unit of measurement for heat is
   a. Fahrenheit.
   b. Celsius.
   c. American thermal unit.
   d. British thermal unit.

3. To maintain a comfortable body temperature, it is recommended to maintain
   a. a temperature of 75°F (24°C) with a relative humidity of 32%.
   b. a temperature of 72°F (22°C) with a relative humidity of 32%.
   c. a temperature of 72°F (22°C) with a relative humidity of 35%.
   d. a temperature of 70°F (21°C) with a relative humidity of 35%.

4. In northeastern United States, it is recommended to install a heating system capable of maintaining a temperature of
   a. 75°F (24°C) when the outside temperature is −10°F (−23.3°C) with a 10-mile-per-hour (32-kilometer-per-hour) wind.
   b. 72°F (22°C) when the outside temperature is −10°F (−23.3°C) with a 10-mile-per-hour (16-kilometer-per-hour) wind.
   c. 70°F (21°C) when the outside temperature is −10°F (−23.3°C) with a 10-mile-per-hour (16-kilometer-per-hour) wind.
   d. 70°F (21°C) when the outside temperature is −10°F (−23.3°C) with no wind.

5. A room becomes cooler because
   a. cold seeps into the heated area.
   b. cold air is radiated through the outside surfaces.
   c. the heat is absorbed by the surrounding cold surfaces.
   d. the heat loses its strength over a period of time.

6. Charts for calculating heat loss may be obtained from
   a. the local utility company.
   b. manufacturers of the heating equipment.
   c. the National Fire Protection Association.
   d. both a and b.

7. A major advantage of electric heat is that
   a. it is quick and clean.
   b. it is very inexpensive.
   c. it never fails.
   d. it is available in quantity.

8. The best location for baseboard heaters is
   a. along the inside walls.
   b. anywhere on an outside wall.
   c. under windows.
   d. beside the door.

9. Electric heating cables installed in the ceiling transfer heat by
   a. conduction.
   b. convection.
   c. radiation.
   d. all of the above.

10. An electric furnace transfers heat by
    a. conduction.
    b. induction.
    c. convection.
    d. radiation.

11. The liquid refrigerant in a heat pump boils at a
    a. very high temperature.
    b. very low temperature.
    c. temperature of 212°F (100°C).
    d. temperature of 200°F (93.3°C).

12. Compression of the vapor within a heat pump causes the temperature of the vapor to
    a. increase.
    b. decrease.
    c. remain the same.
    d. decrease slightly.

13. Two general classifications for heating controls are
    a. operating controls and sensing controls.
    b. safety controls and sensing controls.
    c. operating controls and safety controls.
    d. operating controls and monitoring controls.

14. The operating control that is installed in the plenum chamber of an electric furnace serves
    a. to operate the heaters.
    b. to stop and start the fan motor.

c. to disconnect the heaters from the supply if the temperature exceeds a predetermined value.
   d. all of the above.

15. Common voltage ratings for thermostats are
    a. 24 V, 120 V, 240 V.
    b. 120 V, 208 V, 480 V.
    c. 32 V, 120 V, 277 V.
    d. 16 V, 150 V, 280 V.

16. The room thermostat for an electric furnace controls the current to the
    a. fan.
    b. heaters.
    c. fan and heaters.
    d. entire heating system.

17. Duct heaters are similar to
    a. baseboard heating units.
    b. the electric furnace.
    c. ceiling panels.
    d. electric cables.

18. A heat pump
    a. heats the area to which it is connected.
    b. cools the area to which it is connected.
    c. both a and b.
    d. neither a nor b.

19. A humidity control is used to regulate
    a. the moisture content in the air.
    b. the temperature of the air.
    c. both moisture and temperature.
    d. the temperature of the moisture.

20. For health and comfort, the relative humidity should be between
    a. 25% and 35%.
    b. 30% and 40%.
    c. 35% and 45%.
    d. 40% and 50%.

21. A machine used to add moisture to the air is called a
    a. dehumidifier.
    b. humidifier.
    c. exhaust fan.
    d. humidistat.

22. Dielectric heating is based on the principle of
    a. capacitance.
    b. resistance.
    c. inductance.
    d. conductance.

23. Induction heating is based on the principle of
    a. capacitance.
    b. inductance.
    c. resistance.
    d. conductance.

24. Arc heating occurs when the air between two electrodes of opposite polarity becomes
    a. moistened.
    b. dry.
    c. ionized.
    d. polarized.

*Give Complete Answers*

1. List three methods used to transmit heat.

2. Describe one method to transmit heat.

3. Define the term *British thermal unit.*

4. What temperature is generally recommended to maintain body comfort?

5. What percent humidity is recommended for good health and body comfort?

6. What temperature and wind factor should be used when designing a heating system for a house in northeastern United States?

7. Whom might one contact for assistance in calculating heat loss?

8. List three advantages of electric heating systems.

9. What locations are recommended for electric baseboard units?

10. List three types of electric heating systems that have the advantage of individual control for each room.

11. What is an electric furnace?

12. Describe the operation of an electric furnace.

13. What are duct heaters?

14. What is a heat pump?

15. What is the purpose of the compressor used in conjunction with a heat pump?

16. List two types of controls designed for use with electric heating systems.

17. What is the purpose of the control that is built into every baseboard heating unit?

18. List the controls that must be used in conjunction with an electric furnace.

19. What is the purpose of a humidifier?

20. List two methods used to remove moisture from the air.

21. What is a humidistat?

22. Describe dielectric heating.

23. What is inductive heating?

24. What method of heating is used for spot welding?

25. List two uses for electric arc heating.

# DC Generators

## OBJECTIVES

After studying this chapter, the student will be able to:

- Explain the theory of electromagnetic induction.
- Describe the basic operation of the AC generator.
- Describe the construction and operation of various types of DC generators.
- Describe the uses and operating characteristics of various types of DC generators.
- Discuss the methods of connecting DC generators and the basic troubleshooting procedures.

# ELECTROMAGNETIC INDUCTION

In Chapter 6, we discussed the electron theory of magnetism and explained that a magnetic field is produced whenever an electric current flows. Another relationship between electricity and magnetism lies in a phenomenon called **electromagnetic induction**. Electromagnetic induction takes place whenever a conductor moves across a magnetic field (cuts the lines of force) or when a magnetic field moves across a conductor. When magnetic lines of force are cut by a conductor, a voltage is induced into the conductor. This voltage is called *electromotive force* (emf) because it is the force that causes the current to flow in a circuit.

When magnetism is used to produce electricity, it makes no difference whether the conductor moves through the field or the magnetic field moves across the conductor. It is the interaction between the two that develops electrical pressure (emf) (Figures 16–1A and 16–1B).

The direction of the emf can be determined by the use of the left-hand rule for a generator. Using the left hand, extend the thumb, index finger, and middle finger so they form right angles to one another. When the index finger points in the direction of the magnetic field (from north to south) and the thumb points in the direction of motion of the conductor, the middle finger will point in the direction of the induced emf (Figure 16–2).

# GENERATOR CONSTRUCTION

It is frequently stated that a generator is a machine that changes mechanical energy into electrical energy. This statement, although true, is somewhat misleading. A generator actually converts both mechanical and magnetic energy into electrical energy.

There are many ways to drive generators. For example, in northeastern United States, many generators are driven by steam turbines. The chemical energy of the fuel is converted into heat energy to make steam. The steam pressure drives the prime mover, which in turn drives the generator rotor. In large generators, the rotor contains the electromagnets. These magnets are rotated past coiled conductors wound on the **stator** (the stationary part of the generator). As the magnetic field moves across the stator coils, it induces an emf into them, producing electrical energy.

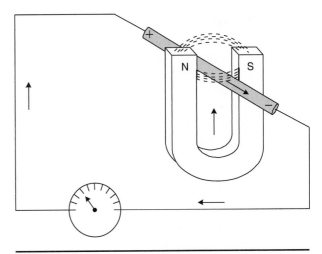

**FIGURE 16–1A** Moving a conductor up through a magnetic field (electromagnetic induction).

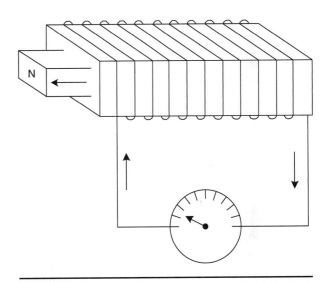

**FIGURE 16–1B** Moving a magnetic field through a coil (electromagnetic induction).

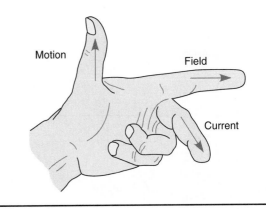

**FIGURE 16–2** Left-hand rule for a generator.

## Basic Generator

The simplest generator (the magneto) consists of a permanent magnet mounted on a frame (called the *yoke*). A coil of insulated wire mounted on a laminated iron core (called the **rotor** or **armature**) is positioned between the poles of the magnet. The armature is arranged so that it can rotate through the magnetic field (Figure 16–3). In small generators, such as those used in megohmmeters, the armature is rotated by hand.

The amount of emf produced by a generator depends upon the following factors:

- The strength of the main field flux (the number of lines of force per square inch).
- The number of loops of wire on the armature.
- The angle at which the armature coils move across the lines of force.
- The speed at which the coils rotate through the magnetic field.

A strong magnetic field produces a large emf. Many turns of wire on the armature also help to produce a large emf. When the coils move at right angles to the magnetic lines of force, they produce the most voltage. Coils moving parallel to the lines of force produce no voltage. The faster the coil rotates through the magnetic field, the greater the emf.

Because the armature revolves through the magnetic field, it does not produce a steady value of voltage. This can be understood by examining a single loop of wire as it rotates in a uniform magnetic field. Figure 16–4A illustrates the loop at the instant it is moving parallel with the lines of force. The two circles represent the ends of a single-loop coil. The meter indicates zero voltage. This is shown on the graph.

As the coil rotates and begins to cut across the field, a small emf is induced into the coil, and the meter indicates a low voltage (Figure 16–4B). Notice

that the graph also indicates a low voltage. When the coil is at a 36-degree angle to the lines of force (Figure 16–4C), the emf is 5.88 volts. At a 54-degree angle, the emf is 8.09 volts (Figure 16–4D), and at a 72-degree angle the emf is 9.51 volts, as seen in Figure 16–4E. At 90 degrees, the value is 10 volts (Figure 16–4F). At this point, the angle begins to decrease and the voltage decreases (Figures 16–4G through 16–4J), until the coil is again moving parallel with the lines of force and the voltage is zero (Figure 16–4K). Conductor *b* now begins to move down through the field and conductor *a* moves up. The meter indicates that the voltage has reversed direction (Figure 16–4L). The electrical pressure is in the opposite direction to what it was in Figures 16–4A through 16–4J. The graph indicates this change of direction by showing the values of voltage below the horizontal line.

In Figures 16–4K through 16–4U, it can be seen that the emf repeats the same values as before, but in the opposite direction. Because this force is first in one direction and then in the opposite direction, it is called an *alternating voltage*. The current that flows as a result of this voltage is called an *alternating current*. All rotating types of generators produce an alternating emf. Figure 16–4V shows the typical graph of an alternating voltage.

It may be said that an alternating emf is an electrical pressure that reverses its direction of force periodically. An alternating current is the flow of electrons first in one direction and then in the opposite direction for equal periods of time.

In order to increase the amount of voltage produced by the aforementioned generator, it is necessary to add more loops to the coil. Each added loop produces an emf equal to that of the first loop. If three loops are used, they form a series circuit and their voltages add together to give a maximum value of 30 volts.

From the graph it can be seen that the values of emf vary with the angle at which the coil moves across the field. It has also been stated that the number of loops in series on the armature determines the value of voltage. If an electromagnet is used in place of the permanent field magnet, the field strength can be increased. This causes a further increase in the generated emf (Figure 16–5).

The type of current obtained from a rotating coil depends upon the method of connecting the load circuit to the generator (the takeoff system). Figure 16–6 shows a method used to obtain alternating current. The ends of the coil from the armature are connected to solid brass or copper rings. These rings, called **slip rings**, are mounted on and insulated from the rotor shaft. They rotate with the armature. Stationary carbon brushes ride on the slip

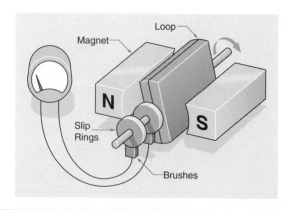

**FIGURE 16–3**   Parts of a generator.

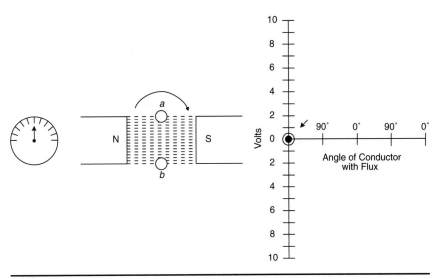

**FIGURE 16–4A** Loop of wire moving parallel to magnetic flux. No voltage is induced into the loop.

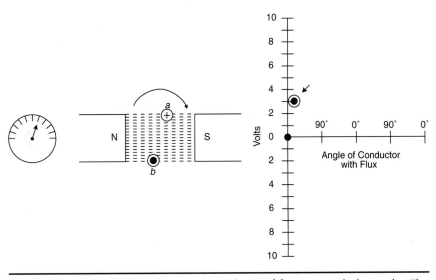

**FIGURE 16–4B** Loop of wire cutting lines of force at a slight angle. The induced emf is 3 V.

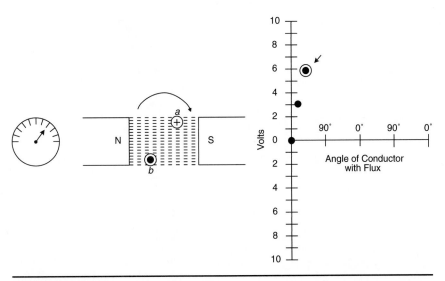

**FIGURE 16–4C** Loop of wire cutting lines of force at a 36-degree angle. The induced emf is 5.88 V.

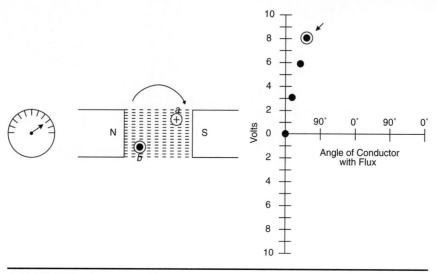

**FIGURE 16–4D** Loop of wire cutting lines of force at a 54-degree angle. The induced emf is 8.09 V.

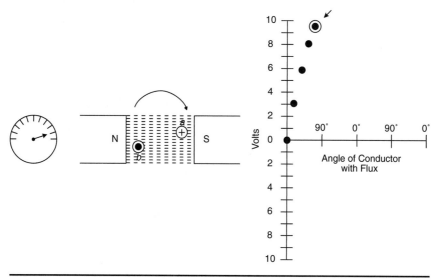

**FIGURE 16–4E** Loop of wire cutting lines of force at a 72-degree angle. The induced emf is 9.51 V.

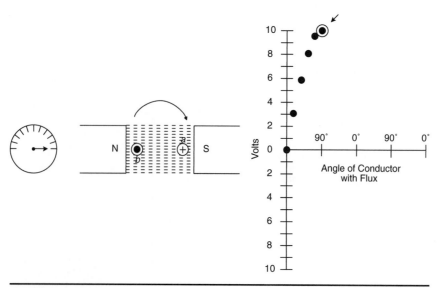

**FIGURE 16–4F** Loop of wire cutting lines of force at a 90-degree angle. The induced emf is 10 V.

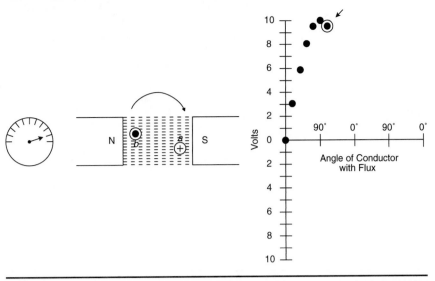

**FIGURE 16–4G** Loop of wire cutting lines of force at a 72-degree angle. The induced emf is 9.51 V.

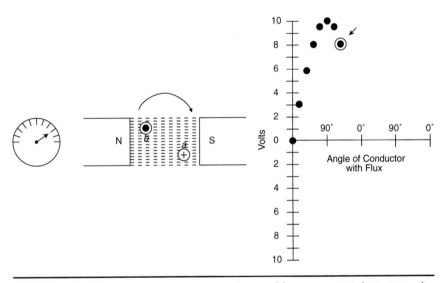

**FIGURE 16–4H** Loop of wire cutting lines of force at a 54-degree angle. The induced emf is 8.09 V.

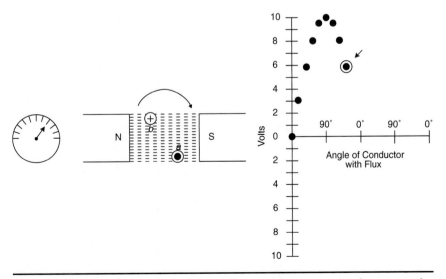

**FIGURE 16–4I** Loop of wire cutting lines of force at a 36-degree angle. The induced emf is 5.88 V.

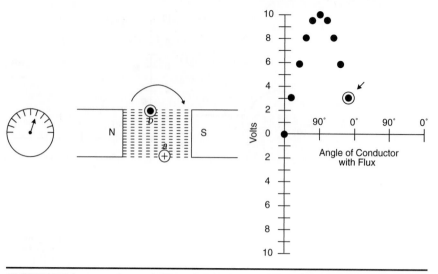

**FIGURE 16–4J** Loop of wire cutting lines of force at a slight angle. The induced emf is 3 V.

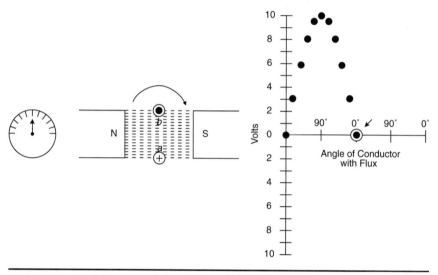

**FIGURE 16–4K** Loop of wire moving parallel to magnetic flux. No voltage is induced into the loop.

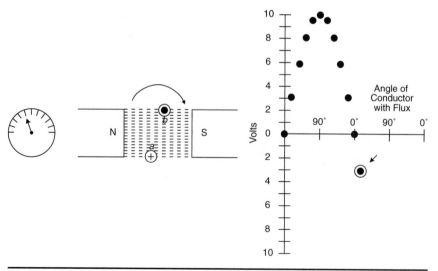

**FIGURE 16–4L** Loop of wire cutting lines of force at a slight angle. The induced emf is 3 V.

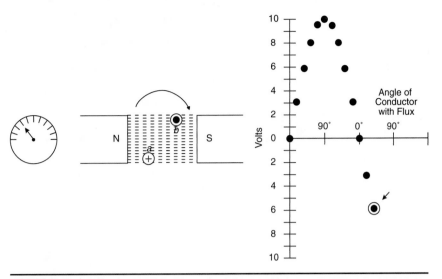

**FIGURE 16–4M** Loop of wire cutting lines of force at a 36-degree angle. The induced emf is 5.88 V.

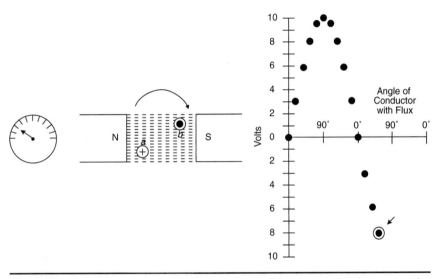

**FIGURE 16–4N** Loop of wire cutting lines of force at a 54-degree angle. The induced emf is 8.09 V.

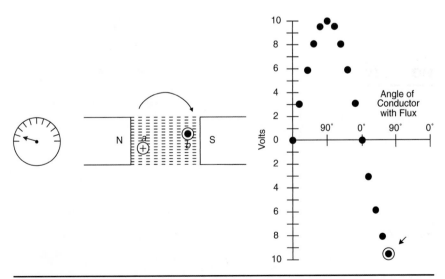

**FIGURE 16–4O** Loop of wire cutting lines of force at a 72-degree angle. The induced emf is 9.51 V.

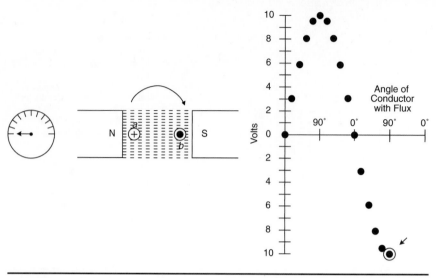

**FIGURE 16–4P**  Loop of wire cutting lines of force at a 90-degree angle. The induced emf is 10 V.

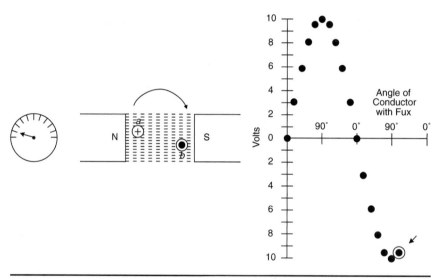

**FIGURE 16–4Q**  Loop of wire cutting lines of force at a 72-degree angle. The induced emf is 9.51 V.

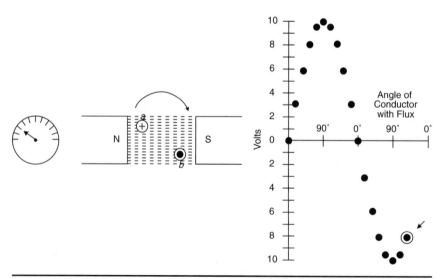

**FIGURE 16–4R**  Loop of wire cutting lines of force at a 54-degree angle. The induced emf is 8.09 V.

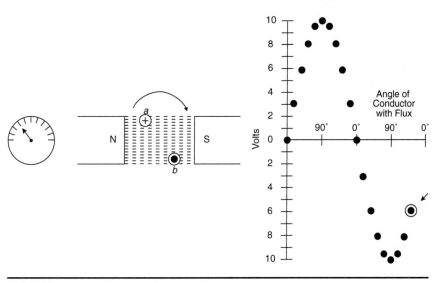

**FIGURE 16–4S**  Loop of wire cutting lines of force at a 36-degree angle. The induced emf is 5.88 V.

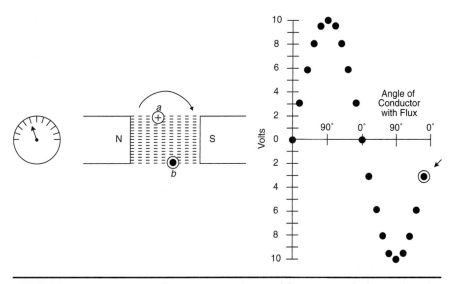

**FIGURE 16–4T**  Loop of wire cutting lines of force at a slight angle. The induced emf is 3 V.

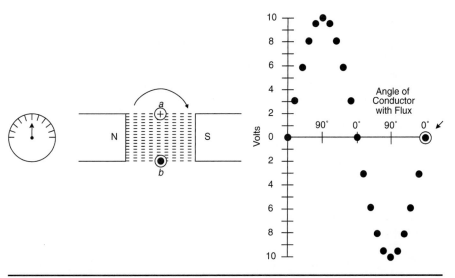

**FIGURE 16–4U**  Loop of wire moving parallel to magnetic flux. No voltage is induced into the loop.

rings and make contact with the load circuit. As the armature rotates, an alternating emf is produced that causes an alternating current to flow through the load. This type of generator is frequently called an *alternator*.

If direct current (current that flows only in one direction) is desired, a different takeoff device is needed. A commutator is used in place of the slip rings. A **commutator** is a type of rotating switch that changes the connections to the load circuit at the same instant as the emf reverses in the armature. Figure 16–7A illustrates the takeoff device for a DC generator.

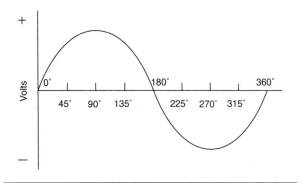

**FIGURE 16–4V** Typical graph of an alternating voltage.

Figure 16–7B shows a graph of the emf produced by the generator shown in Figure 16–7A. Even though the voltage and current reverse direction in the armature, the current continues to flow in the same direction in the load circuit. Note that the pressure is always in the same direction, but it is not a steady pressure. For most DC apparatuses, it is necessary to maintain a reasonably constant value of voltage.

One method used to reduce the pulsation of voltage is to add more coils of wire to the armature and more segments to the commutator. For every coil added to the armature, two segments are added to the commutator. The ends of each coil are connected to a pair of commutator segments. The more coils and segments there are, the smoother the DC output will be (see Figures 16–8A and 16–8B).

In order to maintain a steady direct voltage, the following conditions are necessary:

1. The main field flux must be uniform and steady.

2. The armature must rotate at a constant speed.

3. The armature must contain many coils; each coil must be connected to a pair of commutator segments.

Figure 16–9 shows a typical armature, including the commutator, for a DC generator.

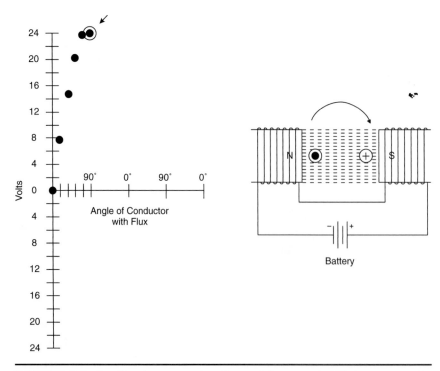

**FIGURE 16–5** Conductor cutting lines of force at a 90-degree angle. The induced emf is 25 V.

**FIGURE 16-6** AC generator (alternator) takeoff system.

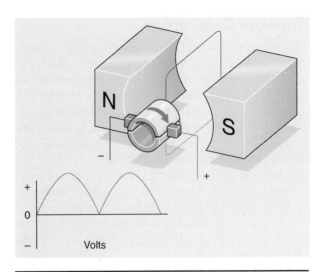

**FIGURE 16-8A** Single-coil generator and graph of DC output.

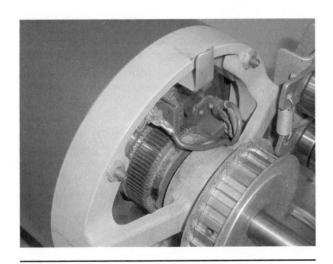

**FIGURE 16-7A** DC generator takeoff system.

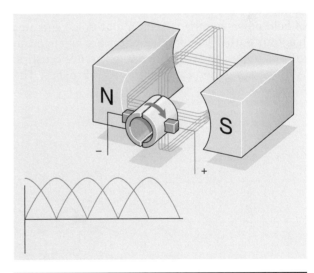

**FIGURE 16-8B** Two-coil generator and graph of DC output.

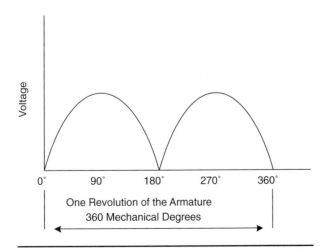

**FIGURE 16-7B** Graph of output voltage of a single-loop DC generator.

**FIGURE 16-9** Armature with commutator for a DC generator.

## Armature Construction

The armature coils of all rotating-type generators are wound on a core made of iron or mild steel. The core is mounted on a shaft set in bearings, which are inserted into the end bells. The entire assembly is arranged to rotate in the magnetic field of the generator. The core also serves as a path of low reluctance for the main field flux. Slots are cut lengthwise into the core and the windings are placed in the slots. The windings are arranged so that when one side of a coil is passing the north pole of the main field, the opposite side of the same coil is passing the south pole.

Two basic types of windings are in general use for DC generator armatures: the *lap winding* and the *wave winding*.

A lap winding is shown in Figure 16–10. Note that the ends of each winding are connected to adjacent commutator segments. The coils are arranged to form parallel paths, thereby allowing for more ampacity without increasing the voltage. On an armature of this type, there are two paths for each set of field poles. In other words, a four-pole generator has four paths. The parallel paths are arranged in groups of two and connected to one set of brushes. A four-path generator contains four brushes. The voltage across any one set of brushes is equal to the voltage across any other set and is equal to the output voltage. The maximum current output of the generator is equal to the sum of the ampacity of each path. Figure 16–11 shows a lap-wound armature and the field poles of a four-pole generator.

The wave-wound armature contains only two paths. If it contained the same size, number, and type of coils as a lap-wound armature, it would produce twice as much voltage. However, the maximum current output would be one half that of the lap winding. The induced emf for each coil of a single path adds up to the total voltage of that path.

Notice in Figure 16–12 that the coil ends are not connected to adjacent segments. Their position on the commutator is determined by the number of field poles. The distance between the ends of each coil is equal to the distance between field poles of the same polarity.

In summary, it can be said that a lap winding is used to obtain high current capacity, and a wave winding is used to obtain high voltage output. Various combinations of the two windings can be used to obtain the desired current and voltage combinations.

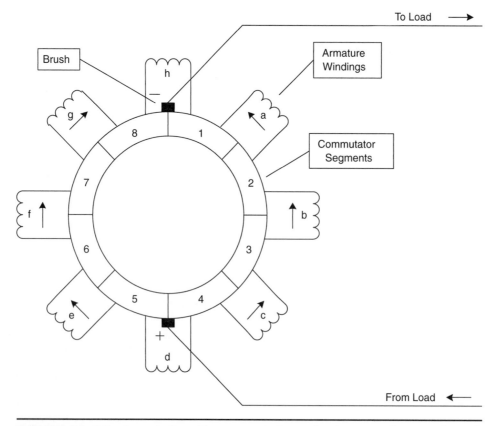

**FIGURE 16–10** Lap winding. The arrows indicate the direction of current through the armature coils.

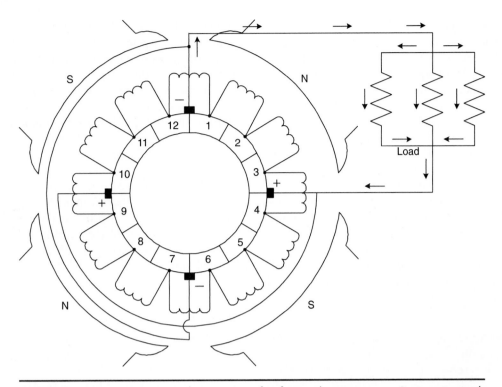

**FIGURE 16–11** Lap-wound armature of a four-pole generator. Current in each path adds to equal load current.

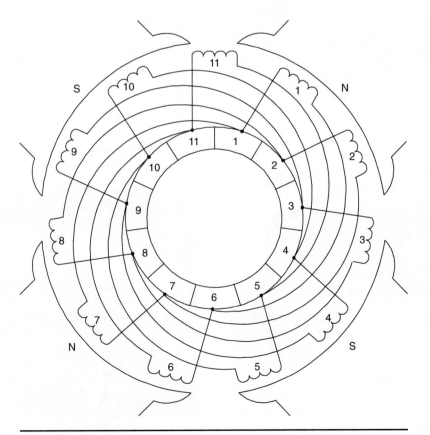

**FIGURE 16–12** Wave-wound armature of a four-pole generator.

The armature core is made of soft iron or mild steel disks. Each disk is about 0.025-inch (0.0635-centimeter) thick. Figure 16–13A shows a typical disk. In Figure 16–13B, the disks are mounted on the rotor shaft. These disks are commonly called *laminations*.

Prior to assembly of the disks on the shaft, they are dipped into an insulating varnish. The varnish insulates them from one another and from the shaft. This type of construction is necessary because the core is a good conductor of electricity. If the core is solid, rotating it in a magnetic field will result in large currents that circulate within the metal. These currents, called *eddy currents,* cause the core to generate enough heat to melt the insulation on the windings, causing a short circuit within the armature.

The armature core is slotted to hold the windings. The slots are lined with an insulation called *fish paper*, which serves as an additional insulation between the core and the windings. It also helps to prevent damage to the insulation on the windings. The windings expand and contract as the load is increased and decreased. If it were not for the fish paper, this continued rubbing against the metal core would wear away the insulation. The windings are held in the slots with fiber or plastic wedges. The sections of the coils not in the slots are held in place by band wires. Figure 16–14 shows a complete armature assembly.

The commutator is made of copper segments that are insulated from one another, and from the supporting rings, with mica. These segments are held firmly in place by clamping rings. The leads from the armature coils are soldered to the commutator segments.

## Brushes

The **brushes** connect the commutator to the load conductors. Brushes are usually made of graphite and carbon and placed in holders similar to the holder shown in Figure 16–15. The brushes

**FIGURE 16–13A** Typical disk (lamination) that makes up the rotor core.

**FIGURE 16–13B** Disks assembled and mounted on a shaft to form the armature core.

**FIGURE 16–14** Complete armature assembly for a DC generator.

**FIGURE 16–15** Brush holder for a DC generator.

should slide freely in the holders so as to follow the irregularities in the commutator. The brush pressure should be 1.75 to 2.5 pounds per square inch (0.124 to 0.176 kilogram per square centimeter). The edge of the brush holder should be 0.0625 to 0.125 inch (0.158 to 0.3175 centimeter) from the commutator. A spring is provided on the brush holder and arranged to adjust the brush pressure (Figure 16–15).

To decrease the electrical resistance of the brush, the upper portion is frequently plated with copper. The brush and plating are connected to the brush holder by a flexible copper wire. The brush holder is insulated from the frame of the machine.

## Frame and Field Poles

The *frame*, or *yoke*, of a generator serves as the mechanical support for the machine and forms part of the magnetic circuit. The field cores are attached to the frame. They are made of steel laminations and have a rectangular shape. The ends near the armature are flared in order to hold the coils in place (Figure 16–16).

The field coils are usually wound with cotton-covered wire. After the coil is formed, it is wrapped with cotton tape and immersed in an insulating enamel. It is placed in an oven, and the enamel is baked to a hard finish. The coil is then installed on the field cores between the frame and the armature.

## Field Excitation

In all generators, except the very small ones called **magnetos**, the field flux is produced by current flowing in coils placed on the pole pieces (field cores). The voltage for the field coils may be obtained from a separate source, such as batteries

**FIGURE 16–16** DC generator frame (yoke).

or another generator, or from the armature of the generator itself. When the field is excited from a separate source, the machine is called a *separately excited generator*. When the current is obtained from the machine's own armature, the machine is called a *self-excited generator*.

# GENERATOR OPERATION

The basic DC generator consists of coils of insulated wire rotating in a uniform magnetic field. An emf is produced as the coils rotate through the field. The emf is transmitted to the electrical system by way of the commutator and brushes.

## Effect of Armature Current

A magnetic field is produced whenever an electric current flows. The emf induced into the armature coils of a generator provides the voltage for the load current. Because the emf originates in the armature, the resulting load current must also flow through the armature. Current flowing in the armature produces a magnetic field, which reacts with the main field flux. This action causes the main field to become distorted. The result of the interaction between the two fields is known as **armature reaction** (Figures 16–17A and 16–17B).

When there is no load on the generator, the main field flux forms a direct path from the north pole to

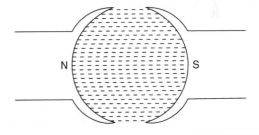

**FIGURE 16–17A** Magnetic field of a two-pole generator when the armature current is zero.

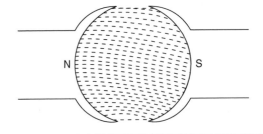

**FIGURE 16–17B** Magnetic field of a two-pole generator with current flowing in the armature coils.

the south pole (Figure 16–17A). As a load is applied, the armature flux causes the main field to bend (Figure 16–17B). The degree of bend depends upon the amount of current flowing in the armature. The more the current flowing, the greater the degree of bend.

## Neutral Plane

When the armature coils are moving parallel with the lines of force, no emf is induced within them. Figure 16–18A illustrates this phenomenon. The two circles (*a* and *b*) represent the ends of a single-loop coil. At the instant shown in Figure 16–18A,

the coil is moving parallel with the main field flux. If a line is drawn connecting these two points (Figure 16–18B), it will indicate the area known as the **neutral plane** of the generator. Therefore, it can be said that when the armature coils are moving parallel with the lines of force, they are moving through the neutral plane.

The neutral plane is as indicated in Figure 16–18B when there is no load on the generator. However, when a load is applied and current flows in the armature coils, the main field flux becomes distorted. This distortion causes the neutral plane to shift. The shift is always in the direction in which the armature

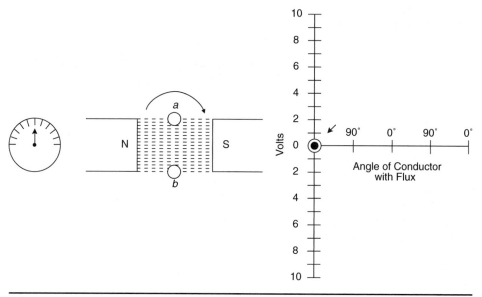

**FIGURE 16–18A**  Coil moving parallel with field flux. No emf is induced into the coil.

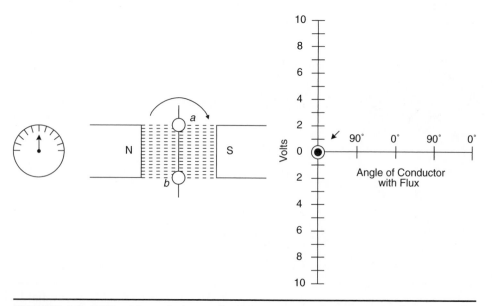

**FIGURE 16–18B**  Neutral plane of a DC generator. No current is flowing in the armature.

is rotating. The value of the current determines the degree of shift (Figure 16–18C).

Shifting of the neutral plane can be a serious problem if the generator is not designed correctly. The following situation illustrates the problems that may arise. Figure 16–19A represents a simple, two-pole generator connected to a load. (The armature connections indicate two current paths.) If the rotation is in a clockwise direction, the current flow is as indicated by the arrows. (This can be determined by using the left-hand rule for a generator.) The current flows out brush A, through the load, and back to brush B. At brush B, it divides equally, with 50 amperes flowing through the coils on the right and 50 amperes flowing through the coils on the left. Both currents return to brush A. At brush A, the two currents join and flow back out through the load.

It is important that the two paths in the armature be identical. The currents must divide equally, and the total emf induced into each path must be equal.

In Figure 16–19A, all coils are cutting the lines of force. Therefore, an emf is induced into all six windings. In Figure 16–19B, however, coils 3 and 6 are moving parallel with the lines of force, and no emf is produced. Coils 3 and 6 are short-circuited by the brushes. If an emf is induced into these coils, short-circuit currents will flow. The value of current depends upon the amount of emf and the resistance of the windings. If the current is large enough, overheating occurs.

Another important point to note is that each brush makes contact with two segments. Brush A is making contact with segments 3 and 4 while brush B is making contact with segments 1 and 6. As the armature rotates, the brushes will break contact with segments 1 and 4 and make contact with 3 and 6. If currents are flowing in coils 3 and 6 at the instant the contacts are broken, arcing will take place. The arcing will cause pitting of the commutator and excessive wear on the brushes. It is of extreme importance that the coils being short-circuited by the brushes always be in the neutral plane.

One way to eliminate this arcing is to move the brushes around the commutator into the new neutral plane (Figure 16–20). This method is successful only if the generator is supplying a constant load. The load on most generators increases and

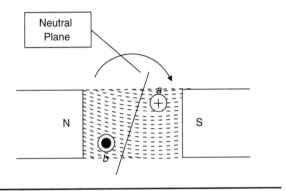

**FIGURE 16–18C** Neutral plane shifts in the direction of armature rotation.

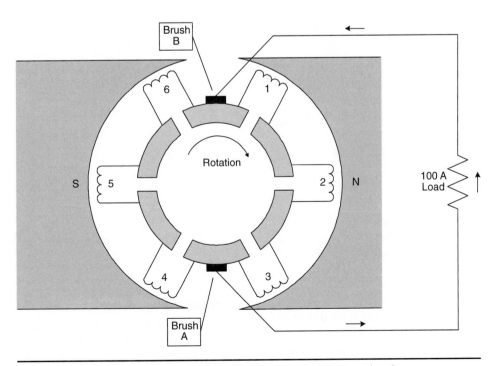

**FIGURE 16–19A** Two-pole generator supplying power to a load.

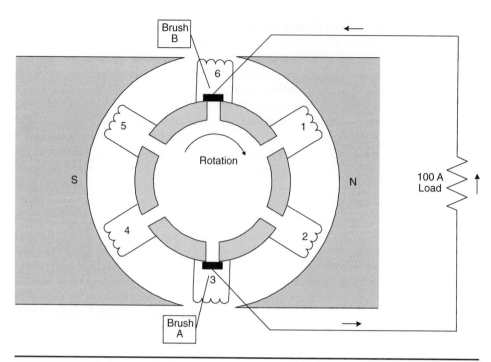

**FIGURE 16–19B** Two-pole generator: brushes short-circuiting coils 3 and 6.

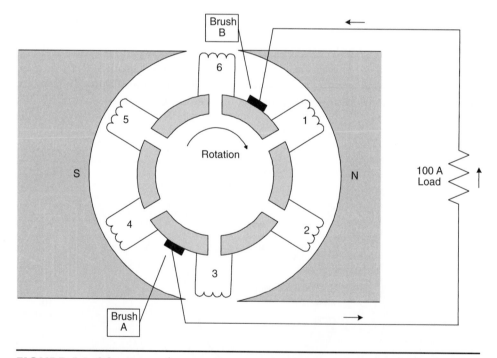

**FIGURE 16–20** Two-pole generator supplying power to a load. The brushes have been shifted to the new neutral plane.

decreases according to the demand on the system. Under these conditions, a generator needs a means of shifting the brushes with every change in load. A way to eliminate the problem is discussed later in this chapter.

## Armature Self-Induction

**Self-induction** is another phenomenon that takes place when current flows in a coil. When the current increases, the magnetic field caused by the current also increases. The field expands and moves across the loops on the coil. In other words, relative motion takes place between the conductors and the magnetic field. The coil does not move, but the magnetic field does. This relative motion induces an emf into the conductors of the coil. Therefore, two voltages exist across the coil:

1. The generated voltage, which is causing the current to flow
2. The voltage of self-induction (which is opposite to the applied voltage)

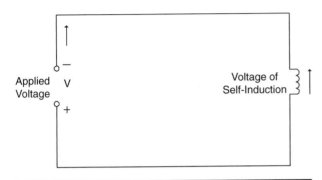

**FIGURE 16–21A** Current flow through a straight wire and load.

The end result is that it takes the current a longer period of time to reach a steady value (Figures 16–21A and 16–21B).

When the coil current decreases, the magnetic field collapses. This means that the field again moves across the conductors, but this time in the opposite direction. The induced emf caused by the collapsing magnetic field is in the same direction as the applied voltage. This tends to cause the current to flow even after the applied voltage has dropped to zero.

The current decreases as the armature coils of a generator move through the neutral plane. Theoretically, the current drops to zero. Because of self-induction, the current does not reach zero value until after the coil has moved beyond the neutral plane.

It can be said that a generator has three neutral planes (Figure 16–22). The no-load neutral plane, called the mechanical neutral plane, is midway between the main field poles. The magnetic neutral plane, caused by current flowing in the armature, is halfway between the mechanical neutral plane and the commutating plane. The electrical neutral plane, referred to as the commutating plane, is ahead of the magnetic plane by a distance equal to the distance between the mechanical plane and the magnetic plane.

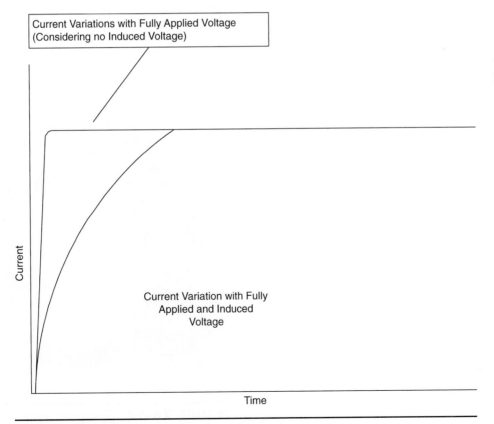

**FIGURE 16–21B** Current flow through a coil.

The brushes must be set in the commutating plane in order to obtain good commutation (to eliminate arcing). The commutating plane shifts with changes in the load. Therefore, this method of reducing arcing at the brushes is practical only for a constant load. To set the brushes to the neutral plane, refer to Figure 16–23 and:

1. Connect an AC voltmeter across the shunt field winding.
2. Obtain a low-voltage AC source.
3. Connect the AC source across the armature.
4. Carefully move the brush assembly back and forth, noting the change in voltage as indicated on the AC voltmeter.
5. Place the brush assembly in the location that produces the lowest indicated voltage on the AC voltmeter.
6. The brushes are now located at the neutral plane.

## Interpoles (Commutating Poles)

Because most generators supply power to varying loads, some means other than shifting the brushes must be provided. One method is to install small poles midway between the main poles. These poles, called **interpoles** or *commutating poles*, are connected in series with the armature. As the load changes, the current through the interpoles changes by the same amount. Figure 16–24 illustrates a two-pole generator with interpoles.

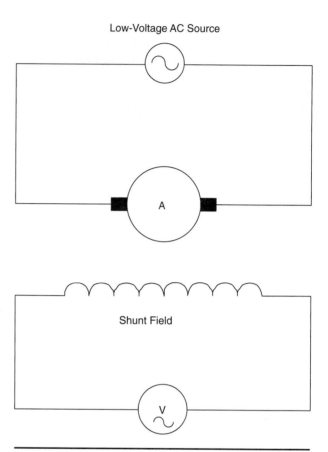

**FIGURE 16–23** Connections for setting the neutral plane.

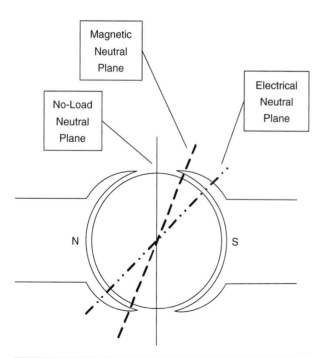

**FIGURE 16–22** Neutral planes in a DC generator.

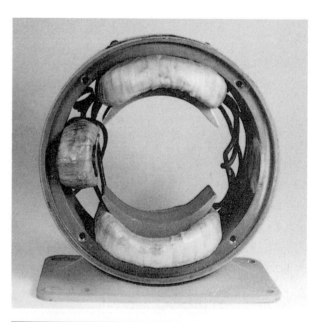

**FIGURE 16–24** Two-pole generator with one interpole.

When interpoles are properly designed and installed, they produce a flux that neutralizes the flux set up by current flowing in the armature. Besides canceling the flux produced by armature reaction, the interpoles also induce an emf into the armature coils that are undergoing commutation. This emf is equal and opposite to the emf of self-induction. The final result is that no current flows in the coils that are short-circuited by the brushes. Thus, there is no arcing at the brushes.

## Compensating for Armature Reaction

Interpoles provide good commutation, but they do not eliminate armature reaction. Another method (although more costly) is to install **compensating windings**. Compensating windings are placed in the main pole faces and are connected in series with the armature. They are arranged so that each conductor embedded in the field pole carries a current equal and opposite to the adjacent armature conductor. Figure 16–25 shows a generator with compensating windings.

Although the compensating windings eliminate armature reaction, they do not solve the problem of self-induction. To eliminate arcing, the brushes must be set slightly ahead of the mechanical neutral plane (in the direction of rotation). If the brushes are set in the proper location, the coils undergoing commutation will be cutting enough flux in the opposite direction to generate a voltage equal and opposite to that of self-induction. It is almost impossible to eliminate arcing completely, but with careful engineering and proper brush placement, it can be reduced to a minimum.

Generators with interpoles are more common than machines with compensating windings. Interpole machines are much less expensive to build. Compensating windings are generally found on large generators that operate at high speeds and produce high voltages. Some machines that serve very heavy loads under wide load variations and that operate at high speeds use both interpoles and compensating windings.

## Other Effects of Armature Current

The magnetic fields set up by current flowing in the armature distort the main field flux and also produce a magnetomotive force (mmf) that opposes the main field flux. The result is a weakening of the overall flux and a decrease in the generated emf. Therefore, there are two factors to consider that affect the output voltage of a generator: the resistance of the wire on the armature causes a voltage drop, and armature reaction reduces the generated emf.

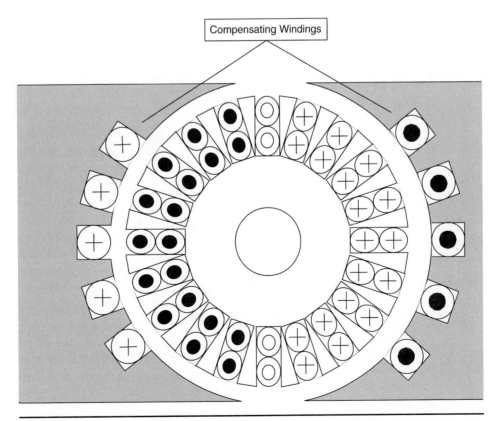

**FIGURE 16–25** DC generator with compensating windings.

It can be said that as the load increases on a generator, the output voltage tends to decrease.

Reversed torque developed in the rotor is another result of armature reaction. In Figure 16–26, notice that the rotor is being driven in a clockwise direction. The polarity of the armature is such that the north pole of the armature is driven toward the north pole of the main field, and the south pole of the armature is driven toward the south pole of the main field. Because like poles repel, the prime mover (the machine driving the generator) must overcome this repelling force. Therefore, as the load is increased on the generator, the prime mover must develop more torque in order to maintain a constant rotor speed. This phenomenon is frequently called *motor action in a generator*.

## GENERATOR VOLTAGE

The amount of emf induced into a conductor depends upon the speed at which it is cutting the flux. When a conductor cuts flux at the rate of $10^8$ lines of force per second, 1 volt is induced. By utilizing this fact, we find it possible to derive an equation that will give the average emf produced by a generator. The equation is as follows:

$$E_G = \frac{PZ\Phi N}{10^8(60b)} \qquad \text{(Eq. 16.1)}$$

where  $E_G$ = generated emf, which at no load is the same value as the terminal voltage

$P$ = number of poles in the main field

$\Phi$ = number of lines of force per pole

$N$ = armature speed, in revolutions per minute (r/min)

$b$ = number of parallel paths through the armature

$Z$ = total number of conductors (inductors) on the armature. Because there are 2 inductors per turn, the total number of inductors is equal to 2 times the number of turns.

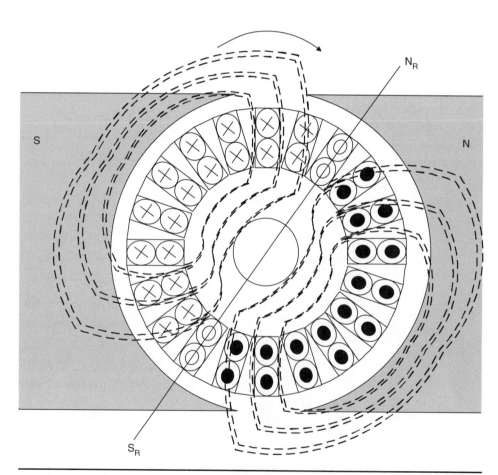

**FIGURE 16–26**  Motor action in a generator.

## Example 1

Determine the average voltage generated in a 6-pole machine, running at 900 r/min, if the armature has 300 conductors cutting the field. The flux per pole is $5 \times 10^6$ magnetic lines. The armature has 6 parallel paths.

$$E_G = \frac{PZ\Phi N}{10^8(60b)}$$

$$E_G = \frac{6 \times 300 \times 5 \times 10^6 \times 900}{10^8 \times 60 \times 6}$$

$$E_G = 225 \text{ V}$$

For a specific generator, all of the factors in the equation are fixed values with the exception of the speed ($N$) and the flux per pole ($\Phi$). Therefore, the letter $K$ may be substituted for all values except $N$ and $\Phi$. The equation now becomes

$$E_G = K\Phi N \qquad \text{(Eq. 16.2)}$$

where   $K$ = combined result of all the fixed values

## Example 2

Determine the value of $K$ for Example 1.

$$K = \frac{PZ}{10^8(60b)}$$

$$K = \frac{6 \times 300}{10^8 \times 60 \times 6}$$

$$K = 5 \times 10^{-8}$$

Proof:

$$E_G = K\Phi N$$

$$E_G = 5 \times 10^{-8} \times 5 \times 10^6 \times 900$$

$$E_G = 225 \text{ V}$$

The formula $E_G = K\Phi N$ can be useful for designing generators. In practice, however, generators usually operate at a fairly constant speed, and the only variable is the field flux ($\Phi$). A rheostat, or a similar device, is frequently installed in the field circuit to provide a means of controlling the field current. Varying the field current will vary the flux and thus the generated emf.

## Saturation Curve

The equation $E_G = K\Phi N$ indicates that the induced emf is proportional to the flux per pole and the revolutions per minute of the generator. At a constant speed, the generated emf is proportional to the field strength. In a given machine, the flux depends upon the field current. It is not, however, directly proportional to the field current, because beyond a certain number of ampere-turns, all electromagnets become saturated. Figure 16–27A is a graphic example of this feature. Because of the residual magnetism, the curved part at point *a* does not start at zero. Between points *a* and *b*, the curve is almost a straight line, indicating that the flux in this area is proportional to the field current. At point *b*, the line begins to curve sharply, indicating that the magnetic circuit is reaching saturation.

Because the generated emf varies directly with the field flux, the voltage curve is the same as the flux curve (Figure 16–27B). Between points *a* and *b*, the voltage increases rapidly with a given change in field current. Beyond point *b*, a large increase in current causes only a slight increase in voltage. The voltage curve will vary somewhat with the design of the machine.

## SELF-EXCITED GENERATOR

As discussed previously, a self-excited generator receives the current for the field from its own armature. This can be accomplished because of residual magnetism in the field cores.

There are three types of self-excited generators, depending upon how the armature and the field windings are connected. These types are the shunt generator, the series generator, and the compound generator. Figure 16–28 illustrates the connections for a self-excited shunt generator. It can be observed that there is no external source of power for the field. As the prime mover drives the generator, the armature rotates in a very weak magnetic field. This field is produced by residual magnetism in the field cores. The armature conductors, cutting the residual field, produce a small emf. Because the shunt field windings are connected directly across the armature, a small current flows through the shunt field. This current causes an increase in the field flux. A stronger field flux produces a larger emf, which forces even more current through the field coils. The flux density again increases, which induces a greater emf into the armature, again increasing the field current. This process continues until the field cores are saturated. At this point, the generator has reached its no-load voltage. The entire building-up process takes about 20 seconds to 40 seconds.

If a self-excited generator is being driven by the prime mover but fails to build up a voltage, there are several possible causes.

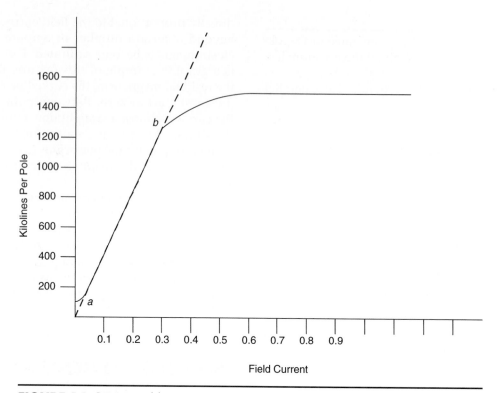

**FIGURE 16–27A**  Field current versus lines of force.

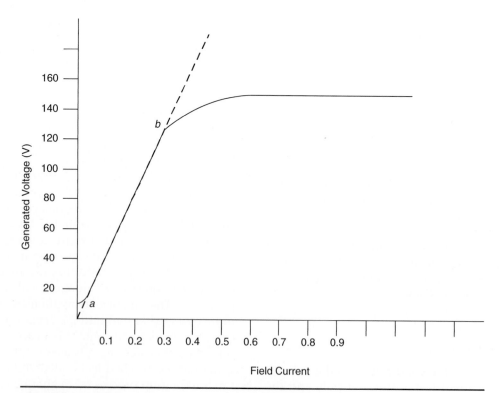

**FIGURE 16–27B**  Field current versus generated voltage.

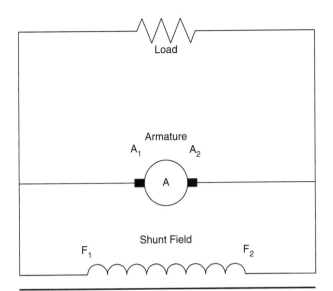

**FIGURE 16–28** Schematic diagram of a shunt generator.

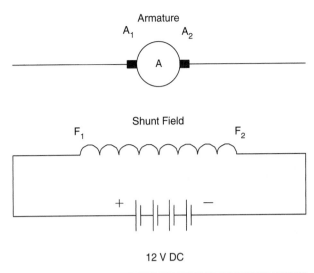

**FIGURE 16–29** Connections for flashing the field.

## Causes

1. Loss of residual magnetism may be a cause. If the generator sits idle for a long period of time, it may lose its magnetism. If it has been moved, the jarring it received may have caused the loss of magnetism. Overloads and current surges could also be a cause.

2. If the voltage induced into the armature forces the current through the field in a direction that produces an mmf in opposition to the residual magnetism, it will not build up a voltage.

3. A break or opening in the field or armature circuit will prevent a voltage buildup.

4. Loose brush connections or contacts may be a cause.

5. A dirty or severely pitted commutator frequently prevents voltage buildup.

6. A short circuit in the armature or field may be the problem.

7. If the speed of rotation of the armature is too slow, the generator will not produce its rated voltage.

8. If the armature is rotating in the wrong direction, the magnetic fields will oppose each other.

9. If the field circuit resistance is too great, the voltage will not build up.

The problems listed can be eliminated as follows:

## Solutions

1. Loss of residual magnetism. Disconnect the field from the armature, being sure to note the field polarity. Connect a DC source across the field as seen in Figure 16–29. It is important to maintain the same polarity that the field had before it lost its residual magnetism. It is best to use a low-voltage DC source for this purpose. For a 120-volt to 600-volt generator, a 12-volt automobile battery would be sufficient. Allow the DC source to remain connected for approximately 10 minutes. This is called "flashing the field."

Disconnect the DC power and reconnect the armature and the field. Be sure to maintain correct polarity. The generator should now build up a voltage.

> ⚠ **CAUTION** If a voltage greater than 0.1 times the rating of the generator is used, discharge resistors must be connected in parallel with the field. The purpose of these resistors is to absorb the power dissipated by the collapsing magnetic field when the DC supply is removed. The rating of this resistance bank depends upon the value of voltage used, the resistance of the field, and the inductive effect. Never use a voltage greater than the rating of the generator.

If the polarity of the residual magnetism has been reversed, the polarity of the generated emf will be reversed. Reversal of polarity can damage some types of equipment.

> ⚠ **CAUTION** It is important to check the polarity before restoring the generator to the line. If the residual magnetism is reversed, it will be necessary to recharge the residual field. After ensuring the correct polarity, apply the DC and allow it to remain connected for about 10 minutes.

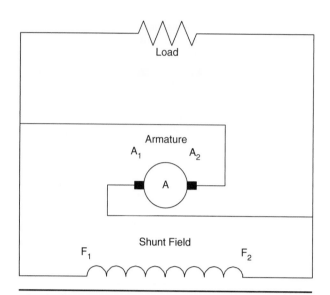

**FIGURE 16-30** Interchange the armature leads to change the polarity.

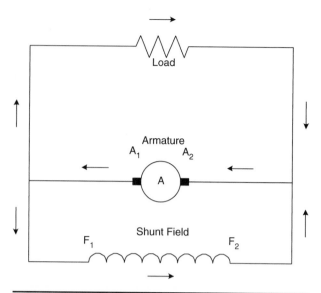

**FIGURE 16-31** Shunt generator. The arrows show the direction of current flow.

2. The generator fails to build up a voltage because of reversed polarity on the armature. Reversing the armature connections will correct this condition. Figure 16–30 illustrates the process.

3. Open circuits in the field or armature. This can be corrected by locating the open circuit and taking necessary steps to remedy it.

4. Poor brush contact or loose brush connections. Check the brushes for excessive wear, and replace them if necessary. Check the commutator for pitting. If necessary, "turn down" the commutator. This is accomplished by using a lathe to remove a small amount of copper, just enough to obtain a smooth surface. If necessary, undercut the mica so that it is slightly below the commutator segments. Always clean the commutator when poor brush contact is discovered. Check the brush tension and readjust it if necessary. Tighten any loose connections.

5. A dirty and/or pitted commutator. In this case, follow the same procedure as outlined in Step 4.

6. A short circuit in the armature or field. Short circuits should be located and cleared. A thorough inspection should be performed to determine the cause.

7. The armature is rotating too slowly. Increase the speed of the prime mover.

8. The armature rotation is reversed. The prime mover must be adjusted to drive it in the correct direction.

9. The field circuit resistance is too great. This can usually be corrected by adjusting the rheostat in the field circuit.

## Shunt Generators

A *shunt generator* is a type of self-excited generator in which the field and armature are connected in parallel. (Figure 16–28 illustrates the connections for this type of generator.) The building-up process for a shunt generator is the same as described in the section on self-excited generators. When the magnetic circuit of the field has become saturated, the field coils will receive full armature voltage.

The field coils are constructed of many turns of small wire. The use of small wire and many turns produces a strong field while keeping the current to a minimum. This results in a more economical and compact construction, and it improves the efficiency of the machine.

Because the armature supplies the emf for both the load and the field, all the current must flow though the armature. Figure 16–31 is a schematic drawing for a shunt generator. The arrows indicate the current flow.

The armature current increases as a load is added to a shunt generator. An increase in the armature current causes an increase in the effects of armature reaction. In other words, as the load current increases, the generated emf decreases. This decrease in the generated emf results in a lower output voltage. Another factor that affects the output voltage of a generator is the IR (voltage) drop in the armature.

Because the field receives its voltage from the armature, any decrease in terminal voltage results in less field current. The reduction in field current causes a weakening of the main field flux. The weaker field results in a further decrease in the output voltage.

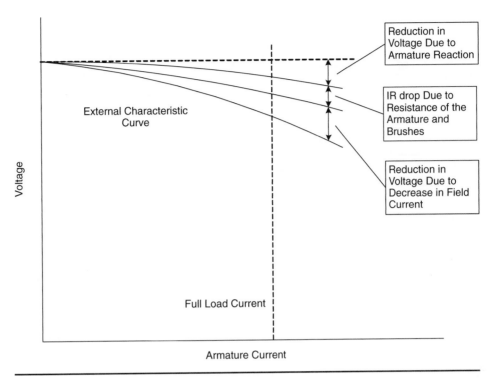

**FIGURE 16-32** Load/voltage characteristics of a shunt generator.

It may appear from this information that the generated emf would quickly drop to zero. This does not occur, however, due to the internal design of the generator. Figure 16–32 is a graphic illustration of the load/voltage characteristics of a shunt generator.

It is generally desirable that the voltage across the load be a constant value. However, shunt generators still are used in industry because they are economical and provide efficient means for supplying power to constant loads. Well-designed generators do not have more than an 8 percent decrease in voltage from no load to full load. These machines are suitable when slight voltage fluctuations are not a problem.

One advantage of the load/voltage characteristics of the shunt generator is self-protection. A short circuit on the load will cause the output voltage to drop rapidly (Figure 16–33). This sharp drop in voltage reduces the current to a minimum and, thus, prevents overheating.

## Series Generators

A *series generator,* as the name implies, has the field connected in a series with the armature. Figure 16–34 shows these connections. The arrows indicate the current flow. Notice that the entire load current flows through the field windings. For

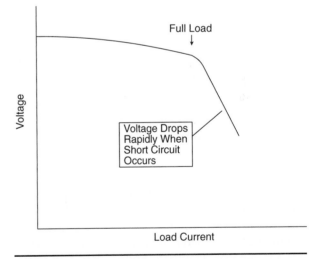

**FIGURE 16-33** Voltage curve of a shunt generator when a short circuit occurs across the armature.

this reason, the field coils must be wound with wire large enough to carry the full load current of the generator.

In a series generator, an increase in load causes a similar increase in the field current. This results in both an increase in the generated emf and a higher output voltage. The increase in voltage with increases in load will continue until the magnetic circuit of the field is saturated. An increase in

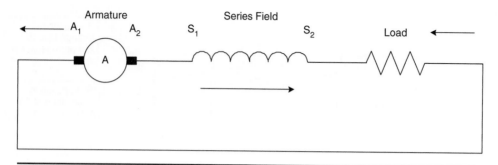

**FIGURE 16-34** Schematic diagram of a series generator.

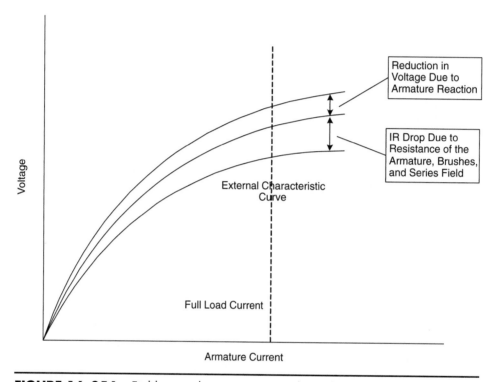

**FIGURE 16-35A** Field strength versus output voltage for a series generator.

load beyond this point will cause the output voltage to decrease. The decrease is a result of IR drop and armature reaction. Figures 16–35A and 16–35B demonstrate the load/voltage characteristics of a series generator. The load/voltage curve in these figures can be compared to the saturation curve.

In the past, series generators were used extensively for series arc lighting, particularly for street lighting. Later, when arc lamps were replaced by series tungsten lamps, the series generator still supplied the power. Most tungsten lamps used for this purpose have been replaced by mercury vapor or high-pressure sodium lamps. These lamps require alternating current. Modern technology has practically phased out the series generator. However, it is sometimes found in remote areas or in applications where it is supplying constant loads.

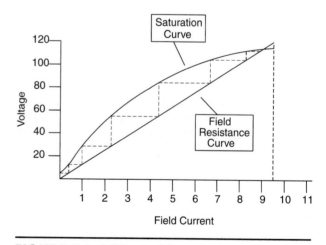

**FIGURE 16-35B** Field strength versus output voltage for a series generator (continued from Figure 16–35A).

## Compound Generators

In industry, most loads require a constant value of voltage, but the load on the generator fluctuates considerably when a plant is in operation. This type of demand makes both the shunt and series generators undesirable for most conditions of use. However, it is possible to maintain a constant voltage under varying loads by combining certain features of the shunt and series generators into one machine. This type of machine is called a *compound generator* (Figure 16–36).

If the series winding is connected so that its magnetic field will aid the shunt magnetic field, it is called a cumulative-compound generator. When the load current increases in a shunt machine, the main field flux is weakened as a result of armature reaction and IR drop. In a cumulative-compound generator (with the series field aiding the shunt field), the results are somewhat different. The load current flowing through the series field causes an increase in the flux. If the increase in flux is equal to the decrease caused by armature reaction and IR drop, the output voltage remains constant.

The load/voltage characteristics of a compound generator depend upon the number of turns on the series field. With many turns on the series field, the machine begins to assume the characteristics of a series generator. Few turns result in predominantly shunt characteristics.

The number of turns on the series field determines the amount of compounding. A machine with many turns on the series field is called an overcompounded generator. A generator with just enough series turns to maintain a steady voltage from no load to full load is a flat-compounded generator. One with fewer turns is said to be an undercompounded generator. Most compound generators are designed for overcompounding. The degree of compounding is determined by a *diverter* (a resistance of specific value) connected in parallel with the series field (Figure 16–37). The load/voltage curves for over-, under-, and flat-compounded generators are depicted in Figure 16–38. In practice, flat compounding does not actually produce a flat curve. The nearest approach is to adjust the machine so that the terminal voltage rises slightly and then drops again, reaching the same value at full load as at no load.

Flat-compounded generators are used when the load is located near the generator. If the load is located some distance away, an overcompounded machine is used. The overcompounding will compensate for the voltage drop in the line wires. The amount of overcompounding used depends upon the type of service. For generators in which the load is a great distance away, 10 percent overcompounding is common. Generators supplying power to street railway, monorail, and subway systems fall into this category.

The differential-compound generator is another type of compound generator. This machine is connected in such a way that the series field mmf opposes the shunt field mmf. In a machine of this type, the voltage drops off sharply as load is added. This machine is used most commonly in arc welding. When a heavy load occurs, the series field flux tends to neutralize much of the shunt field flux. The result is a decrease of the overall flux. The induced emf drops, reducing the armature current and preventing the armature from overheating.

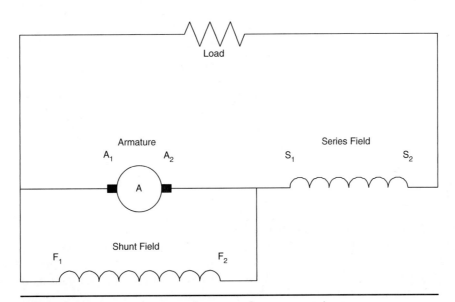

**FIGURE 16–36** Schematic diagram of a compound generator.

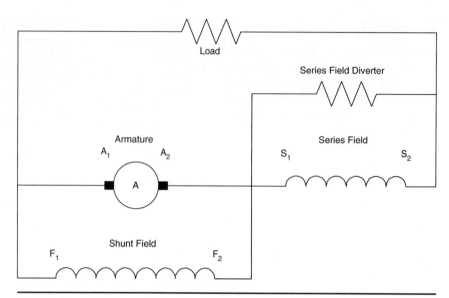

**FIGURE 16–37** Compound generator with series field diverter.

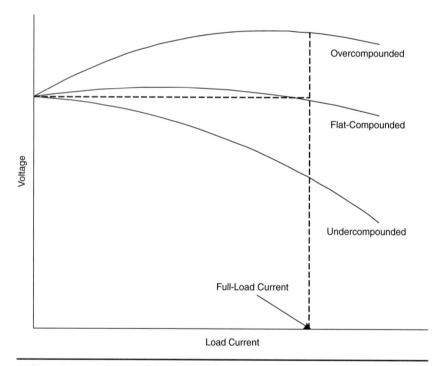

**FIGURE 16–38** Load/voltage characteristics of compound generators.

# SEPARATELY EXCITED GENERATOR

In a separately excited generator, the magnetization current for the field coils is supplied from a source outside of the generator. This supply may be another DC generator, batteries, or a rectifier. Figure 16–39 illustrates the connections for this type of generator.

If a separately excited generator is operated at a constant speed and with a constant field voltage, the terminal voltage at no load will be equal to the generated emf. When a load is applied, terminal voltage will be less than the generated emf. This decrease in voltage is caused by armature reaction and IR drop. Figure 16–40 depicts the load/voltage characteristics of the separately excited generator.

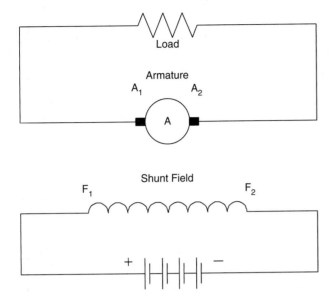

**FIGURE 16-39** Schematic diagram of a separately excited generator.

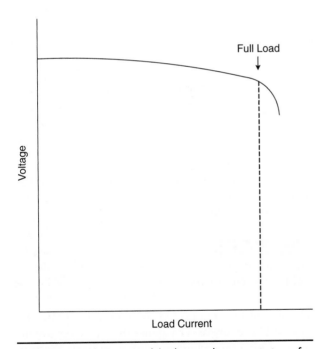

**FIGURE 16-40** Load/voltage characteristics of a separately excited generator.

Because the field current is independent of the armature emf, the magnetic flux of the field is less affected by load changes than it is in the self-excited generator. One method for controlling the terminal voltage of the separately excited generator is to control the field current. This may be accomplished by installing a rheostat in the field circuit. If a constant voltage is required for all loads, the voltage source must be great enough to maintain the rated output voltage at full load.

Separately excited generators are used only for special installations because of the high cost of construction and their physical size.

## VOLTAGE CONTROL VERSUS VOLTAGE REGULATION

The load/voltage curves for a generator illustrate the ability of the machine to regulate its output voltage with changes in load. A generator that maintains a nearly constant output voltage from no load to full load has excellent regulation. Voltage regulation, therefore, is determined by the design of the machine.

**Voltage control** takes place outside of the generator. It is generally accomplished by controlling the current through the shunt field. One method of achieving this is to install a rheostat in the field circuit. This method is used on separately excited generators, shunt generators, and compound generators. On series generators, a variable diverter is connected in parallel with the series field.

Hand control of generators is not satisfactory for all conditions of use. In most installations, an automatic control device is preferable. A sensing device is installed to sense any changes in voltage and/or current output. This device actuates relays or electronic circuits that control the amount of current in the main field of the generator.

## PARALLEL OPERATION OF GENERATORS

More than one generator is used when a large amount of DC power is required, such as in metal-refining plants, paper mills, electric railway systems, and marine work. In such applications, the generators are connected in parallel. Parallel operation provides a very efficient system. During reduced loading periods, some generators can be removed from the line. All generators perform at maximum efficiency when operating at full load. Preventive maintenance can be performed on the generators not in use. If a generator breaks down, it can be removed from the line and another one can be placed in service with minimum downtime.

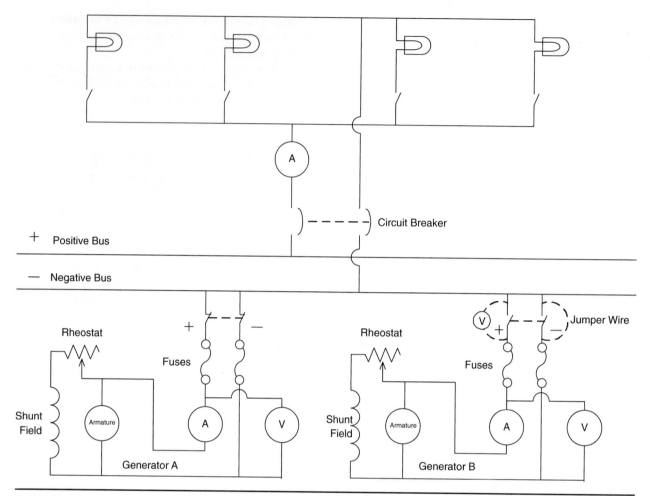

**FIGURE 16–41** Shunt generators connected in parallel.

## Shunt Generators in Parallel

Certain procedures must be performed in order to connect generators in parallel. The following procedure is recommended for shunt generators (Figure 16–41; generator circuit breakers have been omitted in order to simplify the drawing.) Generator A is supplying power to the load. It is operating at nearly full capacity. Additional equipment must be put into operation, requiring more DC power. This requires placing another generator on the line.

Power is supplied to the prime mover of generator B. When generator B has reached its rated speed, the output voltage is adjusted to the same value as generator A. A polarity test is performed to ensure that the positive terminal is connected to the positive bus and the negative terminal is connected to the negative bus.

⚠️ **CAUTION** Reversed polarity can cause serious hazards. Therefore, a polarity test is very important.

A polarity test is accomplished as follows. With the switch for generator B open, connect a jumper wire from the line side to the load side of one pole.

⚠️ **CAUTION** Take great care in making this connection, because both the load and line sides of the switch are energized.

The line side is energized by generator B and the load side is energized by generator A. Select a voltmeter with a range high enough to indicate the sum of the output voltage of both generators. Connect the voltmeter from the line to the load side of the other pole. The meter should indicate zero. If it indicates the sum of the output voltages of A and B, the polarity is reversed. To correct this condition, remove the jumper wire and voltmeter, and shut down the prime mover for generator B. Then, interchange the positive and negative lines from generator B. Start the prime mover and bring generator B up to its rated speed, adjusting the output voltage to the same value as A. With the values of voltage equal, reconnect the jumper wire and voltmeter. The voltmeter should now indicate zero. If the meter indicates a low value of voltage, adjust the rheostat of B until the meter indicates zero.

With the output voltage and polarity correct, close the switch to generator B. Disconnect the jumper wire and the voltmeter. Generator B is now floating on the line; it is not supplying power to the load.

In order to have generator B supply power to the load, it is necessary to increase the voltage of B while decreasing the voltage of A. During this procedure, the bus voltage must be kept at a constant value. This is accomplished by decreasing the voltage of A by the same amount as B is increased, and at the same time. By way of this procedure, the load can be gradually shifted from A to B in any desired amount. The additional load can thus be added to the line without overloading either generator.

To remove a generator from the line, first check the load to be sure the remaining generator(s) can carry it. Assume that generator A (Figure 16–41) is to be removed. Increase the voltage of generator B while decreasing the voltage of A. Be sure that the bus voltage remains constant. Check the ammeters to observe the load shift, and continue adjusting the voltages until the ammeter for A indicates zero. Generator A is now floating on the line, and the switch can be opened.

## Compound Generators in Parallel

When connecting compound generators in parallel, one must give special consideration to the series field. Reaction of the series field to momentary changes in voltage and/or load currents can cause shifting of the load from one machine to the other. The load shifting becomes cumulative, and, eventually, one machine carries all the load, usually resulting in an overload.

In order to prevent the possibility of unwanted **load shifting**, connect the series field of each machine to a common **equalizer bus**. In effect, the series fields are connected in parallel with each other (Figure 16–42). The result is that the series field currents divide proportionally between the generators.

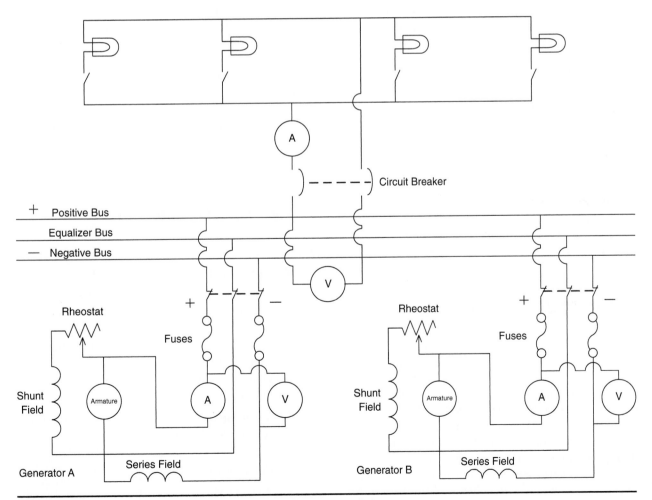

**FIGURE 16–42** Compound generators connected in parallel.

> **⚠ CAUTION** All the safety precautions for connecting shunt generators in parallel should be followed when connecting compound generators in parallel.

The general procedure is as follows. Generator A (Figure 16–42) is supplying power to the load. Bring generator B up to speed and adjust the output voltage to match the bus voltage. Connect a jumper wire across the positive side of the three-pole switch for generator B. Select a voltmeter with a range high enough to indicate the sum of the output voltages of both generators.

Connect the voltmeter across the negative side of the three-pole switch. The meter should indicate zero. If it indicates the sum of the output voltages of A and B, the polarity of generator B is reversed. To correct this condition, remove the jumper wire and voltmeter, and shut down the prime mover for B. Then, interchange the positive and negative lines from B. Start the prime mover and bring generator B up to its rated speed, adjusting the output voltage to the same value as A. With the values of voltage equal, reconnect the jumper wire and voltmeter. The voltmeter should now indicate zero. If the voltmeter indicates a low value (less than the output voltage), adjust the rheostat of generator B until the voltmeter indicates zero.

With the voltage and polarity correct, close the three-pole switch for generator B. Remove the voltmeter and jumper wire. Generator B is now floating on the line. Following the same procedure outlined for the shunt generators divides the load between the generators as needed.

If the generators have interpoles, the equalizer bus must be connected between the series field and the interpoles.

To remove a generator from the line, follow the same procedure outlined for shunt generators.

# GENERATOR EFFICIENCY

It is important to conserve sources of energy. Therefore, generators must be designed for maximum efficiency. A well-designed generator can be a very efficient machine. The efficiency of industrial generators is often as high as 90 percent.

Within the generator there are three major losses: mechanical, electrical, and magnetic. Friction is the major cause of these losses.

Mechanical losses include friction at the bearings and between the brushes and the commutator, and the wind resistance (windage losses) to the rotating parts. A well-balanced rotor and the use of correct bearings and proper bearing lubrication aid in reducing the mechanical losses. The brushes are made of carbon and graphite and are plated with copper. The graphite serves as a lubricant. No other lubricant should be used between the brushes and the commutator. Proper design of the vent openings and the shape of the armature can reduce the windage loss.

Electrical losses are caused by the resistance of the field and armature conductors. Using materials of low resistance and adequate size can keep these losses to a minimum.

Magnetic losses are a result of reluctance in the magnetic circuit. The two major losses are caused by eddy currents and hysteresis. The reluctance can be kept to a minimum by the use of good magnetic materials. It is also important to keep the **air gap** between the armature and the stator field to a minimum.

Eddy currents (currents set up in the armature and field cores) are caused by voltages induced into the cores. The best way to minimize eddy currents is to laminate the cores and insulate the laminations from one another.

Hysteresis losses are caused by fluctuating magnetic fields, which are produced by changes of current flowing in the coils. **Hysteresis** is a result of friction within the armature and field cores. Changes in magnetic strength and/or polarity cause a movement of the molecules and magnetic stresses within the cores.

Hysteresis can be reduced by selecting the proper materials for the core. Generally, materials that have good permeability produce less hysteresis loss.

# REVIEW QUESTIONS

*Multiple Choice*

1. Electromagnetic induction takes place
   a. whenever an electric current flows.
   b. when there is relative motion between a conductor and a magnetic field.
   c. when a steady direct current flows through a conductor.
   d. when a conductor is moved parallel with the lines of magnetic force.

2. In northeastern United States, many generating stations depend upon
   a. water power.
   b. steam power.
   c. solar power.
   d. wind power.

3. Increasing the number of turns of wire on the armature coil causes the induced emf to
   a. decrease.
   b. increase.
   c. become more constant.
   d. none of the above.

4. When it is desired to take alternating current from a generator,
   a. pickup rings are used.
   b. a commutator is used.
   c. slip rings are used.
   d. compensating poles are used.

5. One method used to reduce the pulsation of the output voltage of a DC generator is to
   a. add more coils of wire to the armature.
   b. add more coils of wire to the field.
   c. increase the speed of the rotor.
   d. decrease the speed of the rotor.

6. The armature of a DC generator contains a lap winding. This means that the ends of each winding are connected to
   a. every other segment.
   b. segments on opposite sides of the commutator.
   c. every third segment.
   d. adjacent segments.

7. An armature constructed with a wave winding has
   a. four current paths.
   b. three current paths.
   c. two current paths.
   d. one current path.

8. The insulation separating the segments of the commutator is made of
   a. mica.
   b. silk.
   c. PVC.
   d. rubber.

9. The brushes used on a generator are usually made of
   a. graphite and carbon.
   b. graphite and copper.
   c. aluminum.
   d. aluminum and copper.

10. The neutral plane of a generator is located
    a. in the armature windings.
    b. in the field winding.

c. at a point where the coils cut the maximum number of lines of force.
d. at a point where the armature coils do not cut lines of force.

11. When a load is placed on a generator, the neutral plane shifts
    a. in the direction of armature rotation.
    b. in the direction opposite to the armature rotation.
    c. first in one direction and then in the other direction.
    d. 90°.

12. As the armature coils move through the neutral plane, the current
    a. increases.
    b. decreases.
    c. remains the same.
    d. fluctuates.

13. In order to obtain good commutation, it is necessary that the brushes be set in the
    a. current neutral plane.
    b. mechanical neutral plane.
    c. magnetic neutral plane.
    d. electrical neutral plane.

14. The purpose of interpoles is to
    a. improve commutation.
    b. reduce flux leakage.
    c. strengthen the main field flux.
    d. develop a greater emf.

15. Compensating windings compensate for
    a. motor action in the generator.
    b. armature reaction.
    c. flux losses.
    d. field reversals.

16. The generated emf varies directly with the
    a. armature current.
    b. strength of the field flux.
    c. resistance of the armature windings.
    d. resistance of the field.

17. A self-excited shunt generator has the field winding connected
    a. across the armature.
    b. to a separate DC source.
    c. in parallel with the interpoles.
    d. in series with the armature.

18. A shunt generator has the field connected
    a. in parallel with the armature.
    b. in series with the armature.
    c. to a separate source.
    d. in parallel with the interpoles.

19. A compound generator has
    a. two types of field windings.
    b. interpoles connected in series with the field.
    c. twin armatures.
    d. only one field.

20. Voltage regulation of a generator is
    a. using a rheostat to regulate the output voltage.
    b. varying the speed to regulate the output voltage.
    c. the ability of a generator to regulate its output voltage with changes in load.
    d. using a transformer to regulate the output voltage.

*Give Complete Answers*

1. Describe electromagnetic induction.

2. What do the letters *emf* stand for? What is the unit of measurement of emf?

3. Write the left-hand rule for a generator.

4. What is a generator?

5. Describe the types of energy conversion that take place in a generating station.

6. List four factors that determine the amount of emf induced into a generator armature.

7. Why does a single-loop generator not produce a steady voltage?

8. What type of voltage is produced in the armature of all rotating-type generators?

9. What determines the type of current taken from a generator?

10. How does one obtain AC from a generator?

11. What is the purpose of brushes on a generator?

12. Define *commutator,* and explain its purpose on a generator.

13. What provisions are made in the construction of a generator in order to obtain a smooth DC?

14. Identify two types of windings that are used on a DC generator armature.

15. What is the least number of current paths in a DC generator armature?

16. List the main parts of a DC generator, and explain the purpose of each part.

17. Explain how an alternating voltage is developed in the armature of a generator.

18. A generator is being constructed to produce a high voltage with minimum current. Will the plans call for a lap winding or a wave winding?

19. Describe the construction of the armature core of a DC generator.

20. Define *eddy currents,* and explain what causes them.

21. Why must eddy currents be kept to a minimum?

22. How are the armature and the field cores of generators constructed in order to keep eddy currents to a minimum?

23. Of what materials are brushes made, and why are these materials used?

24. Describe the yoke of a generator, and explain its purpose.

25. Describe field coils, and explain how they are installed on a generator.

26. Explain the meaning of the term *field excitation.*

27. Describe two methods commonly used to excite the field of a DC generator.

28. What is a prime mover?

29. What causes armature reaction in a generator?

30. Describe the effects of armature reaction in a generator.

31. Define the term *neutral plane.*

32. What causes the neutral plane to shift?

33. As the load is increased on the generator, in which direction does the neutral plane shift?

34. Describe the phenomenon known as self-induction.

35. Explain the purpose of interpoles, and describe how they are connected.

36. What are compensating windings, and what is their purpose in a generator?

37. Describe motor action in a generator, and explain what causes it.

38. What is meant by the saturation point of a generator?

39. Draw a schematic diagram of a separately excited generator.

40. Draw a schematic diagram of a shunt generator.

41. Describe the building-up process of a shunt generator.

42. List nine reasons why a self-excited generator may fail to build up a voltage.

43. Draw a schematic diagram of a series generator.

44. Draw a schematic diagram of a compound generator.

45. What does the load/voltage curve of each of the following generators indicate?
    a. Shunt generator
    b. Series generator
    c. Flat-compounded generator

46. What is the most common type of DC generator in use today? Why?

47. Explain the procedures to be followed when connecting shunt generators in parallel.

48. Describe the procedure for connecting compound generators in parallel.

49. Explain the difference between over-, under-, and, flat-compounded generators.

50. What is the difference between a cumulative-compound generator and a differential-compound generator?

*Extended Study*

1. In constructing a compound generator, how is the amount of compounding regulated?

2. Where are overcompounded generators usually used?

3. Name one use for a differential-compound generator, and explain why it is used for this purpose.

4. Define the term *voltage control*.

5. Define the term *voltage regulation*.

6. Describe one method used to control the output voltage of a shunt generator.

7. Describe one method used to control the output voltage of a series generator.

8. How is the output voltage of a compound generator controlled?

9. Define *hysteresis*.

10. List the three major losses that occur in a generator, and explain how to keep the losses to a minimum.

# DC Motors

## OBJECTIVES

After studying this chapter, the student will be able to:

- Explain the principles upon which DC motors operate.
- Describe the construction of DC motors.
- Discuss the different types of DC motors and their operating characteristics.
- Describe basic motor maintenance procedures.

The electric motor utilizes electrical energy and magnetic energy to produce mechanical energy. The purpose of a motor is to produce a rotating force (torque).

The basic construction of a DC motor is very similar to that of a DC generator. Each has a frame, end bells, field poles, an armature, and a commutator. There is, however, a vast difference in their use and operation.

A generator is driven by some type of mechanical machine (steam or water turbine, gasoline engine, or electric motor). It requires mechanical energy for its operation. The goal is to convert mechanical energy into electrical energy. A motor, in contrast, requires electrical energy for its operation. An electric current flowing through the motor windings causes the armature to rotate. The goal is to produce mechanical motion, thus converting electrical energy into mechanical energy.

# BASIC MOTOR OPERATION

The operation of a motor depends upon the interaction between two magnetic fields. One magnetic field is stationary, and the other is free to rotate.

All current-carrying conductors produce a magnetic field. This magnetic field is circular in shape, and the direction of the flux depends upon the direction of current (Figure 17–1).

In Chapter 6, we described the action of two magnetic forces within magnetic reach of each other. Figure 17–2 depicts the results of this condition. This phenomenon can be used to demonstrate the theory of motor action.

If a current-carrying conductor is looped through a magnetic field, it will cause the main flux to become distorted (Figure 17–3).

As seen in Figure 17–2, the crowding of the magnetic lines causes a pressure, which tends to force the conductors apart. A similar condition exists during the distortion of the magnetic flux, as in Figure 17–3. The lines of force are crowded below the conductor on the left and above the conductor on the right. This produces an upward force on the left-hand conductor and a downward force on the right-hand conductor. If the loop is free to rotate, it revolves in a clockwise direction until it is midway between the two poles. In this position, the forces are equal and opposite to each other (Figure 17–4). If the loop is forced beyond this point, by either centrifugal force or other means, the magnetic action will be as shown in Figure 17–5. An oscillating force is produced rather than a rotating force. In order to develop a rotating force, it is necessary to reverse the direction of current through the loop every time the loop passes the midpoint between the poles.

A single-loop motor does not produce a steady torque, nor is it practical. In order to construct a motor that will develop a reasonably steady and usable torque, it is necessary to have many coils on the armature. The armature now becomes a practical electromagnet with a north pole and a south pole. A cutaway end view of the armature is illustrated in Figure 17–6.

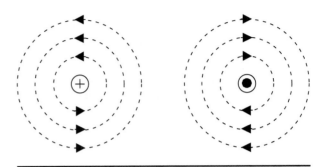

**FIGURE 17–1** Magnetic field around a conductor carrying current.

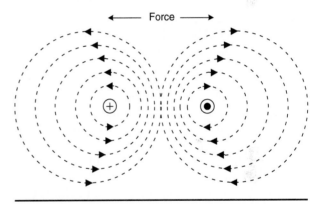

**FIGURE 17–2** Two current-carrying conductors within magnetic reach of each other.

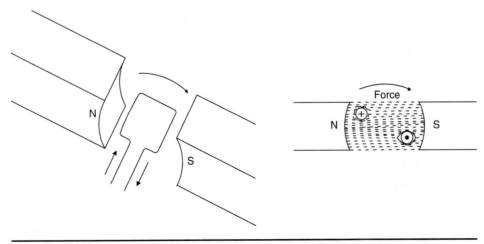

**FIGURE 17–3** Current-carrying conductor looped through a magnetic field.

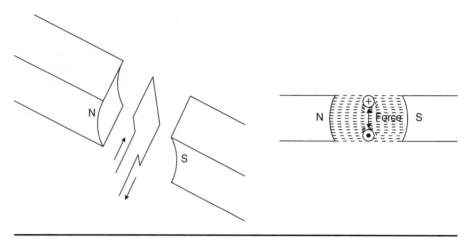

**FIGURE 17-4** Current-carrying conductor in the neutral plane.

**FIGURE 17-5** Magnetic reaction between two magnetic fields produces an oscillating force.

When a voltage is applied to the armature coils, a current will flow, causing the magnetic reaction shown in Figure 17–7. Notice that the north pole of the armature is attracted to the south pole of the main field, and the south pole of the armature is attracted to the north pole of the main field. When the opposite poles are aligned, there is no longer a rotating force. In order to have continuous torque, the current must be reversed in the armature at the instant the unlike poles are aligned. This reversal of current is accomplished through the action of the commutator.

When the current flow is reversed, the magnetic polarity of the armature reverses. The south pole of the armature is near the south pole of the main field; the north pole of the armature is near the north pole of the main field. Because like poles repel, the armature will continue to rotate. This action continues as long as the motor is in operation.

The operating principle of a motor depends upon the following laws of magnetism:

1. When a current-carrying conductor is placed in a magnetic field, it will move at right angles to the field.

2. Like magnetic poles repel each other and unlike magnetic poles attract each other.

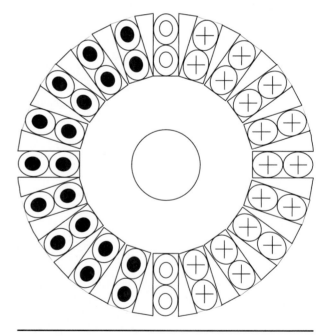

**FIGURE 17-6** End view of motor armature. The wire is wound in the slots to form a coil.

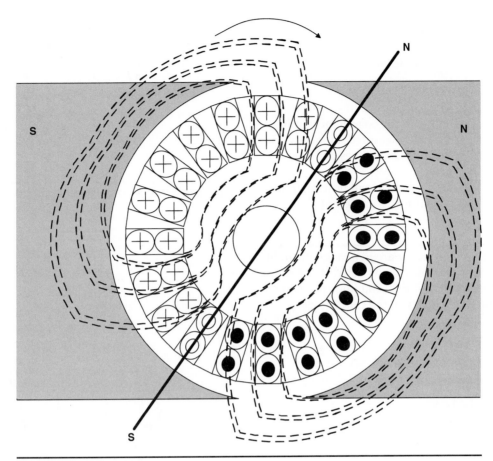

**FIGURE 17–7**  Magnetic reaction in a DC motor.

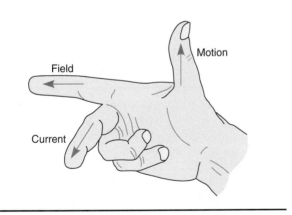

**FIGURE 17–8**  Right-hand rule for a motor.

## Right-Hand Rule for a Motor

To determine the direction of force on a conductor, use the right-hand rule for a motor. Hold the thumb, index finger, and middle finger at right angles to one another. By pointing the index finger in the direction of the main field flux (north to south) and pointing the middle finger in the direction of current flow through the conductor, the thumb will point in the direction of the force (Figure 17–8).

Another method is to use the left hand. Place the open palm of the left hand on the conductor in question. The fingers should point in the direction of the main field flux (north to south), and the thumb should point in the direction of current flow. The force will cause the conductor to move in the direction away from the palm of the hand.

The direction of rotation of a DC motor may be reversed by interchanging either the armature or the field connections. Reversing both connections will not change the direction of rotation. Because of the many different field connections, it is recommended that the armature connections be interchanged.

## Force Exerted on a Conductor

The amount of force acting on a current-carrying conductor in a magnetic field is directly proportional to the following factors:

- The strength of the main field flux
- The length of the conductor within the field
- The amount of current flowing in the conductor

The following equation can be used to calculate this force:

$$F = \frac{8.85Bl\ell}{10^8} \qquad \text{(Eq. 17.1)}$$

where  $F$ = force, in pounds (lb)
$\ell$ = length of the conductor within the field measured, in inches (in.)
$B$ = flux density, in lines per square inch
$I$ = amount of current flowing in the conductor, in amperes (A)

## Example 1

Twelve inches of conductor are in a magnetic field. The current in the conductor is 50 A. If the main field flux has a density of 70,000 lines per square inch, how much force acts on the conductor?

$$F = \frac{8.85Bl\ell}{10^8}$$

$$F = \frac{8.85 \times 7 \times 10^4 \times 50 \times 12}{10^8}$$

$$F = 3.717 \text{ lb}$$

## Torque and Power

Because the armature of a DC motor is wound in much the same way as a DC generator, the current from the supply must divide and flow through two or more paths. The currents through the conductors under a specific pole are in the same direction (Figures 17–9A and 17–9B).

The force acting on the conductors causes the armature to rotate in the direction indicated by the arrows. As the conductors move from the influence of one pole into the influence of the next, the current flow in that conductor is reversed. This reversal of current flow maintains a constant torque in one direction.

The force acting on each conductor adds to produce a total force. The effect of these forces in the armature depends upon their magnitude (total pounds exerted). The effect also depends upon the **radial distance** (the distance from the center of the rotor to the center of the conductors) through which they act. This effect is usually expressed as the product of the force and the radial distance. The unit of measurement is pound-feet (lb ft). (The SI unit is newton-meters, or N m.) The equation for torque is:

$$T = FR \qquad \text{(Eq. 17.2)}$$

where  $T$ = torque, in pound-feet (lb ft)
$F$ = force, in pounds (lb)
$R$ = radial distance, in feet (ft)

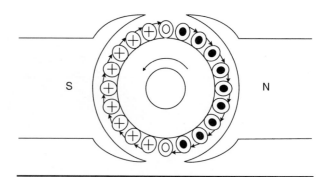

**FIGURE 17–9A**  Two-pole motor.

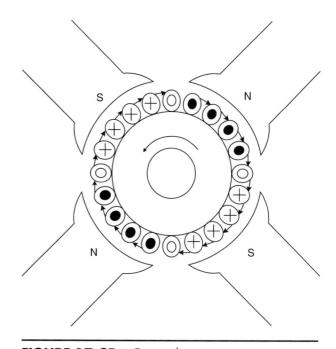

**FIGURE 17–9B**  Four-pole motor.

## Example 2

The motor in Figure 17–9A has 20 conductors within the magnetic field. The length of each conductor within the field is 10 in. (25.4 cm), and the radial distance of the armature is 3 in. (7.62 cm). The current through each conductor is 20 A, and the flux density is 60,000 lines per square inch (1 in.$^2$ = 6.4516 cm$^2$). Find the torque, in pound-feet, developed by the motor. Finally, convert the pound-feet value to newton-meters (1 lb ft = 1.3588 N m).

1. $F = \dfrac{8.85Bl\ell}{10^8}$

$$F = \frac{8.85 \times 6 \times 10^4 \times 20 \times 10}{10^8}$$

$$F = 1.062 \text{ lb}$$

2. $T = FR$

$$T = 1.062 \times \frac{3}{12}$$

$T = 0.2655$ lb ft on one conductor

3. $0.2655$ lb ft $\times$ 20 conductors $= 5.31$ lb ft total torque

4. 1 lb ft $= 1.3588$ N m

5. Therefore, 5.31 lb ft $= 7.215$ N m

## Example 3

A DC motor has 100 conductors within the magnetic field. The length of each conductor within the field is 20 in. (50.8 cm), and the radial distance of the armature is 6 in. (15.24 cm). The current through each conductor is 30 A, and the flux density is 60,000 lines per square inch. Find the torque developed by the motor. If the rated speed of the motor is 1000 r/min, what is the horsepower output?

1. $F = \dfrac{8.85 B I \ell}{10^8}$

$$F = \frac{8.85 \times 6 \times 10^4 \times 30 \times 20}{10^8}$$

$F = 3.186$ lb

2. $T = FR$

$$T = 3.186 \times \frac{6}{12}$$

$T = 1.593$ lb ft

3. 1.593 lb ft $\times$ 100 conductors $= 159.3$ lb ft
$= 216.5$ N m

The total torque is 159.3 lb ft (216.5 N m). From Equation 3.13,

$$hp = \frac{TN}{5252}$$

$$hp = \frac{159.3 \times 1000}{5252}$$

$$hp = 30.33$$

An instrument designed to measure torque is called a **prony brake**. There are several types available to industry. The most common types are the mechanical, the eddy current, and the electrodynameter. Other prony brakes available are the hysteresis, fluidics, and hydraulic types.

# GENERATOR ACTION IN A MOTOR

In the motor, as in the generator, conductors rotate through a magnetic field. If an emf is induced into the armature conductors of a generator because they are cutting flux, the same must be true of motors.

Consider the motor in Figure 17–9A. If the armature is rotating in the direction indicated by the arrow, the emf is in a direction opposite to the current flow. This can be verified by using the left-hand rule for a generator (see Chapter 16).

From this information it can be seen that there are two voltages at work in a motor armature:

1. The applied voltage from the source that causes the current to flow

2. The induced emf that opposes the current flow

The induced emf is called **counter electromotive force** (cemf) because it acts against the applied voltage. The cemf serves a very useful purpose. Because it opposes the current flow, the armature circuit resistance may be kept to a minimum. The lower the resistance of the armature, the fewer the $I^2R$ losses, and the more efficient the motor.

A 5-horsepower, 230-volt motor can be used to illustrate this phenomenon. The resistance of the armature windings of a motor of this size is about 0.1 ohm. According to Ohm's law, the armature current would be $230 \div 0.1 = 2300$ amperes. Such a high current would cause overheating, and the insulation would burn on the armature windings. It may even burn the windings.

Because the induced voltage (cemf) opposes the applied voltage, the formula for current is:

$$I_a = \frac{E_a - E_g}{R_a} \qquad \text{(Eq. 17.3)}$$

where $I_a$ = armature current
$E_a$ = voltage applied to the armature
$E_g$ = voltage induced into the armature (cemf)
$R_a$ = resistance of the armature windings

The cemf of the 5-horsepower, 230-volt motor is about 228 volts. Because of the cemf, the current flowing in the motor armature is only 20 amperes:

$$I_a = \frac{E_a - E_g}{R_a}$$

$$I_a = \frac{230 - 228}{0.1}$$

$$I_a = 20 \text{ A}$$

# COMMUTATION

The armature coil current must periodically reverse its direction of flow in order to produce a constant torque. This is accomplished by the commutator, which is a type of rotating switch. The commutator must change the connections to the armature coils at the instant the coils pass through the neutral plane. At the same instant, the induced voltage in the armature also reverses direction.

## Armature Reaction in a Motor

In a generator, the armature current flows in the same direction as that of the induced emf. In a motor, the armature current flows in the direction opposite to that of the induced emf (cemf). Therefore, for the same direction of armature rotation, the magnetic polarity of the armature in a motor is the reverse of that in a generator. As a result, their neutral planes shift in opposite directions.

In a generator, the neutral plane shifts in the direction of the armature rotation. In a motor, the neutral plane shifts in the direction opposite to the direction of armature rotation. Figure 17–10A shows the field distortion and shifting of the neutral

plane for a generator. Figure 17–10B shows the same conditions for a motor. It should be noted that the directions of rotation and field polarity are the same for both machines. The armature current, however, is in opposite directions. The direction of the current through the armature coils determines the pattern of the main field flux. (As a result, in a generator, the brushes must be shifted in the direction of armature rotation. In a motor, they must be shifted opposite to the direction of rotation.)

Self-induction takes place in a motor just as it does in a generator. It can be said that a motor also has three neutral planes:

1. The no-load neutral plane
2. The magnetic neutral plane
3. The commutating plane

Figures 17–11A, 17–11B, and 17–11C illustrate the neutral planes for a motor.

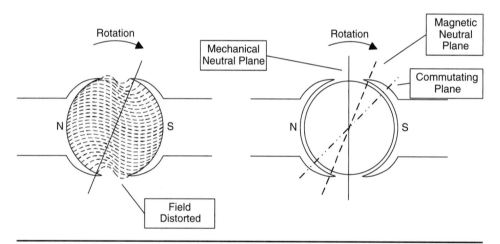

**FIGURE 17–10A**  Armature reaction in a generator. The result is a distorted field flux and a shifting of the neutral plane.

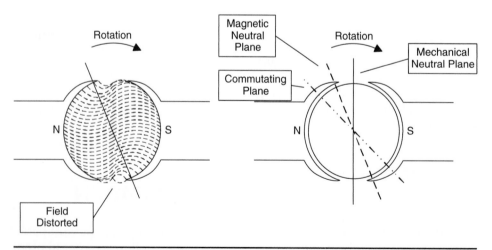

**FIGURE 17–10B**  Armature reaction in a motor. The result is a distorted field flux and a shifting of the neutral plane.

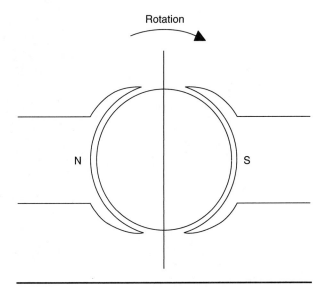

**FIGURE 17–11A**   No-load neutral plane.

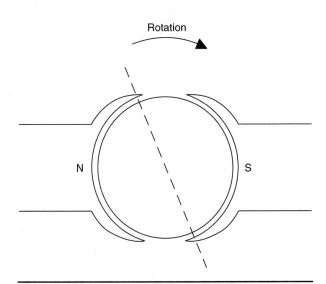

**FIGURE 17–11B**   Magnetic neutral plane.

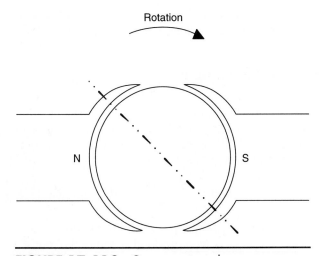

**FIGURE 17–11C**   Commutating plane.

## Motor Interpoles (Commutating Poles)

Interpoles are as important in motors as they are in generators. If a motor were not equipped with some means of compensating for armature reaction, the brushes would have to be shifted with every change in load.

Interpoles serve the same purpose in motors and generators. The one difference is their polarity. For motors, the interpoles must have the same polarity as the main poles directly behind them. For generators, the polarity of the interpoles must be the same as that of the main poles just ahead of them. Interpole polarity is very important. If it is reversed, severe arcing will occur. Figures 17–12A through 17–12D illustrate the correct polarity.

## Compensating Windings in Motors

Compensating windings are seldom used in motors, because of their cost. When they are used, it is generally in conjunction with interpoles. Large machines that are subject to very heavy loads or extreme variations in load and speed may require both interpoles and compensating windings.

The compensating winding is embedded in the face of the main pole piece and then wound around the interpole. It has the same polarity as the interpole it surrounds.

## | MOTOR SPEED

The speed of a DC motor is proportional to the cemf. A weakening of the main field flux reduces the cemf. The lower cemf allows more current to flow in the armature circuit. This increase in the armature current provides a stronger magnetic field in the armature, which causes an increase in the armature speed. The speed increases until the cemf can limit the armature current to a new value. This value is determined by the main field strength. At this point, the motor drives the load at a constant speed.

Decreasing the armature current also affects the motor speed. Assume that the motor is supplying a constant load. A decrease in the armature current results in a decrease in armature reaction. The decrease in armature reaction allows the main field flux to increase, and the armature slows down.

A motor is rated at its maximum horsepower and speed at a constant load. In other words, a 10-horsepower motor rated at 1200 revolutions per

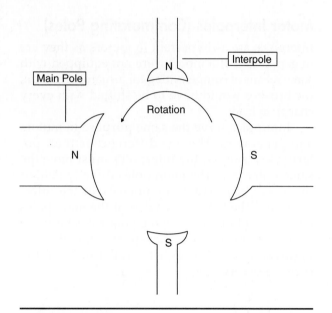

**FIGURE 17–12A**   Interpoles in a two-pole generator.

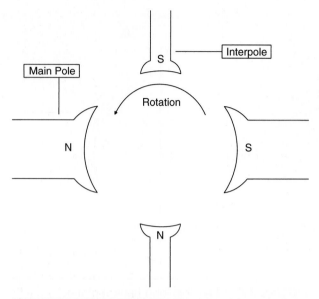

**FIGURE 17–12C**   Interpoles in a two-pole motor.

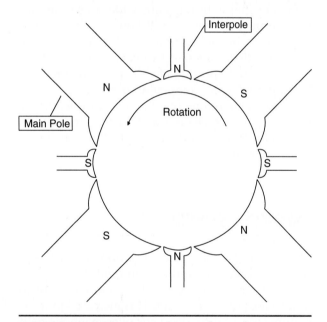

**FIGURE 17–12B**   Interpoles in a four-pole generator.

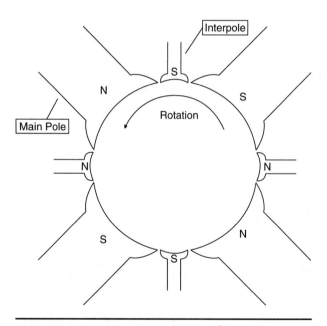

**FIGURE 17–12D**   Interpoles in a four-pole motor.

minute can deliver a maximum of 10 horsepower at a speed of 1200 revolutions per minute. From Equation 3.13, hp = $TN/5252$ it can be seen that at a specific horsepower, an increase in speed requires a similar decrease in torque in order to maintain the same horsepower. Therefore, to operate over a wide range of speeds, a motor does not generally drive its maximum load. It should, however, operate as closely as possible to its maximum load. Motors are most efficient when operating at full load. Efficiency, as well as speed, should be of major concern when selecting motors, particularly for those requiring wide variations in speed.

## Speed Regulation

**Speed regulation** refers to the manner in which a motor adjusts its speed to changes in load. It is the ratio of the loss in speed, between no load and full load, to the full-load speed. The formula for the percent of speed regulation is:

$$\% \text{Reg} = \frac{N_1 - N_2}{N_2} \times 100 \qquad \text{(Eq. 17.4)}$$

where   % Reg = percent of regulation
        $N_1$ = no-load speed of the motor
        $N_2$ = full-load speed of the motor

## Example 4

A motor is rated at 1725 r/min. What is the percentage of speed regulation if the no-load speed is 1775 r/min?

$$\%\,Reg = \frac{N_1 - N_2}{N_2} \times 100$$

$$\%\,Reg = \frac{1775 - 1725}{1725} \times 100$$

$$\%\,Reg = \frac{50}{1725} \times 100$$

$$\%\,Reg = 2.9\%$$

# TYPES OF DC MOTORS

The three major types of DC motors are the series, shunt, and compound motors. In recent years, permanent-magnet motors have gained in popularity. They are extremely efficient.

## Series Motors

The *series motor*, like the series generator, has the field connected in series with the armature (Figure 17–13A). This method of connecting the field and armature has a very definite effect on the operating characteristics of the motor. Because all the current that flows through the armature must also flow through the field, the field strength varies with changes in the load.

A series motor has a very high starting torque. In some motors, it may be as high as 5 times the full-load torque. The speed of the motor also varies with the load.

For this reason, series motors are always connected directly to the load. Belt drives are never used with series motors. As the load increases, the speed decreases. Figure 17–13B shows a graph of the load/speed characteristics of a series motor. Figure 17–13C shows the effects of the load on the speed, torque, and efficiency.

> ⚠️ **CAUTION** This type of motor should never be operated without a load. The only field flux present at no load is that caused by residual magnetism; therefore, the field is very weak. Operating the motor without a load allows the rotor to reach such high speeds that the centrifugal force so generated causes the windings to tear free.

Figure 17–13C demonstrates that when the armature current is 4 amperes, the motor develops a torque

of 10 pound-feet (13.588 newton-meters). At 8 amperes, the torque is 40 pound-feet (108.704 newton-meters). It can be seen that when the armature current doubles, the torque becomes 4 times as great. Thus, the torque increases rapidly near and above full load. Such characteristics make the use of series motors desirable when it is necessary to supply a large torque with a moderate increase in current.

Series motors are used chiefly for wide variations in load when extreme speed changes are not objectionable. They are used extensively for cranes, hoists, electric railway cars, and electric automobiles as well as for starting gasoline engines. In such applications, variations in speed with load are not objectionable. Occasionally, the variations are desirable. For instance, when a crane is being used to lift a heavy load, it is generally desirable to proceed

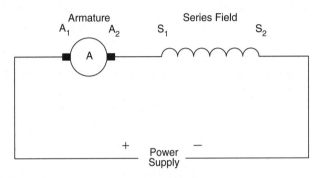

**FIGURE 17–13A**  Series motor.

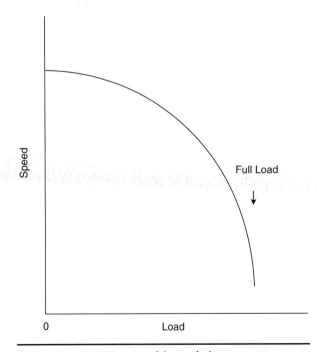

**FIGURE 17–13B**  Load/speed characteristic curve for a series motor.

slowly. The enormous torque of a series motor makes it very suitable for work that demands frequent acceleration under heavy loads.

Series motors are also used extensively in portable tools because of their light weight in comparison to the horsepower they deliver. Because of their characteristics, the use of series motors is restricted to machines that require the presence of an operator.

> ⚠️ **CAUTION** It is most important to remember that a series motor should never be operated without a load.

## Shunt Motors

In a *shunt motor*, the field is connected directly across the line (Figure 17–14). As a result, the field current and the field flux are constant. When a shunt motor is operating without a load, the retarding torque is small since it is due only to the windage and the friction. Because of the constant field, the armature will develop cemf that will limit the current to the value needed to develop only the required torque.

Shunt motors are classified as constant speed motors. In other words, there is very little variation in the speed of the shunt motor from no load to full load. Equation 17.5 may be used to determine the speed of a motor at various loads:

$$N = \frac{E_a - IR}{K_1 \Phi} \qquad \text{(Eq. 17.5)}$$

where   $N$ = speed of the armature, in revolutions per minute (r/min)

$E_a$ = applied voltage

$I$ = armature current at a specific load

$R$ = armature resistance

$\Phi$ = lines of force per square centimeter

$K_1$ = a constant value for the specific motor

In a shunt motor, $E_a$, $R$, $K_1$, and $\Phi$ are practically constant values and $I$ is the only variable. When no load is applied to the motor, the value of $I$ is small because the speed and cemf are both at a maximum. In the equation for $N$, $IR$ (the voltage drop in the armature coils) is negligible when compared to $E_a$. At full load, $IR$ is generally 5 percent of $E_a$. The actual value depends upon the size and design of the motor. Consequently, at full load the speed is about 95 percent of the no-load value. This decrease in speed is reduced slightly by armature reaction, which causes a decrease in flux and a corresponding increase in speed. In some cases, armature reaction is sufficient to cause the speed to remain constant from no load to full load.

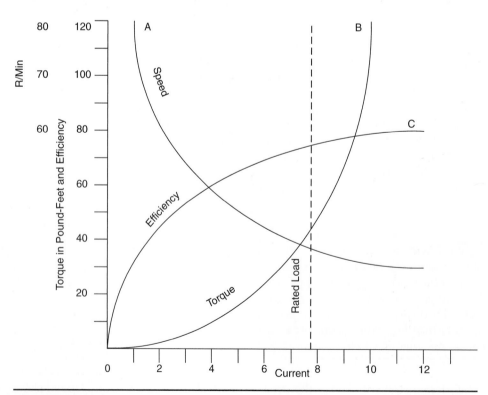

**FIGURE 17–13C** Characteristics of a series motor.

The starting torque of a shunt motor is about 2.75 times the full-load torque. The shunt motor does not have as high a starting torque as the series motor, but it has much better speed regulation.

The graph in Figure 17–15 illustrates the speed, torque, and efficiency of a shunt motor under varying loads.

The shunt motor is best suited for constant speed drives. It meets the requirements of many industrial applications. Some specific applications are machine tools, printing presses, blowers, and motor–generator sets.

> ⚠ **CAUTION** When working with a shunt motor, never open the field circuit when it is in operation. The residual field will cause the rotor speed to increase. At light loads, this speed could become dangerously high.

## Compound Motors

The *compound motor* is a combination of the series motor and the shunt motor. It has two fields. One field is connected in parallel with the armature; the other field is connected in series with the armature (Figure 17–16).

Most compound motors are connected for cumulative compounding. The cumulative-compound motor has the combined characteristics of the shunt and the series motors (Figure 17–17). The cumulative-compound motor has a definite no-load speed and, therefore, may be operated without a load. As the load is increased, the speed decreases more rapidly than it does in shunt motors. This is caused by the increase in the series field flux as the armature current increases.

The speed/torque characteristics of the compound motor may resemble those of either the series motor or the shunt motor, depending upon the strength of both fields. If the motor has only a few turns on the series field, it will have a better starting

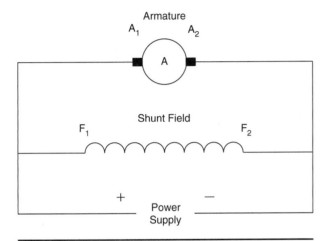

**FIGURE 17–14** Shunt motor.

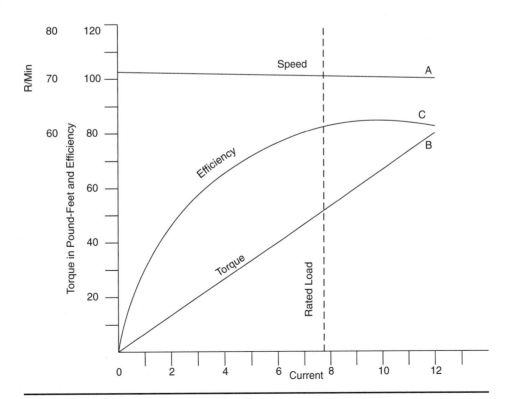

**FIGURE 17–15** Speed, torque, and efficiency curves for a shunt motor.

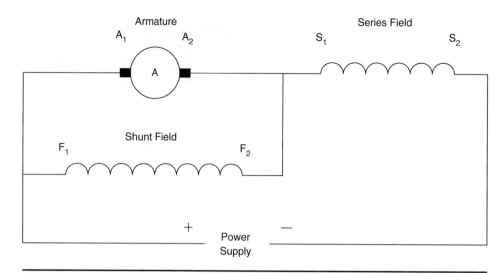

**FIGURE 17–16** Compound motor.

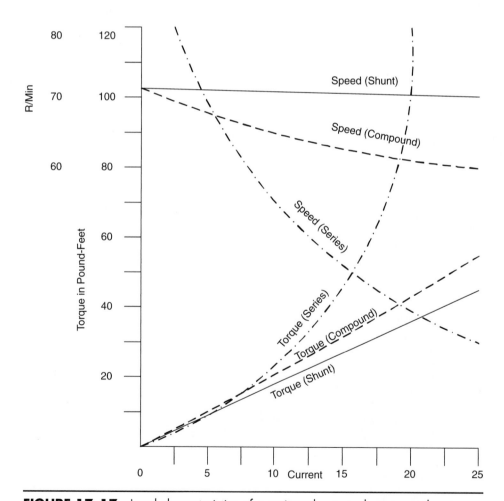

**FIGURE 17–17** Load characteristics of a series, shunt, and compound motor.

and running torque than the shunt motor. It will still retain the good speed regulation and efficient speed control of the shunt motor. If the motor has fewer turns on the shunt field (sufficient to limit the no-load speed to a safe value) and many turns on the series field, it will have most of the characteristics of the series motor.

Compound motors are used to drive machines that require a relatively constant speed under varying loads. They are frequently used on machines that require application of heavy loads for very short periods of time, such as presses, shears, compressors, reciprocating tools, and elevators. Compound motors are also used when it is desired to protect the motor by causing it to decrease in speed under heavy loads.

> **⚠ CAUTION** Never open the shunt field of a compound motor when the motor is operating at light load.

## Flywheel Effect

The power required by some machine tools is very irregular. For example, in a punch press or stamping machine, almost no power is required until the punch or die comes in contact with the material. If the moving parts in such a machine are not very heavy the current taken by the driving motor will vary widely. Figure 17–18 shows this current curve. The motor selected to drive such a machine must be capable of carrying the greatest value of current without overheating or excessive arcing at the brushes. If a considerably overcompounded motor is used (about 30 percent) with a heavy **flywheel**, the armature current will vary less, as indicated by the dotted curve in Figure 17–18. From this curve, it can be observed that the driving motor may be much smaller when a flywheel is used.

The effect of the flywheel can be explained as follows. At point *a* in Figure 17–18, the load on the motor and flywheel suddenly increases, causing them to slow down. The centrifugal force of the flywheel supplies the energy to help overcome the opposition of the load. This reduces the demand on the motor. In other words, with an increase in load, the flywheel assists the motor in driving the load. Although the motor current increases, it does not increase as much as it would without the flywheel.

At point *b*, the power required to drive the load is less than the input to the motor. As a result, there is an increase in rotor speed. The increased speed produces a greater cemf, causing a decrease in current. This process continues as long as the machine is in operation.

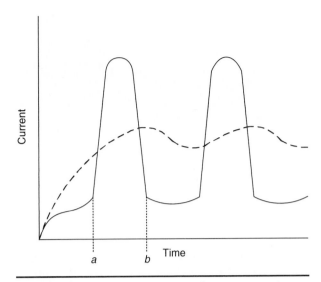

**FIGURE 17–18** Current curve of a DC motor driving a punch press.

## Permanent-Magnet Motors

A *permanent-magnet motor* is a motor in which the main field flux is produced by permanent magnets. An electromagnet is used for the secondary field or armature flux.

Because of the constant field flux, the standard permanent-magnet motor has many of the same characteristics of the shunt motor. Variations in design, however, can change these characteristics considerably.

These motors are frequently used for servomotors, torque motors, and industrial drive motors. They are used on machines requiring exact positioning of an object or component, where high starting and operating torque are required, and where a constant torque is required. Some examples are the opening of a valve under pressure, precise positioning of dampers and three-way valves, and other specific operations in various control systems.

The stationary permanent-magnet motor consists of permanent magnets mounted on a frame with a rotating armature placed between them. Electrical energy is supplied to the rotor by way of a commutator and brushes. This is the conventional construction, and it is suitable for many uses. Figure 17–19 shows a motor of this type.

The revolving permanent-magnet and the wound-stator types of motors have several advantages over the wound-rotor type of motor. Because the windings are on the stator, they can dissipate heat more rapidly. In addition, there is less stress on the windings because of the lack of centrifugal force. Another advantage is easy access to the windings.

**FIGURE 17–19**   Permanent-magnet servomotor.

**FIGURE 17–20**   Commutator and slip rings in the stator of a permanent-magnet motor.

This permits easier and more frequent insulation checks, monitoring of temperature, and use of **static control**.

The method used for commutation (reversing the current flow in the stator windings) is quite different from the methods for motors discussed previously. A specially designed commutator and slip rings are mounted inside the stator (Figure 17–20). Rolling contacts are mounted on the rotor. As the rotor revolves, one roller makes contact with a slip ring. Half of the rollers contact one slip ring for one polarity; the other half contact the other slip ring for the opposite polarity. The rollers are positioned so as to energize the stator at the correct instant and with the proper polarity to maintain a constant torque.

Figure 17–21 shows a rotor used on permanent-magnet motors.

The development of the rotating magnet motor has led to improvements in the magnetic materials used. One of the materials that has received wide acclaim is the cobalt rare earth magnet. These magnets are made thin in the direction of magnetization to allow a large number of poles to be installed in the rotor, where they will funnel maximum flux into the air gap. This greater flux density results in a higher torque.

This type of construction yields a more efficient motor than those previously used. The new design has approximately the same dimensions as that of the conventional one. However, for the same physical size, it provides 50 percent higher continuous rating. Its peak torque is 55 percent greater than that of the conventional design. Its accelerating ability is almost twice as good as that of the standard machine.

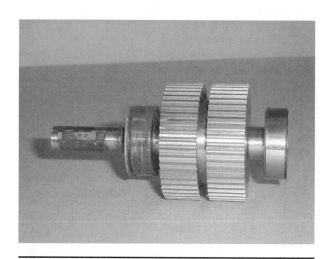

**FIGURE 17–21**   Rotor of a permanent-magnet torque motor using cobalt rare earth magnets.

## Brushless DC Motors

A **brushless DC motor** is a DC motor that does not contain or need brushes. The brushless DC motor is a permanent-magnet motor that uses an electronic circuit to perform the commutation of the applied DC.

Figure 17–22 shows the simplified construction of a brushless DC motor. Notice that the rotor is constructed of permanent magnets, while the field is constructed of windings. Since the rotor is a permanent magnet, there is no need for an electrical connection to the rotor. Therefore, brushes and a commutator are no longer required.

To provide commutation, a special circuit must be employed. This can take the form of one of two types of circuits. One type of circuit is known as an optical encoder. This type of encoder uses an electronic device known as a **phototransistor**. Phototransistors will be studied in Chapter 21. The other type of circuit is known as a magnetic encoder. This type of circuit uses an electronic device known as a **Hall-effect device**.

Figure 17–23 shows a simplified drawing of a brushless DC motor using an optical encoder for commutation. Notice the shutter that blocks phototransistors $Q_2$ and $Q_3$. This means that only $Q_1$ would be conducting, since $Q_1$ is the only phototransistor to be exposed to light. Since $Q_1$ is conducting,

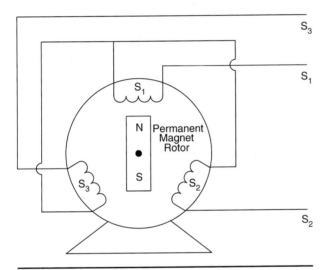

**FIGURE 17–22**  Brushless DC motor.

current flows into the base of **transistor** $Q_4$, causing $Q_4$ also to conduct. With $Q_4$ conducting, current will flow through field winding $S_1$. This causes the permanent-magnet rotor to rotate into alignment with field winding $S_1$.

As the permanent-magnet rotor revolves in a clockwise fashion, the shutter on the end of the motor shaft also rotates. This causes the light to be blocked from phototransistor $Q_1$, and simultaneously allow the light to strike phototransistor $Q_2$. Since light is blocked from phototransistor $Q_1$, $Q_1$ is turned off. This causes transistor $Q_4$ to turn off as well, stopping any current from flowing through field winding $S_1$. However, since phototransistor $Q_2$ is now conducting, transistor $Q_5$ will now be turned on. With $Q_5$ turned on, current will flow through field winding $S_2$. This will cause the permanent-magnet rotor to once again turn in a clockwise fashion and align itself with field winding $S_2$.

As the permanent-magnet rotor aligns with field winding $S_2$, the shutter on the end of the motor shaft also rotates. This will cause the light to be blocked from phototransistor $Q_2$, and simultaneously allow the light to strike phototransistor $Q_3$. Since light is blocked from phototransistor $Q_2$, $Q_2$ is turned off. This causes transistor $Q_5$ to turn off as well, stopping any current from flowing through field winding $S_2$. However, since phototransistor $Q_3$ is now conducting, transistor $Q_6$ will now be turned on. With $Q_6$ turned on, current will flow through

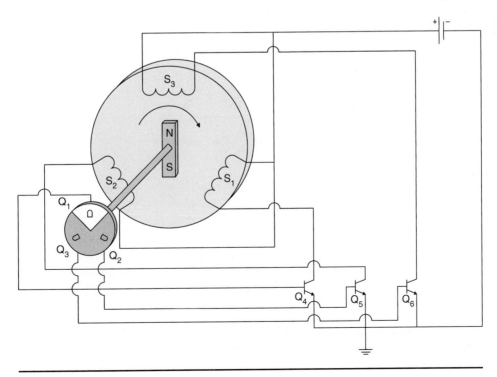

**FIGURE 17–23**  Optical rotor position sensing.

field winding $S_3$. This will cause the permanent-magnet rotor to once again turn in a clockwise fashion and align itself with field winding $S_3$.

A magnetic-encoder-type brushless DC motor is shown in Figure 17–24. Notice that this type of encoder does not require a shutter on the end of the motor shaft. This encoder uses two Hall-effect devices to sense the magnetic field of the permanent-magnet rotor.

Notice in Figure 17–24 that the south pole of the permanent-magnet rotor is positioned under Hall-effect sensor $H_1$. Hall-effect sensor $H_1$ detects the presence of a magnetic field and supplies a voltage to the base of transistor $Q_1$. This voltage causes $Q_1$ to conduct a current, which flows through field winding $S_1$. The magnetic field created by this current in field winding $S_1$ will repel the permanent-magnet rotor in a clockwise direction. As the permanent-magnet rotor rotates, a magnetic field is no longer sensed by Hall-effect device $H_1$. This turns off transistor $Q_1$, removing the current from field winding $S_1$.

As a result of the clockwise rotation of the permanent-magnet rotor, the south pole is now aligned with Hall-effect device $H_2$. Hall-effect sensor $H_2$ detects the presence of a magnetic field and supplies a voltage to the base of transistor $Q_2$. This voltage causes $Q_2$ to conduct a current, which will flow through field winding $S_2$. The magnetic field created by this current in field winding $S_2$ will repel the permanent-magnet rotor in a clockwise direction. As the permanent-magnet rotor rotates, a magnetic field is no longer sensed by Hall-effect device $H_2$. This turns off transistor $Q_2$, removing the current from field winding $S_2$.

Hall-effect sensor $H_1$ detects the presence of a magnetic field from the north pole of the permanent-magnet rotor and supplies a voltage to the base of transistor $Q_3$. Since the north pole of the permanent-magnet rotor is aligned with Hall-effect device $H_1$, the voltage produced will be seen at the second terminal of $H_1$. This voltage causes $Q_3$ to conduct a current, which will flow through field winding $S_3$. The magnetic field created by this current in field winding $S_3$ will repel the permanent-magnet rotor in a clockwise direction. As the permanent-magnet rotor rotates, a magnetic field is no longer sensed by Hall-effect device $H_1$. This turns off transistor $Q_3$, removing the current from field winding $S_3$.

Hall-effect sensor $H_2$ detects the presence of a magnetic field from the north pole of the permanent-magnet rotor and supplies a voltage to the base of transistor $Q_4$. Since the north pole of the permanent-magnet rotor is aligned with Hall-effect device $H_2$, the voltage produced will be seen at the second terminal of $H_2$. This voltage causes $Q_4$ to conduct a current, which will flow through field winding $S_4$.

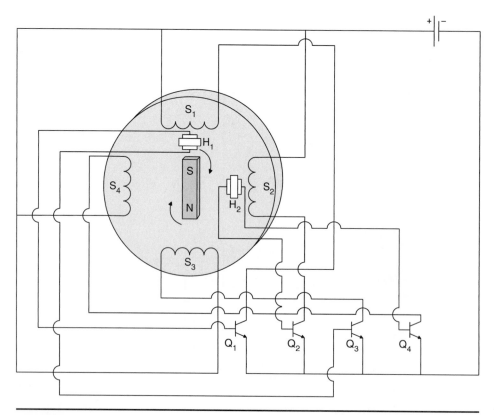

**FIGURE 17–24** Magnetic rotor position sensing using Hall-effect devices.

The magnetic field created by this current in field winding $S_4$ will repel the permanent-magnet rotor in a clockwise direction. As the permanent-magnet rotor rotates, a magnetic field is no longer sensed by Hall-effect device $H_4$. This turns off transistor $Q_4$, removing the current from field winding $S_4$.

## Stepping Motors

The stepping motor is similar in design to the permanent-magnet motor. Because stepping motors are designed to be used for jobs requiring minimum torque, their stators are connected directly to the power source.

Figure 17–25 shows a simplified diagram of the magnetic polarity arrangement of the stator and rotor. The stator windings are arranged so that the direction of current through them can be reversed at the proper instant to obtain the desired direction of rotation. Also, they are arranged to provide for controlling the number of degrees of rotation. Figure 17–26 illustrates the basic circuitry for the stator.

In actual practice, the design is more complex. The stator is designed to have a large number of poles, possibly as many as 40. The rotor is designed to have more poles than the stator. The relationship between the number of poles on the rotor and stator determines the amount of rotation, namely, the *step angle*. The step angle is measured in the number of degrees of rotation.

An electric pulse fed into the stator winding sets up a magnetic field, causing the permanent magnets in the rotor to align themselves with the field of the stator. A series of pulses causes the stator field to rotate, and the rotor follows. A single pulse may advance the rotor as little as 0.5 degree.

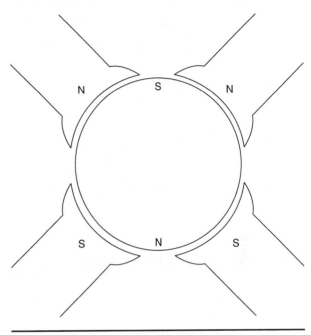

**FIGURE 17–25** Simplified diagram of a stepping motor stator and rotor.

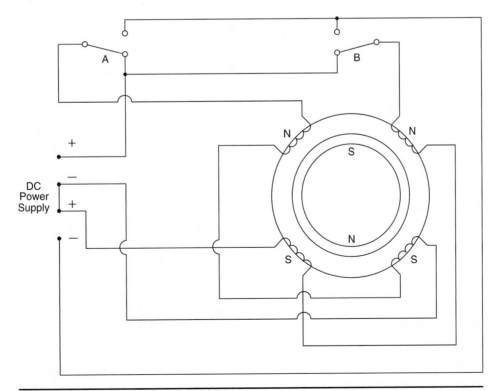

**FIGURE 17–26** Basic circuitry for a permanent-magnet stepping motor.

This can be better understood by referring to Figure 17–26. With switches A and B in the positions shown, the rotor field is in the vertical position. By moving switch A to connect with the top contact, the polarity reverses in one set of poles, and the rotor advances in a clockwise direction. Various switching sequences determine the direction of rotation. In this illustration, each switching operation causes the rotor to move 1/4 turn (90 degrees).

The ability to control the rotation to less than 1 degree allows for many applications in industry. Stepping motors are frequently used in printing shops and machine shops. They are often controlled by computers to obtain fast and accurate pulsing.

Some stepping motors are designed with non–permanent-magnet rotors. These rotors are made of materials that have high permeability. When the stator poles are energized, the rotor aligns itself with the magnetic field. This type of construction is referred to as a *variable reluctance stepping motor*.

# MOTOR MAINTENANCE

A motor should start, deliver its rated load, and run within its speed ranges without excessive vibration, noise, heating, or arcing at the brushes. Failure to start may be a result of one or more of the following:

- A ground
- An open circuit
- A short circuit
- Incorrect connections
- Improper voltage
- Frozen bearings
- An overload

Arcing at the brushes may be traced to the brushes, commutator, armature, or an overload. Arcing may be the result of:

- Insufficient brush contact
- Worn or improper brushes
- Not enough spring tension
- An incorrect brush setting
- A dirty commutator
- High mica
- A rough or pitted commutator surface
- Eccentricity (commutator off center)
- An opening in the armature coil
- A short in the armature or field coils

Vibration and pounding noises may be caused by:

- Worn bearings
- Loose parts
- Rotating parts hitting stationary parts
- Armature unbalance
- Improper alignment of the motor with the driven machine
- Loose coupling
- Insufficient end play
- The motor and/or driven machine loose on the base
- An intermittent load

Overheating is frequently caused by:

- Overload
- Arcing at the brushes
- A wet or shorted armature or field coils
- Too frequent starting or reversing
- Poor ventilation
- Incorrect voltage

Overheated bearings may be a result of:

- A lack of lubricant
- Too much grease
- A dirty lubricant
- Too tight a fitting at the bearings
- Oil rings not rotating
- Too much belt tension or gear thrust
- Insufficient end play
- A rough or bent shaft
- A shaft out of round

To ensure good operating conditions, all motors should be checked periodically. Keep a permanent record of test results, replacement of parts, and general maintenance tasks. The log should indicate the date, time of day, ambient temperature, humidity, and other pertinent data that may affect the test results.

The interior and exterior of the motor should be kept clean and dry. Depending upon the atmospheric conditions, periodically disassemble the motor and clean the interior thoroughly. All loose dirt should be vacuumed away. Clean the commutator and its contacts with a nontoxic, nonabrasive cleaner. Bearings should be lubricated as needed. (Do not overlubricate.) Replace worn bearings immediately.

Insulation tests should be performed on a regular basis. Check for grounds, shorts, open circuits, and leakage currents. A good insulation test can be performed with a megohmmeter (see Chapter 4).

When cleaning or testing motors or doing both, always check the terminal connections to ensure that they are made tight. Any sign of overheating at the terminals is an indication of poor connections. If the connections appear to be tight but there is an indication of overheating, disconnect the lead from the terminal. Clean all the surface area and reconnect it. Vibration or changes in temperature, or both, are frequently the cause of loose connections.

Perhaps the parts in a DC motor that require highest maintenance are the brushes and the commutator. A good commutator will have a light film of oxide. This film will have a different color depending upon the condition of the commutator and brushes. A light, golden bronze colored film indicates good brush performance. It is also an indication that the motor is lightly loaded. If the film is a dark, golden bronze, it means that the brushes are performing perfectly and the motor is properly loaded. Should the film have a very dark brown appearance, the indication is that the brushes are performing poorly and the motor is heavily loaded. There may also be some contamination on the commutator. In all of the aforementioned situations, the color of the film on the commutator should be uniform throughout.

More serious problems in the brush/commutator area are evidenced by the following:

Non-uniform appearance or streaking of the commutator film.

This can be caused by a rough brush face surface. There may be metal contamination between the brush face and commutator, or weak brush spring tension, or contamination of the commutator surface.

Threading of the commutator surface.

If the streaking condition noted in preceding text is not corrected, it can progress to threading. Threading has the appearance of small grooves or scratches that are worn into the commutator surface. This is a result of the rough face surface of the brush scratching or wearing on the surface of the commutator.

Grooving of the commutator surface.

The threading condition, if left unchecked, can progress to the grooving stage. Grooving appears as wide, deep groves (the width of the brush) worn into the commutator surface. This could result from contamination between the brush face and the commutator. It can also be caused by replacing brushes with a more abrasive grade of material. Another cause may be excessive brush arcing due to light spring tension.

# REVIEW QUESTIONS

*Multiple Choice*

1. An electric motor is used to
   a. produce electrical energy.
   b. change mechanical energy into electrical energy.
   c. change electrical energy into mechanical energy.
   d. produce an alternating motion.

2. The operation of a motor depends upon
   a. static electricity.
   b. interaction between two magnetic fields.
   c. interaction between two electrostatic fields.
   d. interaction between an electrostatic field and a magnetic field.

3. When a current-carrying conductor is placed in a magnetic field, it will
   a. remain stationary.
   b. move parallel with the lines of force.
   c. move at right angles to the lines of force.
   d. rotate.

4. The direction of rotation of a DC motor armature may be reversed by
   a. interchanging either the armature or the field connections.
   b. interchanging both the armature and field connections.
   c. neither a nor b.
   d. interchanging the supply connections.

5. The amount of force acting on a current-carrying conductor in a magnetic field is proportional to
   a. the strength of the main field flux.
   b. the length of the conductor within the field.
   c. the amount of current flowing in the conductor.
   d. all of the above.

6. The effect of the forces in the armature of a motor depends upon
   a. the total force acting on the armature conductors.
   b. the radial distance through which the force acts.
   c. both a and b.
   d. neither a nor b.

7. An instrument designed to measure torque is called a
   a. torque meter.
   b. prony brake.
   c. torque scale.
   d. ohmmeter.

8. The voltage induced into a motor armature is called the
   a. voltage drop.
   b. applied voltage.
   c. counter electromotive force.
   d. IR drop.

9. The current in a motor armature flows in
   a. the same direction as the applied voltage.
   b. the opposite direction to the induced voltage.
   c. the same direction as the induced voltage.
   d. both a and b.

10. The polarity of an interpole in a DC motor is
    a. the same polarity as the main pole directly behind it.
    b. the same polarity as the main pole directly ahead of it.
    c. the opposite polarity of the main pole directly behind it.
    d. none of the above.

11. The speed of a DC motor is proportional to the
    a. applied voltage.
    b. counter electromotive force.
    c. armature reaction.
    d. IR drop in the armature.

12. Speed regulation in a motor refers to
    a. the manner in which a motor adjusts its speed to changes in load.
    b. a means of varying the speed of a motor.
    c. limiting the maximum speed of a motor.
    d. all of the above.

13. The series motor has
    a. low starting torque.
    b. high starting torque.
    c. low starting current.
    d. both b and c.

14. Series motors are generally used where
    a. good speed control is essential.
    b. they are subjected to wide variations in load.
    c. a constant speed is essential.
    d. a constant torque is needed.

15. A series motor
    a. is frequently operated without a load.
    b. should never be operated without a load.
    c. develops a constant speed at all loads.
    d. develops a constant torque at all loads.

16. Shunt motors are classified as
    a. variable speed motors.
    b. constant speed motors.
    c. constant torque motors.
    d. variable torque motors.

17. A shunt motor should never be operated
    a. without a load.
    b. with the field circuit open.
    c. with the armature circuit open.
    d. without an operator in attendance.

18. A compound motor has
    a. only a series field.
    b. only a shunt field.
    c. two fields.
    d. three fields.

19. Compound motors are used to drive machines that require a
    a. relatively constant speed under varying loads.
    b. relatively constant speed at a constant load.
    c. constant load and a constant speed.
    d. wide variations of loads and speed.

*Give Complete Answers*

1. Explain the principle of operation of a DC motor.

2. Write two laws of magnetism upon which the operation of a motor depends.

3. Describe the right-hand rule for a motor.

4. Describe a method used to reverse the direction of rotation of a DC motor armature.

5. List three factors that determine the amount of force exerted on a conductor that is carrying current while in a magnetic field.

6. Define *torque.*

7. Identify the unit of measurement for torque.

8. List six types of prony brakes.

9. Define the term *counter electromotive force.*

10. List the factors that determine the value of current through the armature of a DC motor.

11. As the load is increased on a motor, does the neutral plane shift in the direction of rotation or opposite to the direction of rotation?

12. How are compensating windings installed in a motor?

13. Define *speed regulation* with regard to DC motors.

14. List three major types of DC motors and describe how their armatures and fields are connected.

15. What type of DC motor is used to drive machines with wide variations in load, such as hoists, electric railway cars, and electric automobiles?

16. For what type of machines are shunt motors generally used?

17. What type of DC motor is usually used to drive machines that require relatively constant speeds and where sudden application of heavy loads occurs?

18. What is the advantage of a flywheel on a machine?

19. Name two types of permanent-magnet motors.

20. What is the advantage of using cobalt rare earth magnets for the rotor of a permanent-magnet motor?

21. Explain the purpose of a commutator in a conventional DC motor.

22. Describe how commutation is accomplished in a brushless DC motor.

23. Describe the purpose of the shutter in an optical-encoder-type brushless DC motor.

24. Give seven reasons why a DC motor may fail to start.

25. List 10 problems that may cause arcing at the brushes on a DC motor.

26. List nine causes of motor vibration or pounding noises or both.

27. Give six reasons why a DC motor may overheat.

28. List nine causes of overheated bearings.

# AC Generators (Alternators)

## OBJECTIVES

After studying this chapter, the student will be able to:

- Describe the construction and operating characteristics of various types of alternating current generators.

- Discuss the methods for controlling the output voltage and frequency of alternating current generators.

- Discuss the methods for producing single-phase and multiphase voltages.

- Describe the procedures for connecting alternating current generators in parallel.

## AC GENERATORS VERSUS DC GENERATORS

As previously stated, alternating current is the primary source of electrical energy. It is less expensive to produce and transmit than direct current. These advantages, however, do not rule out the use of DC. There are many applications for which direct current is preferred or required. DC generators are often used to provide power for operations such as metal refining, electroplating, and battery charging.

DC generators, however, have certain built-in limitations that restrict their power output. For this reason, and because AC voltage is induced into the armature of all types of generators, AC generators are generally more practical.

## ALTERNATOR CONSTRUCTION

The commutator on a DC generator converts alternating current to direct current. Figure 18–1 shows a single-loop DC generator. If the commutator is replaced with slip rings (Figure 18–2), alternating current will be supplied to the external circuit.

In the two-pole alternator, one cycle is produced for each revolution of the armature. If four poles are used, two cycles are produced for each revolution; six poles produce three cycles and so on. Figure 18–3 illustrates a four-pole machine.

In order to maintain a uniform magnetic field, it is necessary to provide direct current for the field circuit. A separate DC generator is usually mounted on the same drive shaft as the alternator rotor. This generator, called an *exciter*, provides the DC for the alternator field. Except for the slip rings, the construction of many alternators is similar to that of DC generators. Large alternators, however, have a stationary armature and a rotating field. The field receives its energy from the DC exciter, through slip rings. The exciter voltage is usually 250 volts. At this voltage, there is no serious difficulty with insulation breakdown or arcing. Because the armature is stationary, it is called the *stator*, and the revolving field is called the *rotor*.

The rotating field of a low-speed alternator consists of a number of laminated poles dovetailed or bolted to a cast-iron form called a *spider*. The slip rings are attached to and insulated from the spider. The ends of the field windings are connected to the slip rings. Brushes riding on the slip rings connect the DC exciter to the alternator field windings. The field coils for small machines are wound with wire; rectangular copper strips are used for large machines.

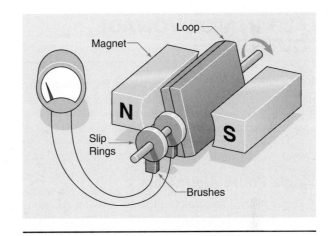

**FIGURE 18–2** Single-loop AC generator.

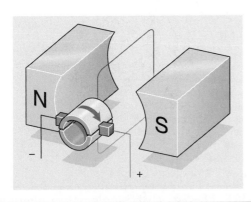

**FIGURE 18–1** Single-loop DC generator.

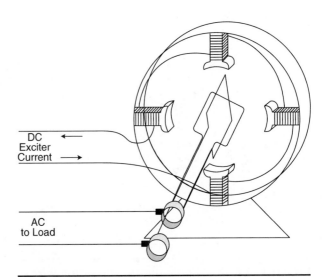

**FIGURE 18–3** Four-pole alternator.

When steam turbines are used to drive alternators, the rotors must be constructed for high-speed operation. Alternators of this type generally have only two, four, or six poles. The poles are wound on a cylindrical form. The windings are embedded in the rotor core in order to reduce windage loss.

These machines produce large quantities of heat because of their speed and design. This heat must be carried away by air currents forced through the passages in the heated parts. A tunnel-like enclosure controls the direction of the air currents and minimizes the noise.

For cooling large **turbo-type alternators**, hydrogen is often used instead of air. Because hydrogen is lighter than air, the windage losses are reduced and the efficiency of the machine is increased. Other advantages of using hydrogen are reduction in noise and less oxidation in the windings.

# ALTERNATOR VOLTAGE OUTPUT

AC generators may be constructed to produce one or more voltages. These voltages and their phase relationship are important factors in the selection and operation of equipment.

## Single-Phase Alternators

A *single-phase alternator* (Figure 18–4) is an AC generator that produces only one voltage. The armature coils are connected in "series-additive." In other words, the sum of the emfs induced into each coil produces the total output voltage. Single-phase

generators are usually constructed in small sizes only. They are used for standby service in case the main power source is interrupted, for supplying temporary power on construction sites, and for permanent installations in remote locations.

## Two-Phase Alternators

The *two-phase alternator* produces two separate voltages 90 degrees out of phase with each other. A two-phase alternator is a combination of two single-phase alternators with their armatures connected so there is a 90-degree phase displacement between the output voltages. Figure 18–5A illustrates this arrangement. The armatures are mounted on the same shaft so they will rotate together. Thus, when generator *a* is producing maximum voltage, generator *b* is producing zero volt. One-quarter turn later, *b* is producing maximum voltage and *a* is producing zero volt.

Figure 18–5B shows the same two windings installed in one generator with a 90-degree phase displacement. The same effect will take place as before; however, now one generator is producing the two separate voltages. Figure 18–6 illustrates a typical two-phase alternator with a rotating field.

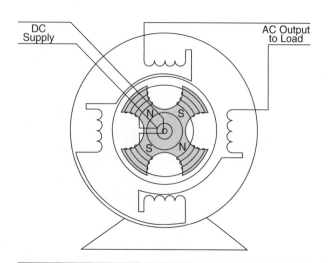

**FIGURE 18–4**  Single-phase alternator with a rotating field.

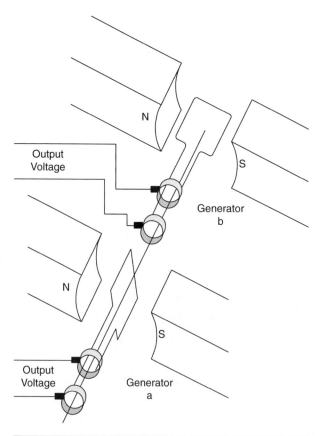

**FIGURE 18–5A**  Two single-phase alternators arranged to produce a two-phase configuration.

There are several different methods for connecting the coils of alternators for a two-phase system. One method is to bring the four wires out separately. The coils are not electrically connected together. This arrangement is known as a *two-phase, four-wire system*. Both single phase and two phase may be supplied from this system. Figure 18–7 illustrates a typical two-phase, four-wire system.

If the two phases of the alternator in Figure 18–6 are connected in series, only three wires are required to supply power to the external load. This arrangement is shown in Figure 18–8 and is known as a two-phase, three-wire system. The voltage between $L_X$ and $L_C$ is equal to the voltage between $L_Y$ and $L_C$. The voltage between $L_X$ and $L_Y$ may be determined by vector addition. Figure 18–9A shows

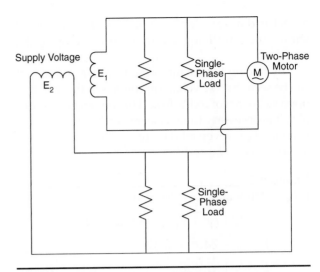

**FIGURE 18–7** Typical two-phase, four-wire system.

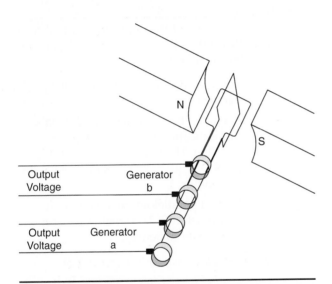

**FIGURE 18–5B** Two-phase alternator with a rotating armature.

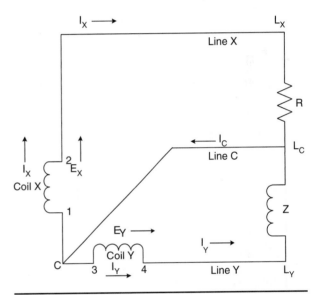

**FIGURE 18–8** Two-phase, three-wire system.

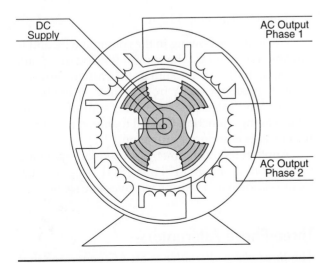

**FIGURE 18–6** Two-phase alternator with a rotating field.

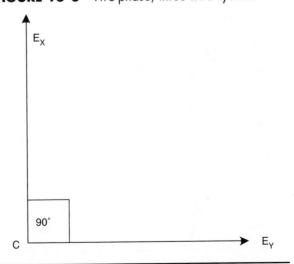

**FIGURE 18–9A** Vector diagram of voltages in a two-phase system.

a vector diagram for the voltages; Figure 18–9B shows the parallelogram method for solving the vectors.

By tracing the circuit formed by coils X and Y, it can be seen that the voltage in coil X is in the direction opposite to that of coil Y. To add these voltages vectorially, it is necessary to reverse one of the two vectors. If vector $E_X$ is reversed (Figure 18–9B), then vector $E$ is the sum of $-E_X$ and $E_Y$.

### Example 1

If $E_X = 240$ V and $E_Y = 240$ V, then:

$$E = \sqrt{-E_X{}^2 + E_Y{}^2} \ (\textit{from Equation 8.4})$$
$$E = \sqrt{-240^2 + 240^2}$$
$$E = \sqrt{115,200}$$
$$E = 339 \text{ V}$$

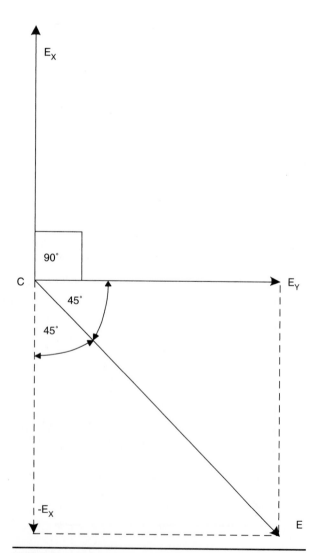

**FIGURE 18–9B** Parallelogram method for calculating voltage in a two-phase, three-wire system.

The currents in lines X and Y (Figure 18–8) are equal to the currents through their corresponding phases. The current in the common wire ($I_C$) is the resultant of the currents in the two phases. These currents must be compared with the voltages that cause them. Current $I_X$ is the result of voltage $E_X$, and $I_Y$ is the result of $E_Y$. Because phase X supplies a pure resistive load, $I_X$ is in phase with $E_X$. Phase Y supplies an inductive load; therefore, $I_Y$ lags $E_Y$ by $\phi$ degrees. The two currents, $I_X$ and $I_Y$, are $(90 + \phi)$ degrees out of phase with each other. The current in the common wire may be determined by adding vectorially the currents through the two phases. Because currents $I_X$ and $I_Y$ are both flowing in the same direction through line C, it is not necessary to reverse one of the current vectors.

### Example 2

Assume that the current through coil X is 10 A and the current through coil Y is 15 A. If $I_Y$ lags $E_Y$ by 30°, what is the current through line C?

Step 1  Lay out current $I_X$ along the voltage vector $E_X$.

Step 2  $I_Y$ lags $E_Y$ by 30°. Therefore, $I_Y$ is laid out indicating the 30° lag (Figure 18–10A). The two currents, $I_X$ and $I_Y$, are 90° + 30° = 120° out of phase with each other.

Step 3  Form a parallelogram as shown in Figure 18–10B. The resultant vector represents current $I_C$.

$$I_C = \sqrt{I_X{}^2 + I_Y{}^2 - 2I_XI_Y \cos\phi} \ (\textit{from Equation 7.5})$$
$$I_C = \sqrt{100 + 225 - (2 \times 10 \times 15 \times 0.5)}$$
$$I_C = \sqrt{175}$$
$$I_C = 13.2 \text{ A}$$

Line C is carrying 13.2 A.

Current $I_C$ is flowing in the common wire and is flowing into point C. The current flowing into point C is equal to the vector sum of the currents flowing through phases X and Y from point C. (Kirchhoff's current law states that the vector sum of the currents leaving a junction is equal to the vector sum of the currents entering the junction.)

Another method sometimes used for two-phase connections is shown in Figure 18–11. This arrangement is known as a two-phase, five-wire system.

## Three-Phase Alternators

The *three-phase alternator* is an AC generator that produces three separate voltages 120 electrical time degrees apart. The coils are arranged around

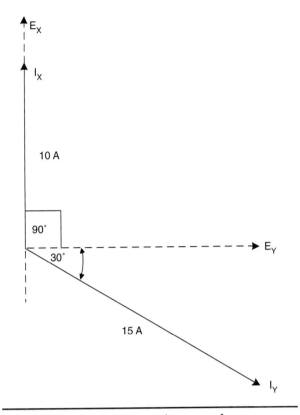

**FIGURE 18–10A** Vector diagram of currents in a two-phase, three-wire system.

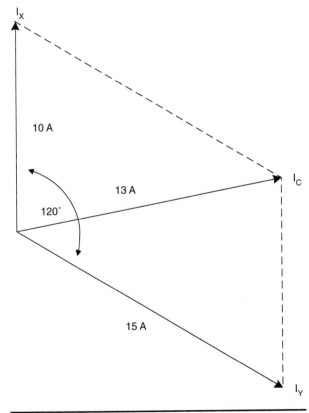

**FIGURE 18–10B** Parallelogram method for calculating the current in the common wire of a two-phase, three-wire system.

the stator in a manner similar to that shown in Figure 18–12A. Mechanically, they are spaced 60 degrees apart. However, the electrical connections are such that the induced voltages are out of phase by 120 electrical time degrees. A graph of the output voltages is shown in Figure 18–12B.

The coils are usually connected together inside the machine. There are two configurations that are generally used. The *wye connection*, sometimes called the star connection, is illustrated in Figure 18–13A. The *delta connection* is shown in Figure 18–13B.

## Wye (Star) Connection

With the wye connection, the beginnings of the coils are connected together, and the ends are brought out to the alternator terminals. Another method used to illustrate the wye connection is shown in Figure 18–14. The common voltages for the three-phase, three-wire wye system are 208 volts and 480 volts. This system provides three single-phase voltages and one three-phase voltage. For example, a 480-volt system can supply 480 volts to a three-phase load and three separate 480-volt circuits for single-phase loads. The single-phase loads are supplied through any two line wires.

Connecting a fourth wire at the junction point of the three phases makes available two values of single-phase voltage. This arrangement is called a three-phase, four-wire wye system. Common voltages obtained from this system are 120 volts and 208 volts, single phase, and 208 volts, three phase. The same type of connection is also used to furnish 277 volts and 480 volts, single phase, and 480 volts, three phase. Figures 18–15A and 18–15B illustrate the load distribution for both the three-wire and four-wire systems.

## Delta Connection

The delta connection is shown in Figure 18–16. In this configuration, the end of each phase coil is connected to the beginning of the next. The line wires are brought out from the points where the connections are made.

With the delta system, only one value of voltage appears for both single phase and three phase. Figure 18–17 depicts the load distribution for a three-phase, three-wire delta system.

## Phase Sequence

The phase sequence is the order in which the voltages follow one another (reach their maximum values). From Figure 18–12B, we can see that phase A reaches its maximum value first, followed

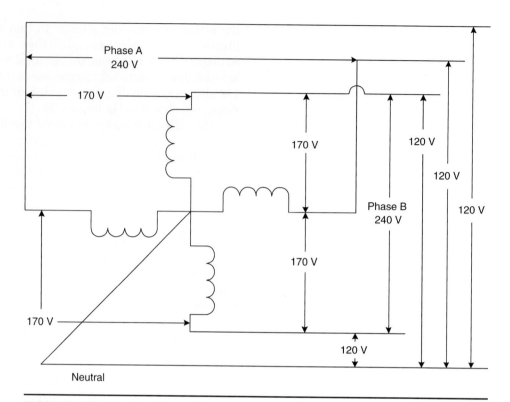

**FIGURE 18–11** Two-phase, five-wire system. Phases A and B supply 240 V, two phase. Single-phase, 120 V, is supplied between the neutral and any one of the phase conductors. The 170 V is not normally used.

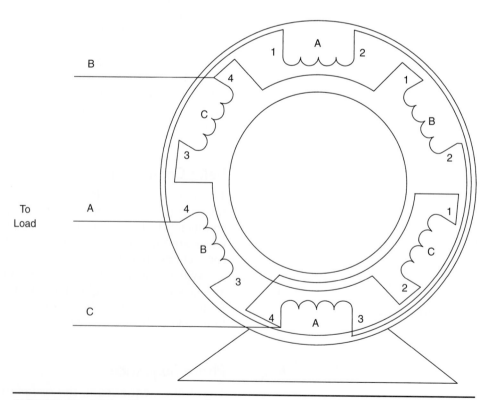

**FIGURE 18–12A** Three-phase alternator stator with the windings connected into a delta configuration.

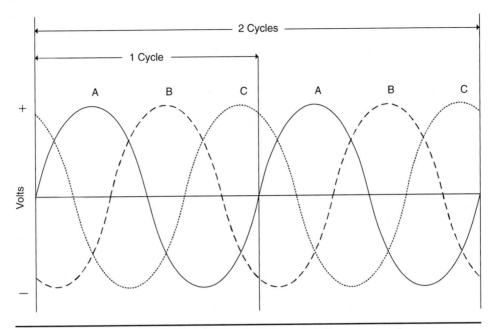

**FIGURE 18–12B**   Graph of the output voltages of a three-phase alternator.

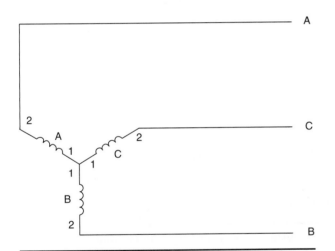

**FIGURE 18–13A**   Wye (star) three-phase connection.

by B and later by C. The phase sequence is an important factor when connecting polyphase generators in parallel and when connecting polyphase motors.

## Voltage and Current in a Delta-Connected Alternator

The voltage induced into the armature coils of a delta-connected alternator is transmitted to the line terminals; therefore the voltage across any one of the armature coils (Figure 18–13B) is equal to the voltage between any two transmission lines.

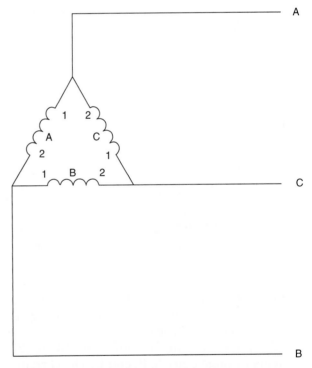

**FIGURE 18–13B**   Delta three-phase connection.

In other words (see Figure 18–13B), the voltage across coil A is equal to the voltage between lines A and B. The voltage across coil B is equal to the voltage between lines B and C, and the voltage across coil C is equal to the voltage between A and C. The voltage between the lines is called the line voltage ($E_L$),

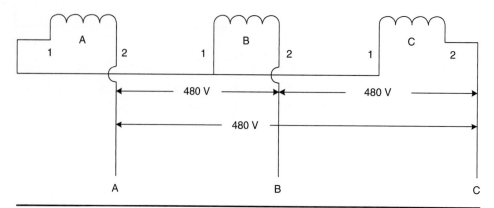

**FIGURE 18-14** Wye (star) three-phase, three-wire connection.

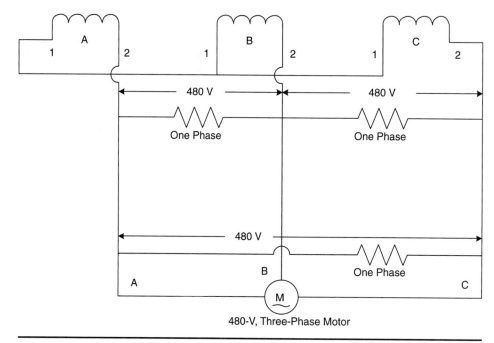

**FIGURE 18-15A** Three-phase, three-wire system supplying single-phase and three-phase loads (wye connection).

and the voltage across the coils is called the phase voltage ($E_p$). For a delta system, the line voltage is equal to the phase voltage ($E_L = E_p$).

The phase currents in Figure 18–13B are the currents through coils A, B, and C. The currents through the line wires are equal to the vector sum of the currents through two of the phases. For example, the current through line A is equal to the vectorial sum of the currents through coils A and C. For a delta system, it has been determined that the vectorial sum of the currents through two of the phases is equal to 1.73 times the current through one phase. Therefore, in a delta-connected alternator, the line voltage is equal to the phase voltage ($E_L = E_p$), and the line current, for balanced loads, is equal to 1.73 times the phase current ($I_L = 1.73I_p$).

## Voltage and Current in a Wye-Connected Alternator

The beginnings of each armature coil of a wye connection are connected together; therefore, the line voltage is equal to the voltage across two of the phase coils. In Figure 18–13A, the voltage between lines A and C is equal to the voltage across coils A and C. If it is assumed that the emf is acting from terminal 1 to terminal 2 in each coil, then the forces are acting in opposition to one another, but

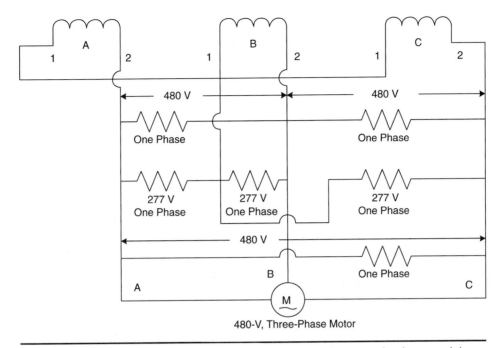

**FIGURE 18–15B** Three-phase, four-wire system supplying single-phase and three-phase loads (wye connection).

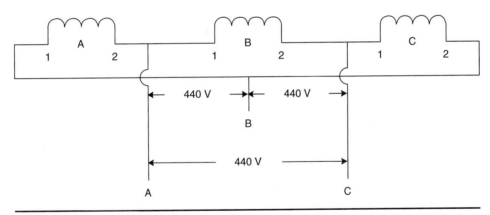

**FIGURE 18–16** Three-phase, three-wire, 440-V delta system.

they are displaced by 120 electrical time degrees. In order to combine the two voltages, they must be subtracted vectorially. It has been determined that the vectorial difference of two voltages in a wye-connected alternator is equal to 1.73 times one phase voltage. Therefore, in a wye-connected alternator, the line voltage is 1.73 times the phase voltage ($E_L = 1.73E_p$).

Assume the current is flowing out coil A (Figure 18–13A) from 1 to 2. It then flows through line A to the load. Therefore, the current through line A is equal to the current through coil A. This is true for each coil and each line in a wye system. Thus, for a wye-connected generator, the line current is equal to the phase current ($I_L = I_p$).

## Power in a Three-Phase System

The power for a balanced three-phase system can be calculated from the formula $P = 1.73IE \cos \phi$. In a balanced three-phase system, the power factor is the cosine of the angle between the phase current and the phase voltage. This formula may also be used for systems that are only slightly unbalanced. The value of current used must be the average of the three values. With only slightly unbalanced systems, this method will provide a reasonably accurate value of power.

If the load is severely unbalanced, a vectorial analysis of each phase must be done, and then the three phases must be combined.

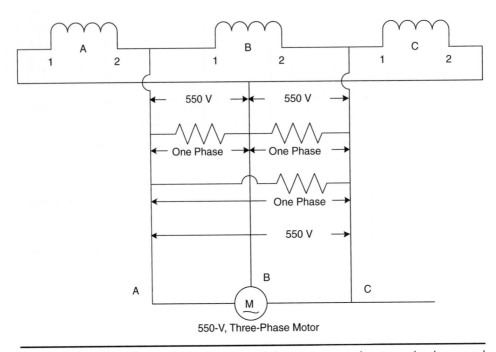

**FIGURE 18–17** Three-phase, three-wire delta system supplying single-phase and three-phase loads.

## VOLTAGE AND FREQUENCY CONTROL

The standard frequency for power distribution in the United States is 60 hertz. Europe, Asia, and South America generally operate on a frequency of 50 hertz, and in some cases 25 hertz.

The frequency of the emf produced by an alternator depends upon the number of field poles and the rotor speed. One cycle is generated whenever one pair of poles passes a coil. A simple formula for calculating the frequency is:

$$f = \frac{PN}{120} \qquad \text{(Eq. 18.1)}$$

where   $f$ = frequency, in hertz (Hz)

  $P$ = number of field poles on the alternator

  $N$ = speed of the rotor, in revolutions per minute (r/min)

Federal regulations require that utility companies maintain a relatively constant frequency. The maximum variation permitted is 3 percent. Therefore, a speed regulator must be installed on the prime mover in order to maintain 60 hertz under varying load conditions.

The voltage supplied by a utility company varies for different sections of the United States and

the world. Some standard voltages supplied by utility companies are 2300 volts, 4800 volts, 6900 volts, and 33,100 volts. Federal regulations also require utility companies to maintain a relatively constant value of voltage. To meet this requirement, utility companies build an automatic voltage regulator into the system. This regulator measures the output voltage of the alternator and adjusts the alternator field current as needed.

## ALTERNATOR CHARACTERISTICS

Three factors affect the output voltage of an alternator.

1. The resistance of the stator windings causes an IR drop within the generator. The value of the voltage drop increases with an increase in load.

2. Self-induction takes place within the stator windings, causing an $IX_L$ drop. This voltage drop also varies with the load.

3. The power factor of the load also affects the output voltage.

Figure 18–18 shows a graph of the effects of power factor on the output voltage. From this information it can be seen that both the amount of load and the type of load affect the output voltage.

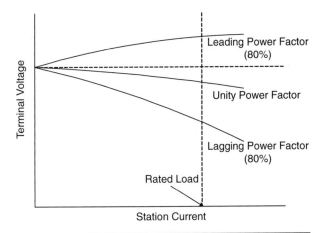

**FIGURE 18-18** Graph of the effects of power factor on the output voltage of an alternator.

# PARALLELING ALTERNATORS

AC power systems generally consist of several alternators connected in parallel to a common distribution bus. Parallel operation of alternators provides for maximum efficiency and allows time for servicing. During the time of day when the load is light, one or more generators can be removed from the line to allow the remaining generators to operate at or near full load. Generators operate at maximum efficiency when fully loaded. Maintenance personnel can take advantage of the time when the generators are not in use to perform preventive maintenance.

When alternators are connected in parallel, certain conditions must be fulfilled. These conditions are as follows:

1. The alternator to be placed on the line must produce a voltage wave of approximately the same shape as the voltage wave across the busses to which it is to be connected.

2. The terminal voltage of the alternator must be equal to the voltage across the busses.

3. The frequency of the output voltage must be equal to that of the bus voltage.

4. With reference to the load, the voltage of the alternator must be in phase with the bus voltage.

5. The phase rotation of the alternator and the distribution bus must be the same.

The procedure to follow in order to connect alternators in parallel is called **synchronizing**. Figure 18-19 shows three alternators connected for parallel operation.

Assume that the third alternator (C) is ready to be connected on the line. The following procedure is recommended. Assume that alternators A and B are supplying 240 volts to the distribution busses. Alternator C is started and brought up to speed. The output voltage is adjusted to equal that of the bus voltage. This is accomplished by varying the **field rheostat**. The frequency of C must be matched to that of the distribution bus. Adjustments in the speed of the rotor accomplish this requirement.

A **synchroscope** is used in order to determine when the frequency of C matches that of the bus. A synchroscope is an instrument that indicates whether it is necessary to increase or decrease the speed of the **prime mover** (Figure 18-20).

For small alternators, the frequency can be matched using incandescent lamps. *Frequency meters may also be used.*

Once the voltage and frequency are correct, it is necessary to check the phase sequence. This operation can be accomplished by using incandescent lamps or a **phase sequence meter**.

If the lamp method is used, the power rating of the lamps must be equal, and the voltage rating must equal the alternator voltage. One lamp is connected across each pole of the three-pole switch (Figure 18-21). If the phase sequence is correct, the three lamps will become bright and dim together. If the lamps become bright and dim one after another, the phase sequence of the alternator is opposite to that of the distribution bus. Interchanging any two of the three leads from the alternator corrects this error.

If the lamps increase and decrease in brightness simultaneously, the sequence is correct but the frequency is incorrect. If the lamps remain dark, the frequency and phase sequence are correct and the alternator may be connected to the distribution bus by closing the switch. For alternators with high voltage outputs, potential transformers can be used to reduce the voltage to that of the lamps.

The *lamp dark method* for synchronizing alternators is generally used only for small operations or for laboratory experiments. Most utility companies have instrument panels equipped with many meters, including synchroscopes and frequency meters. Modern generating stations are equipped with automatic controls that connect or disconnect alternators from the distribution busses with changes in load. All stations, however, must have hand-operated backup equipment in case of automation failure.

## Effect of Varying Field Strength

When all the alternators are **synchronized** and supplying power to the load, they are in phase relative to the load. With reference to one another, they are 180 degrees out of phase. Therefore, no current circulates between the machines.

If the magnetic field of one alternator should increase, the output voltage also increases. This increase in voltage causes a current to circulate between the alternators. Because of the low resistance and high inductance of the alternators, the circulating current is highly reactive and tends to keep the field strength of all the alternators equal. The increase in output voltage is only slight in comparison to the field current, and is generally offset by the impedance drop (IZ) produced by the circulating current.

If two alternators operating in parallel supply equal current to the load and have the same power factor, and the **field excitation** of one machine is

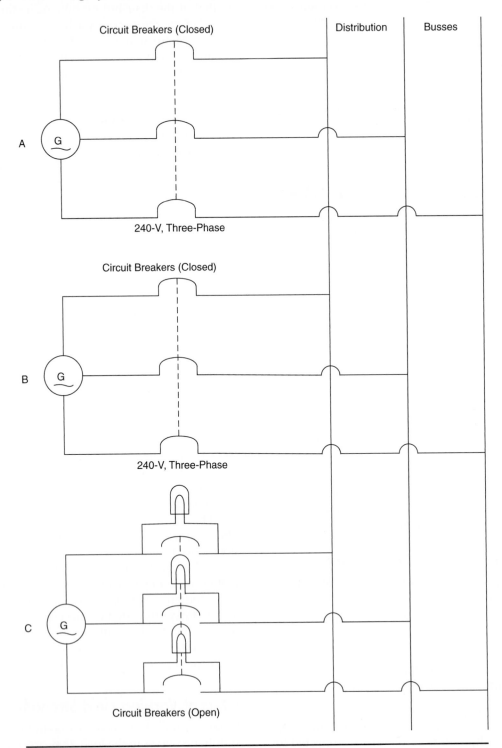

**FIGURE 18–19**  Connecting alternators in parallel.

changed, a circulating current is established. This circulating current may increase the current in one machine and decrease the current in the other. However, the power supplied by each alternator is not greatly affected because the power factor changes inversely with the current. (This means that an alternator cannot be made to take a load merely by increasing the field current. But this method can be used to improve the power factor of the machine, resulting in a decrease in armature current.)

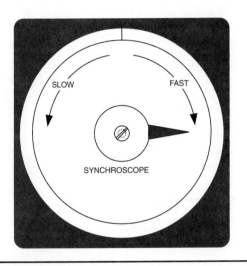

**FIGURE 18–20** Synchroscope.

## Effect of Increased Driving Force

Figure 18–22 illustrates two alternators connected in parallel and supplying power to the load. If the torque of alternator B is increased, the rotor speed increases for a fraction of a revolution until its output voltage, $E_B$, has pulled slightly ahead in phase relationship to $E_A$. The load on B increases, causing a reverse torque, which slows down the rotor to the original speed. Because alternator B has advanced in phase position, the two voltages $E_A$ and $E_B$, no longer neutralize each other, and a resultant voltage exists. The resultant voltage causes a circulating current between the two alternators. Because the resistance of the two alternators is low in comparison to their reactance, the circulating current will lag the resultant voltage by nearly 90 degrees. This current is very close to being in phase with $E_B$ and is nearly 180 degrees out of phase with $E_A$. Because of this phase relationship, alternator B will carry a high current compared to alternator A. (This indicates that if two alternators in parallel supply equal currents to a load, and the driving torque of one machine is increased, the power supplied by this machine is increased and the power supplied by the other machine is decreased. This change occurs without materially affecting the power factor of either machine.)

When one alternator has been synchronized and paralleled with another, it supplies no power to the load. If additional torque is supplied to the

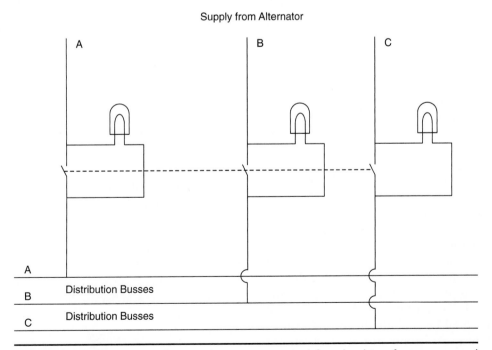

**FIGURE 18–21** Incandescent lamps connected for checking frequency and phase sequence of an oncoming alternator.

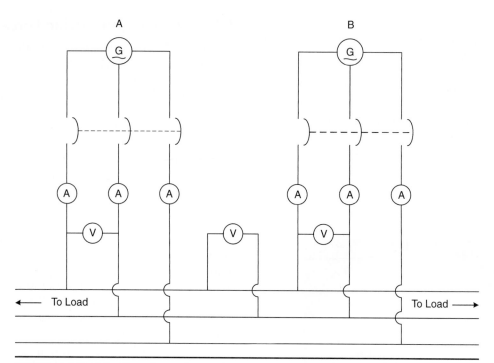

**FIGURE 18-22** Two three-phase alternators connected in parallel.

oncoming machine, its voltage advances a few degrees, causing it to assume part of the load. This increase in load causes the rotor to slow down to its original speed, thereby maintaining a constant frequency.

Operators in power plants can control the steam or water supplied to the prime movers, thus controlling the torque. This permits shifting the load between alternators and adding or removing alternators from the line. These changes are made without interruption of service and with only minor variations of frequency. In modern generating plants, load shifting is accomplished through automated equipment. However, hand-operated backup equipment is installed for emergency purposes.

## MOTOR–GENERATOR SETS

A **motor–generator set** is a combination of two machines. An electric motor is coupled to a generator and arranged so that the motor serves as a prime mover. Motor–generator sets serve many different purposes. For example, an AC motor may drive a DC generator to provide DC power for a specific load. A 600-volt DC motor may drive a 115-volt DC generator. A 60-hertz, 120-volt AC motor may drive a 400-hertz AC generator, or a DC motor may drive an AC generator. These are the most common uses of motor–generator sets.

## ROTARY CONVERTERS (DYNAMOTORS)

A **rotary converter**, sometimes called a *dynamotor*, is a motor–generator set combined into a single housing. The armature for the motor is wound on the same shaft as the generator armature. Rotary converters are used for many of the same purposes as motor–generator sets. They are frequently used to supply DC power in areas where only AC is available. They are also used in communication systems to convert a low-voltage DC to a higher value.

## REVIEW QUESTIONS

*Multiple Choice*

1. Direct current is used for
   a. transformers.
   b. electroplating.
   c. induction motors.
   d. most lighting loads.

2. The voltage induced into the armature of a rotating-type generator is
   a. direct voltage.
   b. alternating voltage.
   c. pulsating voltage.
   d. extremely high voltage.

3. A four-pole alternator will produce
   a. one cycle for each revolution of the rotor.
   b. two cycles for each revolution of the rotor.
   c. three cycles for each revolution of the rotor.
   d. four cycles for each revolution of the rotor.

4. The field circuit of an alternator is supplied with
   a. alternating current.
   b. pulsating current.
   c. direct current.
   d. eddy current.

5. Alternators designed to be driven by steam turbines are built for
   a. high-speed operation.
   b. low-speed operation.
   c. medium-speed operation.
   d. both a and b.

6. Single-phase alternators are generally used to supply
   a. very large loads.
   b. small loads.
   c. any type of load.
   d. only lighting loads.

7. Two-phase alternators produce two separate voltages that are
   a. 60 electrical time degrees apart.
   b. 90 electrical time degrees apart.
   c. 120 electrical time degrees apart.
   d. 180 electrical time degrees apart.

8. Three-phase alternators produce three separate voltages that are
   a. 60 electrical time degrees apart.
   b. 90 electrical time degrees apart.
   c. 120 electrical time degrees apart.
   d. 180 electrical time degrees apart.

9. Another name for the three-phase wye connection is
   a. delta.
   b. alpha.
   c. star.
   d. beta.

10. The order in which the voltages of a three-phase system follow one another is called
    a. phase order.
    b. phase sequence.
    c. phase pattern.
    d. phase system.

11. The line voltage of a delta-connected system is equal to
    a. 1.73 times the phase voltage.
    b. the phase voltage.
    c. the phase voltage divided by 1.73.
    d. the phase voltage plus 1.73.

12. The line current of a delta system is equal to
    a. 1.73 times the phase current.
    b. the phase current.
    c. the phase current divided by 1.73.
    d. the phase current plus 1.73.

13. The line voltage for a wye-connected system is equal to
    a. 1.73 times the phase voltage.
    b. the phase voltage.
    c. the phase voltage divided by 1.73.
    d. the phase voltage plus 1.73.

14. The line current of a wye-connected system is equal to
    a. 1.73 times the phase current.
    b. the phase current.
    c. the phase current divided by 1.73.
    d. the phase current plus 1.73.

15. The most common frequency used in the United States is
    a. 50 Hz.
    b. 60 Hz.
    c. 80 Hz.
    d. 120 Hz.

16. The most common frequency in use in Europe and Asia is
    a. 50 Hz.
    b. 60 Hz.
    c. 80 Hz.
    d. 120 Hz.

17. Federal regulations require that utility companies maintain a relatively constant frequency. The maximum variation permitted is
    a. 1%.
    b. 3%.
    c. 5%.
    d. 7%.

18. AC power systems generally consist of
    a. one large alternator.
    b. several alternators connected in series.
    c. several alternators connected in parallel.
    d. several alternators connected in series-parallel.

19. Generators operate at maximum efficiency when carrying
   a. 75% of their rated load.
   b. 100% of their rated load.
   c. 25% of their rated load.
   d. 125% of their rated load.

20. A synchroscope is an instrument that indicates differences in
   a. voltage.
   b. phase.
   c. speed.
   d. current.

21. When two alternators are connected in parallel, one alternator will assume more load if
   a. its field strength is increased.
   b. its speed is increased.
   c. both its speed and field strength are increased.
   d. its armature current is increased.

22. A motor–generator set is
   a. a generator supplying power to a motor.
   b. an electric motor supplying power to a generator.
   c. both a and b.
   d. neither a nor b.

23. The cost of producing and transmitting alternating current, compared to direct current, is
   a. less expensive.
   b. more expensive.
   c. about the same.
   d. slightly more expensive.

24. A rotary converter is sometimes called a
   a. dynamotor.
   b. rotating generator.
   c. phase shifter.
   d. rectomotor.

25. Rotary converters are frequently used in
   a. air-conditioning systems.
   b. electric railway systems.
   c. communication systems.
   d. heating systems.

## Give Complete Answers

1. Why is AC the primary source of electrical energy?

2. List five applications in which DC is preferred or required over AC.

3. Why is DC necessary for the alternator field circuit?

4. What name is given to the generator that provides DC for the alternator field?

5. Describe the construction of the rotating field of a low-speed alternator.

6. Why is hydrogen used for cooling large turbo-type alternators?

7. Explain how the armature coils are connected on a single-phase alternator.

8. List three uses of single-phase alternators.

9. List three methods for connecting the stator coils on a two-phase alternator.

10. What is a two-phase alternator?

11. What is a three-phase alternator?

12. List two common connections that are used with three-phase alternators.

13. Can single-phase loads be connected to three-phase systems?

14. List the common voltages supplied by a three-phase, four-wire wye system.

15. Describe how the coils are connected for a three-phase, three-wire delta system.

16. List some common transmission voltages used in the United States.

17. Why do power companies prefer to use several smaller alternators connected in parallel rather than one very large alternator?

18. List five conditions that must be fulfilled in order to connect alternators in parallel.

19. Describe the procedure for connecting alternators in parallel.

20. What is meant when it is said that the alternators are being synchronized?

*Solve each problem, showing the method used to arrive at the solution.*

1. The voltage across each phase of a two-phase alternator is 440 V. If the coils are connected to form a two-phase, three-wire system, what is the voltage across the combination?

2. If the current through phase X in the alternator in Problem 1 is 25 A, and the current through phase Y is 30 A, how much current flows in the common wire? Assume that the power factor of both loads is 100%.

3. If the power factor for load Y in Problem 2 is 86.6% lagging, how much current flows in the common wire?

4. On a two-phase, five-wire system, the phase voltage is 440 V. Calculate the voltage between each line wire and the neutral.

5. A four-pole alternator is operating at a speed of 1000 r/min. What is the frequency of the output voltage?

6. If the phase voltage of a wye-connected generator is 277 V, what is the line voltage?

7. If the line voltage of a wye-connected system is 208 V, what is the phase voltage?

8. The line current of a delta-connected system is 600 A. What is the phase current?

9. The current flowing through one phase of a delta-connected generator is 800 A. What is the line current?

10. How much power is taken from a three-phase alternator if the current is 200 A per line and the line voltage is 480 V? The power factor of the load is 90%.

# AC Motors

## OBJECTIVES

After studying this chapter, the student will be able to:

- Describe the construction of various types of AC motors.

- Explain the principle of operation of various types of three-phase motors and single-phase motors.

- Discuss the reasons for the difference in the values of starting and running currents in AC motors.

# AC MOTOR CONSTRUCTION

The **induction motor** is the most common type of AC motor. Its simple, rugged construction makes it relatively inexpensive to manufacture, and it meets most industrial requirements. Its two main components are the stator and the rotor. The stator consists of electromagnets secured to the frame, spaced equal distances apart. The rotor is made of steel lamination in the shape of a cylinder. Windings are placed in slots on the rotor surface. The stator of an AC motor is shown in Figure 19–1A; Figure 19–1B shows the rotor.

**FIGURE 19–1B**   AC motor rotor.

## Squirrel-Cage Rotor Winding

A *squirrel-cage rotor* (Figure 19–2) consists of heavy copper bars connected together at each end by copper or brass end rings. The bars are welded to the end rings.

Some manufacturers use a casting process to construct the rotor. The entire rotor, including the bars, is placed in a mold, and the bar ends are cast to the copper end rings. For small squirrel-cage rotors, the bars, end rings, and fan blades are cast in one piece. Generally, aluminum is used for this process.

## Wound-Rotor Motor

Some industrial applications require a wire-wound rotor. Copper wire is wound into the slots of the rotor as shown in Figure 19–3. For the three-phase motor, the windings are generally connected in wye, and the open ends are connected to slip rings mounted on the shaft.

**FIGURE 19–2**   Squirrel-cage rotor.

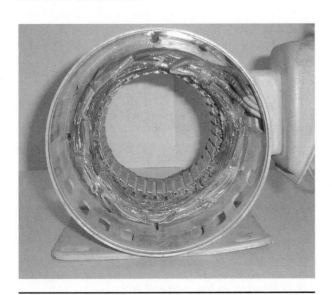

**FIGURE 19–1A**   AC motor stator.

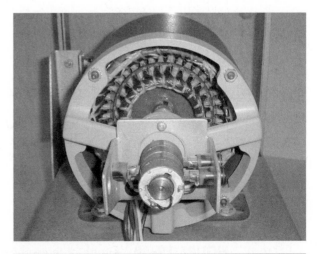

**FIGURE 19–3**   Rotor of a wound-rotor motor.

# THREE-PHASE MOTOR THEORY

The stator of an induction motor has no projecting poles. The windings are embedded in slots (Figure 19–1A). On the three-phase motor, the windings are arranged to produce a rotating magnetic field when connected to a three-phase source.

Figure 19–4A shows a two-pole, three-phase winding. When this winding is energized from a three-phase source, the three-phase currents vary as shown in Figure 19–4B. The currents are 120 electrical time degrees apart and are continuously increasing and decreasing in value and changing in direction. The effect of this variation in strength and change of direction produces a rotating field. Figure 19–4C illustrates one complete revolution.

At instant 1 in Figures 19–4B and 19–4C, the current in phase X is zero, and the currents in Y and Z are equal and opposite. In other words, the current is flowing into winding Y and out winding Z. Figures 19–5A and 19–5B illustrate the current flow at this instant. The magnetic field established by these currents is shown in Figure 19–5C. At instant 2 in Figure 19–4B, the current in phase Y is a negative maximum value, and the currents in X and Z are 50 percent of the positive maximum value. This change in current value per phase causes the flux to shift 30 degrees in a clockwise direction. Figures 19–6A and 19–6B illustrate the current flow at instant 2, and Figure 19–6C shows the magnetic field. At instant 3 current in phase Z is zero, and phases X and Y are equal and opposite. The currents for instant 3 are

shown in Figures 19–7A and 19–7B. Under these conditions, the magnetic field has shifted another 30 degrees clockwise (Figure 19–7C). At instant 4, the current in phase X is the maximum positive value, and the currents in Y and Z are 50 percent of the maximum negative value. Note that the current in

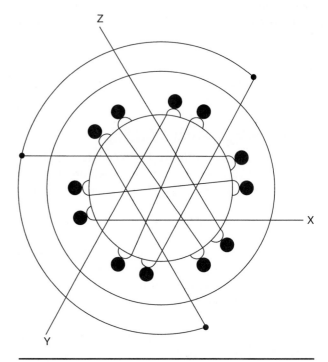

**FIGURE 19–4A**   Two-pole, three-phase stator winding.

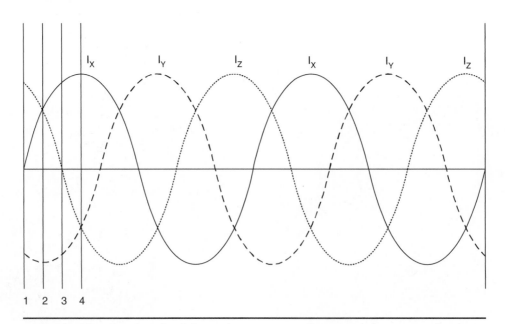

**FIGURE 19–4B**   Graph of three-phase currents flowing in the stator of a three-phase motor.

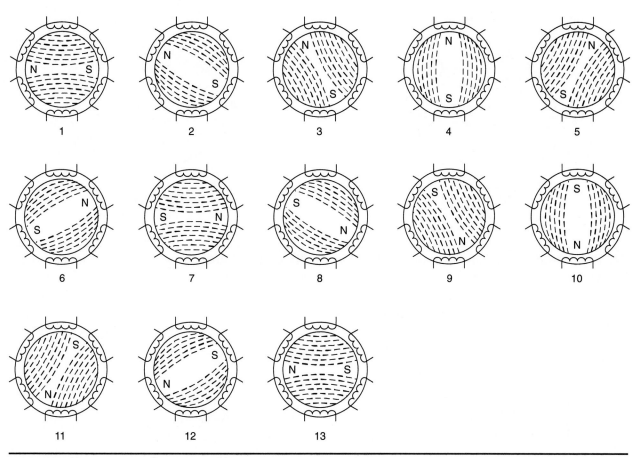

**FIGURE 19–4C** Rotating magnetic field produced by three-phase currents flowing in the stator of a three-phase motor.

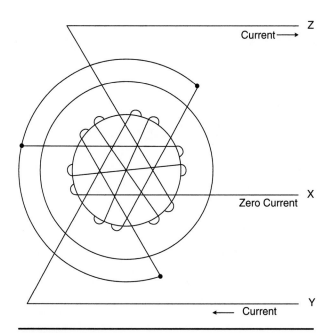

**FIGURE 19–5A** Current flow through stator windings Y and Z at instant 1.

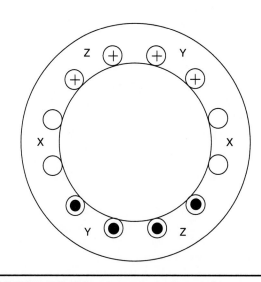

**FIGURE 19–5B** Stator windings showing the direction of current at instant 1.

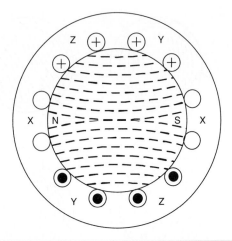

**FIGURE 19–5C** Magnetic field established by three-phase currents at instant 1.

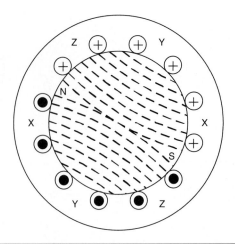

**FIGURE 19–6C** Magnetic field established by three-phase currents at instant 2.

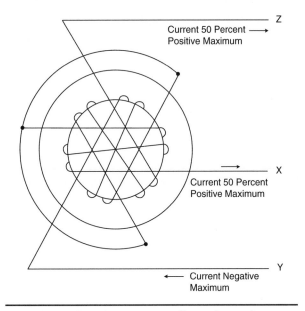

**FIGURE 19–6A** Current flow through stator windings at instant 2.

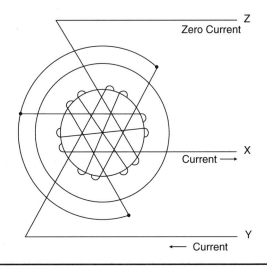

**FIGURE 19–7A** Current flow through stator windings at instant 3.

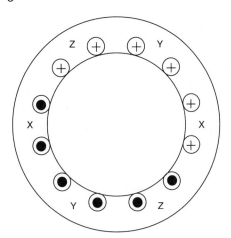

**FIGURE 19–6B** Stator windings showing the direction of current at instant 2.

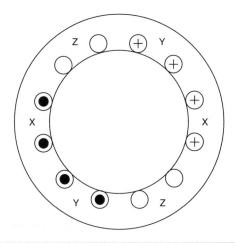

**FIGURE 19–7B** Stator windings showing the direction of current at instant 3.

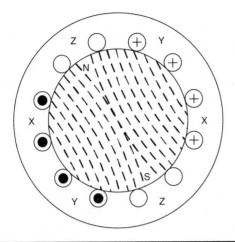

**FIGURE 19–7C** Magnetic field established by three-phase currents at instant 3.

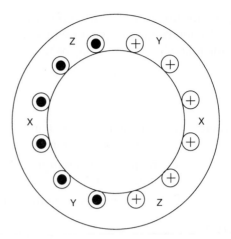

**FIGURE 19–8B** Stator windings showing the direction of current at instant 4.

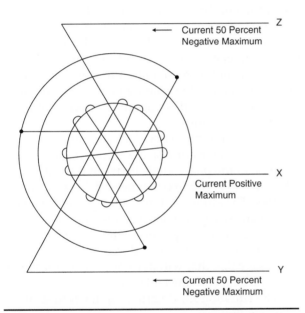

**FIGURE 19–8A** Current flow through stator windings at instant 4.

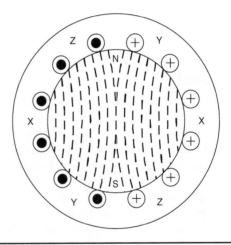

**FIGURE 19–8C** Magnetic field established by three-phase currents at instant 4.

phase Z has reversed direction. Figures 19–8A and 19–8B illustrate this condition, and Figure 19–8C shows the magnetic field, which has again shifted 30 degrees clockwise. Following the procedure for one complete cycle shows that the magnetic field rotates 360 degrees, or one complete revolution, in a clockwise direction. At this point, the cycle of current begins again and the field continues to rotate in a clockwise direction.

As the magnetic field rotates around the stator, the lines of force move across the conductors of the rotor. This action induces a voltage into the rotor conductors. The induced voltage causes a current to flow in the rotor conductors, and the current establishes a magnetic field in the rotor. The poles of the rotor are attracted by the poles of the stator, and a rotation is produced.

## Speed of the Rotating Magnetic Field

The magnetic field of a two-pole induction motor completes 60 revolutions per second (r/s), or 3600 revolutions per minute (r/min), when supplied with 60 hertz. This speed is referred to as the **synchronous speed** of the motor. To obtain lower speeds, it is necessary to increase the number of poles on the stator. The synchronous speed of a four-pole motor is 1800 revolutions per minute;

a six-pole motor produces 1200 revolutions per minute. To determine the synchronous speed of a motor, the following formula may be used:

$$n_1 = \frac{120f}{P} \qquad \text{(Eq. 19.1)}$$

where  $n_1$ = synchronous speed, in revolutions per minute (r/min)

$f$ = frequency of the supply current, in hertz (Hz)

$P$ = number of poles on the stator

### Example 1

What is the synchronous speed of a 12-pole, 60-Hz, squirrel-cage induction motor?

$$n_1 = \frac{120f}{P}$$

$$n_1 = \frac{120 \times 60}{12}$$

$$n_1 = 600 \text{ r/min}$$

## Rotor Speed

The speed of the rotor of an induction motor depends upon the synchronous speed and the load it must drive. The rotor does not revolve at synchronous speed but tends to slip behind. The amount of slip at full load may be as great as 5 percent. At no load, the amount of slip may be as low as 1 percent. For example, the synchronous speed of a two-pole, 60-hertz induction motor is 3600 revolutions per minute. At no load, the rotor revolves at approximately 3564 revolutions per minute, and at full load with 5 percent slip, it revolves at 3420 revolutions per minute.

## Direction of Rotor Rotation

The direction of rotation of the stator field of a three-phase induction motor depends upon the phase sequence. The rotor field is attracted by the field of the stator and therefore revolves in the same direction as the stator field. Interchanging any two of the three-phase leads that supply current to the stator reverses the phase sequence in the motor and the direction of the stator magnetic field. Reversing the direction of the stator field causes the rotor to reverse direction.

## Torque

The torque produced by an induction motor varies with the strength of the stator and rotor fields. The phase relationship between the two fields also affects the torque.

When a three-phase current is first supplied to the stator, the rotor is at a standstill. The three-phase current produces a rotating field in the stator. At this instant, there is 100-percent slip and the stator field is sweeping across the rotor conductors at maximum speed. The voltage induced into the rotor is maximum, and the frequency is the same as that of the stator. Under this condition, the inductive reactance of the rotor is high compared to its resistance. Thus, the rotor current lags the voltage by a large amount. There is only a slight lag between the stator voltage and current, which results in the rotor's current lagging the stator current by a considerable amount. The magnetic fields produced by these currents are also considerably out of phase, resulting in a low starting torque.

As the rotor begins to revolve, the slip decreases, and the rate at which the stator field sweeps across the rotor conductors decreases. This causes a decrease in the rotor voltage and frequency. Further increases in the rotor speed cause further reductions in the frequency and voltage of the rotor. If the rotor speed could increase until it revolved at exactly the same speed as the rotating magnetic field, there would be no induced voltage, no rotor current, and, therefore, no torque.

As the rotor frequency decreases, the inductive reactance of the rotor also decreases, resulting in a decrease in the phase angle between the stator and rotor currents. The two magnetic fields pull closer together. The attractive force between the magnetic fields increases, producing a greater torque.

When a load is applied to the motor, the rotor speed decreases until the induced emf in the rotor reaches a value that will develop enough torque to drive the load at a constant speed.

The equation for determining the torque of an induction motor is:

$$T = K_T \phi_S I_R \cos \phi_R \qquad \text{(Eq. 19.2)}$$

where  $T$ = torque, in pound-feet (lb ft)

$K_T$ = torque constant

$\phi_S$ = stator flux

$I_R$ = rotor current

$\cos \phi_R$ = rotor power factor

## Slip

The difference between the synchronous speed and the rotor speed is called the *slip* of the rotor and may be stated in revolutions per minute or as a percentage. The percentage of slip may be calculated as follows:

$$\%S = \frac{n_1 - n_2}{n_1} \times 100 \qquad \text{(Eq. 19.3)}$$

where  % $S$ = percentage of slip

$n_1$ = synchronous speed, in revolutions per minute (r/min)

$n_2$ = rotor speed, in revolutions per minute (r/min)

Slip in revolutions per minute is calculated as follows:

$$S = n_1 - n_2 \qquad \text{(Eq. 19.4)}$$

where  $S$ = slip, in revolutions per minute (r/min)

$n_1$ = synchronous speed, in revolutions per minute (r/min)

$n_2$ = rotor speed, in revolutions per minute (r/min)

The rotor frequency is directly proportional to the slip. Therefore:

$$f_r = Sf_s \qquad \text{(Eq. 19.5)}$$

where  $f_r$ = rotor frequency

$S$ = slip percentage

$f_s$ = slip frequency

## Example 2

Calculate the percentage of slip and the rotor frequency of a 60-Hz, 8-pole motor operating at 840 r/min.

1.  $n_1 = \dfrac{120f}{P}$

$n_1 = \dfrac{120 \times 60}{8}$

$n_1 = 900 \ r/min$

2.  $\% S = \dfrac{900 - 840}{900} \times 100$

$\% S = 0.067 \times 100$

$\% S = 6.7\%$

3.  $f_r = Sf_s$

$f_r = 0.067 \times 60$

$f_r = 4.01 \ Hz$

# THREE-PHASE MOTOR STARTING AND RUNNING CURRENT

The induction motor is basically a transformer in which the stator is the primary and the rotor is a short-circuited secondary. When a three-phase voltage is applied to the stator (primary), the rotating magnetic field induces a voltage into the rotor (secondary). The rotor current develops a flux that opposes and, therefore, weakens the stator flux. This allows more current to flow in the stator windings, just as an increase in the current in the secondary of a transformer results in a corresponding increase in the primary current. Because of the air gap between the stator and the rotor of an induction motor, sufficient flux leakage occurs to limit the starting current to a value approximately 4 to 6 times the full-load current.

## Loading a Squirrel-Cage Motor

Because of its rugged construction, the three-phase squirrel-cage motor is capable of handling the starting current without damage to itself. Very large motors, however, may require a value of starting current that will cause line drops, which may affect other equipment operating from the same system. For such installations, **reduced voltage starters** are used. The reduced voltage limits the starting current to a lower value.

Starting induction motors under reduced voltage will reduce the starting torque considerably. It is sometimes necessary to start the motor without the load. In this case, once the motor has reached its rated speed, the load is applied. A 50-percent reduction in voltage will reduce the torque to only 25 percent of its normal value.

A torque/current curve of a standard squirrel-cage motor is depicted in Figure 19–9. The maximum torque is reached at about 25 percent slip. Beyond this value, the power factor of the rotor decreases faster than the current increases, causing the torque to decrease. If the motor is loaded beyond maximum torque, that is, its **breakdown torque**, it will quickly slow down and stop. In Figure 19–9, the value of $T$ at starting (at 100 percent slip) is about 150 percent of the full-load torque. The starting current is about 5 times the full-load current. This motor is essentially a constant-speed machine with speed characteristics similar to those of a DC motor. The performance curves for a three-phase, squirrel-cage motor are shown in Figure 19–10. This motor is generally used where average starting torque and relatively constant speed are required.

## Double Squirrel-Cage Rotor

One type of induction motor with excellent operating characteristics has a rotor with two squirrel-cage windings. Figure 19–11 illustrates this type of winding. The bars of the inner winding are made of a low-resistance metal, usually copper surrounded by iron/steel (except for the space between the two

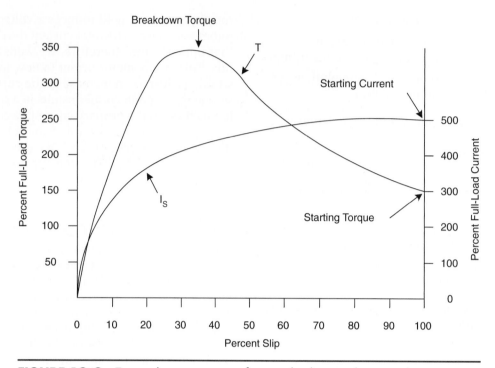

**FIGURE 19-9**  Torque/current curves of a standard squirrel-cage induction motor.

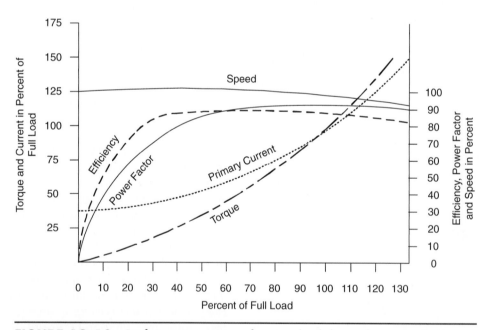

**FIGURE 19-10**  Performance curves of a standard three-phase, squirrel-cage induction motor.

windings). This construction results in a winding with low resistance and high inductance. The bars of the outer winding are made of small copper or aluminum strips having a high resistance when compared with the inner winding. Only the sides of the outer winding of the rotor core are made of iron/steel. This results in a winding with high resistance and low inductance.

When the rotating magnetic field of the stator sweeps across the two windings of the rotor, an equal amount of emf is induced into each winding. At the instant of starting, the rotor frequency is the same as the line frequency. As a result, the reactance of the inner winding is much greater than that of the outer winding. Thus, there is a large current and a high power factor in the outer winding.

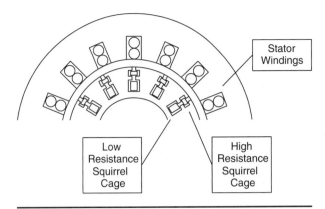

**FIGURE 19–11** Section of a double squirrel-cage winding.

This combination produces a high starting torque. As the rotor approaches synchronous speed, its frequency decreases rapidly, and the current division between the two windings is governed by their resistance. This results in most of the current flowing through the inner winding, and the machine operates as a standard squirrel-cage motor.

# TYPES OF THREE-PHASE MOTORS

Special types of three-phase motors have been designed to provide for variances in speed, frequency, and power factor. The type of motor selected depends upon the service requirements.

## Multispeed Squirrel-Cage Motors

The speed of the standard squirrel-cage induction motor is inherently constant. However, special squirrel-cage motors are manufactured with stator windings in which the number of poles may be changed by changing the external connections. For example, a stator may be wound so that one connection will provide four poles and another connection will provide eight poles. The end of these windings may be connected to a switching device and arranged so that one position will provide the low speed and the other position will provide the high speed. If the motor is supplied with 60-hertz power, the synchronous speeds will be 900 revolutions per minute and 1800 revolutions per minute. If more variations of speed are required, more windings may be placed on the stator. One set of windings may provide connections for 4 or 8 poles; the other set may provide connections for 6 or 12 poles. This arrangement will provide synchronous speeds of 600, 900, 1200, and 1800 revolutions per minute.

## Wound-Rotor Induction Motors

A *wound-rotor induction motor*, sometimes called a *slip-ring motor*, is an induction motor with a wire-wound rotor. The windings are usually connected in wye, and the open ends are connected to slip rings. Figure 19–12A illustrates this connection. Connecting variable resistances between the slip rings inserts resistance into the rotor circuit during the starting period and removes it from the circuit as the rotor accelerates to its rated speed. This method of starting produces a high starting torque at a low current value. When the resistance is removed from the circuit, the slip rings are short-circuited, and the motor operates like a squirrel-cage motor.

When the motor is operating under a constant load, the speed can be varied by varying the resistance of the rotor circuit. Figure 19–12B shows a wound-rotor connected to a three-phase rheostat.

The wound-rotor induction motor has the following advantages compared to the squirrel-cage motor:

- High starting torque with low starting current
- No abnormal heating during the starting period
- Smooth acceleration under heavy loads
- Good speed adjustment when operating under a constant load

The disadvantages of the wound-rotor induction motor are as follows:

- Greater initial cost and maintenance costs than those of the squirrel-cage motor
- Poor speed regulation when operating with resistance in the rotor circuit

## Adjustable-Speed Induction Motor (Brush-Shifting Motor)

One type of polyphase induction motor that operates reasonably well as an adjustable-speed motor is shown in Figures 19–13A and 19–13B. The primary winding is on the rotor and is energized through slip rings. A supplementary winding is wound in the same slots with the primary winding and is called the *regulating winding*. This winding is connected to a commutator. The secondary winding is installed on the stator, and the ends of each phase winding are connected to brushes that slide on the commutator. Figure 19–13B illustrates this *brush-shifting motor*. The brush holders are secured to a yoke and arranged so that they can be shifted around the commutator. When the brushes marked "A" are shifted in one direction, the brushes marked "B" are moved the same distance in the opposite direction.

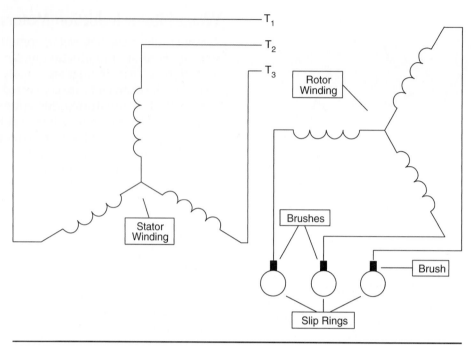

**FIGURE 19–12A**   Connections for a wound-rotor, three-phase induction motor.

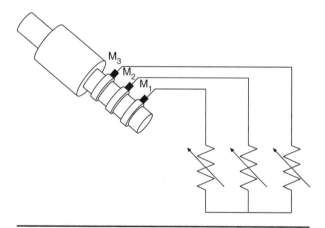

**FIGURE 19–12B**   Wound-rotor motor connected to a three-phase rheostat.

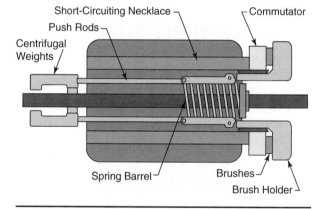

**FIGURE 19–13A**   Cutaway view of an adjustable-speed induction motor (brush-shifting type).

When the primary winding is energized, it develops a rotating magnetic field. This field sweeps across the secondary winding and induces a voltage into the secondary coils. A voltage is also induced into the regulating winding. However, with the brushes positioned as shown in Figure 19–13B, the voltage has very little effect on the motor operation. Current flowing in the secondary winding establishes a rotation flux, which reacts with the primary flux, causing the rotor to turn.

When brushes A and B of each phase are moved apart, the voltage induced into the regulating winding is impressed across the secondary winding. The greater the distance between A and B, the greater the emf across the brushes and across the secondary windings. The voltage of the regulating winding

is of the same frequency as that of the secondary. Depending upon the brush arrangement, this voltage may be in phase or 180 degrees out of phase with the secondary emf. With the brushes located as shown in Figure 19–13C, it can be assumed that the voltages are of the opposite polarity. Under this condition, the secondary current will decrease, causing a decrease in the magnetic flux of the secondary, an increase in the slip, and reduced speed.

If brushes A and B in each phase are shifted to the position shown in Figure 19–13D, the emf across the brushes will be in phase with the secondary emf, and the secondary current will increase. The result is a stronger secondary magnetic field and an increase in the rotor speed.

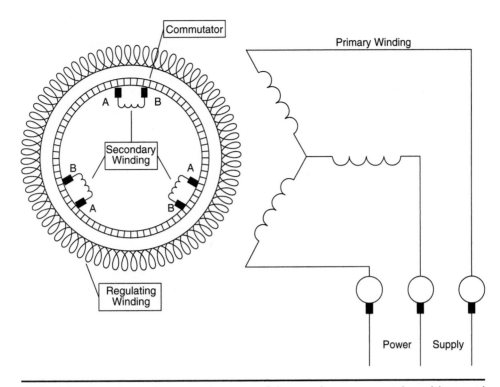

**FIGURE 19–13B** Schematic diagram of the windings on an adjustable-speed induction motor (brush-shifting type).

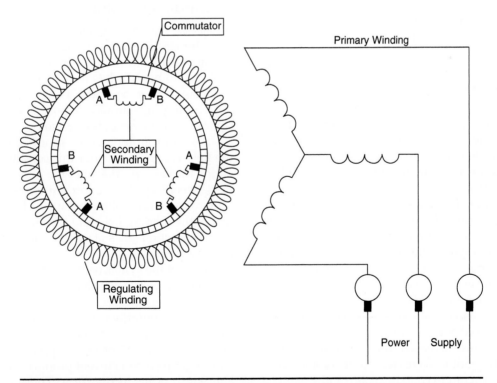

**FIGURE 19–13C** Adjustable-speed induction motor with the brushes set for below-normal speed.

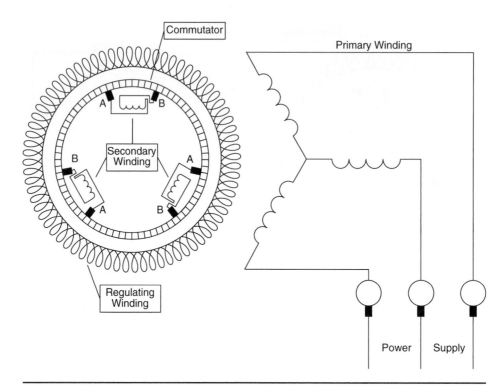

**FIGURE 19–13D**  Adjustable-speed induction motor with the brushes set for above-normal speed.

The adjustable-speed induction motor can provide a wide range of speeds, depending upon the position of the brushes.

## High-Frequency Motors

Certain installations require motors that will operate at speeds greater than those obtainable from a 60-hertz supply. One method used to meet this requirement is to construct motors to operate on higher frequencies. The next higher standard frequency is 180 hertz.

The synchronous speed of the standard 60-hertz, two-pole motor is 3600 revolutions per minute. The same motor constructed to operate on 180 hertz has a synchronous speed of 10,800 revolutions per minute. Lower speeds are obtainable by increasing the number of poles.

Motors operating at frequencies above 60 hertz are generally used where it is necessary to reduce the size or avoid the use of transmission gears. As the frequency is increased, the physical size and weight of the motor decrease. Motors of 400 hertz are used on aircraft to conserve space and weight.

Industrial plants that use 180-hertz motors must convert their 60-hertz supply to an 180-hertz supply. This can be accomplished through the use of motor–generator sets or rotary converters.

## Synchronous Motors

A *polyphase synchronous motor* is an AC motor in which the rotor revolves at the same speed as the rotating magnetic field. The rotor generally has two windings: an AC winding, which may be of the squirrel-cage or the wound-rotor type, and a DC winding. The stator windings are similar to those of the polyphase, squirrel-cage, or wound-rotor motors.

A synchronous motor cannot be started with the field energized. Under this condition, an alternating torque is produced in the rotor. As the stator field sweeps across the rotor, it tends to cause the rotor to try to turn—first in the direction opposite to that of the rotating field and then in the same direction. This action takes place so rapidly that the rotor remains stationary.

To start the synchronous motor, the rotor is left de-energized and the motor is started in the same manner as the squirrel-cage or wound-rotor motor, depending upon the rotor construction. When the rotor reaches approximately 95 percent of synchronous speed, direct current is applied to the exciter winding. The direct current produces definite north and south poles in the rotor. These poles are attracted to, and lock in with, the opposite poles of the stator. Because the rotor is turning at synchronous speed, the magnetic field of the stator is no longer sweeping across the conductors of the rotor.

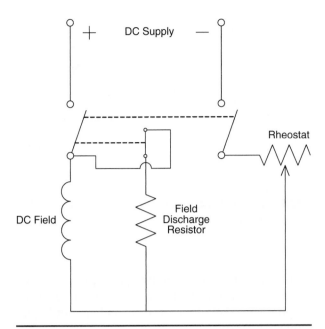

**FIGURE 19–14** Schematic diagram of the DC field of a synchronous motor. A field discharge resistor (FDR) is connected across the DC field when the supply switch is open.

Thus, the only current flowing in the rotor is the DC exciter current. If the rotor momentarily slips below synchronous speed, an induced current will flow and the rotor current will increase. This brings the rotor back to synchronous speed.

During the starting period, the rotating field induces a voltage into both the AC winding and the exciter (DC) winding. Because of the speed ratio and the large number of turns on the exciter winding, this voltage can be very high. The voltage may be so high that an insulation breakdown may occur. A high voltage is also induced when the DC is removed from the exciter winding. For this reason, a low resistance is connected across the exciter field whenever the DC supply switch is open. This resistance is called a *field discharge resistor*. Figure 19–14 shows a schematic diagram of this arrangement.

The synchronous motor has two very important features:

- It operates at a constant speed from no load to full load.
- The power factor of the motor can be controlled by varying the amount of current in the exciter winding.

In plants where the load consists chiefly of induction motors, the power factor is usually very low. Synchronous motors are frequently used to correct this problem. If the motors are operated without a load, they can be adjusted to have a leading power factor, which may be in the vicinity of 10 percent.

Motors operated in this manner are referred to as *synchronous capacitors*. When operated without a load, the synchronous motor requires very little effective power (watts) but develops a high leading reactive power. This leading reactive power compensates for the lagging reactive power of the induction motors. Synchronous motors operating under load can also develop a leading power factor, but not to the extent that they can when operating without a load.

It is not advisable to use synchronous motors when the application requires sudden and heavy loads. Such operation can cause the rotor speed to decrease and remain out of step with the rotating stator field. Synchronous motors are generally used for driving loads requiring constant speeds and infrequent starting and stopping. Some common types of loads are DC generators, blowers, and compressors.

## SINGLE-PHASE MOTORS

Figure 19–15A shows a *single-phase induction motor* with a squirrel-cage-type rotor and a single winding on the stator. When this motor is energized by a single-phase, 60-hertz source, alternating current will flow. At the instant that the current is flowing from $L_1$ to $L_2$ and increasing, a flux is established, which induces an emf into the rotor winding and causes a rotor current as indicated. The current will develop poles in the rotor, as illustrated by $N_R$ and $S_R$. The poles are lined up with the stator poles $N_S$ and $S_S$. No torque is developed because the force is in a straight line. This condition is true for any instant of the AC cycle. It is apparent that this type of single-phase motor is not self-starting.

If a means is provided for turning the rotor by hand, such as by a rope-and-pulley arrangement, a different condition will exist. Because of the rotating action, the rotor conductors will now cut the stator flux. This results in an emf being induced into the rotor conductors, causing a current to flow (Figure 19–15B). Because the rotor is highly inductive, the rotor current lags the rotor voltage by approximately 90 degrees. Therefore, the rotor flux is nearly 90 degrees out of phase with the stator flux (Figure 19–15B). Because the two fields are nearly 90 degrees out of phase, a rotating force is developed.

After a single-phase motor reaches its rated speed, its performance is nearly the same as that of a polyphase motor. The single-phase motor, however, does not develop as smooth a torque as the polyphase motor. Other advantages of the polyphase motor are that it is generally smaller in physical size, is more efficient, and has a higher operating power factor than a single-phase motor of the same horsepower.

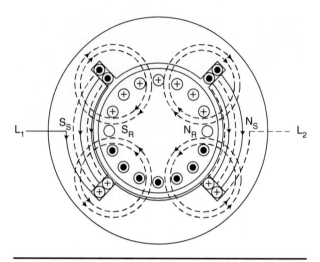

**FIGURE 19–15A**  Single-phase, squirrel-cage induction motor.

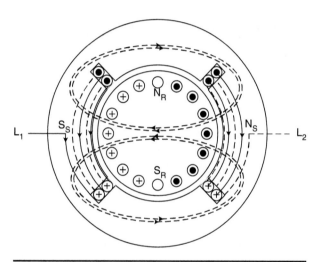

**FIGURE 19–15B**  Change in rotor flux caused by rotor rotation.

## STARTING SINGLE-PHASE MOTORS

One method of making a single-phase motor self-starting is to apply the *phase-splitting principle.* Figure 19–16A shows an arrangement in which the stator has two windings. The main winding, which is connected across the line in the usual manner, has low resistance and high inductance. The auxiliary winding (starting winding) has high resistance and low inductance. The result is that the currents in the two windings are out of phase with each other.

For maximum starting torque to be produced, the two currents should be 90 degrees out of phase. This condition is not possible with the arrangement shown. Because the starting winding has some

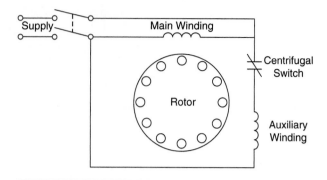

**FIGURE 19–16A**  Single-phase motor with two stator windings.

inductance, the current will lag the applied voltage by a small amount. Because the running winding has some resistance, the current cannot lag the voltage by 90 degrees.

If the current in the starting winding lags the voltage by 25 degrees and the current in the running winding lags the voltage by 70 degrees, the phase displacement is 45 degrees. This difference in phase relationship is enough to provide a weak starting torque, which is adequate for some applications.

The auxiliary winding is generally designed to remain in the circuit for only a short time. If energized too long, overheating will result and will damage the insulation. A centrifugal switch connected in series with this winding (Figure 19–16A) is designed to open when the rotor reaches approximately 75 percent of synchronous speed. The motor then operates as a single-phase induction motor.

### Resistance Split-Phase Motor

In order to obtain better starting torque, *resistance split-phase motors* are designed with a resistance connected in series with the auxiliary winding. This increases the resistance of the starting circuit and produces a greater phase difference between the stator and rotor currents. The results are a greater displacement between the two magnetic fields and better starting torque.

### Capacitor Split-Phase Motors

Some single-phase motor installations require higher starting torque than is available from the resistance split-phase motor. For this purpose, a capacitor-start, induction-run motor, called a *capacitor split-phase motor*, is frequently used.

In the standard split-phase motor, the current in the running winding may lag the supply voltage

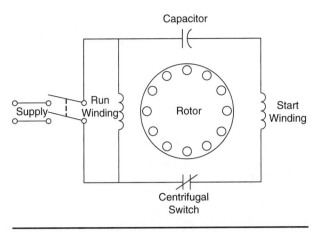

Capacitor

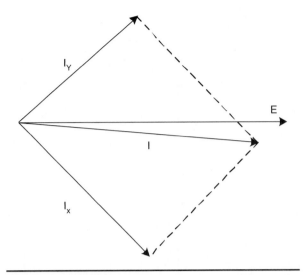

**FIGURE 19–16B** Capacitor-start, split-phase induction motor with the centrifugal switch in the run position.

**FIGURE 19–17B** Vector diagram illustrating the phase relationship of the currents in a permanent capacitor, split-phase motor during the starting period.

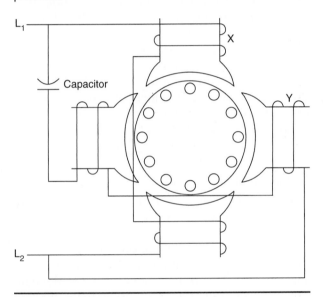

**FIGURE 19–17A** Permanent capacitor, split-phase motor.

main line. The current in coil Y leads the current in coil X by 90 degrees. The total current, $I$, is the vector sum of currents $I_x$ and $I_y$ and is nearly in phase with the supply voltage. This results in a power factor of almost unity (1.0) and a high starting torque.

Capacitor motors, as described, develop high starting torque but require much larger capacitors than are necessary during normal operation. To solve this problem, a way must be provided to reduce the capacitance once the rotor attains 75 percent of full speed. There are two methods frequently used to accomplish this. One method is shown in Figure 19–18A. Two capacitors are connected in parallel and the pair is connected in series with the auxiliary winding. A centrifugal switch is connected in series with one capacitor. When the motor reaches approximately 75 percent of its rated speed, the centrifugal switch opens, disconnecting the larger of the two capacitors. The motor now operates with only the smaller capacitor in the circuit. This arrangement provides excellent starting torque and a high power factor under normal operation.

The arrangement shown in Figure 19–18B is another method that provides high starting torque. During the starting period, the centrifugal switch is closed and the auxiliary winding is connected to point A on the autotransformer. The result is a voltage across the capacitor that is from 2.5 times to 5 times the supply voltage. This high transformer ratio provides a current through the auxiliary winding that is about 20 times as great as the current that would flow if the capacitor were connected directly in series with the auxiliary winding. The high current produces flux strong enough to develop a high starting torque.

by 70 degrees. A properly sized capacitor connected in series with the starting winding can cause a leading current of 20 degrees. With this arrangement, a 90-degree phase displacement is obtained and maximum starting torque is produced. Once the motor has reached approximately 75 percent of synchronous speed, the centrifugal switch opens and the motor operates as a single-phase induction motor (Figure 19–16B).

Figure 19–17A illustrates one type of permanent capacitor motor. This type of motor has both windings connected directly across the line, but the auxiliary winding has a capacitor connected in series with it.

A vector diagram for the starting conditions is shown in Figure 19–17B. In coil X, the current lags the voltage by approximately the same angle as in the

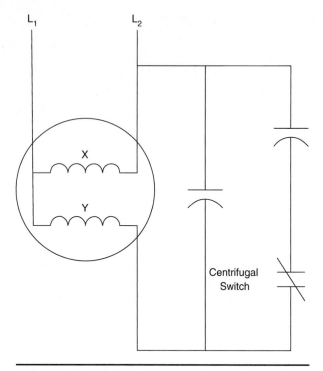

**FIGURE 19–18A** Permanent capacitor, split-phase motor with two capacitors.

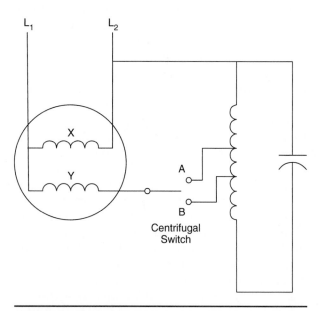

**FIGURE 19–18B** Permanent capacitor, autotransformer split-phase motor.

When the rotor reaches approximately full speed, the centrifugal switch connects the auxiliary winding to point B of the autotransformer. The transformer ratio then becomes about 1 to 2. The current through the auxiliary winding becomes about twice the value that would flow if the capacitor were connected directly in series with

the winding. This type of motor has operating characteristics very similar to those of a three-phase, double squirrel-cage motor.

Capacitor motors are generally divided into the following three classes:

1. Low starting torque
2. Capacitor start, induction run (medium to high starting torque)
3. Capacitor start, capacitor run (high starting torque, high power factor)

## Reversing Split-Phase Motors

Some applications require that the direction of rotation of the rotor be reversed. The direction of rotation depends upon the instantaneous polarities of the main field flux and the flux produced by the auxiliary winding. Therefore, reversing the polarity of one of the fields reverses the torque.

The standard direction of rotation of the rotor on single-phase motors when viewed from the shaft end is counterclockwise. To reverse the direction, interchange the connections to one of the windings.

## Shaded-Pole Motors

Single-phase *shaded-pole motors* are started by means of a low-resistance short-circuited coil placed around one tip of each pole (Figure 19–19). When the current and, therefore, the field flux are increasing, a portion of the flux sweeps across the shading coil. This induces a voltage into the coil, causing a current to flow. The current in the shading coil establishes a flux that opposes the main field flux. During this instant, lines of force are flowing only in the unshaded portion of the pole pieces.

When the main field flux reaches maximum value, the magnetic field is stationary, and zero volt is induced into the shading coil. With zero current flowing in the shading coil, there is no opposing flux, and the lines of force from the main field also flow through the shaded pole.

At the instant when the main field flux is decreasing, an emf is again induced into the shading coil but this time in the opposite direction. The current now flowing establishes a flux that is in the same direction as the main field flux, causing a higher flux in the shaded pole.

The overall effect is that the flux in the shaded pole is always out of phase with the main field flux, resulting in a weak rotating force.

This type of motor has poor starting torque and is manufactured only in very small sizes.

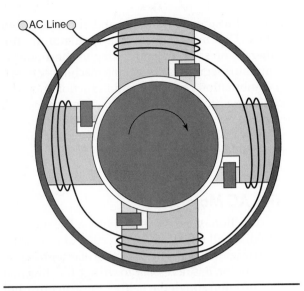

**FIGURE 19–19** Shaded-pole motor.

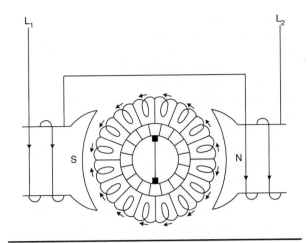

**FIGURE 19–20B** Repulsion motor: the induced emf is equal and opposite. No torque is developed.

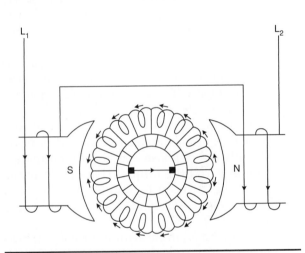

**FIGURE 19–20A** Repulsion motor: the induced emf is additive.

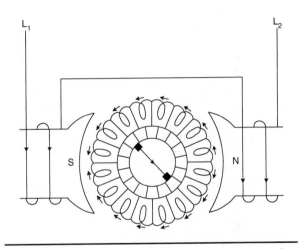

**FIGURE 19–20C** Repulsion motor: an emf is developed across the brushes.

## Repulsion Motors

The *repulsion-start, induction-run motor* is a single-phase motor that starts based on the principle that like poles repel. The rotating member contains windings similar to those of a DC motor. The windings are connected to commutator segments that are secured to, and insulated from, the rotor shaft. The brushes riding on the commutator do not connect to the line but are connected to one another to form a complete circuit through one set of rotor coils.

The method for starting this motor is shown in Figures 19–20A through 19–20D. Although the stator current is AC, it can be assumed that at the instant shown, it is rising from zero to a maximum in the positive direction. The flux produced will induce

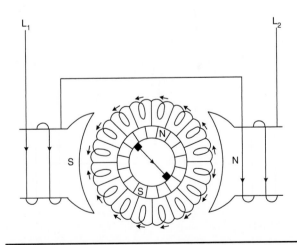

**FIGURE 19–20D** Repulsion motor: current flowing in the rotor coils develops a torque in a counterclockwise direction.

an emf into the rotor conductors, as indicated in Figure 19–20A. The voltages are additive on each side of the brushes. Therefore, a high current is forced through the armature and the short-circuited brushes. No torque will be developed, however, because one half of the conductors under each pole carry current in one direction, and one half carry current in the opposite direction.

If the brushes are shifted 90 degrees (Figure 19–20B), the emf induced into each path is equal and opposite. Therefore, zero current flows in the rotor and no torque is developed.

Shifting the brushes to the position shown in Figure 19–20C causes a resultant emf across the brushes. Current will flow through the rotor and brushes (Figure 19–20D). With this arrangement, all the conductors under one pole carry current in one direction, and all the conductors under the other pole carry current in the opposite direction. Torque is now developed.

Shifting the brushes to the position shown in Figure 19–21 causes the motor to develop torque in the opposite direction.

The machine starts as a repulsion motor, which develops a very high starting torque. As soon as the rotor reaches approximately 75 percent of the rated speed, a centrifugal device causes a short-circuiting ring to connect all the segments of the commutator. This converts the machine to an induction motor. The centrifugal device also lifts the brushes off the commutator, thus decreasing the brush wear.

The *repulsion-induction motor* differs from the repulsion-start, induction-run motor in that it has two windings on the rotor. One winding is either a squirrel-cage-type or a wound-rotor-type winding. The other winding is the repulsion-type and is connected to the commutator and short-circuited brushes. This motor starts as a repulsion motor, but there is no short-circuiting ring to short out the commutator segments. Thus, this machine operates as a combination repulsion and induction motor. It is considered to be a constant-speed motor, which develops very good starting and running torque.

The repulsion motor is expensive and requires considerable maintenance; thus, it is rarely used.

## Series AC Motors

If the direction of current supplying a DC motor is reversed, it will not affect the direction of rotor rotation. In order to reverse the direction of rotor rotation of a DC motor, it is necessary to change the direction of the current through the field or the armature, but not both. Even if a way were developed to reverse the DC supply very rapidly, torque would continue to be developed in only one direction. Thus, if a DC motor is supplied with AC, a unidirectional torque will be developed.

With an *AC shunt motor*, only a very low current will flow because of the high resistance of the field winding coupled with a high inductive reactance caused by alternating current. As a result, a very weak magnetic field is produced. In addition, the inductive effect causes the field flux to be considerably out of phase with the armature flux, resulting in very little torque. These undesirable effects can be reduced to some extent in small motors. Therefore, AC shunt motors are built only in small sizes.

In a *series motor*, the armature current and field current are in phase, resulting in a machine that can develop very high torque. The ordinary DC series motor, however, does not function satisfactorily on AC for the following reasons:

1. The alternating flux produces large eddy currents in the unlaminated parts of the machine, resulting in excessive heating.

2. The high field reactance establishes a large voltage drop across the field winding. This reduces the input current and power factor to such an extent that it makes the motor impractical.

3. The alternating flux develops high currents in the coils, which are short-circuited by the brushes. As a result, there is excessive arcing when the brushes break contact with the commutator segments.

It is obvious from these facts that modifications must be made in order for the series motor to operate from an AC supply. To reduce eddy currents,

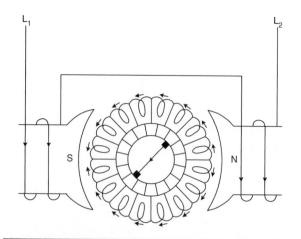

**FIGURE 19–21** Repulsion motor: the current flowing through the rotor coils develops a torque in a clockwise direction.

all metal parts that are within the magnetic circuit must be laminated, and the laminations must be insulated from one another. The laminated field poles and yoke are then supported in a cast-steel housing.

A common method of reducing armature reaction is to make use of a compensating winding. The winding is embedded in the field pole faces (Figure 19–22). It is arranged to supply a magnetizing action that is equal and opposite to that of the armature coils, regardless of the load. This is accomplished either by connecting the compensating winding in series with the armature or by short-circuiting it on itself. In the latter case, the magnetizing action of the compensating winding is obtained by transformer action.

The excessive arcing at the brushes, caused by flux sweeping across the short-circuited coils, can be eliminated by using high-resistance leads to connect the coils to the commutator segments. Each coil short-circuited by the brushes has two resistance leads, but for the main armature current there are two leads in parallel at each brush. Only those resistance leads connected to the commutator segments in contact with the brushes

carry current at any instant. Thus, the resistance leads do not affect the resistance of the armature as a whole.

Because of these modifications, AC series motors are more complex in structure and are heavier per horsepower. They are also more expensive than DC motors of the same rating. The operating characteristics of AC series motors are very similar to those of DC motors.

## UNIVERSAL MOTORS

A *universal motor* is a series motor that will operate on both AC and DC. This motor combines the features of both the AC and DC series motors. The uncompensating winding must be connected in series with the series field. Because of modification problems, it is generally manufactured only in small sizes, usually with only fractional horsepower ratings.

Universal motors are frequently used for vacuum cleaners, fans, portable electrical tools, and other small household and office appliances.

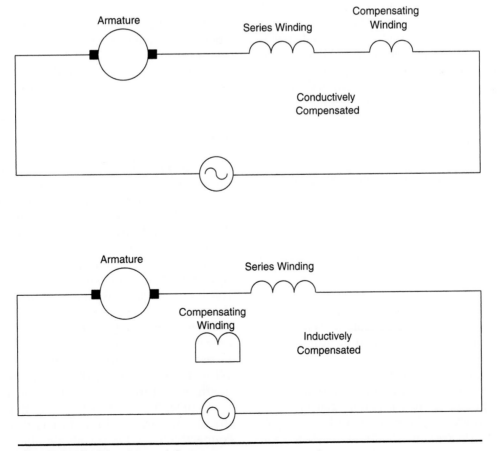

**FIGURE 19–22** Series AC motor.

## SELECTION OF SINGLE-PHASE MOTORS

Small fans, phonographs, and measuring instruments require very little starting torque. The shaded-pole motor is satisfactory for these machines. Larger machines that start without a load, such as small lathes, mills, grinders, and drills, can be started and operated most economically by split-phase motors.

Machines that are required to be started under load must use motors that develop high torque. Two common types of high starting-torque motors are the capacitor motor and the repulsion motor.

## TORQUE MOTORS

Motors used to open and close valves, dampers, doors, gates, windows, drive tool returns, and chuck devices must be designed to stall at a fixed torque. These motors, called *torque motors*, must have their greatest torque output when stalled. Large torque motors are made to operate on three phase, and small ones are usually versions of the universal motor. These motors are not intended for continuous duty. The nameplate of such a motor usually lists the running time, full-load speed, stalling torque, and power supply. Most torque motors can remain stalled for brief periods of time without overheating. Specially designed motors can remain stalled for long periods of time without overheating.

## DUAL-VOLTAGE WINDINGS

It is common practice to manufacture motors that can be operated on either of the two voltages. This can be accomplished, using either a wye or a delta connection, by dividing each phase into two sections.

Figure 19–23A illustrates windings that may be connected either in series or in parallel wye. If the series connection is suitable for 440 volts, then the parallel connection will be suitable for 220 volts.

The series connection is formed by joining $T_4$ to $T_7$, $T_5$ to $T_8$, and $T_6$ to $T_9$. $T_1$, $T_2$, and $T_3$ are connected to the three-phase supply. The parallel connection is obtained by connecting $T_1$ to $T_7$, $T_2$ to $T_8$, and $T_3$ to $T_9$. $T_4$, $T_5$, and $T_6$ are connected together to form a separate junction. The three-phase supply conductors are connected to the junctions of $T_1$ and $T_7$, $T_2$ and $T_8$, and $T_3$ and $T_9$.

Figure 19–23B illustrates the connections for a two-voltage delta. The terminal block diagrams shown below the schematic diagrams illustrate the method of identifying the connections on the nameplate of the motor.

## MULTISPEED INDUCTION MOTORS

The synchronous speed of an induction motor depends upon the supply frequency and the number of poles. Changing the speed by varying the frequency requires the use of variable frequency drives and a means of adjusting the motor current to meet the change in inductive reactance.

This can be accomplished with a solid-state **controller**. The alternating voltage supply is changed to direct voltage and then converted back to alternating voltage at a variable frequency. Varying the frequency provides a smooth speed variation similar to that obtained from a DC motor.

Changing the number of poles provides definite speeds that correspond to the number of poles selected.

Squirrel-cage motors, with windings that may be connected for different numbers of poles, offer an economical and simple means for obtaining definite speeds with minimal additional equipment. These motors are generally manufactured for two, three, and four speeds.

Two-speed motors with single windings are in general use because they require only a few leads and a very simple control. The control serves to change the connections of the stator windings. The two speeds are obtained by producing twice as many poles in the stator for the low-speed operation as are needed for the high-speed operation. To understand how this is accomplished, consider a single-phase motor with only two stator coils (Figures 19–24A and 19–24B). In Figure 19–24A, terminal $T_3$ is left open and the current enters $T_1$ and flows through the coils to $T_2$ to produce one north pole and one south pole. With this connection, the flux passes from the north pole, through the rotor, and through the south pole, then returns to the north pole through the stator core. If the motor is designed to be connected to a 60-hertz, single-phase supply, the

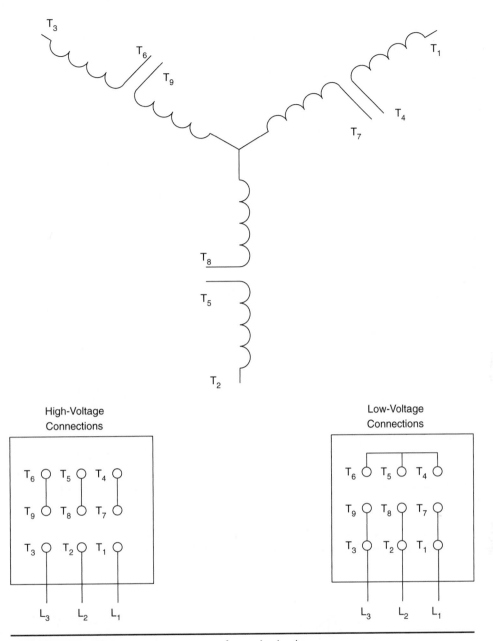

**FIGURE 19–23A** Wye connection for a dual-voltage motor.

two poles will produce a synchronous speed of 3600 revolutions per minute.

If $T_1$ and $T_2$ are connected together and the current enters $T_3$ and flows back through $T_1$ and $T_2$, a parallel connection results. The current flows through the lower coil in the direction opposite to that shown in Figure 19–24A. This change in current direction in the lower coil results in two north poles being established in the stator (Figure 19–24B). Under this condition, the flux cannot follow the same

path as before, but must form a new magnetic path (Figure 19–24B). The overall result is to establish two north poles and two south poles in the stator, or a total of four poles. The synchronous speed for 60 hertz now becomes 1800 revolutions per minute.

Variations in the design of a multispeed motor result in different operating characteristics. Some provide the same maximum horsepower at all speeds; others will produce the same maximum torque at all speeds.

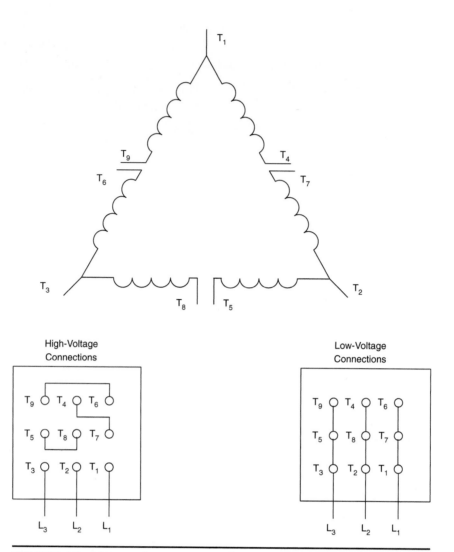

High-Voltage Connections

Low-Voltage Connections

**FIGURE 19-23B** Delta connections for a dual-voltage motor.

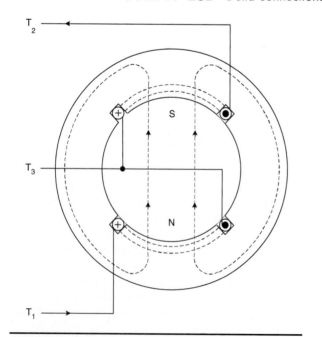

**FIGURE 19-24A** Two-speed motor connected for high speed.

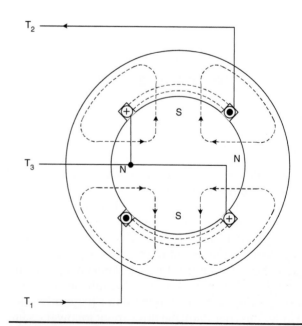

**FIGURE 19-24B** Two-speed motor connected for low speed.

# REVIEW QUESTIONS

*Multiple Choice*

1. A squirrel-cage rotor has windings that consist of
   a. copper wire wound on a core similar to a DC shunt motor.
   b. copper bars placed in slots around the surface of a cylinder.
   c. projecting poles similar to an AC alternator.
   d. either a or b.

2. On a three-phase motor, the stator windings are arranged to produce
   a. an oscillating magnetic field.
   b. a rotating magnetic field.
   c. a fluctuating magnetic field.
   d. a stationary magnetic field.

3. The synchronous speed of an induction motor refers to the speed of the
   a. rotor.
   b. armature.
   c. rotating magnetic field.
   d. rotor flux.

4. A three-phase, two-pole induction motor operating from a 60-Hz supply will produce a synchronous speed, in revolutions per minute, of
   a. 3600.
   b. 1800.
   c. 2400.
   d. 1200.

5. The direction of rotation of the rotor of a three-phase induction motor is determined by the
   a. frequency.
   b. phase sequence.
   c. number of stator poles.
   d. number of phases.

6. When a load is applied to a three-phase, squirrel-cage induction motor, the rotor speed will
   a. increase.
   b. decrease slightly.
   c. remain the same.
   d. fluctuate.

7. The starting current of an induction motor is
   a. equal to the full-load current.
   b. less than the full-load current.
   c. greater than the full-load current.
   d. one-half of the full-load current.

8. Starting an induction motor under reduced voltage will
   a. reduce the starting torque.
   b. increase the starting torque.
   c. have no effect on the starting torque.
   d. increase the horsepower.

9. The speed of a squirrel-cage motor can be varied by
   a. varying the voltage.
   b. using a specially manufactured motor with multiple winding connections.
   c. varying the current.
   d. varying the wattage.

10. High-frequency motors are used to obtain
    a. greater speeds than are available from 60 Hz.
    b. lower speeds than are available from 60 Hz.
    c. more torque per horsepower.
    d. greater horsepower.

11. A polyphase synchronous motor requires
    a. both AC and DC.
    b. two values of voltage.
    c. only AC.
    d. only DC.

12. The field discharge resistor for a synchronous motor is connected into the circuit during the
    a. starting period.
    b. running period.
    c. entire operation.
    d. shutdown period.

13. A synchronous motor is frequently used
    a. to improve the power factor.
    b. to drive variable speed loads.
    c. to maintain a steady frequency.
    d. none of the above.

14. In order for a single-phase induction motor to develop maximum starting torque, the stator field should have a phase displacement of
    a. 30°.
    b. 45°.
    c. 60°.
    d. 90°.

15. A resistance, split-phase induction motor has resistance
    a. added to the rotor circuit.
    b. connected in series with the running winding.
    c. connected in series with the starting winding.
    d. connected in series with the motor circuit.

16. Capacitors are used on split-phase motors
    a. to increase the starting torque.
    b. to improve power factor.
    c. both a and b.
    d. neither a nor b.

17. The direction of rotation on a split-phase motor can be reversed by interchanging the connections to
    a. the main winding.
    b. the auxiliary winding.
    c. either a or b.
    d. both a and b.

18. Shaded-pole motors are used to drive
    a. compressors.
    b. printing presses.
    c. small fans.
    d. pumps.

19. A repulsion motor is a
    a. three-phase motor.
    b. single-phase motor.
    c. DC motor.
    d. universal motor.

20. A universal motor is designed to operate from
    a. an AC single-phase supply.
    b. a DC supply.
    c. both a and b.
    d. a three-phase supply.

*Give Complete Answers*

1. Why is the squirrel-cage induction motor the most common type of motor used in industry?

2. Describe the rotor construction of a squirrel-cage induction motor.

3. Explain how a rotating magnetic field is developed in the stator of a three-phase induction motor.

4. What causes current to flow in the rotor of a three-phase induction motor?

5. What factors affect the speed of the rotating field that is produced in the stator of a three-phase induction motor?

6. Define the term *synchronous speed*.

7. How does the full-load rotor speed compare with the synchronous speed of a three-phase induction motor?

8. What determines the direction of rotation of the stator field of a three-phase induction motor?

9. Describe the operating principle of a three-phase, squirrel-cage induction motor.

10. How does one reverse the direction of rotation of the rotor of a three-phase, squirrel-cage induction motor?

11. What determines the amount of torque developed by a three-phase, squirrel-cage induction motor?

12. Define the term *slip* as it pertains to an induction motor.

13. Describe the operating characteristics of a three-phase, squirrel-cage induction motor.

14. Define *starting current* and *full-load current*.

15. Describe the causes and effects on the supply current for a three-phase, squirrel-cage induction motor from the initial start to the rated speed at full load.

16. Why is it sometimes necessary to start a three-phase induction motor with reduced voltage?

17. How does starting an induction motor with reduced voltage affect the torque?

18. Define the term *breakdown torque*.

19. Describe a double squirrel-cage rotor.

20. What is the purpose of a double squirrel-cage rotor?

21. List two methods used to vary the speed of a three-phase, squirrel-cage induction motor.

22. Which of the methods listed in the answer to Problem 21 provides a smooth variation of speed?

23. Describe a wound-rotor induction motor.

24. List the advantages of a three-phase, wound-rotor induction motor compared to the three-phase, squirrel-cage induction motor.

25. List two disadvantages of the three-phase, wound-rotor induction motor.

26. Describe the three-phase, brush-shifting induction motor.

27. What is the purpose of the brush-shifting arrangement on a brush-shifting induction motor?

28. What advantages are gained by using induction motors that operate on frequencies greater than 60 Hz?

29. Identify two uses for induction motors that operate on frequencies greater than 60 Hz.

30. Describe a three-phase, AC synchronous motor.

31. Explain the operation of a three-phase synchronous motor.

32. List two advantages of a three-phase synchronous motor compared to a three-phase, squirrel-cage induction motor.

33. Name two disadvantages of the three-phase synchronous motor.

34. Why would a three-phase synchronous motor be operated without a load?

35. Describe the procedure to follow when starting a three-phase synchronous motor.

36. What is the purpose of a field discharge resistor when used in conjunction with a three-phase synchronous motor?

37. Define *synchronous capacitor*.

38. Explain why a single-phase induction motor with a squirrel-cage-type rotor and a single winding on the stator does not develop a starting torque.

39. Describe the construction and operation of a standard split-phase motor.

40. Describe the construction and operation of a capacitor-start, induction-run, split-phase motor.

41. What is the advantage of the capacitor-start, induction-run, split-phase motor as compared to the standard split-phase motor?

42. Describe the construction and operation of a capacitor-start, capacitor-run induction motor.

43. What are the advantages of the capacitor-start, capacitor-run induction motor compared to the capacitor-start, induction-run motor?

44. Name two types of capacitor-start, capacitor-run induction motors, and describe the difference between the two.

45. Describe the capacitor, autotransformer, and split-phase induction motor and explain their operation.

46. Define the *centrifugal switch*, and describe its operation.

47. Explain how to reverse the direction of the rotor rotation on a split-phase induction motor.

48. Describe the construction and operation of a shaded-pole motor.

49. Describe the construction and operation of a repulsion induction motor.

50. Describe the construction and operation of a universal motor.

*Extended Study*

1. Describe the difference between the repulsion-start, induction-run motor, and the repulsion induction motor.

2. Explain how to reverse the direction of rotor rotation of a repulsion-start, induction-run motor.

3. Describe the difference between a series AC motor and universal motor.

4. Define the term *torque motor*.

5. List five common uses of a torque motor.

6. Describe a three-phase, double-voltage, squirrel-cage induction motor.

# Motor Control Devices and Circuits

## OBJECTIVES

After studying this chapter, the student will be able to:

■ Discuss several different types of control devices.

■ Identify the proper usage of different types of control devices.

■ Discuss the different types of controlled devices.

■ Identify the proper usage of different types of controlled devices.

■ Describe different methods of starting, controlling speed, reversing, braking, and stopping motors.

■ Develop control circuits to perform the function of starting, controlling speed, reversing, braking, and stopping motors.

When speaking of motor control, we are concerned with controlling the motor, that is, starting, stopping, braking, changing direction, varying or maintaining speed, and so on. There are many different types of devices that are used in the control of a motor, and they are referred to as control and controlled devices. Control devices can be a push button, a thermostat, a switch, or more. Controlled devices can be a relay, motor starter, alarm, or more.

# MOTOR CONTROL DEVICES

## Push Buttons

Perhaps the most common type of control device used in motor control is the *push button*. Figure 20–1 shows two types of common push buttons. Push buttons are manual devices. This means that human intervention is needed in order to operate the push button.

Push buttons come in many different shapes and styles. However, they all perform similar functions: they either open one or more circuits and close one or more circuits, or open and close two or more circuits simultaneously.

Push buttons may operate on a *momentary* basis. A **momentary contact push button** is spring loaded and will remain in its operated state for as long as the button is depressed. Upon releasing the button, the contacts will return to their normal state.

Push buttons may also be *mechanically held*. This type of push button contains a mechanical latching mechanism and will remain in its operated state after being depressed. Typically, another push button, mechanically linked to the first push button, is depressed to release the first push button to its original state. Sometimes a mechanically held push button may be of the push-on push-off design. When this type of push button is depressed, its contacts change state and remain in that condition until the push button is depressed again. This returns the contacts to their original condition. Another type of mechanically held push button uses a locking head design. This is typically found on emergency stop push buttons. When this push button is depressed, the button head is mechanically locked in the depressed position. The button head is returned to the normal position either by pulling it straight back or by twisting and pulling.

This is a safety feature intended to keep the button and equipment in the off state when operated during an emergency shutdown.

Probably the most common push button is a momentary push button that contains a *normally closed (N.C.)* and a *normally open (N.O.)* set of contacts, as seen in Figure 20–2. A design feature of most push buttons allows additional contact blocks to be added to the switch body. This allows additional N.O. and/or N.C. contacts to be added, depending on the circuit requirements. In addition, there are different operator styles available. Push buttons typically have a large surface on which to push. Again refer to Figure 20–2. This is especially helpful when operating a push button while wearing work gloves. Other push buttons may have large, mushroom-shaped heads, as seen in Figure 20–3. These are typically found in push buttons used for emergency stop functions. The large, mushroom-shaped head allows easy operation of the push button by slapping the button head with the palm of the hand. Quick and sure operation is essential in an emergency situation.

The frequency of the operation and the circuit-operating parameters determine the rating of the push button to be used. Push buttons are rated as *standard duty* and *heavy duty*. Standard-duty-rated switches are designed for *pilot duty*. This means that the switch will operate circuits that carry low voltage and low current. A standard duty switch is also

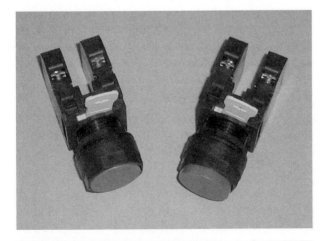

**FIGURE 20–1**  Push buttons.

**FIGURE 20–2**  N.O. and N.C. contact blocks.

**FIGURE 20–3**   Mushroom head (E-stop) push button.

operated on a regular, but somewhat infrequent, basis. Heavy-duty-rated switches are designed to switch higher voltages and currents. These switches are also operated very frequently and are, therefore, subject to hard and rough usage.

In addition to the contacts being N.O. or N.C., contacts may also be *break-before-make* and *make-before-break*. Figure 20–4A shows a schematic representation of a break-before-make contact before the push button is depressed. As the push button is depressed, the moveable contact breaks the connection between terminals A and B before making the connection with terminals C and D, as seen in Figure 20–4B. For a brief period, the circuit will be open as the moveable contact transitions from the top to the bottom set of contacts. Usually, this is not a problem because the transition occurs fairly quickly.

Figure 20–5A shows a schematic representation of a make-before-break contact before the push button is depressed. As the push button is depressed, the lower moveable contact makes the connection between terminals C and D before the upper moveable contact breaks the connection between terminals A and B, as seen in Figure 20–5B. As the push button is depressed fully, the upper moveable contact breaks the connection between terminals A and B, as seen in Figure 20–5C. This type of switch is used in circuits that cannot tolerate the brief open circuit condition created by the break-before-make switch described previously. It should be noted, however, that using a make-before-break switch in the wrong application could cause problems. Imagine the results if this switch were used to reverse the direction of a motor. While the switch transitions, the motor may try to run in both directions simultaneously. This could cause damage to the motor or equipment that the motor is driving. Always be sure you are using the correct switch for the application.

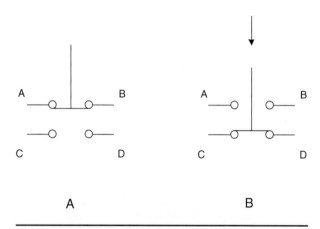

A                                               B

**FIGURE 20–4**   Break-before-make contact.

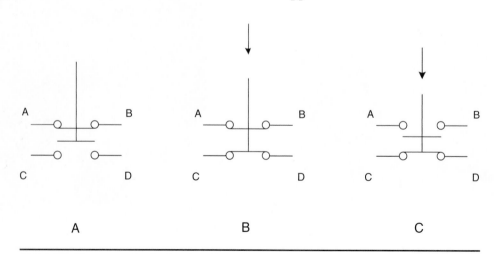

A                         B                         C

**FIGURE 20–5**   Make-before-break contact.

Push buttons are typically used to control a motor by performing the functions of start, stop, and emergency stop. In addition, push buttons can be used to alter the operation of a running motor. In this manner, push buttons may also be used to cause a motor to run forward, run in reverse, jog, jog forward, jog in reverse, run slow, and run fast.

## Rotary Switches

*Rotary switches* are manually operated switches that require a turning motion to operate. Figure 20–6 shows several different types of rotary switches. Rotary switches are sometimes known as *selector switches.* A rotary switch is used to either open one or more circuits, close one or more circuits, or open and close one or more circuits simultaneously.

Rotary switches may be spring loaded, making them operate in a momentary fashion. The contacts will change state as long as a turning force is held against the shaft of the rotary switch. When the turning force is removed, the contacts revert to their original state. Rotary switches may also be mechanically held. This is perhaps the most common type of rotary switch. When a turning force is applied to the rotary switch shaft, the shaft is rotated to a detent. The detent holds the shaft in the new position until it is rotated back to the original position.

Perhaps one of the biggest benefits of using a rotary switch is the almost limitless arrangement of contact configurations. The most basic rotary switch consists of a shaft, the switch deck or wafer, and a set of contacts (N.O. or N.C.) as seen in Figure 20–7. However, it is possible to add additional switch decks to the same shaft, as seen in Figure 20–8, which can increase the number of contact sets available. In addition, the rotary switch can be configured for 2 positions, 3 positions, and so on, up to 12 positions. Often, the rotary switch is equipped with an adjustable stop that allows the user to take a 12-position switch and adjust the stop

so that the switch has only 8 positions, 5 positions, or as many positions between 2 and 12 as desired.

As in the push button switch, rotary switch contacts may also be break-before-make and make-before-break. Figure 20–9A shows a break-before-make contact from a rotary switch. Notice the width of the pole (moveable contact). As the shaft is turned, the moveable contact breaks the connection from terminal A before making the connection with terminal B, as seen in Figure 20–9B. For a brief period, the circuit will be open as the moveable contact transitions from terminal A to terminal B. Usually, this is not a problem because the transition occurs fairly quickly.

Figure 20–10A shows a diagram of a make-before-break contact from a rotary switch. Again, notice the width of the pole piece. The pole piece is wider in the make-before-break-type contact.

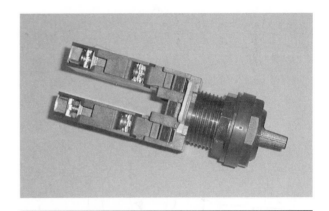

**FIGURE 20–7** Rotary switch with 1-N.O. and 1-N.C. set of contacts.

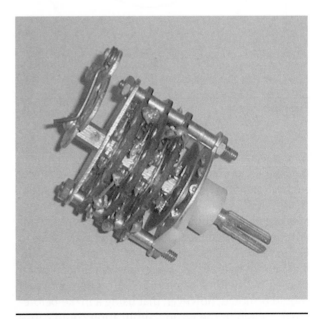

**FIGURE 20–8** A multideck rotary switch.

**FIGURE 20–6** Rotary switches.

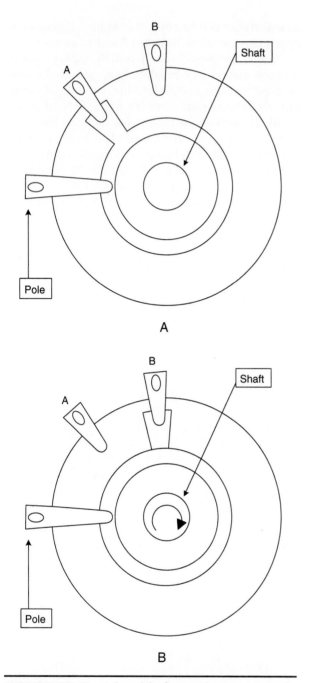

**FIGURE 20–9** A break-before-make rotary switch.

As the shaft is turned, the moveable contact makes the connection with terminal B before breaking the connection with terminal A, as seen in Figure 20–10B. As the shaft is rotated further, the moveable contact breaks the connection with terminal A, while maintaining the connection with terminal B, as seen in Figure 20–10C. This type of switch is used in circuits that cannot tolerate the brief open circuit condition created by the break-before-make switch above. It should be noted, however, that using a make-before-break switch in the wrong application could cause problems. Imagine the results if this switch were

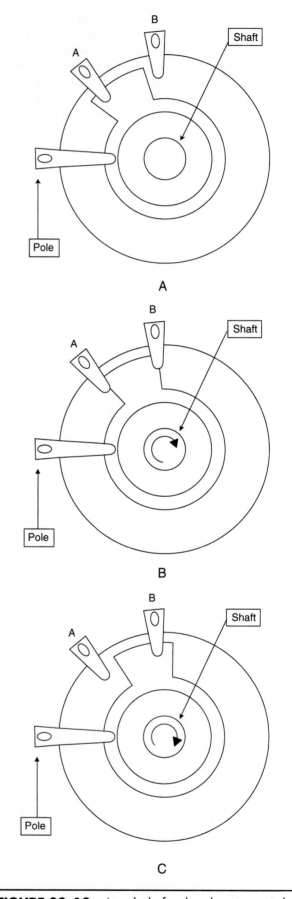

**FIGURE 20–10** A make-before-break rotary switch.

**FIGURE 20-11** Limit switches.

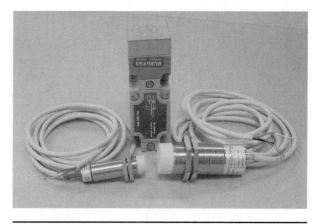

**FIGURE 20-12** Proximity switches (inductive proximity switch on left and rear, capacitive proximity switch on right).

used to switch between two power sources. While the switch transitions, the two power sources will be connected together. This could cause damage to one or both of the power sources. Always be sure you are using the correct switch for the application.

Rotary switches are commonly used to make a selection with regard to the operation of a machine or process. Some typical functions performed by a rotary switch are hand-off-automatic selection, jog-run selection, speed selection, rotation direction selection, and more.

## Limit Switches

Several different types of *limit switches* are shown in Figure 20–11. Limit switches are *automatic* devices that consist of an actuator, a body, and one or more contact blocks. Limit switches will have different actuators, depending on their intended use. Some actuators are simple levers, and some levers may have a smaller roller at the end for smoother operation. Other actuators may appear as a rod, which can move in a full 360-degree range of motion (called a wobble stick). Essentially, the actuator is what makes one limit switch different from another. All limit switches use the mechanical motion of the actuator to operate one or more sets of contacts.

Limit switches are used to sense the presence or absence of an item or to verify that a component of a machine is in the correct position. A box moving along a conveyor could be sensed with a limit switch. The limit switch would be located on the conveyor in such a fashion that the box would press against the lever of the limit switch. This, in turn, causes the switch contacts within the limit switch body to change state, indicating the presence of the

box. Another example would be an access door to a piece of equipment. Due to the existence of dangerous voltages behind the access door, the machine must automatically shut down should the door be removed. A limit switch could be mounted behind the door, so that the presence of the door depresses the lever on the limit switch. This would allow a set of normally open contacts to be closed when the door is in place, and the power for the machine could flow through these contacts. Should the door be removed, the lever moves, causing the contacts to open. This breaks the power to the machine, shutting the machine down.

## Proximity Switches

*Proximity switches* are noncontact–type switches. The typical use for a proximity switch is to sense the presence or absence of an object without actually contacting the object. Proximity switches combine high-speed switching with small physical size. Figure 20–12 shows several different types of proximity switches, the two basic types of which are inductive and capacitive.

### Inductive Proximity Switches

*Inductive proximity switches* contain a coil and an oscillator circuit. The oscillator circuit produces an electromagnetic field, which surrounds the coil. The electromagnetic field extends outward from the target area of the inductive proximity switch, as seen in Figure 20–13A. When a metal object enters the electromagnetic field, the magnetic field is reduced in intensity. The weakened magnetic field is sensed by the electronic circuit contained within the inductive proximity switch, which causes the output of the inductive proximity switch to change state, as seen in Figure 20–13B. When the metal object moves out of the electromagnetic field, the field returns to

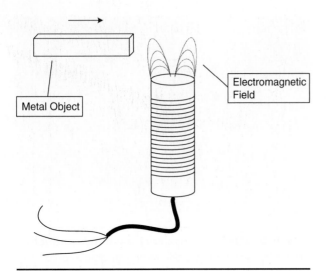

**FIGURE 20–13A**  Inductive proximity switch with electromagnetic field. Object not sensed (before passing switch).

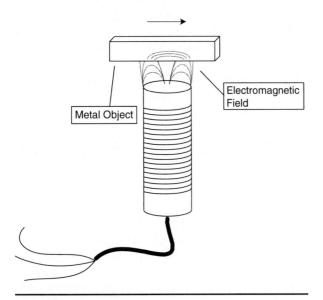

**FIGURE 20–13B**  Inductive proximity switch with electromagnetic field. Object sensed.

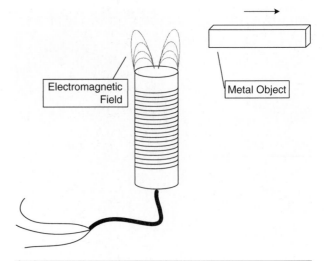

**FIGURE 20–13C**  Inductive proximity switch with electromagnetic field. Object not sensed (after passing switch).

its original intensity. This increase in electromagnetic field strength is sensed by the electronic circuit within the inductive proximity switch. The output of the inductive proximity switch will revert to its original state, as seen in Figure 20–13C. Inductive proximity switches only detect conductive materials, which makes them ideal for applications where the presence of metal must be detected.

There are two versions of the inductive proximity switch—the *inductive shielded* and the *inductive unshielded*. The shielded version is either cylindrical or shaped like a limit switch. It is designed to be flush mounted. The shielded type can detect ferrous and nonferrous metal. Its maximum sensing distance is 0.4 inches. The unshielded version may

be cylindrical, shaped like a limit switch, a small block, or a flat rectangle. It must have clearance around the target end; therefore, it cannot be flush mounted. As a result, the body of the switch will protrude. This means that the sensor may be subject to physical damage. In addition, because the switch protrudes and is unshielded, the maximum sensing distance is 0.7 inches. The unshielded version can detect ferrous and nonferrous metal.

### Capacitive Proximity Switches

Unlike an inductive proximity switch, a *capacitive proximity switch* senses conductive and nonconductive material. As found in an inductive proximity switch, the capacitive proximity switch contains an oscillator circuit, although of a different design. The oscillator circuit produces an *electrostatic field*, which extends from the target area of the capacitive proximity switch, as seen in Figure 20–14A. When an object enters the electrostatic field, the capacitance of the proximity switch's circuit is altered. The capacitance of the circuit will increase when an object is within the target area and decrease when no object is detected. The change in capacitance will cause the output of the capacitive proximity switch to change state, as seen in Figure 20–14B and 20–14C.

## Photoelectric Switches

*Photoelectric switches* are noncontact sensors. This means that a photoelectric switch can be used to sense the presence or absence of an item without having to make physical contact with the item. Figure 20–15 shows several types of photoelectric switches. Photoelectric switches consist of three

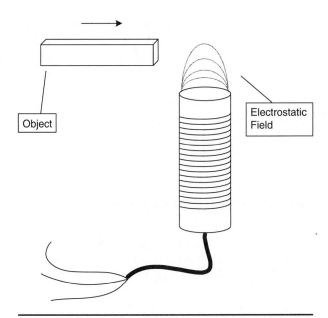

**FIGURE 20-14A** Capacitive proximity switch with electrostatic field. Object not sensed (before passing switch).

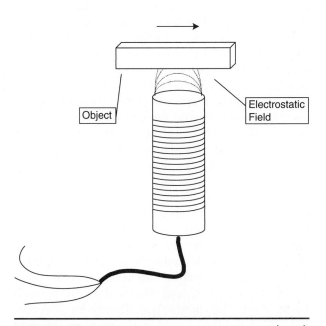

**FIGURE 20-14B** Capacitive proximity switch with electrostatic field. Object sensed.

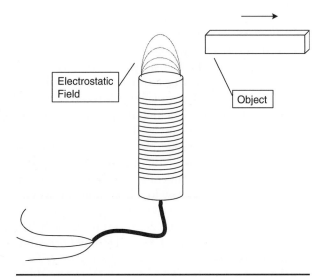

**FIGURE 20-14C** Capacitive proximity switch with electrostatic field. Object not sensed (after passing switch).

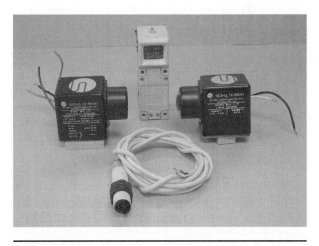

**FIGURE 20-15** Photoelectric switches. Retro-reflective type in front and back. Thru-beam type on left and right (receiver on left, transmitter on right).

basic components: the *transmitter*, the *receiver*, and the *switching device*. Let us look at each of these components in more detail.

### The Transmitter

The transmitter may also be called the *source, light source,* or *emitter.* Transmitters have used incandescent light, or **light-emitting diodes** (**LEDs**), to provide the light source. Incandescent light is identified by the visible white light that is emitted by the transmitter. Unfortunately, incandescent lights have a short life span and high failure rate. They do not operate well in areas of high vibration. Quite frequently, the repair costs of an incandescent-type transmitter exceed the cost of a new transmitter. For these reasons, incandescent transmitters are practically obsolete today.

Most transmitters today use LED technology for the light source because they offer several advantages. LED transmitters are less prone to false activation because of ambient light pollution, LEDs are very inexpensive and have a long life expectancy, and because an LED is a solid-state device, it can withstand operation in areas of high vibration. LEDs do present some problems, however. Heat is the primary cause of failure for an LED. To minimize heating, the LED is rapidly cycled on and off. Because of this cycling, the transmitter must be matched to a specific receiver, and the receiver must be synchronized to the switching frequency of the transmitter in order to "see" the transmitted signal. This prevents one manufacturer's transmitter from being used with a different manufacturer's receiver.

A benefit of the switching frequency is that one transmitter–receiver pair will not typically "see" the signal from another transmitter, because the second transmitter is switching at a different frequency. This allows transmitter–receiver pairs to be located in close proximity to one another. However, if the transmitters are in close enough proximity to one another, interference or cross-talk may occur, and the transmitters will need to be repositioned to eliminate the interference.

There are two basic types of LED transmitters. One type produces visible light, while the other emits an *infrared light* that cannot be seen with the naked eye. This can make troubleshooting difficult because you cannot look at the transmitter and see the light. However, there are devices available from local electronic supply stores that resemble a small card. The card has a material on it that is sensitive to infrared light. Placing the card in the infrared light beam will cause the material to glow. If you do not have one of these cards or are unable to locate one, you can use a camcorder that utilizes CCD (charge coupled device) technology. Most digital cameras use CCD devices. Pointing an operating camera at an operating infrared LED will allow you to "see" the infrared light in the camera's viewfinder. A drawback to the use of infrared LEDs occurs when using fiber-optic cable. Plastic fiber-optic cable will not work with infrared LEDs. You must use glass fiber-optic cable.

### The Receiver

The receiver may also be called the **photocell**, *photodiode*, or *photodetector*. The function of the receiver is to detect the presence or absence of the light transmitted by the light source and then activate or deactivate the switching device.

### The Switching Device

Many different methods are used to provide the switching function. The main difference between the various methods is the response time. The response time is the time interval between the instants when the light is detected (or not detected) and a change in the state of the switching device occurs. Another difference in switching devices lies in their classification as a source or a sink device. A source device requires current to flow from the positive (+), through the output, through the load, to the negative (−). A sink device requires current to flow

from the positive (+), through the load, through the output, to the negative (−). Source and sink devices must always be paired with their opposite partner. For example, when using a PLC with an input module that is a source type, you would use a sink-type photo switch. *Always remember that source does not go with source, and sink does not go with sink.* Figure 20–16 shows several different switching devices and typical characteristics.

## OPERATING METHODS

There are three basic methods of operating a photo switch. They are *thru-beam*, *retro-reflective*, and *diffuse*. We will look at each of these in more detail.

### Thru-Beam

The thru-beam mode is also known as *transmitted beam* or *opposed pair sensing*. With this form of sensing, the transmitter is housed in one unit, and the receiver is housed in another unit. The transmitter and receiver are located some distance apart. The item to be sensed is allowed to pass between the transmitter and receiver, thus breaking the light beam, as seen in Figure 20–17. An alternative approach would be to have the object constantly block the light beam. Should the object be removed, the beam is allowed to pass from the transmitter to the receiver, changing the state of the switching device.

Thru-beam sensing may be used over distances up to 700 feet. However, environmental conditions may limit the distance to a considerably lesser value. If the surrounding air is very dirty or has a high moisture content, the functioning distance between the transmitter and receiver will be restricted. Because there is a potential to cover distances up to 700 feet, another problem can occur when the environment is relatively clean. That is, it may be possible for the transmitted light to pass through the object being sensed and be detected by the receiver. This may occur when the transmitter and receiver are in close proximity to one another and the object being sensed is somewhat opaque. On the other hand, this may be a desired effect. For example, you may wish to determine if a product has been placed within a container. The light beam will pass through an empty container but will be blocked by a full container.

### Retro-Reflective

Retro-reflective devices are also called *reflective sensing* or *reflex sensing*. A retro-reflective sensor houses the transmitter and receiver in a single unit.

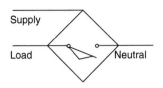

## NPN Output
(Sink or Open Collector Output)

Response Time: 500 μSec
Leakage Current: 10 μA
Maximum Current: 100 μA
Voltage: DC only
Notes:

Fastest transition of DC outputs
Ideal for PLCs
Ideal for high-speed counting
Parallel connected only

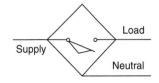

## PNP Output
(Source or Open Emitter Output)

Response Time: 500 μSec
Leakage Current: 10 μA
Maximum Current: 100 μA
Voltage: DC only
Notes:

Fastest transition of DC outputs
Ideal for PLCs
Ideal for high-speed counting
Series or Parallel connected

## Relay Output
(Electromechanical Relay)

Response Time: 10-20 mSec
Leakage Current: None
Maximum Current: 1 A @ 120 V
Voltage: AC or DC
Notes:

Slowest transition time
Mechanically noisy
Shortest life expectancy
Series or Parallel connected

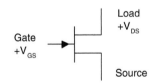

## FET
(Field Effect Transistor)

Response Time: 1 mSec
Leakage Current: Low
Maximum Current: 30 mA
Voltage: AC or DC
Notes:

Faster than the TRIAC
Used with PLCs
Used with solid-state controls
Parallel connected only

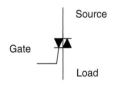

## TRIAC
(Triode AC Switch )

Response Time: 8 mSec
Leakage Current: Some
Maximum Current: 0.75 A @ 120 V
Voltage: AC only
Notes:

Faster than the relay output
Longer life expectancy
Series or Parallel connected

**FIGURE 20-16** Different switching devices and characteristics.

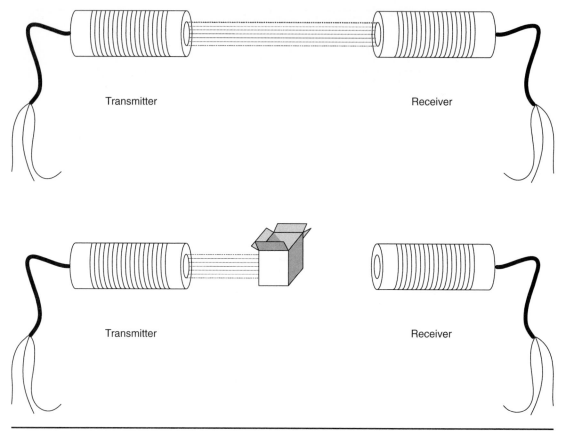

**FIGURE 20–17**   Thru-beam photoelectric switch.

In essence, a retro-reflective device transmits the light to a reflector and detects the reflected light at its receiver (see Figure 20–18). There are two basic types of retro-reflective sensors: the *standard retro-reflector* and the *polarized retro-reflector*.

### Standard Retro-Reflective Sensor

The standard retro-reflective sensor uses an industrial reflector to return the light from the transmitter to the receiver. An industrial-type reflector is used because it reflects a higher percentage of light than a typical, commercial reflector. This enhances the performance of the sensor. A standard retro-reflective sensor will have trouble when used in an environment with high moisture content in the air. The moisture will scatter the light, allowing less light to be returned to the receiver. This may result in the sensor becoming blind.

### Polarized Retro-Reflective Sensor

The polarized retro-reflector is also called an *anti-glare sensor*. A polarized lens is placed over the receiver. The lens allows only light beams that are oriented properly to pass through to the receiver. This is similar to the effect of polarized sunglasses. The benefit of polarized retro-reflective sensing is that shiny objects can be reliably sensed.

### Diffuse

Diffuse sensors use the reflectivity of the object being sensed to return the light from the transmitter to the receiver, as seen in Figure 20–19. There are several different types of diffuse sensors: the *standard diffuse sensor*, the *long-range diffuse sensor*, the *fixed-focus diffuse sensor*, the *wide-angle diffuse sensor*, and the *background suppression sensor*.

### Standard Diffuse Sensor

As mentioned previously, diffuse sensors sense the reflectivity of an object. This can create a problem when the object is no longer present. Diffuse sensors may trigger on the reflectivity of the background that was behind the object being sensed. This results in a false indication that the object is still present. To minimize the possibility of this occurring, the distance between the sensor, the object being sensed, and the background must be kept proportional. The recommended proportion is 1 to 3. This means that if the object to be sensed is located 1 foot away from the sensor, the background should be 3 feet away. Another challenge for sensing objects with a diffuse sensor is the physical appearance or characteristics of the object to be detected. The texture and color will determine the

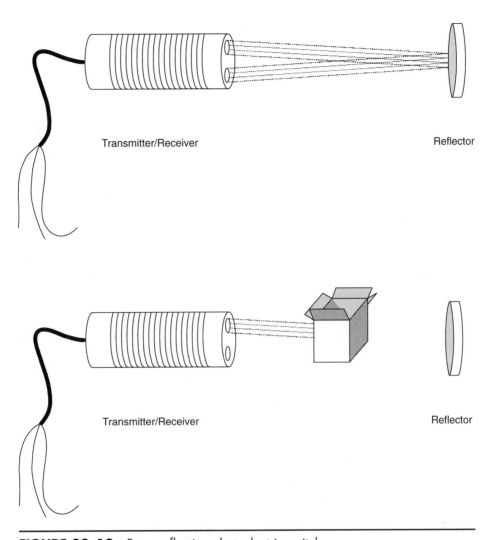

**FIGURE 20–18**  Retro-reflective photoelectric switch.

accuracy and range of detection. A smooth, light-colored object will be detected more readily than a rough, dark-colored object.

### Long-Range Diffuse Sensor

As its name implies, a long-range diffuse sensor is capable of detecting objects over a longer distance than the standard diffuse sensor. The long-range diffuse sensor works best when there is no background present or when the background is a considerable distance away from the object being detected.

### Fixed-Focus Diffuse Sensor

The fixed-focus diffuse sensor utilizes a focusing lens to create a focal point. Light is transmitted from the transmitter, reflects off the object to be detected, and is sensed by the receiver. Should the object be too close or too far away, the reflected light will be out of focus, and the object will not be sensed by the sensor.

### Wide-Angle Diffuse Sensor

The wide-angle diffuse sensor also utilizes a special lens. This lens, however, widens the field of view of the sensor, which allows the sensor to sense objects in a wider area. While this may be desired, it can also present new problems. Insects, dust, and other airborne objects may cause false detections to occur.

### Background Suppression Sensor

As its name implies, a background suppression sensor will sense an object but ignore the background. This type of sensor works best when sensing dark objects against a shiny background. While background suppression offers improved sensing reliability, it does so at the price of reduced sensing range. Typical background suppression sensors are only effective when sensing objects that are located between 3 and 18 inches from the sensor.

Figure 20–20 shows a comparison of the advantages and disadvantages between the different types of photo switches.

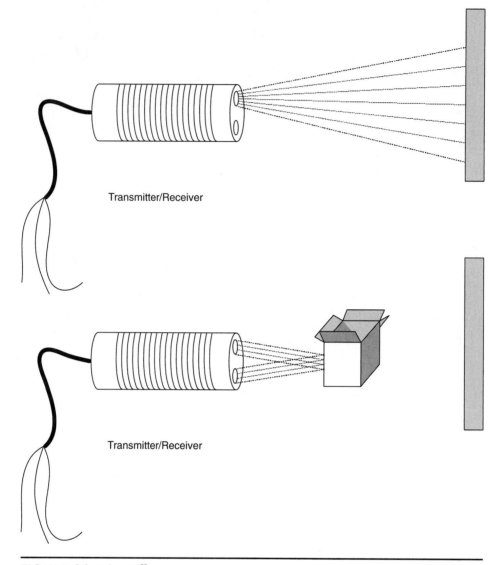

Transmitter/Receiver

Transmitter/Receiver

**FIGURE 20–19**   Diffuse sensor.

# CONTROLLED DEVICES

## Relays

A relay is an *electromechanical switch*. Figure 20–21 shows several different types of relays. The construction of a typical relay is shown in Figure 20–22. Notice that a relay contains a coil of wire wound around an iron core. Attached to the frame of the relay is a set of contacts. One contact is spring loaded and moveable (the pole), and the other contact is fixed in place. The contacts are electrically isolated from the coil.

Relays are manufactured to operate from a DC source or from an AC source. This will be noted on the relay. It is important to use the correct relay for the appropriate voltage source. When a relay is energized, a magnetic field is created around the coil of the relay. The magnetic field attracts the moveable contact, causing the contact to move. When the contact moves, the circuit between the moveable contact and the fixed contact is closed. Essentially, this is an electrically operated switch. More precisely, it is an electromagnetically operated switch. Figure 20–23 shows the relay from Figure 20–22 in its energized state. When the energizing current is removed from the relay, the spring causes the moveable contact to move, opening the circuit with the fixed contact.

Relays are used in many different circuits to perform many different functions. Perhaps the most widely used function of a relay is to allow the switching of one type or magnitude of voltage with a different type or magnitude of voltage. For example, a relay rated at 12 VDC may be used to switch 240 VAC. Recall that the contacts of a relay are electrically isolated from the coil. This allows 12 VDC

| | Thru-beam | Retro-Reflective | | Diffuse | | | | |
|---|---|---|---|---|---|---|---|---|
| | | Standard | Polarized | Standard | Long-Range | Fixed-Focus | Wide-Angle | Background Suppression |
| **Benefits** | Senses over great distances | Lower cost than thru-beam | Reliably senses shiny objects | Reliably senses shiny objects | Can sense over long distances | Senses small object in precise location | Good sensor to detect absence of object | Can sense various colors and textures at different distances |
| | Senses through most dirt and moisture | Easy alignment | Easy alignment | Best without a background | Can sense multiple colored objects at close range | Easily aligned | Senses small objects easily | Will not sense close items |
| | Can sense through some objects | Works well on vibrating equipment | | Works best with reflective object and dark background | Easy sensing of shiny objects | | Ignores most backgrounds | Will not falsely sense shiny backgrounds |
| | Most reliable | Easily senses large items on conveyers | | | Best without a background | | Immune to vibration | |
| **Weakness** | Unwanted sensing through objects | Cannot sense over long distances | Requires special reflectors | Sensing distance is affected by surface characteristics of object | Can falsely sense surrounding objects and backgrounds at great distances | Can sense various colors and textures at different distances | Can sense various colors and textures at different distances | Limited sensing range |
| | Expensive to buy and install | Difficulty in sensing shiny subjects | Slightly less sensing distance than the standard retro-reflective | Can falsely sense overworking objects and background | | Short-sensing range | Senses everything within its sensing area | Complicated installation |
| | Difficult to align at great distances | Does not work well in moisture or dirty environments | | Shiny backgrounds may cause false sensing | | Sensing distance is not adjustable | Does not work in moisture or dirty environments | |
| | | Reflector quality and condition affect performance | | Shorter sensing distance than retro-reflective | | False sensing from vibration | | |

**FIGURE 20–20** Comparison of different types of photoelectric switches.

**FIGURE 20–21**   Various types of relays.

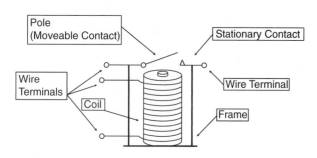

**FIGURE 20–22**   Relay construction (de-energized).

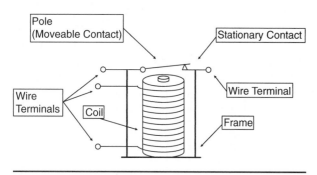

**FIGURE 20–23**   Relay construction (energized).

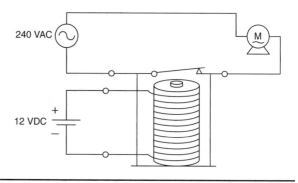

**FIGURE 20–24**   Low-voltage control for high-voltage switching.

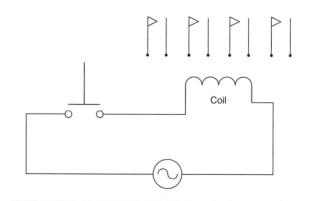

**FIGURE 20–25**   A four-pole relay.

to be applied to the coil without interfering with the voltage switched by the contacts (the 240 VAC), as seen in Figure 20–24.

Another common use for relays is to extend the number of contacts for a device. For example, a push button switch may have only one normally open contact. However, the circuit operation requires four circuits to be closed when the push button is pressed. A relay will solve the dilemma. The single normally open contact from the push button is used to energize a relay that has four sets of normally open contacts. When the push button is pressed, the relay energizes, closing all four contacts simultaneously, as seen in Figure 20–25.

There are many different varieties of relays. As mentioned earlier in this chapter, there are multipole relays. These relays will have two or more sets of contacts. These contacts may be N.O., N.C., or a combination. In addition, the contacts may be single-throw or double-throw.

**Time-delay relays** are an example of a different variety of relay. There are on-delay time-delay relays and off-delay time-delay relays. On-delay relays provide an adjustable time delay between the time the relay is energized and when the contacts change state. Off-delay relay contacts change state immediately when the relay is energized. However, there is an adjustable time delay between the time the relay is de-energized and when the contacts change to their initial state.

Relays may also be of the latching type. These relays consist of two separate coils and a mechanical linkage. One relay coil is called the latch, or set, coil and the other relay coil is called the unlatch, or reset, coil. When the latch coil is energized, the contacts change state. The contacts will remain in this condition even after the relay is de-energized. This is a result of the mechanical linkage. To change the state of the contacts back to their original condition, the unlatch coil is energized. The contacts will now remain in their original condition after the unlatch coil is de-energized. Again, the mechanical linkage maintains the contacts in their last state.

Some relays may have heavy-duty contacts. These contacts are physically larger and, therefore, can handle higher currents. A relay with heavy-duty contacts is called a *contactor*.

## MOTOR STARTERS

A motor starter is an electromechanical switch similar to a relay. A motor starter also contains heavy-duty contacts similar to a contactor. The difference between a motor starter and a relay or contactor is that a motor starter also contains overload protection for the motor. Figure 20–26 shows several different types of motor starters.

Motor starters are available for operation from AC or DC voltage. Different coil voltage ratings are available to match the motor starter to the control circuit voltage. In addition, motor starters are available in different sizes as determined by the motor that they are controlling. Figure 20–27 shows an example of a motor starter selection chart.

There are essentially two different types of motor starters, the standard motor starter and the *reversing motor starter*. The standard motor starter allows a control voltage to be used to energize the motor starter and subsequently cause the motor to run. Removing the control voltage from the motor starter causes the motor to stop. The motor was allowed to run in one direction only. The reversing motor starter, as its name implies, allows the direction of rotation of a motor to be changed. This motor starter consists of two coils, a forward coil and a reverse coil. When control voltage is applied to

**FIGURE 20–26** Various motor starters.

the forward coil, the forward set of contacts close, causing the motor to start and run in the forward direction. When the control voltage is applied to the reverse coil, the reverse set of contacts close, causing the motor to start and run in the reverse direction. Typically, reversing motor starters contain some type of mechanical interlock, which prevents both the forward and reverse sets of contacts from closing simultaneously. In addition, and as a further precaution, most forward/reverse control circuits utilize an electrical interlock. This electrical interlock may prevent the energizing of both the forward and reverse coils simultaneously, or it may require the motor to be stopped before the direction of rotation can be reversed.

Motor starters are also available with an assortment of *auxiliary contacts*. These contacts operate when the main contacts operate. Auxiliary contacts are available in N.O., N.C., and N.O.N.C. arrangements.

## ANNUNCIATORS

An *annunciator* is a signaling apparatus that may be audible or visual. Examples of an audible annunciator would be a bell, klaxon, chime, horn, loudspeaker, and siren. An example of a visible annunciator would be an indicator light, strobe light, and rotating beacon. Annunciators are often used as a warning device to alert personnel of an event or change in status. Annunciators may also be used to indicate the current condition of a machine or process.

| NEMA Size | Continuous Current Rating | Motor Volts | Maximum HP | Coil Voltage |
|-----------|---------------------------|-------------|------------|--------------|
| 00 | 9 A | 200 V | 1½ | 208 V |
|  |  | 230 V | 1½ | 240 V |
|  |  | 460 V | 2 | 480 V |
|  |  | 575 V | 2 | 600 V |
| 0 | 18 A | 200 V | 3 | 208 V |
|  |  | 230 V | 3 | 240 V |
|  |  | 460 V | 5 | 480 V |
|  |  | 575 V | 5 | 600 V |
| 1 | 27 A | 200 V | 7½ | 208 V |
|  |  | 230 V | 7½ | 240 V |
|  |  | 460 V | 10 | 480 V |
|  |  | 575 V | 10 | 600 V |
| 2 | 45 A | 200 V | 10 | 208 V |
|  |  | 230 V | 15 | 240 V |
|  |  | 460 V | 25 | 480 V |
|  |  | 575 V | 25 | 600 V |
| 3 | 90 A | 200 V | 25 | 208 V |
|  |  | 230 V | 30 | 240 V |
|  |  | 460 V | 50 | 480 V |
|  |  | 575 V | 50 | 600 V |
| 4 | 135 A | 200 V | 40 | 208 V |
|  |  | 230 V | 50 | 240 V |
|  |  | 460 V | 100 | 480 V |
|  |  | 575 V | 100 | 600 V |
| 5 | 270 A | 200 V | 75 | 208 V |
|  |  | 230 V | 100 | 240 V |
|  |  | 460 V | 200 | 480 V |
|  |  | 575 V | 200 | 600 V |
| 6 | 540 A | 200 V | 150 | 208 V |
|  |  | 230 V | 200 | 240 V |
|  |  | 460 V | 400 | 480 V |
|  |  | 575 V | 400 | 600 V |
| 7 | 810 A | 200 V | — | 208 V |
|  |  | 230 V | 300 | 240 V |
|  |  | 460 V | 600 | 480 V |
|  |  | 575 V | 600 | 600 V |

**FIGURE 20–27**  Motor starter size chart.

# MOTOR CONTROL CIRCUITS

The design possibilities are endless when it comes to developing circuits that will start, control the speed, reverse, brake, and stop motors. Following are some basic designs of motor control circuits. These are by no means the only circuits that will perform these functions. Likewise, there are other functions of motor control that are not covered in this unit. Each section will present one or more circuits and a brief description of the circuit operation. Remember, there are other possible designs that will accomplish the same task. The circuits presented here are perhaps the most common.

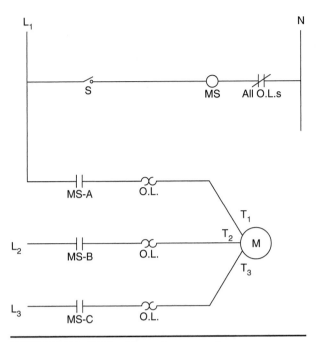

**FIGURE 20–28A** Two-wire control.

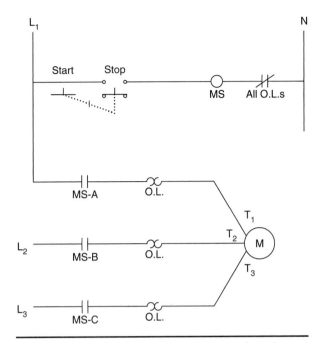

**FIGURE 20–28B** Two-wire control with two mechanically linked push buttons.

## Two-Wire Controls

A *two-wire control* is a method of motor control in which manual devices are used to start and stop the motor. As a result, only two wires are needed to connect the control device into the circuit. Because the control device is mechanically held in the "on" position, the motor will restart automatically when power is restored after a power failure.

Figure 20–28A shows a two-wire control using a snap-action toggle switch. Closing the toggle switch causes the motor to run. Opening the toggle switch causes the motor to stop. Figure 20–28B shows a two-wire control using two push buttons that are mechanically linked together. Pressing the start push button causes the motor to run. The start button will be mechanically held in the depressed position, while the stop push button will mechanically move to the extended position. Depressing the stop push button will stop the motor, while causing the start push button to move to the extended position. The stop push button will be mechanically held in the depressed position.

## Three-Wire Controls

As the name implies, *three-wire controls* need three wires to connect the control device into the circuit, as seen in Figure 20–29A. Three-wire controls also use some type of electromechanical switch to provide a latching, seal-in, memory, or holding function. This function is what causes the motor to continue to run, even after the start push button is

released. Let us follow the operation of this circuit, step-by-step.

1. The circuit is shown in its stopped mode (see Figure 20–29A).
2. Pressing the start push button energizes the motor starter coil, MS.
   A. This causes the holding contacts, MS-1, to close, maintaining a current path to the MS coil.
   B. This also causes the main motor contacts, MS-A, MS-B, and MS-C, to close.
3. The motor runs.
4. Releasing the start push button will have no effect on the circuit because the holding contacts, MS-1, maintain current flow to the motor starter, MS.
5. Pressing the stop push button interrupts the current flow to the motor starter coil, MS.
   A. This causes the holding contacts, MS-1, to open.
   B. This also causes the main motor contacts, MS-A, MS-B, and MS-C, to open.
6. The motor stops.

Should an overload condition occur, the overload contacts would open, interrupting the current path of the motor starter coil, MS. This would cause the motor starter to de-energize, stopping the motor. A power interruption will also cause the motor starter to de-energize, stopping the motor. However, unlike the two-wire control,

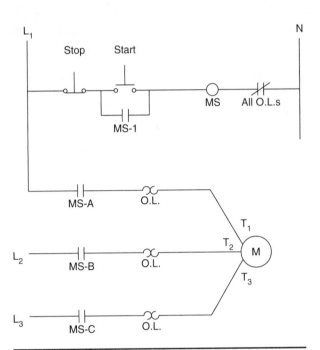

**FIGURE 20–29A**  Three-wire control.

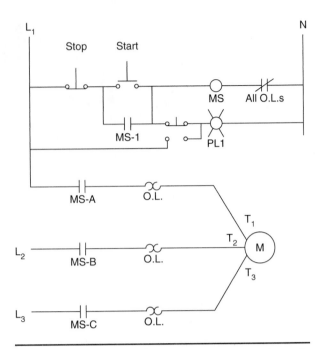

**FIGURE 20–29C**  Three-wire control with push-to-test motor running indicator light.

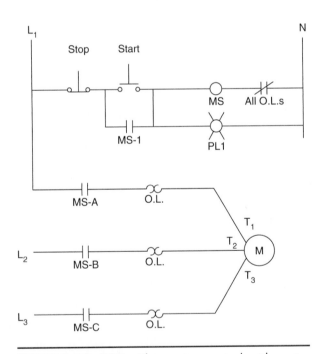

**FIGURE 20–29B**  Three-wire control with motor running indicator light.

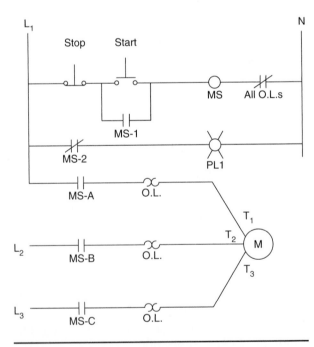

**FIGURE 20–29D**  Three-wire control with motor stopped indicator light.

when power is restored, the motor will not restart automatically.

Figures 20–29B, 20–29C, and 20–29D show the same basic circuit as in Figure 20–29A with some minor modifications. Figure 20–29B has a pilot light added, which indicates that the motor is running. The pilot light is lit when the motor is running and extinguished when the motor is stopped.

Figure 20–29C includes a motor running, push-to-test pilot light. Depressing the lens of the pilot light causes the light to light (if it is working). This test can be performed whether or not the motor is running. In this manner, a defective pilot light can be found.

The circuit found in Figure 20–29D shows a pilot light that is used to indicate when the motor is stopped. In this circuit, the pilot light will be extinguished while the motor is running and lit when the motor is stopped.

## MULTIPLE START/STOP CONTROLS

In some instances, it is necessary to control two motors from one location. An example might be a control panel. This would require two separate start/stop stations and two separate motors. Figure 20–30 shows a circuit that would perform this function. Notice that each motor has its own start/stop station.

This allows each motor to be started and stopped independently of the other. However, notice the master stop push button. Depressing the master stop push button interrupts the current flowing to both start/stop circuits. This will stop both motors simultaneously. Notice, however, that should an overload condition occur on one motor, the other motor will not be affected.

In some instances, it might be necessary to control one motor from several different locations. For example, with a long conveyor belt, it may be desirable to be able to start and stop the conveyor at either end and in the middle. This would require three separate start/stop stations and one motor. Figure 20–31 shows the control circuit that could

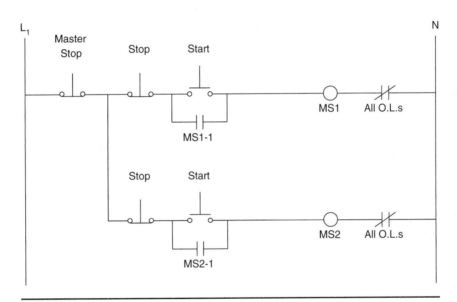

**FIGURE 20–30** Two separate start/stop controls with a master stop.

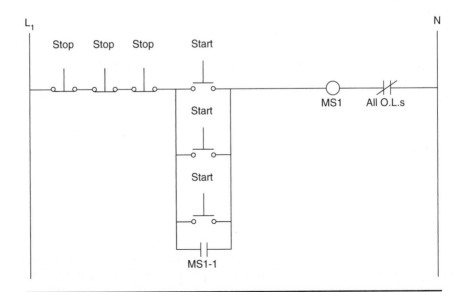

**FIGURE 20–31** Multiple start/stop controls controlling a single motor.

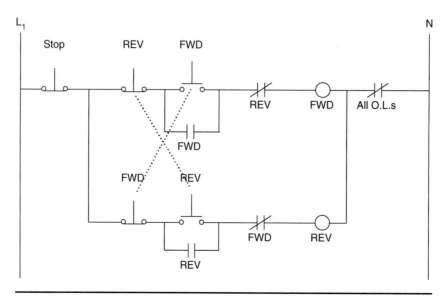

**FIGURE 20–32A** Forward/reverse control with push button and electrical interlocks.

be used to accomplish this feat. Notice that the stop push buttons are connected in series. The result is that depressing any one of the stop push buttons interrupts the current to the remainder of the control circuit. Notice also that the start push buttons are wired in parallel. The result is that depressing any one of the start push buttons causes current to flow to the motor starter, MS1. This means that an operator can start the conveyor by depressing any one of the three start push buttons located along the length of the conveyor. Likewise, the operator can stop the conveyor by depressing any one of the stop push buttons located with the start push buttons.

## Forward/Reverse Controls

Often, it is necessary to change the direction of rotation of a motor. Figure 20–32A shows a circuit that can be used to allow a motor to run in either the forward or the reverse directions. Let us step through the operation of this circuit:

1. The circuit is shown in the stopped mode (see Figure 20–32A).
   A. Notice that there are two sets of contacts (one N.O. and one N.C.) associated with the forward push button. These contacts are mechanically linked together (indicated by the dotted line).
   B. Notice that there are two sets of contacts (one N.O. and one N.C.) associated with the reverse push button. These contacts are mechanically linked together (indicated by the dotted line).

2. The forward push button is depressed.
   A. This energizes the forward motor starter, FWD.
      1. This causes the FWD holding contacts to close.
      2. This causes the FWD N.C. contacts in front of the reverse coil, REV, to open.
3. The motor is now running in the forward direction.
4. Depressing the reverse push button interrupts the current flow to the FWD coil.
   A. This de-energizes the FWD coil.
      1. This causes the FWD holding contacts to open.
      2. This causes the FWD N.C. contacts in front of the reverse coil, REV, to reclose.
   B. This allows the REV coil to energize.
      1. This causes the REV holding contacts to close.
      2. This causes the REV N.C. contacts in front of the forward coil, FWD, to open.
5. The motor is now running in the reverse direction.
6. Depressing the stop push button interrupts the current to both the FWD and REV circuits, causing the motor to stop.

Notice that it was not necessary to stop the motor before changing the direction of rotation. This may be a desirable characteristic. Changing direction without stopping the motor allows for a rapid change of direction. However, this may be hard on the motor and may shorten its life expectancy.

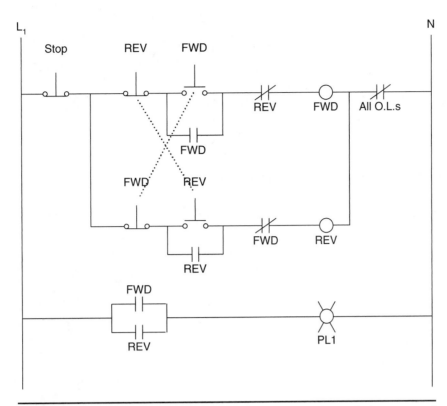

**FIGURE 20-32B** Forward/reverse control with push button and electrical interlocks. Also includes a motor running indicator light.

The requirements of the application will dictate whether the motor should be designed for this type of operation, or whether a circuit that requires stopping be used.

Figures 20–32B and 20–32C show the same essential circuit as Figure 20–32A but with some slight modifications. Figure 20–32B shows a forward/reverse circuit with a pilot light that indicates the motor is running. The pilot light does not indicate in which direction the motor is running, only that the motor is running.

Figure 20–32C shows a circuit in which the direction of rotation is indicated. This circuit uses two pilot lights. One pilot light, PL1, indicates that the motor is running in the forward direction when illuminated. Pilot light PL2 indicates that the motor is running in reverse when illuminated.

Another type of direction control circuit is shown in Figure 20–32D. This circuit uses a control device known as a *drum switch*. A drum switch is a mechanically held device that allows the direction of rotation of a motor to be changed. Notice the switching arrangement as shown in the small diagrams at the bottom of the drawing in Figure 20–32D. Essentially, when the drum switch is placed in the forward position, $L_1$ is connected to $T_1$, $L_2$ is connected to $T_2$, and $L_3$ is connected to $T_3$. Placing

the drum switch in the reverse position connects $L_1$ to $T_3$, $L_2$ to $T_2$, and $L_3$ to $T_1$. You may recall that to reverse the direction of rotation of a three-phase motor, you must interchange the connections to two of the three motor leads. The drum switch accomplishes this function by switching the connections to $T_1$ and $T_3$.

## Speed Control

Another function required of motor control circuits is to control the speed of the motor. This can be accomplished by a circuit that resembles the forward/reverse control seen previously. Figure 20–33A shows a basic two-speed control circuit. We will step through the sequence of operation:

1. The circuit is shown in the stopped mode.
   A. Notice that there are two sets of contacts (one N.O. and one N.C.) associated with the slow push button. These contacts are mechanically linked together (indicated by the dotted line).
   B. Notice that there are two sets of contacts (one N.O. and one N.C.) associated with the fast push button. These contacts are mechanically linked together (indicated by the dotted line).

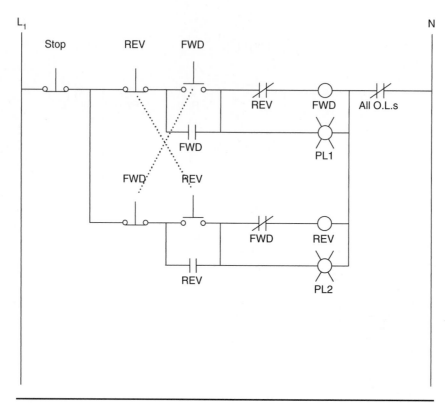

**FIGURE 20–32C** Forward/reverse control with push button and electrical interlocks. Also includes indicator lights for direction of rotation.

2. The slow push button is depressed.
   A. This energizes the slow motor starter, SLOW.
      1. This causes the SLOW holding contacts to close.
      2. This causes the SLOW N.C. contacts in front of the fast coil, FAST, to open.
3. The motor is now running at low speed.
4. Depressing the fast push button interrupts the current flow to the SLOW coil.
   A. This de-energizes the SLOW coil.
      1. This causes the SLOW holding contacts to open.
      2. This causes the SLOW N.C. contacts in front of the fast coil, FAST, to reclose.
   B. This allows the FAST coil to energize.
      1. This causes the FAST holding contacts to close.
      2. This causes the FAST N.C. contacts in front of the slow coil, SLOW, to open.
5. The motor is now running at high speed.
6. Depressing the stop push button interrupts the current to both the SLOW and FAST circuits, causing the motor to stop.

Notice that it was not necessary to stop the motor before changing the speed of the motor. This

allows a smooth transition from low to high speed. However, it should be noted that it is possible to start the motor in the high-speed mode. This may not be desirable. Often, a motor starting at the higher of two speeds will not develop sufficient starting torque to start under load. It is, therefore, necessary to start the motor at a low speed, then switch to full speed.

Figures 20–33B and 20–33C show the same essential circuit as in Figure 20–33A, but with some slight modifications. Figure 20–33B shows a two-speed circuit with a pilot light that indicates the motor is running. The pilot light does not indicate whether the motor is running in low- or high-speed mode, only that the motor is running.

Figure 20–33C shows a circuit in which the speed of rotation is indicated. This circuit uses two pilot lights. One pilot light, PL1, indicates that the motor is running at low speed when illuminated. Pilot light PL2 indicates that the motor is running at high speed when illuminated.

## Jog Control

*Jogging* is a type of motor control in which the motor may be bumped or jogged slightly. This is helpful when using a motor to position an object or piece of machinery (such as a table on a milling

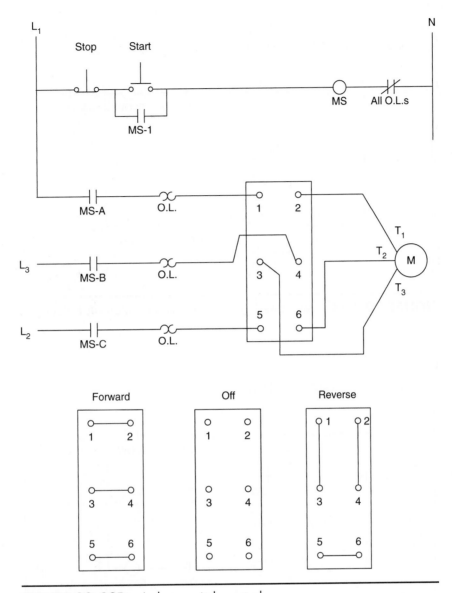

**FIGURE 20–32D**  A drum switch control.

machine). Essentially, the jog function is performed by using a momentary switch that applies power to the motor for as long as the switch is closed. Imagine using a spring-loaded push button. The motor would run for as long as the push button is depressed. Releasing the push button de-energizes the motor. Figure 20–34A shows a basic start/stop motor control with a jog button. Notice that the motor can be started and stopped normally. However, by depressing the jog push button, the motor will run for as long as the jog push button is depressed.

Figure 20–34A shows a jog circuit that required a separate push button to perform the jog function. Figure 20–34B shows a modification to this circuit where a selector switch is used to select between the normal (run) function and jog. Notice that the

jog function is accomplished by depressing the start button when the selector switch is in the jog position. When the selector switch is in the run position, the start push button performs the normal start function.

Figure 20–34C shows a forward/reverse control circuit with jog function added to both forward and reverse directions. Again, a selector switch is used to select between normal (run) function and jog. With the selector switch in the run position, the circuit operates as a standard forward/reverse control. With the selector switch in the jog position, the holding contact circuits for both forward and reverse are opened. This allows the forward push button and the reverse push button to perform the functions of jog forward and jog reverse, respectively.

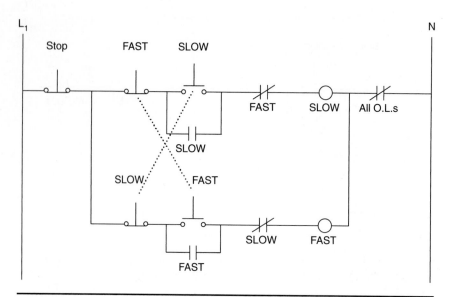

**FIGURE 20–33A** Two-speed control with push button and electrical interlocks.

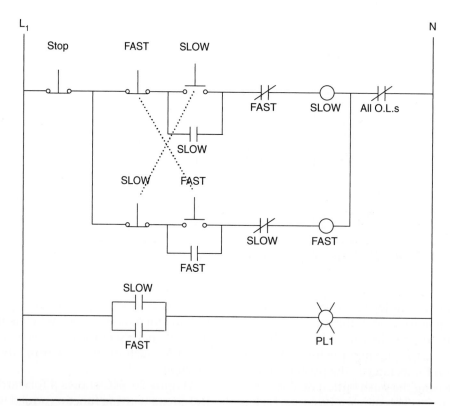

**FIGURE 20–33B** Two-speed control with push button and electrical interlocks. Also includes a motor running indicator light.

## Hand-Off-Automatic Control

Figure 20–35 also uses a selector switch; however, this circuit does not perform a jog function. The purpose of the circuit shown in Figure 20–35 is known as a *hand-off-automatic control*. Hand-off-automatic, or H-O-A, control allows a circuit to function under the control of either an automatic device or a manual control. Notice the thermostat control. This is an automatic device. With the H-O-A switch in the "automatic" position, and depending upon the temperature sensed, the motor is turned on or off automatically based on the state of the

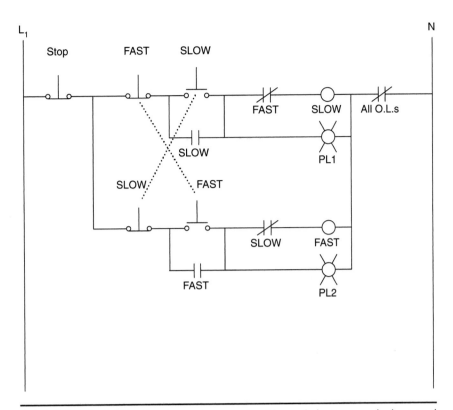

**FIGURE 20–33C** Two-speed control with push button and electrical interlocks. Also includes indicator lights for operating speed.

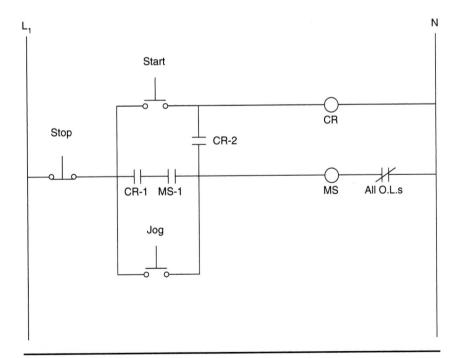

**FIGURE 20–34A** Start/stop control with separate jog push button.

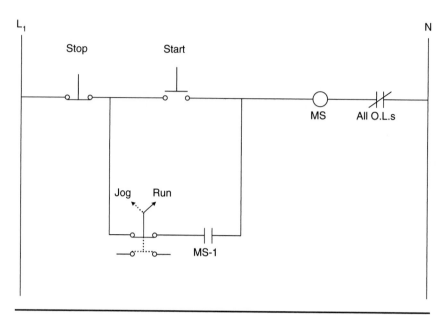

**FIGURE 20-34B**  Jog/run selector switch control.

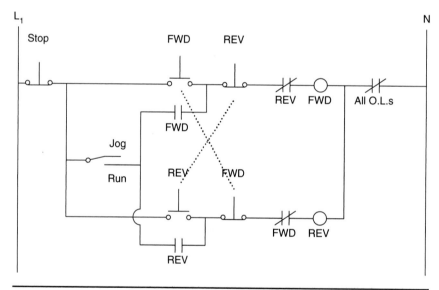

**FIGURE 20-34C**  Forward/reverse control with jog-forward and jog-reverse feature.

thermostat. However, if the H-O-A switch is placed in the "hand" position, the motor will turn on, regardless of the state of the thermostat. The high-temperature cutout switch prevents the motor from operating, regardless of the position of the H-O-A switch, should the temperature rise to too high a value.

Control circuits often use ungrounded power supply conductors. Should one of these conductors become grounded, failure of the control circuit will occur. To sense a grounded conductor, the circuit in Figure 20-36 is used. Under normal conditions, pilot light PL1 and pilot light PL2 are connected

in series across the power supply conductors. As a result, both lights will glow dimly, as they each will have one-half of the supply voltage available to them. Should a ground fault occur on $L_1$, PL1 will be shorted by the ground fault. This effectively removes PL1 from the circuit but connects PL2 directly between $L_1$ and $L_2$. This will cause PL2 to burn brightly. Should a ground fault occur on $L_2$, PL2 will be shorted by the ground fault. This effectively removes PL2 from the circuit but connects PL1 directly between $L_1$ and $L_2$. This will cause PL1 to burn brightly. Push-to-test push buttons are used to verify that both lights are operational.

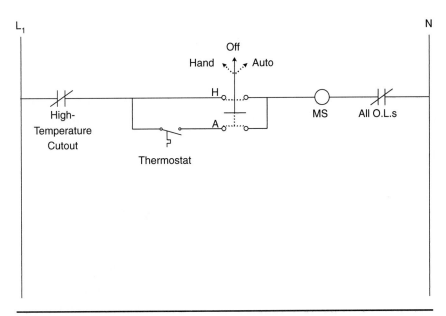

**FIGURE 20–35** Hand-off-automatic (H-O-A) control.

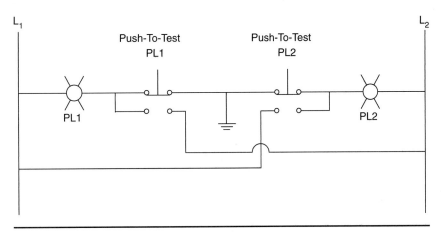

**FIGURE 20–36** Ground fault detection circuit.

# MULTIPLE MOTOR STARTER CONTROL

Let us assume that we need a control circuit that will allow us to start three motors simultaneously. However, should the first motor not run, the second and third motors will not run either. Furthermore, should the first motor start, and the second motor fail to run, the third motor will not be allowed to run. Figure 20–37 shows a control circuit that will perform this function.

Notice that a set of N.O. contacts from MS1 control current to the second motor starter, MS2. Notice also that a set of N.O. contacts from MS2 control current to the third motor starter, MS3. This type of circuit is known as a sequential starting circuit.

MS1 must be energized before the MS1-1 N.O. contact can close to provide current to the MS2 motor starter. Likewise, MS2 must be energized before the MS2-1 N.O. contact can close to provide current to the MS3 motor starter.

Should motor starter MS1 not energize, motor starters MS2 and MS3 will not energize. Should motor starter MS1 energize and motor starter MS2 not energize, motor starter MS3 will not energize.

An overload on MS3 will shut down motor starter MS3 only. An overload on MS2 will shut down motor starters MS2 and MS3, but not MS1. An overload on motor starter MS1 will shut down all motor starters.

Figure 20–38 also shows the control of three separate motors. This circuit provides independent

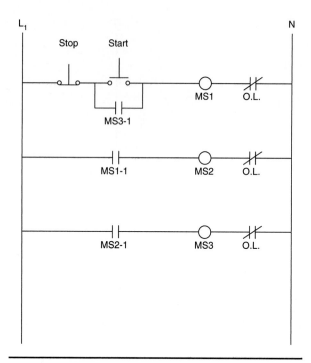

**FIGURE 20-37** Simultaneous starting of three motors.

control of the three motors, however. Notice that each motor has its own start/stop station. Each motor can be started and stopped independently of the others. However, there is a master stop push button that will stop all motors simultaneously. In addition, notice that all overloads are located on the first rung. This means that if an overload condition occurred on any of the motors, all motors will be shut down.

## SEQUENTIAL STARTING CONTROL

Under certain conditions, it is desirable to start one motor, and then after a short time delay, start a second motor. An example might be a cutting tool. Initially, a coolant pump is started to pump cutting fluid to the cutting tool. After the coolant starts flowing, the cutting tool motor is allowed to run. This insures that cutting fluid is available before the cutting operation can begin. Figure 20–39 shows a

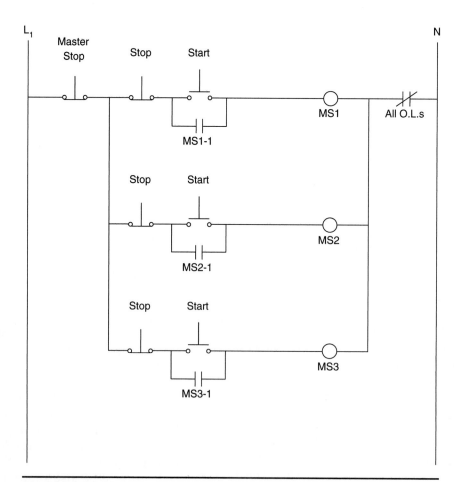

**FIGURE 20-38** Three separate start/stop controls with one master stop.

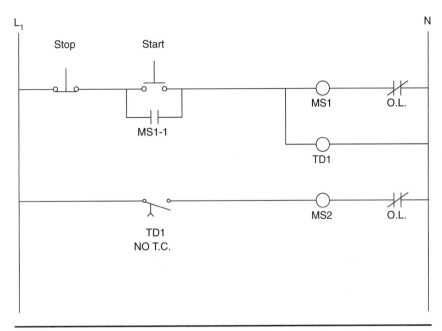

**FIGURE 20-39** Time-delayed sequential start.

circuit that could perform this action. Here is how this circuit works:

1. The circuit is shown in the initial de-energized state.
2. The start push button is depressed.
   A. This energizes MS1 coil.
      1. This causes the MS1-1 holding contacts to close.
   B. TD1, the time-delay relay coil also energizes.
3. After a short delay, the N.O. timed close (N.O. T.C.) contacts, TD1, close.
   A. This energizes MS2 coil.
4. Pressing the stop push button stops the complete circuit.

It should be noted that should an overload occur on the MS2 motor, only the MS2 motor will stop. The MS1 motor will continue to operate; however, should an overload occur on the MS1 motor, the entire circuit will shut down.

Figure 20-40A shows a different type of sequential control. In this circuit, depressing the start push button, which is connected to MS2, has no effect until MS1 is started. This circuit could be used for the same application as the circuit seen in Figure 20-38; however, the second motor must be started manually, not automatically.

A sequential control is also shown in Figure 20-40B. This circuit also uses automatic controls. However, one **automatic control** is in the form of

a pressure switch, while the other automatic control is a time-delay relay. Let us step through the operation of this circuit:

1. The circuit is shown in the de-energized state.
2. The start push button is depressed.
   A. This energizes MS1 coil and PL1.
      1. This causes the MS1-1 holding contacts to close.
      2. PL1 indicates that MS1 is energized.
3. If MS1 were controlling a pump, pressure switch PS1 would close after MS1 started and built up pressure.
   A. This causes MS2, PL2, and TD1 to energize.
   B. PL2 indicates that MS2 is energized.
4. After a short time delay, the N.O. T.C. contact of TD1 closes.
   A. This energizes MS3 and PL3.
   B. PL3 indicates that MS3 is energized.
5. Pressing the stop push button will stop the entire process.

Notice that an overload condition on MS3 only de-energizes MS3. MS2 and MS1 will not be affected by an overload on MS3. An overload on MS2 will de-energize MS2 and MS3. MS1 will not be affected by the MS2 overload. Should an overload occur on MS1, the entire process will be halted.

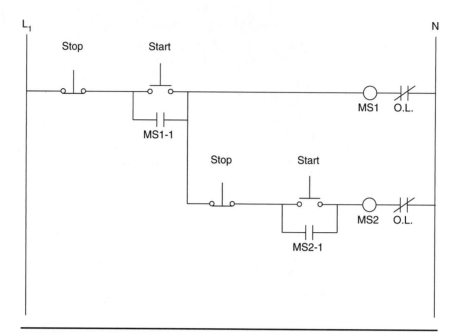

**FIGURE 20–40A** Sequential start. MS1 must be started before MS2 can be started.

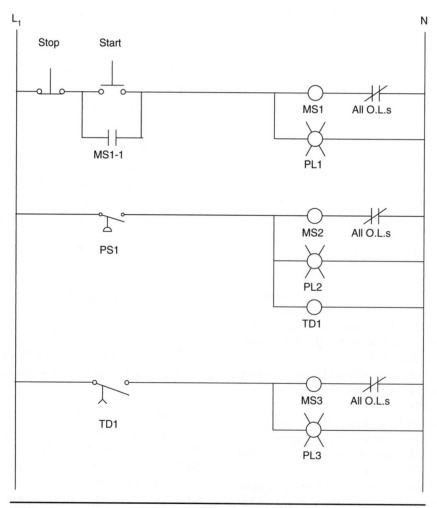

**FIGURE 20–40B** Sequential start with automatic control.

## VARIOUS STARTING METHODS

The circuits that we have seen so far are called **across-the-line starter** circuits. This is because the motors are connected directly across the power lines, providing full voltage and current to the motor. Sometimes it is desirable to limit the voltage or current that the motor receives on start-up. The following circuits are examples of different methods of starting.

Figure 20–41A shows a circuit called a *primary resistor starter*. When the motor is initially started, resistors are connected in series with the motor. These resistors limit the inrush of current to the motor windings. After a certain amount of time, the time-delay relay, TD1, energizes control relay CR1. The CR1 contacts CR1-1, CR1-2, and CR1-3 close around the resistors, effectively shorting out the resistors. This allows full current to be applied to the motor. The result is a smooth acceleration to the rated speed.

Figure 20–41B shows a circuit known as an *autotransformer starter*. Notice there are two control relays labeled CR-S and CR-R. These are the start relay (CR-S) and run relay (CR-R). When the circuit is started, the start relay is energized. This con-

nects the autotransformer windings to the power lines through the CR-Sa, CR-Sb, CR-Sc, and CR-Sd contacts. The center taps of the windings feed power to the motor. This lowers the voltage applied to the motor. After a certain amount of time, the time-delay relay contacts (TD1-A and TD1-B) change state. Contact TD1-A closes while contact TD1-B opens. This de-energizes the start relay CR-S and energizes the run relay CR-R. This causes the start relay contacts CR-Sa, CR-Sb, CR-Sc, and CR-Sd to open. At the same time, the run relay contacts CR-Ra and CR-Rb close. The effect of this is that the autotransformer is electrically removed from the circuit and the motor is connected directly to the power lines. This applies full voltage and current to the motor.

The circuit in Figure 20–41C uses a *nine-lead motor* because it uses one winding of the motor for starting and parallels a second winding of the motor for running. At start-up, MS1 is energized, closing contacts MS1-A, MS1-B, and MS1-C. This allows the motor to start on windings $T_1$, $T_2$, and $T_3$. After a preset time delay, the N.O. T.C. contact of TD1 closes. This energizes MS2, closing contacts MS2-A, MS2-B, and MS2-C. Windings $T_7$, $T_8$, and $T_9$ are now connected in parallel with $T_1$, $T_2$, and $T_3$. This circuit is known as a two-step, part-winding starting circuit.

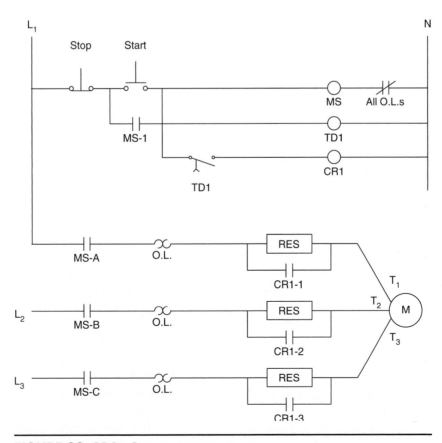

**FIGURE 20–41A** Primary resistor starter.

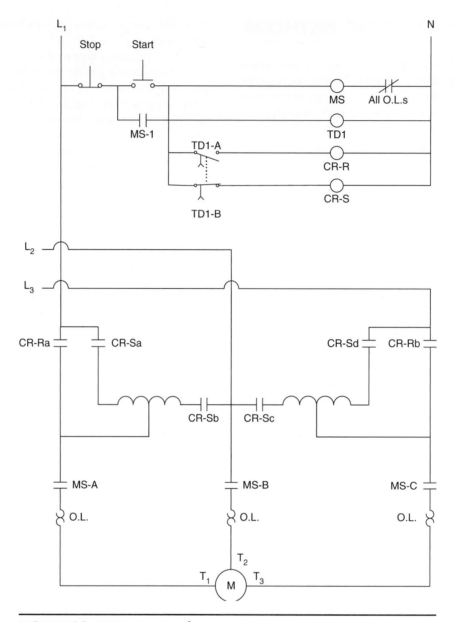

**FIGURE 20–41B**   Autotransformer starter.

The circuit shown in Figure 20–41D is known as a *wye-delta starting circuit.* A wye-delta motor is required for this type of circuit. This is a motor that contains two sets of windings. However, the connections for each end of the winding must extend to the motor terminal box. This is known as a *twelve-lead motor.* Let us follow through the operation of this circuit:

1. The circuit is shown in its de-energized state.
2. The start push button is depressed.
   A. This energizes C1, CR, MS1, and TD1 and keeps C2 from energizing.
      1. C1 and C2 are contactors; CR is a control relay.

2. Contacts C1-1, C1-2, and C1-3 will close.
   a) Contact C1-1 provides part of the holding circuit for the start push button.
   b) Contacts C1-2 and C1-3 connect one end of each of the motor windings together to form the middle (common point) of the wye circuit.
3. Contact MS-1 forms the remaining part of the holding circuits for the start push button.
4. Contacts MS1-A, MS1-B, and MS1-C will close.
   a) These contacts connect motor terminals $T_1$, $T_2$, and $T_3$ to the power lines.

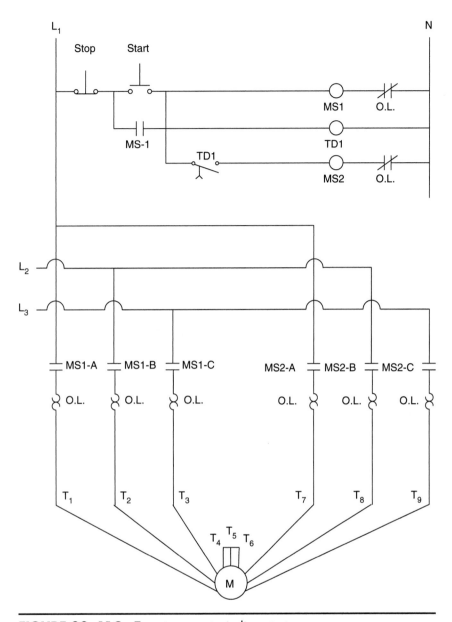

**FIGURE 20–41C** Two-step, part-winding starter.

B. The motor has now started as a wye-connected motor.

C. After a short time delay, the N.C. T.O. contact, TD1-1, opens.

   1. This de-energizes C1 and CR1.

     a) Contact C1-1 opens. The holding circuit is still maintained by the MS-1 contact.

     b) Contacts C1-2 and C1-3 open. This removes the wye connection from the motor windings.

     c) Contact CR-1 closes energizing C2.

       1. This causes contacts C2-A, C2-B, and C2-C to close.

   2. These contacts connect the motor windings in a delta configuration.

D. The motor is now running as a delta-connected motor.

Figure 20–41E shows a circuit that provides *automatic three-point starting* for a wound-rotor induction motor. Let us see how this circuit operates:

1. The circuit is shown in its de-energized state.

2. Pressing the start push button energizes MS1 and TD1.

   A. This causes MS-1 contact to close.

     1. This provides a holding contact for the start push button.

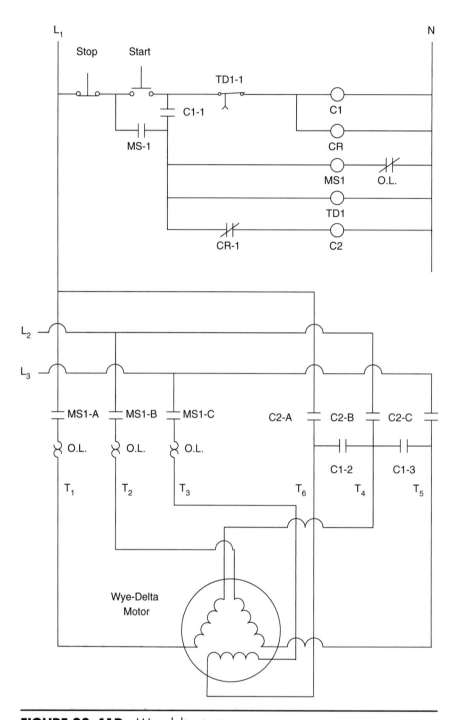

**FIGURE 20–41D**  Wye-delta starter.

B. Contacts MS1-A, MS1-B, and MS1-C will close.

  1. These contacts connect motor terminals $T_1$, $T_2$, and $T_3$ to the power lines.

  2. This causes the motor to begin to rotate.

    a) All six resistors are now connected to the rotor circuit.

    b) This greatly limits the amount of rotor current.

C. After a short delay, N.O. T.C. timer contact TD1-1 will close.

  1. This energizes contactor C1 and time-delay relay TD2.

    a) Contacts C1-1 and C1-2 will close.

    b) This electrically shorts out the right-most set of resistors in the rotor circuit.

    c) This allows an increase in rotor current resulting in an increase in rotor speed.

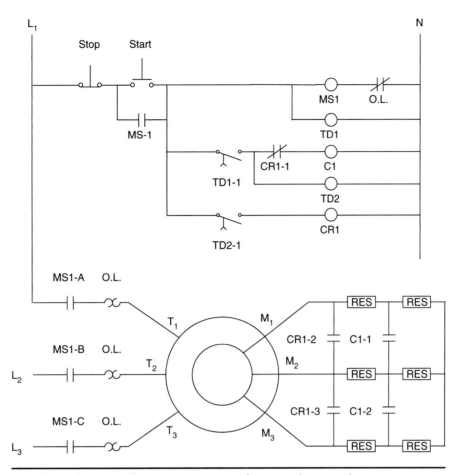

**FIGURE 20–41E** Three-point starting of a wound-rotor induction motor.

D. After a short delay, N.O. T.C. timer contact TD2-1 will close.

   1. This energizes control relay CR1.

      a) This causes the N.C. CR1-1 contact to open.

         1. This de-energizes contactor C1.

            a. This causes contacts C1-1 and C1-2 to open.

      b) This causes contacts CR1-2 and CR1-3 to close.

         1 This electrically shorts out all of the resistors in the rotor circuit.

         2. Full current is now applied to the rotor and the motor accelerates to full speed.

Figure 20–42 shows a circuit that can be used to provide across-the-line starting for a DC motor. Essentially, this is the same circuit as seen for an AC motor with only a few minor modifications. Pressing the start push button energizes the control relay CR1 and the contactor C1. The control relay contact CR1-1 provides the holding contact for the start push button. The energized contactor, C1, will cause the three contacts C1-1, C1-2, and C1-3 to close. This connects the armature and the shunt field directly across the DC supply. The motor is now running. Pressing the stop push button de-energizes control relay CR1 and contactor C1, causing the motor to stop.

You may notice that the control relay appears slightly different in Figure 20–42 than in ones seen previously because it is a DC control relay. These relays have two internal coils and one internal contact. One coil is much larger than the other coil. This large coil is the start winding and is used only when the coil is first energized. It assists the other coil in causing the contacts to move. Once energized, the internal contacts open, removing the start winding from the circuit. Now, only the smaller holding coil remains energized to maintain the contacts in their energized state. You may wonder about contactor contacts C1-1 and C1-2. You might ask why are there two contacts connected in series as seen in this circuit. This arrangement of the contacts is necessary to suppress the arcing that occurs when switching DC circuits.

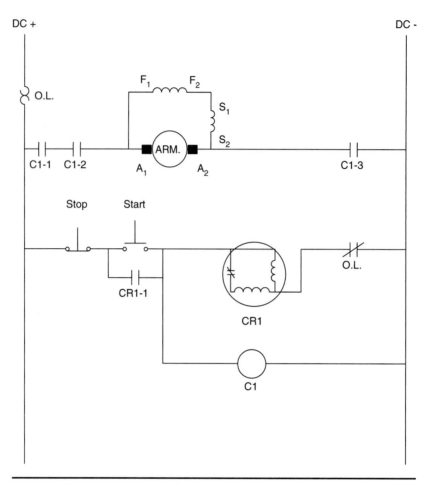

**FIGURE 20–42** Across-the-line starting of a DC motor.

# BRAKING

*Plugging* is a method of braking a motor in which a motor is brought to a rapid stop by forcing it to rotate in the direction opposite to the direction in which it was running. Imagine driving your car at 70 mph down the interstate and throwing the transmission into reverse. You will come to a quick halt, but you will possibly do some damage to your vehicle. The same principle is applied to a motor. Plugging reverses the direction of a running motor. The braking action is very quick. However, plugging should be reserved for emergency stop applications only. Repeated plugging of a motor will cause damage.

Figure 20–43A shows a plugging circuit. Here is how this circuit works:

1. The circuit is shown in its de-energized state.
   A. Notice timer contact TD1-1. This is a N.O. T.O. contact. While the contact is drawn in the open state, the contact will be closed when the circuit is started.

2. The run push button is pressed.
   A. This energizes control relay CR1, time-delay relay TD1, and the FWD motor starter.
      1. CR1-1 contacts close to provide a holding function around the run push button.
      2. CR1-2 contacts open.
      3. Timer contact TD1-1 closes.
   B. The motor is now running.

3. The emergency stop push button is pressed.
   A. This de-energizes control relay CR1, time-delay relay TD1, and the FWD motor starter.
      1. The holding contacts, CR1-1, around the run push button open.
      2. CR1-2 contacts reclose.
      3. Time-delay relay TD1 begins timing.
      4. Control relay CR2 and the REV motor starter energize.
   B. The motor reverses its direction of rotation.

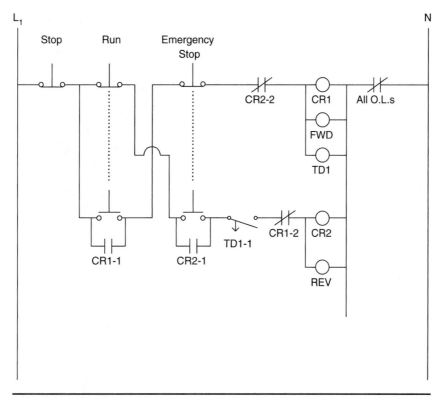

**FIGURE 20–43A** Plugging.

4. Time-delay relay TD1 times out.
   A. Timer contacts TD1-1 reopen.
      1. This de-energizes CR2 and the REV motor starter.
   B. The motor has now stopped.

It should be noted that the setting of the time delay of timer TD1 is critical. If the time is set too short, the motor will not come to a full stop. If the time is set too long, the motor will continue to run in the opposite direction. The setting of the time delay is also dependent upon the load that the motor is driving. A setting made with the motor lightly loaded will not be correct should the load increase. Likewise, a setting made with a heavy load will not be correct should the load decrease.

Figure 20–43B shows a braking circuit for a DC motor. This braking circuit uses a form of braking known as *dynamic braking*. When power is removed from an operating motor, the motor will coast to a stop. While coasting, the motor acts like a generator. There is a residual magnetic field remaining in the iron core of the motor. The coasting motor provides relative motion of the motor windings, which causes the motor to generate a small voltage. This voltage

will oppose the applied voltage to the motor. If this voltage could be reapplied to the motor, the motor would try to rotate in the opposite direction. This would have the effect of braking the motor. This action is similar to dynamic braking of a three-phase induction motor. In the circuit shown in Figure 20–43B, resistors are used to provide a path for the generated voltage back to the motor. Let us see how this circuit works:

1. The circuit is shown in its de-energized state.
2. The start push button is pressed.
   A. This causes contactor C1, control relay CR1, and pilot light PL2 to energize.
      1. Contactor contacts C1-1 and C1-2 close.
         a) The motor is now running.
      2. Contactor contact C1-3 opens.
      3. Control relay contact CR1-1 closes to provide a holding function for the start push button.
      4. Pilot light PL2 indicates that the motor is running.
3. The stop push button is pressed.
   A. This causes contactor C1, control relay CR1, and pilot light PL2 to de-energize.

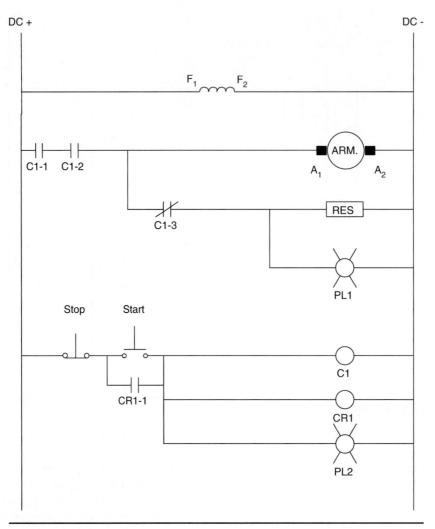

**FIGURE 20–43B** Dynamic braking of a DC motor.

1. Contactor contacts C1-1 and C1-2 reopen.
2. The motor is now coasting.
3. Contactor contact C1-3 recloses.
    a) This connects the dynamic braking resistor and pilot light PL1 across the armature of the motor.
    1. This circuit provides the braking effect to the motor.
    2. Pilot light PL1 will initially be bright and gradually grow dim as the braking energy from the motor bleeds off.
B. The motor is now stopped.

One disadvantage of **dynamic breaking** is that the braking effect diminishes as the motor slows down. It is, therefore, necessary to provide some means of auxiliary braking if a complete stop is required. This is usually accomplished by the addition of a mechanical brake.

## REVIEW QUESTIONS

*Multiple Choice*

1. A motor controller
    a. allows the motor to start and stop.
    b. allows the motor speed to be controlled.
    c. allows the direction of rotation to be controlled.
    d. all of the above.

2. When using multiple start/stop stations,
    a. the stop push buttons are wired in series and the start push buttons are wired in parallel.
    b. the stop push buttons are wired in parallel and the start push buttons are wired in series.
    c. the stop and start push buttons are wired in parallel.
    d. the stop and start push buttons are wired in series.

3. The start button on a momentary push-button station completes the circuit to the control coil
   a. only while the button is being pressed.
   b. continually once the button has been pressed.
   c. only after the button has been released.
   d. none of the above.

4. The stop button on a momentary push-button station completes the circuit to the main control coil
   a. only while the button is being pressed.
   b. continually once the button has been pressed.
   c. only after the button has been released.
   d. none of the above.

5. A manual starter is operated by hand
   a. from a remote location.
   b. at the starter location.
   c. at a central location.
   d. from many locations.

6. Jogging refers to
   a. a motor that is not capable of developing smooth torque.
   b. a method used to stop a motor for exact positioning of components.
   c. a motor that starts and stops periodically.
   d. none of the above.

7. An interlocking arrangement on a forward/reverse controller
   a. prevents short circuits.
   b. ensures proper operating sequence.
   c. provides overload protection.
   d. prevents the motor from overheating.

8. Dynamic braking on three-phase induction motors is accomplished
   a. in the same manner as on DC motors.
   b. by energizing the rotor with DC.
   c. by energizing the stator with DC.
   d. by reversing the phase sequence.

9. Plugging on a three-phase induction motor is accomplished by
   a. energizing the stator with DC.
   b. reversing the phase sequence to the stator.

   c. reversing the phase sequence to the rotor.
   d. energizing the rotor with DC.

10. A wye-delta controller is used for starting three-phase induction motors. When in the start position, the motor windings are connected in
    a. wye.
    b. delta.
    c. open delta.
    d. open wye.

11. A primary resistance starter provides
    a. sudden spurts of torque as the motor accelerates.
    b. smooth acceleration to the rated speed.
    c. sudden spurts of speed as the motor accelerates.
    d. rough acceleration but good speed control.

*Give Complete Answers*

1. What is a motor controller?

2. List two methods used to limit the amount of starting current of a motor.

3. What is an across-the-line starter?

4. What is the difference between a manual across-the-line starter and a remotely operated across-the-line starter?

5. Describe a momentary push-button station.

6. What safety precaution is provided by a three-wire control that cannot be provided with a two-wire control?

7. What is the purpose of jogging control?

8. Describe how jogging can be accomplished.

9. Explain how the direction of rotation of a three-phase motor is accomplished.

10. Describe the construction and operation of a reversing starter.

11. What is the purpose of interlocks on a forward/reverse controller?

12. What is meant by the term *dynamic braking*?

13. How is dynamic braking accomplished with a three-phase induction motor?

14. Define the term *plugging* and explain how it is accomplished with a three-phase induction motor.

# Basic Industrial Electronics

**OBJECTIVES**

After studying this chapter, the student will be able to:

- Identify common symbols used in solid-state devices.
- Explain the operation of common solid-state devices.
- Identify common operational amplifier circuits.
- Identify common symbols used in digital logic.
- Explain the operation of common digital logic gates.
- Perform basic go-no-go tests on some common devices.

This chapter introduces you to some of the more common electronic devices that are found in industrial applications today. Not too long ago, maintenance technicians pulled wire and ran **conduit**. The electronic technicians were the individuals who were responsible for understanding, installing, and troubleshooting devices containing electronic components. With the advent of programmable logic controllers (PLCs) and electronic variable speed/frequency drives, the field of **electronics** has entered the world of the maintenance technician. It is, therefore, important that you achieve a basic understanding of electronic devices and their applications.

# SEMICONDUCTORS

Electronic circuits include materials known as semiconductors. A **semiconductor** is a material that is neither a good conductor nor a good insulator.

Two of the most common materials used to make semiconductors are germanium and silicon. These materials, in their pure state, are not much use in the electronic industry; however, by adding an **impurity** such as arsenic or indium, they take on very different characteristics. This process is called *doping*.

If arsenic is added to pure silicon, the latter's atomic structure is altered. A silicon atom contains 14 electrons, 4 of which are in the outer shell (**valance**) (Figure 21–1).

Atoms with 1 to 4 electrons in the outer shell are generally good conductors of electricity. Atoms with 5, 6, or 7 electrons in the outer shell are classified as poor conductors. Those with 8 electrons in the outer shell are insulators. The structure of silicon, however, presents a different phenomenon.

Figure 21–2 illustrates a small section of silicon at the atomic scale. To simplify the drawing, only the valance electrons are shown for each atom. As the electrons orbit around the nucleus of their respective atoms, their valances overlap. This arrangement allows a sharing of valance electrons. A close examination

shows that each atom has 4 of its own electrons in the valance and 4 electrons from the atoms surrounding it. This makes a total of 8 valance electrons. Pure silicon might then be classified as an insulator. This arrangement is called **covalent bonding**.

An arsenic atom has 5 electrons in its valance. When arsenic is combined with silicon, 4 of the 5 valance electrons share valances with the adjacent silicon atoms. This leaves 1 electron free to drift. In this semiconductor there are many silicon and arsenic atoms that share valance electrons, leaving many free electrons. Because the material contains free electrons, it is called an *N-type material* (N for negative) (Figure 21–3).

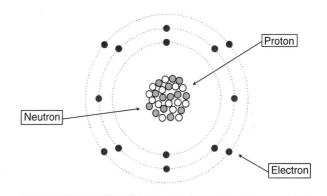

**FIGURE 21–1**  Silicon atom.

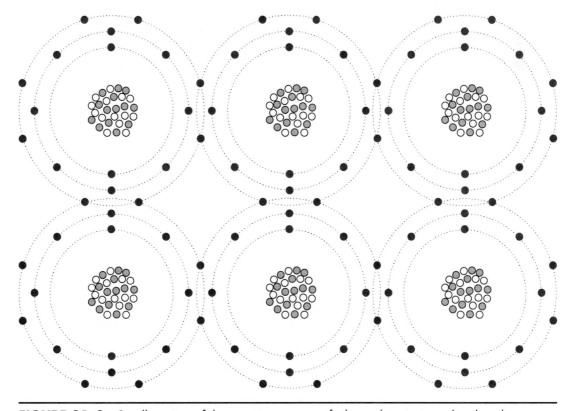

**FIGURE 21–2**  Small section of the atomic structure of silicon showing covalent bonding.

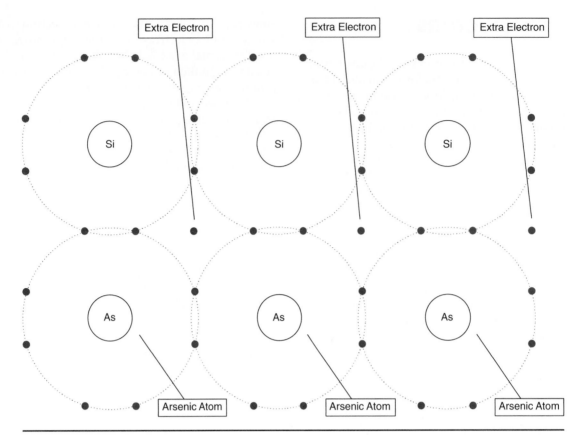

**FIGURE 21–3**  Silicon doped with arsenic to form an N-type material.

Other materials that may be used for doping are indium and gallium. The atoms of these impurities have only 3 valance electrons. When indium is added to silicon, the 3 valance electrons share valances with the 4 adjacent silicon atoms. This arrangement provides 7 valance electrons, leaving a hole where the 8th electron should be (Figure 21–4). Because of the absence of electrons, the material now has a positive charge. Therefore, it is called a *P-type material.*

The mixing of impurities (arsenic, indium, or gallium) with silicon or germanium is a chemical process called doping. An impurity that causes the material to contain extra electrons is called a *donor.* Arsenic is a donor. An impurity that causes a shortage of electrons is called an *acceptor.* Indium and gallium are acceptors.

Diffusion occurs when P and N materials are joined together (Figure 21–5). Some electrons in the N material, near the junction, are attracted to the holes in the P material, thus leaving holes in the N material. This diffusion of electrical charges produces a potential difference in a small area near the junction. As a result, the material will conduct in one direction but not in the opposite direction. For this reason, the area in which this emf exists is called a *barrier.*

A PN device is known as a *diode.* The diode is a two-terminal device. Several diode styles are shown in Figure 21–6. The schematic symbol and a physical drawing of a diode are shown in Figure 21–7. One terminal is called the **anode**. This is the terminal represented by the arrowhead symbol. The other terminal is called the **cathode**. The cathode is represented by the T-shaped symbol.

Diodes are placed in a circuit in such a fashion that the cathode has a more negative voltage applied to it with respect to the anode. When a diode is connected in this fashion, the diode will conduct an electrical current. The diode is said to be *forward biased.* Figure 21–8 shows a diode connected in the forward-biased condition. If the polarities of the diode are reversed, that is, if the cathode is more positive with respect to the anode, the diode will block the current. The diode is now said to be *reverse biased.* Figure 21–9 shows a diode connected in the reverse-biased condition.

When a diode is reverse biased, it is possible to cause the diode to breakdown and conduct a current in the reverse direction. Diodes have a specification called the *peak inverse voltage* rating (or PIV). This rating is a measure of how much voltage—the peak voltage—a diode can withstand in the reverse direction. For example, if you have a diode that is

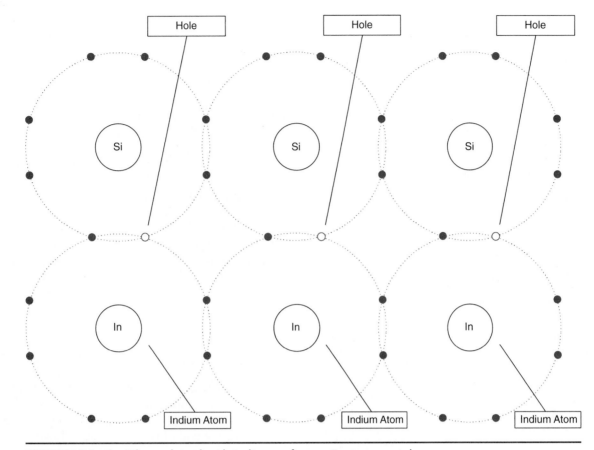

**FIGURE 21–4** Silicon doped with indium to form a P-type material.

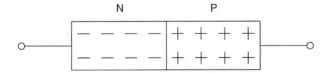

**FIGURE 21–5** PN junction.

**FIGURE 21–6** Various diodes.

**FIGURE 21–7** Schematic symbol of a diode. Note the cathode and anode.

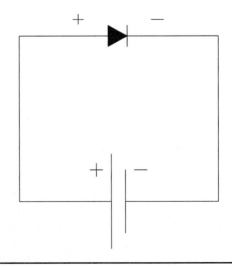

**FIGURE 21–8** Forward biasing a diode.

rated at 400 volts PIV, this means that the diode can withstand a peak reverse voltage of up to 400 volts before it will breakdown and conduct in the reverse direction. At this point, the diode is destroyed and must be replaced. It is important to realize that when replacing a defective diode, you must always equal or exceed the PIV rating of the defective diode. For example, if the defective diode in your circuit has a

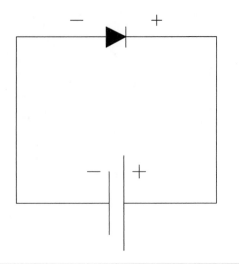

**FIGURE 21-9**   Reverse biasing a diode.

PIV rating of 200 volts, you can replace it with a diode rated at 200 V PIV, 400 V PIV, or more. However, you cannot replace it with a diode rated at 100 V PIV. The replacement diode will probably be destroyed when the circuit operates.

Another factor to consider when replacing a defective diode is the current rating of the diode. For example, a diode may be rated at 1 ampere of current. This means that the maximum current that can flow through this diode when it conducts in the forward direction is 1 ampere. If the forward current flow exceeds 1 ampere, the diode will be destroyed. When replacing this diode, the replacement diode must have a current rating equal to 1 ampere or greater.

## Rectifier Diodes

As we have learned, a diode will only conduct current when forward biased. If a diode is placed in a circuit that has an alternating current applied to it, the diode will only conduct for half of the 360 degrees of the AC sine wave. Figure 21–10 shows a circuit with a diode and an applied AC voltage. Let's freeze time and analyze this circuit one step at a time.

Let us assume that L1 is positive and L2, therefore, is negative. Current will flow from L2, through the load, through the diode, and return to L1. Notice that the diode was forward biased and allowed a current to flow. This causes the output voltage to appear for this half of the AC sine wave input (Figure 21–11). Now, imagine that the AC sine wave input voltage has reversed its polarity. Now, L1 is negative and L2, therefore, is positive. Current will be blocked by the diode since the diode is now reverse biased. Since the current flow is blocked, there can be no voltage across the load for this half of the AC sine wave input.

Since the output voltage will be either all positive voltage or all negative voltage (depending on how the diode is placed in the circuit), the output voltage is a DC voltage. The diode has been used to convert the AC input voltage into a DC output voltage. This conversion is called *rectification*. When a diode is used to perform rectification, the diode is called a *rectifier diode*, or simply a **rectifier**. Since the output voltage appears for only half of the AC sine wave input voltage, this circuit is called a *half-wave rectifier*.

Suppose we needed a DC voltage that was present for both half cycles of the AC sine wave input voltage. If we use two diodes and a center-tapped transformer with an AC sine wave input voltage, the output voltage will appear during both half-cycles of the AC input. This circuit appears in Figure 21–12. Again, we will freeze time and look at this step by step.

Refer to Figure 21–13. During one-half cycle of the AC input sine wave, the secondary of the transformer will have a negative polarity at one end of the winding and a positive polarity at the other end. The polarity of the center tap will depend on which end we compare it to. For instance, the polarity of the center tap will be negative with respect to the positive end of the secondary. However, the polarity of the center tap will be positive with respect to the negative end of the secondary. Let's consider the center tap as being negative with respect to the positive end of the secondary. Current will flow from the center tap, through the load, and through diode D1. Current is able to flow through diode D1 because its anode is positive with respect to its cathode. Diode D1 is forward biased and will conduct. Current will not be able to flow through diode D2 because its anode is negative with respect to its cathode. This means that diode D2 is reverse biased and cannot conduct.

Now, refer to Figure 21–14. During the next half cycle of the AC input sine wave, the secondary of the transformer polarity will flip. This means that the end of the winding that was negative will now be positive. The end of the winding that was positive will now be negative. The polarity of the center tap will still depend on which end we compare it to. Let's consider the center tap as being negative with respect to the positive end of the secondary. Current will flow from the center tap, through the load, and through diode D2. Current is able to flow through diode D2 because its anode is positive with respect to its cathode. Diode D2 is forward biased and will conduct. Current will not be able to flow through diode D1 because its anode is negative with respect to its cathode. This means that diode D1 is reverse biased and cannot conduct.

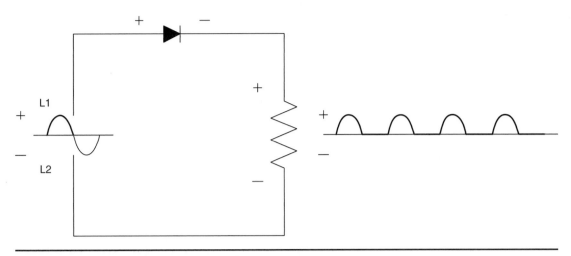

**FIGURE 21-10**   Rectification: diode forward biased.

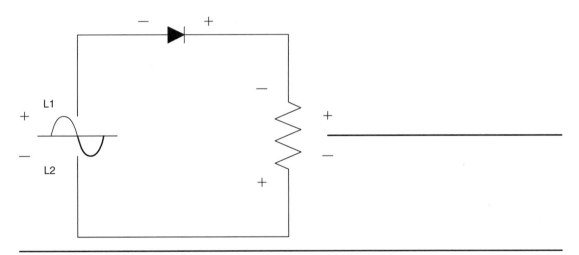

**FIGURE 21-11**   Rectification: diode reverse biased.

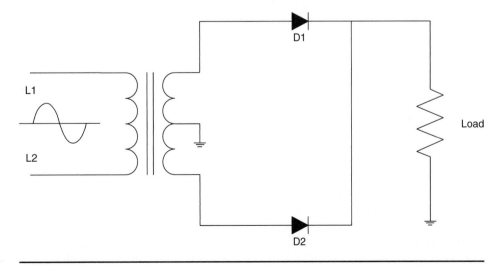

**FIGURE 21-12**   A full-wave rectifier circuit with a center-tapped transformer.

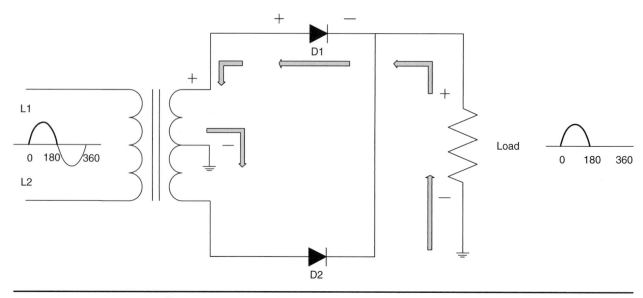

**FIGURE 21-13**　Current flow with diode D1 forward biased and diode D2 reverse biased.

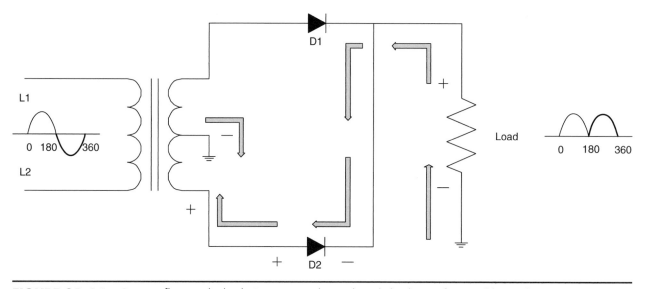

**FIGURE 21-14**　Current flow with diode D1 reverse biased and diode D2 forward biased.

Notice that a DC voltage was present across the load during both half cycles of the AC input sine wave voltage. Notice, too, that the polarity of the DC load voltage was the same for both halves of the AC input voltage. This circuit is called a *full-wave recti-fier* because there is a DC output voltage during both halves of the AC input voltage.

A full-wave rectifier can also be fashioned with a slightly different circuit. The center-tapped transformer is replaced by a non–center-tapped secondary. Four diodes will be used instead of two. Figure 21–15 shows this new circuit, and next we'll look at how this circuit works.

Refer to Figure 21–16. During one-half cycle of the AC input sine wave, the secondary of the transformer will have a negative polarity at one end of the winding and a positive polarity at the other end. Current will flow from the negative end of the secondary winding, through diode D1, through the load, through diode D3, and back to the positive end of the secondary winding. Current is able to flow through diodes D1 and D3 because their anodes are positive with respect to their cathodes. Diodes D1 and D3 are forward biased and will conduct. Current will not be able to flow through diodes D2 and D4 because their anodes are negative with respect

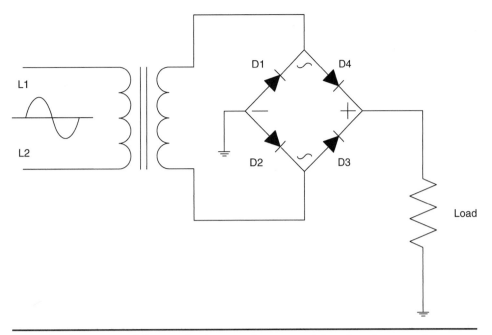

**FIGURE 21-15** A full-wave bridge rectifier circuit. Note the absence of a center-tapped transformer.

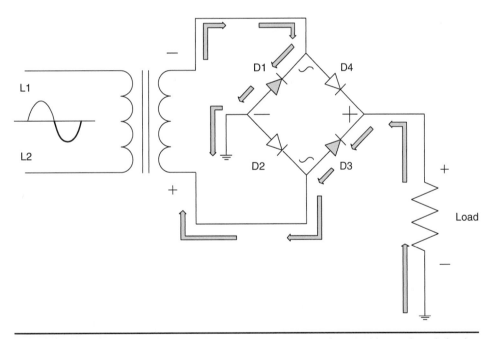

**FIGURE 21-16** Current flow with diodes D1 and D3 forward biased and diodes D2 and D4 reverse biased.

to their cathodes. This means that diodes D2 and D4 are reverse biased and cannot conduct.

Refer to Figure 21–17. During the next half cycle of the AC input sine wave, the secondary of the transformer polarity will flip. This means that the end of the winding that was negative will now be positive, and the end of the winding that was positive will now be negative. Current will flow from the negative end of the secondary winding, through diode D2,

through the load, through diode D4, and back to the positive end of the secondary winding. Current is able to flow through diodes D2 and D4 because their anodes are positive with respect to their cathodes. Diodes D2 and D4 are forward biased and will conduct. Current will not be able to flow through diodes D1 and D3 because their anodes are negative with respect to their cathodes. This means that diodes D1 and D3 are reverse biased and cannot conduct.

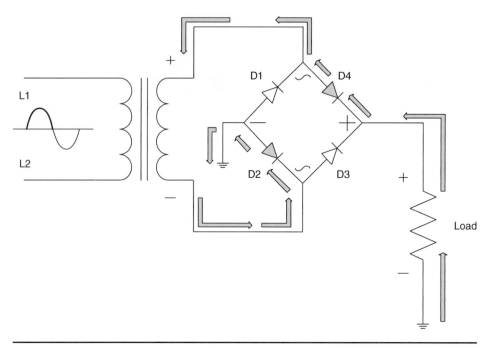

**FIGURE 21-17**   Current flow with diodes D1 and D3 reverse biased and diodes D2 and D4 forward biased.

Notice that a DC voltage was present across the load during both half cycles of the AC input sine wave voltage. Notice, too, that the polarity of the DC load voltage was the same for both halves of the AC input voltage. This circuit is called a *full-wave recti-fier* because there is a DC output voltage during both halves of the AC input voltage. However, due to the configuration of the four diodes in this circuit, this circuit is called a *full-wave bridge rectifier.*

## RECTIFIERS

Direct current has many uses in industry. Chemical applications that require DC are charging batteries, electroplating, and separating metal from ore. Applications that require a constant electric charge include dust and smoke precipitation, spray painting, and voltage testing of insulation. Mechanical applications using DC include some types of speed control, magnetic chucks, magnetic brakes, and magnetic clutches. Electromagnetic hoists also often require DC. Batteries, generators, and rectifiers are used to obtain DC for these applications.

Batteries are too bulky for certain applications, and they require maintenance and recharging. For some applications, the life of the battery is too limited. Generators are mechanical-moving equipment that require frequent maintenance. In addition, their noise during operation may be objectionable. If DC is distributed over long distances, losses in voltage and power may occur.

Rectifiers for converting AC to DC have many advantages. As a power source, AC is universally available. Any desired value of direct voltage can be obtained from an AC source by using transformers in combination with rectifiers. Further, electrical and electronic rectifiers operate with no noise and require minimum maintenance.

### Ripple

The fluctuating output voltages obtained from a rectifier are actually a combination of AC and DC voltages. The AC component that causes the pulsating voltage is called the *ripple.* The magnitude of ripple voltages is measured in the percentage of output voltage. The percent of ripple and the frequency of the pulses are two factors that determine the selection of a rectifier for a specific job.

Two cycles of input voltage to a rectifier are shown in Figure 21–18. Figure 21–19 shows that the output waveform of a half-wave, single-phase rectifier has the same ripple frequency as the input voltage because the distance from one peak to the next is the same. A full-wave, single-phase rectifier, however, has an output ripple frequency that is twice the input frequency because there are twice as many peaks in the output wave. Thus with a 60-hertz input, the ripple in the output of a half-wave, single-phase rectifier is also 60 hertz. The ripple frequency in the output of a full-wave, single-phase rectifier is 120 hertz (Figure 21–20).

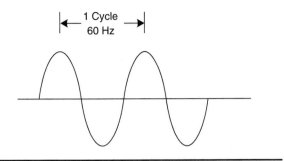

**FIGURE 21–18** Two cycles of input voltage to a rectifier.

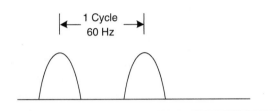

**FIGURE 21–19** Output voltage waveform of a half-wave, single-phase rectifier.

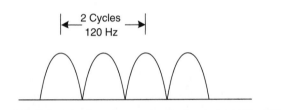

**FIGURE 21–20** Output waveform of a full-wave, single-phase rectifier.

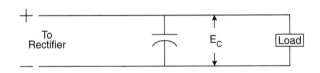

**FIGURE 21–21** Filtering capacitor connected to the output of a rectifier.

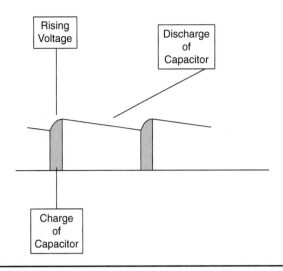

**FIGURE 21–22** Illustration of a filtering capacitor charge and discharge.

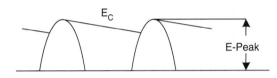

**FIGURE 21–23** Waveform of the output voltage of a half-wave rectifier with a single-filtering capacitor.

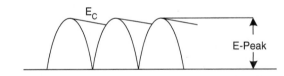

**FIGURE 21–24** Waveform of the output voltage of a full-wave rectifier with a single-filtering capacitor.

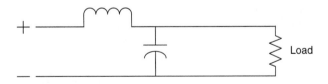

**FIGURE 21–25** Single LC filter circuit supplied from a rectifier (choke input).

## Filters

When DC voltages must have little or no ripple, a filter is used to smooth out the pulse. A filter usually consists of capacitance and inductance or resistance. The number of capacitors and inductors depends upon the degree of filtration required.

Figure 21–21 shows a filtering capacitor connected to the output of a rectifier. As illustrated in Figure 21–22, when the secondary voltage rises, the capacitor charges. When the output voltage of the rectifier starts to decrease, the capacitor begins to discharge, thus maintaining a current through the load. If the resistance of the load is high, there is

little flow from the capacitor, and it maintains almost full charge. However, if the load resistance is relatively low, there is a greater discharge from the capacitor, resulting in a pulsing voltage. The curves in Figures 21–23 and 21–24 show waveforms from half-wave and full-wave rectifiers with a single filtering capacitor. The output obtained from the full-wave rectifier is steadier and, therefore, more desirable for most applications than the output from a half-wave rectifier.

The use of a capacitor and **choke** (Figure 21–25) further decreases the pulsing and produces a

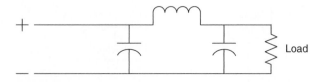

**FIGURE 21–26** Double-capacitor, single-choke filter circuit supplied from a rectifier (capacitor input).

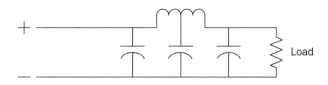

**FIGURE 21–27** Triple-capacitor, double-choke filter circuit supplied from a rectifier (capacitor input).

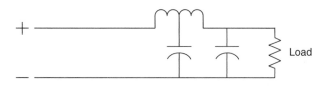

**FIGURE 21–28** Double-capacitor, double-choke filter circuit supplied from a rectifier (choke input).

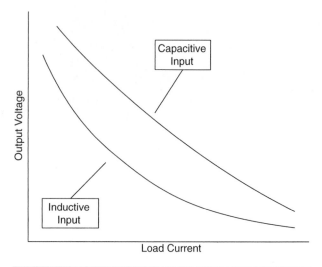

**FIGURE 21–29** Characteristic curves of the output voltage for choke and capacitor filter circuits.

**FIGURE 21–30** Zener diode: schematic and physical appearance. Note that the band indicates cathode.

smoother load current. The addition of still more inductance and capacitance (LC) increases the filtering effect to the point where the voltage and current have practically no ripple. Figures 21–26 and 21–27 show circuit diagrams of capacitor input filters. Figures 21–25 and 21–28 show circuit diagrams of choke input filters. Comparative characteristics of the output voltage from the choke (inductive) and the capacitor (capacitive) filters are shown in Figure 21–29. With little or no load current, the output voltage is practically the peak voltage. When the load is high, the choke-input-type filter has better regulation.

## Zener Diodes

**Zener diodes** are a special type of diode used to provide voltage regulation. Figure 21–30 shows the schematic symbol and the physical drawing of a zener diode. A zener diode has a physical appearance very similar to that of a regular diode. You cannot tell the two apart by simply looking at them. You must note the identifying number imprinted on the body of the diode. By contacting the manufacturer of the diode, you will be able to determine whether the diode is a regular diode or a zener diode.

When used in a circuit, the zener diode operates in the reverse breakdown region. When operated in this manner, a zener diode will maintain a constant

voltage drop across it. This means that any load, connected in parallel with the zener, will have a constant voltage as well. Even if the load varies, the voltage across the zener will remain constant (within design limits). Zener diodes are available with voltage ratings from 2 volts to 200 volts and practically any voltage in between. Let's look at an example to see how a zener diode functions as a voltage regulator.

Refer to Figure 21–31. Notice the manner in which the zener diode is connected in this circuit. The anode is connected to the negative portion of the circuit, while the cathode is connected to the positive portion. This is how a zener diode is installed so that it operates in the reverse breakdown region. When the circuit is energized, the zener diode will not conduct until the voltage drop across the zener is slightly more than the voltage rating of the zener. For example, a 12-volt zener diode will not conduct if the voltage across the zener is 11.5 volts. The load, at this point, will see 11.5 volts as well. However, if the voltage drop across the zener should increase to 12.1 volts, the zener diode will conduct. When the zener conducts, the voltage drop across the zener will be a constant 12 volts. The load will see 12 volts as well. Should the input voltage to the zener diode increase to 15 volts, the voltage drop across the zener will remain at 12 volts. This means that the load will still see 12 volts as well. This is how a zener diode is used to provide voltage regulation. Notice, however, that should the voltage fall below the voltage rating of the zener, the load voltage will fall as well. In this respect, the zener diode only provides regulation for excess voltage, not for under-voltage conditions.

## Light-Emitting Diodes

*Light-emitting diodes*, or LEDs, are diodes that have been doped in such a fashion that, when forward biased, emit light. Figure 21–32 shows several LED styles. LEDs have a cathode, which must be negative, and an anode, which must be positive, in order for the LED to conduct. Figure 21–33 shows the schematic symbol of an LED. LEDs require a higher voltage and more current to conduct. Typically, LEDs require approximately 1.4 volts to forward bias the junction. When LEDs conduct, there usually

will be a current of approximately 20 milliamperes flowing through the junction.

LEDs are available in a wide variety of colors. Some LEDs have a clear lens, while others have a colored lens. LEDs are available in red, yellow, green, blue, and purple. In addition, infrared LEDs (IRLEDs) are available. IRLEDs are used in remote controls for TVs and VCRs.

An interesting adaptation of the LED is the voltage indicator. A manufacturer will package a red and yellow LED in inverse parallel in the same package (see Figure 21–34). When the LED is connected to a source of DC, either the red or the yellow LED will light. Let's suppose that the red LED is lit. Now, if the polarity of the DC is reversed, the other LED will light. In our example, the yellow LED is now lit, which tells us the polarity of the DC. Suppose the LED was connected to a source of AC. The AC alternates in polarity (typically 60 times

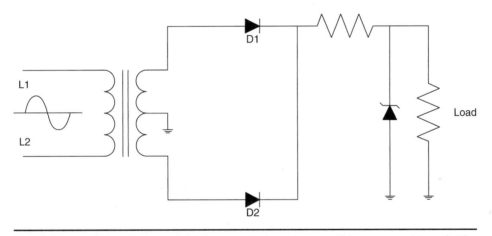

**FIGURE 21–31** A zener diode is installed so that it will be reverse biased. The load is placed in parallel with the zener diode.

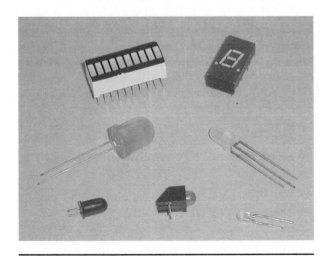

**FIGURE 21–32** Various light emitting diode (LED) packages.

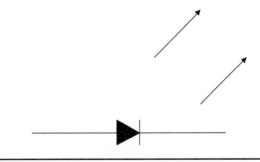

**FIGURE 21–33** Schematic symbol of an LED.

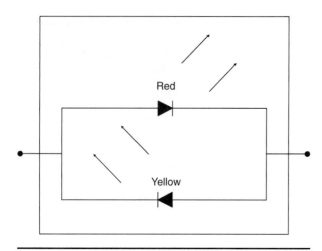

**FIGURE 21–34** A type of voltage and polarity indicator. Glows red with positive DC on the left and negative DC on the right. Glows yellow with positive DC on the right and negative DC on the left. Glows orange with AC applied.

a second in the United States). This will cause the red LED to light, then the yellow LED, then the red LED, and so on. To our eyes, the alternating lighting of the red and yellow LEDs occurs too fast to distinguish. We see the combination of the red and yellow as orange. Now, we have a voltage indicator. We can tell if the voltage is AC (orange light) or DC (red or yellow light) and if DC, we can tell the polarity (red or yellow light).

## Photo Diodes

If the doping of a diode is changed, we can produce yet another type of diode, the *photo diode*. The schematic diagram of the photo diode is shown in Figure 21–35. The photo diode will conduct in the presence of light. The diode package has a small window, which allows light to enter (refer to Figure 21–36). The light enters the window and strikes the PN junction. The light that strikes the PN junction causes the photo diode to conduct in the reverse direction. If more light enters the window, the photo diode will conduct more current. If less light enters the window, the photo diode will conduct less current. Photo diodes are used in applications where we wish to sense the presence or absence of light.

## Testing Diodes

It is possible to perform a simple test on a diode to determine if the diode is functioning properly. The test may be performed either in-circuit or out-of-circuit. However, when testing in-circuit, be certain

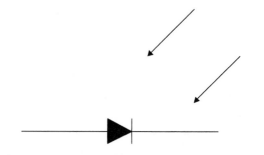

**FIGURE 21–35** Schematic symbol of a photodiode.

that power is removed and any capacitors are fully discharged. Also, if any of the readings are questionable, the diode should be removed from the circuit and retested out-of-circuit.

This test will use nothing more than a simple multimeter. To see how this works, refer to Figure 21–37. Most digital multimeters (DMMs) have a diode-check position, as indicated by the diode symbol. Set the meter to this position, and place the negative lead of the DMM on the cathode of the diode to be tested. Connect the positive lead from the DMM to the anode of the diode under test. Notice the reading on the DMM. If the diode is connected properly (negative to the cathode and positive to the anode), and if the diode is good, the DMM will display a reading of a few tenths of a volt. If the DMM does not display this reading, there can be several explanations. The diode may be defective, the diode may be connected backwards (reverse biased), or the DMM may be defective. You can verify that the DMM is working properly by checking a known good diode. Switch the connections to the diode under test as shown in Figure 21–38. This will check the diode for conduction in the reversed-bias direction. If the DMM now reads zero volt, the diode is reverse biased. This diode is probably good. If the DMM does not produce a reading with the diode connected either way, the diode is probably open and should be replaced. If the DMM produces the same reading with the diode connected either way, the diode is probably shorted and should be replaced. Even if the diode checks good, it is still possible that under circuit conditions the diode may fail. This is a remote possibility but a possibility, nonetheless.

## TRANSISTORS

The **bipolar junction transistor** (BJT) is a three-terminal device and is available in two versions. The schematic symbols for both versions are shown

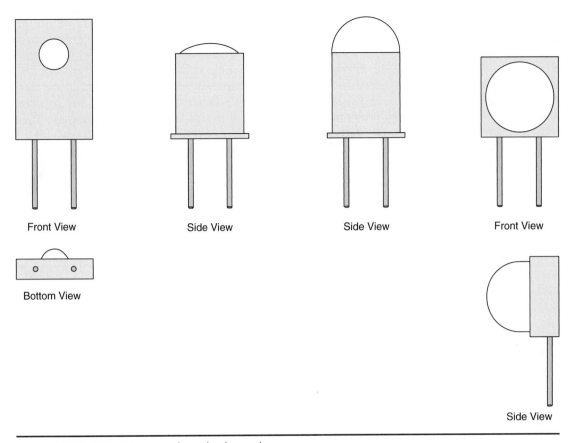

Front View

Side View

Side View

Front View

Bottom View

Side View

**FIGURE 21-36** Various photodiode packages.

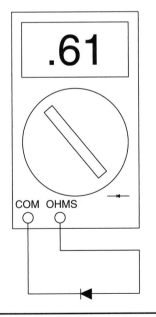

**FIGURE 21-37** Testing a diode in the forward bias direction with an ohmmeter on the diode-check function. Note the low reading.

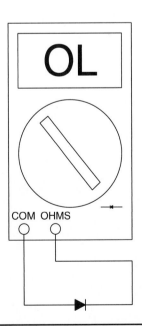

**FIGURE 21-38** Testing a diode in the reverse bias direction with an ohmmeter on the diode-check function. Note the high reading.

in Figure 21–39. Notice that the direction of the arrowhead on one of the terminals is reversed. The symbol with the arrowhead pointing in is known as a *PNP transistor*. The symbol with the arrowhead pointing out is known as an *NPN transistor*. The theory of operation of both types of transistors is identical; only the direction of current flow through the two devices is reversed. The three terminals are known by the following names: the terminal with the arrowhead is called the *emitter*, the terminal that is T-shaped is called the *base*, and the remaining terminal is called the *collector*. Several types of transistors are shown in Figure 21–40. Some transistors have only two leads, as shown in Figure 21–41. In this instance, the collector is usually the metal body of the transistor itself.

To help in understanding of the operation of a transistor, we will use an analogy to a water faucet, which is shown in Figure 21–42. The collector of the transistor acts like the supply pipe to the faucet. The base of the transistor functions like the valve that controls the flow of water (electrons). Lastly, the emitter acts as the part of the faucet where the water (electrons) flows out. In an actual transistor, the current flowing through the base and emitter controls the current flowing through the collector and emitter.

Transistors can be used in switching or amplification functions. When used as a switch, transistors have the capability of turning on and off several thousand times a second. This switching occurs

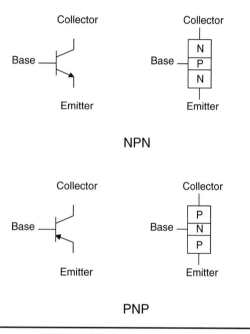

**NPN**

**PNP**

**FIGURE 21–39** Schematic symbols of a transistor. NPN on the top, PNP on the bottom.

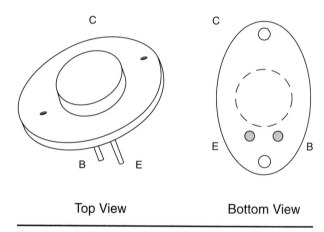

Top View          Bottom View

**FIGURE 21–41** Some transistors use the metal case as the third lead.

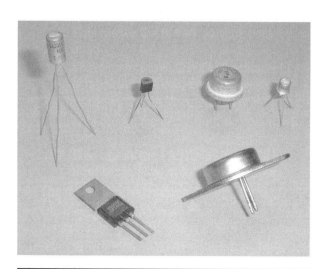

**FIGURE 21–40** Various transistor packages.

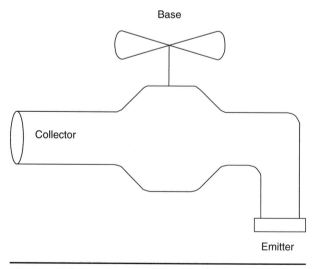

**FIGURE 21–42** The current from collector to emitter is controlled by the base.

without moving parts, making the transistor very efficient and reliable. When used as an **amplifier**, transistors are capable of reproducing and amplifying an input signal. It is not unusual for transistor amplifiers to have a gain of several hundred. In addition, transistors may be connected together in such a fashion as to cause an even greater gain to occur. When transistors are connected in this fashion, a *Darlington amplifier* is formed. Figure 21–43 shows the schematic symbol of a Darlington amplifier. The gain of a Darlington transistor may exceed several thousand.

## Opto-Couplers

Opto-couplers are devices that also provide electrical isolation. Figure 21–44 shows the schematic symbol of an opto-coupler.

An **opto-coupler** consists of a light source (such as an LED) and a photoconductive device (such as a phototransistor) mounted in a single package. These devices are electrically isolated from each other. Figure 21–44 shows an opto-coupler with an LED and a phototransistor. A phototransistor is similar to a regular BJT except that it conducts when its base is

struck by light. In the case of this opto-coupler, the LED is the light source.

In order for the opto-coupler to work, a voltage is applied to the LED, causing the LED to light. This light strikes the base of the phototransistor, causing the phototransistor to conduct. Because light was used to turn on the phototransistor, and since there is no electrical connection between the LED and the phototransistor, there is electrical isolation within the coupler.

## Testing Transistors

The transistor may be tested in-circuit or out-of-circuit. When testing in-circuit, remember to de-energize the circuit and verify that any capacitors are fully discharged. It is important to note that if a transistor is questionable when tested in-circuit, you will need to retest the transistor out-of-circuit to be certain it is defective.

Refer to Figure 21–45. We begin by setting the DMM to a low-resistance range, such as the 200-Ω or 2-kΩ position. Do not use the low-Ω or diode-check position as these will not supply sufficient current to accurately check the transistor under test.

Notice that the NPN transistor has been connected to a DMM. We have connected the negative lead of the DMM to the emitter of the transistor. The positive lead of the DMM is connected to the base of the transistor. If the transistor is good, the DMM will read a low resistance (typically less than

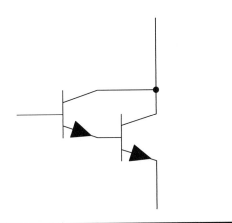

**FIGURE 21–43** Schematic symbol for a Darlington transistor.

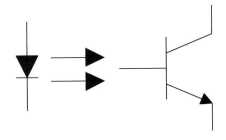

**FIGURE 21–44** Schematic symbol for an opto-coupler.

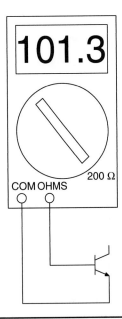

**FIGURE 21–45** Testing the emitter-base junction (forward bias) of an NPN transistor with an ohmmeter on the 200-Ω range. Note the low reading.

1 kiloohm). Now, reverse the DMM connections, as shown in Figure 21–46. Put the positive lead of the DMM on the emitter and the negative lead from the DMM on the base. A good transistor will typically measure in excess of 100 kiloohms. If you measured a low value in both instances, the emitter-base junction is probably shorted. If the measurement was high in both instances, the emitter-base junction is probably open. In both instances, the transistor will need to be replaced.

If the emitter-base junction checked OK, we must then check the collector-base junction (Figure 21–47). Connect the negative lead of the DMM to the collector of the transistor. Connect the positive lead of the DMM to the base of the transistor. If the collector-base junction is good, you should measure less than 1 kiloohm. Now, reverse the connections to the transistor, as shown in Figure 21–48. Connect the positive lead from the DMM to the collector of the transistor. Connect the negative lead of the DMM to the base of the transistor. If the collector-base junction is good, you should measure greater than 100 kiloohms. If you measured a low value in both instances, the collector-base junction is probably shorted. If the measurement was high in both instances, the collector-base junction is probably open. In both instances, the transistor will need to be replaced.

If the collector-base junction checked OK, we will need to perform a final check of the emitter-collector junction (Figure 21–49). We begin by connecting the negative lead of the DMM to the emitter

terminal. Next, we connect the positive lead of the DMM to the collector. If the transistor is good, the DMM will indicate more than 100 kiloohms. Now, reverse the connections to the transistor, as shown in Figure 21–50. Connect the positive lead of the DMM to the emitter and the negative lead of the DMM to the collector. Again, the DMM should indicate more than 100 kiloohms. If the DMM indicated

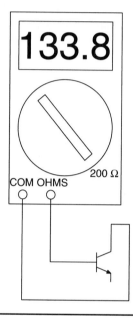

**FIGURE 21–47**  Testing the collector-base junction (forward bias) of an NPN transistor with an ohmmeter on the 200-Ω range. Note the low reading.

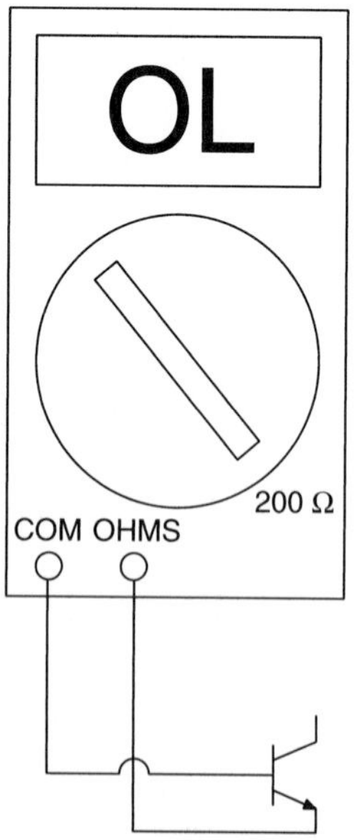

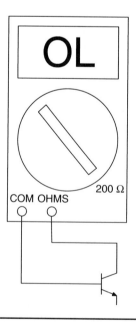

**FIGURE 21–48**  Testing the collector-base junction (reverse bias) of an NPN transistor with an ohmmeter on the 200-Ω range. Note the high reading.

**FIGURE 21–46**  Testing the emitter-base junction (reverse bias) of an NPN transistor with an ohmmeter on the 200-Ω range. Note the high reading.

a low resistance in both tests, the transistor is probably shorted from the emitter to collector and will need to be replaced.

If the transistor passed all of these tests, it is more than likely OK. However, it is possible that under circuit voltages, the transistor may breakdown and fail. It is also common for transistors to fail after they have heated up. Therefore, the DMM test is not 100 percent reliable, but it can be an effective and quick troubleshooting tool.

## Insulated Gate Bipolar Transistors

Another three-terminal device is called the **insulated gate bipolar transistor** (or **IGBT**). There are two different schematic symbols in use for the IGBT. These are shown in Figure 21–51 and Figure 21–52. In Figure 21–51, the terminal with the arrowhead is the emitter, the L-shaped terminal is the gate, and the remaining terminal is the collector. In Figure 21–52, the emitter is the terminal with the arrowhead symbol, the L-shaped lead is the gate, and the remaining lead is the collector.

IGBTs are finding more usage in industry due to their high-speed switching ability. In addition, they have very low internal resistance, which decreases any losses that they inject into the circuit.

Because of the way in which an IGBT is constructed, a reliable check cannot be performed with an ohmmeter. A better understanding of electronic circuits is needed if one is to determine if an IGBT is defective.

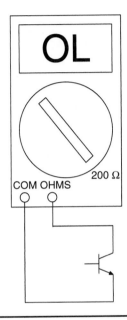

**FIGURE 21–49** Testing the collector-emitter junction (forward bias) of an NPN transistor with an ohmmeter on the 200-Ω range. Note the high reading.

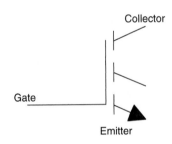

**FIGURE 21–51** One form of a schematic symbol used to represent an IGBT.

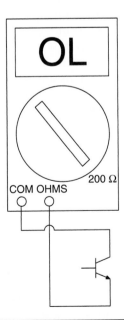

**FIGURE 21–50** Testing the collector-emitter junction (reverse bias) of an NPN transistor with an ohmmeter on the 200-Ω range. Note the high reading.

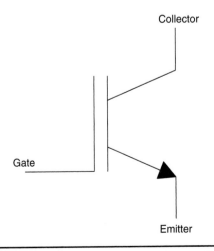

**FIGURE 21–52** A more common schematic symbol used to represent an IGBT.

# FIELD EFFECT TRANSISTORS

There are two types of **field effect transistors**: the **junction field effect transistor** (or **JFET**) and the **metal oxide semiconductor field effect transistor** (or **MOSFET**).

## JFETs

JFETs are further divided into two types: the *n-channel JFET* and the *p-channel JFET*. The three leads of a JFET are labeled *source, gate,* and *drain*. Notice the schematic symbols of an n-channel (Figure 21–53) and p-channel (Figure 21–54) JFET. The difference is the direction of the arrowhead on the gate lead. The arrowhead points inward on the n-channel JFET and points outward on the p-channel JFET. The operation of the n- and p-channel JFET is identical with the exception that the polarities of the voltages are reversed.

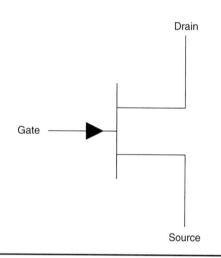

**FIGURE 21–53** N-channel JFET schematic symbol.

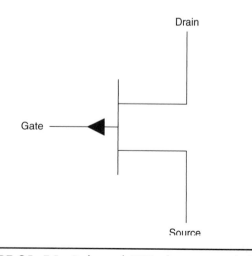

**FIGURE 21–54** P-channel JFET schematic symbol.

Unlike BJTs, *JFETS are voltage-controlled devices.* To control the conduction of current from the source to the drain, the gate voltage must be more negative than the source voltage. Increasing the amount of negative voltage (notice the size of the battery) applied to the gate will reduce the current flow from source to drain, as shown in Figure 21–55. Decreasing the amount of negative voltage (again, notice the size of the battery) applied to the gate will cause the current flow from source to drain to increase, as shown in Figure 21–56. JFETs have a higher internal impedance than BJTs. Typically, the impedance of a JFET is in the neighborhood of 20,000 megaohms.

JFETs can be tested with an ohmmeter with reasonable accuracy, in-circuit or out-of-circuit. When testing in-circuit, remember to de-energize the circuit and verify that any capacitors are fully discharged. It is important to note that if a JFET is questionable when tested in-circuit, you will need to retest the JFET out-of-circuit to be certain it is defective.

Refer to Figure 21–57. We begin by setting the DMM to the diode-check position. Notice that we have taken an n-channel JFET and connected it to a DMM. We have connected the negative lead of the DMM to the source of the JFET. The positive lead of the DMM is connected to the gate of the JFET. If the JFET is good, the DMM will read a few tenths of a volt (typically 0.5 to 0.7 volt). Now, reverse the DMM connections, as shown in Figure 21–58. Place the positive lead of the DMM on the source and the negative lead from the DMM on the gate. A good JFET will typically measure 0 volt. If you measured a low value in both instances, the source-gate junction is probably shorted. If the measurement was high in both instances, the source-gate junction is probably open. In both instances, the JFET will need to be replaced.

If the source-gate junction checked OK, we must then check the drain-gate junction (Figure 21–59). Connect the negative lead of the DMM to the drain of the JFET. Connect the positive lead of the DMM to the gate of the JFET. If the drain-gate junction is good, you should measure a few tenths of a volt (typically 0.5 to 0.7 volt). Now, reverse the connections to the JFET, as shown in Figure 21–60. Connect the positive lead from the DMM to the drain of the JFET. Connect the negative lead of the DMM to the gate of the JFET. If the drain-gate junction is good, you should not measure any voltage. If you measured a low value in both instances, the drain-gate junction is probably shorted. If the measurement was high in both instances, the drain-gate junction is probably open. In both instances, the JFET will need to be replaced.

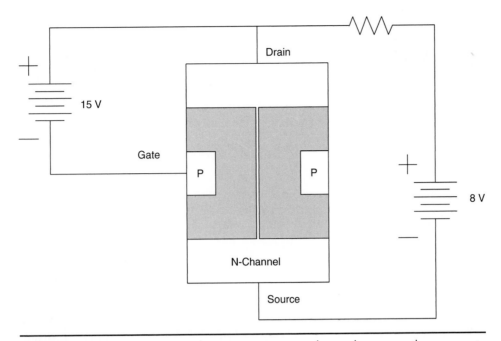

**FIGURE 21–55**  Increasing the gate-to-source voltage decreases the source to drain current.

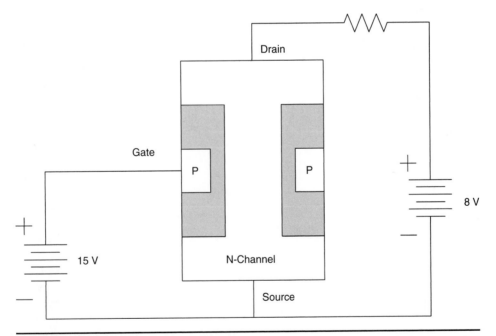

**FIGURE 21–56**  Decreasing the gate-to-source voltage increases the source to drain current.

If the drain-gate junction checked OK, we will need to perform a final check on the source-drain junction (Figure 21–61). We begin by connecting the negative lead of the DMM to the source terminal. Next, we connect the positive lead of the DMM to the drain. If the JFET is good, the DMM will indicate a few tenths of a volt. The actual amount of voltage will vary from JFET to JFET. Now, reverse the connections to the JFET, as shown in Figure 21–62.

Connect the positive lead of the DMM to the source and the negative lead of the DMM to the drain. Again, the DMM should indicate a few tenths of a volt.

If the JFET passed all of these tests, it is more than likely OK. However, it is possible that under circuit-operating voltages, the JFET may have a breakdown and fail. Therefore, the DMM test is not 100 percent reliable, but it can be an effective and quick troubleshooting tool.

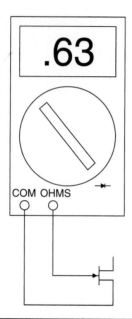

**FIGURE 21-57** Testing the source-gate junction (forward bias) of an N-channel JFET with an ohmmeter on the diode-check function. Note the low reading.

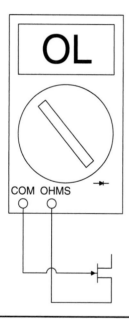

**FIGURE 21-58** Testing the source-gate junction (reverse bias) of an N-channel JFET with an ohmmeter on the diode-check function. Note the high reading.

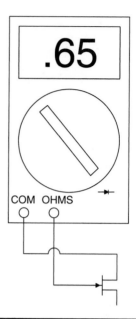

**FIGURE 21-59** Testing the drain-gate junction (forward bias) of an N-channel JFET with an ohmmeter on the diode-check function. Note the low reading.

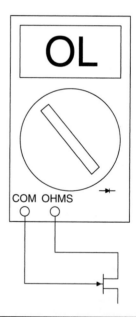

**FIGURE 21-60** Testing the drain-gate junction (reverse bias) of an N-channel JFET with an ohmmeter on the diode-check function. Note the high reading.

## MOSFETs

MOSFETs are divided into two types: depletion-enhancement MOSFET (**DE-MOSFET**) and enhancement only MOSFET (**E-MOSFET**). Within each type, there are n-channel and p-channel devices. Figures 21–63 through 21–66 show the different schematic symbols for the DE-MOSFET n-channel and p-channel, and the E-MOSFET n-channel and p-channel, respectively.

MOSFETs are a three-terminal device. One of the terminals is called the source, which is the terminal with the arrowhead. Another terminal is the gate, which is the terminal that is L-shaped on the schematic. The remaining terminal is the drain.

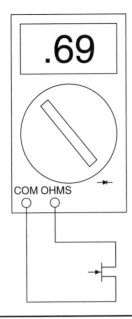

**FIGURE 21-61** Testing the source-drain junction (forward bias) of an N-channel JFET with an ohmmeter on the diode-check function. Note the low reading.

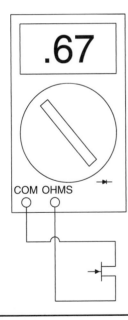

**FIGURE 21-62** Testing the source-drain junction (reverse bias) of an N-channel JFET with an ohmmeter on the diode-check function. Note the low reading.

MOSFETs have a much higher input resistance than the JFETs. Typically, the input resistance of a MOSFET is around 2 gigaohms As a result, MOSFETs are considered by some to be a perfect switch. This means that when the MOSFET is turned on, there is very little internal resistance. When the MOSFET is turned off, its input resistance is very high.

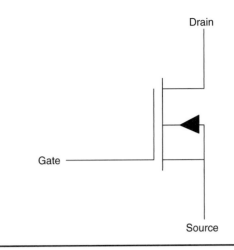

**FIGURE 21-63** DE-MOSFET N-channel schematic symbol.

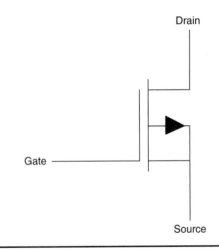

**FIGURE 21-64** DE-MOSFET P-channel schematic symbol.

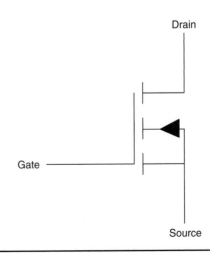

**FIGURE 21-65** E-MOSFET N-channel schematic symbol.

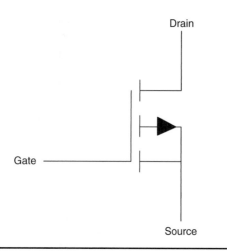

**FIGURE 21–66** E-MOSFET P-channel schematic symbol.

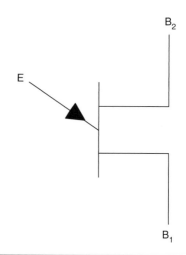

**FIGURE 21–67** Schematic symbol of a unijunction transistor (UJT).

Due to the nature of a MOSFET, a reliable check cannot be performed with an ohmmeter. A better understanding of electronic circuits is needed to determine if a MOSFET is defective. Also, most MOSFETs are sensitive to static electricity. Special handling precautions are required so that these devices are not damaged by static discharge.

# THYRISTORS

The family of devices known as thyristors includes **unijunction transistors (UJTs)**, **silicon-controlled rectifiers (SCRs)**, **diacs**, and **triacs**.

## Unijunction Transistors

The unijunction transistor (UJT) has one emitter and two bases, as seen in Figure 21–67. UJTs are considered digital devices because they have only two states: off or on. To turn on a UJT, a voltage must be applied to the emitter. This voltage must be approximately 10 volts more positive than the voltage applied to base $B_1$. Base $B_2$ must be more positive than the emitter for the UJT to conduct. Current will flow from base $B_1$ to base $B_2$ until the emitter voltage drops to a value that is 3 volts more positive than base $B_1$. At this point, the UJT turns off. The UJT is a device that is very useful in producing pulses.

UJTs can be tested with a digital multimeter with reasonable accuracy, in-circuit or out-of-circuit. When testing in-circuit, remember to de-energize the circuit and verify that any capacitors are fully discharged. It is important to note that if a UJT is questionable when tested in-circuit, you will need to retest the UJT out-of-circuit to be certain it is defective.

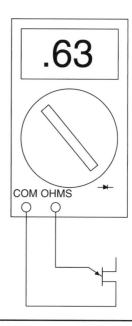

**FIGURE 21–68** Testing the emitter-B1 junction (forward bias) of a UJT with an ohmmeter on the diode-check function. Note the low reading.

Refer to Figure 21–68. We begin by setting the DMM to the diode-check position. We have connected the negative lead of the DMM to base $B_1$ of the UJT. The positive lead of the DMM is connected to the emitter of the UJT. If the UJT is good, the DMM will read a few tenths of a volt (typically 0.5 to 0.7 volt). Now, reverse the DMM connections, as shown in Figure 21–69. Place the positive lead of the DMM on base $B_1$ and the negative lead from the DMM on the emitter. A good UJT will typically measure 0 volt. If you measured a low value in both instances, the $B_1$-emitter junction is probably shorted. If the measurement was high in both

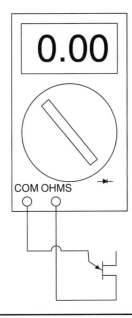

**FIGURE 21–69** Testing the emitter-B1 junction (reverse bias) of a UJT with an ohmmeter on the diode-check function. Note the 0 reading.

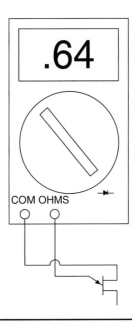

**FIGURE 21–70** Testing the emitter-B2 junction (forward bias) of a UJT with an ohmmeter on the diode-check function. Note the low reading.

instances, the $B_1$-emitter junction is probably open. In both instances, the UJT will need to be replaced. If the $B_1$-emitter junction checked OK, we must then check the $B_2$-emitter junction.

Refer to Figure 21–70. Connect the negative lead of the DMM to base $B_2$ of the UJT. Connect the positive lead of the DMM to the emitter of the UJT. If the $B_2$-emitter junction is good, you should measure a few tenths of a volt (typically 0.5 to 0.7 volt). Now, reverse the connections to the UJT, as shown in Figure 21–71. Connect the positive lead from the DMM to base $B_2$ of the UJT. Connect the negative lead of the DMM to the emitter of the UJT. If the $B_2$-emitter junction is good, you should not measure any voltage. If you measured a low value in both instances, the $B_2$-emitter junction is probably shorted. If the measurement was high in both instances, the $B_2$-emitter junction is probably open. In both instances, the UJT will need to be replaced.

If the $B_2$-emitter junction checked OK, we will need to perform a final check on the $B_1$-$B_2$ junction. Refer to Figure 21–72. We begin by connecting the negative lead of the DMM to base $B_2$ terminal. Next, we connect the positive lead of the DMM to base $B_1$. If the UJT is good, the DMM will indicate a few tenths of a volt. The actual amount of voltage will vary from UJT to UJT. Now, reverse the connections to the UJT, as shown in Figure 21–73. Connect the positive lead of the DMM to base $B_2$ and the negative lead of the DMM to base $B_1$. Again, the DMM should indicate a few tenths of a volt.

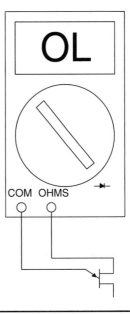

**FIGURE 21–71** Testing the emitter-B2 junction (reverse bias) of a UJT with an ohmmeter on the diode-check function. Note the high reading.

If the UJT passed all of these tests, it is more than likely OK. However, it is possible that under circuit-operating voltages, the UJT may breakdown and fail. Therefore, the DMM test is not 100 percent reliable, but it can be an effective and quick troubleshooting tool.

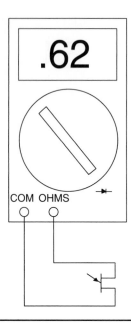

**FIGURE 21-72** Testing the B1-B2 junction (forward bias) of a UJT with an ohmmeter on the diode-check function. Note the low reading.

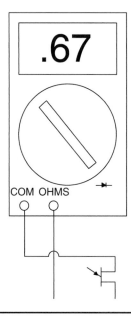

**FIGURE 21-73** Testing the B1-B2 junction (reverse bias) of a UJT with an ohmmeter on the diode-check function. Note the low reading.

## Silicon-Controlled Rectifiers

Figure 21–74 shows the schematic symbol of a silicon-controlled rectifier or SCR. You will notice that the SCR looks very similar to a diode except that there is an extra terminal. The terminals of the SCR are known as the anode (which is the arrowhead symbol), the cathode (which is the T-shaped symbol), and the gate or trigger (which is the remaining terminal).

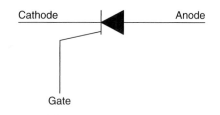

**FIGURE 21-74** Schematic symbol of a silicon-controlled rectifier (SCR).

SCRs function similarly to a diode, with the exception of the gate circuit. As we have seen, a diode will only conduct current when forward biased. The current flow will be blocked when the diode is reverse biased. The same is true for the SCR with one exception—the SCR will conduct current when forward biased (cathode negative with respect to the anode). However, the SCR will not conduct current in the forward direction until the proper gate voltage is applied. This means that we can control the condition under which the SCR conducts.

In order for the SCR to conduct, we must apply a negative voltage to the cathode and a positive voltage to the anode. At this point, the SCR does not conduct. By applying a positive voltage to the gate, the SCR is triggered into conduction. If the gate voltage is removed, the SCR will continue to conduct as long as current flows through the cathode–anode junction. Should the current drop below the minimum holding current rating of the SCR, the SCR will turn off. The SCR will not turn on again until another positive gate voltage is applied. Different SCRs have different values of minimum holding current.

It is possible to perform a simple check of an SCR with a DMM. This test may be performed either in-circuit or out-of-circuit. However, when testing in-circuit, be certain that power is removed and any capacitors are fully discharged. Also, if any of the readings are questionable, the SCR should be removed from the circuit and retested out-of-circuit.

Refer to Figure 21–75. We will begin by testing the diode portion of the SCR. Set the DMM to either the 200-Ω or the 2-kΩ position. Do not use the low-Ω or diode-check position as these will not supply sufficient current to accurately check the SCR under test. Connect the negative lead from the DMM to the cathode of the SCR under test. Connect the positive lead from the DMM to the anode of the SCR. A good SCR will typically measure in excess of 100 kiloohms. Now, reverse your connections, as shown in Figure 21–76. Connect the negative lead from the DMM to the anode of the SCR, and connect the positive lead from the DMM to the cathode of the SCR. You should still measure in excess of 100 kiloohms if the SCR is good.

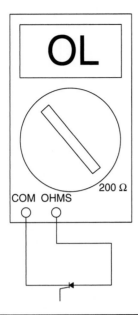

**FIGURE 21–75** Testing the cathode–anode junction (forward bias untriggered) of an SCR with the ohmmeter on the 200 Ω range. Note the high reading.

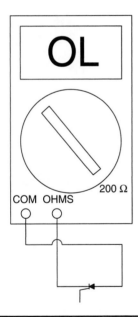

**FIGURE 21–76** Testing the cathode–anode junction (reverse bias untriggered) of an SCR with the ohmmeter on the 200 Ω range. Note the high reading.

Reverse your connections again so that the negative lead from the DMM is connected to the cathode of the SCR and the positive lead from the DMM is connected to the anode of the SCR, as shown in Figure 21–75. Now, take another test lead and, while leaving the DMM connected, connect the positive lead of the DMM to the gate terminal of the SCR, as shown in Figure 21–77. (The positive lead of the

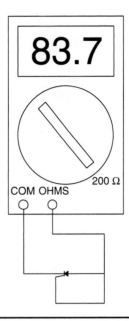

**FIGURE 21–77** Testing the cathode–anode junction (forward bias triggered) of an SCR with the ohmmeter on the 200-Ω range. Note the low reading.

DMM is now connected to both the anode and the gate of the SCR.) If the SCR is good, the resistance reading should drop to less than 1 kiloohm. If there is no change in the resistance reading, the SCR is probably defective and will need to be replaced. If you now remove the connection to the gate terminal, the resistance reading should remain at less than 1 kiloohm. If the reading returns to a high resistance value, the SCR is not necessarily defective. Some DMMs do not supply sufficient holding current for the SCR to maintain conduction. This is especially true with larger, high-current SCRs. Breaking the connection to either the anode or the cathode will turn off the SCR. The test can now be repeated.

## Diacs

The diac (sometimes called a bilateral trigger diode) is a bidirectional diode. This means that a diac can operate in an AC circuit. The diac has two terminals, called main terminal 1 (MT 1) and main terminal 2 (MT 2), as shown by the schematic symbol in Figure 21–78. The diac functions like an AC version of a UJT. That is, a diac does not conduct until a threshold voltage is reached. The diac continues to conduct until the voltage falls below its minimum conduction voltage. At this point, the diac turns off. For example, a diac may have a turn-on voltage rating of 32 volts and a turn-off voltage rating of 25 volts. This means that when power is first applied to the circuit, the diac will not conduct. Should the

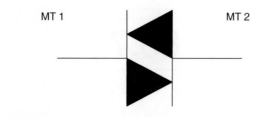

**FIGURE 21–78A**  Diac schematic symbol.

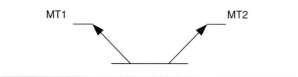

**FIGURE 21–78B**  Alternate diac schematic symbol.

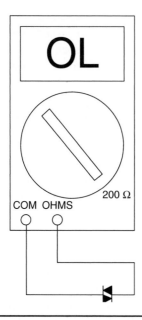

**FIGURE 21–79**  Testing a diac with the ohmmeter on the 200-Ω range. Note the high reading.

voltage rise to ±32 volts, the diac turns on and allows current flow. Should the voltage drop to ±30 volts, the diac will remain on, and if the voltage drops below ±25 volts, the diac will turn off and current flow will stop. Diacs are available with different threshold voltages. You can think of the diac as a voltage-sensitive AC switch.

The only test that can be performed on a diac with an ohmmeter is to check for a shorted diac. Under normal conditions, a good diac will indicate an open circuit when tested with an ohmmeter, as shown in Figures 21–79 and 21–80. A low resistance reading in both directions indicates that the diac is shorted and will need to be replaced.

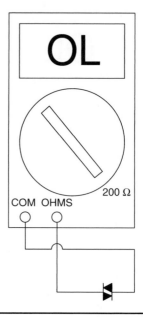

**FIGURE 21–80**  Testing a diac with the ohmmeter on the 200-Ω range. Note that the connections have been reversed, but the reading remains high.

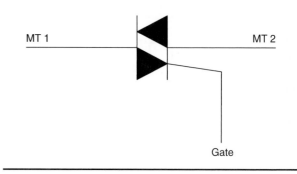

**FIGURE 21–81**  Schematic symbol of a triac.

## Triacs

A triac is a device that operates like an SCR for an AC circuit. The triac will conduct during both halves of the AC waveform. This means that the output of a triac is AC, not DC, as is the case with an SCR. The schematic symbol of a triac is shown in Figure 21–81. The triac is a three-terminal device. One terminal is called the gate. The terminal closest to the gate is called main terminal 1 (or MT 1), while the remaining terminal is called main terminal 2 (or MT 2). The gate voltage must be of the same polarity as MT 2 to turn on the triac. The triac will continue to conduct until the current flowing through the triac decreases below the minimum holding current level for the particular triac.

It is possible to perform a simple check of a triac with a DMM. This test may be performed either in-circuit or out-of-circuit. However, when testing in-circuit, be certain that power is removed and any

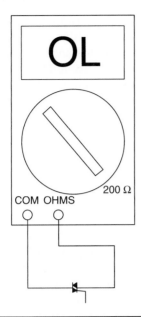

**FIGURE 21–82** Testing the MT 1-to-MT 2 junction (untriggered) of a triac with the ohmmeter on the 200-Ω range. Note the high reading.

capacitors are fully discharged. In addition, if any of the readings are questionable, the triac should be removed from the circuit and retested out-of-circuit.

Refer to Figure 21–82. We will begin by testing one of the diode portions of the triac. Set the DMM to either the 200-Ω or the 2-kΩ position. Do not use the low-W or diode-check position as these will not supply sufficient current to accurately check the triac under test. Connect the negative lead from the DMM to MT 1 of the triac under test. Connect the positive lead from the DMM to MT 2 of the triac. A good triac will typically measure in excess of 100 kiloohms. Now, reverse your connections, as shown in Figure 21–83. Connect the negative lead from the DMM to MT 2 of the triac, and connect the positive lead from the DMM to MT 1 of the triac. You should still measure in excess of 100 kiloohms if the triac is good.

Now, take another test lead and, while leaving the DMM connected, connect the negative lead of the DMM to the gate terminal of the triac, as shown in Figure 21–84. (The negative lead of the DMM is now connected to both MT 2 and the gate of the triac.) If the triac is good, the resistance reading should drop to less than 1 kiloohm. If there is no change in the resistance reading, the triac is probably defective and will need to be replaced. If you remove the connection to the gate terminal, the resistance reading should remain at less than 1 kiloohm. If the reading returns to a high resistance value, the triac is not necessarily defective. Some DMMs do not supply sufficient holding current for the triac to maintain conduction. This is especially true with larger,

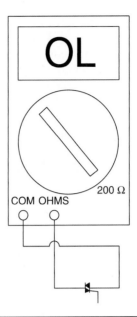

**FIGURE 21–83** Testing the MT 1-to-MT 2 junction (untriggered) of a triac with the ohmmeter on the 200-Ω range. Note that the connections have been reversed, but the reading remains high.

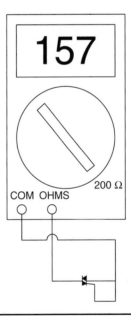

**FIGURE 21–84** Testing the MT 1-to-MT 2 junction (triggered) of a triac with the ohmmeter on the 200-Ω range. Note the low reading.

high-current triacs. Breaking the connection to either MT 1 or MT 2 will turn off the triac.

The test can now be repeated on the other half of the triac. Refer to Figure 21–85. Connect the negative lead from the DMM to MT 1, and connect the positive lead of the DMM to MT 2. Also, connect the positive lead of the DMM to the gate terminal of the triac. (The positive lead of the DMM is now connected to both MT 2 and the gate of the triac.) If the triac is good, the resistance reading should drop to less than 1 kiloohm. If there was no change in the resistance reading, the triac is probably defective and will need to be replaced. If you remove the

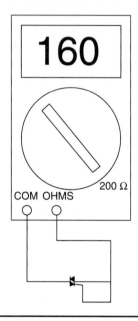

**FIGURE 21-85** Testing the MT 1-to-MT 2 junction (triggered) of a triac with the ohmmeter on the 200-Ω range. Note that the connections have been reversed, but the reading remains low.

connection to the gate terminal, the resistance reading should remain at less than 1 kiloohm. If the reading returns to a high resistance value, the triac is not necessarily defective. Some DMMs do not supply sufficient holding current for the triac to maintain conduction. This is especially true with larger, high-current triacs. Breaking the connection to either MT 1 or MT 2 will turn off the triac.

## Photoconductive Cells

A *photoconductive cell* changes its internal resistance when its surface is struck by light. Most of these devices are made of a material known as either cadmium sulfide or cadmium selenide. For this reason, these devices are sometimes referred to simply as *cad cells*.

Under low light-level conditions (or no light for that matter), the resistance of the cad cell is several hundred thousands to several million ohms of resistance. If this cad cell were installed in the coil circuit of a relay, as shown in Figure 21–86, the coil would not receive sufficient current to energize.

When the light increases to a sufficient level, the internal resistance of the cad cell lowers. Typically, a cad cell in the presence of light will have an internal resistance in the order of several thousand ohms. This allows current to flow through the cad cell and the relay coil, energizing the relay.

Figure 21–86 represents a circuit used in an oil pump control circuit for an oil-burning furnace.

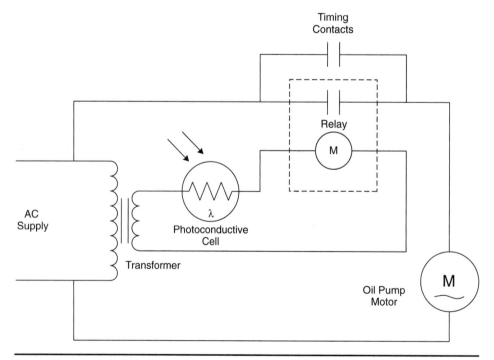

**FIGURE 21-86** Simplified diagram of an oil pump controlled by a photoconductive cell.

Normally, the resistance of the photoconductive cell is quite high. This blocks the current to the relay, keeping the relay contacts open. When the burner flame ignites, the cad cell reacts to the light from the flame. The internal resistance of the cad cell lowers, allowing the current to increase to the relay coil. As a result, the relay coil energizes, closing the relay contacts. With the relay contacts closed, the oil pump motor turns on. As long as the burner flame is burning, the oil pump motor pumps. When the burner flame is extinguished, the cad cell resistance increases, blocking current to the relay. This de-energizes the relay, opening the relay contacts and shutting down the oil pump motor.

This is the same principle used in the operation of street lamps. By using a cad cell and a relay, the street lamps are turned off when sufficient light is present and turned on during low light-level conditions.

## Hall-Effect Devices

A Hall-effect device is a device that detects a magnetic field. A Hall-effect device generally has four terminals. Refer to Figure 21–87. Two of the terminals are used to supply the operating current to the Hall-effect device. The remaining two terminals supply an output voltage.

Figure 21–87 shows a Hall-effect device as it would appear without the presence of a magnetic field. Notice that the current flow through the device is undistorted. Now, refer to Figure 21–88. In this drawing, a magnetic field has been brought into close proximity of the Hall-effect device. Notice how the current has been deflected to one side of the device. This causes a voltage to appear across the output terminals of the Hall-effect device.

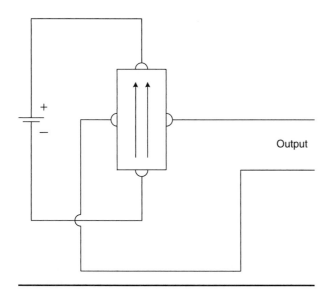

**FIGURE 21–87** Hall-effect sensor.

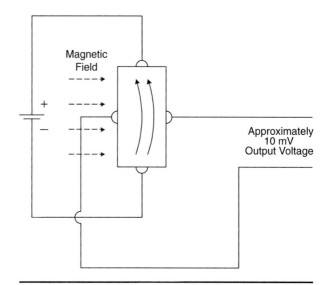

**FIGURE 21–88** Hall-effect sensor in the presence of a magnetic field.

Typically, Hall-effect devices produce approximately 10 millivolts of voltage at their output terminals. The actual amount of output voltage is dependent upon the strength of the magnetic field. For this reason, a voltage amplifier is often used in conjunction with a Hall-effect device.

## THE 555 TIMER

The *555 timer* is a member of a family of devices known as *integrated circuits* or ICs. An integrated circuit is a device that contains anywhere from a few components to several thousand components in one package. These components are connected to form complete circuits. The circuits are miniaturized and contained within a device known as an integrated circuit.

The most common package for an integrated circuit is known as a dual-in-line pin, or DIP, package. The 555 timer is typically found in an 8-pin DIP package. However, a 555 timer may also appear in a TO-5 package. In addition, there is a 556 timer that has two 555 timers in a 14-pin DIP package. There is also a 558 timer, which contains four 555 timers in a 14-pin DIP package. Because the 8-pin package is the most common, we will show how the pins of the 8-pin DIP and TO-5 package are numbered. Figure 21–89 shows the pin numbers for an 8-pin DIP package, while Figure 21–90 shows the pin numbers for an 8-pin TO-5 package.

555 timers are used to perform many types of timing, pulsing, and delaying operations. As we will see, by changing the connections and components associated with the 555 timer, we will be able

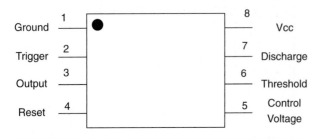

**FIGURE 21–89**  Pin-outs of a 555 timer, DIP package.

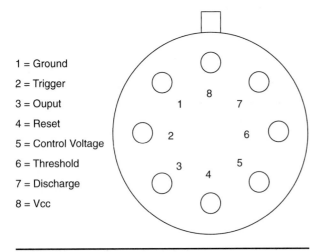

1 = Ground

2 = Trigger

3 = Ouput

4 = Reset

5 = Control Voltage

6 = Threshold

7 = Discharge

8 = Vcc

**FIGURE 21–90**  Pin-outs of a 555 timer, TO-5 package.

to create different timing circuits to meet different needs. After we have learned the function of each of the pins of a 555 timer, we will study a few of the many timing circuits that are possible.

## Pin Assignments

Each pin of the 555 timer has a function. Here are the functions of these pins:

Pin #1   **Ground**. This pin is connected to the most negative potential of the external circuit. This is typically the circuit ground when the circuit uses a positive voltage supply.

Pin #2   **Trigger**. The trigger is used to fire or start the timing operation. The 555 will be triggered when the voltage on pin #2 is lowered to less than one-third of the supply voltage to the 555 timer, or if the voltage on pin #2 is one-half of the voltage present on pin #5. Typically, pin #2 is connected to the circuit ground to trigger the 555. It is important to

realize that this connection must be momentary. If the connection to ground is maintained, the 555 will not operate.

Pin #3   **Output**. This is the output pin for the 555 timer. When the trigger (pin #2) voltage is lowered, the output will be on. To turn the output off, either the voltage at pin #6 must be raised above two-thirds of the supply voltage to the 555 timer, or the voltage at pin #4 must be lowered to less than 0.7 volt.

Pin #4   **Reset**. Lowering the voltage at pin #4 to less than 0.7 volt will reset or turn off the output (pin #3). This will occur regardless of the state or condition of the trigger (pin #2), the threshold (pin #6), or the discharge (pin #7). If the reset function is not used, pin #4 should be connected to the 555 supply voltage to avoid false resetting.

Pin #5   **Control Voltage**. Two-thirds of the 555 supply voltage will appear at this pin. By connecting a variable resistor from pin #5 to the 555 supply voltage, the *on time* can be varied. This will have no effect on the *off time*. By connecting a variable resistor from pin #5 to ground, the off time can be varied. This will have no effect on the on time. If the control voltage function is not used, a 0.01-microfarad capacitor should be connected between pin #5 and ground. This will filter any electrical noise from entering pin #5.

Pin #6   **Threshold**. When the voltage applied to pin #6 of the 555 is raised above two-thirds of the 555 supply voltage, the output (pin #3) of the 555 will turn off.

Pin #7   **Discharge**. The state of pin #7 will be the opposite of the state of pin #3. That is, when the output (pin #3) is on, the discharge (pin #7) will be off, and vice versa. Pin #7 is sometimes used as an auxiliary output, with its state being the compliment of the output (pin #3).

Pin #8   **Vcc or V+**. This is the positive supply voltage terminal for the 555 timer. Typically, a 555 can operate within a supply voltage range of +4.5 volts to +16 volts; however, the accuracy of the timing function is assured when operated between +5 volts and +15 volts.

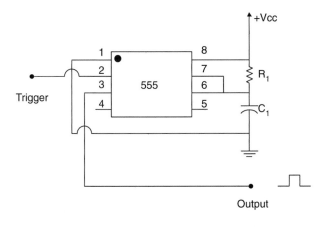

**FIGURE 21-91**  *555 timer connected in the mono-stable mode.*

## Monostable Mode

555 timers are operated in one of two modes. These modes are *monostable* (or one-shot) and *astable*. In the monostable mode, the 555 has a single, stable state. In the astable mode, the 555 will generate a constant output of pulses. Let's see how a 555 timer works in the monostable mode.

Refer to Figure 21–91. The single, stable state of a 555 is the off state. When triggered, the 555 switches to its on state. The 555 will remain in the on state for a length of time determined by the values of a resistor–capacitor network. After the time has expired, the 555 will return to its off state. This results in the generation of an output pulse at pin #3 of the 555. The on time of the output pulse is determined by the values of the resistor–capacitor network. The 555 must be manually retriggered to generate another output pulse.

## Astable Mode

A 555 timer, operating in the astable mode, is also called a *multivibrator*. When used in this mode, the 555 will produce a string of pulses at the output (pin #3). To produce the string of pulses, the 555 timer must be constantly retriggered. This is done automatically by connecting the trigger (pin #2) to the threshold (pin #6), as shown in Figure 21–92. If a variable resistor is connected between the supply voltage and the control voltage (pin #5) of the 555, the on time can be made longer or shorter. The variable resistor will not affect the off time or the time between pulses. If a variable resistor is connected between the control voltage (pin #5) of the 555 and ground, the off time, or time between pulses, can be made longer or shorter. The variable resistor will not affect the on time. A resistor–capacitor network is used to control the frequency of the pulses.

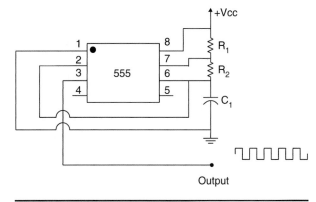

**FIGURE 21-92**  *555 timer connected in the astable mode.*

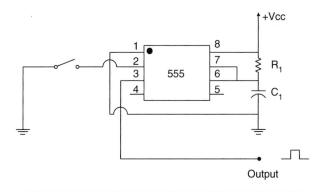

**FIGURE 21-93**  *555 timer connected as a one-shot.*

## 555 CIRCUITS

The 555 timer is a very versatile and useful device. There are many circuits and applications for the 555. In the following three topics, we will show a few of these applications. The circuit diagram will be presented as well as an explanation of the circuit and example of use.

## One-Shot

Figure 21–93 shows a 555 timer configured as a one-shot. Notice the RC network consisting of $R_1$ and $C_1$. This network is connected between the supply voltage and ground. Notice also that the junction of $R_1$ and $C_1$ is connected to pins #6 and #7 of the 555. The trigger source is applied to pin #2 of the 555. The output pulse will be seen at pin #3. Here is how the 555 operates in the one-shot mode.

Initially, the trigger source is a *high* condition (approximately one-third of V+). With a positive trigger signal at pin #2, the output of the 555 will be practically 0 volt. When a negative-going trigger pulse is applied to pin #2, the output of the 555 will change states. That is, the output of the 555 will start to produce a positive pulse. At this time, capacitor $C_1$ begins to charge through resistor $R_1$. The length of time it takes for $C_1$ to charge is dependent upon the RC time constant of $R_1$ and $C_1$. When $C_1$ charges to approximately two-thirds of the supply voltage, the output of the 555 will switch back to practically 0 volt. This discharges capacitor $C_1$. The 555 is now reset and "awaiting" the next negative-going trigger pulse to pin #2.

The length of time taken for the output pulse to be achieved can be varied by changing the value of either $C_1$ or $R_1$. This will affect the RC time constant. In this manner, a longer- or shorter-duration output pulse can be generated. It is possible to produce output pulses from as short a time period as 10 microseconds to as long as practical, based on available values for $R_1$ and $C_1$.

A 555 timer used as a one-shot can be used as a triggering device. This is helpful when a well-defined pulse is needed for triggering purposes.

## Oscillator

Figure 21–94 shows a 555 timer configured as an oscillator. Notice that in this circuit pins #2 and #6 are connected together. Also, notice that a second resistor, $R_2$, has been added. The RC network now consists of $R_1$, $R_2$, and $C_1$. This network is connected between the supply voltage and ground. The junction of $R_2$ and $C_1$ is connected to pins #2 and #6 of the 555, and the junction of $R_1$ and $R_2$ is connected to pin #7 of the 555. Since this 555 circuit operates in the astable mode, there is no separate trigger source. The output pulse will be seen at pin #3. Here is how the 555 operates as an oscillator.

$C_1$ will be uncharged when power is initially applied to the circuit. The output of the 555 at pin #3 will be high. The high output condition will cause capacitor $C_1$ to begin charging through $R_1$ and $R_2$. When the charge on $C_1$ reaches two-thirds of the supply voltage, the output of the 555 will switch to the low state. $C_1$ will now begin to discharge.

When the voltage on $C_1$ drops to approximately one-third of the supply voltage, the output of the 555 will switch to the high state. This causes $C_1$ to begin charging again and causes the entire cycle to repeat. The result is a continuous stream of rectangular pulses at the output of the 555 at pin #3.

The frequency of the output pulses is dependant upon the values of $R_1$, $R_2$, and $C_1$. The time that the output of the 555 is on is dependant upon the values of $R_1$ and $R_2$. The time that the output of the 555 is off is dependant upon the value of $R_2$ only.

A 555 timer used as an oscillator can produce various sounds when connected to an amplifier stage. The timer could then be used to provide an audible indication of continuity or some other application requiring an audible alert.

## On-Delay Timer

Figure 21–95 shows a 555 timer connected to provide an on-delay timer function. A 555 timer used in this fashion can be used to energize a relay after a preset time delay. This is the basis of most electronic time-delay relays. Let's see how this works.

First, notice that a capacitor, $C_2$, has been added from pin #5 of the 555 to ground. This capacitor provides electrical noise immunity to the 555. Notice also that the 555 is configured for astable operation. However, notice that pin #4 is connected to the collector of $Q_2$. This will be used as to reset the 555. Diode $D_1$ is called a *free-wheeling diode*. It is used to dissipate the voltage spike produced by the collapsing magnetic field in the coil of relay $K_1$ when the relay de-energizes. We will assume that before power is first applied to the circuit, all capacitors are in a discharged state.

When power is applied, capacitors $C_1$ and $C_3$ begin to charge. The charge time for capacitor $C_3$ is dependant on the RC time constant of $C_3$ and $R_4$. This time constant provides a short time delay before $C_3$ becomes charged. During this time delay, the output of the 555 is in the on state. Since the output of the 555 is on, transistor $Q_1$ will be turned on as well. As a result of $Q_1$ being on, $Q_2$ will be off. Because $Q_2$ is off, the relay, $K_1$, will be de-energized.

The charge time of capacitor $C_1$ is determined by the values of $C_1$, $R_1$, and $R_2$. When $C_1$ charges to approximately two-thirds of the supply voltage, the output of the 555 will change from on to off. When the output of the 555 changes to the off state, transistor $Q_1$ will turn off as well. Because transistor $Q_1$ has turned off, transistor $Q_2$ will turn on. This allows current to flow through the coil of relay $K_1$, energizing the relay. Notice that relay $K_1$ energized after a time delay from when power was applied to the circuit.

By replacing resistor $R_1$ with a variable resistor, we can adjust the time for $C_1$ to charge. By varying $R_1$, we

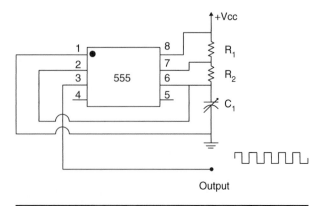

**FIGURE 21–94**   *555 timer connected as an oscillator.*

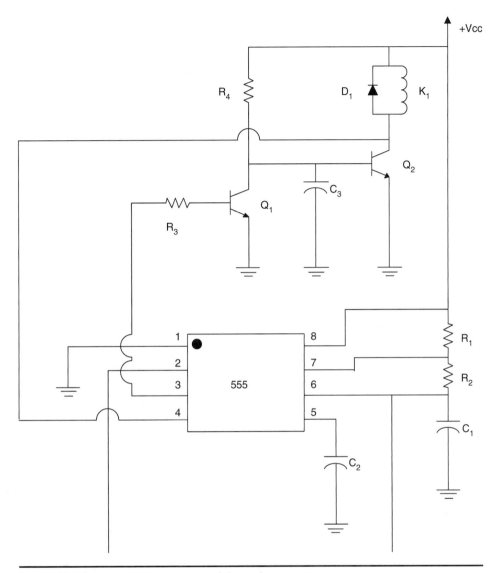

**FIGURE 21–95**  *555 timer time delay relay circuit.*

could vary the delay before relay $K_1$ energizes. Thus, we would have a variable on-delay, time-delay relay.

As mentioned earlier, notice that the collector of $Q_2$ is connected to pin #4 of the 555. This connection keeps the 555 from operating when the voltage at pin #4 is less than two-thirds of the supply voltage. This occurs when transistor $Q_2$ is turned on. The result of this is that power must be removed from the circuit to reset the timer and allow the delay function to be available again.

# OPERATIONAL AMPLIFIERS

Another popular type of integrated circuit is the **operational amplifier**, or **op-amp**. The schematic symbol of an op-amp is shown in Figure 21–96. An op-amp is also a very useful and widely used device. Op-amps can perform many different

functions depending on the manner in which they are connected in a circuit. Perhaps the most common op-amp is the 741 op-amp. The 741 is available in both the 8-pin DIP package as well as the TO-5 package. Since the 7-pin DIP package is most common, we will focus our attention on it. We will begin our study of op-amps by learning about the function of the various pins.

## Pin Assignments

Pin #1    **Offset Null**. This pin is used in conjunction with pin #5 and a potentiometer to eliminate an offset output voltage condition.

Pin #2    **Inverting Input**. A signal applied to the inverting input of the op-amp will appear inverted at the output of the op-amp.

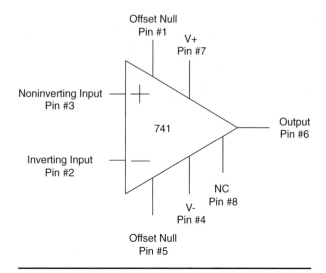

**FIGURE 21–96**   741 op-amp pin-outs.

Pin #3   **Noninverting Input**. A signal applied to the noninverting input of the op-amp will not be inverted at the output of the op-amp.

Pin #4   **V−**. This pin may be connected in one of two methods. One method is to connect a negative supply voltage to this pin. The other method is to connect a circuit ground to this pin.

Pin #5   **Offset Null**. This pin is used in conjunction with pin #1 and a potentiometer to eliminate an offset output voltage condition.

Pin #6   **Output**. The output signal appears at this pin.

Pin #7   **V+**. This pin may be connected in one of two methods. One method is to connect a positive supply voltage to this pin. The other method is to connect a circuit ground to this pin.

Pin #8   **NC**. There is no connection to pin #8. This is a dummy or dead pin and is not used.

## Noninverting Amplifiers

Figure 21–97 shows an op-amp circuit that is configured so that the op-amp acts as a **noninverting amplifier**. The input signal is applied to the non-inverting input, pin #3, of the op-amp. The output signal, at pin #6, will be of the same polarity as the input signal.

Resistors $R_1$ and $R_2$ determine the voltage gain of the circuit. The formula to determine the voltage gain of a noninverting op-amp is:

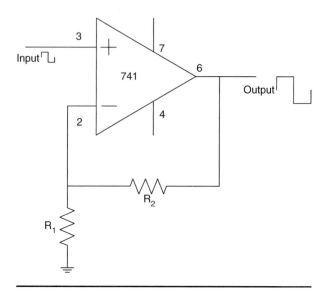

**FIGURE 21–97**   Noninverting amplifier.

$$A_v = 1 + \frac{R_2}{R_1}$$

where:   $A_v$ = voltage gain
   $R_1$ = the resistance from the inverting input, pin #2, to ground, measured in ohms
   $R_2$ = the feedback resistance from the output, pin #6, to the inverting input, pin #2, measured in ohms

For example, what is the voltage gain of a non-inverting op-amp with $R_1$ = 33 kΩ, and $R_2$ = 1 MΩ? We will use the following formula to determine the voltage gain of this circuit:

$$A_v = 1 + \frac{R_2}{R_1}$$

Here is what we know:

$$R_1 = 33 \text{ k}\Omega$$
$$R_2 = 1 \text{ M}\Omega$$

Here is what we don't know:

$$A_v = ?$$

Now, plug the known values into the formula and solve for $A_v$:

$$A_v = 1 + \frac{R_2}{R_1}$$
$$A_v = 1 + \frac{1 \text{ M}\Omega}{33 \text{ k}\Omega}$$
$$A_v = 30.3$$

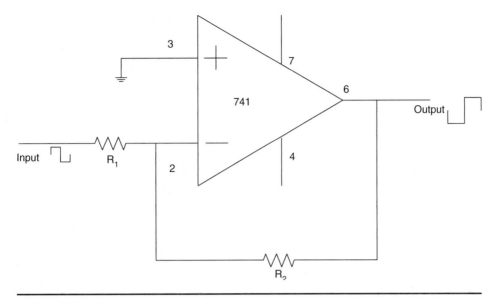

**FIGURE 21–98**   Inverting amplifier.

Therefore, our noninverting op-amp circuit has a voltage gain of 30.3. This means that a 1-volt signal applied to the noninverting input, pin #3, of the op-amp produces a 30.3-volt signal at the output, pin #6, of the op-amp.

## Inverting Amplifiers

Figure 21–98 shows an op-amp circuit that is configured so that the op-amp acts as an **inverting amplifier**. The input signal is applied to the inverting input, pin #2, of the op-amp. The output signal, at pin #6, will be of the polarity opposite to that of the input signal.

Resistors $R_1$ and $R_2$ determine the voltage gain of the circuit. The formula to determine the voltage gain of an inverting op-amp is:

$$A_v = \frac{R_2}{R_1}$$

where:   $A_v$ = voltage gain
$R_1$ = the input resistance to the inverting input, pin #2, measured in ohms
$R_2$ = the feedback resistance from the output, pin #6, to the inverting input, pin #2, measured in ohms

For example, what is the voltage gain of a noninverting op-amp with $R_1$ = 22 kΩ, and $R_2$ = 150 kΩ? We will use the following formula to determine the voltage gain of this circuit:

$$A_v = -\frac{R_2}{R_1}$$

Here is what we know:

$$R_1 = 22 \text{ k}\Omega$$
$$R_2 = 150 \text{ k}\Omega$$

Here is what we don't know:

$$A_v = \text{?}$$

Now, plug the known values into the formula and solve for $A_v$:

$$A_v = -\frac{R_2}{R_1}$$
$$A_v = -\frac{150 \text{ k}\Omega}{22 \text{ k}\Omega}$$
$$A_v = -6.82$$

Therefore, our inverting op-amp circuit has a voltage gain of −6.82. This means that a 1-volt signal applied to the inverting input, pin #2, of the op-amp produces a −6.82-volt signal at the output, pin #6, of the op-amp. Notice that the output voltage is of an opposite polarity from the input voltage. Hence the name, inverting op-amp.

## Buffer Amplifiers

Figure 21–99 shows an op-amp circuit that is configured so that the op-amp acts as a **buffer amplifier**. The input signal is applied to the noninverting input, pin #3, of the op-amp. The output signal, at pin #6, will be of the same polarity and magnitude as the input signal. You might ask, "Why use this circuit if the output is exactly the same as the input?" This circuit is used to provide impedance matching

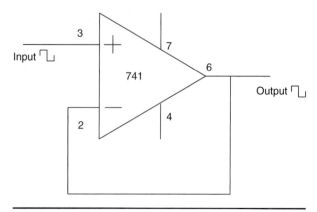

**FIGURE 21–99**  Buffer amplifier.

between the output of a previous circuit and the input to the following circuit. An op-amp buffer amplifier will typically have a high impedance input and a low impedance output. The voltage gain of an op-amp buffer amplifier is 1.

## Difference or Differential Amplifiers

Figure 21–100 shows an op-amp circuit that is configured so that the op-amp acts as a *difference amplifier*. This means that the difference between the two input signals will be amplified and results in the output signal being proportional to the difference between the input signals. One input signal is applied to the noninverting input, pin #3, of the op-amp. Another input signal is applied to the inverting input, pin #2, of the op-amp. The difference between these two input signals is amplified and appears at the output, pin #6, of the op-amp.

Resistor $R_3$ is the feedback resistor, resistors $R_1$ and $R_2$ are the input resistors, and $R_1$ and $R_2$ will be of equal value. The formula to determine the voltage gain of a differential op-amp is:

$$A_v = \frac{R_f}{R_i}(E_1 - E_2)$$

where:  $A_v$ = voltage gain

$R_i$ = the resistance value of one of the input resistors, measured in ohms

$R_f$ = the feedback resistance, measured in ohms

$E_1$ = the voltage applied to the inverting input, pin #2, of the op-amp

$E_2$ = the voltage applied to the noninverting input, pin #3, of the op-amp

For example, what is the voltage gain of a differential op-amp where $R_i$ = 5 kΩ, $R_f$ = 80 kΩ, $E_1$ = 1.6 V, and $E_2$ = 1.2 V? We will use the following formula to determine the voltage gain of this circuit:

$$A_v = \frac{R_f}{R_i}(E_1 - E_2)$$

Here is what we know:

$$R_f = 80 \text{ k}\Omega$$
$$R_i = 5 \text{ k}\Omega$$
$$E_1 = 1.6 \text{ V}$$
$$E_2 = 1.2 \text{ V}$$

Here is what we don't know:

$$A_v = ?$$

Now, plug the known values into the formula and solve for $A_v$:

$$A_v = \frac{R_f}{R_i}(E_1 - E_2)$$

$$A_v = \frac{80 \text{ k}\Omega}{5 \text{ k}\Omega}(1.6 \text{ V} - 1.2 \text{ V})$$

$$A_v = \frac{80 \text{ k}\Omega}{5 \text{ k}\Omega}(0.4 \text{ V})$$

$$A_v = 6.4$$

Therefore, our differential op-amp circuit has a voltage gain of 6.4.

## Summing Amplifiers

Figure 21–101 shows an op-amp circuit that is configured so that the op-amp acts as a *summing amplifier*. This means that the sum of several input signals will be amplified and results in the output signal being the algebraic sum of all input signals. If the input signals are applied to the noninverting input, pin #3, of the op-amp, the output signal will be of the same polarity. If the input signals are applied to the inverting input, pin #2, of the op-amp, the output signal will be of the opposite polarity.

## Comparators

Figure 21–102 shows an op-amp circuit that is configured as a *comparator*, or *level detector*. In this figure, the inverting input of the op-amp is connected to a fixed-reference voltage. The noninverting input of the op-amp is connected to a varying signal. If the varying signal on the noninverting input is less than the reference voltage applied to the inverting input of the op-amp, the output of the op-amp will

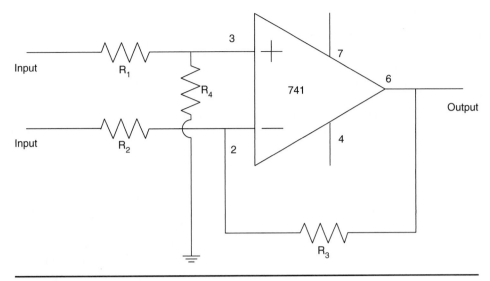

**FIGURE 21-100** Difference amplifier.

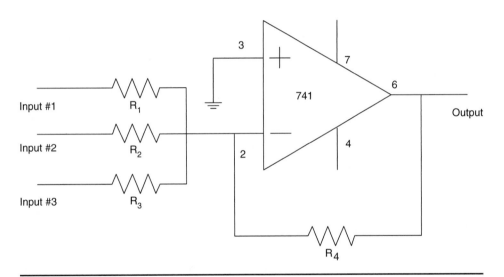

**FIGURE 21-101** Summing amplifier.

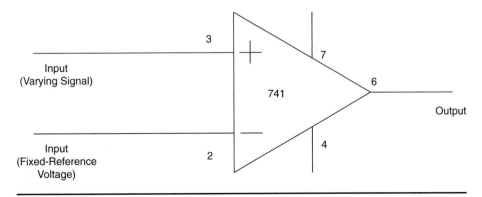

**FIGURE 21-102** Comparator with fixed-reference voltage.

be positive. When the varying input signal rises to a level above the reference voltage that is applied to the inverting input of the op-amp, the output of the op-amp will be driven negative. In this fashion, a comparator can provide an indication when a varying input signal exceeds a reference voltage level.

Figure 21–103 shows a variation of the comparator shown in Figure 21–102. The circuit in Figure 21–103 shows a comparator with a variable resistor connected to the inverting input of the op-amp. This allows the reference voltage to be varied. The varying input signal will then be compared to whatever voltage is set by the potentiometer.

## Integrators

An op-amp *integrator* is shown in Figure 21–104. Notice the resistor that is connected to the inverting input of the op-amp. Also, notice the capacitor that is connected between the inverting input and the output of the op-amp. An integrator will produce an output that is proportional to the *integral* of the input. This means that the output voltage will be proportional to the area of the input waveform.

Refer to Figure 21–105. If we were to apply a square-wave signal to the inverting input of the integrator, the output will be a triangular wave. *An integrator is a square-wave-to-triangular-wave converter.* A triangular-output wave is produced because of the charge/discharge time of the capacitor. Now, consider the area under the positive portion of the square wave. As this area varies, the output will vary proportionally. For example, if this area becomes smaller (smaller integral), the output must change faster. Should the area become larger (larger integral), the output must change more slowly. Figure 21–106A shows the relationship between the square-wave input signal and the triangular-wave output signal. Figure 21–106B shows the relationship with a smaller integral, and Figure 21–106C shows the relationship with a larger integral.

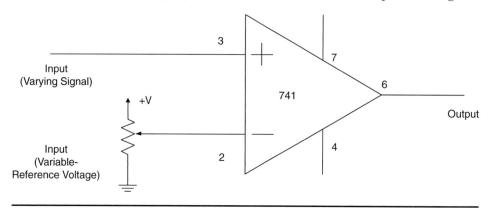

**FIGURE 21–103** Comparator with variable-reference voltage.

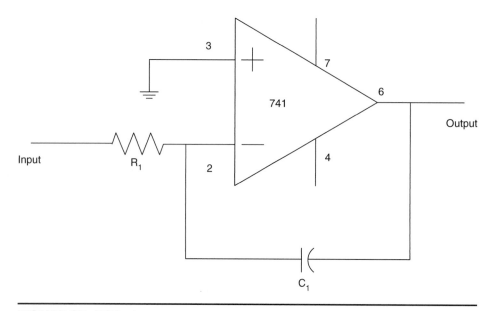

**FIGURE 21–104** Integrator.

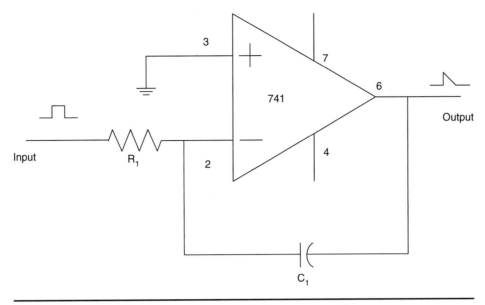

**FIGURE 21-105** Square-wave input produces a triangular-wave output.

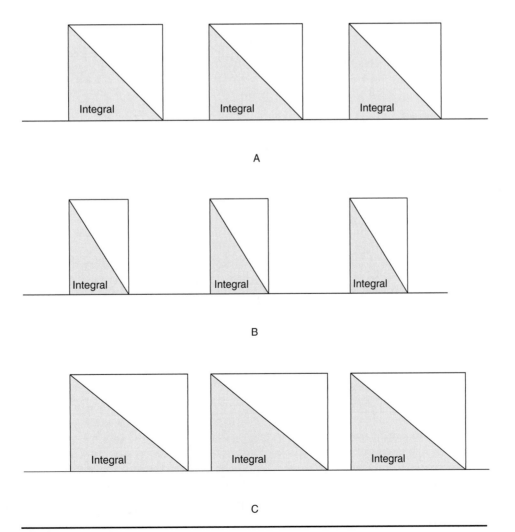

**FIGURE 21-106** The integral varies as the width of the square-wave varies.

## Differentiators

Figure 21–107 shows a *differentiator* op-amp circuit. Compare the placement of the resistor and capacitor in this circuit with those in Figure 21–104. A differentiator will produce an output that is proportional to the rate of change of the input signal. This means that an output will be present only while the input signal is changing.

Refer to Figure 21–108. If we were to apply a triangular-wave signal to the inverting input of the differentiator, the output will be a square wave. *A differentiator is a triangle-wave-to-square-wave converter.* A square-wave output is produced because of the switching action of the op-amp. Figure 21–109A shows the relationship between the triangular-wave input signal and the square-wave output signal. Notice the rate of change in the triangular-input signal between time $T_0$ and time $T_1$. This rate of change is constant and rising in the positive direction. During this time, the output of the differentiator is negative (because the input signal is applied to the inverting input of the op-amp). When the triangular-input signal begins to fall in the negative direction, the output of the op-amp switches from negative to positive. Since the rate of change between time $T_1$ and $T_2$ is constant and falling in the negative direction, the output of the differentiator is positive (again because the input signal is applied to the inverting input

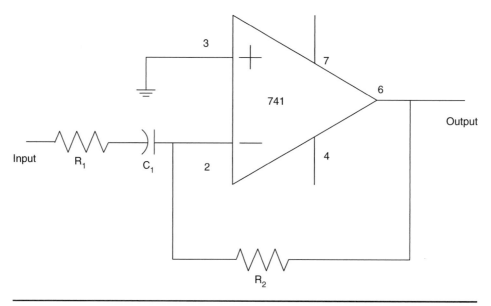

**FIGURE 21–107**   Differentiator.

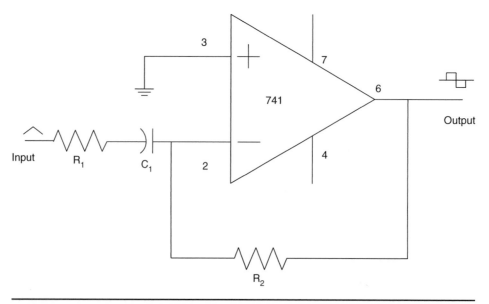

**FIGURE 21–108**   Triangular-wave input produces a square-wave output.

of the op-amp). Notice that the width of the square-wave output pulses is proportional to the rise and fall times of the triangular-input signal. If the rise and fall times of the triangular-input signal were more rapid, the output pulses would be narrower. If the rise and fall times of the triangular-input signal were longer, the output pulses would be wider. This can be seen in Figures 21–109B and 21–109C.

## DIGITAL LOGIC

A *digital device* is a device that has only two conditions: off and on. A light switch on the wall is an example of a digital device. Either it is off, or it is on. Increasingly, more electronic devices are using digital components to perform *logic functions*. By logic functions, we

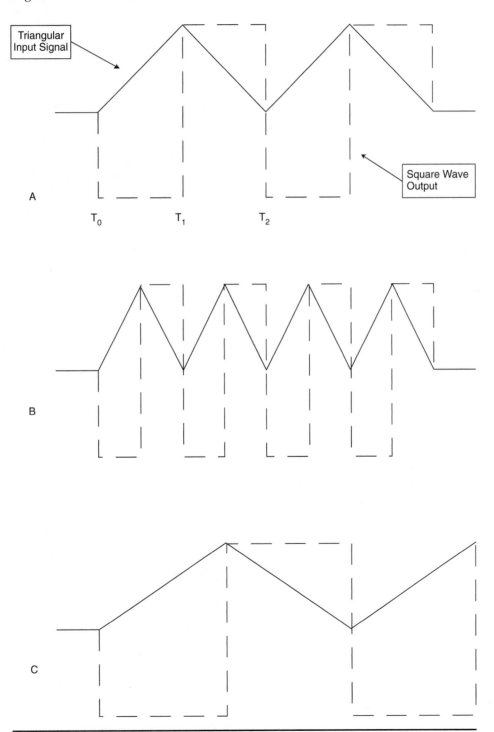

**FIGURE 21-109** The width of the output square wave varies as the rate of change of the triangular-input wave varies.

mean circuits that make decisions based on certain input conditions. For example, if an input voltage is present, a relay may be energized. The condition is the presence or absence of a voltage. The logical decision is whether or not to energize the relay.

As mentioned earlier, digital devices operate in one of two states, off or on. We use a 0 (zero) to represent the off condition. We use a 1 (one) to represent an on condition. Typically, an off condition is indicated by 0 volt, while the on condition is represented by +5 volts. There are different families of digital devices that operate at different logic voltage levels. For the purposes of this topic, we will use 0 and 1 to represent off and on, respectively. This means that the logic will be consistent regardless the family of devices that is used.

We will now study a family of integrated circuits known as *digital logic gates*. These gates make logical decisions based upon certain input conditions. However, since these are digital devices, the inputs can only be one of two states, off (0) or on (1). Likewise, the output can only be one of two states, off (0) or on (1). We will also learn how to use a truth table to indicate all possible input and output conditions.

## The Inverter Gate

The simplest logic gate is the **inverter**. The inverter is also called a NOT gate. An inverter gate will invert the input signal. We sometimes refer to the output of an inverter as the *compliment* of the input. This means that if the input signal is 0, the output will be 1. Likewise, if the input signal is 1, the output will be 0. The output is inverted when compared to the input, or we can say that the output is the compliment of

the input. An easy way to remember the function of an inverter is to say that you will have an output when you do *not* have an input.

An inverter has only one input terminal and one output terminal. Figure 21–110A shows the schematic symbol of an inverter. (Figure 21–110B shows the IEC symbol for an inverter.) The schematic symbol only shows the symbol and input and output terminals of the inverter. It does not show the supply voltage terminal or the supply ground terminal for the integrated circuit. These are found by identifying the device and obtaining the specifications and terminal connections, or pin-outs, for the particular IC that is being used.

Figure 21–111 shows a truth table for an inverter. A truth table is a handy tool when evaluating a logic gate's response to different input conditions. A truth table will list all possible input conditions and then show all possible output conditions for those inputs. We will use the letter *A* to represent the input terminal and the letter *Y* to represent the output terminal. Since the inverter has only one input terminal, there can be only two possible input conditions, 0 or 1. Since the inverter has only one output terminal, there can be only two possible output conditions, 0 or 1. Notice that when the input is a 0, the output is a 1. Likewise, when the input is a 1, the output is a 0. Now, you can see how a truth table can be used to show all possible output conditions based on all possible input conditions.

We can relate the inverter to the operation of a switch in a control circuit. Since the inverter provides an output when there is no input, the inverter functions similar to a normally closed switch. Figure 21–112 shows a ladder diagram with a

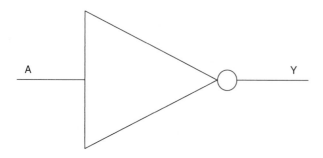

**FIGURE 21–110A**  An inverter.

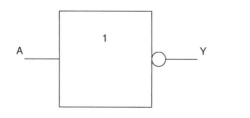

**FIGURE 21–110B**  IEC symbol of an inverter.

| Inputs | Output |
|:------:|:------:|
| **A** | **Y** |
| 0 | 1 |
| 1 | 0 |

**FIGURE 21–111**  Inverter truth table.

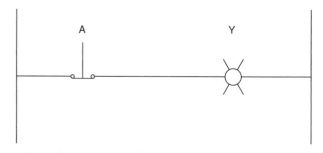

**FIGURE 21–112**  Ladder diagram circuit performing an inverter function.

normally closed push button switch and a pilot light. Notice that if the push button switch is *not* pressed, the pilot light is on. If the push button switch is pressed, the pilot light is off. Figure 21–113 shows a truth table for the operation of this circuit. Notice that the truth table for this circuit is identical to the truth table for an inverter. This means that an inverter performs the same logical operation as a normally closed switch.

## The AND Gate

The *AND* gate has a minimum of two input terminals but will always have one output terminal. AND gates are available with more than two inputs—some have three inputs, four, or more. We will study a two-input AND gate and a three-input AND gate. Figure 21–114A shows the schematic symbol of a two-input AND gate on the left, while the schematic symbol of a three-input AND gate is shown on the right (Figure 21–114B shows the IEC symbol for a two-input AND gate). Once you have studied these, you should be able to determine the truth table for an AND gate with additional inputs.

The function of an AND gate is to produce an output (1) only when all inputs are 1s. This means that should one or more of the inputs be

a 0, the output will be a 0. An easy way to remember the function of an AND gate is to say you will have an output when input A *AND* input B are 1s. Figure 21–115 shows the truth table of a two-input AND gate. Notice that the only time there is a 1 at the output is when input A *AND* input B are 1s. Figure 21–116 shows the truth table for a three-input AND gate. Again, notice the only time there is a 1 at the output is when inputs A *AND* B *AND* C are 1s.

We can relate the operation of a two-input AND gate to the operation of two switches in a control circuit. Since the AND gate provides an output only when both inputs are 1s, the two-input AND gate functions similarly to two normally open switches connected in series. Figure 21–117 shows a ladder diagram with two normally open push button switches, connected in series, and a pilot light. Notice that if

| Inputs | | Output |
|---|---|---|
| **A** | **B** | **Y** |
| 0 | 0 | 0 |
| 0 | 1 | 0 |
| 1 | 0 | 0 |
| 1 | 1 | 1 |

**FIGURE 21-115** Two-input AND gate truth table.

| Inputs | | | Output |
|---|---|---|---|
| **A** | **B** | **C** | **Y** |
| 0 | 0 | 0 | 0 |
| 0 | 0 | 1 | 0 |
| 0 | 1 | 0 | 0 |
| 0 | 1 | 1 | 0 |
| 1 | 0 | 0 | 0 |
| 1 | 0 | 1 | 0 |
| 1 | 1 | 0 | 0 |
| 1 | 1 | 1 | 1 |

**FIGURE 21-116** Three-input AND gate truth table.

| Inputs | Output |
|---|---|
| **A** | **Y** |
| 0 (Not Depressed) | 1 (On) |
| 1 (Depressed) | 0 (Off) |

**FIGURE 21-113** Ladder diagram truth table for the inverter function.

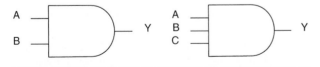

**FIGURE 21-114A** AND gates: two-input AND gate on the left, and three-input AND gate on the right.

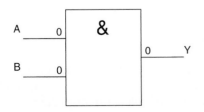

**FIGURE 21-114B** IEC symbol of an AND gate.

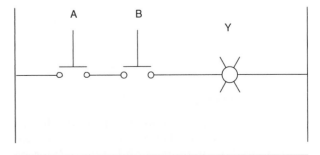

**FIGURE 21-117** Ladder diagram circuit performing a two-input AND function.

| Inputs | | Output |
|---|---|---|
| **A** | **B** | **Y** |
| 0 (Not Depressed) | 0 (Not Depressed) | 0 (Off) |
| 0 (Not Depressed) | 1 (Depressed) | 0 (Off) |
| 1 (Depressed) | 0 (Not Depressed) | 0 (Off) |
| 1 (Depressed) | 1 (Depressed) | 1 (On) |

**FIGURE 21–118** Ladder diagram truth table for the two-input AND function.

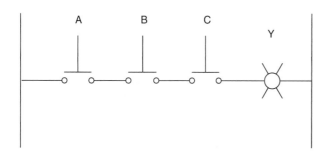

**FIGURE 21–119** Ladder diagram circuit performing a three-input AND function.

none of the push buttons is pressed, the pilot light is off. If only switch A is pressed, the pilot light is off. Likewise, if only switch B is pressed, the pilot light is off. However, if both switch A *AND* switch B are pressed, the pilot light is on. Figure 21–118 shows a truth table for the operation of this circuit. Notice that the truth table for this circuit is identical to the truth table for a two-input AND gate. This means that a two-input AND gate performs the same logical operation as two normally open switches wired in series.

Figure 21–119 shows a ladder diagram with three normally open push button switches, connected in series, and a pilot light. Notice that if none of the push buttons is pressed, the pilot light is off. If only switch A is pressed, the pilot light is off. Likewise, if only switch B is pressed, the pilot light is off. If only switch C is pressed, the pilot light will be off. However, if switch A *AND* switch B *AND* switch C are pressed, the pilot light is on. Figure 21–120 shows a truth table for the operation of this circuit. The truth table for this circuit is identical to the truth table for a three-input AND gate. This means that a three-input AND gate performs the same logical operation as three normally open switches wired in series.

## The NAND Gate

Logic gates can be combined to provide different logical operations. For example, if an inverter is added to the output of an AND gate, a different set of logical conditions will be created. This new

circuit is called a *NAND* gate. The word *NAND* is a combination of NOT and AND. This means that this gate functions as an AND gate with an inverted output. The left side of Figure 21–121A shows a two-input AND gate with an inverter added to the output, and the right side of Figure 21–121A shows the more common schematic for a two-input NAND gate. (Figure 21–121B shows the IEC symbol for a two-input NAND gate.) Notice the inverter is now represented by the small circle at the output of the AND gate. The small circle will be used in future schematics to represent the inverter function when added to another gate.

Figure 21–122 shows the truth table for a two-input NAND gate. Notice that the only time there is not an output is when both inputs are 1s. Compare this truth table to that of a two-input AND gate in Figure 21–115. The truth table for a two-input NAND gate is the exact opposite of a two-input AND gate. This is because a NAND gate is an AND gate with an inverted output. While our discussion has focused on a two-input NAND gate, you should realize that NAND gates may have three, four, or more inputs, just like the AND gate.

Figure 21–123 shows a ladder diagram with two normally open push button switches, a relay, and a pilot light. The push buttons are both normally open switches and are wired in series with the relay coil. Notice that if none of the push buttons is pressed, the pilot light is on. If only switch A is pressed, the pilot light is on. Likewise, if only switch B is pressed, the pilot light is on. However, when both switch A *AND* switch B are pressed, the pilot light is off. Figure 21–124 shows a truth table for the operation of this circuit. Notice that the truth table for this circuit is identical to the truth table for a two-input NAND gate.

## The OR Gate

An *OR* gate will have a minimum of two input terminals but will always have one output terminal. OR gates are available with more than two inputs— some have three inputs, four, or more. We will

| Inputs | | | Output |
|---|---|---|---|
| **A** | **B** | **C** | **Y** |
| 0 (Not Depressed) | 0 (Not Depressed) | 0 (Not Depressed) | 0 (Off) |
| 0 (Not Depressed) | 0 (Not Depressed) | 1 (Depressed) | 0 (Off) |
| 0 (Not Depressed) | 1 (Depressed) | 0 (Not Depressed) | 0 (Off) |
| 0 (Not Depressed) | 1 (Depressed) | 1 (Depressed) | 0 (Off) |
| 1 (Depressed) | 0 (Not Depressed) | 0 (Not Depressed) | 0 (Off) |
| 1 (Depressed) | 0 (Not Depressed) | 1 (Depressed) | 0 (Off) |
| 1 (Depressed) | 1 (Depressed) | 0 (Not Depressed) | 0 (Off) |
| 1 (Depressed) | 1 (Depressed) | 1 (Depressed) | 1 (On) |

**FIGURE 21-120** Ladder diagram truth table for the three-input AND function.

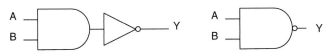

**FIGURE 21-121A** NAND gate: AND gate with inverter, AND gate with inverted output.

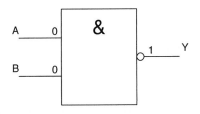

**FIGURE 21-121B** IEC symbol of a NAND gate.

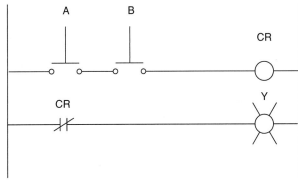

**FIGURE 21-123** Ladder diagram circuit performing a two-input NAND function.

| Inputs | | Output |
|---|---|---|
| **A** | **B** | **Y** |
| 0 | 0 | 1 |
| 0 | 1 | 1 |
| 1 | 0 | 1 |
| 1 | 1 | 0 |

**FIGURE 21-122** Two-input NAND gate truth table.

study two-input and three-input OR gates. Figure 21–125A shows the schematic symbol of a two-input OR gate on the left and the schematic symbol of a three-input OR gate on the right. (Figure 21–125B shows the IEC symbol for a two-input OR gate.) Once you have studied these, you should be able to determine the truth table for an OR gate with additional inputs.

The function of an OR gate is to produce an output (1) when any input is a 1. This means that when both inputs are a 0, the output will be a 0. An easy way to remember the function of an OR gate is to say you will have an output when input A *OR*

input B is a 1. Figure 21–126 shows the truth table of a two-input OR gate. Notice that the only time there is a 1 at the output is when input A *OR* input B is a 1. Figure 21–127 shows the truth table for a three-input OR gate. Again, notice the only time there is a 1 at the output is when input A *OR* B *OR* C is a 1.

We can relate the operation of a two-input OR gate to the operation of two switches in a control circuit. Since the OR gate provides an output only when either input is a 1, the two-input OR gate functions similarly to two normally open switches connected in parallel. Figure 21–128 shows a ladder diagram with two normally open push button switches, connected in parallel, and a pilot light. Notice that if none of the push buttons is pressed, the pilot light is off. If only switch A is pressed, the pilot light is on. Likewise, if only switch B is pressed, the pilot light is on. If both switch A and switch B are pressed, the pilot light is on. Figure 21–129 shows a truth table for the operation of this circuit. Notice that the truth table for this circuit is identical to the truth table for a two-input OR gate. This means that a two-input OR gate performs the same logical operation as two normally open switches wired in parallel.

| Inputs | | Output |
|---|---|---|
| **A** | **B** | **Y** |
| 0  (Not Depressed) | 0  (Not Depressed) | 1  (On) |
| 0  (Not Depressed) | 1  (Depressed) | 1  (On) |
| 1  (Depressed) | 0  (Not Depressed) | 1  (On) |
| 1  (Depressed) | 1  (Depressed) | 0  (Off) |

**FIGURE 21–124**  Ladder diagram truth table for the two-input NAND function.

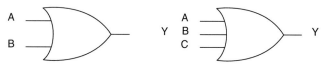

**FIGURE 21–125A**  OR gates: two-input OR gate on the left, and three-input OR gate on the right.

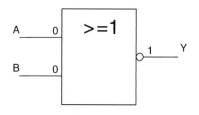

**FIGURE 21–125B**  IEC symbol of an OR gate.

| Inputs | | Output |
|---|---|---|
| **A** | **B** | **Y** |
| 0 | 0 | 0 |
| 0 | 1 | 1 |
| 1 | 0 | 1 |
| 1 | 1 | 1 |

**FIGURE 21–126**  Two-input OR gate truth table.

| Inputs | | | Output |
|---|---|---|---|
| **A** | **B** | **C** | **Y** |
| 0 | 0 | 0 | 0 |
| 0 | 0 | 1 | 1 |
| 0 | 1 | 0 | 1 |
| 0 | 1 | 1 | 1 |
| 1 | 0 | 0 | 1 |
| 1 | 0 | 1 | 1 |
| 1 | 1 | 0 | 1 |
| 1 | 1 | 1 | 1 |

**FIGURE 21–127**  Three-input OR gate truth table.

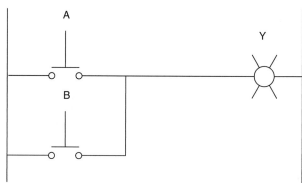

**FIGURE 21–128**  Ladder diagram circuit performing a two-input OR function.

Figure 21–130 shows a ladder diagram with three normally open push button switches, connected in parallel, and a pilot light. If none of the push buttons is pressed, the pilot light is off. If only switch A is pressed, the pilot light is on. If only switch B is pressed, the pilot light is on. If only switch C is pressed, the pilot light will be on. Pressing switches A and B will turn the pilot light on. Likewise, pressing switches B and C will turn the pilot light on, and pressing switches A and C will turn the pilot light on. Finally, if all three switches, A, B, and C, are pressed, the pilot light will be on. Figure 21–131 shows a truth table for the operation of this circuit. Notice that the truth table for this circuit is identical to the truth table for a three-input OR gate. This means that a three-input OR gate performs the same logical operation as three normally open switches wired in parallel.

## The NOR Gate

If an inverter is added to the output of an OR gate, a different set of logical conditions will be created. This new circuit is called a *NOR* gate. The word *NOR* is a combination of NOT and OR. This means that this gate functions as an OR gate with an inverted output. Figure 21–132A shows a two-input NOR gate.

| Inputs | | | | Output | |
|---|---|---|---|---|---|
| **A** | | **B** | | **Y** | |
| 0 | (Not Depressed) | 0 | (Not Depressed) | 0 | (Off) |
| 0 | (Not Depressed) | 1 | (Depressed) | 1 | (On) |
| 1 | (Depressed) | 0 | (Not Depressed) | 1 | (On) |
| 1 | (Depressed) | 1 | (Depressed) | 1 | (On) |

**FIGURE 21-129** Ladder diagram truth table for the two-input OR function.

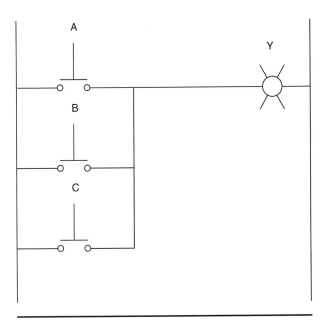

**FIGURE 21-130** Ladder diagram circuit performing a three-input OR function.

(Figure 21–132B shows the IEC symbol for a two-input NOR gate.) Recall that the inverter is represented by the small circle at the output of the OR gate.

Figure 21–133 shows the truth table for a two-input NOR gate. Notice that the only time there is an output is when both inputs are 0s. Compare this truth table to that of a two-input OR gate in Figure 21–126. The truth table for a two-input NOR gate is the exact opposite of a two-input OR gate. This is because a NOR gate is an OR gate with an inverted output. While our discussion has focused on a two-input NOR gate, you should realize that NOR gates may have three, four, or more inputs, just like the OR gate.

Figure 21–134 shows a ladder diagram with two normally open push button switches, a relay, and a pilot light. The push buttons are both normally open switches and are wired in parallel with the relay coil. Notice that if none of the push buttons is pressed, the pilot light is on. If only switch A is pressed, the pilot light is off. Likewise, if only switch B is pressed, the pilot light is off. When both

switch A *and* switch B are pressed, the pilot light is off. Figure 21–135 shows a truth table for the operation of this circuit. Notice that the truth table for this circuit is identical to the truth table for a two-input NOR gate.

## The Exclusive OR Gate

Figure 21–136A shows the schematic symbol of our final logic gate, the *Exclusive OR gate*. (Figure 21–136B shows the IEC symbol for an Exclusive OR gate.) The Exclusive OR gate is also called an *XOR gate*. The XOR gate will produce an output when either input is a 1, but not when both inputs are 0s or 1s. The truth table for an XOR gate is shown in Figure 21–137.

We can relate the operation of an XOR gate to the operation of two switches in a control circuit. Since the XOR gate provides an output only when either input is a 1, the XOR gate functions similarly to two switches, each with a normally open and a normally closed contact, connected as shown in Figure 21–138. Notice that if none of the push buttons is pressed, the pilot light is off. If only switch A is pressed, the pilot light is on. Likewise, if only switch B is pressed, the pilot light is on. If both switch A and switch B are pressed, the pilot light is off. Figure 21–139 shows a truth table for the operation of this circuit. Notice that the truth table for this circuit is identical to the truth table for an XOR gate, as shown in Figure 21–137.

## REPLACING SOLID-STATE DEVICES

When replacing a defective solid-state device, you should identify the device by the number that is imprinted on the device body. You can then contact the manufacturer for a replacement. There are also several companies that manufacture replacement or substitute devices that meet or exceed the original

| Inputs | | | Output |
|---|---|---|---|
| **A** | **B** | **C** | **Y** |
| 0 (Not Depressed) | 0 (Not Depressed) | 0 (Not Depressed) | 0 (Off) |
| 0 (Not Depressed) | 0 (Not Depressed) | 1 (Depressed) | 1 (On) |
| 0 (Not Depressed) | 1 (Depressed) | 0 (Not Depressed) | 1 (On) |
| 0 (Not Depressed) | 1 (Depressed) | 1 (Depressed) | 1 (On) |
| 1 (Depressed) | 0 (Not Depressed) | 0 (Not Depressed) | 1 (On) |
| 1 (Depressed) | 0 (Not Depressed) | 1 (Depressed) | 1 (On) |
| 1 (Depressed) | 1 (Depressed) | 0 (Not Depressed) | 1 (On) |
| 1 (Depressed) | 1 (Depressed) | 1 (Depressed) | 1 (On) |

**FIGURE 21-131**   Ladder diagram truth table for the three-input OR function.

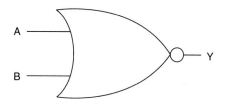

**FIGURE 21-132A**   Two-input NOR gate.

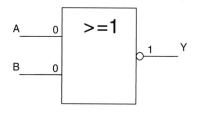

**FIGURE 21-132B**   IEC symbol of a NOR gate.

| Inputs | | Output |
|---|---|---|
| **A** | **B** | **Y** |
| 0 | 0 | 1 |
| 0 | 1 | 0 |
| 1 | 0 | 0 |
| 1 | 1 | 0 |

**FIGURE 21-133**   Two-input NOR gate truth table.

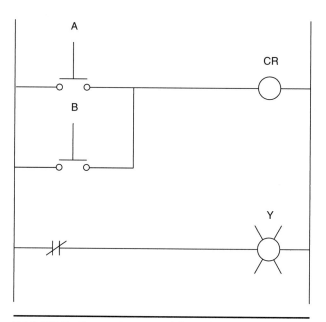

**FIGURE 21-134**   Ladder diagram circuit performing a two-input NOR gate function.

| Inputs | | Output |
|---|---|---|
| **A** | **B** | **Y** |
| 0 (Not Depressed) | 0 (Not Depressed) | 1 (On) |
| 0 (Not Depressed) | 1 (Depressed) | 0 (Off) |
| 1 (Depressed) | 0 (Not Depressed) | 0 (Off) |
| 1 (Depressed) | 1 (Depressed) | 0 (Off) |

**FIGURE 21-135**   Ladder diagram truth table for the two-input NOR gate function.

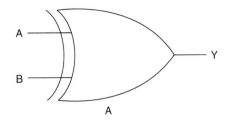

**FIGURE 21-136A**  Exclusive OR (XOR) gate.

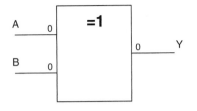

**FIGURE 21-136B**  IEC symbol of an XOR gate.

| Inputs | | Output |
|---|---|---|
| A | B | Y |
| 0 | 0 | 0 |
| 0 | 1 | 1 |
| 1 | 0 | 1 |
| 1 | 1 | 0 |

**FIGURE 21-137**  XOR gate truth table.

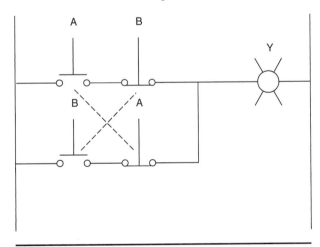

**FIGURE 21-138**  Ladder diagram circuit performing an XOR gate function.

| Inputs | | Output |
|---|---|---|
| A | B | Y |
| 0 (Not Depressed) | 0 (Not Depressed) | 0 (Off) |
| 0 (Not Depressed) | 1 (Depressed) | 1 (On) |
| 1 (Depressed) | 0 (Not Depressed) | 1 (On) |
| 1 (Depressed) | 1 (Depressed) | 0 (Off) |

**FIGURE 21-139**  Ladder diagram truth table for the XOR gate function.

manufacturer's specifications. These companies typically publish a cross-reference guide where you simply look up the original manufacturer's part number in a listing of all replacements available. Once you have found your component, you will find the part number of the substitute part. Usually, you will also find the specifications for the substitute part. This helps you ensure that the substitute will work as a replacement for the defective manufacturer's device.

Some devices are printed with the manufacturer's own unique number. These numbers cannot be cross-referenced, which makes it impossible to find a suitable substitute part. In this situation, you must contact the manufacturer for a replacement. Quite often, the manufacturer will not sell just the component, and your only option is to replace the entire assembly.

## REVIEW QUESTIONS

*Multiple Choice*

1. A semiconductor is
   a. a good conductor.
   b. a poor conductor.
   c. neither a good conductor nor a good insulator.
   d. a type of resistor.

2. Two of the most common materials used to make semiconductors are
   a. copper and aluminum.
   b. copper and magnesium.
   c. lead and silicon.
   d. germanium and silicon.

3. Covalent bonding is
   a. electrons moving from atom to atom.
   b. valance electrons sharing outer shells with adjacent atoms.
   c. the electrons in the atom that are bonded together.
   d. electrons bonded to protons.

4. The mixing of impurities (arsenic, indium, or gallium) with silicon or germanium is a chemical process called
   a. doping.
   b. bonding.
   c. building.
   d. purifying.

5. A semiconductor material that has an excess of electrons is called
   a. P material.
   b. N material.
   c. V material.
   d. E material.

6. Diffusion occurs when P and N materials are
   a. separated.
   b. joined together.
   c. mixed.
   d. bonded.

7. A PN device is known as a
   a. diode.
   b. triode.
   c. biode.
   d. pentode.

8. A diode
   a. causes current flow.
   b. blocks current flow.
   c. increases current flow.
   d. blocks current flow in one direction.

9. Diodes conduct when they are connected in
   a. reverse bias.
   b. forward bias.
   c. neutral bias.
   d. none of the above.

10. A rectifier is a device that changes
    a. AC to DC.
    b. DC to AC.
    c. the value of current.
    d. the value of voltage.

11. A half-wave rectifier produces a
    a. smooth DC.
    b. pulsating DC.
    c. square-wave AC.
    d. sawtooth-wave DC.

12. A three-element transistor has an/a:
    a. emitter, collector, and base.
    b. emitter, collector, and grid.
    c. collector, grid, and base.
    d. emitter, collector, and resistor.

13. What two functions can a transistor perform?
    a. Amplification and blocking.
    b. Amplification and switching.
    c. Blocking and switching.
    d. Isolation and switching.

14. A type of transistor amplifier with a gain in excess of several thousand is called a:
    a. Dakota amplifier.
    b. Denmark amplifier.
    c. Darlington amplifier.
    d. Davenport amplifier.

15. A solid-state device that provides electrical isolation is called:
    a. an IGBT.
    b. an opto-coupler.
    c. a JFET.
    d. a diac.

16. The emitter-base junction of a good NPN transistor will measure:
    a. low resistance in both directions.
    b. low resistance with the negative ohmmeter lead connected to the emitter and the positive ohmmeter lead connected to the base.
    c. high resistance in both directions.
    d. high resistance with the negative ohmmeter lead connected to the emitter and the positive ohmmeter lead connected to the base.

17. You can think of a diac as:
    a. a current-sensitive AC switch.
    b. a voltage-sensitive AC switch.
    c. a current-sensitive DC switch.
    d. a voltage-sensitive DC switch.

18. Triacs conduct in
    a. one direction only.
    b. both directions.
    c. neither direction.
    d. specific periods of time.

19. SCR stands for
    a. solid-state control rectifier.
    b. silicon-controlled rectifier.
    c. solid-state current rectifier.
    d. silicon-controlled relay.

20. The three terminals of an SCR are:
    a. anode, cathode, and gate.
    b. emitter, gate, and collector.
    c. source, gate, and drain.
    d. MT1, MT2, and gate.

21. An SCR will:
    a. only conduct in one direction when properly biased and triggered.
    b. only conduct in one direction at any time.
    c. conduct in both directions when properly biased and triggered.
    d. conduct in both directions at any time.

*Give Complete Answers*

1. What is a transistor?
2. What are semiconductors?
3. Describe the action that occurs when an impurity is added to silicon to produce N material.
4. How is P material developed?
5. What is meant by *covalent bonding*?
6. List three groups of semiconductors.
7. What is a diode?
8. Describe the action that takes place when P and N materials are joined together.
9. Describe a half-wave rectifier.
10. Describe a full-wave rectifier.
11. Describe the difference in operation between a diode and a zener diode.
12. What occurs when an LED is forward biased?
13. Describe a PNP transistor.
14. Describe an NPN transistor.
15. Describe how a BJT and a JFET differ.
16. What is the difference in construction between a JFET and a UJT?
17. Describe how you would connect a diode to a DC source so that the diode is forward biased.
18. What is the name of the solid-state device that provides electrical isolation?
19. Hall-effect devices are used to detect what?
20. What is a thyristor?
21. Describe a diac.
22. Describe a triac.
23. What is the purpose of the SCR?
24. Name the three basic types of circuit configurations of an op-amp.
25. Explain the procedure you would follow to determine if an NPN transistor is defective.
26. List three circuit configurations for an operational amplifier.
27. Which two-input digital logic gate will only produce an output when both inputs are 1s?
28. Which two-input digital logic gate will produce an output when either input is a 1 but not when both inputs are a 1 or a 0?

# DC Electronic Variable Speed Drives

## OBJECTIVES

After studying this chapter, the student will be able to:

■ Name the various sections of a DC drive.

■ State the general operating principle of a DC drive.

■ Describe the operation of a switching amplifier field current controller.

■ Describe the operation of an SCR armature voltage controller.

■ Explain the operation of buck and boost choppers.

■ Identify problems and troubleshoot a DC drive.

Motors have been used in industry for many years. Typically, these motors have been operated at full speed or slightly below their full speeds. When necessary, the speed of the motor may be varied, usually by adding series resistance. Although this practice works well and is easy to accomplish, it is inefficient, wasting a large amount of energy in the form of heat given off by the resistors. Electronic variable speed drives are now used to vary the speed of these motors with greater efficiency and more precise speed control than resistors offer. This chapter will introduce you to DC electronic variable speed drives.

# DC DRIVES

An electronic variable speed drive (see Figure 22–1) has two basic sections: the *control section* and the *power section*. The control section governs or controls the power section while the power section supplies controlled power to the DC motor

The control section allows us to control not only the motor's speed but also its torque. (Recall that torque is the turning force that a motor produces.) Motor speed and torque control can be accomplished by one of two methods: we can vary either the voltage to the armature of the DC motor or the current to the field. When we vary the armature voltage, the motor produces full torque, but the speed is varied. However, if the field current is varied, both the motor speed and the torque will vary. Because of the need to vary the armature voltage or the field current, a separately excited DC motor, which allows very precise control over speed and torque, is the most commonly used type of motor.

Attaining precise control over motor speed and torque requires a means of evaluating the motor's performance and automatically compensating for any variations from the desired levels. The control section of a DC drive uses three types of signals to evaluate motor performance: *the command, feedback,* and *error signals.*

A command signal, sometimes called the set point or reference signal, is programmed into the DC drive and sets the desired operating speed of a DC motor. While the motor is operating, a feedback signal from the motor indicates the motor's performance. The feedback signal can originate either from the *counter electromotive force* produced by the motor or from a *tachometer-generator* or *encoder* mounted on the motor's shaft. Counter electromotive force (cemf) is the voltage produced in the motor's rotating armature, which cuts the magnetic lines of force in the field as it revolves. This armature voltage is called counter emf because it opposes the applied voltage from the electronic variable speed drive. As a result of this opposing voltage, the amount of armature current will be limited. A difference exists between the control signal and the feedback signal. This difference is the error signal. The electronic variable speed drive's controller automatically adjusts the motor's performance until the error signal is reduced practically to zero. This function is an ongoing process.

A controller's *regulation* determines how well it responds to changes in motor performance and is usually expressed as a percentage. Different types of feedback signals result in variable degrees of regulation. For instance, controllers using counter emf feedback, also called *armature voltage feedback,* typically have a regulation of 5 to 8 percent. This means that the speed of a DC motor set to operate at 1800 rpm can vary from 1944 rpm to 1656 rpm (1800 rpm ± 8 percent). On the other hand, when shaft-coupled encoders are used to provide the feedback signal, regulation is much tighter. A typical shaft-coupled encoder produces a regulation of 0.01 percent. Thus, the speed variation of the same motor operating at 1800 rpm would drop to between 1799.92 rpm and 1800.18 rpm (1800 rpm ± 0.01 percent).

In addition to managing motor speed and torque, the control section of a DC drive determines the direction of motor rotation and controls motor braking as well.

**FIGURE 22–1** Eurotherm model 590SP DC drive with programming panel.

# SWITCHING AMPLIFIER FIELD CURRENT CONTROLLER

Before you can understand how a switching amplifier field current controller works, you must understand what is meant by *open-loop control* and *closed-loop control*. In open-loop control, also called

**manual control** any variations in motor speed must be compensated by a manual adjustment. As you can imagine, paying someone to monitor and adjust motor speed constantly under varying load conditions would be very costly and inefficient. Therefore, open-loop control is limited to applications where the motor load is fairly constant.

When the motor load varies considerably or frequently, closed-loop control is used. Closed-loop control, also called *automatic control*, uses feedback information to monitor the performance of a motor. The information that is fed back automatically causes the control circuit to adjust the motor speed to varying load conditions.

A switching amplifier field current controller, as shown in Figure 22–2, receives and responds to a feedback signal from a DC motor that provides information about the speed of the motor. The controller first compares the feedback signal to a reference signal. Depending on the result of this comparison, the controller automatically increases or decreases the motor speed until it reaches the level indicated by the reference signal. To accomplish this adjustment, the controller switches the shunt field current on and off at varying rates.

To understand the following detailed explanation, refer to Figure 22–2. We begin with the feedback device.

In the upper right corner of Figure 22–2 is a schematic view of a tachometer-generator (tach. gen.), a DC generator attached to the DC shunt motor shaft. As the DC shunt motor turns, the DC tachometer-generator turns and produces a positive DC voltage as a result. The faster the DC shunt motor turns, the more positive DC voltage the DC tachometer-generator produces. Thus, the output of the DC tachometer-generator is proportional to the speed of the DC shunt motor.

Next, we will consider the feedback section, shown in detail in Figure 22–3. The positive voltage produced by the DC tachometer-generator and fed through R1 into buffer amplifier U1 is the feedback signal. Resistor R1 is known as a scaling resistor. The value of R1 is adjusted to provide the proper amount of DC voltage from the tachometer-generator for a given rpm. The output of U1 is fed to the inverting input of U2, which inverts the polarity of the tachometer-generator voltage from positive to negative. The resulting negative voltage is then fed through R7 to the noninverting input of U3 in the preamplifier section. Note that one end of R8 is connected to the noninverting input of U3 and the other end of R8 is connected to variable resistor R6. Resistor R6 allows us to set the reference voltage level, which determines at what speed the DC shunt motor will run. This reference voltage may

also be called the command signal or set point. The reference voltage (positive) and the tachometer-generator voltage (negative) are added together at the junction of R7 and R8, called the *summing point*. The difference of these two voltages, which appears at the noninverting input of U3, is the error signal.

To understand how this process works, assume that we wish the DC shunt motor to turn at 1800 rpm. We find that R6 must be set to 10 volts, the reference voltage required to attain a motor speed of 1800 rpm. Assume also that at a speed of 1800 rpm, the DC tachometer-generator produces a positive output of 5 volts. This value will be inverted by U2. Thus, at the junction of R7 and R8, we will have a positive reference voltage of 10 volts and a negative DC tachometer-generator feedback voltage of 5 volts. The sum of these two voltages is a positive error voltage of 5 volts $(10 + (-5) = 5)$.

If the speed of the DC shunt motor decreases below the reference speed, the tachometer-generator produces less positive DC voltage. Now, assume that the DC tachometer-generator produces a positive voltage of 3 volts DC, which results in a negative voltage of 3 volts at R7 from the output of U2. Because the reference voltage does not change, we still have a positive voltage of 10 volts at R8 from R6. The sum of these two voltages at the junction of R7 and R8 is a positive error voltage of 7 volts $(10 + (-3) = 7)$. Therefore, we can conclude that decreases in motor speed produce higher positive voltage at the noninverting input of U3. Likewise, increases in motor speed produce lower positive voltage at the noninverting input of U3.

Figure 22–4 illustrates U3, a preamplifier stage that takes the voltage at its noninverting input and provides an amplified positive output voltage of the proper level for the comparator to work with. Before examining the comparator stage, we will first look at the sawtooth generator stage.

Inverters U6 and U7, together with R20 and C2, form an oscillator circuit, as shown in Figure 22–5. Assume that the circuit's frequency of oscillation is 3 kilohertz. This oscillator circuit produces a rectangular pulse that is applied to U10, which is a pulse-shaping circuit. The output of U10, along with that of Q3, produces a 3-kilohertz sawtooth ramp voltage that is applied to the inverting input of U4 in the comparator stage.

In the comparator, shown in Figure 22–6, the noninverting input of U4 receives a positive voltage from the output of the preamplifier stage. Simultaneously, the inverting input of U4 receives a positive 3-kilohertz sawtooth ramp voltage from the sawtooth generator stage. This 3-kilohertz sawtooth voltage causes U4 to switch on and off. Whenever the noninverting input has a higher positive

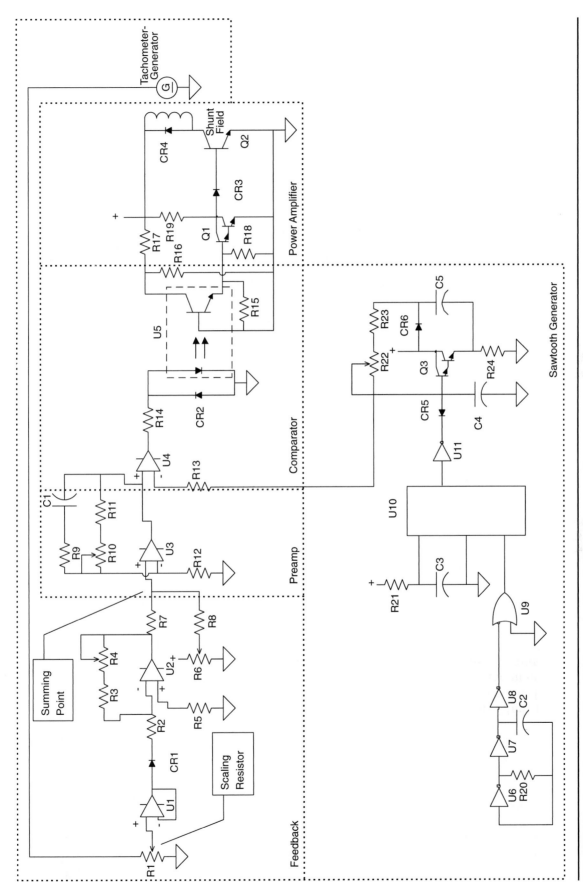

**FIGURE 22-2** Schematic of a switching amplifier field current controller.

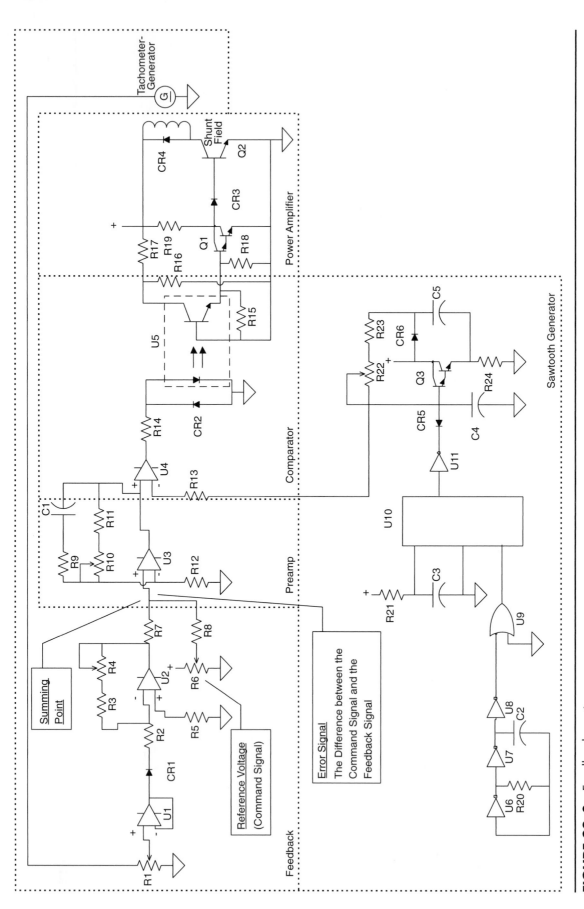

**FIGURE 22–3**   Feedback section.

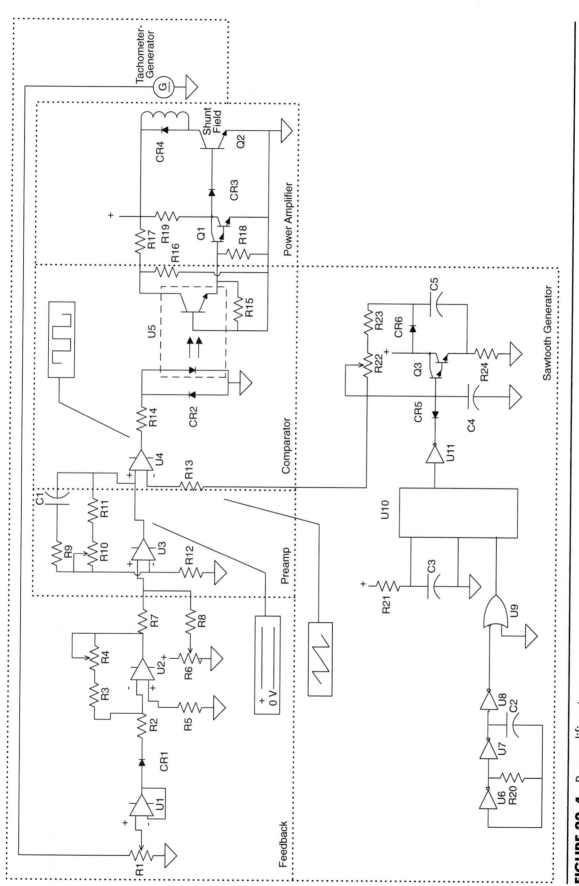

**FIGURE 22–4** Preamplifier stage.

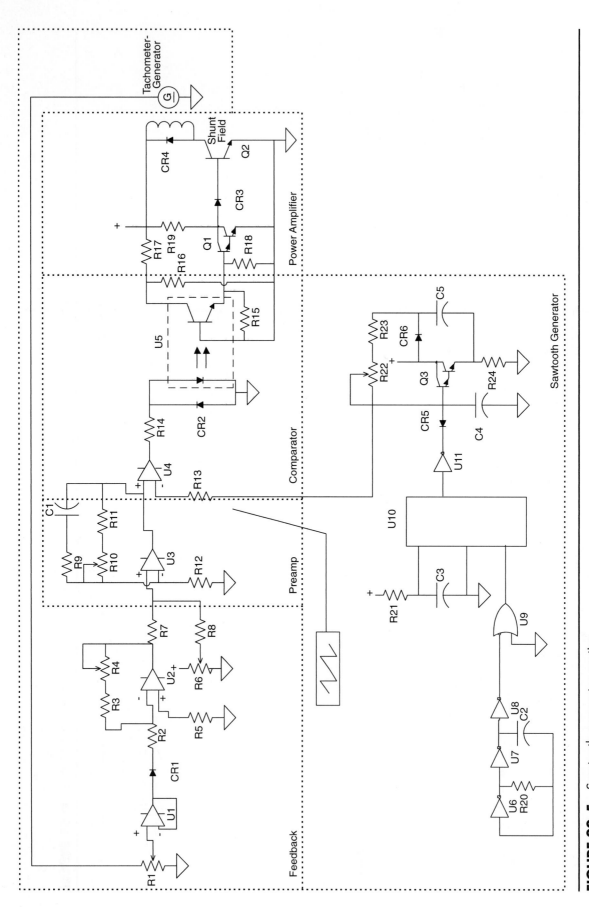

**FIGURE 22–5**   Sawtooth generator section.

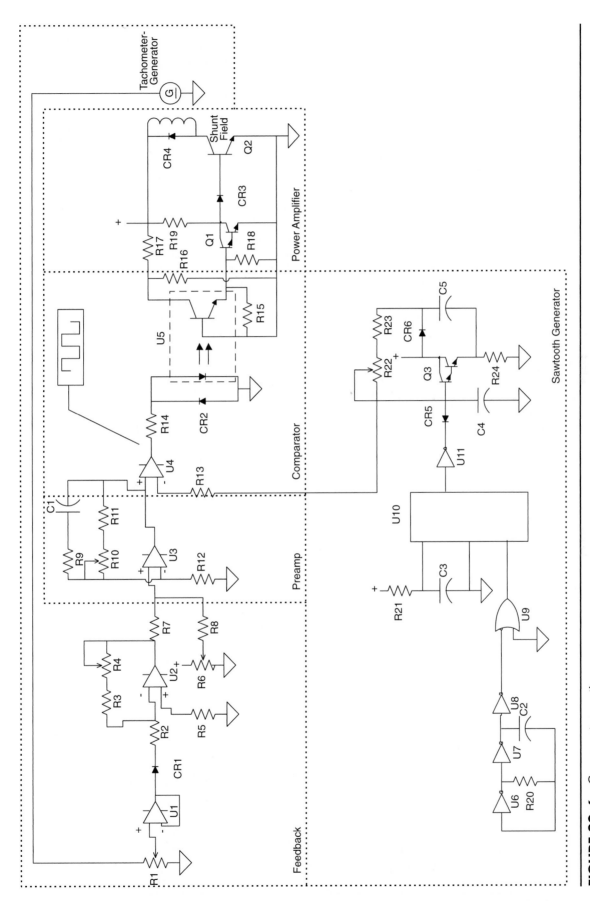

**FIGURE 22-6** Comparator section.

value than the inverting input from the sawtooth generator, U4 will produce an output voltage. Conversely, when the inverting input from the sawtooth generator has a higher positive value than the noninverting input, the output of U4 will be turned off and a rectangular pulse will be produced at the output of U4 as a result, as shown in Figure 22–7. The width of U4's output pulse is controlled by the reference voltage from R6. If R6 is adjusted for a higher DC reference voltage, then U3 will produce a higher positive voltage for the noninverting input of U4. In turn, the sawtooth ramp voltage must increase to a higher positive value at the inverting input of U4. Since the inverting input takes more time to attain a higher positive value than the noninverting input does, the output pulse of U4 becomes wider, as shown in Figure 22–8. Likewise, if R6 is adjusted for a lower DC reference voltage, U3 will produce a lower positive DC voltage at the noninverting input of U4. Consequently, the sawtooth ramp voltage does not need to reach as high a positive value, and U4 turns off sooner. In this case, U4's output pulse will be narrower, as shown in Figure 22–9. The effect is called pulse-width modulation, or PWM. The output pulse of U4 is applied to opto-coupler (or opto-isolator) U5. When the output of U4 is on, opto-coupler U5 is on. When the output of U4 is off, opto-coupler U5 is also off. Opto-coupler

U5 isolates the low-voltage logic circuits from the higher-voltage power circuits in the power amplifier stage and also prevents electrical noise that may be induced onto the power circuits from entering the control section of the drive.

Figure 22–10 illustrates how the output of opto-coupler U5 is fed to the base of Q1 in the power amplifier stage. The power amplifier stage consists primarily of Darlington transistor Q1 and power transistor Q2. Transistor Q1 receives its input signal from opto-coupler U5. Whenever U5 turns on, Q1 will conduct. Likewise, whenever U5 is turned off, Q1 will not conduct. The collector of Q1 drives the base of Q2. Whenever Q1 is conducting, Q2 is turned off, and no current flows through the shunt field of the DC shunt motor. As a result, the DC shunt motor speeds up. Conversely, whenever Q1 is turned off, Q2 does conduct a current, which flows through the shunt field of the DC shunt motor, causing the motor to slow down.

Reconsider our earlier example of a motor set to a speed of 1800 rpm. Recall that we adjusted R6 for this speed and that this produced 10 volts DC at R8. Because the DC shunt motor is turning at 1800 rpm, the tachometer-generator produces a negative voltage of 5 volts DC at R7. Therefore, at the junction of R7 and R8 we have a positive voltage of 5 volts DC. This voltage is applied to the noninverting input

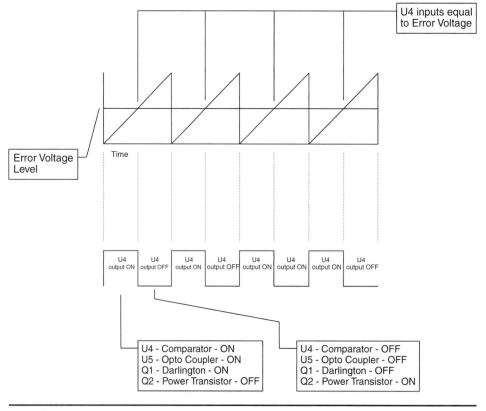

**FIGURE 22–7**   Output of the comparator section.

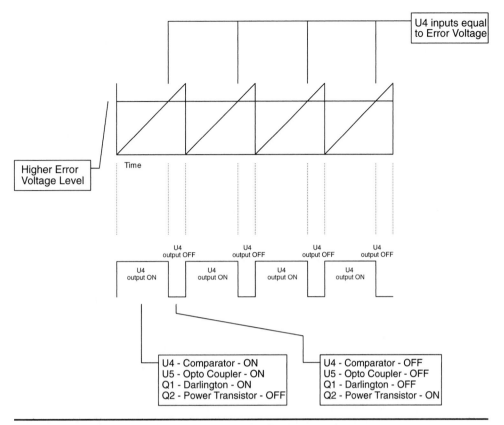

**FIGURE 22-8** Comparator output with higher error voltage.

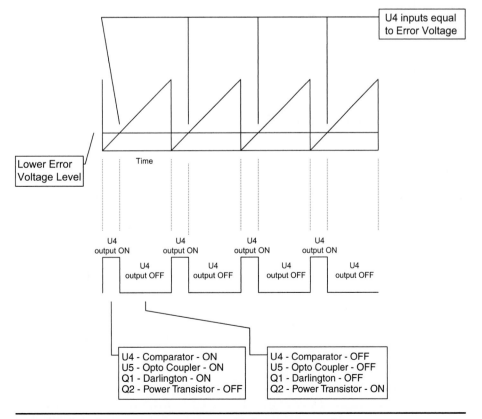

**FIGURE 22-9** Comparator output with lower error voltage.

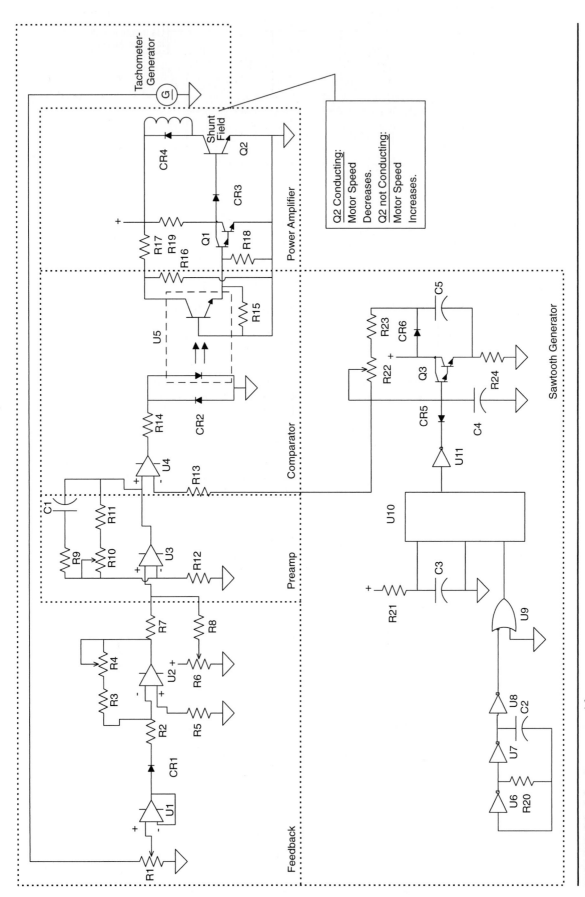

**FIGURE 22-10** Power amplifier section.

of the comparator, U4, and then compared with the sawtooth ramp voltage at the inverting input of U4. Assume that as a result of this comparison, the output of U4 is turned on for 166 microseconds. Consequently, U5 and Q1 are also turned on for 166 microseconds. However, Q2 is turned off for 166 microseconds. Because 166 microseconds represents a 50 percent duty cycle at 3 kilohertz, the outputs of U4, U5, and Q1 will all be switched on and then off every 166 microseconds (% duty cycle = time of "on" pulse/time of one cycle). Consequently, Q2 will also be turned on and off every 166 microseconds. Therefore, the current through the shunt field of the DC motor pulsates at an average value that keeps the motor turning at 1800 rpm.

Applying an increased load to the DC shunt motor slows the motor down. As the DC shunt motor turns more slowly, the tachometer-generator produces less output voltage. This reduction in voltage results in a lower negative voltage at R7. Suppose that the tachometer-generator voltage drops from 5 volts to 3 volts. The resulting voltage at the junction of R7 and R8 will be positive 7 volts (10 + (−3) = 7). Consequently, a higher voltage (7 volts) is applied to the noninverting input of U4. When this voltage is compared to the sawtooth ramp voltage, the value of the sawtooth ramp voltage must increase to a higher value before U4 is turned off. Thus, the output of U4 will be turned on for a longer period of time than it was previously. Assume that U4 is now turned on for 233 microseconds. This implies that the output of U4 will be turned off for a shorter period of time (100 microseconds). The longer U4 is turned on and Q2 is turned off, the longer the time during which no current flows through the shunt field of the DC shunt motor. Therefore, the average value of the current flowing through the shunt field will be lower. Reducing the shunt field current causes the DC shunt motor to speed up. The reverse is true if the load on the DC shunt motor is lessened. In this way, an increase or decrease in load on a DC shunt motor is automatically compensated. The process by which this compensation is accomplished is closed-loop control, or automatic control.

## SCR ARMATURE VOLTAGE CONTROLLER

An SCR armature voltage controller is another method of DC motor speed control. The speed of a DC motor may be varied by controlling either the shunt field current or the armature voltage. Armature voltage control is the type most commonly used in DC drives. By controlling the firing of SCRs, we can control the voltage of the armature of a DC motor. A typical SCR armature voltage controller is shown in Figure 22–11.

As you look at Figure 22–11, try to recognize some of its similarities in relation to Figure 22–2. Both units use the same type of circuit for closed-loop feedback control. We will begin by focusing on the lower portion of Figure 22–11, where the circuit that consists of the null detector, pulse shaper, and sawtooth generator are located.

Begin by looking at the null detector in Figure 22–12. Note that the AC voltage is rectified by the full-wave bridge circuit consisting of CR7–CR10. This circuit causes an unfiltered, pulsating DC voltage to appear across zener diode Z1, R28, and the LED of opto-coupler U10.

First, we will consider what happens as this pulsating DC voltage rises from 0 volt to its **peak value**. The LED of U10 will not conduct until the DC voltage reaches the turn-on voltage of the LED. Assume that this voltage is approximately 1.5 volts. As the DC voltage increases from 0 volt to 1.5 volts, U10 is turned off. Therefore, the output of U10 will be a positive pulse. This positive pulse at the output of U10 causes Q2 to turn on. When Q2 conducts, a negative pulse will appear at the collector of Q2. This negative pulse resets the output of the sawtooth generator Q3 back to 0.

Now, we will look at what happens when the LED of U10 turns on. As the pulsating DC voltage rises, eventually it reaches a value of 1.5 volts. The voltage across the LED of U10 will never exceed the voltage rating of zener diode Z1. At this point, the LED of U10 turns on and conducts, causing the output of U10 to supply a negative pulse that turns Q2 off. When Q2 is turned off, the sawtooth generator outputs a sawtooth waveform to the inverting input of U5 (the comparator). The sawtooth generator will continue to output this waveform until it is reset when Q2 turns on, that is, when the pulsating DC voltage falls below 1.5 volts and continues to decrease to 0 volt. In this way, the pulsating DC voltage causes the sawtooth generator to provide the comparator with a reference signal that is used to vary the firing angle of the SCRs that control the armature voltage.

Next, look at U5, the comparator in Figure 22–13. The comparator has two inputs. One, which comes from the feedback circuit, is a measure of the speed of the armature. The other, as we just learned, is a ramp signal from the sawtooth generator. The noninverting input from the feedback amplifier will be constant if the load on the armature is constant. The inverting input is a rising amplitude signal from the sawtooth generator. Initially, the noninverting

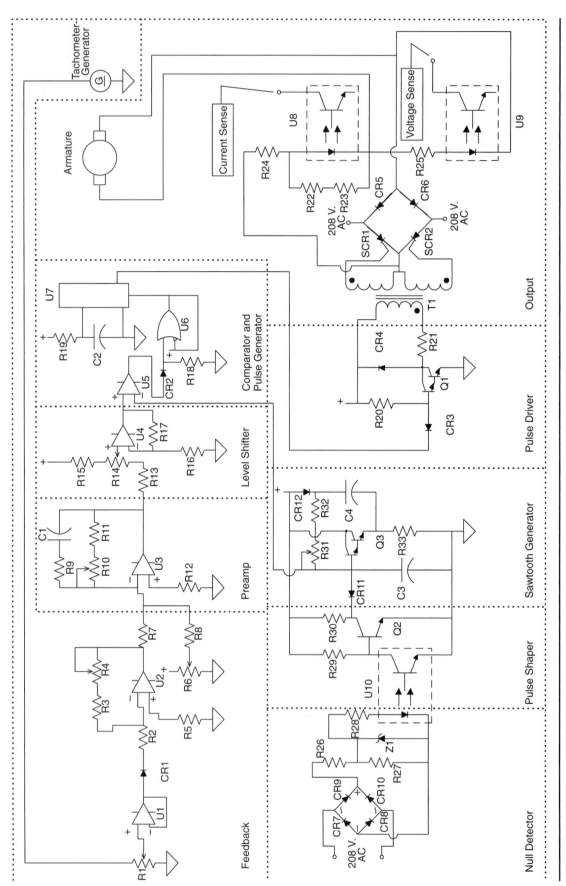

**FIGURE 22-11** Schematic of an SCR armature voltage controller.

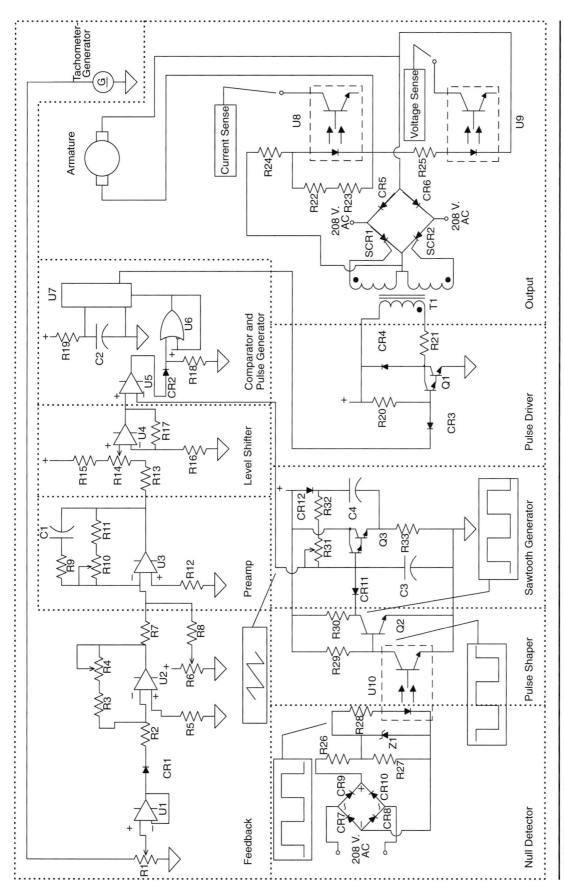

**FIGURE 22–12** Null detector, pulse shaper, and sawtooth generator sections.

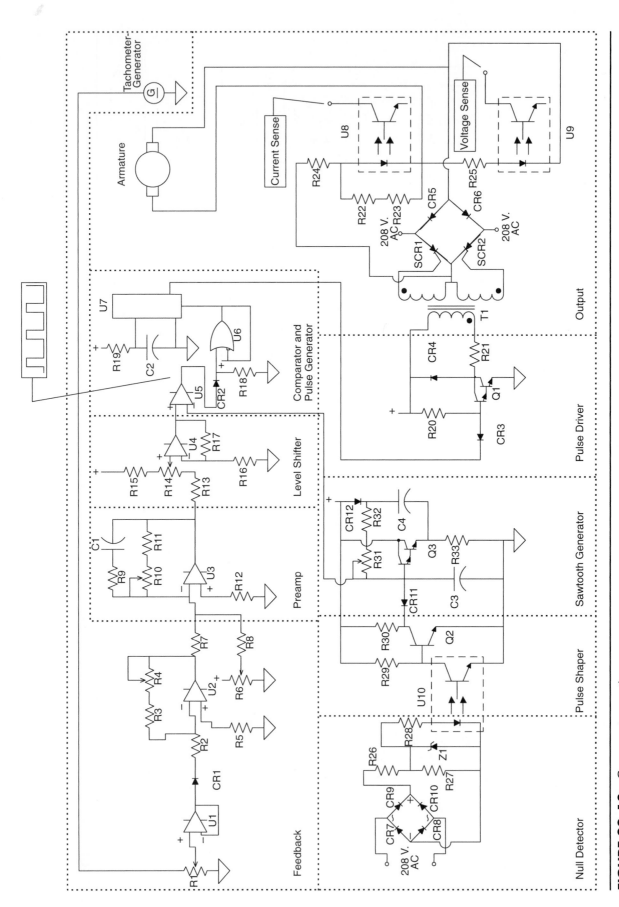

**FIGURE 22-13**   Comparator section.

input causes the output of U5 to be positive. However, as the inverting input ramp signal climbs, at a given point the amplitudes of both inputs become equal. At this point, the output of U5 switches off and becomes a rectangular pulse. The width of this pulse is determined by the length of time it takes for both inputs to become equal. This is what determines whether the SCR conducts earlier or later within the positive half of the cycle.

The pulse generator, shown in Figure 22–14, is an integrated circuit that contains a one-shot, or monostable, multivibrator. The purpose of the pulse generator is to trigger on the output of the comparator and, thus, provide a narrow pulse to Q1, the pulse driver.

Referring to Figure 22–15, recall that Q1 will not conduct until a positive pulse appears at its base. When this occurs, current will flow through the primary of T1. Q1 will only conduct for the duration of the pulse applied to its base. The current flowing through the primary of T1 will cause a current flow in the secondary of T1. Therefore, the gates of SCR1 and SCR2 receive a trigger pulse simultaneously. Figure 22–16 shows some of the common types of SCRs in use. However, note that the anodes of SCR1 and SCR2 are connected to an AC source. As a result, when the anode of SCR1 is positive, the anode of SCR2 is negative, and vice versa. Thus, when SCR1 and SCR2 are triggered, only the SCR with the positive anode conducts, so the SCRs conduct alternately. Consequently, DC current flows through the armature of the DC motor. Figure 22–17 shows the circuitry used to control the SCRs, while Figure 22–18 shows the circuitry used to rectify the AC with SCRs.

Now, let's bring this all together. Figure 22–19 shows the waveforms at various points in the comparator and pulse generator, pulse driver, and output stages. These waveforms have been lined up vertically so that you understand the timing of the waveforms that must occur in order for the output SCRs to trigger at the appropriate time. Here's how it works. The waveform in Figure 22–19 waveform (a) shows the sawtooth waveform at the inverting input of U5, the comparator. At the same time, Figure 22–19 waveform (a) also shows the DC voltage that is applied to the noninverting input of comparator U5. Recall that this voltage is the feedback voltage from the tachometer-generator and is an indication of the armature speed. The DC voltage level shown is representative of a motor speed of 1800 rpm. These two inputs cause the output of the comparator to produce a pulsating DC, as shown in Figure 22–19 waveform (b). Notice that the pulse is initially positive and is switched off when the two inputs of the comparator become equal. The output of U5 remains off until the sawtooth generator resets. At this point, the output of U5 switches on

and remains on until the sawtooth signal is equal to the feedback voltage.

The pulsating output of U5 is fed through diode CR2 to the input of U6. Because of CR2, the input signal to U6 will be inverted compared to the output of U5. This is shown in Figure 22–19 waveform (c). The positive pulses from the output of U6 trigger the one-shot, U7. The output of the one-shot, U7, will appear in phase with the input pulses, as seen in Figure 22–19 waveform (d).

The output of the one-shot, U7, is fed through diode CR3 to the base of the Darlington transistor, Q1. Notice in Figure 22–19 waveform (e) that the collector of Q1 is 180 degrees out of phase with the output of the one-shot. When Q1 conducts, current will flow through the pulse transformer, T1. When current flows through pulse transformer T1, a pulse appears at the gates of SCR1 and SCR2. This is shown in Figure 22–19 waveform (f). Recall that only the SCR that is properly biased will conduct. The SCR that is triggered and properly biased will conduct until the AC voltage drops to the zero-crossing point. The next pulse from T1 will cause the other SCR to conduct. Therefore, the SCRs will conduct alternately, as shown in Figure 22–19 waveform (g).

Let's assume that an increase in load has caused the motor speed to decrease to 1000 rpm. Refer to Figure 22–20. The wave pattern in Figure 22–20 waveform (a) shows the sawtooth waveform at the inverting input of U5, the comparator. At the same time, Figure 22–20 waveform (a) shows the DC voltage that is applied to the noninverting input of comparator U5. The DC voltage level shown is representative of a motor speed of 1000 rpm. Notice that this level is lower than that shown in Figure 22–19 waveform (a). This is a result of the lower motor speed and lower output voltage from the tachometer-generator. These two inputs cause the output of the comparator to produce a pulsating DC, as shown in Figure 22–20 waveform (b). Notice that the pulse is initially positive and is switched off when the two inputs of the comparator become equal. The output of U5 remains off until the sawtooth generator resets. At this point, the output of U5 switches on and remains on until the sawtooth signal is equal to the feedback voltage. Compare the width of the positive pulses in Figure 22–20 waveform (b) with those in Figure 22–19 waveform (b). Notice that the positive pulses are narrower in Figure 22–20 waveform (b).

The pulsating output of U5 is fed through diode CR2 to the input of U6. Because of CR2, the input signal to U6 will be inverted compared to the output of U5. This is shown in Figure 22–20 waveform (c). Again, compare the positive pulses in Figure 22–20 waveform (c) with those shown in

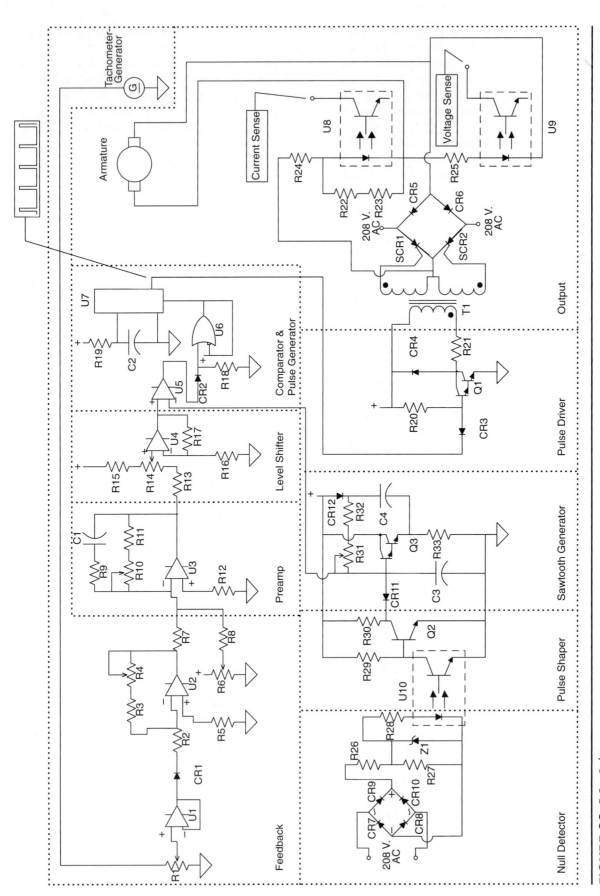

**FIGURE 22-14** Pulse-generator section.

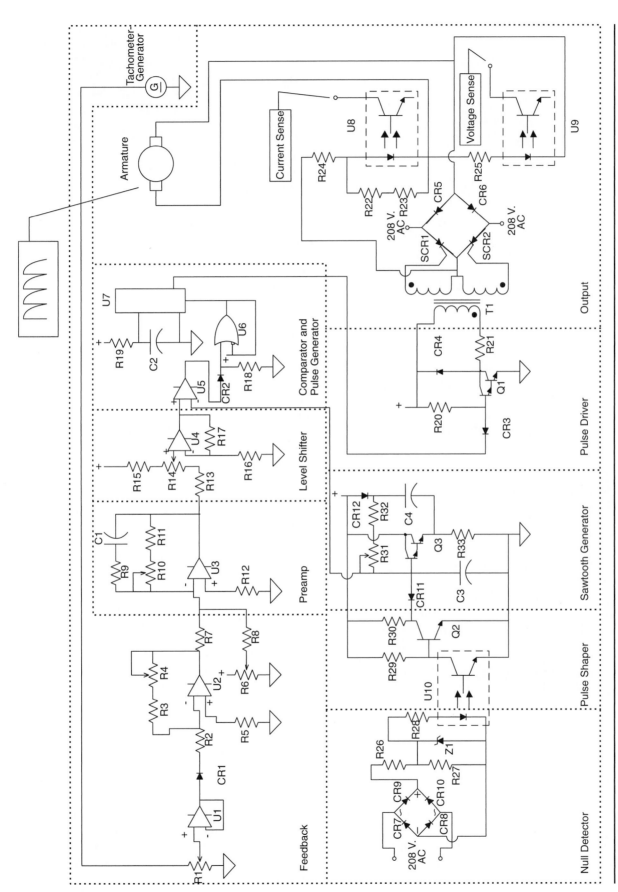

**FIGURE 22-15** Output section.

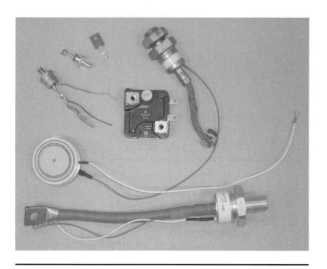

**FIGURE 22–16** Various SCR packages.

**FIGURE 22–18** SCR power rectifier section.

**FIGURE 22–17** Control board assembly.

Figure 22–19 waveform (c). Notice that the positive pulses are wider in Figure 22–20 waveform (c) than those shown in Figure 22–19 waveform (c). The positive pulses from the output of U6 trigger the one-shot, U7. The output of the one-shot, U7, will appear in phase with the input pulses, as seen in Figure 22–20 waveform (d). Notice that the output pulse of the one-shot now occurs earlier than it did in Figure 22–19 waveform (d), which will cause the SCRs to fire sooner and conduct for a longer period

of time. The following example will test if this is what occurs. The output of the one-shot, U7, is fed through diode CR3 to the base of the Darlington transistor, Q1. Notice in Figure 22–20 waveform (e) that the collector of Q1 is 180 degrees out of phase with the output of the one-shot. When Q1 conducts, current will flow through the pulse transformer, T1. When current flows through pulse transformer T1, a pulse appears at the gates of SCR1 and SCR2. This is shown in Figure 22–20 waveform (f). Recall that only an SCR that is properly biased will conduct. The SCR that is triggered and properly biased will conduct until the AC voltage drops to the zero-crossing point. The next pulse from T1 will cause the other SCR to conduct. Therefore, the SCRs will conduct alternately, as shown in Figure 22–20 waveform (g). Notice that the SCRs conduct for a longer amount of time as compared to the conduction of the SCRs in Figure 22–19 waveform (g). This results in a higher average voltage applied to the armature of the DC motor and will cause the speed of the DC motor to increase.

Let's assume that a decrease in load has caused the motor speed to increase to 2600 rpm. Refer to Figure 22–21. The wave pattern in Figure 22–21 waveform (a) shows the sawtooth waveform at the inverting input of U5, the comparator. At the same time, Figure 22–21 waveform (a) also shows the DC voltage that is applied to the noninverting input of comparator U5. The DC voltage level shown is representative of a motor speed of 2600 rpm. Notice that this level is higher than that shown in Figure 22–19 waveform (a). This is a result of the higher motor speed and higher output voltage from the tachometer-generator. These two inputs cause the output of the comparator to produce a pulsating DC, as shown in Figure 22–21 waveform (b). Notice that the pulse is initially positive and is switched off when the two inputs of the comparator become equal. The output of

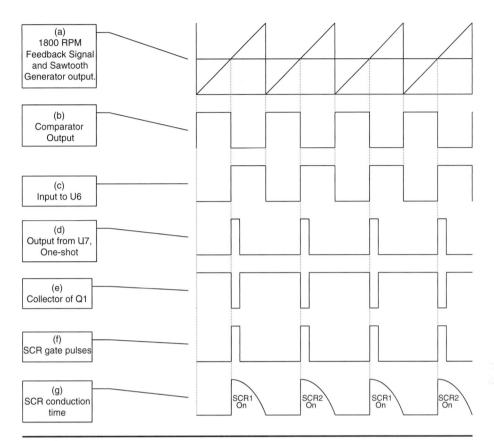

**FIGURE 22-19** Waveform timing and SCR conduction time at 1800 rpm.

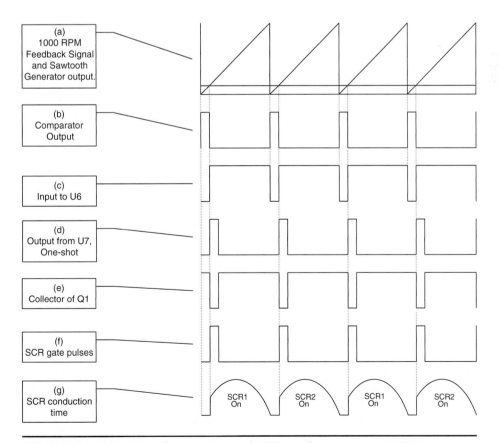

**FIGURE 22-20** Waveform timing and SCR conduction time at 1000 rpm.

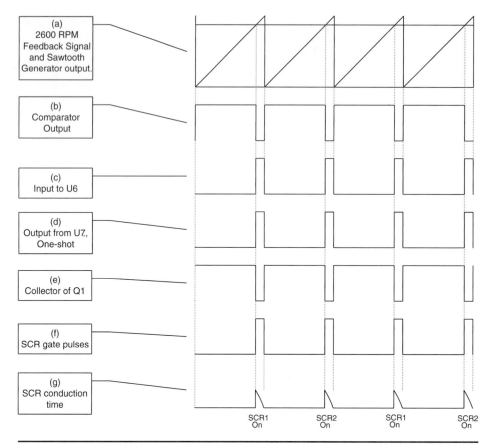

**FIGURE 22–21** Waveform timing and SCR conduction time at 2600 rpm.

U5 remains off until the sawtooth generator resets. At this point, the output of U5 switches on and remains on until the sawtooth signal is equal to the feedback voltage. Compare the width of the positive pulses in Figure 22–21 waveform (b) with those in Figure 22–19 waveform (b). Notice that the positive pulses are wider in Figure 22–21 waveform (b).

The pulsating output of U5 is fed through diode CR2 to the input of U6. Because of CR2, the input signal to U6 will be inverted compared to the output of U5. This is shown in Figure 22–21 waveform (c). Again, compare the positive pulses in Figure 22–21 waveform (c) with those shown in Figure 22–19 waveform (c). Notice that the positive pulses are narrower in Figure 22–21 waveform (c). The positive pulses from the output of U6 trigger the one-shot, U7. The output of the one-shot, U7, will appear in phase with the input pulses, as seen in Figure 22–21 waveform (d). Notice that the output pulse of the one-shot now occurs later than it did in Figure 22–19 waveform (d), which will cause the SCRs to fire later and conduct for a shorter period of time. The following example will test if this is what occurs. The output of the one-shot, U7, is fed through diode CR3 to the base of the Darlington transistor, Q1. Notice in Figure 22–21 waveform

(e) that the collector of Q1 is 180 degrees out of phase with the output of the one-shot. When Q1 conducts, current will flow through the pulse transformer, T1. When current flows through pulse transformer T1, a pulse appears at the gates of SCR1 and SCR2, as shown in Figure 22–21 waveform (f). The SCR that is triggered and properly biased will conduct until the AC voltage drops to the zero-crossing point. The next pulse from T1 will cause the other SCR to conduct. Therefore, the SCRs will conduct alternately, as shown in Figure 22–21 waveform (g). Notice that the SCRs conduct for a shorter amount of time as compared to the conduction of the SCRs in Figure 22–19 waveform (g). This results in a lower average voltage applied to the armature of the DC motor and will cause the speed of the DC motor to decrease.

## CHOPPERS

A *chopper* is a circuit that uses very fast electronic switches. These switches consist of transistors, SCRs, or metal oxide semiconductor field effect transistors (MOSFETs). MOSFETs switch the applied DC on and off more rapidly. By using them, we can create a DC output level that will be either lower or higher than the applied DC. If the chopper's output is lower than the applied DC, the chopper is called a *buck chopper*, or *step-down chopper*. If the chopper output is higher than the applied DC, the chopper is called a *boost chopper*, or *step-up chopper*.

In Figure 22–22, notice that the gate of MOSFET Q1 is connected to a control signal source. The control signal emitted by this source is a *pulse-width modulated* (PWM) signal. We will begin by setting the duty cycle of the PWM signal to 50 percent. (Remember that the duty cycle is the ratio of the time of the "on" pulse to the time of one cycle. Therefore, a pulse set at a 50 percent duty cycle is on for the duration of one-half cycle.)

When the control signal pulse is on, it is applied to the gate of the MOSFET, and Q1 conducts as a result. However, Q1 will only conduct for half of one cycle. While Q1 conducts, the armature of the DC motor is connected to the DC supply. This connection causes diode CR1, a free-wheeling diode, to appear as an open circuit because it is reverse biased. Therefore, the current flowing through the armature of the DC motor increases. During the next half-cycle, this pulse to the gate of Q1 is switched off, so Q1 is also turned off. As a result, the armature of the DC motor is no longer connected to the DC supply, and diode CR1 provides a discharge path for the collapsing magnetic field of the armature.

The output voltage of a buck chopper is proportional to the duty cycle of the control signal. If we

increase the duty cycle (thus making the "on" pulse wider), the average value of the chopper's output voltage also increases. Likewise, if we decrease the duty cycle (making the "on" pulse narrower), the average value of the chopper's output voltage decreases as well. Varying the frequency of the control signal has no effect on the level of output voltage and current. However, a higher frequency control signal produces less ripple frequency in the output voltage and current, and thus results in smoother motor operation. Because of losses in the MOSFET, at no time will the output voltage of the buck chopper exceed the DC supply voltage. That is the reason why this particular chopper is known as a buck, or step-down, chopper.

The boost chopper shown in Figure 22–23 again uses a PWM signal to turn the gate of MOSFET Q1 on and off. Let's begin with a 50-percent duty cycle and a charge on capacitor C1. When the signal pulse is on, it is applied to the gate of MOSFET Q1, and Q1 is also turned on. With Q1 turned on, a very low voltage will develop across the source to drain leads of Q1, which in turn allows current to flow through inductor L1 and, thus, builds up a magnetic field. Because diode CR1 is now reverse biased, no current will flow through it. Therefore, capacitor C1 discharges through the armature of the DC motor, causing the armature current, voltage, and motor speed to decrease as the charge on C1 decreases.

When the signal pulse to Q1 is switched off during the next half-cycle, Q1 is also turned off, causing the source-to-drain voltage of Q1 to increase. As a result, CR1 now becomes forward biased, allowing the magnetic field of inductor L1 to collapse. The discharge current from L1, along with the supply current, recharges C1 and also increases the amount of current flowing through the armature of the DC motor. In turn, this increased armature current causes corresponding increases in armature voltage and motor speed.

To sum up, if Q1 is turned on for a longer period of time, motor speed decreases. If, on the other hand, Q1 is turned on for a shorter period of time, motor speed increases. Again, varying the

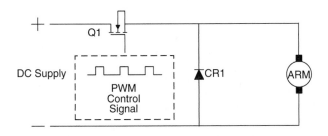

**FIGURE 22–22** Buck or step-down chopper.

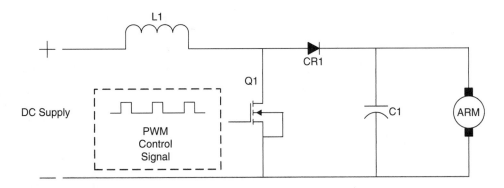

**FIGURE 22–23** Boost or step-up chopper.

frequency of the control signal has no effect on the level of output voltage and current. However, a higher frequency control signal produces less ripple frequency in the output voltage and current, resulting in smoother motor operation. Because the inductor voltage boosts, or supplements, the supply voltage, the output voltage can be higher than the input voltage. That is the reason why this circuit is known as a boost, or step-up, chopper.

We must realize that current will only flow in one direction, from source to load, in both of these drives. Thus, both of these controls have some shortcomings. Most notable is the fact that speed control is difficult with high-inertia loads, which may cause *overhauling*.

Overhauling occurs when the load has enough momentum to cause the motor rotor to continue rotating even after the motor has been disconnected. A flywheel is an example of this type of load. The type of drive in which this phenomenon occurs is a one-quadrant drive.

## THE FOUR QUADRANTS OF MOTOR OPERATION

First, look at the four quadrants shown in Figure 22–24. Note the polarities of the current and voltage in each quadrant. In quadrant 1, the current and the voltage are both positive. When a motor is operating in quadrant 1, we say that the motor is motoring, that is, driving a load and turning in the forward direction.

Look at quadrant 3 in Figure 22–25. Note that in this quadrant, the current and the voltage are both negative. A motor operating in quadrant 3 is still motoring, but it is now operating in reverse.

Sometimes a motor must operate in both quadrants 1 and 3. This is necessary when a motor must drive a load both forward and in reverse. Examples are the motor in an electric vehicle that runs forward and backward or in a crane that must raise and lower a load.

In Figure 22–26, look first at quadrant 2. Note that although the voltage is still positive, the

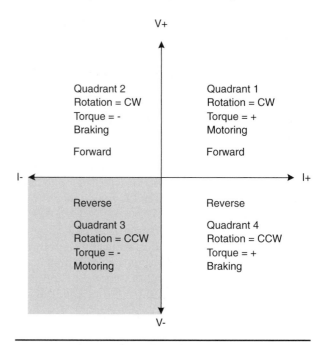

**FIGURE 22–25**   Quadrant 3: motoring in reverse.

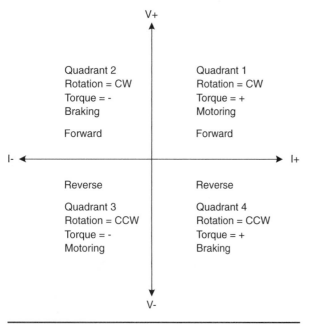

**FIGURE 22–24**   The four quadrants of motor operation.

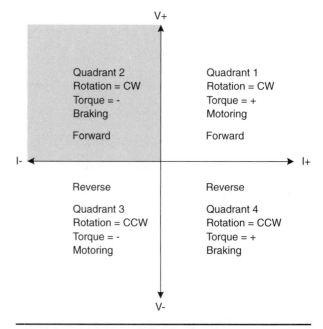

**FIGURE 22–26**   Quadrant 2: braking in forward.

current is negative. We say that a motor operating in quadrant 2 is *regenerating*. During this time, the motor provides energy back to the power source while rotating forward. In other words, the motor acts like a generator. This regenerated energy provides braking for the motor, affording us better control over the motor's speed. However, operating in quadrant 2 does not allow for braking when the motor turns in reverse.

Now, look at quadrant 4 in Figure 22–27. Here, the voltage is negative, and the current is positive. Motor operation in quadrant 4 produces **regenerative braking** action when the motor turns in reverse. Think about the electric vehicle example. We want the vehicle to be able not only to travel forward and backward, but also to stop in either direction of travel. Therefore, a DC motor for an electric vehicle must be able to operate in all four quadrants. Likewise, in the example of a crane, not only must we be able to lift and lower a load, we must also be able to control the speed at which it raises and lowers the load. Again, the DC motor must operate in all four quadrants.

As its name implies, a four-quadrant chopper operates in all four quadrants. That is, the motor operates both forward and in reverse, and it also has regenerative braking capabilities in both directions.

Regenerative braking occurs when the motor still rotates even after power has been removed. When power is removed from a motor, the motor continues to rotate until friction and windage slow it to a stop. While the motor is rotating, the magnetic fields in

the armature and stator are collapsing. The motion of the armature through these collapsing magnetic fields generates a voltage and a current. The polarity of this generated voltage and current is the opposite of the polarity of the voltage and current that were originally applied to the DC motor. This opposite polarity tries to make the motor turn in the opposite direction, in effect braking the motor. Because the braking force is a result of the generated voltage and current, it is called regenerative braking.

In Figure 22–28, notice the four MOSFETs: Q1, Q2, Q3, and Q4. Also, notice that each MOSFET has an associated free-wheeling diode: CR1, CR2, CR3, and CR4. The gates of the four MOSFETs are controlled by a switching control circuit. The switching control circuit switches the MOSFETs on and off in pairs: Q1 with Q4 and Q2 with Q3. When one pair is turned on, the other is turned off.

We will begin by assuming that Q1 and Q4 are turned on. Therefore, current flows through the armature of the DC motor and through Q1 and Q4. We will also assume that at this time the motor is turning clockwise, that is, the motor is operating in quadrant 1 and is now motoring.

Look at Figure 22–29, and assume that we now turn off Q1 and Q4. The armature of the motor will continue to rotate within the now-collapsing magnetic field. As this rotation continues, magnetic lines of force are cut by the armature. This produces a voltage that opposes the applied voltage. This voltage will cause a current to flow from the armature, through CR3, to the supply, through CR2, and back to the armature. Since this voltage is opposite to the applied voltage, a counter torque is produced. This counter torque provides a braking action to the motor. Notice, however, that the generated energy is now being applied to the power source. This is known as regenerative braking, and the motor is now operating in quadrant 2.

A motor operating in quadrant 3, as shown in Figure 22–30, must rotate in the opposite direction from a motor operating in quadrant 1. In our example, therefore, the motor now rotates counterclockwise. Turning on Q2 and Q3 in this case allows current flow in the reverse direction compared to quadrant 1, which causes the motor to turn counterclockwise when it operates in quadrant 3.

Let us now turn off Q2 and Q3, as shown in Figure 22–31. As before, the armature of the motor will continue to rotate in the reverse direction within the now-collapsing magnetic field. As this rotation continues, magnetic lines of force are cut by the armature, which produces a voltage that opposes the applied voltage. This voltage will cause a current to flow from the armature, through CR4, to the supply, through CR1, and back to the armature.

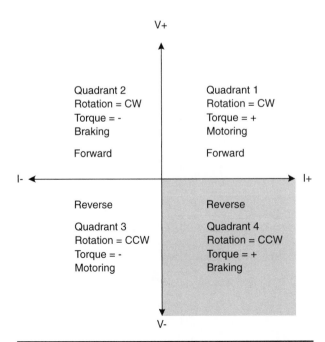

**FIGURE 22–27** *Quadrant 4: braking in reverse.*

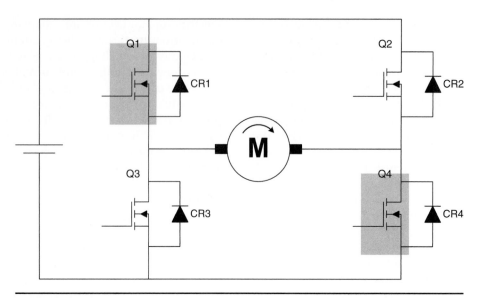

**FIGURE 22-28**    Four-quadrant chopper.

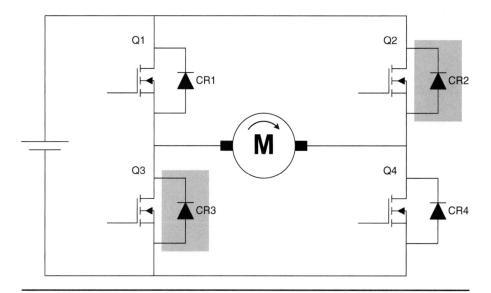

**FIGURE 22-29**    CR2 and CR3 conducting.

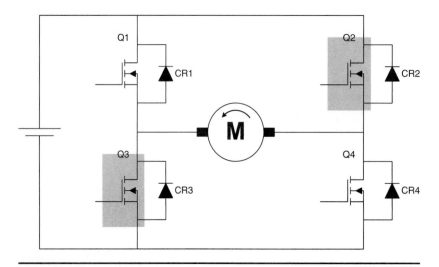

**FIGURE 22-30**    Q2 and Q3 conducting.

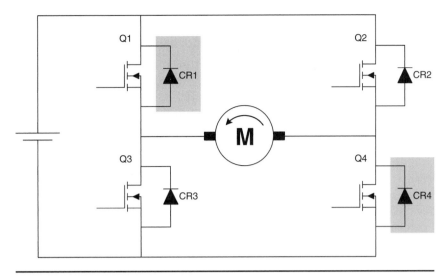

**FIGURE 22–31** CR1 and CR4 conducting.

Since this voltage is opposite to the applied voltage, a counter torque is produced. This counter torque provides a braking action to the motor. Notice, however, that the generated energy is now being applied to the power source. This provides regenerative braking from the reverse direction, and the motor is now operating in quadrant 4.

We can control a motor's speed quite easily in all four quadrants. Remember that each MOSFET is controlled by a pulse-width modulation signal applied to its gate. If the width of the "on" pulse is wider, the MOSFET will conduct for a longer period of time, and the motor will operate at a higher average speed. If we make the "on" pulse narrower, the MOSFET will conduct for a shorter period of time, and the average speed of the motor will be lower. Varying the frequency of the control signal has no effect on the level of output voltage and current. However, a higher frequency control signal produces less ripple frequency in the output voltage and current, resulting in smoother motor operation. Therefore, a four-quadrant chopper controls not only the quadrant of operation, but also the speed of the motor.

# TROUBLESHOOTING DC DRIVES

All too often, a maintenance technician who is introduced to something new will automatically assume that any problems are located in the new item. As we will see, this is not always true.

*Assumption #1*: If the motor doesn't run, the problem must be the drive.

First, note that in a DC drive system there are four main areas to check for possible problems:

1. The electrical supply to the motor and the drive
2. The motor and/or its load
3. The feedback device and/or the sensors that provide signals to the drive
4. The drive itself

Even though the problem may be in any one or several of these areas, the best place to begin troubleshooting is at the drive unit itself because most drives have some type of display that aids in troubleshooting. This display may be simply an LED that illuminates to indicate a specific fault condition, or it could be an error or fault code that may be deciphered by looking in the operator's manual. For this reason, it is strongly suggested that a copy of the fault codes be made and fastened to the inside of the drive cabinet where it will be readily accessible to the maintenance technician. The original should be placed in a safe location, such as the maintenance supervisor's office.

*Assumption #2*: After checking the drive and reading the code, it's time to make the repairs.

Wrong! Before going any further, you must consider safety factors. First, stop and think about what you are doing! Before working on any electrical circuit, disconnect all power. At times when this is not possible or permissible, work carefully and wear appropriate safety equipment. Do not rely on safety interlocks, fuses, or circuit breakers to provide personal protection. Always use a voltmeter to verify that the equipment is de-energized and tag and lock out the circuit!

Even when the power has been disconnected, you are still likely to be subjected to shock and burn hazards. Most drives have high-power resistors inside, and these can and do get hot! Give them time to cool down before touching them. Most drives also have large electrolytic capacitors that can and do store an electrical charge. Usually, the capacitors have a bleeder circuit to dissipate this charge. However, be aware that this circuit may have failed. Always verify that electrolytic capacitors are fully discharged by carefully measuring any voltage present across the capacitor terminals. If voltage is present, use an approved shorting device to discharge the capacitor completely.

Having followed these steps, what should you do next? Use your senses! Most problems can be identified by using your senses.

Look!  Are there any charred or blackened components? Have you noticed any arcing? Do fuses or circuit breakers appear to have been blown or tripped? Do you see any discoloration around wires, terminals, or components? A good visual inspection can save a lot of troubleshooting time.

Listen!  Did you hear any funny or unusual noises? A "frying" or "buzzing" sound may indicate arcing. A "hum" may be normal or an indication of loose laminations in a transformer core. A "rubbing" or "chafing" sound may indicate a cooling fan is not rotating freely.

Smell!  Do you notice any unusual odors? Burnt components and wires give off a distinctive odor when overheated. Metal will smell hot if there is too much friction.

Touch!  (But very carefully!) If components feel cool, perhaps no current is flowing through the device. If components feel warm, chances are that everything is normal. If components feel hot, everything may be normal too, although it is more likely that the current is too high or the cooling device is not working properly. In any event, there may be a problem worth further investigation.

The point is that by being observant, you have a good chance of discovering the problem or problem area. We will now discuss in more detail the four main areas mentioned earlier and some of the problems that we may encounter in each.

## The Electrical Supply to the Motor and the Drive

Most maintenance technicians believe that the power distribution in an industrial environment is reliable, stable, and free of interference. Nothing could be further from reality. Power outages, voltage spikes and sags, and electrical noise are frequent occurrences. The effect of these phenomena is not as detrimental to motor performance as it can be to the operation of the drive itself. Most DC drives are designed to operate within a range of variation of supply voltages. Typically, the incoming power can vary as much as ±10 percent with no noticeable change in drive performance. However, in the real world, it is not unusual for power line fluctuations to exceed 10 percent. These fluctuations may occasionally cause a controller to "trip," and if they occur repeatedly, a power line regulator may be required to hold the power at a constant level. A power line regulator will be of little use, however, should the power supply to the controller fail. In this situation, an uninterruptible power supply (UPS) will be needed. Several manufacturers produce complete power line conditioning units that combine a UPS with a power line regulator.

Quite often controllers are connected to an inappropriate supply voltage. For example, it is not unusual to find a drive rated 208 volts connected to a 240-volt supply. Likewise, a drive rated 440 volts may be connected to a 460-volt or even a 480-volt source. Usually, the source voltage should not exceed the voltage rating of the drive by more than 10 percent. For a drive rated 208 volts, the maximum supply voltage should not exceed 229 volts $(208 \times 10\% = 20.8 + 208 = 229)$.

Obviously, a 208-volt drive connected to a 240-volt power supply is overvoltaged and should not be used under such circumstances. For a 440-volt drive, the maximum supply voltage should not exceed 484 volts $(440 \times 10\% = 44 + 440 = 484)$. Furthermore, although this value is within acceptable limits, a potential problem still exists. Suppose that the power line voltage fluctuates by 10 percent. If the 440-volt source suffers a 10 percent spike, the voltage will increase to 484 volts. This value falls within the permissible design parameters of the drive. But consider what can happen if we connect a 440-volt drive to a 460-volt or 480-volt power line. If we experience that same 10 percent spike, the voltage will increase to 506 volts in the 460-volt line $(460 \times 10\% = 46 + 460 = 506)$ and to 528 volts in the 480-volt line $(480 \times 10\% = 48 + 480 = 528)$.

Exceeding the voltage rating of the drive to this extent will probably damage some internal drive components. Most susceptible to excessive voltage and spikes or transients are the SCRs, MOSFETs, and power transistors. Premature failure of capacitors can also occur. As you can see, it is very important to match the line voltage to the voltage rating of the drive.

An equally serious problem occurs when the phase voltages are unbalanced. Typically, during construction, care is taken to balance the electrical loads on the individual phases. As time goes by and new construction and remodeling occur, it is not unusual for the loading to become unbalanced. This imbalance will cause intermittent tripping of the controller, which can result in premature failure of certain components.

To determine if phase imbalance exists, you must do the following:

1. Measure and record the phase voltages (L1 to L2, L2 to L3, and L1 to L3).

2. Add the three voltage measurements obtained in step 1 and record the sum of the phase voltages.

3. Divide the sum obtained in step 2 by 3 and record the resulting average phase voltage.

4. Now, subtract the average phase voltage obtained in step 3 from each phase voltage measurement taken in step 1 and record the results. (Treat any negative answers as positive values.) These values are the individual phase imbalances.

5. Add the individual phase imbalances obtained in step 4 and record the resulting total phase imbalance.

6. Divide the total phase imbalance obtained in step 5 by 2 and record the adjusted total phase imbalance.

7. Next, divide the adjusted total phase imbalance from step 6 by the average phase voltage found in step 3 and record the resulting calculated phase imbalance.

8. Finally, multiply the calculated phase imbalance from step 7 by 100 and record this percent of total phase imbalance.

Let's work through an example involving a 440-volt, three-phase supply to a DC drive to see how this procedure works:

1. Assume that L1 to L2 = 437 V, L2 to L3 = 443 V, and L1 to L3 = 444 V.

2. The sum of these phase voltages equals 437 V + 443 V + 444 V, or 1324 V.

3. The average phase voltage equals 1324 V ÷ 3, or 441.3 V.

4. To find the individual phase imbalances, we subtract the average phase voltage from the individual phase voltages and treat any negative values as positive. Therefore
   L1 to L2 = 437 V − 441.3 V, or 4.3 V;
   L2 to L3 = 443 V − 441.3 V, or 1.7 V; and
   L1 to L3 = 444 V − 441.3 V, or 2.7 V.

5. Now, we find the total phase imbalance by adding together these individual phase imbalances: 4.3 V + 1.7 V + 2.7 V = 8.7 V.

6. To find the adjusted total phase imbalance, we divide the total phase imbalance by 2, therefore, 8.7 V ÷ 2 = 4.35 V.

7. Next, we divide the adjusted total phase imbalance by the average phase voltage to find the calculated phase imbalance: 4.35 V ÷ 441.3 V = 0.0099.

8. Finally, we multiply the calculated phase imbalance by 100 to find the percentage total phase imbalance: 0.0099 × 100 = 0.99%.

In this example, the values are within tolerances, and the differences in the phase voltages should not cause any problems. In fact, as long as the percentage total phase imbalance does not exceed two percent, we should not experience any difficulties as a result of the differences in phase voltages.

## What to Check When the Motor or the Load is the Suspected Problem?

Probably the most common cause of motor failure is heat. Excess heat can be simply a result of the motor's operating environment. Many motors are operated in areas of high ambient temperature. If steps are not taken to keep the motor within its operating temperature limits, the motor will ultimately fail.

Some motors have an internal fan to provide cooling. If such a motor is operated at reduced speed, the internal fan may not turn fast enough to cool the motor sufficiently. In these instances, an external fan may be needed to provide additional cooling to the motor. Typically, these fans are interlocked with the motor operation in such a way that the motor will not operate unless the fan operates as well. Therefore, it is possible for a fault in an external fan control to prevent a motor from operating.

The temperature sensors used in motors generally consist of a nonadjustable thermostatic switch that is normally closed and opens only when the temperature rises beyond an acceptable level. Therefore, it may be necessary to wait for an overheated motor to cool down sufficiently before the temperature sensor can be reset and the motor restarted.

Periodic inspection of the motor and any external cooling fans is strongly recommended. The fans should be checked for missing or bent vanes. All openings in the motor's and fan's housing intended to promote cooling should be kept free of obstructions. Any accumulation of dirt, grease, or oil there or elsewhere should be removed. Any filters used in the motor or fan must be cleaned or replaced on a routine schedule.

Heat may also cause other problems. When motor windings become overheated, the insulation on the wires may break down, causing a short circuit that may lead to an "open" condition. A common practice used to find shorts or opens in motor windings is to "Megger" the windings with a megohmmeter. Extreme caution must be taken when using a Megger. When using a Megger on the motor leads, be certain that you have disconnected the leads from the drive. Failure to do so will cause the Megger to apply a high voltage into the output section of the drive, and damage to the power semiconductors will result. You may decide to Megger the motor leads at the drive cabinet. Again, be certain that you have disconnected the leads from the drive unit, and Megger the motor leads and motor winding. Never Megger the output of the drive itself!

Because the motor drives a load of some kind, the load may also create problems in motor operation. The drive may trip out if the load causes the motor to draw an excessive amount of current for too long a time. Most drives display some type of fault indication when this phenomenon occurs. This problem may be a result of excessive motor operating speed. Quite often, a minor reduction in speed is all that is necessary to prevent the repetitive tripping of the drive. The same effect occurs if the motor is truly overloaded by too large a load. Obviously, in this case, either the motor size must be increased or the size of the load decreased to prevent the drive from tripping.

Some loads have a high inertia. They require not only a lot of energy to move them, but once moving, a lot of energy to stop them. If the drive cannot provide sufficient braking action to match the inertia of the load, the drive may trip, or overhaul. A drive with greater braking capacity is needed to prevent the occurrence of overhauling.

## Problems Associated with Feedback Devices and Sensors

Mechanical vibration may loosen the mounting or alter the alignment of feedback devices. Periodic inspections are necessary to verify that these devices are aligned and mounted properly.

It is also important to verify that the wiring to these devices is in good condition and that the terminations are clean and tight. Another consideration related to the wiring of feedback devices is electrical interference. Feedback devices produce low-voltage, low-current signals that are applied to the drive. If the signal wires from these devices are routed next to high-power cables, interference can occur. This interference may result in improper drive operation. To eliminate the possibility of interference, you must follow several steps. First, make certain that the signal wires from the feedback device are installed in a separate conduit. Do not install power wiring and signal wiring in the same conduit. The signal wires should consist of shielded cable, and the shield wire should be properly grounded at the drive cabinet only. Do not ground both ends of a shielded cable. When routing the shielded signal cable to its terminals in the drive cabinet, do not run or bundle the signal cable parallel to any power cables. The signal cable should be routed at a right angle to power cables. Also, do not route the signal cable near any high-power contactors or relays. When the coils of a contactor or relay are energized and de-energized, a spike is produced, which can also create interference in the drive. To suppress this spike, it may be necessary to install a free-wheeling diode across the DC coils or a snubber circuit across the AC coils.

## Problems that Exist in the Drive Itself

First, look for any fault codes or fault indicators. Most drives provide some form of diagnostics, and this can be a great time-saver. The operator's manual interprets the fault codes and gives instructions for clearing the fault condition. However, there are other problems related to the drive that we should be aware of.

Heat can produce problems in the drive unit and elsewhere. The drive cabinet may have one or more cooling fans, with or without filters. These fans are often interlocked with the drive power in such a way that the fan must operate in order for the drive to operate. Make certain that the fans are operational and the filters are cleaned or replaced regularly. The drive's power semiconductors are typically mounted on heat sinks. A small thermostat may be mounted on a heat sink to detect excessive temperatures in the power semiconductors. If the heat sink becomes too hot, the thermostat opens and the drive trips. Usually, these thermostats reset themselves. You must wait for them to cool down and reset before the drive will operate. If overheating occurs repeatedly, a more serious problem exists that will require further investigation.

If the drive is newly installed, problems are often the result of improper adjustments. On the other hand, if the drive has been in operation for some time, it is unlikely that readjustments are needed. All too often, an untrained individual will try to readjust a setting to see if that fixes it. Usually, such readjustments only make things worse! This is not to say that adjustments are never needed. For example, changes in the process being controlled or replacement of some component of the equipment will probably require changes in the drive settings. It is, therefore, very important to record the current drive settings and any changes made to them over the years. This record should be placed in a safe location, and a copy of it should be made and placed in the drive cabinet for easy access by maintenance personnel.

Consider the more common adjustments that may be performed on typical DC drives. Not all drives permit all of these adjustments. Some drives require small adjustments performed with a screwdriver, whereas others allow you to program adjustments using a keypad or jumpers:

- **Acceleration/Deceleration Rates, or Ramp**. This adjustment controls how rapidly the motor speeds up or slows down, as shown in Figure 22–32. If the motor must respond more quickly than its given load allows, the drive will trip.

- **Field Voltage**. This adjustment sets the field voltage when the motor is not running. It not only saves energy, but also lowers the winding temperature and increases motor life expectancy.

- **IR Compensation**. Used to sense cemf from the motor as an indication of motor speed, this adjustment matches the motor's characteristics to the drive. Therefore, readjustment should not be needed unless the motor or the drive is replaced. This adjustment is more commonly found on older drives and is rarely seen on drives that use feedback devices to sense motor speed.

- **Jogging or Inching Speed**. This is the speed of the motor expressed in small increments, usually 10 percent of the motor's full speed.

- **Maximum Current**. This setting allows the motor to draw 150 to 300 percent more current than the motor's maximum rating for a short period of time. The higher the maximum current setting, the shorter the period of time that the motor can draw this current.

- **Minimum Current**. This setting prevents the DC motor from overspeeding in the event that the shunt field circuit opens or the shunt field current becomes too small to produce a magnetic field strong enough to generate sufficient torque.

- **Overspeeding**. Typically, this setting trips the drive if the motor speed exceeds the desired speed by more than 10 percent.

- **Watchdog Circuit**. Adjusted to detect certain levels of electrical interference or "noise," this setting also trips the drive in the event of a voltage sag, spike, or single phasing.

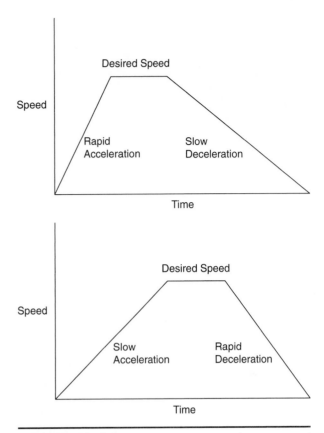

**FIGURE 22–32** Variations in acceleration and deceleration rates.

Remember, you should never adjust these settings unless you have been properly trained and know their effects.

## What to Do if the Drive Still Doesn't Work

If the drive still doesn't work, you might need a new one. Replacing the drive should clearly be the last choice. More often than not, even if the drive is defective, something external to the drive is the reason for the drive's failure. Replacing the drive without determining the cause of its failure may cause damage to the replacement drive. However, whenever a drive fails, regardless of the cause, some possibility exists that you can get it to work again.

Most drive failures occur in the power section, where you will find power SCRs, transistors, MOSFETs, and so on. In some drives, these devices will be individual components. You can test these devices with reasonable accuracy using nothing more than an ohmmeter. In other drives, SCRs, transistors, and MOSFETs are contained within a power module.

An SCR power module is shown in Figure 22–33. One of these modules would be used for each phase of the three-phase AC. Notice that this module

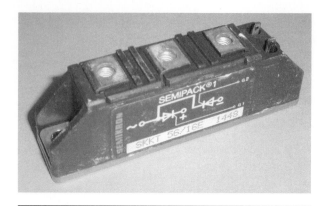

**FIGURE 22–33** SCR power module.

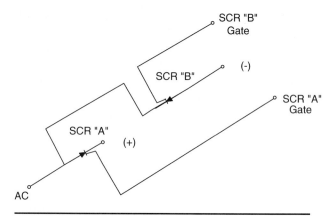

**FIGURE 22–34** SCR power module schematic.

contains two SCRs. We will call the SCR to the left "A" and the SCR to the right "B." You can use an ohmmeter to test an SCR module with reasonable accuracy. To do so, you must understand what each terminal of the module represents. A schematic diagram of the SCR module is shown in Figure 22–34.

Notice in Figure 22–33 that there are a total of five terminals on the SCR module. Referring to Figure 22–34, you will see the same five terminals. Two of the terminals (the small terminals in Figure 22–33) represent the gate connections for each SCR. The gate terminal toward the back of the module is connected to the gate of SCR B, while the gate terminal at the front of the module is connected to the gate of SCR A. The large terminal closest to the gate terminals is the negative terminal of the module, which is also the anode of SCR B. The large terminal farthest from the gate terminals is the AC terminal for the module, which is also the anode of SCR A. Notice that the anode of SCR A is internally connected to the cathode of SCR B. The large terminal in the center is the positive terminal of the module, which is also the cathode of SCR A.

To test the module with an ohmmeter, follow these steps:

1. Place the ohmmeter in the "Ohms $\Omega$" position, if using a digital multimeter. (Use the "200 $\Omega$" position if using an analog multimeter.)

2. Place the black (negative) lead of your ohmmeter on the anode (terminal farthest from the gate terminals) of SCR A.

3. Place the red (positive) lead of your ohmmeter on the cathode (center terminal) of SCR A.

4. Your meter should indicate a very high or infinite resistance. A low-resistance reading indicates a faulty SCR. Replace the module.

5. Reverse your meter connections. Place the black (negative) lead of your ohmmeter on the cathode (center terminal) of SCR A.

6. Place the red (positive) lead of your ohmmeter on the anode (terminal farthest from the gate terminals) of SCR A.

7. Your meter should indicate a very high or infinite resistance. A low-resistance reading indicates a faulty SCR. Replace the module.

8. While leaving your ohmmeter connected as in steps 5 and 6, use a clip lead to connect the gate of SCR A (small terminal at the front of the module) to the red (positive) lead of your ohmmeter.

9. You should notice a drop in the resistance reading. This is a result of triggering SCR A into conduction. If your resistance reading does not drop, the SCR may be faulty and you should replace the module. However, it is possible that your ohmmeter is not supplying sufficient current to trigger the SCR into conduction, and the SCR may be functioning normally. The old saying, "If in doubt, change it out!" would apply.

If SCR A appears normal, repeat the same process to check SCR B:

1. Place the black (negative) lead of your ohmmeter on the anode (terminal closest to the gate terminals) of SCR B.

2. Place the red (positive) lead of your ohmmeter on the cathode (terminal farthest from the gate terminals) of SCR B.

3. Your meter should indicate a very high or infinite resistance. A low-resistance reading indicates a faulty SCR. Replace the module.

4. Reverse your meter connections. Place the black (negative) lead of your ohmmeter on the cathode (terminal farthest from the gate terminals) of SCR B.

5. Place the red (positive) lead of your ohmmeter on the anode (terminal closest to the gate terminals) of SCR B.

6. Your meter should indicate a very high or infinite resistance. A low-resistance reading indicates a faulty SCR. Replace the module.

7. While leaving your ohmmeter connected as in steps 5 and 6, use a clip lead to connect the gate of SCR B (small terminal at the back of the module) to the red (positive) lead of your ohmmeter.

8. You should notice a drop in the resistance reading. This is a result of triggering SCR B into conduction. If your resistance reading does not drop, the SCR may be faulty and you should replace the module. However, it is possible that your ohmmeter is not supplying sufficient current to trigger the SCR into conduction, and the SCR may be functioning normally. Again, "If in doubt, change it out!"

While we are on the subject of power modules, there is another type of power module that you may find in the drive upon which you are working. This module is technically not part of the power output section, although it does have a power function in the drive. The module is a three-phase bridge rectifier module. It is used to convert three-phase AC into rectified DC. A picture of this module appears in Figure 22–35, and a schematic of this module appears in Figure 22–36. Notice that this module has five terminals. The two horizontal terminals at the left end of the module are the (+) and (–) DC connections. The three vertical terminals are the connections for the three-phase AC (L1, L2, and L3).

To test the module with an ohmmeter, follow these steps:

1. Place the ohmmeter in the "Diode Test" position, if using a digital multimeter. (Use the "200 Ω" position if using an analog multimeter.)

2. Place the black (negative) lead of your ohmmeter on the (–) terminal of the module.

3. Place the red (positive) lead of your ohmmeter on AC terminal L1 (this is actually the cathode of diode CR4).

4. Your meter should indicate a very high or infinite resistance. A low-resistance reading indicates a faulty diode. Replace the module.

5. Move the red (positive) lead of your ohmmeter from AC terminal L1 to AC terminal L2 (this is actually the cathode of diode CR5).

6. Your meter should indicate a very high or infinite resistance. A low-resistance reading indicates a faulty diode. Replace the module.

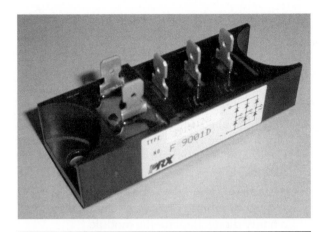

**FIGURE 22–35**  Three-phase bridge rectifier module.

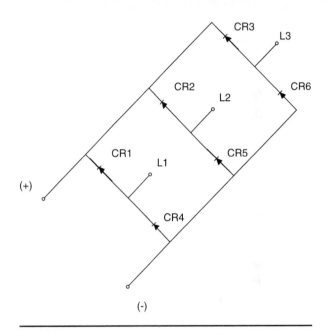

**FIGURE 22–36**  Three-phase bridge rectifier module schematic.

7. Move the red (positive) lead of your ohmmeter from AC terminal L2 to AC terminal L3 (this is actually the cathode of diode CR6).

8. Your meter should indicate a very high or infinite resistance. A low-resistance reading indicates a faulty diode. Replace the module.

9. Reverse your meter connections. Place the red (positive) lead of your ohmmeter on the (–) terminal of the module.

10. Place the black (negative) lead of your ohmmeter on AC terminal L1 (this is actually the cathode of diode CR4).

11. Your meter should indicate a low resistance. A high- or infinite-resistance reading indicates a faulty diode. Replace the module.

12. Move the black (negative) lead of your ohmmeter from AC terminal L1 to AC terminal L2 (this is actually the cathode of diode CR5).

13. Your meter should indicate a low resistance. A high- or infinite-resistance reading indicates a faulty diode. Replace the module.

14. Move the black (negative) lead of your ohmmeter from AC terminal L2 to AC terminal L3 (this is actually the cathode of diode CR6).

15. Your meter should indicate a low resistance. A high- or infinite-resistance reading indicates a faulty diode. Replace the module.

16. Place the red (positive) lead of your ohmmeter on the (+) terminal of the module.

17. Place the black (negative) lead of your ohmmeter on AC terminal L1 (this is actually the anode of diode CR1).

18. Your meter should indicate a very high or infinite resistance. A low-resistance reading indicates a faulty diode. Replace the module.

19. Move the black (negative) lead of your ohmmeter from AC terminal L1 to AC terminal L2 (this is actually the anode of diode CR2).

20. Your meter should indicate a very high or infinite resistance. A low-resistance reading indicates a faulty diode. Replace the module.

21. Move the black (negative) lead of your ohmmeter from AC terminal L2 to AC terminal L3 (this is actually the anode of diode CR3).

22. Your meter should indicate a very high or infinite resistance. A low-resistance reading indicates a faulty diode. Replace the module.

23. Reverse your meter connections. Place the black (negative) lead of your ohmmeter on the (+) terminal of the module.

24. Place the red (positive) lead of your ohmmeter on AC terminal L1 (this is actually the anode of diode CR1).

25. Your meter should indicate a low resistance. A high- or infinite-resistance reading indicates a faulty diode. Replace the module.

26. Move the red (positive) lead of your ohmmeter from AC terminal L1 to AC terminal L2 (this is actually the anode of diode CR2).

27. Your meter should indicate a low resistance. A high- or infinite-resistance reading indicates a faulty diode. Replace the module.

28. Move the red (positive) lead of your ohmmeter from AC terminal L2 to AC terminal L3 (this is actually the anode of diode CR3).

29. Your meter should indicate a low resistance. A high- or infinite-resistance reading indicates a faulty diode. Replace the module.

Once you have completed testing the module, and you determine that the module is defective, you can usually obtain a substitute part from a local electronics parts supplier. If the part is not available, you will have to return the drive to the manufacturer for service or call a service technician for on-site repairs.

If the problem is not in the drive's power section, then it must be located in its control section. The electronics in the control section are more complex; therefore, troubleshooting is not recommended. In this event, you should definitely return the drive to the manufacturer for repair or call a service technician for on-site repairs.

## REVIEW QUESTIONS

*Multiple Choice*

1. Torque is best defined as
   a. an opposing voltage produced in a motor's armature.
   b. the difference between the actual speed and the desired speed of a motor.
   c. an undesirable phenomenon, and steps should be taken to eliminate it.
   d. the turning force produced by a motor.

2. An opto-coupler
   a. protects low-voltage circuits from high-voltage circuits.
   b. provides electrical noise immunity.
   c. is also called an opto-isolator.
   d. a and b only.
   e. a, b, and c.

3. The component responsible for producing the higher output levels in a boost chopper is
   a. a capacitor.
   b. a MOSFET.
   c. an inductor.
   d. a and c.
   e. a, b, and c.

4. In which operating quadrant does motoring in reverse occur?
   a. First
   b. Second
   c. Third
   d. Fourth

5. In which quadrant(s) must a motor operate to achieve both forward and reverse operation with braking capabilities in both directions?
   a. First
   b. Second
   c. Third
   d. Fourth

*True or False*

1. T or F   A higher percentage of regulation is better than a lower percentage or regulation.
2. T or F   Counter emf is undesirable.
3. T or F   The summing point is that point in a circuit where the output from the sawtooth generator is joined with the output from the preamplifier.
4. T or F   The SCR armature voltage controller is the most commonly used method for controlling the speed of a DC motor.
5. T or F   Pulse-width modulation is used with the MOSFETs in a chopper.
6. T or F   Varying the frequency of the control signal does not affect the output voltage level of a buck chopper.
7. T or F   Varying the frequency of the control signal does not affect the output voltage level of a boost chopper.
8. T or F   Motor speed increases as the "on" pulse becomes wider.
9. T or F   Increasing the frequency of the "on" pulse will cause the motor's speed to increase.
10. T or F   Increasing the frequency of the "on" pulse will cause the motor to operate more smoothly.
11. T or F   An inoperable cooling fan can prevent a drive from operating.
12. T or F   When using shielded cable on feedback devices, you must ground the shield wire at both ends of the cable.

*Give Complete Answers*

1. Name the four parameters of a DC motor that a DC drive controls.
2. Name at least two sources of the feedback signal.
3. Explain the importance of the feedback signal. Why is it needed?
4. What does the term *error signal* mean?
5. What is meant by counter emf (cemf)?
6. Describe the operation of a tachometer-generator.
7. Is the voltage generated by a tachometer-generator directly proportional or inversely proportional to motor speed?
8. Explain the function of the comparator stage.
9. Explain what is meant by closed-loop control.
10. What is another name for closed-loop control?
11. If the load on a DC motor varies frequently, would you use a DC drive with open- or closed-loop control? Why?
12. Explain the effect of increasing the armature voltage on the speed of a DC motor.
13. List some of the differences between the SCR armature voltage controller and the switching amplifier field current controller.
14. List some of the similarities between the SCR armature voltage controller and the switching amplifier field current controller.
15. Explain the process by which the SCR armature voltage controller adjusts the speed of a DC motor.
16. Explain what a chopper does.
17. What is the purpose of the free-wheeling diode in the buck chopper?
18. Is the length of time that a MOSFET is switched on directly proportional or inversely proportional to the output voltage level of a buck chopper? Explain your answer.
19. Is the length of time that a MOSFET is switched on directly proportional or inversely proportional to the output voltage level of a boost chopper? Explain your answer.
20. Describe each of the four quadrants as they affect motor operation.
21. Explain the term *pulse-width modulation* (PWM).
22. Name the four main areas to check when a problem with a DC drive system occurs.
23. Where is the best place to begin troubleshooting a DC drive system problem? Why?
24. List some safety steps that you should follow prior to working on a DC drive system.

25. Is it permissible to connect a DC drive to a supply voltage that is higher than the nameplate rating of the drive? Why or why not?

26. Explain phase imbalance.

27. Calculate the phase imbalance for a supply to a DC drive that has the following voltages: L1 to L2 = 209 V; L2 to L3 = 205 V; and L1 to L3 = 210 V.

28. Describe the cooling problem created by using a variable speed drive to operate a motor with a shaft-mounted fan.

29. Describe the dangers of using a Megger on a DC drive system.

30. Explain how the motor load can affect the performance of a DC drive.

31. Describe any precautions that you should observe when routing the feedback device cable outside the drive cabinet.

32. Describe any precautions that you should observe when routing the feedback device cable inside the drive cabinet.

33. Explain when it is permissible to readjust drive settings.

34. Describe the purpose of the IR compensation adjustment.

35. Which two adjustments inhibit DC motor overspeeding?

36. Explain when a DC drive should be returned to the manufacturer for repair.

# AC (Inverter) Drives

## OBJECTIVES

After studying this chapter, the student will be able to:

- Name the various sections of an AC drive.
- State the general operating principle of an AC drive.
- Describe the operation of a variable voltage inverter.
- Describe the operation of a pulse-width modulated variable voltage inverter.
- Describe the operation of a current source inverter.
- Discuss the operation of a flux vector drive.
- Identify problems and troubleshoot an AC drive.

# INVERTER DRIVES

For many years, the mainstay motor of industry has been the three-phase, squirrel-cage induction motor. This motor has the advantages of low cost and low maintenance. Its biggest disadvantage has been its fixed operating speed. If you needed a three-phase induction motor with variable speed, you had to use a wound-rotor induction motor with a potentiometer. This configuration entailed added expense and increased maintenance.

Fortunately, we now have AC drives, more commonly called *inverters* (Figure 23–1). Inverters are now used not only to vary the speed of the squirrel-cage motor, but also to vary its torque, start the motor slowly and smoothly, and increase the motor's efficiency.

An AC drive has three basic sections: the *converter*, the *DC filter*, and the inverter. Figure 23–2 shows the inside view of an inverter drive. The converter rectifies the applied AC into DC. The DC filter (also called the *DC link* or *DC bus*) provides a smooth, rectified DC. The inverter switches the DC on and off so rapidly that the motor receives a pulsating DC that appears similar to AC. Because this DC switching is controlled, we can vary the frequency of the artificial AC that is applied to the motor, something that we are normally unable to do.

The AC line frequency is set by the electric utility companies across the United States at a standard rate of 60 hertz. Because the AC line frequency could not normally be varied, the speed of a squirrel-cage motor was, for the most part, fixed. The number of the motor's poles could be increased or decreased, causing the motor to slow down or speed up, respectively. However, this could only be done as the motor was being built. We could also vary the stator voltage. However, doing so reduces the motor's ability to drive a load at low speeds, so this method has limited usefulness.

Today, we can use inverters to vary the frequency of the "AC" (the pulsating DC) applied to the squirrel-cage motor, and thus vary the motor's speed, while maintaining constant torque. As we will see, in today's inverter drives, both the stator voltage and the frequency are variable.

The different methods used to create the pulsating DC constitute the basic differences among the various types of inverter drives. Another area of design variation is that of the ratio of *volts to hertz* (V/Hz). The V/Hz ratio should be maintained at a constant value. This means that a motor turning

**FIGURE 23–1**  Carotron Vista II digital inverter drive.

**FIGURE 23–2**  Inside view of a digital inverter drive.

at 1800 rpm, operating from 208 volts at 60 hertz, would have to operate from 104 volts at 30 hertz to attain a speed of 900 rpm, because these values do not change the V/Hz ratio from its original value.

There are two major types of inverter drives: the *voltage source inverter*, also known as the *voltage fed inverter* (VSI), and the *current source inverter* (CSI). The voltage source inverter may be further subdivided into the categories of the *variable voltage inverter* and the *pulse-width modulated inverter*.

## Variable Voltage Inverter

A variable voltage inverter (VVI) is one of two categories of adjustable frequency, variable speed AC drives. In a VVI, the DC voltage is controlled, and the DC is free to respond to the motor needs. The VVI uses a converter, a DC link, and an inverter to vary the frequency of the applied AC voltage.

A converter is a circuit that changes the incoming AC power (fixed voltage, fixed frequency) into DC power. The converter circuit is simply a rectifier circuit that produces an unfiltered, pulsating DC. The converter can be either single-phase or three-phase, depending on the type of power that the AC induction motor you are using needs.

A typical DC link, also called a DC filter or DC bus, is simply a filter circuit composed of an inductor and a capacitor. The purpose of the DC link is to filter or smoothen the AC ripple from the output of the converter stage. The filtered DC from the DC link is fed via the DC bus to the input of the inverter stage.

An inverter converts the applied DC voltage to a pulsating DC voltage. Because we can vary the magnitude and the frequency of this pulsating DC, we can, therefore, control the speed of an AC induction motor. This pulsating DC power acts as artificial AC power on the induction motor.

The VVI shown in the schematic diagram in Figure 23–3 is divided into three basic parts: the converter, the DC link, and the inverter. Basically, three-phase AC is applied to the converter stage of the VVI. Naturally, this AC has both fixed amplitude and fixed frequency. The converter rectifies the AC into DC, which is then smoothed by the DC link. This filtered DC is applied to the inverter, which chops the DC into pulsating AC. This AC is then applied to the motor. The chopping rate applied to the DC varies the frequency of the AC applied to the motor.

We can look at this in more detail in Figures 23–4A through 23–4D, which are simplified schematics of the inverter stage of a six-step VVI. The inverter stage functions in the same way regardless of whether you use the phase control or the chopper control method, both of which we will study later. In the first one-sixth of the cycle, transistors Q1, Q5, and Q6 conduct. In the second one-sixth of the cycle, transistors Q1 and Q6 continue to conduct; however, transistor Q5 turns off, and transistor Q2 turns on. During the third one-sixth of the cycle, transistors Q2 and Q6 continue to conduct, but transistor Q1 turns off, and transistor Q4 conducts. During the fourth one-sixth of the cycle, transistors Q2 and Q4 remain on, while transistor Q6 turns off and Q3 turns on. After two more steps, the motor will have completed one revolution. This six-step cycle is the reason for the inverter's name.

By turning the transistors on at a lower or higher rate, we can vary the frequency of the voltage applied to the motor. As the frequency varies, the inductive reactance of the motor windings also varies proportionally. An increase in reactance causes a decrease in current at higher speeds, and as a result the motor will not function properly. Therefore, the voltage must be increased proportionally to the

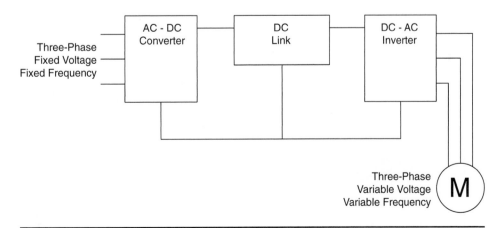

**FIGURE 23–3** Variable voltage inverter block diagram.

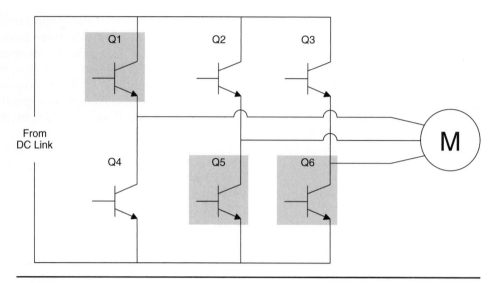

**FIGURE 23–4A**   Six-step inverter with Q1, Q5, and Q6 conducting.

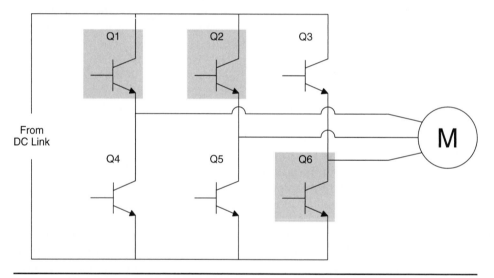

**FIGURE 23–4B**   Six-step inverter with Q1, Q2, and Q6 conducting.

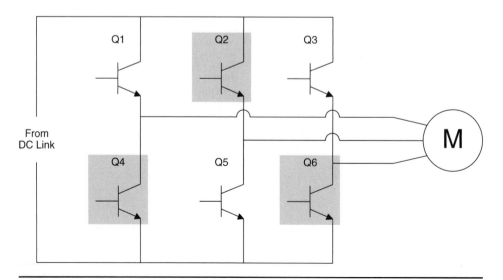

**FIGURE 23–4C**   Six-step inverter with Q2, Q4, and Q6 conducting.

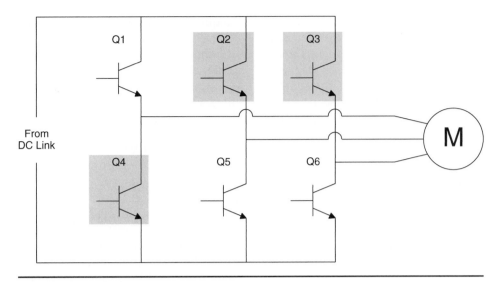

**FIGURE 23–4D**  Six-step inverter with Q2, Q3, and Q4 conducting.

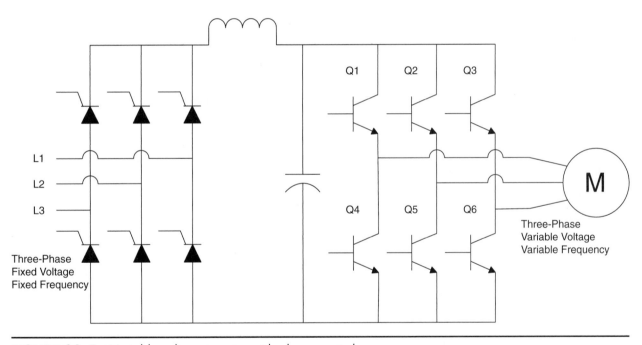

**FIGURE 23–5**  Variable voltage inverter with phase control.

frequency, in a ratio known as the volts/hertz ratio, or V/Hz. The V/Hz ratio must be kept constant, and this can be accomplished by one of two methods: phase control or chopper control.

In Figure 23–5, the converter stage is composed of a three-phase SCR bridge circuit. Remember that we can control when and for how long an SCR conducts. If we turn the SCR on early in the cycle, we can provide a higher average voltage to the inverter because the SCR will conduct for a longer portion of the cycle. If the SCR is turned on late in the cycle, the SCR will conduct for a shorter period of time; therefore, the inverter will receive a lower average

voltage. By varying the DC bus voltage level in this way and simultaneously varying the frequency of the output from the inverter section, we can maintain a constant V/Hz ratio.

The V/Hz ratio (also called the V/f ratio) is very important for a number of reasons. Remember that we need to be concerned with many characteristics when dealing with AC and inductors. We not only have to be aware of voltage, current, and resistance, but we must also recognize the importance of reactance, hysteresis, and eddy currents. We will see what can happen if we fail to maintain a constant V/Hz ratio.

Suppose that we could increase the voltage applied to the motor without adjusting the frequency of the applied voltage. What would happen? The increased voltage would produce increased magnetic flux, which in turn would saturate the iron components of the motor. This flux would cause increased iron losses in the form of hysteresis and eddy currents. It would also increase the stator current and possibly damage the motor windings as a result.

If we examine what happened to the ratio of volts to hertz (V/Hz) in the preceding example, we would see that the ratio increased because the frequency remained constant as the voltage increased. The same effect occurs if the voltage is kept constant as the frequency is reduced. Excessive current will flow, which will result in more heat produced. As we saw in the example, the net results of such a change to the V/Hz ratio are the disruption of normal motor operation and the potential for serious damage to the motor. That is why it is so important for continued and proper motor operation that we maintain a constant V/Hz ratio.

In Figure 23–6, we begin again with the same inverter section discussed earlier. Here we have returned to a simple three-phase diode bridge rectifier circuit. However, we have now added a chopper, which is nothing more than an electronic switch. This switch may be a transistor or, more commonly, a MOSFET.

Here is how the chopper works. Three-phase AC is applied to the three-phase diode bridge. The bridge rectifies the AC into a fixed DC voltage. Capacitor C1 filters the DC for the chopper circuit. The smoothed DC is then "chopped" at a fixed

frequency by electronic switches; however, the ratio of the chopper's "on" time to "off" time is varied. If the chopper is turned on for a longer period of time, a higher DC voltage is applied to the inverter. If the chopper is turned off for a longer period of time, the DC applied to the inverter will be lower. Varying these on and off times allows us to maintain a constant V/Hz ratio.

All types of VVIs provide open-circuit protection, can handle multiple motor applications and undersized motors, and do not require high-speed switching devices. These drives are quite light in weight, utilize relatively simple control circuits, and exhibit good efficiency at low speeds.

On the negative side, all VVIs lack short-circuit protection, and they cannot handle oversized motors. Another disadvantage that both types of VVIs share is that they do not operate smoothly at low frequencies (<6 hertz). Regenerative braking is not possible with a chopper-controlled VVI, but is possible with a phase-controlled VVI. On the other hand, a chopper-controlled VVI can be operated from batteries, whereas a phase-controlled VVI cannot.

## Pulse-Width Modulated Variable Voltage Inverter

Another form of variable voltage inverter uses pulse-width modulation (PWM) to vary the voltage applied to an AC induction motor. Many similarities exist between this type of inverter and the DC drives discussed earlier.

Notice in Figure 23–7 that the feedback device used is a tachometer-generator. (You will find it

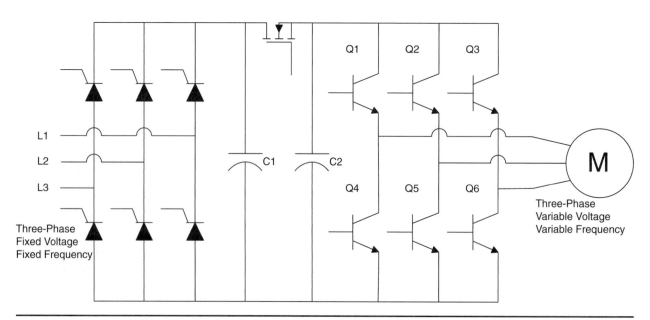

**FIGURE 23–6**  Variable voltage inverter with chopper control.

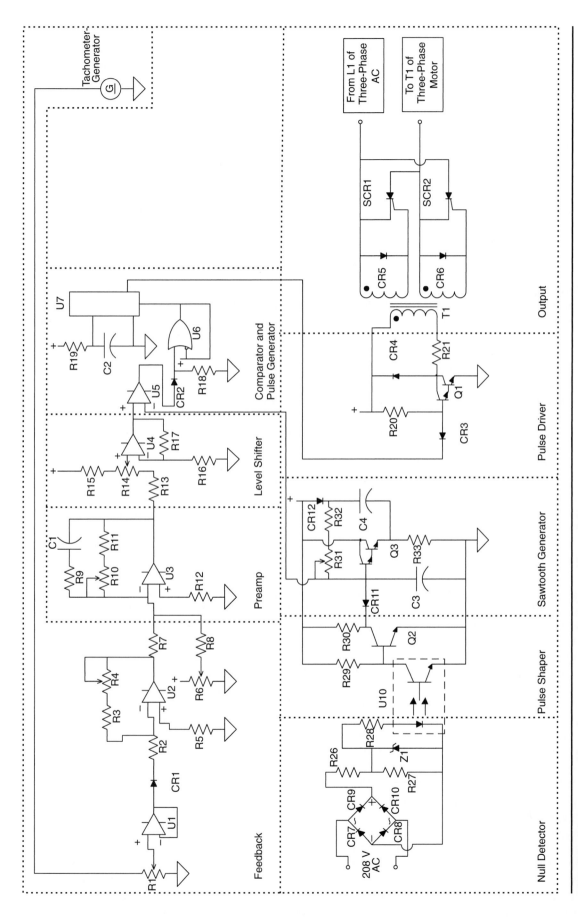

**FIGURE 23-7** Variable voltage inverter with pulse-width modulation.

located in the upper right corner of the schematic.) Recall that the tachometer-generator is a DC generator attached to an AC induction motor's shaft. As the AC induction motor turns, the DC tachometer-generator also turns, producing a positive DC voltage. The faster the AC induction motor turns, the more positive DC voltage the DC tachometer-generator produces. In other words, the output of the DC tachometer-generator is proportional to the speed of the AC induction motor.

Now, we will consider the feedback section, shown in Figure 23–8. The positive voltage, or feedback signal, produced by the DC tachometer-generator is fed through R1 into buffer amplifier U1. The output of U1 is then fed to the inverting input of U2, which inverts the polarity of the tachometer-generator voltage from positive to negative. The resulting negative voltage is then fed through R7 to the noninverting input of U3 in the preamplifier section. Notice that one end of R8 is also connected to the noninverting input of U3 and that the other end of R8 is connected to variable resistor R6. Resistor R6 allows us to set the reference voltage level, sometimes called the command signal, which is the voltage necessary for the AC induction motor to run at a certain speed. The reference voltage (positive) and the tachometer-generator voltage (negative) are added together at the junction of R7 and R8, which is known as the summing point. The difference of these two voltages is the value of the voltage that will appear at the noninverting input of U3. This value is known as the error signal. Now let's leave this section for a moment and move on to the null detector and sawtooth generator circuits.

Notice that in the null detector shown in Figure 23–9, the applied AC voltage is rectified by the full-wave bridge circuit consisting of CR7–CR10. This process causes an unfiltered, pulsating DC voltage to appear across zener diode Z1 and across R28 and the LED of opto-coupler U10.

What happens as the pulsating DC voltage rises from 0 volt to its peak value? The LED of U10 does not conduct until the DC voltage reaches the turn-on voltage of the LED. Assume that this value is approximately 1.5 volts. As the DC voltage increases from 0 volt to 1.5 volts, U10 is off. Therefore, the output of U10 is positive, causing Q2 to be on. When Q2 conducts, a negative pulse appears at the collector of Q2. This negative pulse is used to reset the output of the sawtooth generator back to 0.

What happens when the LED of U10 turns on? As the pulsating DC voltage rises, eventually it will reach 1.5 volts. The voltage across the LED of U10 will never exceed the voltage rating of zener diode Z1. At this point, the LED of U10 turns on and conducts, causing the output of U10 to decrease to a level that turns Q2 off. With Q2 turned off, the sawtooth generator outputs a sawtooth waveform to the inverting input of U5 (the comparator). The sawtooth generator continues to output this waveform until it is reset when Q2 turns on, that is, when the pulsating DC voltage falls below 1.5 volts and decreases to 0 volt. Therefore, the pulsating DC voltage causes the sawtooth generator to provide a reference signal to the comparator. This reference signal is used to vary the firing angle of the SCRs that control the stator voltage.

Consider U5, the comparator, in Figure 23–10. The comparator has two inputs. One comes from the feedback circuit and is a measure of the speed of the armature. The other input, as we just learned, is a ramp signal from the sawtooth generator. The noninverting input from the feedback amplifier will be constant if the load on the armature is constant. In contrast, the inverting input is a rising amplitude signal from the sawtooth generator. Initially, the noninverting input will cause the output of U5 to be positive. However, as the inverting input ramp signal climbs, a point is reached where the amplitudes of both inputs are equal. At this point, the output of U5 will switch off, becoming a rectangular pulse. The width of this pulse is determined by the length of time it takes for both inputs to become equal. This is what determines whether the SCRs conduct earlier or later within each half cycle.

The pulse generator shown in Figure 23–11 is an integrated circuit that contains a one-shot, or monostable, multivibrator. Its purpose is to trigger on the output of the comparator and, in turn, provide a narrow pulse to Q1, the pulse driver.

Referring to Figure 23–12, recall that Q1 does not conduct until a positive pulse appears at its base. When this occurs, current flows through the primary of T1. Furthermore, Q1 conducts only for the duration of the pulse applied to its base. The current flowing through the primary of T1 causes a current flow in the secondary of T1. Therefore, the gates of SCR1 and SCR2 simultaneously receive a trigger pulse.

However, note that the anode of SCR1 and the cathode of SCR2 are connected to motor lead T1. The cathode of SCR1 and the anode of SCR2 are connected to L1 of the AC source. As a result, when the cathode of SCR1 is positive, the anode of SCR2 is positive. When the cathode of SCR1 is negative, the anode of SCR2 is negative. Therefore, when SCR1 and SCR2 are triggered, only the SCR with the positive anode or negative cathode conducts. This causes the SCRs to conduct alternately. Thus, AC flows through the stator of the AC induction motor. For simplicity's sake, we have considered only one phase of a three-phase controller here. The same circuitry is duplicated for each phase.

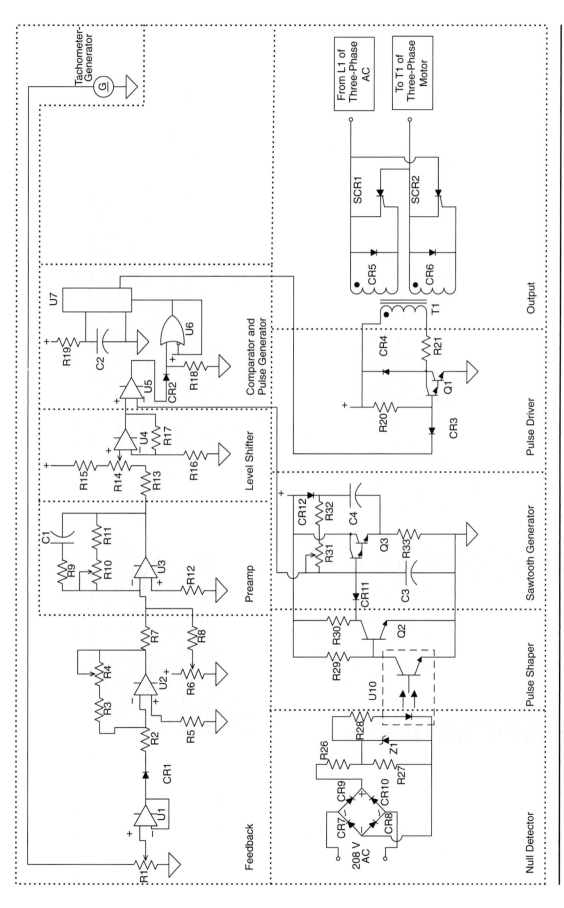

**FIGURE 23-8** Variable voltage inverter with pulse-width modulation feedback section.

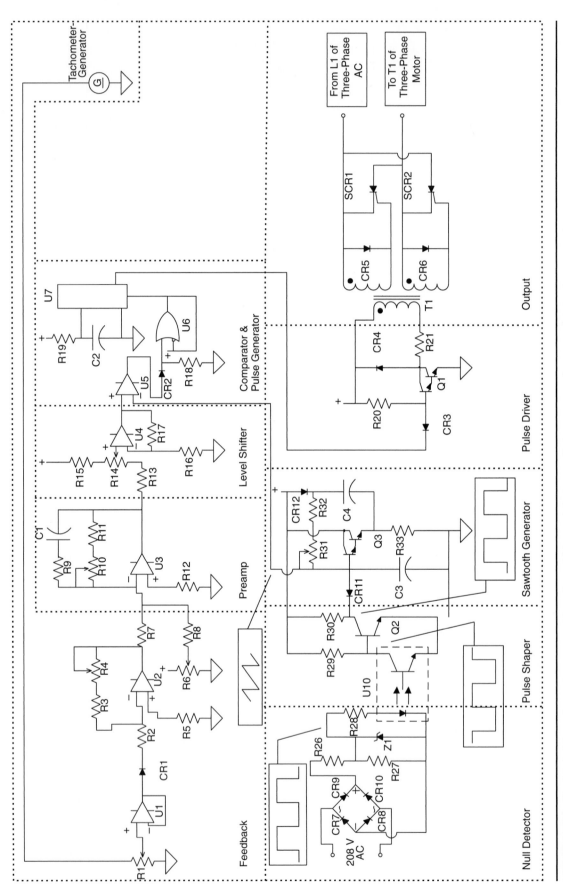

**FIGURE 23-9** Variable voltage inverter with pulse-width modulation null detector and sawtooth generator section.

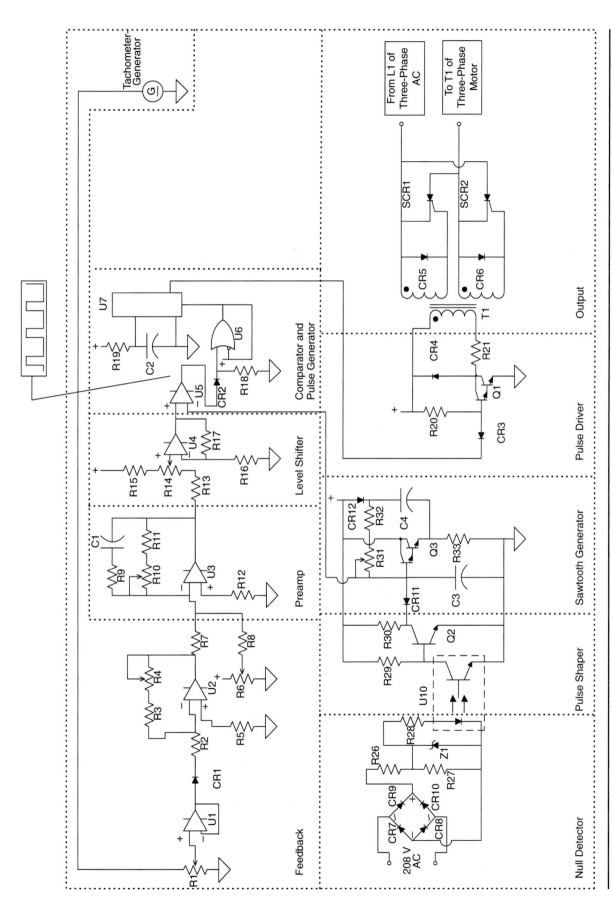

**FIGURE 23-10** Variable voltage inverter with pulse-width modulation comparator section.

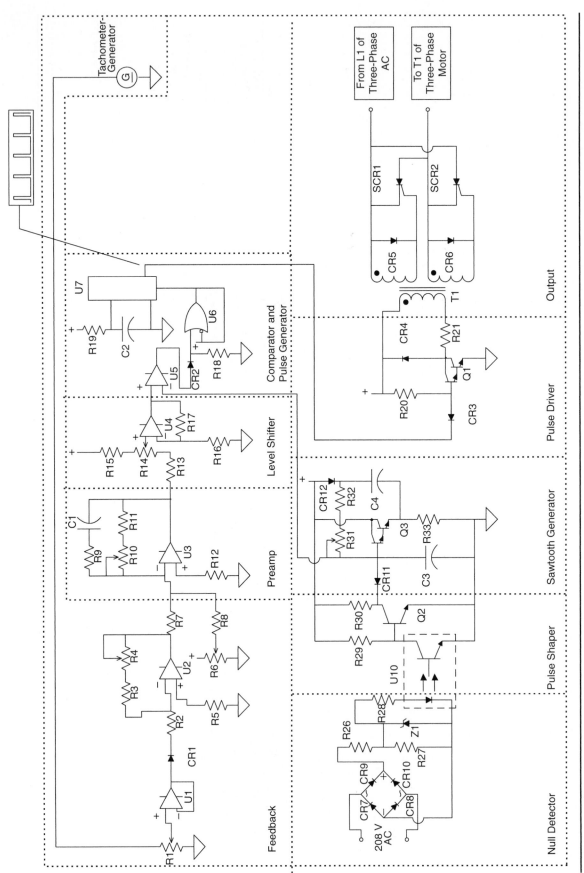

**FIGURE 23–11** Variable voltage inverter with pulse-width modulation pulse generator section.

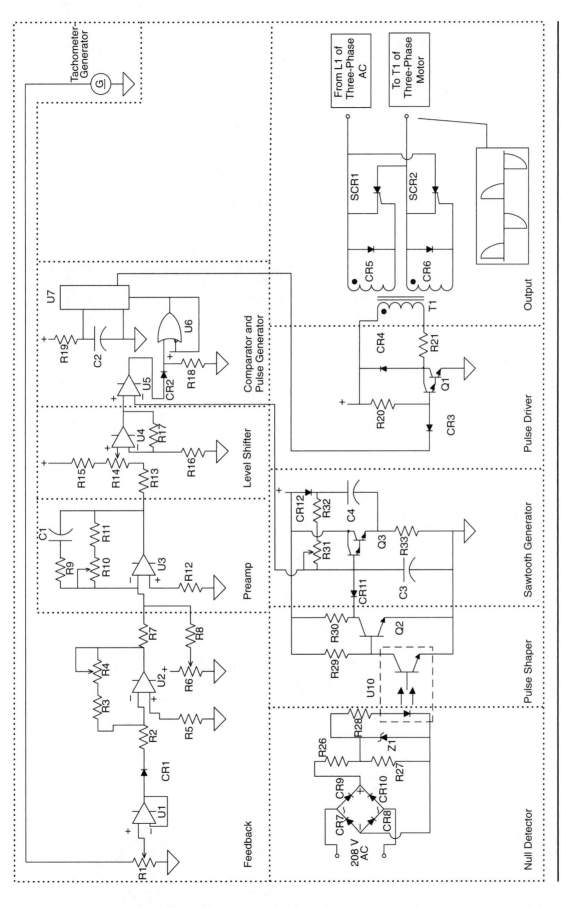

**FIGURE 23–12** Variable voltage inverter with pulse-width modulation output section.

Now, let's bring this all together. Figure 23–13 shows the waveforms at various points in the comparator and pulse generator, pulse driver, and output stages. These waveforms have been lined up vertically so that you understand the timing of the waveforms that must occur in order for the output SCRs to trigger at the appropriate time. Here's how it works. The wave pattern in Figure 23–13 waveform (a) shows the sawtooth waveform at the inverting input of U5, the comparator. At the same time, Figure 23–13 waveform (a) shows the DC voltage that is applied to the noninverting input of comparator U5. Recall that this voltage is the feedback voltage from the tachometer-generator and is an indication of the armature speed. The DC voltage level shown is representative of a motor speed of 1800 rpm. These two inputs cause the output of the comparator to produce a pulsating DC, as shown in Figure 23–13 waveform (b). Notice that the pulse is initially positive and is switched off when the two inputs of the comparator become equal. The output of U5 remains off until the sawtooth generator resets. At this point, the output of U5 switches on and remains on until the sawtooth signal is equal to the feedback voltage.

The pulsating output of U5 is fed through diode CR2 to the input of U6. Because of CR2, the input signal to U6 will be inverted compared to the output of U5. This is shown in Figure 23–13 waveform (c). The positive pulses from the output of U6 trigger the one-shot, U7. The output of the one-shot, U7, will appear in phase with the input pulses, as seen in Figure 23–13 waveform (d).

The output of the one-shot, U7, is fed through diode CR3 to the base of the Darlington transistor, Q1. Notice in Figure 23–13 waveform (e) that the collector of Q1 is 180 degrees out of phase with the output of the one-shot. When Q1 conducts, current

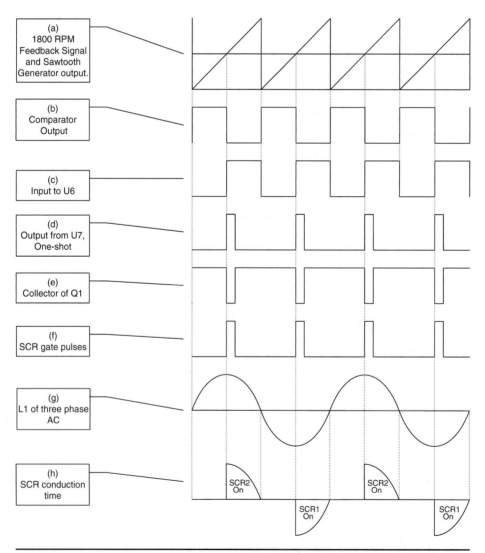

**FIGURE 23–13**  Waveform timing and SCR conduction time at 1800 rpm.

will flow through the pulse transformer, T1. When current flows through pulse transformer T1, a pulse appears at the gates of SCR1 and SCR2, as shown in Figure 23–13 waveform (f). Recall that only an SCR that is properly biased will conduct. The SCR that is triggered and properly biased will conduct until the AC voltage drops to the zero-crossing point. [The AC voltage waveform is shown in Figure 23–13 waveform (g)]. The next pulse from T1 will cause the other SCR to conduct. Therefore, the SCRs will conduct alternately, as shown in Figure 23–13 waveform (h).

Let's assume that an increase in load has caused the motor speed to decrease to 1000 rpm. Refer to Figure 23–14. The wave pattern in Figure 23–14 waveform (a) shows the sawtooth waveform at the inverting input of U5, the comparator. At the same time, Figure 23–14 waveform (a) also shows the DC

voltage that is applied to the noninverting input of comparator U5. The DC voltage level shown is representative of a motor speed of 1000 rpm. Notice that this level is lower than that shown in Figure 23–13 waveform (a) as a result of the lower motor speed and lower output voltage from the tachometer-generator. These two inputs cause the output of the comparator to produce a pulsating DC, as shown in Figure 23–14 waveform (b). Notice that the pulse is initially positive and is switched off when the two inputs of the comparator become equal. The output of U5 remains off until the sawtooth generator resets. At this point, the output of U5 switches on and remains on until the sawtooth signal is equal to the feedback voltage. Compare the width of the positive pulses in Figure 23–14 waveform (b) with those in Figure 23–13 waveform (b). Notice that the positive pulses are narrower in Figure 23–14 waveform (b).

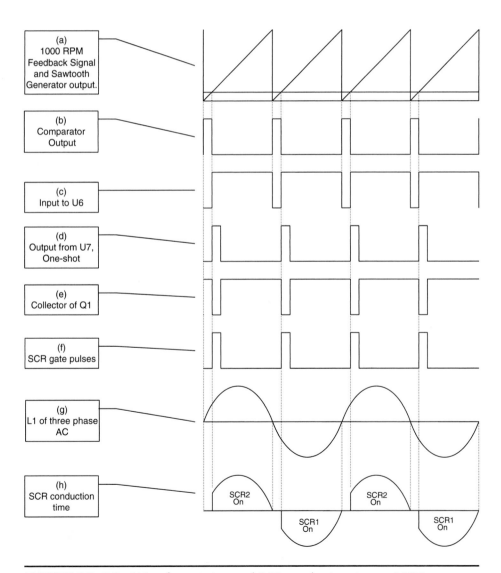

**FIGURE 23–14**  Waveform timing and SCR conduction time at 1000 rpm.

The pulsating output of U5 is fed through diode CR2 to the input of U6. Because of CR2, the input signal to U6 will be inverted compared to the output of U5, as shown in Figure 23–14 waveform (c). Again, compare the positive pulses in Figure 23–14 waveform (c) with those shown in Figure 23–13 waveform (c). Notice that the positive pulses are wider in Figure 23–14 waveform (c) than those shown in Figure 23–13 waveform (c). The positive pulses from the output of U6 trigger the one-shot, U7. The output of the one-shot, U7, will appear in phase with the input pulses, as seen in Figure 23–14 waveform (d). Notice that the output pulse of the one-shot now occurs earlier than it did previously [see Figure 23–13 waveform (d)]. This will cause the SCRs to fire sooner, causing them to conduct for a longer period of time. Let's see if this is what occurs.

The output of the one-shot, U7, is fed through diode CR3 to the base of the Darlington transistor, Q1. Notice in Figure 23–14 waveform (e) that the collector of Q1 is 180 degrees out of phase with the output of the one-shot. When Q1 conducts, current will flow through the pulse transformer, T1. When current flows through pulse transformer T1, a pulse appears at the gates of SCR1 and SCR2, as shown in Figure 23–14 waveform (f). Recall that only an SCR that is properly biased will conduct. The SCR that is triggered and properly biased will conduct until the AC voltage drops to the zero-crossing point. The next pulse from T1 will cause the other SCR to conduct. Therefore, the SCRs will conduct alternately, as shown in Figure 23–14 waveform (g). Notice that the SCRs conduct for a longer amount of time as compared to the conduction of the SCRs in Figure 23–13 waveform (g). This results in a higher average voltage applied to the AC motor. This will cause the speed of the AC motor to increase.

Let's assume that a decrease in load has caused the motor speed to increase to 2600 rpm. Refer to Figure 23–15. The wave pattern in Figure 23–15 waveform (a) shows the sawtooth waveform at the

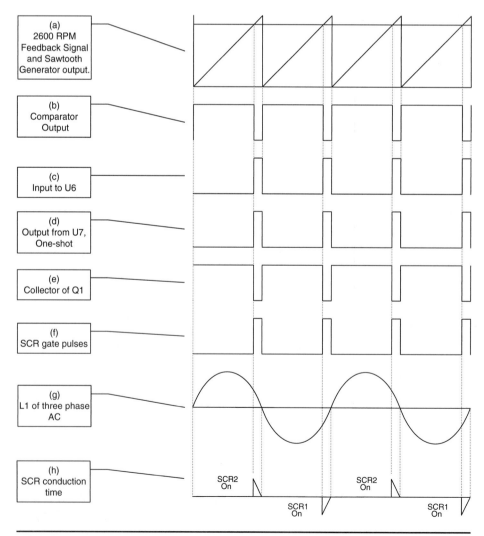

**FIGURE 23–15** Waveform timing and SCR conduction time at 2600 rpm.

inverting input of U5, the comparator. At the same time, Figure 23–15 waveform (a) also shows the DC voltage that is applied to the noninverting input of comparator U5. The DC voltage level shown is representative of a motor speed of 2600 rpm. Notice that this level is higher than that shown in Figure 23–13 waveform (a). This is a result of the higher motor speed and higher output voltage from the tachometer-generator. These two inputs cause the output of the comparator to produce a pulsating DC, as shown in Figure 23–15 waveform (b). Notice that the pulse is initially positive and is switched off when the two inputs of the comparator become equal. The output of U5 remains off until the sawtooth generator resets. At this point, the output of U5 switches on and remains on until the sawtooth signal is equal to the feedback voltage. Compare the width of the positive pulses in Figure 23–15 waveform (b) with those in Figure 23–13 waveform (b). Notice that the positive pulses are wider in Figure 23–15 waveform (b).

The pulsating output of U5 is fed through diode CR2 to the input of U6. Because of CR2, the input signal to U6 will be inverted compared to the output of U5, as shown in Figure 23–15 waveform (c). Again, compare the positive pulses in Figure 23–15 waveform (c) with those shown in Figure 23–13 waveform (c). Notice that the positive pulses are narrower in Figure 23–15 waveform (c) than those shown in Figure 23–13 waveform (c). The positive pulses from the output of U6 trigger the one-shot, U7. The output of the one-shot, U7, will appear in phase with the input pulses, as seen in Figure 23–15 waveform (d). Notice that the output pulse of the one-shot now occurs later than it did previously [see Figure 23–13 waveform (d)]. This will cause the SCRs to fire later, causing them to conduct for a shorter period of time. Let's see if this is what occurs.

The output of the one-shot, U7, is fed through diode CR3 to the base of the Darlington transistor, Q1. Notice in Figure 23–15 waveform (e) that the collector of Q1 is 180 degrees out of phase with the output of the one-shot. When Q1 conducts, current will flow through the pulse transformer, T1. When current flows through pulse transformer T1, a pulse appears at the gates of SCR1 and SCR2, as shown in Figure 23–15 waveform (f). Recall that only an SCR that is properly biased will conduct. The SCR that is triggered and properly biased will conduct until the AC voltage drops to the zero-crossing point. The next pulse from T1 will cause the other SCR to conduct. Therefore, the SCRs will conduct alternately, as shown in Figure 23–15 waveform (g). Notice that the SCRs conduct for a shorter amount of time as compared to the conduction of the SCRs

shown in Figure 23–13 waveform (g). This results in a lower average voltage applied to the AC motor. This will cause the speed of the AC motor to decrease.

## Current Source Inverter

The *current source inverter* (CSI), or *current fed inverter*, is a type of inverter drive in which the current is controlled while the voltage is varied to satisfy the motor's needs. To accomplish these tasks, a large inductor is used in the DC link section. Recall that in the variable voltage inverter a large capacitor was used to keep the DC voltage constant. A capacitor opposes a change in voltage. In the current source inverter, a large inductor is used to keep the DC constant. Remember that an inductor opposes a change in current.

A CSI can operate in two basic ways. One is by means of a *phase-controlled bridge rectifier* circuit; the other is via a diode bridge rectifier circuit and chopper control. We will examine the phase-controlled bridge rectifier type shown in Figure 23–16.

You should recognize the bridge rectifier circuit, constructed with SCRs. By varying the firing angle, we can cause the SCRs to conduct later in their cycle, thus producing a lower average output current to the DC bus. Conversely, if we fire the SCRs earlier in their cycle, we will produce a higher average output current to the DC bus. The variable DC is fed to the large inductor in the DC bus section. This inductor provides a constant current to the inverter section. The inverter then provides either six-step control or pulse-width modulation control.

The other operating method used in a current source inverter is a diode bridge rectifier with chopper control. Figure 23–17 shows how this method works. Notice that a standard, run-of-the-mill diode bridge is used in the converter stage to rectify the AC into DC. The DC is then fed into a chopper circuit. Recall that a chopper is basically an electronic switch that turns on and off rapidly, chopping the DC. In this way, the chopper circuit can vary the DC. If the chopper is closed (or conducting) longer than it is open (not conducting), a higher average DC flows. If the chopper is open (not conducting) longer than it is closed (or conducting), a lower average DC flows. This variable DC is fed to the large inductor in the DC bus section. The inductor provides the inverter stage with a constant current. As before, the inverter can be either a six-step inverter or a pulse-width modulated inverter.

Current source inverters have several distinct advantages over variable voltage inverters. They

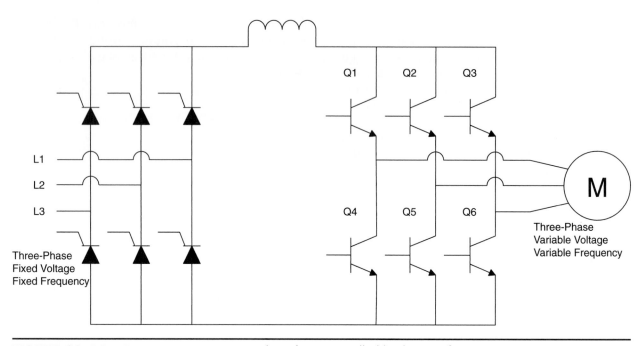

**FIGURE 23–16**   Current source inverter with a phase-controlled bridge rectifier.

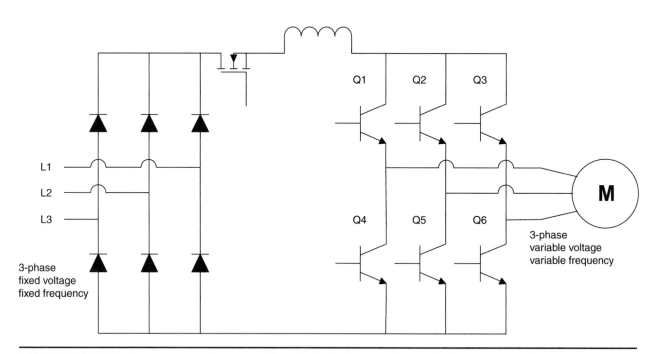

**FIGURE 23–17**   Current source inverter with a diode bridge rectifier and chopper control.

provide protection against short circuits in the output stage, and they can handle oversized motors. In addition, CSIs have relatively simple control circuits and good efficiency. As to their disadvantages, CSIs produce torque pulsations at low speed, cannot handle undersized motors, and are large and heavy. The phase-controlled bridge rectifier

CSI is less noisy than its chopper-controlled counterpart, does not need high-speed switching devices, and cannot operate from batteries. The chopper-controlled CSI can operate from batteries and produces more noise as a result of its need for high-speed switching devices. Refer to the chart in Figure 23–18 for a quick comparison between

| Features | Variable Voltage Inverter with Phase Control | Variable Voltage Inverter with Chopper Control | Current Source Inverter with Phase Control | Current Source Inverter with Chopper Control | Pulse-Width Modulation |
|---|---|---|---|---|---|
| Open Circuit Protection | YES | YES | | | YES |
| Short Circuit Protection | | | YES | YES | |
| Ability to Handle Oversized Motors | | | YES | YES | |
| Ability to Handle Undersized Motors | YES | YES | | | YES |
| Multiple Motor Applications | YES | YES | | | YES |
| Low-Speed Torque Pulsations | YES | YES | YES | YES | |
| Requires High-Speed Switching Devices | | YES | | YES | YES |
| Battery Operation | | YES | | YES | YES |
| Regenerative Operation | YES | | YES | | |
| Low-Speed Efficiency | GOOD | GOOD | GOOD | GOOD | MEDIUM |
| Complex Control Circuit | MEDIUM | MEDIUM | MEDIUM | MEDIUM | HIGH |
| Size and Weight | MEDIUM | MEDIUM | HIGH | HIGH | LOW |

**FIGURE 23–18**  Inverter comparison chart.

the various types of variable voltage and current source inverters.

# FLUX VECTOR DRIVES

Because this text was written with the maintenance technician in mind, care has been taken to approach the theory behind the operation of electronic variable speed drives in a simple, straightforward manner. However, one rather complicated type of drive, the flux vector drive, is gaining popularity. Figure 23–19 shows a SECO model VR flux vector drive. Figure 23–20 shows the control section of a SECO VR flux vector drive, and Figure 23–21 shows the power section of a SECO VR flux vector drive. Due

to its complexity, the theory behind this drive is somewhat beyond the scope of this text. However, for the individual exposed to flux vector drives, this section provides a basic understanding of their operation. We will begin our discussion of flux vector drives with a slightly different look at general motor theory.

To explain the operation of a typical, three-phase, AC induction motor, we always speak in terms of the rotating magnetic field. In order to understand flux vector drives, we must delve a little deeper into the characteristics of the rotating magnetic field.

Figure 23–22 shows a typical, three-phase sine wave. Notice also the vectors drawn to represent the balanced three-phase currents. (Recall that a vector represents both magnitude and direction.)

**FIGURE 23–19** SECO model VR flux vector drive.

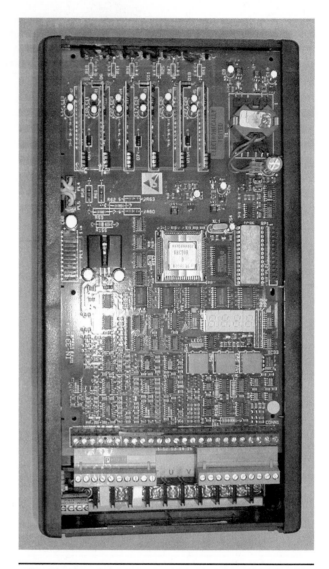

**FIGURE 23–20** Control section of a SECO VR flux vector drive.

Observe that these vectors are drawn 120 degrees apart to represent the normal phase shift in a three-phase system. Note, too, that at 30-degree vectors, L1 and L2 are drawn with their arrowheads pointing outward, away from the center, or neutral point. Vector L3 is drawn with its arrowhead pointing inward, or toward the neutral point. The direction of these arrowheads corresponds to the polarity of the instantaneous current. If the current is in the positive portion of the sine wave, the arrowhead is drawn pointing outward or away from the neutral. If the current is in the negative portion of the sine wave, the arrowhead will be drawn pointing inward or toward the neutral point. Also, the length of the vectors varies to represent the magnitude of the current.

Now, look at Figure 23–23, which includes another set of drawings that show the addition of the current vectors and the resulting magnetic flux. (Recall that vectors are added by positioning them head to tail.) Notice that, as we advance from 30 degrees to 90 degrees, the current vectors appear to rotate in a counterclockwise direction, causing the flux to rotate in the same direction. As we continue to advance from 90 degrees to 150 degrees, the current

vectors and the flux continue to rotate counterclockwise. If we continue to advance to 360 degrees, the current vectors and the flux will complete one full revolution. All of this occurs in a predictable manner because the three-phase supply is balanced and sinusoidal, and it has a constant amplitude and a constant frequency. What happens when we introduce some changes to this configuration?

Before we discuss changes to the frequency, amplitude, or phase rotation, be aware that the AC sine wave supplied to the motor is artificially produced by an inverter stage in the flux vector drive. This inverter stage allows us to exercise control over the synthetic AC and to make the changes described in the remainder of this chapter.

In Figure 23–24, we have caused the devices that produce the artificial AC to "jump ahead."

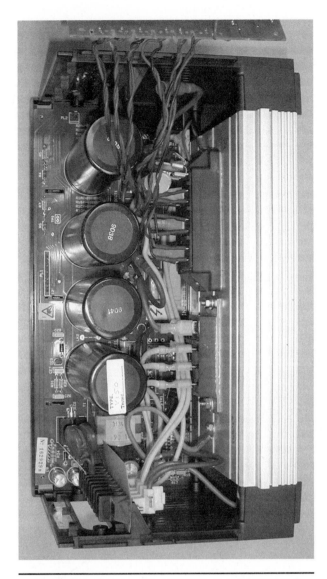

**FIGURE 23–21**   Power section of a SECO VR flux vector drive.

In other words, the AC was advanced from 60 degrees to 90 degrees instantaneously. This change has no effect on the direction of rotation or on the strength of the magnetic flux. This does, however, cause the motor suddenly to advance 30 degrees in rotation. Because we can control the firing sequence of the devices that produce the synthetic AC, we can also control the amount of advancement of the flux field. This jump is known as a step change.

Given this control over the devices that produce the synthetic AC, what happens when these same devices are held on for an extended period of time? Referring to Figure 23–25, notice that at the 60-degree point, the step change discussed previously does not occur, nor is the current allowed to advance normally. In this instance, the current has been held constant for 30 degrees. As a result, the

flux rotation has been stopped or held stationary for the 30-degree period. You should also recognize that by taking advantage of our control over the AC, we can change the phase rotation and, thus, the direction of rotation of the motor as well.

Flux vector drives operate by monitoring rotor position and changing the rotor's position as compared to its orientation to the stator field. To know the rotor's position, encoders are typically used as feedback devices. In addition, current feedback from two of the three phases is used. This amount and variety of feedback is necessary to achieve the variety of control mechanisms that a flux vector drive offers. For instance, if we vary the amplitude of the current, we will vary the magnitude of the current vectors. If we vary the phase rotation, we will vary the direction of rotation. Causing a step change to occur will result in an instantaneous advance in motor position. Finally, if we hold the phase current constant, we will cause the motor to hold a constant position.

The flux vector drive must constantly monitor the motor's performance and allow for high-speed corrections to occur to maintain accuracy in motor positioning and performance. For this reason, flux vector drives are matched to the particular motor that they will operate. Although this makes flux vector drives more expensive, the resulting improved performance in the areas of rapid response and precise control makes these drives well worth their price.

## TROUBLESHOOTING INVERTER DRIVES

In an AC inverter drive, the four main areas of possible problems are the same as those in a DC drive:

1. The electrical supply to the motor and the drive
2. The motor and/or its load
3. The feedback device and/or sensors that provide signals to the drive
4. The drive itself

Even though the problem may be in any one or several of these areas, the best place to begin troubleshooting is the drive unit itself. The reason is that most drives have some type of display that aids in troubleshooting. This display may be simply an LED that illuminates to indicate a specific fault condition, or it may be an error or fault code that can be looked up in the operator's manual. For this reason, it is strongly suggested that a copy of the fault codes be made and fastened to the inside of

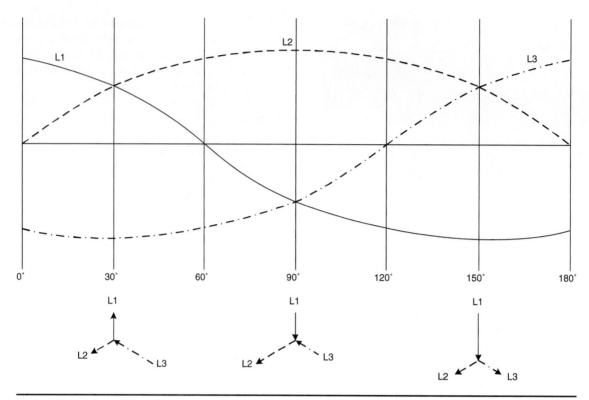

**FIGURE 23–22** Typical three-phase AC with current vectors.

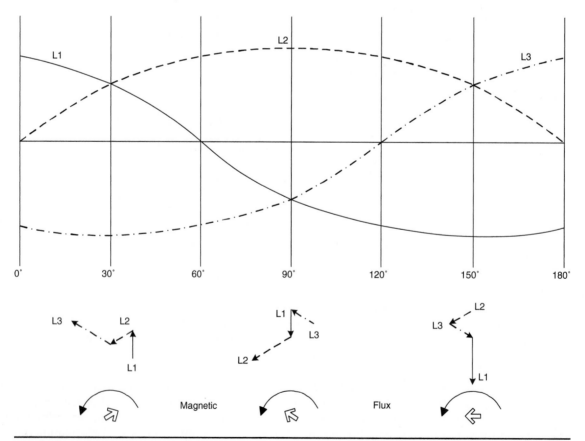

**FIGURE 23–23** Typical three-phase AC showing addition of the current vectors and the resulting magnetic flux.

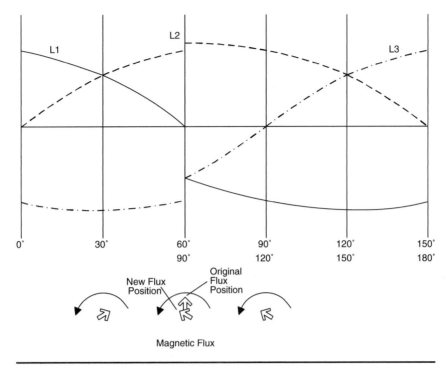

**FIGURE 23–24**  Change in magnetic flux position due to a step change.

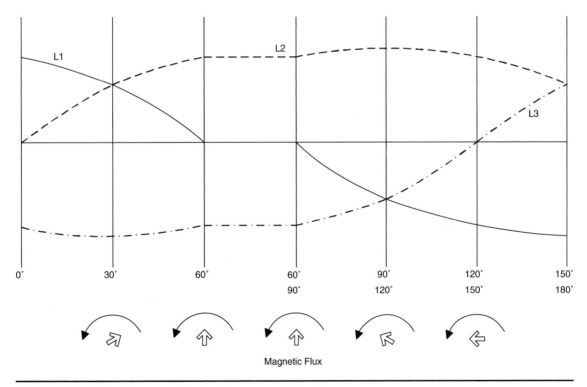

**FIGURE 23–25**  Magnetic flux held constant for 30°.

the drive cabinet, where it will be readily available to the maintenance technician. The original should be placed in a safe location such as the maintenance supervisor's office.

Before you proceed, stop and think about what you are doing. Before working on any electrical circuit, remove all power. Sometimes removing the power is not possible or permissible. In these instances, work carefully and wear appropriate safety equipment. Do not rely on safety interlocks, fuses, or circuit breakers to provide personal protection. Always use a voltmeter to verify that the equipment is de-energized, and tag and lock the circuit out!

Even when the power has been removed, you are still subject to shock and burn hazards. Most drives have high-power resistors inside, and these resistors can and do get hot! Give them time to cool down before touching them. Most drives also have large electrolytic capacitors. These capacitors can and do store an electrical charge. Usually, the capacitors have a bleeder circuit that dissipates this charge. However, this circuit may have failed. Always verify that electrolytic capacitors are fully discharged. Do this by measuring carefully any voltage present across the capacitor terminals. If voltage is present, use an approved shorting device to discharge the capacitor completely.

Now, slow down, take your time, and use your senses, even though production has stopped, and you are under pressure to fix the equipment. Sometimes a little extra initial time to take stock of the situation can save considerable time later. So do the following:

Look!    Do you see any charred or blackened components? Have you noticed any arcing? Do fuses or circuit breakers appear blown or tripped? Do you see any discoloration around wires, terminals, or components? A good visual inspection can save a lot of troubleshooting time.

Listen!  Did you hear any funny or unusual noises? A frying or buzzing sound may indicate arcing. A hum may be normal or an indication of loose laminations in a transformer core. A rubbing or chafing sound may indicate that a cooling fan is not rotating freely.

Smell!   Do you notice any unusual odors? Burnt components and wires give off a distinctive odor. Metal will smell hot if subjected to too much friction.

Touch!   (Be very careful with this!) If components feel cool, that may indicate that no current is flowing through the device. If components feel warm, chances are that everything is normal. If components feel hot, things may be normal, although more than likely too much current is flowing, or too little cooling is taking place. In any event, there may be a problem worth further investigation.

The point of all this is that by being observant you have a good chance of discovering the problem or problem area. Before we consider in more detail the four main areas mentioned earlier, let us review the proper techniques for installing an inverter.

1. Make the AC power connections. Be careful to use the proper voltage and connect the power source to the correct terminals on the inverter. If the supply voltage is different than the voltage indicated on the inverter, you may have to move some jumpers on the inverter to reconfigure it for the proper voltage levels.

2. After the AC supply has been connected, make the motor connections. Again, be careful to connect the motor leads to the proper terminals on the inverter. You should also verify that the voltage rating of the motor matches the voltage rating of the inverter.

3. Next, you must program or otherwise set the proper operating parameters for the inverter. We will now examine how to accomplish this in more detail.

4. Depending on the model, you may need to set the inverter for remote control (start/stop/adjust speed from a remote location) or local control (start/stop/adjust speed from the inverter's control panel).

5. If you are using remote control, you will need to wire the control circuit. You will also need to wire the speed/torque control potentiometer.

The preceding instructions are general guidelines for installing an inverter. Check the manual that comes with the particular inverter that you are using and follow the procedures presented there precisely.

Now, let's examine some of the parameters that can be programmed into an inverter. Keep in mind that not all inverters share these parameters. Some manufacturers list the same parameters in different order, while certain manufacturers have other names for the settings discussed here. Also, keep in mind that some inverters will have more parameters and others will have fewer parameters than those mentioned here. Again, familiarize yourself with the particular inverter that you are using. In general, programmable inverter parameters include the following:

| | |
|---|---|
| Analog input select | Sets either a 0–10-volt or 4–20-milliampere reference that is proportional to the frequency input. |
| Analog output select | Provides a 4–20-milliampere output signal that is proportional to the motor speed. |
| Basic set-up | Can be set for constant torque or constant speed. |
| Current | The output current of the inverter. |
| Current limit | Sets the maximum current available to the motor. If the setting is at the maximum permissible value, the motor will have maximum starting torque. This value can be on the order of 160 percent of nominal motor current. If the current limit is set too low, the inverter can trip out. |
| DC brake time | If selected, this parameter will provide additional braking torque at low motor speeds. |
| Digital input select | If selected, this value bypasses the ramp time down setting, and the inverter decelerates the motor in the shortest possible amount of time as a result. |
| Frequency | The output frequency of the inverter. |
| Jogging speed | Sets the speed of the motor when jogging. |
| Local/remote | Can be set for local control from the inverter's control panel or remote control from a start/stop station located away from the inverter cabinet. |
| Maximum speed | Depending on the setting, it may be possible to attain a speed higher than the rated speed of the motor. This parameter must be set higher than the minimum speed setting. If it is set lower than the minimum speed, the motor will not run. |
| Minimum speed | Lowest speed setting at which the motor will run. |
| Motor magnetization | Set to the no load current rating on the motor's nameplate. |
| Motor nominal current | Set to the full load current rating on the motor's nameplate. |
| Motor nominal frequency | The rated frequency of the motor from the motor nameplate. This value should be set as closely as possible to the specified value. |
| Motor nominal voltage | The rated line voltage of the motor from the motor nameplate. This value should be set as closely as possible to the specified value. |
| Motor power | This is the motor's power rating expressed either in horsepower (hp) or in kilowatts (kW). Some drives will accept the value in either unit of measurement while others require converting the rating from one unit to the other. |
| Ramp time down | This is the deceleration time (the time required to get from maximum speed to minimum speed) expressed in seconds. If this time is set too short, the inverter can trip out. |
| Ramp time up | This is the acceleration time (the time required to get from minimum speed to maximum speed) expressed in seconds. If this time is too short, the inverter can trip out. |
| Relay output select | Provides a contact closure when the inverter is placed in the run mode. |

| | |
|---|---|
| Slip compensation | Typically, the factory setting for this value should be adequate. This setting is affected by the motor power, motor nominal voltage, and motor nominal frequency values. |
| Start compensation | Typically, the factory setting for this value should be adequate. This setting is affected by the motor power, motor nominal voltage, and motor nominal frequency values. |
| Start/stop mode | This value is programmed for the various types of start/stop circuits used. For example, a two-wire start/stop, a three-wire start/stop, a three-wire start/stop with a jog, and so on. |
| Start voltage | Typically, the factory setting for this value should be adequate. This setting is affected by the motor power, motor nominal voltage, and motor nominal frequency values. |
| Thermal motor protect | Depending on the setting chosen, this parameter will either flash the display when the motor's critical temperature is reached or trip the inverter. |
| Torque | Calculated motor torque. Dependent on the programmed settings of the motor nominal current and the motor magnetization. |
| Trip reset mode | If selected, this parameter will prevent the inverter from restarting automatically after a trip. |
| V/f ratio | Typically, the factory setting for this value should be adequate. This setting is affected by the motor power, motor nominal voltage, and motor nominal frequency values. |
| Voltage | The output voltage of the inverter. |

In order to begin troubleshooting, we need to understand the general, step-by-step sequence of events that most inverter drive systems follow on start-up and shutdown. We begin with the AC power applied to the inverter. We must adjust the reference or set point for the desired speed and torque characteristics of the motor. Next, the start/stop circuit is placed in the start or run mode. Instantly, the main control components begin a diagnostic routine. We will examine some of the routine's fault codes shortly. If no faults are detected, the driver activates the power semiconductors. These produce the output frequency and V/f ratio programmed into the inverter to match the speed and torque settings. The programmed setting for the ramp time up will control how long it takes the motor to reach the desired speed. While the motor speed is ramping up, the motor current is monitored. Should the current exceed the programmed current limit setting, the inverter may (depending on the manufacturer) automatically adjust the ramp-time-up program or simply trip. If tripping occurs repeatedly, the ramp-time-up program may need to be modified to accommodate a longer acceleration time. The ramp-time-up setting will cause the inverter output voltage and frequency to increase, accelerating the motor to the desired speed set point.

When it becomes necessary to stop the motor, most inverters offer several options. One option is basically to let the motor coast to a stop. This can take a very long time for high-inertia loads. The ramp down time setting allows the inverter to slow the motor gradually. This deceleration is accomplished by allowing the motor to feed its self-generated energy back into the inverter. The inverter will use large resistors to absorb this energy. This process, called dynamic braking, is usually insufficient to bring the motor to a controlled and rapid stop because as the motor slows, less energy is self-generated. Therefore, it is common practice to use a mechanical brake in conjunction with dynamic braking to provide additional braking action at low speeds. Another method of stopping the motor without using a mechanical brake is plugging, or DC injection. When the operator wishes to stop the motor, DC is fed into the motor winding. This current replaces the rotating magnetic field with a fixed magnetic field. The rotor soon becomes locked with this fixed magnetic field, which, in effect, stops the motor rotation. This method should not be used repeatedly because heat can build up in the motor and damage the windings. Now that we have a basic understanding of the processes that an inverter follows on start-up and shutdown, let's investigate the types of problems that can be encountered in the four main areas of an inverter drive system.

## The Electrical Supply to the Motor and the Drive

Most maintenance technicians believe that the power distribution in an industrial environment is reliable, stable, and free of interference. Nothing could be further from reality! Frequent outages, voltage spikes and sags, and electrical noise are normal operating occurrences. The effect of these is not as detrimental to motor performance as it can be to the operation of the drive itself. Most AC inverters are designed to operate despite variations in supply voltages. Typically, the incoming power can vary as much as ±10 percent with no noticeable change in drive performance. However, in the real world, it is not unusual for power line fluctuations to exceed 10 percent. These fluctuations may occasionally cause a controller to trip. If tripping occurs repeatedly, a power line regulator may be required to hold the power at a constant level.

A power line regulator will be of little use, however, should the power supply to the controller fail. In this situation, a UPS (uninterruptible power supply) is needed. Several manufacturers produce a complete power line conditioning unit. These units combine a UPS with a power line regulator.

Quite often, controllers are connected to an inappropriate supply voltage. For example, it is not unusual for a drive rated 208 volts to be connected to a 240-volt supply. Likewise, a 440-volt-rated drive may be connected to a 460-volt or even a 480-volt source. Usually, the source voltage should not exceed the voltage rating of the drive by more than 10 percent. For a drive rated at 208 volts, the maximum supply voltage is 229 volts ($208 \times 10\% = 20.8 + 208 = 229$). Obviously, the 208-volt drive, when connected to the 240-volt supply, is receiving excess voltage and should not be used. For our 440-volt drive, the maximum supply voltage is 484 volts ($440 \times 10\% + 440 = 484$). Although this value appears to fall within permissible limits, another potential problem exists. Suppose that the power line voltage fluctuates by 10 percent. If the 440-volt source suffers a 10 percent spike, the voltage will increase to 484 volts. This value is within the design limits of the drive. But what happens when we connect the 440-volt drive to a 460-volt or a 480-volt power line? If we experience that same 10 percent spike, the 460-volt line will increase to 506 volts ($460 \times 10\% = 46 + 460 = 506$), and the 480-volt line will increase to 528 volts ($480 \times 10\% = 48 + 480 = 528$). We have thus exceeded the voltage rating of the drive and probably damaged some internal components! Most susceptible to excess voltage and spikes or transients are SCRs, MOSFETs, and power transistors. Premature failure of capacitors can also occur. As you can see, it is very important to match the line voltage to the voltage rating of the drive.

An equally serious problem occurs when the phase voltages are unbalanced. Typically, during construction, care is taken to balance the electrical loads on the individual phases. As time goes by and new construction and remodeling occurs, it is not unusual for the loading to become imbalanced, causing intermittent tripping of the controller and perhaps premature failure of components. To determine if an imbalanced phase condition exists, you will need to do the following:

1. Measure and record the phase voltages (L1 to L2, L2 to L3, and L1 to L3).

2. Add the three voltage measurements from step 1 and record the sum of all phase voltages.

3. Divide the sum from step 2 by 3 and record the average phase voltage.

4. Now, subtract the average phase voltage obtained in step 3 from each phase voltage measurement in step 1 and record the results. (Treat any negative answers as positive answers.) These values are the individual phase imbalances.

5. Add the individual phase imbalances from step 4 and record the total phase imbalance.

6. Divide the total phase imbalance from step 5 by 2 and record the adjusted total phase imbalance.

7. Now, divide the adjusted total phase imbalance from step 6 by the average phase voltage from step 3 and record the calculated phase imbalance.

8. Finally, multiply the calculated phase imbalance from step 7 by 100 and record the percentage of total phase imbalance.

Consider an example involving a 440-V, three-phase supply to an AC inverter drive to see how this process works:

1. L1 to L2 = 432 V; L2 to L3 = 435 V; and L1 to L3 = 440 V.

2. The sum of all phase voltages equals 432 V + 435 V + 440 V, or 1307 V.

3. The average phase voltage is equal to 1307 V ÷ 3, or 435.7 V.

4. To find the individual phase imbalances, we subtract the average phase voltage from the individual phase voltages and treat any negative values as positive. Therefore,

   L1 to L2 = 432 V − 435.7 V, or 3.7 V;

   L2 to L3 = 435 V − 435.7 V, or 0.7 V; and

   L1 to L3 = 440 V − 435.7 V, or 4.3 V.

5. Now, we find the total phase imbalances by adding the individual phase imbalances: 3.7 V + 0.7 V + 4.3 V = 8.7 V.

6. To find the adjusted total phase imbalance, we divide the total phase imbalance by 2: 8.7 V ÷ 2 = 4.35 V.

7. Next, we find the calculated phase imbalance by dividing the adjusted total phase imbalance by the average phase voltage: 4.35 V ÷ 435.7 V = 0.00998.

8. Finally, we multiply the calculated phase imbalance by 100 to find the percentage total phase imbalance: 0.00998 × 100 = 0.998%.

In this example, the values are within tolerances, and the differences in the phase voltages should not cause any problems. In fact, as long as the percentage total phase imbalance does not exceed 2 percent, we should not experience any difficulties as a result of the differences in phase voltages.

## Problems that Can Occur with the Motor and the Load

Probably the most common cause of motor failure is heat. Heat can occur simply as a result of the operating environment of a motor. Many motors are operated in areas of high ambient temperature. If steps are not taken to keep the motor within its operating temperature limits, the motor will fail. Some motors have an internal fan that cools the motor. If the motor is operated at reduced speed, this internal fan may not turn fast enough to cool the motor sufficiently. In these instances, an additional external fan may be needed to provide additional cooling to the motor. Typically, such fans are interlocked with the motor operation in such a way that the motor will not operate unless the fan operates as well. Therefore, a fault in the external fan control may prevent the motor from operating.

The sensors used to sense motor temperatures consist simply of a nonadjustable thermostatic switch that is normally closed and opens when the temperature rises to a certain level. In this event, you must wait for the motor to cool down sufficiently before resetting the temperature sensor and restarting the motor.

Periodic inspection of the motor and any external cooling fans is strongly recommended. The fans should be checked for missing or bent vanes. All openings for cooling should be kept free of obstructions. Any accumulation of dirt, grease, or oil should be removed. If filters are used, these must be cleaned or replaced on a routine schedule.

Heat may also cause other problems. When motor windings become overheated, the insulation on the wires may break down. This breakdown may cause a short, which may lead to an open condition. A common practice used to find shorts or opens in motor windings is to Megger the windings with a megohmmeter. Extreme caution must be taken when using a Megger on the motor leads. Be certain that you have disconnected the motor leads from the drive. Failure to do this will cause the Megger to apply a high voltage to the output section of the drive, resulting in damage to the power semiconductors. You may also decide to Megger the motor leads at the drive cabinet. Again, be certain that you have disconnected the leads from the drive unit, and Megger only the motor leads and motor winding. Never Megger the output of the drive itself!

Since the motor drives a load of some kind, it is also possible for the load to create problems. The drive may trip out if the load causes the motor to draw an excessive amount of current for too long a time. When this occurs, most drives display some type of fault indication. The problem may be a result of the motor operating at too high a speed. Quite often, a minor reduction in speed is all that is necessary to prevent repeated tripping of the drive. The same effect occurs if the motor is truly overloaded. Obviously, in this case, either the motor size needs to be increased or the size of the load decreased to prevent the drive from tripping.

Some loads have a high inertia. They require not only a large amount of energy to move, but once moving, a large amount of energy to stop. If the drive cannot provide sufficient braking action to match the inertia of the load, the drive may trip, or overhaul. A drive with greater braking capacity is needed in such cases to prevent tripping from recurring.

## Feedback Devices or Sensors May Cause Problems

Mechanical vibration may cause the mounting of feedback devices to loosen and their alignment to vary. Periodic inspections are necessary to verify that these devices are aligned and mounted properly.

It is also important to verify that the wiring to these devices is in good condition, and the terminations are clean and tight. Another consideration regarding the wiring of feedback devices is electrical interference. Feedback devices produce low-voltage/low-current signals that are applied to the drive. If the signal wires from these devices are routed next to high-power cables, interference can occur. This interference may result in improper drive operation. To eliminate the possibility that this will occur, several steps must be taken. The signal wires from the feedback device should be installed in their own conduits. Do not install power wiring and signal wiring in the same conduit. The signal wires should be shielded cable, with the shield wire grounded to a good ground at the drive cabinet only. Do not ground both ends of a shielded cable. When the shielded signal cable is routed to its terminals in the drive cabinet, the cable should not be run or bundled parallel to any power cables, but instead at right angles to such power cables. Furthermore, the signal cable should not be routed near any high-power contactors or relays. When the coils of a contactor or relay are energized and de-energized, a spike is produced. This spike can create interference with the drive. To suppress this spike, it may be necessary to install a free-wheeling diode across any DC coils or a snubber circuit across any AC coils.

## Problems Can Occur in the Inverter Drive Itself

First, look for fault codes or fault indicators. Most drives provide some form of diagnostics, and this can be a great time-saver. Next, we will look at some of the fault codes, symptoms, probable causes, and fixes for some common problems:

■ **The inverter is inoperable, and no LED indicators are illuminated**. Possibly no incoming power is present. You can verify this by measuring the voltage at the power supply input terminals in the inverter cabinet. The problem may be caused by a blown fuse, an open switch, an open circuit breaker, or an open disconnect. There are several things to check if the fuses are blown. One item to look for is a shorted metal oxide varistor (MOV), a device that provides surge protection to the inverter. If a significant power surge occurred, the MOV may have shorted to protect the inverter. Because the MOV is located across the power supply lines (to provide protection), a shorted MOV can cause fuses to blow. Another reason why fuses blow is a shorted diode in the rectifier circuit. A shorted or leaky filter capacitor in the power supply may also cause fuses to blow.

■ **The inverter is powered up but does not work**. There are indications of a fault condition. The "watchdog" circuit may have tripped. Remember that a watchdog circuit monitors the power lines for disturbances, and if the disturbance occurs for a long duration, the watchdog circuit trips

the controller to protect it from damage. A heavy starting load may sag the power line voltage to such a point that the inverter receives insufficient voltage. This deficiency can trip the inverter. Likewise, if the load has high inertia, it is possible for the regenerative effect to provide excess voltage to the inverter. Another way that the inverter may receive excess voltage is by use of power factor correction capacitors while the load is removed.

■ If the watchdog circuit has not tripped, other possible reasons exist for this fault condition. It is possible that there are interlocks on the cabinet or cooling fans, and one or more of these may be open. Likewise, a temperature sensor on a heat sink on the power semiconductors in the inverter or in the motor itself may have detected an excessive temperature condition and opened as a result.

■ **The inverter is energized, and a fault is indicated**. The motor does not respond to any control signals. If the load is too high for the motor settings, the motor may fail to rotate. It may be necessary to increase the current limit or voltage boost settings to allow the motor to overcome the load. Another possibility is that the load is overhauling the motor. If this is the situation, it will be necessary to adjust the deceleration time to allow the motor to take longer to brake the load. It may also be necessary to add auxiliary braking in the form of a mechanical brake. If the motor leads have developed a short or the motor itself has a shorted winding or is overloaded or stalled, the current limit sensor may trip the inverter. Tripping may also occur as a result of a shorted power semiconductor.

As you can see, you need to be aware of many areas when dealing with an inoperable inverter. Fortunately, the inverter itself can help a great deal by displaying fault codes. The operator's manual will interpret the fault codes and give instructions for clearing the fault condition. Next, we will examine some fault codes and how they can help in your troubleshooting. Remember that not all manufacturers provide the same fault codes: some provide more, and others provide fewer.

As mentioned earlier, one fault that may be displayed indicates an overcurrent limit condition. This indicator should direct you to examine the motor for mechanical binding, jams, and so forth. To verify whether one of these conditions is the cause of the problem, disconnect the motor from the load and reset the inverter. If the fault clears, then you know that the load is the cause. If the fault reappears, you need to look further, perhaps at the motor itself.

Another fault code, overvoltage, may be the result of a high-inertia load that causes overhauling. This fault code may also be a result of setting the deceleration ramp down parameter for rapid deceleration. Lengthening the deceleration time may clear the fault. If lengthening the time is not possible, additional mechanical braking may be required to bring the load to a rapid stop.

The inverter overload fault code is an indication of electrical problems. Examples of these are shorted or grounded motor leads or windings and/or defective power semiconductors. If the motor is suspect, disconnect it from the inverter. If the fault clears when you reset the inverter, you can assume that the problem lies in the motor and/or its leads. To verify whether the problem is in the power semiconductors, disconnect the gate lead from one of the devices, and reset the inverter. If the fault clears, you have found the problem. If the fault is still present, reconnect the gate lead, move to the next device, and disconnect its gate lead. Reset the inverter. Again, if the fault clears, you have found the problem. If the fault is still present, repeat the preceding steps until you have tested all of the power semiconductors.

Another fault code indicates shorted control wiring. If this code is displayed, simply disconnect the control wiring and reset the inverter. The fault should clear. This result indicates problems in the control wiring. If the fault does not clear, try unplugging the control board and resetting the inverter. A cleared fault condition in this case indicates problems in the control board.

It is a good idea to maintain an inventory of spare pc boards for the various inverters at your plant. That way, if you determine that the problem is caused by a defective pc board, it should be a fairly simple matter to replace the board with a spare. This will minimize downtime and allow you to try to repair the pc board in the shop, under a lot less pressure. If the pc board cannot be repaired, it may be possible to return it to the manufacturer for repair or exchange.

Although fault codes are excellent troubleshooting tools and can save a great deal of time, you should be aware of other potential problems that may or may not show up as fault codes, depending on the manufacturer.

Heat can produce problems in the drive unit. The cabinet may have one or more cooling fans, with or without filters. These fans are often interlocked with the drive power in such a way that the fan must operate in order for the drive to operate. Make certain that the fans are operational and the filters are cleaned or replaced regularly. The power semiconductors are typically mounted on heat sinks. A heat sink may have a small thermostat mounted on it to detect an excess temperature condition in the power semiconductors. If the heat sink becomes too hot, the thermostat will open, and the drive will trip. Usually these thermostats are self-resetting. You must wait for them to cool down and reset themselves before the drive will operate. If tripping occurs repeatedly, a more serious problem exists that requires further investigation.

If the drive is newly installed, problems are often the result of improper adjustments to the drive. On the other hand, if the drive has been in operation for some time, it is unlikely that readjustments are needed. All too often, an untrained individual will try to adjust a setting to "see if this fixes it." Usually such adjustments only make things worse! This is not to say that adjustments are never needed. For example, if the process being controlled is changed or some component of the equipment has been replaced, it probably will be necessary to change the drive settings. For this reason, it is very important to record the initial settings and any changes made over the years. This record should be placed in a safe location and a copy made and placed in the drive cabinet for easy access by maintenance personnel.

Replacing the drive should be the final choice. More often than not, if the drive is defective, something external to the drive is the reason for the drive's failure. Replacing the drive without determining what caused the failure may result in damage to the replacement drive. If the drive has failed, the possibility still exists that you can get it to work again.

Most drive failures occur in the power section, where you will find power SCRs, transistors, MOSFETs, and so on. In some drives, these devices will be individual components. You can test these devices with reasonable accuracy using nothing more than an ohmmeter. In other drives, the SCRs, transistors, and MOSFETs are contained within a power module.

An SCR power module is shown in Figure 23–26. One of these modules would be used for each phase of the three-phase AC. Notice that this module contains two SCRs. We will call the SCR to the left "A" and the SCR to the right "B." You can use an ohmmeter to test an SCR module with reasonable accuracy. To do so, you must understand what each terminal of the module represents. A schematic diagram of the SCR module is shown in Figure 23–27.

Notice in Figure 23–26 that there are a total of five terminals on the SCR module. Referring to Figure 23–27, you will see the same five terminals. Two of the terminals (the small terminals in Figure 23–26),

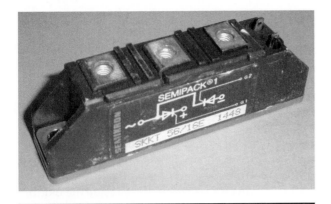

**FIGURE 23–26**   SCR power module.

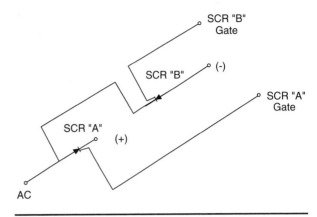

**FIGURE 23–27**   SCR power module schematic.

represent the gate connections for each SCR. The gate terminal toward the back of the module is connected to the gate of SCR B, while the gate terminal at the front of the module is connected to the gate of SCR A. The large terminal closest to the gate terminals is the negative terminal of the module, which is also the anode of SCR B. The large terminal farthest from the gate terminals is the AC terminal for the module, which is also the anode of SCR A. Notice that the anode of SCR A is internally connected to the cathode of SCR B. The large terminal in the center is the positive terminal of the module, which is also the cathode of SCR A.

To test the module with an ohmmeter, follow these steps:

1. Place the ohmmeter in the "Ω" or "Ohms" position, if using a digital multimeter. (Use the "200 Ω" position if using an analog multimeter.)

2. Place the black (negative) lead of your ohmmeter on the anode (terminal farthest from the gate terminals) of SCR A.

3. Place the red (positive) lead of your ohmmeter on the cathode (center terminal) of SCR A.

4. Your meter should indicate a very high or infinite resistance. A low-resistance reading indicates a faulty SCR. Replace the module.

5. Reverse your meter connections. Place the black (negative) lead of your ohmmeter on the cathode (center terminal) of SCR A.

6. Place the red (positive) lead of your ohmmeter on the anode (terminal farthest from the gate terminals) of SCR A.

7. Your meter should indicate a very high or infinite resistance. A low-resistance reading indicates a faulty SCR. Replace the module.

8. While leaving your ohmmeter connected as in steps 5 and 6, use a clip lead to connect the gate of SCR A (small terminal at the front of the module) to the red (positive) lead of your ohmmeter.

9. You should notice a drop in the resistance reading. This is a result of your triggering SCR A into conduction. If your resistance reading does not drop, the SCR may be faulty, and you should replace the module. However, it is possible that your ohmmeter is not supplying sufficient current to trigger the SCR into conduction, and the SCR may be functioning normally. The old saying, "If in doubt, change it out" would apply.

If SCR A appears normal, repeat the same process to check SCR B.

1. Place the black (negative) lead of your ohmmeter on the anode (terminal closest to the gate terminals) of SCR B.

2. Place the red (positive) lead of your ohmmeter on the cathode (terminal farthest from the gate terminals) of SCR B.

3. Your meter should indicate a very high or infinite resistance. A low-resistance reading indicates a faulty SCR. Replace the module.

4. Reverse your meter connections. Place the black (negative) lead of your ohmmeter on the cathode (terminal farthest from the gate terminals) of SCR B.

5. Place the red (positive) lead of your ohmmeter on the anode (terminal closest to the gate terminals) of SCR B.

6. Your meter should indicate a very high or infinite resistance. A low-resistance reading indicates a faulty SCR. Replace the module.

7. While leaving your ohmmeter connected as in steps 5 and 6, use a clip lead to connect the gate of SCR B (small terminal at the back of the module) to the red (positive) lead of your ohmmeter.

8. You should notice a drop in the resistance reading. This is a result of your triggering SCR B into conduction. If your resistance reading does not drop, the SCR may be faulty, and you should replace the module. However, it is possible that your ohmmeter is not supplying sufficient current to trigger the SCR into conduction, and the SCR may be functioning normally. Again, "If in doubt, change it out."

Figure 23–28 shows a transistor power module. Notice that this module has 17 terminals, 5 large and 12 small. Figure 23–29 shows the schematic diagram of the transistor power module. This module may look intimidating at first glance, but notice that the module consists simply of six transistors connected to perform an inverter function.

Looking at both Figure 23–28 and Figure 23–29, notice the row of six small terminals at the top of the module. Three of these terminals (BU, BV, and BW) are connected to the bases of transistors Q1, Q2, and Q3. The remaining three terminals (EU, EV, and EW) are connected to the emitters of the same transistors. Now, notice the row of six small terminals at the bottom of the module. Three of these terminals (BX, BY, and BZ) are connected to the bases of transistors Q4, Q5, and Q6. The remaining three terminals (EX, EY, and EZ) are connected to the emitters of the same transistors. You will also see two large terminals at the left end of the module. The large terminal toward the top of the module is the positive (+) DC input to the module. This terminal is internally connected to the collectors of transistors Q1, Q2, and Q3. The large terminal toward the bottom of the module is the negative (–) DC input to the module. This terminal is internally connected to the

emitters of transistors Q4, Q5, and Q6. Finally, you should see three large terminals in the center of the module (between the top and bottom rows of small terminals). The large terminal (W) at the right side of the module is connected to the emitter of Q3 and the collector of Q6. The next large terminal (V) to the left is connected to the emitter of Q2 and the collector of Q5. The last large terminal (U) to the left is connected to the emitter of Q1 and the collector of Q4.

You can perform a reasonably accurate test of this module with an ohmmeter. You should be aware, however, that the circuitry inside the module often is more complex than what is shown in Figure 23–29. The additional, but unknown, components may cause erroneous or unexpected readings on the ohmmeter. Therefore, you will need to exercise some judgment in interpreting the ohmmeter readings to determine if the module is faulty or not.

To test the module with an ohmmeter, proceed as follows:

1. Place the ohmmeter in the "Ω" position, if using a digital multimeter. (Use the "200-Ω" position if using an analog multimeter.)

2. Place the black (negative) lead of your ohmmeter on the base (terminal BU) of transistor Q1.

3. Place the red (positive) lead of your ohmmeter on the emitter (terminal EU) of transistor Q1.

4. Your meter should indicate a very high or infinite resistance. A low-resistance reading indicates a faulty transistor. Replace the module.

**FIGURE 23–28** Transistor power module.

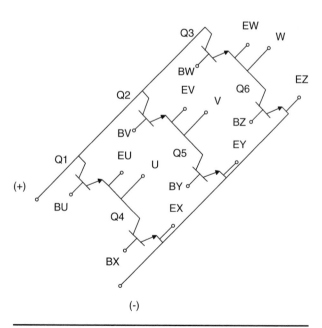

**FIGURE 23–29** Transistor power module schematic.

5. Leave the black (negative) lead of your ohmmeter connected to the base (terminal BU) of transistor Q1.

6. Place the red (positive) lead of your ohmmeter on the (+) terminal of the module. (This connects your red lead to the collector of transistor Q1.)

7. Your meter should indicate a very high or infinite resistance. A low-resistance reading indicates a faulty transistor. Replace the module.

8. Reverse your meter connections. Place the black (negative) lead of your ohmmeter on the emitter (terminal EU) of transistor Q1.

9. Place the red (positive) lead of your ohmmeter on the base (terminal BU) of transistor Q1.

10. Your meter should indicate a low resistance. A high- or infinite-resistance reading indicates a faulty transistor. Replace the module.

11. Leave the red (positive) lead of your ohmmeter connected to the base (terminal BU) of transistor Q1.

12. Place the black (negative) lead of your ohmmeter on the (+) terminal of the module. (This connects your black lead to the collector of transistor Q1.)

13. Your meter should indicate a low resistance. A high- or infinite-resistance reading indicates a faulty transistor. Replace the module.

14. Place the red (positive) lead of your ohmmeter on the emitter (terminal EU) of transistor Q1.

15. Place the black (negative) lead of your ohmmeter on the (+) terminal of the module. (This connects your black lead to the collector of transistor Q1.)

16. Your meter should indicate a very high or infinite resistance. A low-resistance reading indicates a faulty transistor. Replace the module.

17. Reverse your meter connections. Place the red (positive) lead of your ohmmeter on the (+) terminal of the module. (This connects your red lead to the collector of transistor Q1.)

18. Place the black (negative) lead of your ohmmeter on the emitter (terminal EU) of transistor Q1.

19. Your meter should indicate a very high or infinite resistance. A low-resistance reading indicates a faulty transistor. Replace the module.

20. Repeat steps 1 through 19 for each of the remaining five transistors in the module. If any readings are questionable, remove any doubt by checking your measurements against a known good module, or simply replace the questionable module.

There is another type of power module that you may find in the drive upon which you are working. This module is technically not part of the power output section, although it does have a power function in the drive. The module is a three-phase bridge rectifier module. It is used to convert three-phase AC into rectified DC. A picture of this module appears in Figure 23–30 and a schematic of this module appears in Figure 23–31.

Notice that this module has five terminals. The two horizontal terminals at the left end of the module are the (+) and (−) DC connections. The three vertical terminals are the connections for the three-phase AC (L1, L2, and L3).

**FIGURE 23–30** Three-phase bridge rectifier module.

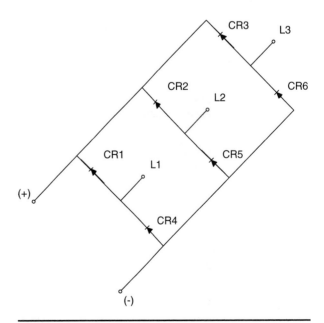

**FIGURE 23–31** Three-phase bridge rectifier module schematic.

To test the module with an ohmmeter, follow these steps:

1. Place the ohmmeter in the "Diode Test" position, if using a digital multimeter. (Use the "200 Ω" position if using an analog multimeter.)

2. Place the black (negative) lead of your ohmmeter on the (−) terminal of the module.

3. Place the red (positive) lead of your ohmmeter on AC terminal L1. (This is actually the cathode of diode CR4.)

4. Your meter should indicate a very high or infinite resistance. A low-resistance reading indicates a faulty diode. Replace the module.

5. Move the red (positive) lead of your ohmmeter from AC terminal L1 to AC terminal L2. (This is actually the cathode of diode CR5.)

6. Your meter should indicate a very high or infinite resistance. A low-resistance reading indicates a faulty diode. Replace the module.

7. Move the red (positive) lead of your ohmmeter from AC terminal L2 to AC terminal L3. (This is actually the cathode of diode CR6.)

8. Your meter should indicate a very high or infinite resistance. A low-resistance reading indicates a faulty diode. Replace the module.

9. Reverse your meter connections. Place the red (positive) lead of your ohmmeter on the (−) terminal of the module.

10. Place the black (negative) lead of your ohmmeter on AC terminal L1. (This is actually the cathode of diode CR4.)

11. Your meter should indicate a low resistance. A high- or infinite-resistance reading indicates a faulty diode. Replace the module.

12. Move the black (negative) lead of your ohmmeter from AC terminal L1 to AC terminal L2. (This is actually the cathode of diode CR5.)

13. Your meter should indicate a low resistance. A high- or infinite-resistance reading indicates a faulty diode. Replace the module.

14. Move the black (negative) lead of your ohmmeter from AC terminal L2 to AC terminal L3. (This is actually the cathode of diode CR6.)

15. Your meter should indicate a low resistance. A high- or infinite-resistance reading indicates a faulty diode. Replace the module.

16. Place the red (positive) lead of your ohmmeter on the (+) terminal of the module.

17. Place the black (negative) lead of your ohmmeter on AC terminal L1. (This is actually the anode of diode CR1.)

18. Your meter should indicate a very high or infinite resistance. A low-resistance reading indicates a faulty diode. Replace the module.

19. Move the black (negative) lead of your ohmmeter from AC terminal L1 to AC terminal L2. (This is actually the anode of diode CR2.)

20. Your meter should indicate a very high or infinite resistance. A low-resistance reading indicates a faulty diode. Replace the module.

21. Move the black (negative) lead of your ohmmeter from AC terminal L2 to AC terminal L3. (This is actually the anode of diode CR3.)

22. Your meter should indicate a very high or infinite resistance. A low-resistance reading indicates a faulty diode. Replace the module.

23. Reverse your meter connections. Place the black (negative) lead of your ohmmeter on the (+) terminal of the module.

24. Place the red (positive) lead of your ohmmeter on AC terminal L1. (This is actually the anode of diode CR1.)

25. Your meter should indicate a low resistance. A high- or infinite-resistance reading indicates a faulty diode. Replace the module.

26. Move the red (positive) lead of your ohmmeter from AC terminal L1 to AC terminal L2. (This is actually the anode of diode CR2.)

27. Your meter should indicate a low resistance. A high- or infinite-resistance reading indicates a faulty diode. Replace the module.

28. Move the red (positive) lead of your ohmmeter from AC terminal L2 to AC terminal L3. (This is actually the anode of diode CR3.)

29. Your meter should indicate a low resistance. A high- or infinite-resistance reading indicates a faulty diode. Replace the module.

Once you have completed testing the module, and you determine that the module is defective, you can usually obtain a substitute part from a local electronics parts supplier. If the part is not available, you will have to return the drive to the manufacturer for service or call a service technician for on-site repairs.

If it is determined that the problem is not in the power section, then it must be located in the control section of the drive. The electronics used in the control section are more complex; therefore, troubleshooting is not recommended. In this event, the drive must be returned to the manufacturer for repair or arrangements made for on-site repair by a factory-trained technician.

# REVIEW QUESTIONS

*Multiple Choice*

1. The DC filter is also known as
   a. the DC bus.
   b. the DC link.
   c. the converter.
   d. pulsating DC.

2. A chopper
   a. changes AC into DC.
   b. changes DC into AC.
   c. changes pulsating DC into steady DC.
   d. changes steady DC into pulsating DC.

3. The inverter(s) that does/do not produce torque pulsations at low speeds is/are the
   a. VVI with phase control.
   b. VVI with chopper control.
   c. CSI with phase control.
   d. CSI with chopper control.
   e. pulse-width modulation.

4. The inverter(s) that can operate from batteries is/are the
   a. VVI with phase control.
   b. VVI with chopper control.
   c. CSI with phase control.
   d. CSI with chopper control.
   e. pulse-width modulation.

5. The inverter(s) that provide(s) short circuit protection is/are the
   a. VVI with phase control.
   b. VVI with chopper control.
   c. CSI with phase control.
   d. CSI with chopper control.
   e. pulse-width modulation.

6. The inverter(s) that does/do not need high-speed switching devices is/are the
   a. VVI with phase control.
   b. VVI with chopper control.
   c. CSI with phase control.
   d. CSI with chopper control.
   e. pulse-width modulation.

7. The inverter(s) that is/are not as efficient at low speeds is/are the
   a. VVI with phase control.
   b. VVI with chopper control.
   c. CSI with phase control.
   d. CSI with chopper control.
   e. pulse-width modulation.

8. The inverter(s) that use(s) more complex control circuitry is/are the
   a. VVI with phase control.
   b. VVI with chopper control.
   c. CSI with phase control.
   d. CSI with chopper control.
   e. pulse-width modulation.

9. The inverter(s) that can handle oversized motors is/are the
   a. VVI with phase control.
   b. VVI with chopper control.
   c. CSI with phase control.
   d. CSI with chopper control.
   e. pulse-width modulation.

10. The inverter(s) that does/do not produce high-frequency noise is/are the
    a. VVI with phase control.
    b. VVI with chopper control.
    c. CSI with phase control.
    d. CSI with chopper control.
    e. pulse-width modulation.

11. The biggest and heaviest inverter(s) is/are the
    a. VVI with phase control.
    b. VVI with chopper control.
    c. CSI with phase control.
    d. CSI with chopper control.
    e. pulse-width modulation.

*True or False*

1. T or F    Another name for an AC drive is a converter.

2. T or F    In a VVI, the DC voltage is controlled.

3. T or F    Flux vector drives must be matched to the particular motor they operate.

4. T or F    Flux vector drives are no more expensive than their counterparts.

5. T or F    Flux vector drives provide rapid response and precise control.

6. T or F    An inoperable cooling fan can prevent an inverter from operating.

7. T or F    When using shielded cable on feedback devices, you must ground the shield wire at both ends of the cable.

*Give Complete Answers*

1. Explain, in general, the principles behind the operation of an AC inverter drive.

2. List the three basic sections of an AC inverter drive and describe the purpose of each section.

3. Explain why we need AC inverter drives.

4. List two main types of inverter drives.

5. What do the letters "VVI" mean?

6. Describe the function of a converter.

7. Describe the function of an inverter.

8. Explain what the V/Hz ratio is and why it is so important.

9. Explain how phase control is accomplished.

10. Which type of VVI, phase- or chopper-controlled, can be operated from batteries?

11. Is regenerative braking possible with a chopper-controlled VVI?

12. Name one disadvantage common to both types of VVI.

13. Does a pulse-width modulated variable voltage inverter drive maintain a constant V/Hz ratio? Explain.

14. Describe the effect on motor speed of varying the frequency of the AC. Describe the effect on torque.

15. Describe the effect on motor speed of varying the stator voltage. Describe the effect on torque.

16. List several ways in which the PWM VVI drive is similar to the SCR armature voltage controller introduced in Chapter 22.

17. Explain how the firing angles of the SCRs are varied.

18. What do the letters "CSI" stand for?

19. What is another name for a CSI drive?

20. Describe the device responsible for maintaining a constant current to the inverter section and explain how it works.

21. List some differences between a CSI and a VVI.

22. Name two different methods of producing a variable current to the DC bus and explain how they work.

23. What is the term used to describe the jumping ahead of the artificial AC in a flux vector drive?

24. What happens when a jump ahead does not occur, nor is a change allowed in the flux field?

25. Name the four main areas to check when a problem occurs with an AC inverter drive.

26. Where is the best place to begin troubleshooting an AC inverter drive? Why?

27. List some safety steps you should follow prior to working on an AC inverter drive.

28. Is it permissible to connect an AC inverter to a supply voltage that is higher than the nameplate rating of the drive? Why or why not?

29. Explain phase imbalance.

30. Calculate the phase imbalance for a supply to an AC inverter that has the following voltages: L1 to L2 = 231 V; L2 to L3 = 241 V; and L1 to L3 = 243 V.

31. Describe the cooling problem caused by using an inverter drive to operate a motor with a shaft-mounted fan.

32. Describe the dangers of using a Megger on an AC inverter drive system.

33. Explain how the motor load can affect the performance of the AC inverter drive.

34. Describe any precautions that should be observed when routing the feedback device cable outside the drive cabinet.

35. Describe any precautions that should be observed when routing the feedback device cable inside the drive cabinet.

36. Explain when it is permissible to readjust the settings on the inverter.

37. Explain when an AC inverter should be returned to the manufacturer for repair.

# CHAPTER 24

# Programmable Logic Controllers

## OBJECTIVES

After studying this chapter, the student will be able to:

- Identify the components of a PLC.
- Describe the function of a PLC.
- Correctly wire a PLC I/O module.
- Develop a simple PLC program.
- Develop a PLC I/O wiring diagram.
- Define various terms used in conjunction with PLCs.

The **programmable logic controller** (PLC) is an assembly of solid-state digital logic elements designed to make logical decisions and provide control. PLCs are used for the control and operation of manufacturing process equipment and machinery. The PLC is an industrially hardened computer, designed to perform control functions in industrial environments. This means that unlike your desktop personal computer (PC), the PLC must be capable of operating in temperature extremes, with poor power conditions (spikes, sags, etc.), in dusty, dirty, corrosive atmospheres, and withstand shock and vibration. In addition, PLCs are designed to be programmed by individuals who are familiar with motor control circuits. Therefore, most PLCs program in a language that resembles ladder diagrams, which makes learning PLCs very easy for most electricians. In addition to typical switching functions, PLCs can also perform counting, calculations, comparisons, processing of analog signals, and more.

## PLC COMPONENTS

PLCs are available in three sizes: small, medium, and large. The size of a PLC is determined by the number of *I/O* (input/output) devices it can handle and the amount of program memory available. Figure 24–1 shows a comparison between small, medium, and large PLCs. Most PLCs consist of an assortment of *modules* (input and output), a *CPU*, a power supply, and a rack in which to mount the aforementioned modules.

Figure 24–2 shows a fundamental block diagram of a PLC. Essentially, the PLC monitors input signals from various sensors and switches through an input module, as seen in Figure 24–3. The status of the input device is compared to the program stored within the PLC. Figure 24–4 shows a complete PLC. Based on the program logic, the PLC activates or deactivates an output, located on an output module, as seen in Figure 24–5. The output is connected to some type of device, such as a relay or a motor starter. The programming device is used to input the program and make modifications to an existing program. The programming device may be a handheld programmer, as seen in Figure 24–6, or it may be a PC (desktop or laptop). Typically, there is also some type of display, which may be as simple as an LCD display located on the programming device, or it may be the monitor of the PC, which is used to program the PLC.

### Input Module

The main purpose of the input module is to take the input signal from the field device (switch or sensor) and convert it to a signal that can be processed by the PLC **central processing unit** (CPU). Typically, this means converting the input signal to a 5 VDC level. In addition, the input module provides electrical isolation between the field device and the PLC.

There are many different types of input modules available. One type is the *digital input module* (Figure 24–3), which accepts input signals that are either off or on. Digital input modules are available for DC inputs, AC inputs, or a mix. Most digital input modules are available for 8 inputs, 16 inputs, or 32 inputs. In addition, digital input modules are available with a mix of inputs and outputs on the same module.

Another type of input module is the *analog input module*, as seen in Figure 24–7. This module will accept input signals that are either 0 to 10 VDC or 4 to 20 milliamperes. These modules are very useful when used with instrumentation and control loops. Analog input modules are available for 4 inputs, 8 inputs, or 16 inputs. In addition, analog input modules are available with a mix of inputs and outputs on the same module.

There are many other types of input modules available. These include *high-speed counter input modules*, *thermocouple input modules*, *RTD input modules*, and *motion control modules*, to name a few.

### Output Module

The main purpose of the output module is to take the signal from the PLC CPU and convert it to a signal for the field device (relay, motor starter, etc.).

| | Small | Medium | Large |
|---|---|---|---|
| Program Memory Size | Up to 18 k | Up to 64 k | Up to 100 k |
| Digital I/O | Up to 84 (shared with Analog I/O) | Up to 4096 | Up to 50176 |
| Analog I/O | Up to 84 (shared with Digital I/O) | Up to 96 | Up to 50176 |
| Timers/Counters | Limited by available memory | Limited by available memory | Limited by available memory |
| Scan time | As fast as 1 μSec | As fast as 0.225 mSec | As fast as 0.5 mSec |
| Expansion | Up to 4 expansion modules with up to 14 I/O points per module | Up to 3 expansion chassis with up to 30 I/O slots per chassis | Up to 125 expansion chassis with up to 16 I/O slots per chassis |

**FIGURE 24–1**　Small, medium, and large PLC comparison chart.

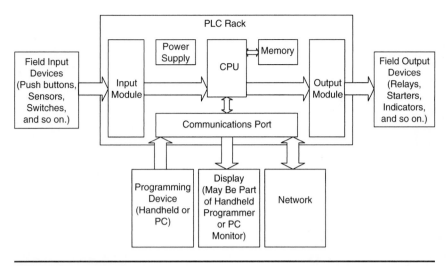

**FIGURE 24-2** PLC block diagram.

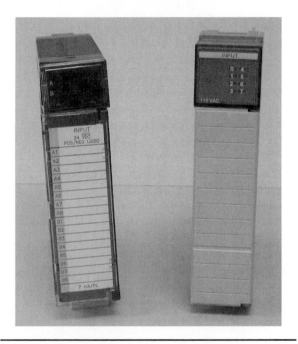

**FIGURE 24-3** PLC input modules.

**FIGURE 24-5** Output modules.

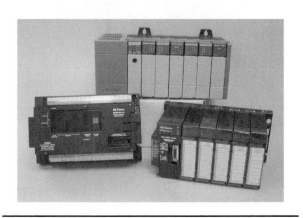

**FIGURE 24-4** Complete PLCs.

**FIGURE 24-6** Handheld programmer (HHP).

**FIGURE 24–7**   Analog input modules.

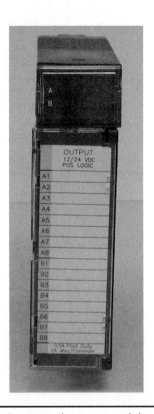

**FIGURE 24–8**   Digital output module.

Typically, this means converting the 5-VDC level from the CPU to the required voltage for the output field device. In addition, the output module provides electrical isolation between the PLC and the field device.

There are many different types of output modules available. One type is the *digital output module* (Figure 24–8), which produces output signals that are either off or on. Digital output modules are available for DC outputs, AC outputs, or a mix. Most digital output modules are available for 8 outputs, 16 outputs, or 32 outputs. In addition, digital output modules are available with a mix of inputs and outputs on the same module.

Another type of output module is the *analog output module*, as seen in Figure 24–9. This module will produce output signals that are either 0 to 10 VDC or 4 to 20 milliamperes. Like the analog input modules, these modules are very useful when used with instrumentation and control loops. Analog output modules are available for 4 outputs or a mix of inputs and outputs on the same module.

There are many other types of output modules available. These include *process control* and *positioning*, to name a few.

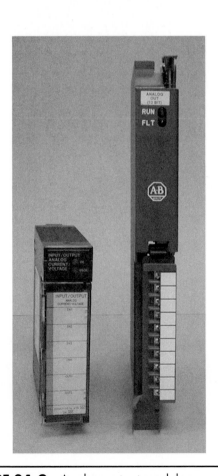

**FIGURE 24–9**   Analog output module.

## I/O WIRING

One big advantage of using a PLC over point-to-point wiring of relay controls is that the wiring is greatly simplified and reduced. Figure 24–10 shows a forward/reverse circuit, as it would typically be wired with relay-type controls. Compare this drawing to the one shown in Figure 24–11. Notice that there is less wiring in Figure 24–11. In addition, the wiring is easier to follow, which is the benefit of using a PLC. The PLC program performs the logical decisions. Therefore, the logic connections are made within the PLC program, *not* with the circuit wiring. In addition, modifications to the operation of the circuit do not require rewiring. Simply edit the PLC program for the desired operation.

Let us look at the wiring of the PLC I/O modules in more detail. Refer again to Figure 24–11. We will begin with the input module, which is a 120-volt AC input module. This means that the

input devices must supply 120 VAC to the input terminals of the input module.

Notice the stop, forward, and reverse push buttons. One side of each of these push buttons is connected to the 120-volt AC line. Notice also that power is applied to the NC contacts of the stop push button, while power is applied to the NO contacts of the forward and reverse push buttons. Each push button is then connected to its own terminal on the 120-volt AC input module.

Before we continue, we must draw attention to a few details. First, recall that the stop push button is wired NC. This is the manner in which stop push buttons are generally wired in point-to-point wiring. Notice, also, that we are only using one set of contacts for the forward and reverse push buttons. This may seem odd at first, but remember that the PLC is capable of performing functions that were previously accomplished through the circuit wiring. Finally, notice the 120-volt AC neutral wire that is connected to the input module. This is necessary for the module to operate properly.

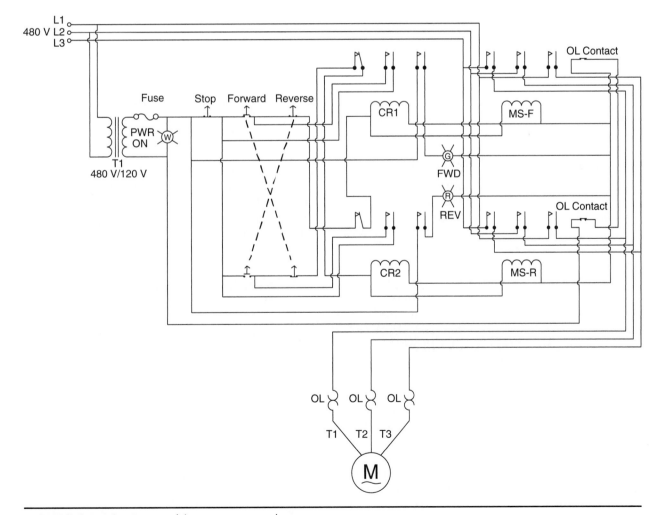

**FIGURE 24–10** Forward/reverse wiring diagram.

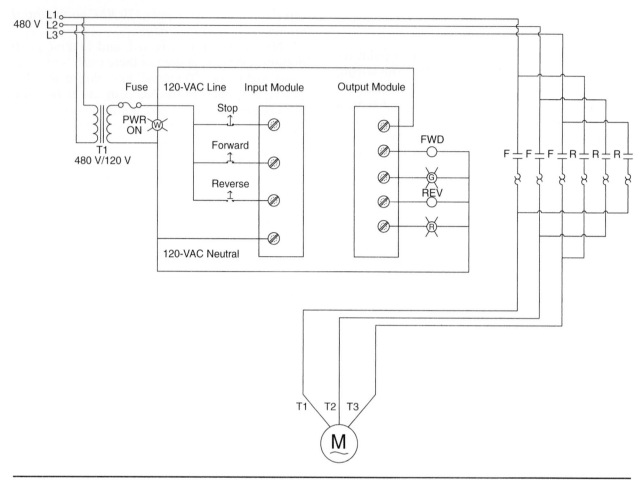

**FIGURE 24–11**  Forward/reverse wiring diagram when using a PLC.

This is all of the wiring needed for our input devices. We will now turn our attention to the output module. This particular output module is a 120-volt AC output module, which means that the module will supply 120 VAC to the output terminals of the output module.

Notice the forward and reverse motor starters. One side of each of these motor starters is connected to 120-volt AC neutral. Each motor starter is then connected to its own terminal on the 120-volt AC output module. The two pilot lights are connected in a similar manner. One side of each pilot light is connected to 120-volt AC neutral, while the other side of each pilot light is connected to its own terminal on the 120-volt AC output module. Notice, also, the 120-volt AC line that is connected to the output module, which is necessary for the module to operate properly.

We have now completed the wiring of our PLC forward/reverse control circuit. Notice how much simpler the wiring is. Figure 24–12 shows the panel for the forward/reverse control as it appeared when

wired in a point-to-point fashion. Figure 24–13 shows the same panel as wired for PLC control. Now, let us learn how to program this circuit. We will begin with a simple three-wire control circuit.

## PROGRAMMING

Over the years, different PLC manufacturers have created different programming languages for their PLCs. This has created confusion and increased training costs. The IEC (International Electrotechnical Commission) has created a standard (IEC 1131–3) for five programming languages for PLCs. These five languages are known as:

1. Function block diagram (FBD)

2. Instruction list (IL)

3. Ladder diagram (LD)

4. Sequential function chart (SFC)

5. Structured text (ST)

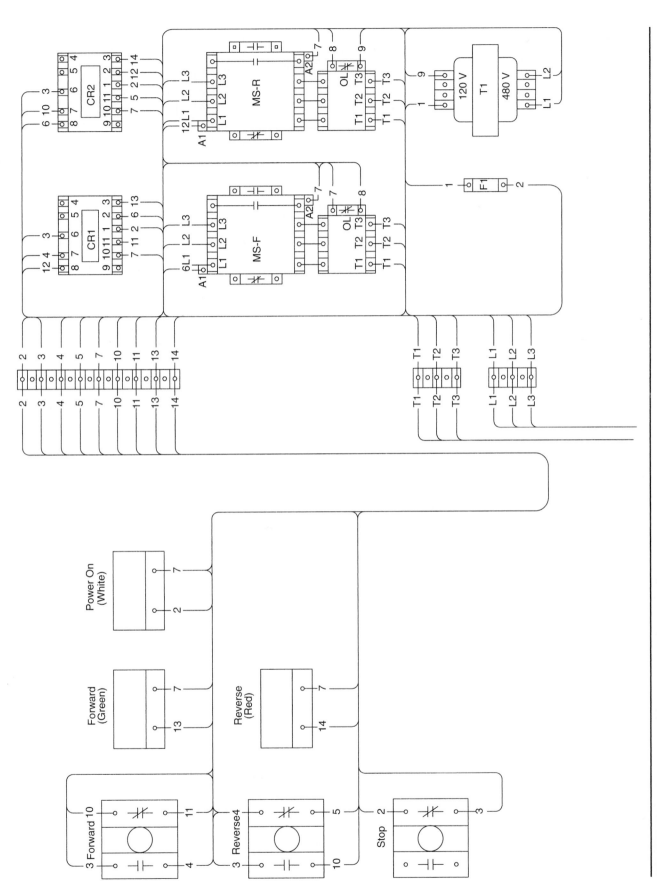

**FIGURE 24-12** Forward/reverse panel wiring diagram.

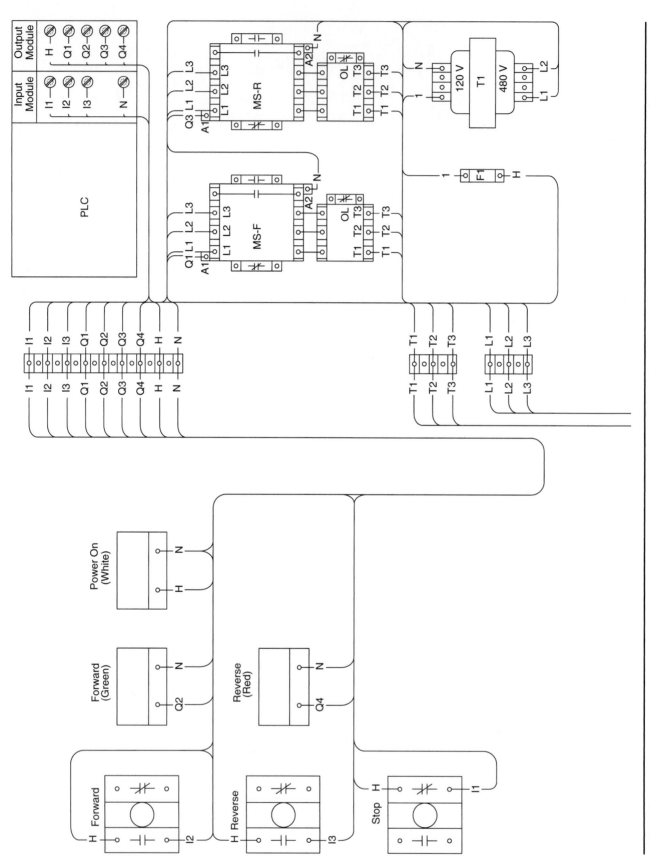

**FIGURE 24-13** Forward/reverse panel wiring diagram when using a PLC.

**631**

## Function Block Diagram (FBD)

This programming language is a *graphic language* that uses a library of functions (math, logic, etc.) in combination with custom functions (modem control, PID control, etc.) to create programs.

## Instruction List (IL)

Instruction List is a *low-level program* best suited for small applications and fast execution.

## Ladder Diagram (LD)

This is perhaps the most popular programming language. Ladder diagrams use graphics, discreet control, interlocking logic, and may incorporate function block instructions. This text will focus on the ladder diagram language.

## Sequential Function Chart (SFC)

SFC is also a *graphical language*. While providing structure and coordination of sequential events, alternative and parallel sequences are supported as well.

## Structured Text (ST)

Structured text programming language is a BASIC-style language. ST is an excellent language for complex processes or calculations that are not graphic friendly.

## PLC Program

Before we begin to develop a PLC program, we need to realize that to a PLC, an input is an input, is an input and an output, is an output, is an output. This means that a PLC does not care what type of device provides the input signal to the input module. The input could come from a push button, a limit switch, a contact from a relay, or whatever. A PLC will treat them all the same. They will be looked upon as either a normally open or a normally closed contact. The same holds true for an output. The output module may provide an output to a motor starter, control relay, pilot light, siren, or whatever. A PLC will treat them all the same, and they will be looked upon as a load. Therefore, a ladder diagram, written for relay logic, will look different when converted to a PLC ladder logic diagram.

Figure 24–14A shows a relay logic ladder diagram for a three-wire control. Figure 24–14B shows the PLC ladder logic program for the same circuit. Notice the different symbols used to represent the start, stop, and motor starter functions. Notice, also, the addressing of the start, stop, and motor

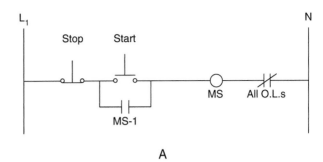

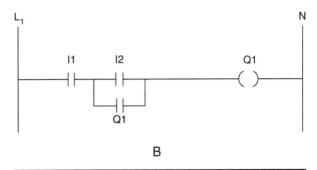

**FIGURE 24–14** (A) Relay logic for a three-wire control. (B) PLC program for a three-wire control.

starter functions. Individual PLC manufacturers have their own methods of addressing functions. You will need to become familiar with the method used by the PLC at your facility. For the purposes of this text, we will take a somewhat generic approach to addressing. Input devices will begin with the letter "I," and output devices will begin with the letter "Q."

Now, look at Figure 24–15. This is the same PLC program as shown in Figure 24–14B, except the input and output module wiring is shown as well. We will now walk through the operation of this circuit.

When the PLC is running, it is constantly scanning the inputs of the input module looking for voltage. At this point, the PLC would detect 120 VAC from the stop push button at terminal I1 of the input module. This is a result of the stop push button wired normally closed, and causing 120-volt AC line voltage to appear at this terminal. However, 120 VAC is not detected at terminal I2 of the input module because terminal I2 is connected to the start push button, which is wired normally open. Therefore, the PLC program sees the stop push button as passing power, but not the start push button. On the PLC ladder logic program seen in Figure 24–16, the contact representing the stop push button is highlighted to show power flow. Notice that the contact representing the start push button is not highlighted because the start push button is not passing power to the input module at this time.

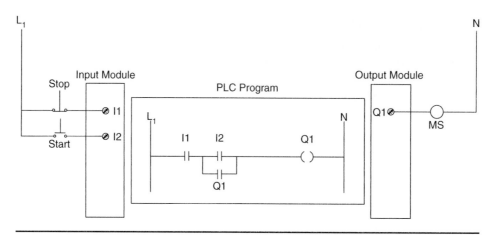

**FIGURE 24-15** I/O wiring diagram and program for a three-wire control.

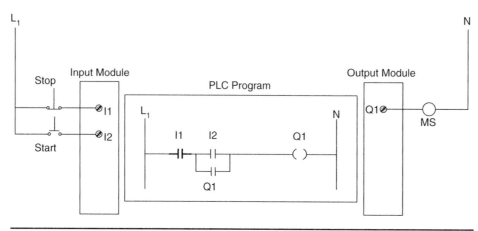

**FIGURE 24-16** Initial conditions: stop push button not pressed, start push button not pressed, motor starter de-energized.

Notice, also, that the coil representing the motor starter is not highlighted. Because the start push button has not been pressed, power cannot flow to the motor starter coil. The motor starter is de-energized. This also means that the motor starter auxiliary contacts, Q1, remain open.

Imagine pressing the start push button. Refer to Figure 24–17. When this happens, the 120-volt AC line is connected to the second input terminal on the 120-volt AC input module. The PLC detects the presence of voltage at terminal I2 and causes the associated rung element, I2, to close. This is indicated by the highlighting of the start push-button contact, I2, and causes the output element, Q1, to energize (highlighted). Now, two things occur simultaneously. The holding contact addressed "Q1" will close (highlighted). This provides a latch or seal-in for the start push button. In addition, terminal Q1 on the output module will provide 120 volts AC to the motor starter coil that is wired to terminal Q1, which will result in the motor starter energizing.

Figure 24–18 shows the effects of releasing the start push button. The PLC detects the absence of voltage at input module terminal I2. Therefore, the start push-button rung element, I2, is no longer highlighted. However, because this circuit contains a latch or seal-in circuit, all other elements in the PLC program remain highlighted. This means that there is power flow from the 120-volt AC line, through the holding contacts, Q1, through the motor starter coil, Q1, to the 120-volt AC neutral. The motor starter will remain energized.

We will now see what happens when we press the stop push button. Refer to Figure 24–19. When the stop push button is pressed, the PLC detects the absence of voltage at input module terminal I1. Therefore, the stop push-button rung element, I1, is no longer highlighted. This means that there is no longer power flow from the 120-volt AC line, through the stop push button. As a result, the motor starter, Q1, de-energizes, causing the holding contacts, Q1, to open. In addition, terminal Q1 on the

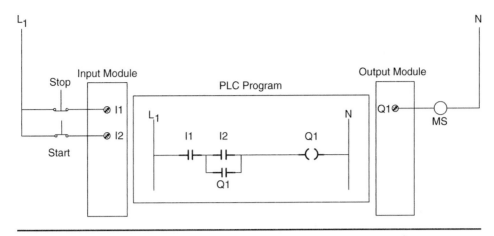

**FIGURE 24–17** Start push button pressed. Motor starter energized.

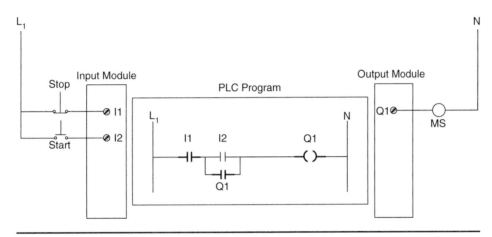

**FIGURE 24–18** Start push button released. Motor starter energized.

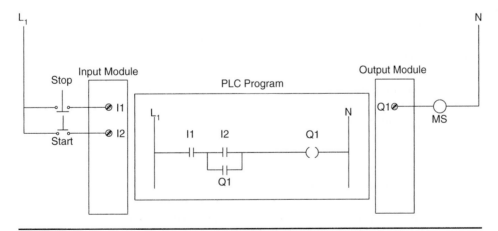

**FIGURE 24–19** Stop push button pressed. Motor starter de-energized.

output module will no longer provide 120 volts AC to the motor starter coil that is wired to terminal Q1, which will result in the motor starter de-energizing. This is indicated in the PLC program by the absence of the highlighting of the stop push button, I1, the start pushbutton, I2, the motor starter coil, Q1, and

the holding contacts, Q1. When the stop push button is released, the circuit is returned to its original condition, as seen in Figure 24–20.

At this point, you may have two questions that are bothering you. First, the start push button is wired as a *normally open* push button to input

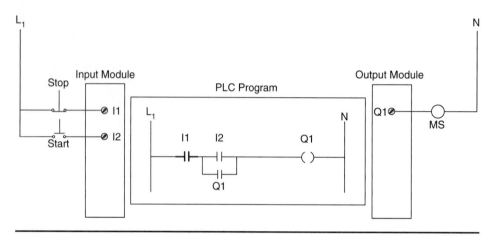

**FIGURE 24-20**   Stop push button released. Motor starter de-energized.

terminal I2, and the program shows a *normally open* contact, I2. However, *the stop push button is wired as a normally closed push button*, wired to input terminal I1, and *the program shows a normally open contact*, I1. Is this correct? Your other question might be, what is the origin of the holding contacts, Q1? They do not appear as a set of contacts that have been wired into the input module. We'll start with the first question.

Earlier, we stepped through the operation of the three-wire control circuit and found that it performs as expected. We will now make a slight modification to the program so that the stop push-button rung element, I1, appears as a normally closed contact. This would appear to be the correct thing to do, since the start push button is wired normally open and programmed normally open. It would make sense to wire the stop push button normally closed and program it normally closed as well. However, observe the results as seen in Figure 24–21.

Recall that the PLC would detect 120 VAC from the stop push button at terminal I1 of the input module. This is a result of the stop push button being wired normally closed and causing a 120-volt AC line voltage to appear at this terminal. Therefore, the PLC program sees the stop push button as passing power. On the PLC ladder logic program seen in Figure 24–21, the stop push button is *not* highlighted. Because the PLC senses voltage at input module terminal I1, the PLC interprets the stop push button as being activated. The PLC will, therefore, *open* the programmed normally closed contact, I1. Let us look at this a different way.

We will again modify the circuit. Figure 24–22 shows the same circuit, except the stop push button has now been wired as a normally open switch. The PLC program has remained unchanged. The stop push-button rung element, I1, has been pro-

grammed as a normally closed contact. Let us see how this affects the operation of our circuit.

The PLC would *not* detect 120 volts AC from the stop push button at terminal I1 of the input module. This is a result of the stop push button being wired normally open. Therefore, the PLC program sees the stop push button as not passing power. On the PLC ladder logic program seen in Figure 24–22, the stop push button *is* highlighted to show power flow due to the stop push-button rung element, I1, being programmed as a normally closed contact. Pressing the stop push button will cause the rung element, I1, to open, breaking the power flow.

This arrangement appears to function as well as the first circuit (Figure 24–16). However, there is one potential problem with the circuit shown in Figure 24–22. Imagine that due to vibration, a wire becomes disconnected from the stop push button. Pressing the stop push button will not apply 120 volts AC to input module terminal I1. Therefore, the PLC never sees the stop push button being depressed. *This means that the circuit cannot be stopped!* The general rule is to wire the stop push button normally closed and program the rung element as a normally open contact. In fact, should this be a critical stop push button, such as an emergency stop push button, the general rule is to wire emergency stop push buttons directly to the circuit and not through the PLC logic. Should the PLC suffer a failure, you would still be able to stop the process or machine. Figure 24–23 shows a typical PLC safety circuit. Notice that when either emergency stop push button is pressed, power is removed from the output devices but not the PLC or input devices. In this fashion, troubleshooting can be performed on the PLC while the machine or process is halted. Notice, also, that the relay that performs this function is called a master control relay (MCR). Do not confuse this

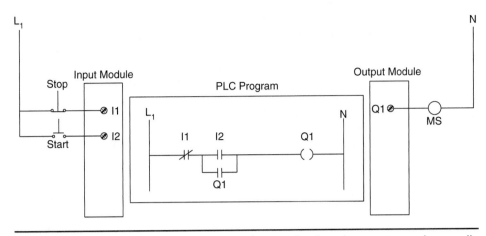

**FIGURE 24-21** Stop push button wired normally closed, programmed normally closed.

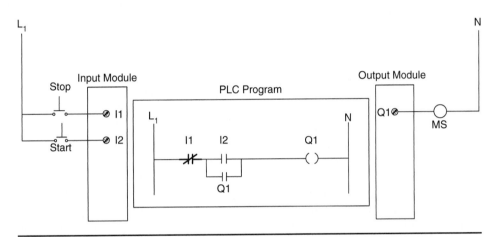

**FIGURE 24-22** Stop push button wired normally open, programmed normally closed.

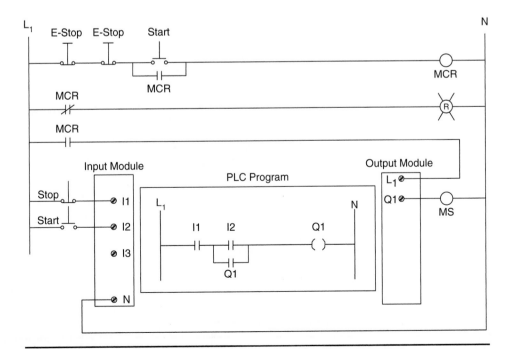

**FIGURE 24-23** PLC safety circuit.

relay with the PLC program instruction MCR. The programmed MCR instruction should never be used in place of a hard-wired master control relay.

Now, we return to the second question, what is the origin of the holding contacts, Q1? They do not appear as a set of contacts that have been wired into the input module. In addition to performing logical decisions, PLCs have within them devices known as internal coils (relays), timers, counters, and more. When we programmed the output element, Q1, we really addressed an internal coil. This coil has practically an unlimited number of normally open and normally closed contacts available. By simply programming a normally open contact and addressing it the same as the coil, Q1, we have instructed the PLC to operate the contact whenever the coil is energized.

This should now draw attention to one of the greatest advantages of using PLCs over relay-type devices. You have available to you a large number of devices with practically unlimited contact arrangements. In addition, there are functions available to you that cannot be performed with relay-type devices.

Now, let us turn our attention back to the forward/reverse control that was seen previously. The circuit is shown again as Figure 24–24, only now we have added the PLC program as well. Here, we are only using one normally open set of contacts from the forward push button. Notice that we are only using one set of normally open contacts from the reverse push button. Looking at the PLC program, notice there are two contacts shown for the forward and reverse push buttons. These are internal

contacts. When an input address is used, there are practically an unlimited number of normally open and normally closed contacts available to be assigned to that input address. This is what helps simplify the wiring in a circuit that uses a PLC.

## TYPES OF INSTRUCTIONS

PLCs are programmed by combining various instructions to form a logical control circuit. Many of these instructions are common to the different PLC manufacturers. There may be variations to the name, how they appear, or how they are programmed, but their basic function will be essentially the same. Again, you will need to familiarize yourself with the particular PLC at your facility. We will now look at some of the common types of instructions used in PLC programming.

### Relay Functions

- Contacts: Refer to Figure 24–25.
  - Contacts are used to indicate the status of a referenced component. This may be a field-wired device (switch, sensor, etc.) or an internal element (coil). Contacts may be normally open or normally closed.
  - A normally open contact acts as a switch to pass power when the status of its referenced component is on.

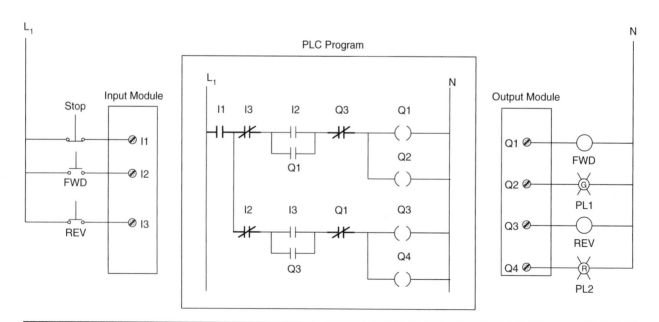

**FIGURE 24–24**   Forward/reverse program and I/O wiring.

■ A normally closed contact acts as a switch to pass power when the status of its referenced component is off.

■ Coils: Refer to Figure 24–26.

■ There are two basic types of coils: *nonretentive* and *retentive*.

■ Nonretentive coils revert to their de-energized state whenever the PLC is cycled from stop to run or whenever the power is cycled.

■ Retentive coils retain their current state whenever the PLC is cycled from stop to run or whenever the power is cycled.

| Type of Contact | Display | Contact Passes Power |
|---|---|---|
| Normally Open | —‖— | When Referenced Component Is ON |
| Normally Closed | —/‖— | When Referenced Component Is OFF |

**FIGURE 24–25** Contact instructions.

| Type of Coil | Display | Power | Result |
|---|---|---|---|
| Normal | —( )— | ON | Set referenced component ON |
| Normal | —( )— | OFF | Set referenced component OFF |
| Negated | —( / )— | ON | Set referenced component OFF |
| Negated | —( / )— | OFF | Set referenced component ON |
| Retentive | —( M )— | ON | Set referenced component ON (retained) |
| Retentive | —( M )— | OFF | Set referenced component OFF (retained) |
| Negated Retentive | —(/M)— | ON | Set referenced component OFF (retained) |
| Negated Retentive | —(/M)— | OFF | Set referenced component ON (retained) |
| SET | —( S )— | ON | Set referenced component ON until reset by —( R )— |
| SET | —( S )— | OFF | Coil state unchanged |
| RESET | —( R )— | ON | Reset referenced component OFF until set by —( S )— |
| RESET | —( R )— | OFF | Coil state unchanged |
| Retentive SET | —(SM)— | ON | Set referenced component ON until reset by —( R )— (retained) |
| Retentive SET | —(SM)— | OFF | Coil state unchanged |
| Retentive RESET | —(RM)— | ON | Reset referenced component OFF until set by —( S )— (retained) |
| Retentive RESET | —(RM)— | OFF | Coil state unchanged |

**FIGURE 24–26** Coil instructions.

- COIL controls a referenced contact. When the coil receives power (becomes energized), a referenced normally open contact will close to pass power, and a referenced normally closed contact will open to block power.

- Negated COIL operates inversely to a standard coil. A negated coil is *energized* when it does not receive power, and is *de-energized* when it receives power. When a negated coil is *not* receiving power (is *energized*), a referenced normally open contact will close to pass power, and a referenced normally closed contact will open to block power.

- Retentive COIL operates in a similar fashion to a coil. However, its condition is not altered by a power failure.

- Negated retentive COIL operates in a similar fashion to a negated coil. However, its condition is not altered by a power failure.

- SET coil is a nonretentive coil used to perform a latch function. When a SET coil receives power (becomes energized), a referenced normally open contact will close to pass power, and a referenced normally closed contact will open to block power. These conditions will remain, even if the SET coil is de-energized. They will not remain, should the PLC lose power. SET coils require the programming of a RESET coil with the same address.

- RESET coil is used to negate a SET coil's condition. The RESET coil must have the same address as the SET coil to perform this function.

- Retentive SET coil is used to perform a latch function. When a retentive SET coil receives power (becomes energized), a referenced normally open contact will close to pass power, and a referenced normally closed contact will open to block power. These conditions will remain, even if the SET coil is de-energized or should the PLC lose power. SET coils require the programming of a RESET coil with the same address.

- Retentive RESET coil is used to negate a SET coil's condition. The negated RESET coil must have the same address as the SET coil to perform this function.

## Timers and Counters

- Timers and counters: Refer to Figure 24–27.

- There are two basic types of time delay functions: *on-delay* and *off-delay*.

  - On-delay timers perform a timing function while receiving power. The state of the associated contact(s) do not change until the timer has timed out.

  - Off-delay timers perform a timing function after becoming de-energized. The state of

| Mnemonic | Function | Is Set When | Remains Set Until |
|---|---|---|---|
| TMR or TON | On-Delay Timer | Rung conditions allow power flow, and accumulated value is equal to or greater than the preset value | Rung conditions break power flow |
| OFDT or TOF | Off-Delay Timer | Rung conditions allow power flow | Rung conditions break power flow, and accumulated value is equal to or greater than the preset value |
| ONDTR or RTO | On-Delay Timer (Retentive) | Rung conditions allow power flow, and accumulated value is equal to or greater than the preset value | Associated RESET element is energized |
| CTU or UPCTR | Up-Counter | Rung element transitions from open to closed, and accumulated value is equal to or greater than the preset value | Associated RESET element is energized |
| CTD or DNCTR | Down-Counter | Rung element transitions from open to closed, and accumulated value is less than or equal to zero | Associated RESET element is energized |

**FIGURE 24–27**  Timer and counter instructions.

the associated contact(s) change immediately upon the timer becoming energized. The contact(s) revert to their original state after the timer has timed out after becoming de-energized.

- There are both nonretentive and retentive timers.
  - Nonretentive timers will reset their timing function upon loss of power.
  - Retentive timers will retain their timed value during a loss of power.
    - Retentive timers require an associated reset control to clear the timed value.
- There are two basic types of counter functions: *up-counters* and *down-counters.*
  - Up-counters perform a counting function when the associated input element transitions from an off to an on state. Up-counters begin at some preset value and increment upward. Up-counters are retentive and require an associated reset element to clear the counted values.
  - Down-counters perform a counting function when the associated input element transitions

from an off to an on state. Down-counters begin at some preset value and decrement downward. Down-counters are retentive and require an associated reset element to clear the counted values.

## Math Functions

- Standard math functions: Refer to Figure 24–28.
- Addition
  - Adds two integers
- Subtraction
  - Subtracts one integer from another
- Multiplication
- Multiplies two integers
- Division
  - Divides one integer by another
- Additional math functions: Refer to Figure 24–28.
- Square Root
  - Finds the square root of an integer

| Mnemonic | Function | Description |
|----------|----------|-------------|
| ADD | Addition | Adds one integer to another |
| SUB | Subtraction | Subtracts one integer from another |
| MUL | Multiplication | Multiplies two integers |
| DIV | Division | Divides one integer by another |
| SOR | Square Root | Finds the square root of an integer |
| SIN | Sine | Finds the sine of an integer |
| COS | Cosine | Finds the cosine of an integer |
| TAN | Tangent | Finds the tangent of an integer |
| ASIN | Arcsine | Finds the arcsine of an integer |
| ACOS | Arccosine | Finds the arccosine of an integer |
| ATAN | Arctangent | Finds the arctangent of an integer |
| LOG | Logarithm | Finds the base 10 logarithm of an integer |
| LN | Natural Logarithm | Finds the natural logarithm of an integer |
| DEG | Degrees | Converts an integer in degree units to radians |
| RAD | Radians | Converts an integer in radian units to degrees |

**FIGURE 24–28** Math instructions.

- Trigonometric Functions
  - Sine
  - Cosine
  - Tangent
  - Arcsine
  - Arccosine
  - Arctangent
- Logarithmic/Exponential Functions
  - Logarithm
  - Natural logarithm
  - Exponent
- Radian Conversion
  - Radians
  - Degrees

## Relational Functions

- Relational Functions: Refer to Figure 24–29.
- Equal
  - Test two integers for equality
- Not Equal
  - Test two integers for non-equality
- Greater Than
  - Test for one integer greater than another
- Greater Than or Equal To
  - Test for one integer greater than or equal to another
- Less Than
  - Test for one integer less than another

- Less Than or Equal To
  - Test for one integer less than or equal to another
- Range
  - Test for one integer to have a value within a minimum and maximum value

## Program Control Instructions

- Program Control Instructions: Refer to Figure 24–30.
  - MCR: Master Control Relay
    - Denotes an area within a program that is controlled by a master control relay. The master control relay is located at the beginning of the area, and an ENDMCR is located at the end of the area. When an MCR is activated, any program elements located within the MCR area are ignored by the PLC.
  - ENDMCR: Master Control Relay END
    - Denotes an area within a program that is controlled by a master control relay. The master control relay is located at the beginning of the area, and an ENDMCR is located at the end of the area. When an MCR is activated, any program elements located within the MCR area are ignored and deactivated by the PLC.
  - JUMP
    - When active, causes the PLC to jump to a specific location within the program as specified by a LABEL.
  - LABEL
    - Designator used with a JUMP command.

| Mnemonic | Function | Description |
|---|---|---|
| EQ | Equal | Test whether two integers are equal |
| NE | Not Equal | Test whether two integers are not equal |
| GT | Greater Than | Test whether one integer is greater than another |
| GE | Greater Than or Equal | Test whether one integer is greater than or equal to another |
| LT | Less Than | Test whether one integer is less than another |
| LE | Less Than or Equal | Test whether one integer is less than or equal to another |
| RANGE or LIMIT | Range or Limit Test | Test whether an integer is within a specified range or limit |

**FIGURE 24–29**　Relational instructions.

There are many other functions available depending upon the manufacturer, model, and size of the PLC. Only by becoming familiar with the PLC at your facility can you fully realize the power and features available.

# REVIEW QUESTIONS

*Multiple Choice*

1. A PLC is designed
   a. the same as an office computer.
   b. to operate in an industrial environment.
   c. the same as a home computer.
   d. to withstand heat.

2. Most PLCs are designed to be programmed using
   a. BASIC programming language.
   b. FORTRAN programming language.
   c. ladder logic programming language.
   d. special computer programming language.

3. The basic components of a PLC are
   a. the power supply and CPU.
   b. the programming terminal or device.
   c. the input/output modules.
   d. all of the above.

4. The CPU performs
   a. only special functions.
   b. all logic functions.
   c. only input functions.
   d. only output functions.

5. The programming terminal or device is used to program the
   a. power supply.
   b. input unit.
   c. output unit.
   d. CPU.

*Give Complete Answers*

1. What is a PLC?

2. List three advantages of a PLC over electromagnetic control systems.

3. List two differences between a home or business PC and a PLC.

4. Name five components that form a PLC.

5. What is the purpose of an input module?

6. What is the most common PLC programming language in use today?

7. Describe the recommended method to wire and program a stop push button.

8. What is the recommended method of wiring and programming an emergency stop push button?

9. List one advantage of using a computer to program and monitor a PLC program as opposed to a handheld programmer.

| Mnemonic | Function | Description |
|---|---|---|
| MCR | Master Control Relay | Creates zone within program for controlling all nonretentive outputs within the zone |
| ENDMCR | End Master Control Relay | Defines the end of a Master Control Relay zone |
| JUMP | Jump to Label | Causes the PLC to jump over designated program rungs to specified label |
| LABEL | Label | Target rung for a JUMP command |
| CALL or JSR | Call or Jump to Specified Subroutine | Causes the PLC to leave the main program and enter a subroutine program |
| RET | Return From Subroutine | Causes the PLC to leave a subroutine program and return to the main program |

**FIGURE 24–30** Program control instructions.

**TABLE A-1**  Conductor properties.

| Size (AWG or kemil) | Area mm² | Area Circular mils | Stranding Quantity | Stranding Diameter mm | Stranding Diameter in. | Overall Diameter mm | Overall Diameter in. | Overall Area mm² | Overall Area in.² | Copper Uncoated ohm/km | Copper Uncoated ohm/kFT | Copper Coated ohm/km | Copper Coated ohm/kFT | Aluminum ohm/km | Aluminum ohm/kFT |
|---|---|---|---|---|---|---|---|---|---|---|---|---|---|---|---|
| 18 | 0.823 | 1620 | 1 | — | — | 1.02 | 0.040 | 0.823 | 0.001 | 25.5 | 7.77 | 26.5 | 8.08 | 42.0 | 12.8 |
| 18 | 0.823 | 1620 | 7 | 0.39 | 0.015 | 1.16 | 0.046 | 1.06 | 0.002 | 26.1 | 7.95 | 27.7 | 8.45 | 42.8 | 13.1 |
| 16 | 1.31 | 2580 | 1 | — | — | 1.29 | 0.051 | 1.31 | 0.002 | 16.0 | 4.89 | 16.7 | 5.08 | 26.4 | 8.05 |
| 16 | 1.31 | 2580 | 7 | 0.49 | 0.019 | 1.46 | 0.058 | 1.68 | 0.003 | 16.4 | 4.99 | 17.3 | 5.29 | 26.9 | 8.21 |
| 14 | 2.08 | 4110 | 1 | — | — | 1.63 | 0.064 | 2.08 | 0.003 | 10.1 | 3.07 | 10.4 | 3.19 | 16.6 | 5.06 |
| 14 | 2.08 | 4110 | 7 | 0.62 | 0.024 | 1.85 | 0.073 | 2.68 | 0.004 | 10.3 | 3.14 | 10.7 | 3.26 | 16.9 | 5.17 |
| 12 | 3.31 | 6530 | 1 | — | — | 2.05 | 0.081 | 3.31 | 0.005 | 6.34 | 1.93 | 6.57 | 2.01 | 10.45 | 3.18 |
| 12 | 3.31 | 6530 | 7 | 0.78 | 0.030 | 2.32 | 0.092 | 4.25 | 0.006 | 6.50 | 1.98 | 6.73 | 2.05 | 10.69 | 3.25 |
| 10 | 5.261 | 10380 | 1 | — | — | 2.588 | 0.102 | 5.26 | 0.008 | 3.984 | 1.21 | 4.148 | 1.26 | 6.561 | 2.00 |
| 10 | 5.261 | 10380 | 7 | 0.98 | 0.038 | 2.95 | 0.116 | 6.76 | 0.011 | 4.070 | 1.24 | 4.226 | 1.29 | 6.679 | 2.04 |
| 8 | 8.367 | 16510 | 1 | — | — | 3.264 | 0.128 | 8.37 | 0.013 | 2.506 | 0.764 | 2.579 | 0.786 | 4.125 | 1.26 |
| 8 | 8.367 | 16510 | 7 | 1.23 | 0.049 | 3.71 | 0.146 | 10.76 | 0.017 | 2.551 | 0.778 | 2.653 | 0.809 | 4.204 | 1.28 |
| 6 | 13.30 | 26240 | 7 | 1.56 | 0.061 | 4.67 | 0.184 | 17.09 | 0.027 | 1.608 | 0.491 | 1.671 | 0.510 | 2.652 | 0.808 |
| 4 | 21.15 | 41740 | 7 | 1.96 | 0.077 | 5.89 | 0.232 | 27.19 | 0.042 | 1.010 | 0.308 | 1.053 | 0.321 | 1.666 | 0.508 |
| 3 | 26.67 | 52620 | 7 | 2.20 | 0.087 | 6.60 | 0.260 | 34.28 | 0.053 | 0.802 | 0.245 | 0.833 | 0.254 | 1.320 | 0.403 |
| 2 | 33.62 | 66360 | 7 | 2.47 | 0.097 | 7.42 | 0.292 | 43.23 | 0.067 | 0.634 | 0.194 | 0.661 | 0.201 | 1.045 | 0.319 |
| 1 | 42.41 | 83690 | 19 | 1.69 | 0.066 | 8.43 | 0.332 | 55.80 | 0.087 | 0.505 | 0.154 | 0.524 | 0.160 | 0.829 | 0.253 |
| 1/0 | 53.49 | 105600 | 19 | 1.89 | 0.074 | 9.45 | 0.372 | 70.41 | 0.109 | 0.399 | 0.122 | 0.415 | 0.127 | 0.660 | 0.201 |
| 2/0 | 67.43 | 133100 | 19 | 2.13 | 0.084 | 10.62 | 0.418 | 88.74 | 0.137 | 0.3170 | 0.0967 | 0.329 | 0.101 | 0.523 | 0.159 |
| 3/0 | 85.01 | 167800 | 19 | 2.39 | 0.094 | 11.94 | 0.470 | 111.9 | 0.173 | 0.2512 | 0.0766 | 0.2610 | 0.0797 | 0.413 | 0.126 |
| 4/0 | 107.2 | 211600 | 19 | 2.68 | 0.106 | 13.41 | 0.528 | 141.1 | 0.219 | 0.1996 | 0.0608 | 0.2050 | 0.0626 | 0.328 | 0.100 |
| 250 | | — | 37 | 2.09 | 0.082 | 14.61 | 0.575 | 168 | 0.260 | 0.1687 | 0.0515 | 0.1753 | 0.0535 | 0.2778 | 0.0847 |
| 300 | | — | 37 | 2.29 | 0.090 | 16.00 | 0.630 | 201 | 0.312 | 0.1409 | 0.0429 | 0.1463 | 0.0446 | 0.2318 | 0.0707 |
| 350 | | — | 37 | 2.47 | 0.097 | 17.30 | 0.681 | 235 | 0.364 | 0.1205 | 0.0367 | 0.1252 | 0.0382 | 0.1984 | 0.0605 |
| 400 | | — | 37 | 2.64 | 0.104 | 18.49 | 0.728 | 268 | 0.416 | 0.1053 | 0.0321 | 0.1084 | 0.0331 | 0.1737 | 0.0529 |
| 500 | | — | 37 | 2.95 | 0.116 | 20.65 | 0.813 | 336 | 0.519 | 0.0845 | 0.0258 | 0.0869 | 0.0265 | 0.1391 | 0.0424 |
| 600 | | — | 61 | 2.52 | 0.099 | 22.68 | 0.893 | 404 | 0.626 | 0.0704 | 0.0214 | 0.0732 | 0.0223 | 0.1159 | 0.0353 |
| 700 | | — | 61 | 2.72 | 0.107 | 24.49 | 0.964 | 471 | 0.730 | 0.0603 | 0.0184 | 0.0622 | 0.0189 | 0.0994 | 0.0303 |
| 750 | | — | 61 | 2.82 | 0.111 | 25.35 | 0.998 | 505 | 0.782 | 0.0563 | 0.0171 | 0.0579 | 0.0176 | 0.0927 | 0.0282 |
| 800 | | — | 61 | 2.91 | 0.114 | 26.16 | 1.030 | 538 | 0.834 | 0.0528 | 0.0161 | 0.0544 | 0.0166 | 0.0868 | 0.0265 |
| 900 | | — | 61 | 3.09 | 0.122 | 27.79 | 1.094 | 606 | 0.940 | 0.0470 | 0.0143 | 0.0481 | 0.0147 | 0.0770 | 0.0235 |
| 1000 | | — | 61 | 3.25 | 0.128 | 29.26 | 1.152 | 673 | 1.042 | 0.0423 | 0.0129 | 0.0434 | 0.0132 | 0.0695 | 0.0212 |
| 1250 | | — | 91 | 2.98 | 0.117 | 32.74 | 1.289 | 842 | 1.305 | 0.0338 | 0.0103 | 0.0347 | 0.0106 | 0.0554 | 0.0169 |
| 1500 | | — | 91 | 3.26 | 0.128 | 35.86 | 1.412 | 1011 | 1.566 | 0.02814 | 0.00858 | 0.02814 | 0.00883 | 0.0464 | 0.0141 |
| 1750 | | — | 127 | 2.98 | 0.117 | 38.76 | 1.526 | 1180 | 1.829 | 0.02410 | 0.00735 | 0.02410 | 0.00756 | 0.0397 | 0.0121 |
| 2000 | | — | 127 | 3.19 | 0.126 | 41.45 | 1.632 | 1349 | 2.092 | 0.02109 | 0.00643 | 0.02109 | 0.00662 | 0.0348 | 0.0106 |

Notes:

1. These resistance values are valid only for the parameters as given. Using conductors having coated strands, different stranding type, and especially, other temperature changes the resistance.

2. Formula for temperature change: $R_2 = R_1 [1 + \alpha (T_2 - 75)]$ where $\alpha_{cu} = 0.00323$, $\alpha_{AL} = 0.00330$ at 75°C.

3. Conductors with compact and compressed stranding have about 9 percent and 3 percent, respectively, smaller bare conductor diameters than those shown. See Table 5A [of the NFPA 70-2008, *National Electrical Code*] for actual compact cable dimensions.

4. The IACS conductivities used: bare copper = 100%, aluminum = 61%.

5. Class B stranding is listed as well as solid for some sizes. Its overall diameter and area is that of its circumscribing circle.

FPN: The construction information is per NEMA WC8-1992 or ANSI/UL 1581-1998. The resistance is calculated per National Bureau of Standards Handbook 100, dated 1966, and Handbook 109, dated 1972.

**TABLE B-1** Allowable ampacities of insulated conductors rated 0 through 2000 volts, 60°C through 90°C (140°F through 194°F), not more than three current-carrying conductors in raceway, cable, or earth (directly buried), based on ambient temperature of 30°C (86°F).

| | Temperature Rating of Conductor (See Table 310.13.) | | | | | | |
|---|---|---|---|---|---|---|---|
| | 60°C (140°F) | 75°C (167°F) | 90°C (194°F) | 60°C (140°F) | 75°C (167°F) | 90°C (194°F) | |
| Size AWG or kcmil | Types TW, UF | Types RHW, THHW, THW, THWN, XHHW, USE, ZW | Types FEP, FEPB, MI, RHH, RHW-2, THHN, TTHHW, THW-2, THWN-2, USE-2, XHH, XHHW, XHHW-2, ZW-2 | Types TW, UF | Types RHW, THHW, THW, TTHWN, XHHW, USE | Types TBS, SA, SIS, THHN, THHW, TTHW-2, THWN2, RHH, RHW-2, USE-2, XHH, XHHW, XHHW-2, ZW-2 | Size AWG or kcmil |
| | Copper | | | Aluminum or Copper-Clad Aluminum | | | |
| 18 | — | — | 14 | — | — | — | — |
| 16 | — | — | 18 | — | — | — | — |
| 14* | 20 | 20 | 25 | — | — | — | — |
| 12* | 25 | 25 | 30 | 20 | 20 | 25 | 12* |
| 10* | 30 | 35 | 40 | 25 | 30 | 35 | 10* |
| 8 | 40 | 50 | 55 | 30 | 40 | 45 | 8 |
| 6 | 55 | 65 | 75 | 40 | 50 | 60 | 6 |
| 4 | 70 | 85 | 95 | 55 | 65 | 75 | 4 |
| 3 | 85 | 100 | 110 | 65 | 75 | 85 | 3 |
| 2 | 95 | 115 | 130 | 75 | 90 | 100 | 2 |
| 1 | 110 | 130 | 150 | 85 | 100 | 115 | 1 |
| 1/0 | 125 | 150 | 170 | 100 | 120 | 135 | 1/0 |
| 2/0 | 145 | 175 | 195 | 115 | 135 | 150 | 2/0 |
| 3/0 | 165 | 200 | 225 | 130 | 155 | 175 | 3/0 |
| 4/0 | 195 | 230 | 260 | 150 | 180 | 205 | 4/0 |
| 250 | 215 | 255 | 290 | 170 | 205 | 230 | 250 |
| 300 | 240 | 285 | 320 | 190 | 230 | 255 | 300 |
| 350 | 260 | 310 | 350 | 210 | 250 | 280 | 350 |
| 400 | 280 | 335 | 380 | 225 | 270 | 305 | 400 |
| 500 | 320 | 380 | 430 | 260 | 310 | 350 | 500 |
| 600 | 355 | 420 | 475 | 285 | 340 | 385 | 600 |
| 700 | 385 | 460 | 520 | 310 | 375 | 420 | 700 |
| 750 | 400 | 475 | 535 | 320 | 385 | 435 | 750 |
| 800 | 410 | 490 | 555 | 330 | 395 | 450 | 800 |
| 900 | 435 | 520 | 585 | 355 | 425 | 480 | 900 |
| 1000 | 455 | 545 | 615 | 375 | 445 | 500 | 1000 |
| 1250 | 495 | 590 | 665 | 405 | 485 | 545 | 1250 |
| 1500 | 520 | 625 | 705 | 435 | 520 | 585 | 1500 |
| 1750 | 545 | 650 | 735 | 455 | 545 | 615 | 1750 |
| 2000 | 560 | 665 | 750 | 470 | 560 | 630 | 2000 |

**Correction Factors**

| Ambient Temp. (°C) | For ambient temperatures other than 30°C (86°F), multiply the allowable ampacities shown above by the appropriate actor shown below | | | | | | Ambient Temp. (°F) |
|---|---|---|---|---|---|---|---|
| 21–25 | 1.08 | 1.05 | 1.04 | 1.08 | 1.05 | 1.04 | 70–77 |
| 26–30 | 1.00 | 1.00 | 1.00 | 1.00 | 1.00 | 1.00 | 78–86 |
| 31–35 | 0.91 | 0.94 | 0.96 | 0.91 | 0.94 | 0.96 | 87–95 |
| 36–40 | 0.82 | 0.88 | 0.91 | 0.82 | 0.88 | 0.91 | 96–104 |
| 41–45 | 0.71 | 0.82 | 0.87 | 0.71 | 0.82 | 0.87 | 105–113 |
| 46–50 | 0.58 | 0.75 | 0.82 | 0.58 | 0.75 | 0.82 | 114–122 |
| 51–55 | 0.41 | 0.67 | 0.76 | 0.41 | 0.67 | 0.76 | 123–131 |
| 56–60 | — | 0.58 | 0.71 | — | 0.58 | 0.71 | 132–140 |
| 61–70 | — | 0.33 | 0.58 | — | 0.33 | 0.58 | 141–158 |
| 71–80 | — | — | 0.41 | — | — | 0.41 | 159–176 |

*See 240.4(D).

**TABLE C–1**  Allowable ampacities of single-insulated conductors rated 0 through 2000 volts in free air, based on ambient air temperature of 30°C (86°F).

| Size AWG or kcmil | Temperature Rating of Conductor | | | | | | Size AWG or kcmil |
|---|---|---|---|---|---|---|---|
| | 60°C (140°F) | 75°C (167°F) | 90°C (194°F) | 60°C (140°F) | 75°C (167°F) | 90°C (194°F) | |
| | Types TW, UF | Types RHW, THHW, THW, THWN, XHHW, USE, ZW | Types TBS, SA, SIS, FEP, FEPB, MI, RHH, RHW-2, THHN, TTHHW, THW-2, THWN-2, USE-2, XHH, XHHW, XHHW-2, ZW-2 | Types TW, UF | Types RHW, THHW, THW, TTHWN, XHHW, USE | Types TBS, SA, SIS, THHN, THHW, TTHW-2, THWN-2, RHH, RHW-2, USE-2, XHH, XHHW, XHHW-2, ZW-2 | |
| | Copper | | | Aluminum or Copper-Clad Aluminum | | | |
| 18 | — | — | 18 | — | — | — | — |
| 16 | — | — | 24 | — | — | — | — |
| 14* | 25 | 30 | 35 | — | — | — | — |
| 12* | 30 | 35 | 40 | 25 | 30 | 35 | 12* |
| 10* | 40 | 50 | 55 | 35 | 40 | 40 | 10* |
| 8 | 60 | 70 | 80 | 45 | 55 | 60 | 8 |
| 6 | 80 | 95 | 105 | 60 | 75 | 80 | 6 |
| 4 | 105 | 125 | 140 | 80 | 100 | 110 | 4 |
| 3 | 120 | 145 | 165 | 95 | 115 | 130 | 3 |
| 2 | 140 | 170 | 190 | 110 | 135 | 150 | 2 |
| 1 | 165 | 195 | 220 | 130 | 155 | 175 | 1 |
| 1/0 | 195 | 230 | 260 | 150 | 180 | 205 | 1/0 |
| 2/0 | 225 | 265 | 300 | 175 | 210 | 235 | 2/0 |
| 3/0 | 260 | 310 | 350 | 200 | 240 | 275 | 3/0 |
| 4/0 | 300 | 360 | 405 | 235 | 280 | 315 | 4/0 |
| 250 | 340 | 405 | 455 | 265 | 315 | 355 | 250 |
| 300 | 375 | 445 | 505 | 290 | 350 | 395 | 300 |
| 350 | 420 | 505 | 570 | 330 | 395 | 445 | 350 |
| 400 | 455 | 545 | 615 | 355 | 425 | 480 | 400 |
| 500 | 515 | 620 | 700 | 405 | 485 | 545 | 500 |
| 600 | 575 | 690 | 780 | 455 | 540 | 615 | 600 |
| 700 | 630 | 755 | 855 | 500 | 595 | 675 | 700 |
| 750 | 655 | 785 | 885 | 515 | 620 | 700 | 750 |
| 800 | 680 | 815 | 920 | 535 | 645 | 725 | 800 |
| 900 | 730 | 870 | 985 | 580 | 700 | 785 | 900 |
| 1000 | 780 | 935 | 1055 | 625 | 750 | 845 | 1000 |
| 1250 | 890 | 1065 | 1200 | 710 | 855 | 960 | 1250 |
| 1500 | 980 | 1175 | 1325 | 795 | 950 | 1075 | 1500 |
| 1750 | 1070 | 1280 | 1445 | 875 | 1050 | 1185 | 1750 |
| 2000 | 1155 | 1385 | 1560 | 960 | 1150 | 1335 | 2000 |

| Ambient Temp. (°C) | Correction Factors | | | | | | Ambient Temp. (°F) |
|---|---|---|---|---|---|---|---|
| | For ambient temperatures other than 30°C (86°F), multiply the allowable ampacities shown above by the appropriate factor shown below. | | | | | | |
| 21–25 | 1.08 | 1.05 | 1.04 | 1.08 | 1.05 | 1.04 | 70–77 |
| 26–30 | 1.00 | 1.00 | 1.00 | 1.00 | 1.00 | 1.00 | 78–86 |
| 31–35 | 0.91 | 0.94 | 0.96 | 0.91 | 0.94 | 0.96 | 87–95 |
| 36–40 | 0.82 | 0.88 | 0.91 | 0.82 | 0.88 | 0.91 | 96–104 |
| 41–45 | 0.71 | 0.82 | 0.87 | 0.71 | 0.82 | 0.87 | 105–113 |
| 46–50 | 0.58 | 0.75 | 0.82 | 0.58 | 0.75 | 0.82 | 114–122 |
| 51–55 | 0.41 | 0.67 | 0.76 | 0.41 | 0.67 | 0.76 | 123–131 |
| 56–60 | — | 0.58 | 0.71 | — | 0.58 | 0.71 | 132–140 |
| 61–70 | — | 0.33 | 0.58 | — | 0.33 | 0.58 | 141–158 |
| 71–80 | — | — | 0.41 | — | — | 0.41 | 159–176 |

*See 240.4(D).

**Across-the-line-starter**—An electromagnetic controller designed to start motors at full voltage. It generally contains an overload protective mechanism, a main circuit, and a control circuit. The relay is usually controlled from a remote switch.

**Air gap**—With reference to a motor, the distance between the rotating component and the stationary component within the magnetic field.

**Alloy**—A mixture of two or more metals.

**Alnico**—A mixture of aluminum, nickel, cobalt, and iron used in the manufacturing of permanent magnets.

**Alternating current (AC)**—The flow of electrons first in one direction and then in the opposite direction, in equal periods of time.

**Alternator**—An alternating current generator.

**Ambient**—Surrounding; for example, the air surrounding an electrical conductor.

**American wire gauge (AWG)**—A standard table or scale used for the measurement of the most common wire conductors in use in electrical work. This scale ranges from No. 50, which is equal to 1 circular mil, to No. 0000, which is equal to 211,660 circular mils.

**Ammeter**—A meter used to measure the amount of current flowing in an electrical system, circuit, or component.

**Ampacity**—The current-carrying capacity of electrical conductors, expressed in amperes.

**Ampere**—The rate of flow of electrons through a circuit. One ampere is equal to the flow of 1 coulomb ($628 \times 10^{16}$ electrons) per second.

**Amplifier**—An electronic device used to increase power, voltage, current, and/or sound signals.

**Anode**—The element in an electron tube, sometimes called the plate, that attracts the electrons emitted by the cathode.

**Apparent power**—The product of the current and voltage in an AC system or part of a system. Its unit of measurement is the volt-ampere.

**Armature**—In general, a piece of magnetic material, sometimes surrounded by coiled conductors, arranged to be acted upon by a magnetic field established by current flow. On a generator, the coil(s) into which the voltage is induced is generally referred to as the armature. On a DC motor the rotating part is generally called the armature. It is the moving part of relays, buzzers, or loudspeakers.

**Armature reaction**—The phenomenon that takes place in a motor or generator. Current flowing in the armature conductors produces a magnetic field that weakens and distorts the main field flux.

**Atom**—The smallest unit of any chemical element. Atoms consist of protons, neutrons, and electrons.

**Automatic control**—A controller that is operated by one or more sensing devices such as a thermostat or pressurestat. This type of controller does not require human intervention.

**Autotransformer**—A transformer in which one winding is common to both the primary and secondary.

**Average value**—With reference to AC, the average of the instantaneous values of current or voltage for one half cycle.

**Balanced system**—A multiwire system in which the resultant current in each ungrounded conductor is equal and the current in the grounded neutral conductor is zero.

**Ballasts**—A component used in fluorescent fixtures to change the value of voltage and to limit the amount of current.

**Battery**—Two or more cells connected to produce a specific voltage.

**Bipolar junction transistor**—A transistor consisting of a material sandwiched by a second material (NPN or PNP).

**BJT**—The abbreviation for a bipolar junction transistor.

**Breakdown torque**—The maximum torque of a motor. This usually occurs at about 25 percent slip. Loading a motor beyond the breakdown torque will stall the rotor.

**Brightness**—A measurement of the amount of light reflected from an object or a light source.

**British thermal unit**—The amount of heat required to raise the temperature of 1 pound of water 1 degree Fahrenheit.

**Brushes**—A device, usually made of carbon, used to complete the connection from a stationary circuit to a rotating circuit. Brushes are generally used on motors and generators.

**Brushless DC motor**—A DC motor that uses electronics instead of a commutator and brush assembly to perform the commutation of the applied DC.

**Buffer amplifier**—An amplifier that has no gain. A buffer amplifier is used to provide electrical isolation and/or impedance matching. A buffer amplifier is sometimes called a *voltage follower.*

**Candlepower**—A measurement of the intensity of light expressed in standard candles.

**Capacitance**—The ability to store an electrostatic charge.

**Capacitor**—A device consisting of conductors and insulators that is capable of storing an electric charge.

**Cathode**—The element of an electron tube that emits electrons.

**Cathode ray tube**—Also called a CRT. The cathode ray tube is a special vacuum tube that uses an electron beam to strike the inside, fluorescent coating of the front screen. The electron beam is used to draw waveforms, which can then be seen by viewing the front of the CRT.

**Central processing unit**—A component of a programmable controller that performs all the logic functions.

**Choke**—An inductor used to limit the flow of AC.

**Circuit**—A circuit consists of a power source, conductors, and a load connected in such a fashion that electrical current leaves the power source, flows through a conductor, flows through the load, flows through a second conductor, and returns to the power source. Some circuits include a control device such as a switch.

**Circuit breaker**—A device designed to open and close a circuit by nonautomatic means. It opens the circuit automatically on a predetermined overcurrent, without damage to itself, when properly applied within its rating (*NEC* 1993). These devices can be operated by thermal means or magnetic means or a combination of both.

**Circular mil**—The area of a circle 1 mil in diameter.

**Clamp-on ammeter**—This is a special type of ammeter with a set of open and closable jaws. The jaws are opened and then placed around a conductor; then the jaws are closed. This allows the current, flowing through the conductor, to be measured without disconnecting the conductor from the circuit.

**Coefficient of utilization**—With reference to lighting systems, the ratio between the light reaching the working plane and the light produced by the luminaire.

**Combination circuit**—A circuit containing groups of loads connected in series and in parallel.

**Common ground**—The grounding conductor that connects both the wiring system and the equipment to the grounding electrode.

**Commutator**—Copper bars mounted side by side on the shaft of a DC generator or motor. They are insulated from one another and from the shaft. They are arranged and connected to form a type of rotating switch. On a generator, the commutator serves to change the connections to the load at the same instant that the emf reverses in the armature, thus providing a unidirectional voltage to the load. On a motor, it serves to reverse the direction of the current in the armature at the precise instant to provide a unidirectional torque.

**Compensating winding**—Windings placed in the main pole faces of a generator or motor and connected in series with the armature. The purpose of these windings is to eliminate the effect of armature reaction.

**Complete path**—In a circuit, when the current can flow from the power source, through a conductor to a load, through the load, through another conductor back to the power source, we say there is a complete path for current flow.

**Component**—A part of a system, circuit, or apparatus.

**Condenser**—A term sometimes used for a capacitor. A device consisting of conductors and insulators that is capable of storing an electric charge.

**Conduction heating**—The transfer of heat energy from one object to another by contact.

**Conductor**—Any material that offers very little resistance to electron flow.

**Conduit**—A tube, pipe, or duct for enclosing electrical conductors.

**Continuity**—A continuous electrical connection between two points.

**Contrast**—With reference to light, either the difference in the amount of illumination from two sources or the difference in the amount of light reflected from various surfaces.

**Controller**—A device or group of devices that serve to govern, in some predetermined manner, the electrical power to the apparatus to which it is connected (*NEC* 1993). As used with a motor, a device used to adjust or vary the rotor speed.

**Convection heating**—The transmission of heat energy through a fluid or air.

**Copper losses**—Power dissipated as heat, caused by current flowing through the resistance of the conductors.

**Coulomb**—A measurement of a quantity of electrons. One coulomb is equal to $628 \times 10^{16}$ electrons.

**Counter electromotive force (cemf)**—A voltage induced into a coil that acts in the direction opposite to that of the applied voltage.

**Covalent bonding**—An arrangement of the atomic structure of a material that allows adjacent atoms to share valance electrons.

**Cycle**—A term used to indicate a complete change in values in both directions.

**Dehumidifier**—A machine that removes moisture from the air.

**Demand factor**—The ratio of the maximum demand of a system, or part of a system, to the total connected load of the system or the part of the system under consideration (*NEC* 1993).

**DE-MOSFET**—The abbreviation for the depletion-enhancement metal oxide semiconductor field effect transistor.

**Diac**—A bi-directional diode used in AC circuits.

**Dielectric heating**—The heating of electrically nonconductive materials by a high-frequency alternating current.

**Differential measurements**—A measurement technique where the difference between two measurement values is found.

**Digital multimeter**—An electrical instrument with a digital display designed to measure more than one electrical parameter, for example, voltage, current, and resistance.

**Direct current (DC)**—The flow of electrons in one direction only.

**Direct measurement**—With reference to an oscilloscope probe, a direct measurement is made when the oscilloscope probe is a 1:1 or 1X type probe.

**Dynamic breaking**—A method of stopping a motor quickly. With DC motors, the armature is disconnected from the supply and connected across a resistance load. With AC motors, the supply voltage is reversed and then the motor is disconnected from the source.

**Eddy currents**—Circulating currents caused by voltages induced into a conducting material.

**Effective value**—The effective value of an alternating current is that value that will produce the same heating effect as a specific value of a steady direct current. It is equal to the maximum value multiplied by 0.707.

**Efficiency**—The ratio of the power output to the power input.

**Electric charge**—An accumulation of electrons on an object or a lack of electrons on an object. When an object has an excessive number of electrons, it is negatively charged. If an object has lost some electrons so that it contains more protons than electrons, it is positively charged.

**Electric current**—The movement of electrons from atom to atom through a conductor.

**Electric discharge**—A type of lighting. Examples of electric discharge lighting is fluorescent and high-intensity discharge (HID), which consists of low-pressure sodium, mercury-vapor, metal-halide, and high-pressure sodium lamps.

**Electromagnet**—A magnet produced by an electric current flowing through a coil of wire.

**Electromagnetic induction**—The production of a voltage in a coil as a result of a magnetic field moving across the coil or the coil moving across the magnetic field.

**Electromotive force (emf)**—The electrical pressure produced by a battery, generator, or other apparatus designed to produce a force to cause current flow. Electromotive force is usually called voltage.

**Electron**—The smallest part of an atom; the negative charge of an atom that orbits around the nucleus.

**Electronics**—A branch of electrical science that pertains to the use and control of electrons in circuits containing gas or vacuum tubes, transistors, and amplifiers.

**E-MOSFET**—The abbreviation for the enhancement metal oxide semiconductor field effect transistor.

**Energy**—The ability to do work.

**Engineering notation**—A variation of the powers of 10, in which each step is in increments of 1000 instead of 10. Typically, letters (for example, *m* and *k*) and words (for example, *milli* and *kilo*) are used as a further shorthand expression.

**Equalizer bus**—A conductor used to connect the series fields of two or more compound generators together when the generators are operating in parallel. The purpose of this connection is to equalize the currents through the series fields.

**Equivalent circuit**—A simplified circuit whose characteristics are equivalent to the original circuit.

**Farad**—A measurement of capacitance. A capacitor has a capacitance of 1 farad when it can be charged to 1 coulomb in 1 second by 1 volt of pressure. The farad is too large a measurement for practical use; for most purposes the microfarad is used.

**Field effect transistor**—A voltage-controlled solid-state device. The field effect transistor has three terminals: the source, the gate, and the drain. Field effect transistors are also known as FETs.

**Field excitation**—Current flowing in the field coils of a generator. This current produces the flux necessary to induce a voltage into the armature coils.

**Field rheostat**—A variable resistor used to control the amount of current flowing through the field circuit of a motor or generator.

**Filament**—The element (usually a wire) through which current flows in an electron tube or electric lamp. The current flow produces heat and light.

**Fitting**—An accessory used in an electrical installation. A fitting is intended primarily to perform a mechanical rather than an electrical function.

**Fluorescent**—A material that glows when exposed to ultraviolet light.

**Fluorescent lamp**—A form of electric discharge lighting source. When a voltage is applied across the lamp, the gas within the tube ionizes, causing a radiation within the tube. The radiation activates a fluorescent material, producing a visible light.

**Flux**—Magnetic field.

**Flux density**—The number of lines of force for a given area, usually per square centimeter or square inch.

**Flywheel**—A wheel having a heavy outer perimeter. It is generally used to maintain a steady speed in machines whose load and/or driving force varies widely.

**Foot-candle**—The intensity of illumination on a plane. One foot-candle is the intensity of illumination on a plane 1 foot away from a lighting source of 1 candlepower and at a right angle to the rays from the source.

**Frequency**—In reference to AC, the number of cycles completed per unit of time, usually 1 second.

**Fuse**—An overcurrent protective device that contains a melting element.

**Generator**—A machine that converts mechanical energy into electrical energy; usually a machine that causes a coil to rotate through a magnetic field or a rotating magnetic field to move across a coil.

**Gilbert**—The unit of measurement of the magnetomotive force of a magnet.

**Glare**—An intense and/or bright light that tends to cause eye strain.

**Glow switch**—A type of fluorescent starter.

**Graticule**—The grid pattern that is visible on the face of the CRT of an oscilloscope.

**Ground-fault circuit interrupter**—A device that senses small currents to ground and opens the circuit when the current exceeds a predetermined value.

**Ground-fault protector**—A device that senses ground faults and opens the circuit when the current to ground reaches a predetermined value.

**Grounding conductor**—A conductor used to connect another conductor or equipment to the grounding electrode.

**Hall-effect device**—A solid-state device that can sense or detect the presence of a magnetic field. Hall-effect devices typically have four terminals: two for the operating voltage, and two for the output voltage. The typical output voltage from a Hall-effect device is approximately 10 millivolts.

**Harmonics**—In an AC system, a voltage or current waveform that is a multiple of the initial frequency.

**Heat pump**—A device used to heat or cool dwellings powered by an electrically driven compressor.

**Heat radiation**—Heat emitted or given off by an object.

**Heat transfer**—The exchange of heat energy from one object to another.

**Henry**—The unit of measurement of inductance. A coil has an inductance of 1 henry when an emf of 1 volt will cause the current to change at a rate of 1 ampere per second.

**Hertz**—The measurement of the number of cycles of an alternating current or voltage completed in 1 second.

**High-pressure sodium lamp**—A type of electric discharge lamp.

**Horsepower**—A unit of measurement of mechanical power. One horsepower is equivalent to work produced at the rate of 33,000 foot-pounds per minute.

**Humidifier**—A machine that adds moisture to the air.

**Humidistat**—A device for controlling the amount of moisture in the air. It is a switch that opens and/or closes an electric circuit as the humidity changes.

**Hypotenuse**—The side of a right triangle opposite to the 90° angle. The longest side of a right triangle.

**Hysteresis**—A lagging effect. In magnetic circuits, hysteresis occurs when a ferromagnetic material is placed under the influence of a varying magnetic field. It is sometimes referred to as molecular friction.

**IGBT**—The abbreviation for an insulated gate bipolar transistor.

**Illumination**—A measurement of the density of light projected on a surface. One lumen of light striking an area of 1 square foot will illuminate it at an intensity of 1 foot-candle.

**Impedance**—The total opposition to the flow of a pulsating or alternating current. Impedance is composed of reactance and resistance.

**Impurity**—A substance added to germanium or silicon to produce a semiconductor.

**Incandescent**—Light produced by heating an object to a high temperature.

**Incandescent lamp**—A lighting source, sometimes referred to as a light bulb, that emits light as a result of intense heat.

**Induction heating**—The heating of an electrically conductive material by a high-frequency alternating current.

**Induction motor**—A motor whose operation depends upon the principle of mutual induction. AC flowing in the stator windings induces a voltage into the rotor windings, developing two magnetic fields that interact to produce torque.

**Inductor**—A device that produces an inductive effect in a circuit.

**In phase**—Alternating currents or voltages are said to be in phase when they reach their maximum and zero values at the same time and in the same direction.

**Instantaneous value**—The value of an alternating current or voltage at a specific instant during the cycle.

**Insulated gate bipolar transistor**—A voltage-controlled device with three terminals: the emitter, the collector, and the gate. The insulated gate bipolar transistor, or IGBT, is capable of high-speed switching at high voltages.

**Insulator**—A device or material that has high electrical resistance. Insulators are used to separate or support conductors and prevent current flow between conductors and/or between conductors and ground.

**Interpoles**—Small poles placed midway between the main poles of a generator or motor and connected in series with the armature. The purpose of these poles is to compensate for armature reaction, thus reducing arcing at the brushes.

**Inverter**—A device that changes DC to AC.

**Inverting amplifier**—An amplifier whose output is opposite to its input. This means that as the input signal rises in a positive direction, the output signal falls in a negative direction.

**Isolated inputs**—Where two or more inputs do not share a common connection so that no electrical path exists between them.

**Isolation transformer**—A transformer designed to isolate a part of a system from the source.

**JFET**—The abbreviation for a junction field effect transistor.

**Junction field effect transistor (JFET)**—A voltage-controlled device with three terminals: the source, gate, and drain. JFETs conduct current through a conduction channel, whose width is controlled by the amount of reverse bias voltage present at the gate. JFETs are available as either an n-channel or a p-channel device.

**Ladder diagram**—Abbreviated LD. A variation of a schematic diagram, the ladder diagram shows the actual electrical connections, but not the actual physical location of the components. Ladder diagrams show how the circuit is designed to work in a more logical fashion than the schematic diagram.

**Lamp**—The part of an electrical fixture that emits the light.

**LED**—The abbreviation for a light-emitting diode.

**Light**—A form of radiant energy that stimulates the visual senses.

**Light-emitting diode (LED)**—A special type of diode that emits light when operated in the forward bias mode. Light-emitting diodes, or LEDs, are available in red, orange, yellow, green, and blue.

**Line drop**—The voltage required to force a specific value of current through the resistance of the conductors between the supply and the load. Line drop may refer to the entire system or to a small part of the system.

**Line loss**—The power dissipated in the form of heat caused by current flowing through the conductors between the supply and the load. Line loss may refer to the entire system or to a small part of the system.

**Liquidtight**—Constructed and installed so as to prevent the entrance of liquids.

**Load**—Any device that draws current from a power source.

**Load shifting**—Removing some or all of the load from one voltage source and applying it to another source.

**Loop distribution system**—A system for transmitting electrical energy through conductors and equipment. The energy begins at a specific point, loops through a specified area, and returns to the point of origin.

**Low-pressure sodium lamp**—A type of electric discharge lamp.

**Lumen**—The unit for total light emitted from a lighting source.

**Luminaire**—A lighting source usually in the form of an electric lighting fixture.

**Magnet**—An object that attracts iron and steel.

**Magnetic circuit**—The path through which the magnetic lines of force flow.

**Magnetic field**—The area surrounding a magnet in which the magnetic force exists.

**Magnetic flux**—The total number of lines of force.

**Magnetic lines of force**—Imaginary lines used to indicate the direction of force within the magnetic field.

**Magnetic saturation**—The condition of a material when it is fully magnetized.

**Magnetizing current**—The current necessary to produce a magnetic field in a machine or in equipment.

**Magnetos**—An AC generator that produces an electromotive force by rotating a coil of wire between the poles of a permanent magnet.

**Magnetomotive force**—The force that produces the flux emitting from a magnet.

**Magnet wire**—An insulated wire generally made of copper and used in the construction of electromagnets. The wire is generally manufactured in sizes No. 36 AWG to No. 12 AWG, and has an exact and uniform diameter.

**Maintenance factor**—With reference to lighting systems, the maintenance factor is the percentage of light to be expected with the usual cleaning and/or painting of the luminaries and reflecting surfaces.

**Manual control**—A form of control that depends upon human intervention for its operation.

**Matter**—Anything that occupies space and has weight, such as air, wood, metal, and water.

**Maximum value**—With reference to AC, the greatest value reached during one cycle.

**Maxwell**—A unit of magnetic flux. One maxwell equals 1 gauss per square centimeter, or 1 magnetic line of force.

**Megohmmeter**—A high-range ohmmeter. Its power is provided by a small generator, often hand driven. It is frequently used to measure the resistance of insulation.

**Mercury-vapor lamp**—A type of electric discharge lamp, consisting of an arc tube sealed inside a glass bulb. Current flowing through mercury causes it to vaporize. The mercury vapor regulates the value of current.

**Metal-halide lamp**—A type of electric discharge lamp.

**Metal oxide semiconductor field effect transistor**—A voltage-controlled device with three terminals: the source, gate, and drain. In a MOSFET, the gate is insulated from the conduction channel. There are four types of MOSFETs available: the enhancement only n-channel, the enhancement only p-channel, the depletion-enhancement n-channel, and the depletion-enhancement p-channel.

**Micrometer**—A tool for precisely measuring small thicknesses such as the diameter of a round wire.

**Mil**—A unit of length equal to 1/1000 inch. It is frequently used to express the diameter of a wire.

**Mil-foot**—A unit of measurement consisting of an object 1 foot long and 1 mil in diameter.

**Molecule**—Two or more atoms joined together by chemical bonds.

**Momentary contact push button**—A spring-loaded switching device containing one or more sets of contacts. The contacts are made only when the button is pressed.

**MOSFET**—The abbreviation for a metal oxide semiconductor field effect transistor.

**Motor**—A machine that converts electrical energy into mechanical energy. A machine that produces a rotating motion.

**Motor–generator set**—An electric motor mechanically coupled to a generator. The motor is used to drive the generator.

**Motor starter**—A device used to regulate the current to a motor during the starting period. It may be used to make and break the circuit and/or limit the starting current. It is usually equipped with an overload protection device.

**Multimeter**—An electrical instrument designed to measure more than one unit; for example, voltage, current, and resistance.

**Mutual induction**—The phenomenon by which a fluctuating current flowing in one coil induces an emf in an adjacent coil.

**National Electrical Code (NEC)**—A set of recommended standards developed for the purpose of the practical safeguarding of persons and property from hazards arising from the use of electricity.

**National Fire Protection Association (NFPA)**—A nonprofit organization established to aid the public in fire prevention. NFPA makes recommendations for manufacturing and installing equipment, devices, and electrical conductors, but it has no power or authority to enforce compliance with their recommendations. The organization works in conjunction with insurance companies and fire departments to develop high standards of fire prevention.

**Neutral conductor**—The conductor of a multiwire circuit or system that is grounded and has an equal voltage between it and each ungrounded conductor of the circuit or system.

**Neutral plane**—An imaginary plane in a motor or generator forming right angles to the main field flux.

**Neutron**—The particle in an atom that is electrically neutral; it is found in the nucleus of the atom.

**Noninverting amplifier**—An amplifier whose output is larger than its input. This means that as the input signal rises in the positive direction, the output signal also rises in the positive direction, but at a more rapid rate.

**Ohm**—The unit of electrical resistance. One ohm of resistance will allow 1 ampere to flow when 1 volt is applied.

**Ohmmeter**—A meter used to measure the value of electrical resistance.

**Ohm's law**—A mathematical law that expresses the relationship between current, voltage, and resistance in a DC circuit. It also expresses the relationship between current, voltage, and impedance in an AC circuit.

**Ohm's law triangle**—A memory device to aid in the memorization of the Ohm's law formulas.

**Op-amp**—The abbreviation for an operational amplifier.

**Operational amplifier**—An integrated circuit that consists of two or more amplifier stages. An operational amplifier, or op-amp, typically has two input terminals (one inverting input terminal and one noninverting input terminal) and one output terminal.

**Opto-coupler**—A solid-state device that consists of an LED and a phototransistor. An opto-coupler is used to provide electrical isolation between the input and the output terminals of the opto-coupler.

**Oscilloscope**—An instrument that uses the cathode ray tube. It is used to determine wave shapes, frequency, values of current and voltage; to compare these values; and to determine phase relationships.

**Overload**—An electrical condition in which an excessive amount of current flows. This condition arises when too much utilization equipment is connected into a circuit or system. When a motor is driving a load that requires more horsepower than its rating, it is overloaded, and excessive current flows.

**Parallel circuit**—A circuit that contains more than one path through which the electrons can flow.

**Peak value**—With reference to AC, the highest value reached during one cycle.

**Permanent magnet**—A magnet made of steel, rare earth, or alloys, which will retain its magnetism for an extremely long period of time.

**Permeability**—The measurement of the ability of a material to conduct magnetic lines of force. A material with high permeability is easily magnetized.

**Phase**—A position on an alternating current or voltage curve. The term three-phase indicates three separate voltages, which reach their maximum and zero values at three different but equidistant positions along the curve.

**Phase angle**—The angle of lead or lag between the current and the voltage in an AC circuit. In multiphase circuits, phase angle can refer to the angle between two voltages.

**Phase sequence**—A term referring to the order in which one voltage reaches maximum value in reference to the other.

**Phase sequence meter**—An instrument that indicates the order in which each voltage of a multiphase system reaches its maximum value.

**Photocell**—A material, generally placed within a tube, that produces an emf when light strikes it.

**Phototransistor**—A transistor with a small "window" in the case. The window allows light to strike the junction of the transistor, causing the transistor to conduct.

**Pictorial diagram**—Also called a wiring diagram, shows the relative physical location of the components, but not the operation of the circuit.

**PIRE wheel**—A memory device that shows all 12 Ohm's law and power law formulas.

**Polarity**—An electrical condition indicating the direction in which DC is flowing through a circuit, instrument, or component.

**Potential difference**—The difference in electrical pressure between two points. Potential difference is measured in volts.

**Power**—The rate of doing work.

**Power factor**—The ratio of the true power to the apparent power. It is equal to the cosine of the phase angle between the current and voltage in an AC circuit.

**Power law triangle**—A memory device to aid in the memorization of the power law formulas.

**Power source**—The source of energy in an electrical circuit. This could be a battery, a generator, a solar cell, etc.

**Primary distribution system**—The electrical conductors and equipment between the generator station and the final distribution point.

**Prime mover**—The machine that drives the rotating member of a generator.

**Programmable logic controller (PLC)**—A special type of computer designed to perform specific control functions in a logical order. It is used to control industrial machinery.

**Prony brake**—An instrument used to measure the torque of a motor.

**Proton**—A part of an atom containing a positive charge. Protons are found in the nucleus of the atom.

**Raceway**—An enclosed channel designed to hold electrical conductors.

**Radial distance**—The distance from a center of a circular path to any point on that path.

**Radial distribution system**—A system for distributing electrical energy. The energy is transmitted in all directions from one central point.

**Radiation**—With reference to heat transfer, the transfer of heat energy by way of electromagnetic waves from an infrared source to the object being heated.

**Ratio**—A mathematical relationship between two quantities usually expressed as a quotient (one quantity divided by the other).

**Reactive power**—The power circulating between the source and the load. This is a result of reactance in the circuit. Reactive power is measured in vars (volt-amperes reactive). If plotted on a graph, it has a negative value.

**Rectifier**—A device used to convert AC to DC.

**Reduced voltage starter**—A motor starter designed to lower the voltage applied to a motor during the starting period.

**Regenerative braking**—A type of electric braking in which the energy generated by the motor is returned to the power supply.

**Relative humidity**—The percentage of moisture in the air.

**Relay**—An electromagnetic switch. It consists of a coil and an armature. A small current through the coil causes the armature to move, opening or closing a switch.

**Reluctance**—The opposition a material presents to magnetic lines of force. A material with high reluctance is difficult to magnetize.

**Residual magnetism**—The magnetism that remains in a material after it has been removed from the magnetic influence.

**Resistance**—The opposition to current flow caused by the material from which the circuit is made.

**Resistance heating**—The act of producing heat by current flowing through a resistance.

**Resistivity**—As used in the electrical industry, the resistivity of a conducting material is the resistance of the material per mil-foot at a temperature of 68°F (20°C).

**Resistor**—A device designed to offer a specific amount of opposition to electron flow. It is constructed of materials that have few free electrons.

**Resonance**—The characteristic of a circuit containing resistance, inductance, and capacitance that results in unity power factor. In a resonant circuit, $X_L = X_C$.

**Resonant circuit**—A circuit in which the inductive reactance is equal to the capacitive reactance. The result is zero reactance.

**Resonant frequency**—The frequency at which the inductive reactance is equal to the capacitive reactance. The frequency at which the resistance of a circuit is equal to the impedance.

**Retentivity**—The ability of a material to retain its magnetism.

**Rheostat**—A variable resistance. The resistance can be varied by turning a knob or moving a handle.

**Right triangle**—A triangle containing one 90° angle.

**rms value**—rms stands for root-mean-square. The value is the same as the effective value. It is calculated by determining the square root of the average of the squares of the instantaneous values for one cycle.

**Rotary converter**—A motor generator set housed in one unit. The motor drives the generator and is used for converting one type of electrical energy to another type. For example, it can change AC to DC, or 60 Hz to 180 Hz.

**Rotor**—The rotating part of a motor or generator.

**Schematic diagram**—An electrical diagram that shows the electrical connections of circuits and/or equipment in a simplified manner.

**Scientific notation**—Using the powers of 10 to express a whole or a decimal number.

**Secondary ties**—Secondary conductors of a loop system to which load conductors are connected between transformers.

**Self-induction**—The inducing of an emf in a coil. It is caused by a varying current flowing in the coil.

**Semiconductor**—Materials that are neither good conductors nor good insulators. Certain combinations of these materials allow current to flow in one direction but not in the opposite direction. Semiconductor devices are rapidly replacing electron tubes.

**Series circuit**—A circuit that contains only one path over which the electrons can flow.

**Short circuit**—A condition that exists in an electric circuit when conductors of opposite polarity come into contact with each other. This condition results in excessive current flow.

**Silicon-controlled rectifier**—A diode with three terminals: anode, cathode, and gate. The gate provides control when the SCR conducts. The SCR will only conduct in one direction.

**Single-line drawing**—A very simple drawing that is used to convey overview information but not a lot of detail. A single-line drawing does not show the actual electrical connections, nor does it show the actual physical locations of components.

**Single phase**—A term applied to a single alternating current or voltage. A single waveform is produced within 360 electrical time degrees.

**Skin effect**—The effect produced by AC that tends to cause the current to flow along the outside of the conductor as opposed to flowing through the entire conductor.

**Slip**—With reference to an induction motor, slip is the difference between the synchronous speed and the rotor speed. It may be stated in revolutions per minute or percentage of the synchronous speed.

**Slip rings**—Copper or brass rings mounted on, and insulated from, the shaft of an alternator. On large alternators, they are used to make the connections from the DC source to the revolving field of the alternator. On small alternators, they are used to complete the connections from the alternator to the external circuit. Slip rings are sometimes used on motors to complete connections from a stationary circuit to a revolving circuit.

**Solenoid**—An electromagnet used to cause mechanical movement of an armature, for example, a solenoid valve.

**Speed control**—The ability to vary the speed of a motor by means of a device connected outside of the motor.

**Speed regulation**—The variation in the speed of a motor caused by a change in the load. A motor whose speed remains practically constant from no load to full load is considered to have good speed regulation.

**Splice**—A point in the wiring system where two or more conductors are joined both electrically and mechanically.

**Square mil**—The area of a square measuring 1 mil on a side.

**Starting current**—The amount of current required by a motor at the instant of starting.

**Starting torque**—The torque developed on the rotor at the instant the motor is connected to the line.

**Static control**—Control devices that have no moving parts. They are made of semiconducting materials.

**Stator**—The stationary part of a motor or generator; the part of the machine that is secured to the frame.

**Step-down transformer**—A transformer used to lower the supply voltage to a value required for the load.

**Step-up transformer**—A transformer used to increase the supply voltage to a value required for transmission or for the load.

**Switch**—A device for making or breaking the circuit or changing the current path.

**Synchronized**—With reference to alternators, all conditions are met so the machines can be connected in parallel.

**Synchronizing**—With reference to alternators, adjusting the phase sequence, phase relationship, frequency, and voltage of two or more alternators so they can be connected in parallel.

**Synchronous speed**—The speed at which the electromagnetic field revolves around the stator of an induction motor. The synchronous speed is determined by the frequency of the supply voltage and the number of poles on the motor stator.

**Synchroscope**—An instrument designed to indicate when the emf of two voltage sources are in phase and of the same frequency.

**Temperature coefficient**—The amount by which the resistance of a material changes per degree change in temperature for each ohm of resistance.

**Terminal**—A fitting for connecting electrical conductors to devices and/or equipment.

**Thermocouple**—Two dissimilar metals in contact with each other at one point. When the junction is heated, an emf is developed across the opposite ends of the metals.

**Thermostat**—A device for controlling temperature. It is a switch that opens and/or closes an electric circuit as the temperature changes.

**Three-phase**—A term applied to three alternating currents or voltages of the same frequency, type of wave, and amplitude. The currents and/or voltages are one-third of a cycle (120 electrical time degrees) apart.

**Time-delay relay**—A relay in which the armature movement is arranged to close or open contacts at a predetermined time after the signal has been received.

**Torque**—A turning or twisting force. A force that produces a rotation about an axis.

**Trace**—Another name for the line that is displayed on the CRT face of an oscilloscope.

**Transformer**—A device that generally contains no moving parts and is designed to increase or decrease the voltage and/or isolate a part of the system from the supply. See also *step-up transformer*, *step-down transformer*, and *isolation transformer*.

**Transformer primary**—The part of a transformer containing the winding to which the supply is connected.

**Transformer ratio**—The ratio of the number of turns of wire on the primary of a transformer to the number of turns on the secondary winding.

**Transformer secondary**—The part of a transformer containing the winding to which the load is connected.

**Transistor**—A solid-state device made of certain semiconductor materials; generally used to control electron flow.

**Triac**—An AC version of an SCR. The TRIAC is a three-terminal device with main terminal 1, main terminal 2, and the gate. The gate controls when the TRIAC conducts. Since the TRIAC is an AC device, it will conduct in both directions.

**Triangle**—A three-sided figure formed by connecting three vectors end to end to form a closed pattern.

**Trigger system**—The section of an oscilloscope that controls the stability of the displayed trace.

**Trigonometry**—The study of mathematics pertaining to the function of angles as ratios of the sides of a triangle. The use of vectors, angles, and trigonometry functions.

**True power**—The power, in watts, utilized by a circuit. If plotted on a graph, the true power is the positive power, sometimes called the effective power.

**Turbo-type alternator**—An alternator driven by a turbine.

**Two-phase**—A term applied to two alternating currents or voltages of the same frequency, type of wave, and amplitude. There is a 90° phase displacement between each current and/or voltage.

**UJT**—The abbreviation for a unijunction transistor.

**Unijunction transistor**—The unijunction transistor is a voltage-controlled device that is used primarily in switching and timing applications. The unijunction transistor, or UJT, has three terminals: the emitter, base 1, and base 2.

**Utility company**—A company that supplies electrical energy to consumers.

**Valance**—The outer shell of an atom that contains electrons.

**Vector**—A line segment that has a definite length and direction.

**Vector diagram**—Two or more vectors joined together to convey information.

**Voltage**—The electrical pressure that forces electrons to flow. Voltage is sometimes called electromotive force.

**Voltage control**—The ability to vary the output voltage of a generator by means of a device connected outside of the generator.

**Voltage curve**—A graph indicating the various values of voltage as produced by a generator or other voltage source.

**Voltage regulation**—The variation in output voltage of a generator, which changes with load. A generator whose voltage remains practically constant from no load to full load is said to have good voltage regulation.

**Volt-ampere**—The measurement of the apparent or reactive power of an AC circuit or system; the product of the current and voltage in an AC circuit.

**Voltmeter**—A meter used to measure the electrical pressure between two points.

**Watt**—A unit of measurement of electrical power; 746 watts is equal to 1 horsepower.

**Wattmeter**—An electrical instrument designed to measure the amount of true power utilized by a component, circuit, or system.

**Work**—The overcoming of resistance through a distance.

**Zener diode**—A diode that is designed to operate in reverse bias. When a zener diode is operating, it will have a constant voltage drop across it. Should a load be connected in parallel with the zener diode, a constant (or regulated) voltage will be maintained across the load.

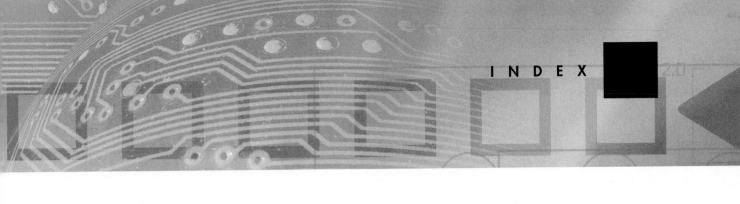